Textbook of
ENGINEERING CHEMISTRY

Textbook of

ENGINEERING CHEMISTRY

C.Parameswara Murthy
M.Sc., Ph.D. Dipl German
Professor of Chemistry
Nizam College, Osmania University,
Hyderabad

C.V. Agarwal
B.Sc. (HONS), M.Sc. Tech, Ph.D.
Professor of Applied Chemistry (Retd.)
Institute of Technology
Banaras Hindu University, Varanasi

Andra Naidu
M.Sc., Ph.D.
Professor of Chemistry,
JNTU College of Engineering
Hyderabad

BSP BS Publications
A unit of **BSP Books Pvt. Ltd.**
4-4-309/316, Giriraj Lane, Sultan Bazar,
Hyderabad - 500 095.
Phone : 040 - 23445605, 23445688

Published by:

BS Publications

A unit of **BSP Books Pvt. Ltd.**

4-4-309/316, Giriraj Lane, Sultan Bazar,
Hyderabad - 500 095.
Phone : 040 - 23445605, 23445688
e-mail : info@bspbooks.net

ISBN : 978-93-85433-86-3 (Hardbound)

*You see things; and you say "**Why?**" But I dream things that never were; and I say "**Why not?**"*

— George Bernard Shaw, Great Dramatist and Nobel Laureate.

Chemistry has an important role to play in achieving a sustainable civilization on earth.

— Dr. Terry Collins, Professor of Chemistry, Carnegie Mellon University.

Most of the environmental problems of past centuries and decades, such as the biological contamination of drinking water, were solved only when the methods of science in general—and chemistry in particular—were applied to them. The phenomenal rise in human life expectancy and in the material quality of life that has come about in recent decades is due in no small measure to chemicals and chemistry.

— Colin Baird, Environmentalist.

Preface

Chemistry is the study of matter, its properties, the changes that it may undergo and how these properties and changes are affected by its composition. It is important for engineers to have knowledge of chemistry, since they can expect to find problems in fields as diverse as the design and development of new materials, quality control and environmental engineering that are basically chemical in nature. The standing and firmness of a structure is decided by the strength of foundation laid, although it is never expressed to the external world. Similarly the applied and engineering sciences derive strength from basic sciences and their growth is decided by inputs given by basic sciences.

Chemistry is the back bone in designing and understanding the nature of various engineering materials. Many advances in engineering either produce a new chemical demand as in the case of polymers or wait upon chemical developments for their application as in the case of implants and alloys. Currently, the electronics and computers engineers are waiting for suitable biopolymers and nano molecules for use in miniature super computers; the electrical engineers are in search of proper conducting polymers; the mechanical engineers are on lookout for micro fluids and the civil engineers are looking for materials that are environment friendly, economical but long lasting.

Engineering Chemistry is a maze of topics of basic chemistry and applied chemistry. In developing this book an attempt is made to mix both the facets of the subject to reach proper consequence and meet the demands of teachers and taught.

This book is intended to be a complete text book that includes both theoretical concepts and practical procedures required for undergraduate Engineering and Technology students of Indian Universities.

This book is an amalgamation of expertise of

Prof. C. Parameswara Murthy, the author of *College Chemistry* three volumes of *University Chemistry* and *Chemistry of Engineering Materials*.

Prof. C. V. Agarwal, the author of celebrated book *Chemistry of Engineering Materials* that received about twelve reprints.

Prof. A. Naidu's teaching experience in Chemistry of Engineering Materials.

We are confident that this book too would get acceptance of teacher and student communities like our previous books.

To give a logical end, the applied topics are preceded by the corresponding basic topics. Thus the fuel related topics are preceded by thermodynamics and thermo chemistry, the corrosion is dealt after giving sufficient information of electrochemistry, solid state and phase rule precede alloys, material science and so on. Several numerical problems, multiple choice questions and short & essay type questions are included at the end of chapters to help the students to prepare for exams and self test their knowledge.

No book is perfect without the inputs of constructive criticisms from the readers. So we would appreciate receiving comments and suggestions for improving the content of text, additions and deletions to be made, errors crept and the same would be acknowledged in the future editions. You can use the email ID *cpmurthy@hotmail.com* for this purpose.

We take this opportunity to express our deep sense of gratitude to the authorities of Osmania University and Jawaharlal Nehru Technological University for permitting us to take up this venture. We are highly thankful to our colleagues for their discussions and suggestions. In particular we thank our colleagues Prof. P. K. Dubey, Prof. A. S. R. Swamy, Prof. S. Sondu, Prof. G. Venkat Reddy, Dr. P. Yadagiriswamy, Dr. B. Satyanarayana, Dr. P. Veera Somaiah, Prof. Arun Kumar, Dr. Gouri shankar for providing various inputs to get a best book.

The good students help in shaping their teachers with their inquisitive questions and observations and their role is acknowledged through this work.

We are thankful to the proprietors of B. S. Publications Mr. Nikhil Shah & Anil Shah for their cooperation. We are also thankful to Mr. Lakshminarayan Jakkani and Mr. Naresh for good DTP work. We are thankful to the management of Tara Publications for their encouragement for writing this book.

Our family members are the under-current of our energies and inspiration and we deeply acknowledge their patience and encouragement.

- Authors

Contents

Chapter 1

Colloids

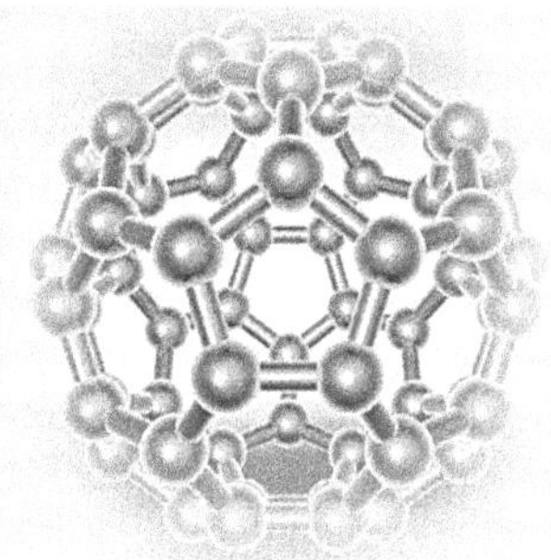

1.1 Introduction

All the macroscopic and microscopic particulate matter may be grouped into two categories (i) pure substances (ii) Mixtures.

(i) **Pure substances :** All the elements and compounds fall in this category. They have a fixed composition and generally have sharp melting and boiling pointsp.

(ii) **Mixtures :** A mixture may be defined as a substance containing two are more pure substances whose particles are in intimate contact but have not lost their individuality. These substances can be separated by simple separation techniques and the properties of mixtures vary according to percentage of components present. They usually do not have sharp melting and boiling points. We find a variety of mixtures in this universe. They can be further grouped into three general categories depending on the sizes of particles that are mixed, (A) Solutions (B) Suspensions and (C) Colloids.

(A) **Solutions :** On adding a spoon of sugar to a glass of water and stirring, we find that it vanishes but leaves a sweet taste to water. The sugar is distributed uniformly and can be regained by evaporating water. Varying amounts of sugar can be accommodated by the water. This mixture is known as a solution of sugar in water. As long as the concentration of sugar is less than that of water, sugar is called the **solute** and water the **solvent**. Properties of solution are different from the properties of solvent. Though the presence of sugar in water could be realized from its taste, attempts to detect the sugar molecules in a sugar solution using a microscope did not yield fruitful results because they assume a size of few angstroms and for the same reason they readily diffuse through parchment membrane. As these substances are generally in crystalline form, Graham called them **crystalloids.**

(B) **Suspensions :** On the other hand by adding a spoon of sand to a glass of water, we find the clarity of water is lost and it becomes turbid. If we keep this mixture aside for a while we find that the sand settles at the bottom under the influence of gravity and the clarity of solvent is restored. Clear water could also be obtained quickly by filtering it. This type of mixtures are called **suspensions**. The particles in a suspension are large enough to be seen either with the naked eye or through a microscope. The suspended particles or **suspensoids** are not distributed uniformly through the medium (here water) and they normally do not have any significant impact on the properties of the medium.

In the laboratory one can prepare a suspension by adding few drops of $AgNO_3$ solution to a solution of NaCl. The white cloud, appears due to formation of water insoluble AgCl, is an example of suspension. Either by filtration or centrifuging the mixture, it is possible to separate the AgCl and get a clear solution of $NaNO_3$.

(C) **Colloids :** This category of mixtures fall intermediate between true solutions and suspensions. The term colloid is derived from the Greek Word '**kolla**' meaning glue. The natural glues are colloidal dispersions of glue in water. Starch solution is a typical example of a colloid. The particles are invisible to naked eye and microscope but their presence could be detected by their scattering effect. The main properties that differentiate colloids from suspensions are their uniform distribution throughout medium or stability towards settling of particles under gravity and passage of particles through the filter paper without getting separated. These two properties resemble that of true solution. The main property that distinguishes a colloidal solution from a true solution is the detection of particles under ultra microscope and their precipitation by the addition of an electrolyte. The terms solute and solvent used in connection with a true solution are replaced by **disperse phase** and **dispersion medium** in case of colloidal solutions.

Colloids particles range from 1 nm, (10 Å), to 1000 nm and they consist of neutral molecules and charged particles like ions. They are not single phase solutions like true solutions but generally contains two phases. This fact could be demonstrated by studying the optical properties of a true solutions, a suspension and a colloidal solutions. For this take the suspension of soil in water in a test tube say (a), the true solution of sodium chloride in water in test tube (b) and a colloidal starch solution in test tube (c). Just by looking at the test tubes with our naked eye, we can identify that test tube (a) is a suspension but test tubes 'b' and 'c' both look fully transparent.

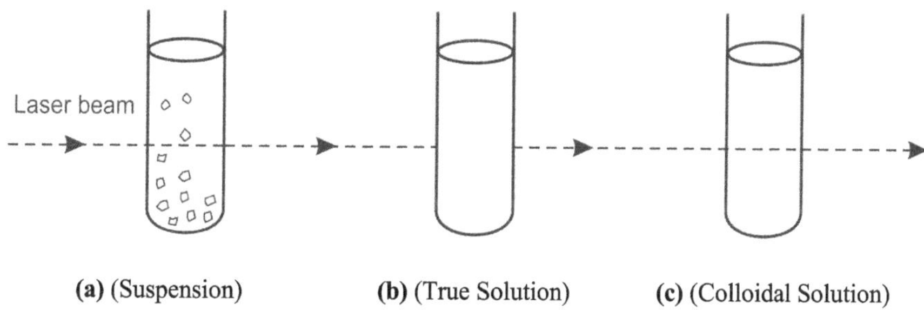

(a) (Suspension) **(b)** (True Solution) **(c)** (Colloidal Solution)

Now pass an intense beam of light through these test tubes. We will be able to trace the path in test tubes (a) and (c) while we cannot locate any ray in the tube (b). This is due to scattering of the light by the suspension particles and colloidal particles. The scattering of light by the colloidal particles was first noticed by Tyndall and is known as Tyndall effect. The distinguishing features of colloids are given in Table 1.1, although a detailed account of properties of colloids will be given later in **section 1.6.**

Table 1.1 Comparative statement of properties of true solutions, colloids and suspensions.

Property	True solutions	Colloids	Suspensions
1. State	Monophasic (Homogeneous)	Biphasic (Heterogeneous)	Biphasic (Heterogeneous)
2. Particle size	<1 nm	1-1000 nm	>1000 nm
3. Visibility of particles	Particles are not visible	Particles are invisible but are discernible under ultra microscope.	Particles are either visible to naked eye or under microscope
4. Settling	Does not settle under gravity	Does not settle under gravity	Settles under gravity
5. Appearance	Clear	Clear	Opaque
6. Filtration	Particles cannot be separated by filtration through filter paper	Particles can not be separated by filtration through filter paper	Particles separate out by passing through filter paper
7. Dialysis	Easily pass through parchment membrane	Passes through parchment membrane but slowly	Does not pass through parchment membrane.
8. Brownian motion	The solute particles have constant kinetic motion	The dispersion phase shows Brownian motion	The motion of the suspension particles is negligible.
9. Diffusion	Diffuse very fast	Diffuse slowly	No diffusion
10. Tyndall effect	Do not scatter light. Tyndall effect not observed	Shows scattering of light. Tyndall effect observed	Shows scattering of light. Tyndall effect observed.

1.2 Types of Colloidal Dispersion

We come across various colloids daily. The milk we drink, the blood flowing through our veins, the protein solutions, the butter, cheese, jelly etc., that we eat, the clouds, the fog, the smoke are all different forms of **colloidal dispersions or colloidal suspensions.**

The colloidal phase and colloidal medium which form the colloidal dispersions can take all the possible combinations of three states of matter viz, solid, liquid and gas. A gas mixes in another gas in all proportions, uniformly at molecular level, this combination only results in a solution. Various colloidal dispersions with suitable example are listed in Table 1.2.

Table 1.2 Types of Colloidal Dispersions.

Disperse Phase	Dispersion Medium	Colloidal Type	Examples
1. Solid	Solid	Solid sol	Colored glasses, pearls many precious stones
2. Solid	Liquid	Sol, Gel	Gold sol, $Fe(OH)_3$ sol starch in water, paint
3. Solid	Gas	Solid aerosol	Smoke (carbon particles in air) Haze (dust in air)
4. Liquid	Solid	Solid emulsion	Cheese, butter (water in fat), jellies (water in fruit pulp)
5. Liquid	Liquid	Emulsion	Milk (fat in water), cream, mayonnaise
6. Liquid	Gas	Liquid aerosol	Fog, cloud (water in air) mist, fog
7. Gas	Solid	Solid foam	Pumice stone (air in silicates), marshmallow
8. Gas	Liquid	Foam	Foam (water in air), froth, shaving cream, whipped cream
9. Gas	Gas	**Always produces a solution**	

1.3 Classification of Colloids

A colloidal suspension has two phases, the *disperse phase* and *dispersion medium*. Basing on the affinity between these two phases colloids are classified into **lyophobic sol and lyophilic sol**. When the dispersion medium is water these are said to be **hydrophobic** and **hydrophilic sols** respectively.

Lyophobic sols : These are also termed as irreversible sols or suspension. This term has its origin in greek and means solvent hating or solvent averting. *In the lyophobic colloids the disperse phase has very little affinity towards the dispersion medium.* This lack of affinity accounts the difficulties encountered in the preparation and their easy precipitation. As the colloids are readily precipitated giving larger visible particles, they are called **suspensoids** and this process of precipitation is known as **coagulation,** irreversibility of this process brought the name **irreversible colloids**. Colloidal dispersions of metals like gold, silver, iron etc., and sparingly soluble salts like As_2S_3, $Fe(OH)_3$ etc., are the examples for this class. They coagulate readily by the addition of even a trace amount of electrolyte. Normally stabilizing agents are added to prevent coagulation.

Lyophilic sols : Lyophilic means solvent loving or solvent liking. These colloids are formed readily by employing simple techniques like heating, shaking a colloidal particle phase in a suitable dispersion medium and called **emulsoids** or **reversible colloids**. In the house we notice the preparation of starch solution, just by adding starch powder to the boiling water. Protein solution, glue (gum), etc., are some typical examples of lyophilic sols. These sols obtained are of reversible type. A coagulated disperse phase could be brought back to the colloidal state just by diluting and warming it again. These sols are self stabilized and require no stabilizer and small amounts of electrolyte have no effect.

The characteristics of lyophobic and lyophilic sols are given in Table 1.3.

Table 1.3 Characteristics of Lyophobic and lyophilic sols.

Property	Lyophobic sol	Lyophilic sol
1. Visibility	The particles are detectable with an ultra microscope	The particle cannot be detected
2. Stability	Less stable	More stable
3. Reversibility	Irreversible	Reversible
4. Surface tension	The surface tension of colloidal dispersion is alsmot of the same order as colloidal medium	The surface tension is less than that of colloidal medium
5. Viscosity	Viscosity is almost of the same order as dispersing medium	Viscosity is higher than dispersing medium.
6. Effect of electrolytes	Small quantities of electrolyte is sufficient to cause coagulation	Large quantities are needed for coagulation

*Contd...*Table 1.3

Property	Lyophobic sol	Lyophilic sol
7. Hydration	Particles are usually hydrated poorly	Particles are heavily hydrated.
8. Charge	Particles carry a characteristic charge	The charge of the particles depends on the pH, may carry negative, positive or even no change.
9. Electrophoresis	They show electro-phoresis	Depending upon the charge on the particle, it may or may not show electrophoresis
10. Concentration	Can have only very low concentration	Higher concentration are possible

1.4 Methods of Preparation of Colloidal Solutions

It is possible to prepare colloidal solutions of all substances by selecting suitable dispersion medium. The particle *size of colloidal dispersion is intermediate to true solutions and suspension.* So we can have two approaches in preparing these dispersions. One approach involves breaking of macroscopic suspension particles into particles of the size of colloidal phase (1 nm - 1000 nm) employing mechanical methods or clectrical methods or by creating conditions for the reversal of coagulation. The methods employed are called **dispersion methods**. The second approach involved bringing the microscopic atoms or molecules of true solutions together and forming colloidal particles. These methods are called **condensation methods**. These two approaches can be represented by the following flow chart.

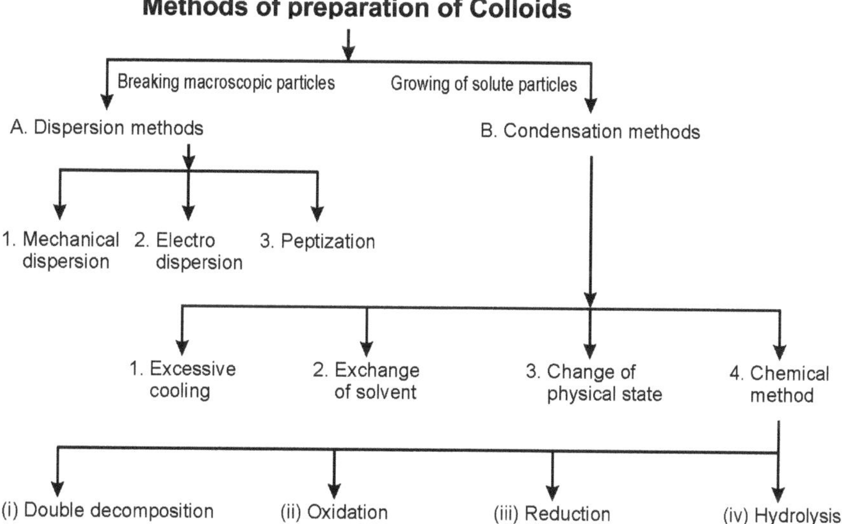

A. Dispersion Methods

1. Mechanical dispersion : Coarser particles of a substance are broken into colloidal size by grinding them between two steels discs having a narrow clearance. For this a suspension along with a suitable stabilizer is charged through a hollow shaft of a colloidal mill (Fig. 1.1), where the steel discs are rotating at a high speed in opposite direction, and the colloidal dispersion is collected.

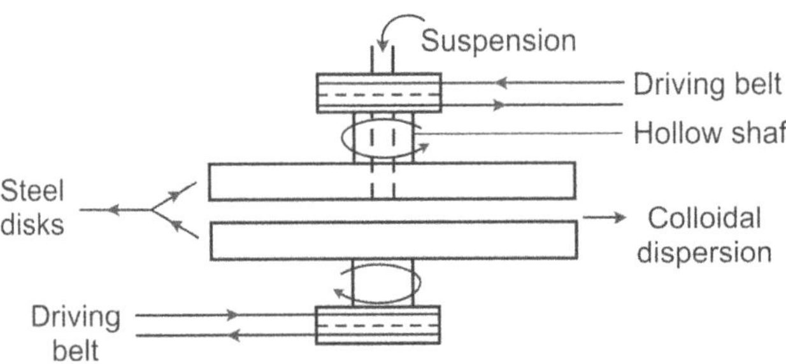

Fig. 1.1 Colloidal mill.

This way indian ink is prepared by grinding lamp black in the colloid mill and stabilizing the dispersion with gum arabic.

2. Electro-dispersion – Bredig's arc method : The colloids dispersion of metals like gold, silver, copper etc., in water are prepared by this method. The metal rods are dipped in the solvent and an arc is set by passing current through the rods. This arc liberates metal particles of colloidal size which are readily absorbed by the medium giving a colloidal suspension (Fig. 1.2). To prevent evaporation of the medium due to heat liberated by the arc the vessel is surrounded by ice and kept cool.

This method is not useful when we take organic solvents as dispersion medium, as the heat generated at the arc chars the medium.

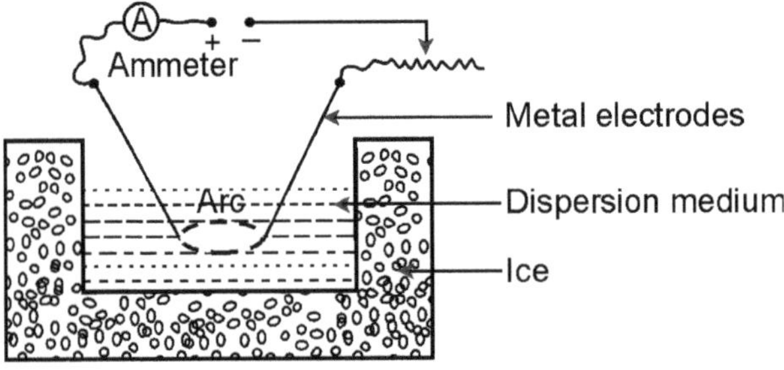

Fig. 1.2 Bredig's setup for electro dispersion.

3. **Peptization :** *Peptization is inverse of coagulation.* As discussed earlier coagulation is the result of lack of affinity between dispersed phase and dispersing medium. An affinity is created by the addition of a stabilizing agent, known as *peptizing agent*, which prompts the coagulated particles back to give colloidal solution and the process is called *peptization.*

A freshly prepared precipitate of ferric hydroxide, $Fe(OH)_3$, is peptized, giving a colloidal dispersion, by the addition of some ferric chloride, $FeCl_3$, solution. This is due to masking of uncharged colloidal particles of $Fe(OH)_3$ by the aquated ferric ions, Fe^{3+} aq, and creating a net positive charge on the colloidal particle and thus bridging the affinity gap between hydrophobic $Fe(OH)_3$, particle and dispersion medium.

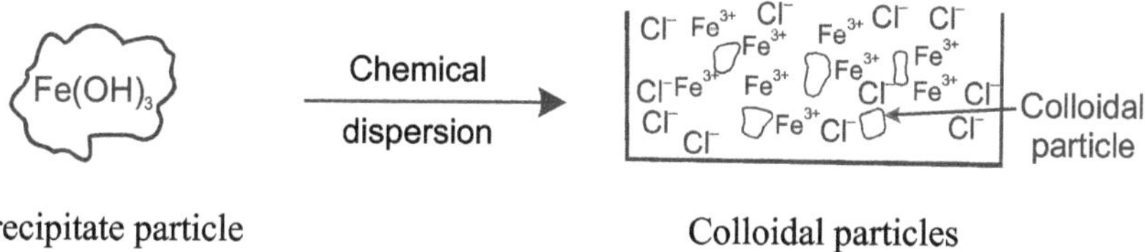

Precipitate particle   Colloidal particles

As these colloidal particles are positively charged, addition of a negatively changed PO_4^{3-} ion to the colloidal solution neutralizes the charge causing destabilization of colloid and precipitation of disperse phase.

We come across many such systems in our day to day life. Milk is a colloidal dispersion of oil in water. As oil is a hydrophobic substance, casein in milk serves the purpose of peptizing agent. Milk coagulates by the addition of common salt. In the formation of indian ink, lamp black is peptized by gum arabic.

B. Condensation Methods

1. **By excessive cooling :** Colloidal solution of ice in organic solvents like chloroform, ether are formed by cooling below 0 °C, a solution of water in these solvents. Water is only sparingly soluble in these solvents and on excessive cooling, the solubility further decreases throwing dissolved water out to form fine colloidal particles of ice.

2. **By solvent exchange :** Change of solvent is another way to lower the solubility of the solute. Thus substances like sulphur, phosphorous, etc., are first dissolved in alcohol and then they are added to excess of water resulting in the formation of hydrosol of these substances. Similarly when we add an alcoholic phenolphthalein indicator to excess of water, we observe a white milky colloidal solution phenolphthalein.

3. **By change of physical state :** In the gaseous state at moderate pressures and temperature, many substances attain a macromolecular size. Hence by passing the vapor, e.g., sulphur and phosphorous, through the dispersion medium containing stabilizer yields a colloid.

4. **Chemical methods :** Chemical reactions that produce partially soluble and insoluble substances lead to the formation of colloids. Depending on the nature of chemical reaction. They can be classified into (i) double decomposition (ii) oxidation (iii) reduction and (iv) hydrolysis.

   (i) **Double decomposition :** On passing H_2S through a dilute solution of arsenious oxide we get a colloidal solution of arsenious sulphide.

   $$As_2O_3 + 3H_2S \longrightarrow As_2S_3 + 3H_2O$$

   Concentrated solutions yield only suspensions. The excess of hydrogen sulphide is removed by boiling the solution which may otherwise lead to precipitation. Similarly on adding few drops of ferric chloride to a dilute solution of potassium ferrocyanide in the presence of a peptizing agent we get a sol of ferri cyanide.

   $$K_4Fe(CN)_6 + FeCl_3 \longrightarrow \underset{\text{Prussian blue colloid}}{KFe[Fe(CN)_6]} + 3\,KCl$$

   (ii) **Oxidation :** H_2S on oxidation gives a colloidal solution of elemental sulphur

   $$2H_2S + O_2 \longrightarrow 2H_2O + \underset{\text{sulphur colloid}}{2S}$$

   A moderate oxidizing agent as bromine or nitric acid could also be used for this purpose.

   $$H_2S + Br_2 \longrightarrow 2HBr + S$$

   (iii) **Reduction :** The colloids of metals such as gold, silver, lead and platinum could be conveniently prepared by reducing the metal salts with reducing agents such as formaldehyde, tannic acid stannous chloride etc

   $$2AuCl_3 + 3SnCl_2 \longrightarrow 3SnCl_4 + \underset{\text{gold sol}}{2\,Au}$$

   Zigmondy prepared the bright red gold sol by reducing the gold chloride ($HAuCl_4.3H_2O$), and phosphorous colloid by reducing a dilute etherial layer of phosphorus.

(iv) **Hydrolysis :** Colloidal dispersions are obtained by the hydrolysis of metal hydroxides, viz., Fe $(OH)_3$ or Al (OH_3) and stabilizing the colloids by the addition of little excess of solutions of corresponding metal chlorides.

$$FeCl_3 + 3H2O \longrightarrow Fe\,(OH)_3 + 3HCl$$

For achieving stability of the colloid it is essential to remove HCl by dialysis.

1.5 Purification of Colloidal Dispersion

The general impurities present in a colloidal dispersion are mainly the suspended particles and the electrolytes. The suspended particles could be easily separated by a simple filtration. The following methods are employed for freeing a colloidal solution from electrolytes.

1. **Dialysis :** The process of separating a colloidal solution from a crystalloid (true solution) is known as dialysis. The property of passing through an animal membrane by the solute and solvent molecules of a crystalloid and inability of colloidal particles was exploited by Graham in designing the apparatus used for purification a colloid. This apparatus is known as dialyzer. The colloidal solution to be dialyzed is taken in a parchment bag and suspended in a running water trough. The soluble impurities pass through the membrane into the running water leaving pure colloidal suspension in the parchment bag (Fig. 1.3). Completion of dialysis could be ensured by monitoring the conductivity of water in the trough. Presence of ions increases the conductivity of pure water.

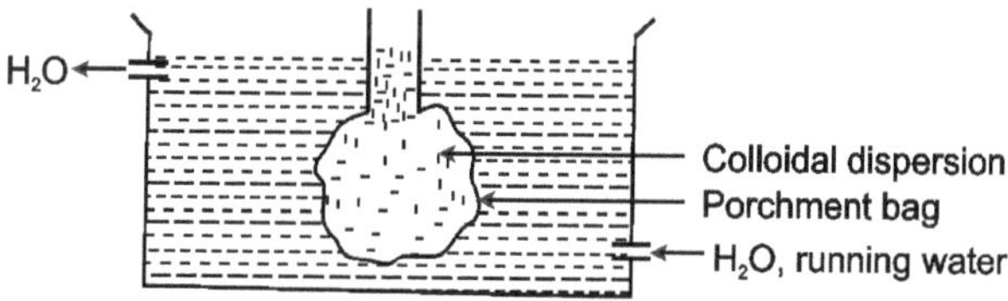

Fig. 1.3 Graham dialyzer.

2. **Electrodialysis :** The principle behind this technique is same as simple dialysis. Normal dialysis is a slow process and usually takes several hours or sometimes days for completion. Colloidal particles normally have a charge due to adsorption or resting of ions over the surface. Hence diffusing of electrolyte through the membrane is hastened by carrying out the dialysis in the presence of electric field and in known as **electrodialysis** (Fig. 1.4).

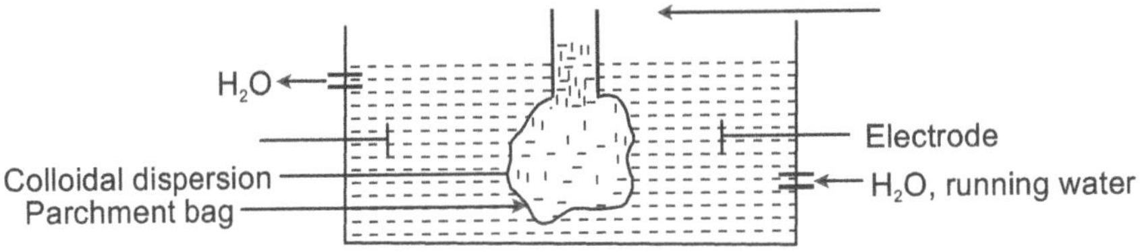

Fig. 1.4 Electrodialysis.

Electrodialysis is obviously not meant for removing uncharged impurities such as sugar, urea etc.

3. **Ultra-filtration :** The pore size of ordinary filter paper is >1000 nm allowing the sol particles, which have a diameter of 10nm - 1000 nm, to pass through them. In the ultra filtration, the filter paper is soaked in a colloidal solution such as gelatin or collodion and dried. This reduced the pore size considerably and the impregnated filter paper, known as ultrafilter, is used in the purification of colloid, similar to a dialysis membrane. By selectively choosing impregnation suspensions of different sizes and preparing corresponding ultra filters, it is also possible to separate a mixture of colloidal solutions having particles of different sizes.

4. **Ultra-centrifuging :** Colloidal particles and crystalloid, both, do not settle under gravitation and exhibit Brownian (zig-zag) motion. The continuous zig zag motion of the colloidal particles prevents the colloidal particles from settling under gravity. But if one considers the velocity of this motion colloids move rather slow in comparison to crystalloids. Hence when a sol is placed in a ultra-centrifuge and rotated at a speed of 15000 or more revolutions per minute, the colloids settle at the bottom of centrifuging tube leaving the electrolyte free. This suspension can be reconverted into a colloidal solution, employing usual dispersion methods of heating, mechanical stirring etc.

1.6 Properties of Colloids

1.6.1 State

Colloidal solutions are heterogeneous mixtures and have two phases, the disperse phase and the dispersion medium.

1.6.2 Particle size

They have a particle size of 1-1000 nm and can be seen only under a ultra microscope, particles are invisible to naked eye and to even for powerful microscopes.

1.6.3 Filtration

Colloidal particles pass through ordinary filter paper. As mentioned in the purification methods specially prepared ultra filter retain the colloidal particles. They also can not pass through animal membranes.

1.6.4 Color

The color of a colloidal dispersion depends on

(a) The size and charge the of the particle

(b) The characteristic absorption, the disperse phase and dispersion medium.

(c) Whether the observer is receiving the transmitted light or reflected light.

(d) The wave length of the light falling.

For example, with the increase in the particle size gold sol changes its color gradually from red to blue. When the particle are small (say 200 nm) the blue component of the white light is reflected and when we observe in the direction of incidence only red color is observed. Similarly with the increase in particle size the gold particles start reflecting even the other wave length thereby start looking blue.

Dilute milk appears blue when observed at right angle to the incident light (reflected rays) and red when observed in the direction of incident light (transmitted light).

A beautiful demonstration of the color properties of colloidal particles is the chemical sunset scene produced by the transmitted light by the addition of hydrochloric acid to sodium thiosulphate. Here a sulphur colloid is produced initially, which gradually coagulates to give suspension particles.

1.6.5 Colligative Properties

The colligative properties viz., osmotic pressure, relative lowering of boiling point depend on the number of solute particles in a given weight of solvent. For a given concentration true solutions have far larger number of solute particles than the colloidal particles. This is due to aggregation of solute particles (about 1000 nm) to give new bigger size colloidal particles. Of the above mentioned colligative properties only osmotic pressure has a measurable value and changes in other properties are insignificant. The change in osmotic pressure of a colloid is a highly useful data in determining the average particle mass of starch particle which was found to be about 32,000 atomic mass units.

1.6.6 Optical Property–Tyndall Effect

As discussed already, when light passing through a true solution is observed at right angles to the path of incident ray, it looks completely dark. The same experiment if we repeat with a colloidal solution, we notice a luminescent path of the ray passing through the solutions. The path of the ray is made visible by the scattering of light by the colloidal particles. The particles of true solution can not scatter light because the wave length of light is larger than the size of the particles (< 10 nm).

This effect was *first noticed by Faraday but detailed studies were made by Tyndall giving it the name* **Tyndall effect** (Fig. 1.5). The appearance of bright path of the projection lamp in a theatre is due to scattering of light by the dust particles.

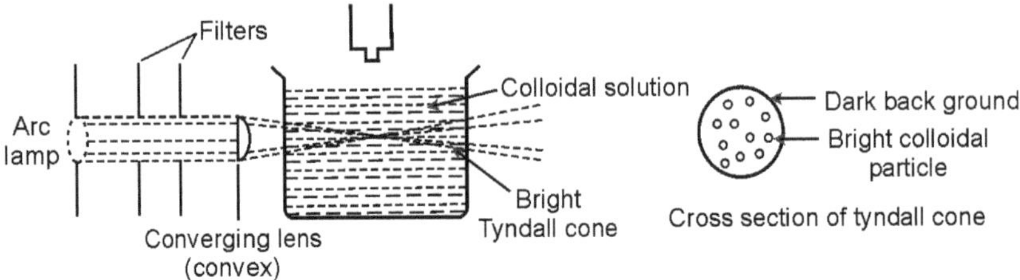

Fig. 1.5 Tyndall effect.

The two conditions that must be satisfied to show Tyndall effect are :

(i) The wave length of light used must not be much larger than the diameter or size of the particle.

(ii) The dispersed phase and dispersed medium should differ greatly in the magnitude of refractive indices.

Condition (i) was already explained and condition (ii) could be understood if we consider the colloidal suspension of freshly prepared albumin. Here the disperse phase, albumin and the dispersion medium water, have a very small difference of refractive indices and the Tyndall effect is not observable. On standing the particle size increases, bringing a change in the refractive index thus showing Tyndall effect.

Tyndall effect is strong for hydrophobic sols due to large difference in the refractive indices of disperse phase and dispersion medium. In case of lyophilic sols the difference in refractive indices is small, resulting in a weak Tyndall effect.

Siedentopf and Zsigmondy made successful attempt to see the colloid particles with the help of a microscope known as **ultra microscope**. Here a bright light beam from an arc is converged with a convex lens and allowed to fall on the colloidal solutions. We find two self-luminous bluish colored cones, called **Tyndall cones**, meeting at the focus of the lens. When a microscope is placed at right angles to the incident light, presence of colloidal particles can be detected as bright moving spots. This is due to initial absorption of light by the colloidal particle before it is scattered.

1.6.7 Kinetic Property – Brownian Movement

The invention of ultra-microscope by Siedentopf and Zsigmondy led to another invention by Robert Brown in 1827. Just like the constant directionless movement of pollen grains in an aqueous suspension, he noticed under microscope that the colloidal particles also move continuously in a zig-zag path without any destiny and this type of motion is called **Brownian movement.**

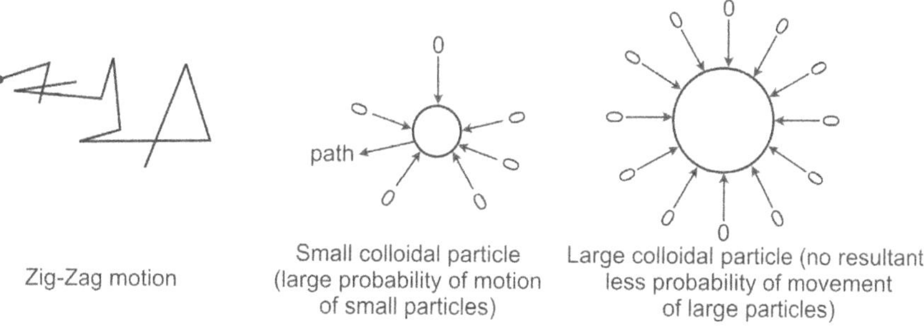

Fig. 1.6 Brownian movement.

This *Brownian movement arises due to constant collisions or striking of the molecules of dispersion medium and colloidal particles*, Fig. 1.6. Owing to the smaller size of the colloidal particles, the striking are limited and the colloidal particles which always move in a resultant direction.

A large particle (e.g., a suspension particle) has a higher inertia (inability of a body to move) due to its heavy mass and in addition due to numerous collisions in all the directions and cancellation of opposing forces the resultant force becomes extremely weak to cause any significant impact. Hence the particles of a suspension show no Brownian movement and settle under gravity.

1.6.8 Electrical Charge

Colloids often carry a charge on their surface. For example, ferric hydroxide sol particles are positively charged while the arsenous sulphide sols are negatively changed. The electrical neutrality of the sol is maintained by the dispersion medium taking equal and opposite charge of the colloidal particle.

A colloidal particle can take both positive and negative charges depending on the conditions employed in the preparation of the colloidal dispersions. The following two examples demonstrate this fact.

Example 1.

If a freshly prepared stannic oxide precipitate is peptized by a few drops of KOH solution the sol assumes a negative charge.

This is due to the adsorption of stannate ions,SnO_3^{2-} formed by the reaction between SnO_2 and KOH.

$$SnO_2 + 2KOH \longrightarrow SnO_3^{2-} + 2K^+ + H_2O$$

This is colloidal particle is represented as $[SnO_2]\ SnO_3^{2-}\ 2K^+$

When the SnO_2 precipitate is peptized with the addition of few drops of HCl, the sol takes a positive charge.

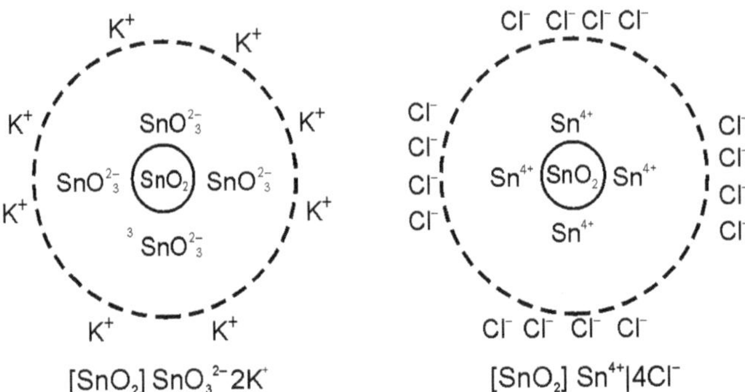

Sn^{4+} absorbed on the sol particle formed by the reaction between SnO_2 and HCl

$$SnO_2 + 4HCl \longrightarrow Sn^{4+} + 4Cl^- + 2H_2O$$

this is represented as $[SnO_2]\ Sn^{4+} * 4Cl^-$

Example 2.

If a dilute solution of potassium iodide is added to a concentrated solution of $AgNO_3$ a positively charged colloid is formed due to adsorption of free Ag^+ ion on the AgI surface and is represented as [Ag I] $Ag^+ * NO^{-3}$. Similarly a negatively charged colloid is produced when dilute silver nitrate is added to a concentrated solution of KI and is represented as [Ag I] $I^- K^+$.

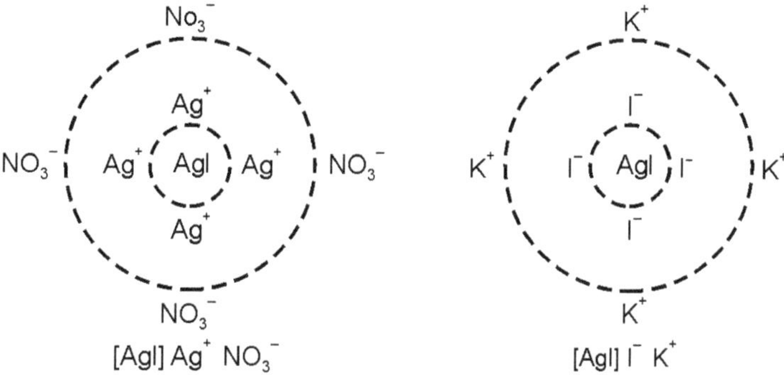

Inspite of co-existence of oppositely charged disperse phase and dispersion medium, the reason for no charge neutralization, and unexpected stability is explained by the **electrical double layer theory.**

1.6.9 Electrical Double Layer Theory

This theory was propounded by Helmholtz. Consider a positively charged fixed layer in contact with a **mobile or movable or diffused, medium.** The negatively charged ions of the movable medium try to approach the fixed layer to neutralize the charge. But the constant thermal agitation or kinetic motion of the ions of the mobile layer create a varying electrical environment

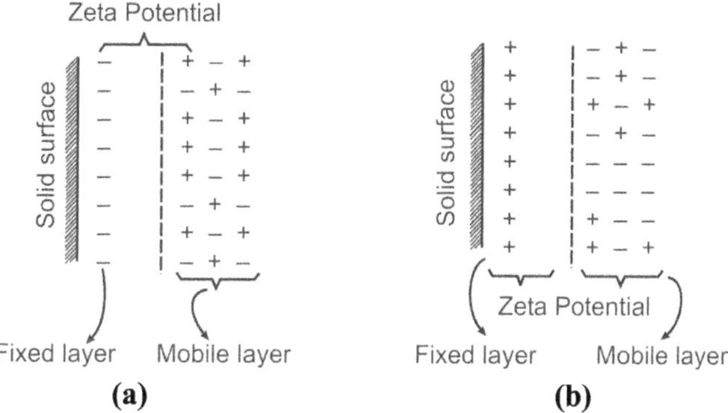

Fig. 1.7 The electrical double layer.

for the positively charged fixed layer. This *constantly shifting environment results in a charge difference between the fixed layer and the mobile layer known as* **Zeta potential or electro kinetic potential.**

The combination of the fixed layer and mobile layer forms an electrical double layer system (Fig. 1.7). The stability of a colloidal dispersion, according to electrical double layer theory, is due to formation of a zeta potential screen between the dispersed particles (fixed layer) and dispersion medium (mobile layer) that restricts the neutralization of electrical charge on the fixed layer. This screen of electrical double layer is broken down when we add multivalent ions carrying the charge of dispersion medium, leading to coagulation of the sol. Thus positively charged ferric hydroxide, $Fe(OH)_3$, sol is coagulated by the addition of trivalently charged PO_4^{3-} ions as shown in Fig. 1.8.

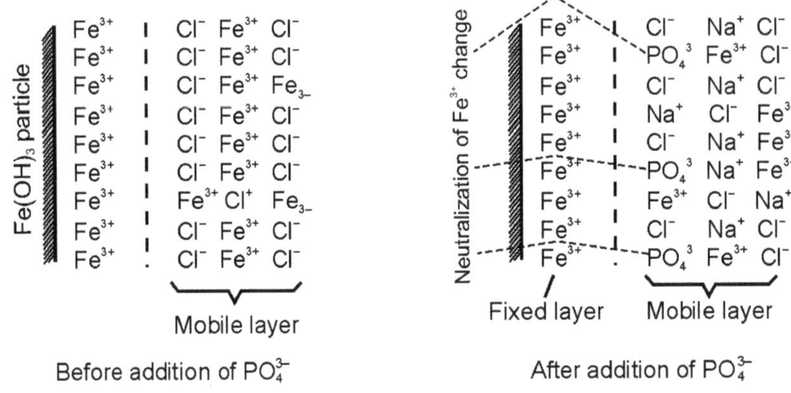

Fig. 1.8 Electrical double layer of Fe (OH) sol.

The phosphate anion, wherever it gets in contact with the colloidal particles, the Zeta potential approaches zero resulting in the coagulation of the sol particle.

1.6.10 Passage of Current

Passage of current through a colloidal solution results in the movement of colloidal particles, as they are charged, towards oppositely charged electrodes. We can design experiments either to allow the movement of the colloidal particles or restrict their motion. The former process is known as **electrophoresis** or **cataphoresis** and the later process is called **electro-osmosis.**

A. Electrophoresis : *The migration of colloidal particles under the influence of applied electric field is known as electrophoresis.* This migration helps us in deciding the charge associated with the colloidal particles to demonstrate electrophoresis, the sol is filled in a U-tube. Two electrodes are fixed, just above the colloidal solutions, in both the open ends of U-tube (Fig. 1.9). Now water is added to the open ends to dip the electrodes. A high electrode potential is applied across the electrodes. Now we mark a movement of sol-water boundary depending on the charge of colloidal particles. For example, arsenic sulphide colloidal particles, charged negatively, move towards the anode causing a raise of level in the anode limb.

As in the first observation migration of boundaries was made with ferric hydroxide colloid, which migrated towards cathode, the phenomena was given a misnomer as **cataphoresis**.

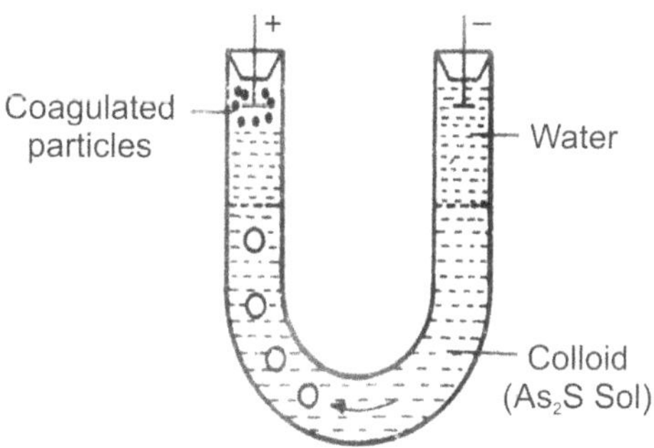

Fig. 1.9 Electrophoresis.

The sol particles move at different velocities depending on the charge, mass and size of the particle. *The distance travelled by the colloidal particles in one second under an applied potential of one volt per centimeter is called* **electrophoretic mobility**. Knowing the electrophoretic mobility enables one to calculate the zeta potential using the formula

$$Z = \frac{4\pi\eta\mu}{D}$$

Where η is the coefficient of viscosity

μ is the electrophoretic mobility

and D is the dielectric constant of the dispersion medium

Knowledge of electrophoretic mobility also helps in separating the sols from a colloidal mixture. Thus from a mixture of metal sol and metal sulphide sol, the lighter colloidal metal sulphide particles are recovered first, followed by the metal sol particles. Biochemists use this technique in fractionating proteins, polynucleic acids and polysaccharides from their colloidal mixtures.

Depending on the Charge carried by colloidal particles they are grouped into positively charged sols and negatively charged sols. Few common sols are listed below Table 1.4:

Table 1.4

Positively charged sols	Negatively charged sols
1. Metal hydroxides: $Fe(OH)_3$, $Cr(OH)_3$, $Al(OH)_3$, etc.	1. Metals : Au, Ag, Cu, Fe, etc
2. Metal oxides : Ti O_2, etc.	2. Metal sulphides : As_2S_3,CdS, Sb_2 S_3, etc.
3. Basic dyestuffs : Bismark brown, methylene blue, etc.	3. Acid dye stuffs : Eosin,congo red, etc.
	4. Organic colloids: Starch, gums, proteins, etc.

Applications of electrophoresis : Electrophoresis has many industrial applications :

(i) Carbon from the colloidal smoke going out of an industrial chimney is coagulated and removed on a cathode.

(ii) Suspended impurities from the dust sensitive assembly units like computer and watch are removed taking advantage of this phenomena.

(iii) Negatively charged latex rubber particles are electroplated on the surface of metal or wood or cloth.

B. Electro-osmosis : The movement of the dispersion medium, when the colloidal particles are restricted from their motion, under the influence of electric field is known as **electroosmosis**. This could be demonstrated by constructing a simple apparatus shown in Fig. 1.10.

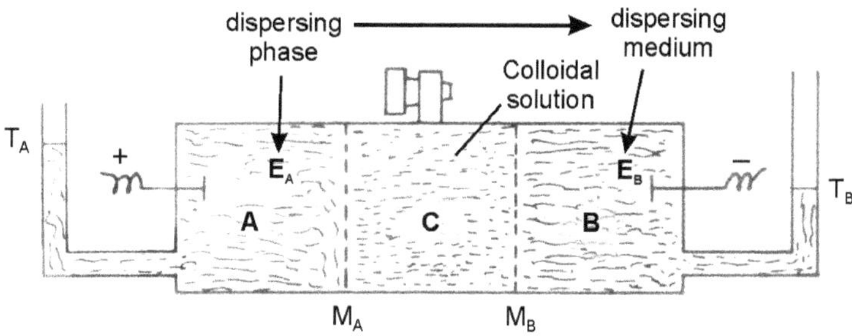

Fig. 1.10 Electroosmosis apparatus (for positive charged particle).

The apparatus consists of three compartments A,B and C separated by two dialyzing membranes M_A and M_B. The central compartment 'C' is filled with the colloidal solution, and the side compartments 'A' and 'B' are fitted with side tubes T_A and T_B respectively.

The side compartments are fitted with platinum electrodes E_A and E_B. An electrode potential (D. C current) is applied across E_A and E_B. The dispersion medium and disperse phase start moving towards oppositely charged electrodes. For example if we take arsenic sulphide sol, the negatively charged As_2S_3 particles start moving towards E_A, the anode and the dispersion medium towards E_A the cathode. But the sol particles are not allowed to get into the anodic compartment 'A' by the dialyzing membrane M_A. The dispersing medium passes into the cathodic compartment 'B' without any hindrance by the membrane M_B. This results in a rise in the level of T_B with a corresponding fall in the level of T_A. In case of positively charged sol particle the rise is in the anodic compartment.

1.6.11 Coagulation

Coagulation is a process of growing the colloidal particle size resulting in the formation of suspension and ultimately leading to the precipitation of the particles of the dispersed medium. It may be spontaneous or accomplished either by the addition or complete removal of electrolyte or by mixing two oppositely charged colloidal solutions.

(i) **Spontaneous coagulation :** Some sols, on long standing the particles tend to come closer and grow in their size leading to precipitation. This spontaneous coagulation is called **ageing**.

(ii) **By the addition of electrolyte :** The colloidal particles are stabilized by assuming a charge cloud on its surface and by forming an electrical double layer or having a zeta potential. Thus as we already discussed As_2S_3 sol particles are negatively charged with positive dispersion medium around them, addition of an electrolyte brings a change in the zeta potential and the colloidal particle is robbed away of its charge leading to coagulation.

Since it is the charge on the ions of added electrolyte which is responsible for the coagulation, it is rational to assume that the valency of the ions is the root cause of precipitation. Hardy and Schulze after rigorous experiments found that *higher the valency of the active ion the greater is its precipitating action* and is known as **Hardy-Schulze law**. The precipitating actions of cations Na^+, Ba^{2+}, Al^{3+} on As_2S_3 sol has been found to be

$$Al^{3+} > Ba^{2+} > Na^+$$

The minimum concentration of an electrolyte which can cause precipitation is called the **flocculation value.**

Table 1.5 Flocculation values of As_2S_3 sol

(milli mole per litre to produce precipitation)

NaCl	51.0	$M_g\ SO_4$	0.81	$Al(NO_3)_3$	0.095
KCl	49.5	$Ba\ Cl_2$	0.69	$Al\ Cl_3$	0.093
		$Ca\ Cl_2$	0.65		

Similarly for the positively charged colloidal particles (e.g., $Fe(OH)_3$ sol) we have to take the anion as the active ion. There the precipitating action is of the order

$$PO_4^{3-} > SO_4^{2-} > NO_3^{+}$$

Table 1.6 Flocculation value of $Fe(OH)_3$ sol

(millie mole per litre required to produce precipitation)

KCl	9.5	K_2SO_4	0.20
NaCl	9.3	$MgSO_4$	0.22
$BaCl_2$	9.6		

When we compare the relative stability of hydrophobic and hydrophilic sols, the hydrophobic sols readily coagulate by the addition of electrolytes. In case of hydrophilic sols even after stripping away the charge present on the colloidal particle by the added electrolyte, the sol is not coagulated. This is due to a shield of water molecules formed around the colloidal particle as a result of aquation. Addition of a few drops of dehydrating agent like alcohol or acetone break opens the shield and coagulation sets in. This is represented in Fig. 1.11.

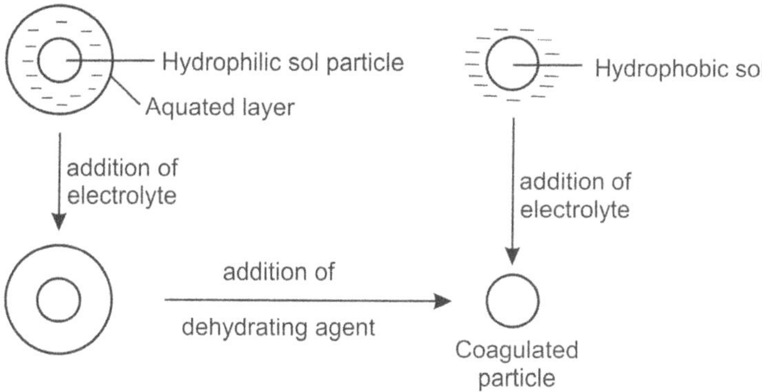

Fig. 1.11 Coagulation of hydrophobic and hydrophilic sols.

(iii) **By the complete removal of electrolyte :** Traces of electrolytes are needed for the stability of a colloid. Prolonged dialysis removes these traces of electrolytes causing precipitation of colloidal dispersion.

(iv) **Mutual coagulation :** When oppositely charged colloidal dispersions are mixed they neutralize their charges mutually causing precipitation known as mutual coagulation. Thus, when As_2S_3 sol and $Fe(OH)_3$ sol are mixed both the colloidal particles coagulate.

1.6.12 Protection

As mentioned above the lyophilic colloids achieve extraordinary stability due to solvation of colloidal particles. When lyophilic sols are added to lyophobic sols depending on their sizes either lyophobic sol is adsorbed in the surface of lyophilic sol or lyophilic sol is adsorbed on the surface of lyophobic sol as represented in Fig. 1.12. In both the cases the hydrophobic sols achieve additional stability. *The process of addition of lyophilic sols by which the lyophobic sols are protected from precipitation on adding an electrolyte is known as protection, and the lyophilic colloid added as the protective colloid.*

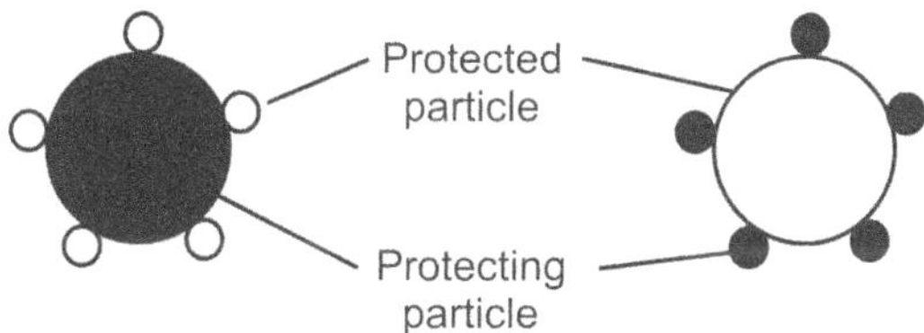

Fig. 1.12 Protection of a colloid.

Thus hydrophilic colloids such as gelatin, starch etc., act as protective colloids for the gold sol and prevent their precipitation by the addition of limited amounts of NaCl.

For a comparative study Zsigmondy introduced a scale of protective action for different protective colloids in terms of **gold number**. *The gold number is the weight in milligrams of a protective colloid which checks the coagulation of 10 ml of a given gold sol on adding 1 ml of 10% sodium chloride. Thus smaller the gold number, greater is the protective action.* The gold numbers of few commonly used protective colloids is given in Table 1.7.

Table 1.7 Gold numbers of some protective colloids.

Gelatin	0.005	Starch	25
Egg Albumin	0.08	Gum arabic	0.10

The data in Table 1.7 suggests that gelatin is a very effective protective colloid.

1.6.13 The Donnan Membrane Equilibrium

The Donnan membrane equilibrium is a special case where a true solution of an electrolyte is separated from a salt which has a non-diffusible ion by a membrane, and at equilibrium the concentration of diffusible electrolyte on both the sides of membrane was found to be different. In case of colloids the nondiffusible ion is the charged colloidal particle.

Thus when we take C_1 moles of sodium chloride as one side of the membrane compartment 'A' and sodium salt of non-diffusible colloidal ion, R^-, on the other side of the membrane (compartment B) after allowing the system to attain an equilibrium, one finds an increase in the concentration of diffusible electrolyte in the compartment 'B'. If x moles/lit of sodium chloride diffused into the compartment 'B' dropping the concentration of electrolyte in compartment 'A' from an initial value of C_1 to $C_1 - X$, then the equilibrium may be represented as

	Compartment 'A' *Sodium chloride*		*Compartment 'B'* *Sodium salt*		
Ions	Na^+	Cl^-	Na^+	R^-	Cl^-
Initial conc. (moles lit^{-1})	C_1	C_1	C_2	C_2	0
At equilibrium (moles lit^{-1})	C_1-X	C_1-X	C_2+X	C_2	X

Donnan (1911) gave explanation for the migration of diffusible ions from one compartment to another compartments separated by a membrane that restricts the movement of non diffusible ions. If no non-diffusible ion is present in the system the electrolyte is present equally in both the compartments. The driving force behind this diffusion is known as **Chemical potential** and for an ideal electrolyte it is equal to the product of concentration of the ions in the electrolyte. At equilibrium the chemical potential in both the compartments is equal.

When non-diffusible electrolyte of concentration C_2 is separated from diffusible electrolyte of concentration C_1 as represented above the chemical potential of diffusible ions at equilibrium are equal

Therefore $(C_1\text{-}X)\,(C_1\text{-}X) = (C_2 + X)\,X$

or $$X/C_1 = C_1/(C_2 + 2C_1)$$

X/C_1 is the fraction of sodium chloride (diffusible electrolyte) that has diffused through the membrane. Higher the concentration of non diffusible electrolyte present low is the fraction of electrolyte diffused. From the above expression it is also clear that in the absence of non-diffusible electrolyte, half of the NaCl migrates from Compartment 'A' to Compartment 'B'.

Total concentration of ions at equilibrium in compartment '*A*' is $2(C_1–X)$ and in compartment 'B' it is $2(C_2 + X)$. Hence the concentration difference across the membrane is

$$= 2(C_2 + X) - 2(C_1 - X)$$

$$= 2(C_2 - C_1 + 2X)$$

As this system is comparable to an osmometer, the osmotic pressure 'B' of the solution is given by

$$\pi = 2(C_2–C_1+2X)\,RT$$

but $$X = \frac{C_1^2}{C_2 + 2C_1}$$

Therefore,

$$\pi = \left\{\frac{2\left(C_2 - C_1 + 2C_1^2\right)}{C_2 + 2C_1}\right\} RT$$

By rearranging we have

$$\pi = 2\,RTC_2\,\frac{(C_1 + C_2)}{2C_1 + C_2}$$

When $C_1 >> C_2$ the above expression reduces to

$$\pi = RTC_2$$

Thus a measurement of osmotic pressure allows us to calculate the concentration of a non-diffusible electrolyte or a charged colloidal particle, there by its molecular weight.

1.7 Emulsion(*l*/*l*)

When liquids form both the disperse phase and dispersion medium of a colloid, then that colloid is said be an ***emulsions***. Any two partially miscible liquids can yield an emulsion. Either liquid can be dispersed in the other, Fig. 1.13.

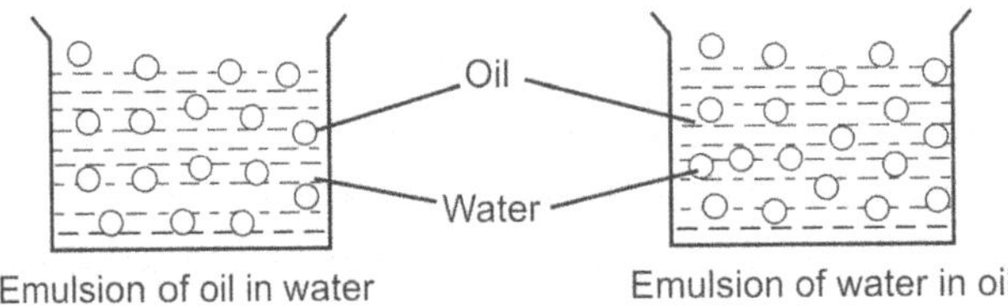

Fig. 1.13 Emulsion of oil in water and water in oil.

Thus we can have an emulsion of oil in water by shaking vigorously few drops of oil in excess water. Similarly an emulsions of water in oil can be obtained by taking few drops of water in oil. These are illustrated in Fig. 1.13.

The emulsions formed entirely of oil and water are quite unstable, as the molecules of the dispersed phase collide with one another and separate out into two layers on standing. Stability of an emulsions is achieved by the addition of a third substance known as **emulsifying agent or emulsifier.** The commonly used emulsifiers are gelatin, soaps and gum arabic. The role of emulsifier, according to Brancraft, *is to surround the particles of dispersed phase and reducing the surface tension on one side of the water oil interphase than the other.* As the spherical surface has always a minimum area the emulsion particle forms a spherical envelope that inhibits the colloidal particles come closer, Fig. 1.14.

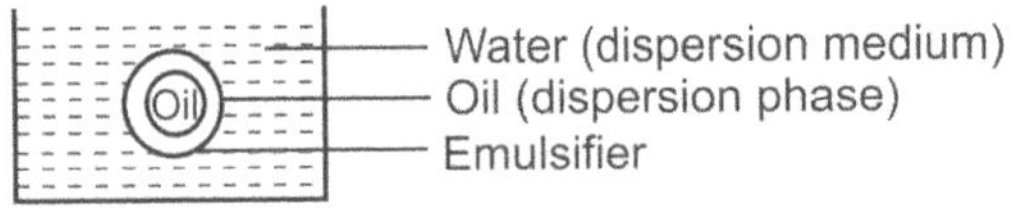

Fig. 1.14 Oil particle dispersed in water.

Certain finely divided solid particles can also act as emulsifiers. They are wetted by both the liquids. The liquid that wets more forms the dispersion medium and the liquid that wets less is protected by the solid particles, Fig. 1.15.

Fig. 1.15 Protective action of soot particles.

We get an emulsions of water in kerosene on shaking them along with some soot particles.

1. **Preparation of emulsions :** Vigorous shaking or stirring of few drops of the disperse phase with dispersion medium in the presence of a liquid emulsifier or finely divided solid particles is sufficient to given an emulsion.

   Emulsions can be obtained by adding a solution to an excess another solvent, which readily dissolves the solvent of the added solution but keeps liquid solute undissolved. Thus an emulsion of oil in water is obtained by adding a dilute solution of oil in alcohol to excess of water.

   Passing a vapor of oil into water containing emulsifying agent also gives an emulsion.

2. **Properties of emulsions :** The properties of emulsions are same as that of colloids. To avoid repetition of electrical, optical, brownian movement etc., only few selective properties are discussed here.

   A. The stability : The stability of an emulsion is determined by two factors (i) the strength of the protective layer of the emulsifying agent (ii) the charge on the emulsoid (dispersed particle).

   B. Coagulation : Coagulation or demulsification takes place by adding a substance that uncovers the emulsifier film. This can be affected by :

   (i) By adding an electrolyte that neutralizes the charge on emulsoid.

   (ii) By excessive heating, cooling and centrifuging.

   (iii) By passing current.

   (iv) By adding dehydrating agent to an emulsion of water in oil.

C. **Vision :** When the liquids participating in the formation of emulsion have refractive indices of the same order, the emulsion looks transparent and when they are different the emulsion is opaque or turbid.

3. **Uses of emulsions :** Emulsions of a substance has more surface area than the pure substance. The large surface area codliver oil emulsions helps in better assimilated by the stomach than pure oil. Similarly phenyl and lysol act as better disinfectants when emulsified in water.

1.8 Gels

Colloidal systems of *liquids dispersed in solids are called* ***gels***. They are obtained by coagulation of the colloidal solutions yielding a semisolid mass. Thus on heating agar-agar with water first we get a colloid and on further boiling this solution agar-agar molecules come closer trapping water in a honey combed like structure and giving a semisolid mass of agar-agar gel, Fig. 1.16.

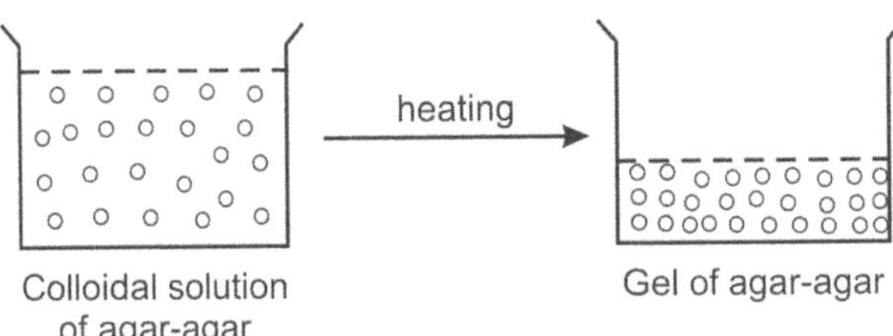

Fig. 1.16 Formation of a gel.

Similarly on adding alcohol to a saturated solution of calcium carbonate in water initially produces a colloidal solutions which coagulate to give a semisolid mass of $CaCO_3$ in which alcohol is trapped. This gel of alcohol in calcium carbonate, called solid alcohol, is used as solid fuel in picnic stoves, military camps etc.

1. **Properties of gels :** On standing gels loose the medium (e.g., water) and shrink. This shrinking of gel with simultaneous exudation (sweating) of the gel is called **synergisis** or weeping of the gel. Gels are classified into three categories.(i) Elastic gels (ii) Rigid gels (iii) Thixotropic gels.

   (i) **Elastic gels :** These gels are defused by the application of a force but retains its original shape on the removal of applied force. e.g., Starch, gelatin, fruit jam and jellies.

(ii) **Rigid gels :** There is no deformation on applying force. The strong covalent bonds between molecules give the rigidity of shape, e.g., silicic acid or commonly known as silica gel.

(iii) **Thixotropic gels :** These become semisolid when kept aside but become liquid when agitated. Kiesel gel forms a thin paste on standing but readily flows on shaking. Solution of high polymers exhibit these properties.

2. **Uses of gel :** Commercially many gels are available in the market. The commonly used boot polishes, polishing waxes, soaps, some brands of tooth pastes and various foodstuffs as petha are all examples of gels.

1.9 Application of Colloid Chemistry

Applications of a class of colloids namely emulsions and gels have already been discussed in subheads 1.7 and 1.8. Application of other forms is given below.

1. **Colloids in medicine :** Medicines in the form of colloids are easily absorbed by the body tissues. Some of the essential metals like silver, gold which can not be absorbed in the metallic form are administered as colloids.

   (i) Argyrol and protargal are protected silver colloids used in the cure of granulations.

   (ii) Colloids of gold, calcium, manganese, etc., are administered as intramuscular injections to enhance the vitality and cure tuberculosis and rickets.

2. **Colloids as pesticides :** Sulphur colloids are sprayed on plants to kill various germs.

3. **Smoke precipitation :** Smoke in the industrial exhausts contain particulate carbon and dust in the colloidal form and these are removed by the application of 30,000 volts and then allowed to escape into the atmosphere through the chimney (Fig. 1.17).

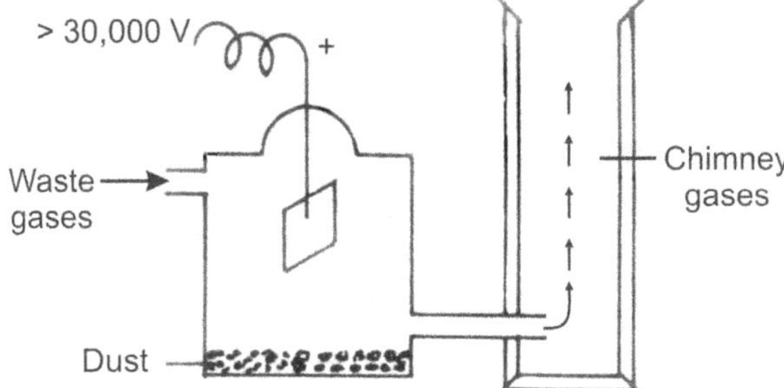

Fig. 1.17 Cotrel smoke precipitator.

4. **Dialysis :** Kidney removes urea and other unwanted electrolytes from the impure blood by passing it through a dialyzing membrane which holds back the colloidal blood. A kidney failure sends these toxic wastes into the blood. Patients having kidney problem are dialyzed externally.

5. **Purification of water :** The clay particles present in the muddy water are removed by the addition of alum (contains Al^{3+} ions) which coagulated them leaving a clear water.

6. **Sewage disposal :** The dirt particles present in the sewage water are in the colloidal form. These are removed in the presence of a high voltage and the clear water free from sewage sludge is let out into the irrigation fields. The sludge is also used as a manure.

7. **Cleaning action of soap :** A cloth gets dirty by the sticking of grease and oil particles which in turn adsorb the dust from the air. Soap is a sodium salt of fatty acid.

It has a polar end and a non polar fatty acid chain. Polar end is water soluble (hydrophilic) while non polar end is oil soluble (hydrophobic). When soap is applied to the cloth the oil soluble end of soap emulsifies with the grease along with the dirt particles sticking to it. Polar ends are present away from the cloth and are water soluble. Hence after the application of soap to the cloth, washing with excess water removes the emulsified grease, dirt particles by solubilizing the polar end of the emulsified soap, Fig. 1.18.

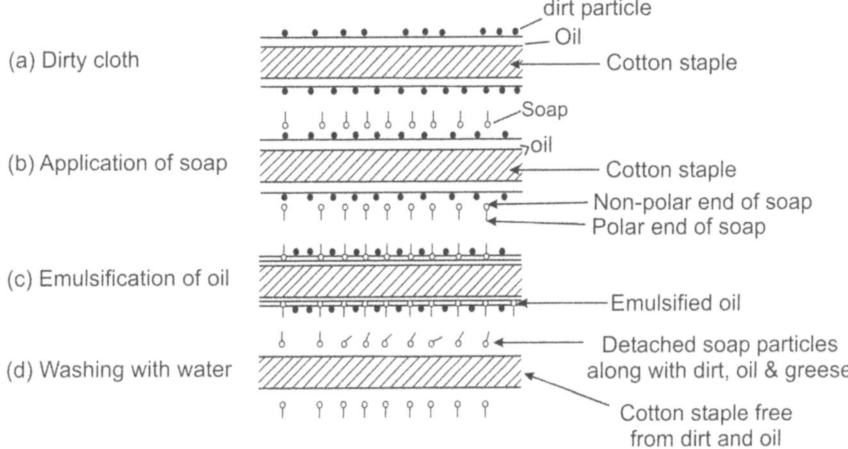

Fig. 1.18

8. Other Miscellaneous Uses :

(a) **Photographic plates :** Silver bromide emulsions are coated on the photographic plates, films etc.,

(b) **Formation of deltas :** Rivers carry silt and other mud particles with them which are negatively charged, when the river water meets sea, the electrolytes in the sea water coagulate the silt forming fertile delta lands.

(c) **Industrial products :** Various industrial products as inks, boot polish, creams, paints, rubber lubricants are all colloidal particles.

(d) **Rubber industry :** Latex is a negatively charged colloidal solution. Coagulation of latex yields rubber.

(e) **Dairy industry :** Dairy products as creams, butter, cheese, milk are all colloidal solutions.

(f) **Leather Industry :** Skins and hides have proteins in them in the colloidal state. Skins are first soaked and put in a colloidal solution of tannin, a negatively changed colloid. Tannin enters the skins and mutual coagulation of tannin and protein takes place giving hardness to the leather.

(g) **Stopping bleeding :** Blood is a colloid. So shaving cuts and other bleedings are stopped using alum which coagulates the blood forming a clot.

9. Explanation of some interesting observations

(a) **Blue color of sky :** Sun light falling on the earth is partially reflected back by the earth's surface. On its way back this light hits the dust particles present in the atmosphere. The colloidal particles of dust are quite small and are capable of reflecting only blue radiation while the radiations of other colors escape through the sky. This given the sky a blue color.

(b) **Tail of comets :** Comets are celestial bodies moving at tremendous speeds. During their motion they leave behind tiny solid particles suspended in air. The light falling on them is scattered leaving a Tyndall cone like structure which we call the tail of comet.

(c) **Artificial rains :** Clouds are colloids of tiny water droplets suspended in air. Artificial rains are caused by spraying a sol or an electrolyte that coagulates these water droplets into rain and this process is called seedling of clouds.

Short and Essay Type Questions

1. What are distinguishing features of the colloidal state: Describe the optical properties of colloids

2. Write a brief notes on

   (i) Protective colloids

   (ii) Zeta potential

3. Write a brief notes on

   (i) Emulsion and

   (ii) Gels. Give two examples each ?

4. What do you understood by the terms protection of colloids? Define the term 'Gold Number'. Three organic substances M.N,P have their gold numbers as 0.5 0.05 and 0.005 respectively which one of these substances has the greatest protective action

5. Write short notes on

   (a) Tyndall effect (b) Cataphoresis (c) Hardy -Schultz rule

6. Explain fully what happens when a colloidal solution of gold is treated as follows.

   (a) an electrolyte is added to it.

   (b) A beam of light is passed through it.

   (c) it is brought under the influence of an electric field.

7. Distinguish between a true solution and a colloidal solution. Describe briefly any two methods of preparing colloidal solutions. Explain the optical and electrical properties of colloids.

8. What is a protective colloid ? Describe its working ?

9. Discuss the origin of the charge on colloidal particle. How would you measure the charge and how is it influenced by addition of electrolyte ?

10. Give an account of methods of preparation and purification of colloidal solutions ?

11. Write a short notes on

    (a) Ultra Microscope

    (b) Ultra centrifuge

    (c) Charge of a colloidal and particle and its stability

    (d) Bredig's arc method for the preparation of a colloid

12. Explain the following;

    (a) A sulphur sol is coagulated by adding a little electrolyte, where as a gelatin sol is apparently unaffected ?

    (b) A colloidal solution is stabilized by to addition of a gel.

13. Give a detailed account on the applications of colloidal chemistry ?

14. Give an account for the stability of colloids interms of electric double layer.

15. How is colloidal sulphur prepared ? Write a note on the optical and kinetic properties of colloids.

16. What are lyophobic and lyophilic colloids.

Self Test Questions

1. The charge on colloidal particles can be distributed by
   (A) Tyndall effect (B) Brownian movement
   (C) Osmosis (D) Cataphoresis
2. The precipitation of colloidal solution is difficult in the presence of
   (A) Nacl (B) $AlPO_4$
   (C) Aerosol (D) Gelatin
3. Fog is a colloidal solution of
   (A) Liquid dispersed in gas (B) Gas dispersed in liquid
   (C) Solid dispersed in liquid (D) Liquid dispersed in liquid.
4. Smoke is an example of
   (A) Emulsion (B) Gel (C) Aerosol (D) Jelly
5. The Brownian movement increases as
   (A) The size of the colloidal particle reduces
   (B) The size of the colloidal particle increases
   (C) In lyopholic solution
   (D) In lyophillic solution
6. Cheese is an example of
   (A) Gel (B) Emulsion
   (C) Aerosol (D) Hydrophil
7. In Coagulating a positively charged colloidal particle the coagulating power of PO_4^{3-} is
   (A) Greater than SO_4^{2-} (B) Less than SO_4^{2-}
   (C) Less than Br^- (D) Less than HCO_3
8. Gold number is minimum in case of
   (A) Nacl (B) Silica gel (C) Starch (D) Gelatin
9. Coagution 10ml of gold solution by 10% Nacl is just prevented by the presence of 0.25 g of starch. The gold number of starch is
   (A) 0.25 (B) 2.5 (C) 250 (D) 0.025
10. Example of positively charged solution
   (A) $Fe(OH)_3$ (B) Ag_2S_3
   (C) Protein (D) Gold colloid

Chapter 2

Environmental Chemistry

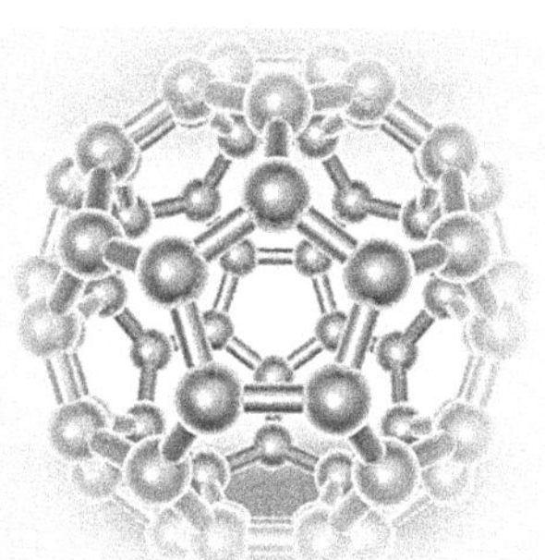

2.1 Introduction

Consequences of introduction of any substance that disturbs / destroys the purity or ecological balance of the nature is the subject matter of environmental chemistry. Study of environmental chemistry covers the sources, transport, absorption, biological control and remedies of various chemical species pumped into air, water, soil etc., out of human activity. Universe was never free from pollution but was in a low profile till the advent of industrialization, started just 300 years ago. In these three centuries, especially in the last century, earth was dumped with several hazardous chemicals, radioactive waste, concrete jungles, smoking chimneys, vehicles, etc., and polluted to such an extent that billions of yesteryears could not do.

The mankind has now marched to a point of no return where problems connected with population explosion and materialistsic comforts of the modern civilization cannot be sacrificed and human race is left with the only option of using their means judiciously and change their life styles to cause minimum pollution. This awareness is the heart of environmental chemistry.

We notice four main reasons for the present difficulties;

(i) The world population has increased enormously in the last 100 years. For example; since our independence the population of our country has tripled. Increase in population, demands more food, more clothing, and more housing. So first step is to arrest the population growth at the present level.

(ii) Increasing energy requirements : The present state of technology demands the burning of fossil fuels, viz., coal, petrol, diesel etc., which in turn inject different hazardous gases into atmosphere. Nuclear fuels are no replacement for fossil fuels as they are also limited and pose the problems connected with disposal of nuclear waste. So we must strive hard to make non-conventional, non-polluting and non-depleting energy sources, like solar energy, wind energy, etc., technologically viable.

(iii) Urbanization increases concentration limits of the pollutants like CO, N_2O, CO_2, dust, human facie, to a level that nature cannot handle in its normal recourse. Hence the big cities must be discouraged and smaller towns should be encouraged.

(iv) The human race should be taught of the pollution hazards, use of biodegradable materials, need to restrain human activities leading to indiscriminate release of pollutants.

2.2 Pollutant and Contaminant

We have a small difference between a pollutant and contaminant.

Pollutant : *Pollutant is a naturally existing substance introduced in greater than natural abundance into environment as a result of human activity, and has a detrimental affect on the living world*, e.g., Toxic metals, carbon monoxide, sulphur dioxide, carbon dioxide, dust, etc.

Contaminant : *Contaminant is a substance of synthetic origin or not present in a described surroundings and is introduced into environment as a result of human activity*. Example : Oil spills that often take place due to wreckage of ships or tankers and leakage of harmful gases like MIC (Methyl isocyanate) that claimed a number of lives in Bhopal tragedy are the cases of few harmful contaminants. Polyethylene textiles, plastic carry bags, disposable syringes that are made out of biologically nondegradable materials are a menace in the environment but not directly harmful contaminants. When a contaminant assumes a detrimental effect on the environment or human health then the contaminant is termed as a pollutant.

2.3 Types of Pollution

Depending upon the place of occurrence, pollutants are classified into (i) Air pollution, (ii) Water pollution, (iii) Soil pollution.

2.3.1 Air Pollution

In the high school chemistry itself we are clear that air is composed of 78.09% of N_2, 20.94% of O_2, 0.93% of Ar and the remaining trace volume fractions of gases like CO_2, CH_4, He, Ne, etc. Introduction of any new toxic gas or imbalancing the natural composition may be termed as **air pollution**. The most important and most frequently occurring air pollutants, arranged according to their physical state, are given in Table 2.1.

Table 2.1 Air pollutants and their physical states.

Pollutant	Physical state
1. Carbon monoxide, CO	Gaseous phase
2. Sulphur oxides, SO_X	Gaseous phase
3. Nitrogen oxides, NO_X	Gaseous phase
4. Natural gas	Gaseous phase
5. Mercury vapor	Gaseous phase
6. Solvent vapors	Gaseous phase

Contd... Table 2.1

Pollutant	Physical state
7. H_2SO_4, HCl, etc., acids	Liquid phase
8. Higher hydrocarbons	Liquid phase
9. Smog	Liquid phase
10. Dust (metallic, mineral, cement, pollen, etc)	Solid phase
11. Soot	Solid phase
12. Fly ash	Solid phase
13. Asbestos	Solid phase

Out of the big list of pollutants given in Table 2.1, the five primary pollutants which contribute to 90% of the global air pollution are (i) CO, (ii) NO_X, (iii) SO_X, (iv) Hydrocarbon and (v) Particulates.

A detailed account of the nature of above pollutants, their sources, sinks and detrimental effects are discussed in the following sections.

Man made air pollution in urban areas in often referred to as smog. The major source of air pollutions comes from automobiles and contribute for approximately two third of CO, one half of the hydrocarbons and nitrogen oxides. In addition engineering industries, chemical industries and power plants are responsible one fourth of the total air pollution. The air pollutants mainly cause lung and throat irritation and damage to sensitive membranes like eyes.

Most of the air pollutants are destroyed/removed by the microorganism present in the soil and some find their way into biological cyclic path way. Bhopal MIC gas leakage in 1984 which affected nearly 2,00,000 Bhopal residents, and Chermobyl nuclear disaster in Russia, on April 26, 1986 which shook entire Europe are two unforgettable tragedies in the history of mankind that always remind us the consequences of overlooking measures to curb the air pollution.

Preventive measures for Air Pollution

1. Urbanization should be discouraged.
2. Industries should be scattered and surrounded by sufficient green belt.
3. Stringent air pollution control measures should be imposed on chemical industries to avoid any leakages of volatile substances and gases. Public should be taught to combat any undesired consequences and alarmed as and when any untoward incident occurs.

4. Nuclear reactors should be either completely banned or fool proof measures be taken in running them.
5. A total ban on the usage of carcinogenic pesticides and fungicides should be imposed.
6. Automobiles should be fitted with catalysts and constant vigilance should be kept on their exhausts.
7. Non-conventional and safe energy sources should be harnessed and fossil-fuel technology should soon be replaced.
8. Workers working in industries like cement, steel, mining and asbestos are highly prone for lung disorders due to regular exposure to dust and should be provided with proper gas masks. Similarly chemical industry workers should be provided with suitable gas masks to fight any untoward incident.
9. The masses should be educated on the dire consequence of air pollution and other types of pollution and dealing with them.

2.3.2 Water Pollution

Human civilizations have formed, grown and flourished only on the banks of water resources like rivers and lakes. Although major portion of the earth is covered with water (70.84% of total surface) the mankind is still struggling hard for adequate supply of safe drinking water (barely 0.4% of total earths surface is covered with fresh water) and general purpose water. Urban and industrial growth creates complex problems of water pollution which seriously undermine the quality of life. Realizing the importance of effective management of water, United Nation has declared 1981-1990 as the International Drinking Water Supply and Sanitation Decade.

Classification of Water Pollutants

Addition of any undesired and harmful substance into water resources of biosphere is termed as ***water pollution***. The large number of water pollutants may be broadly classified under the following categories.

(i) Organic pollutants

(ii) Inorganic pollutants

(iii) Radioactive wastes

(iv) Sediments

(v) Thermal pollutants.

1. **Organic Pollutants :** Organic matter, industrial wastes from food processing industries, tanneries, paper industries, sugar industries, sewage, pesticides, agricultural runouts (fertilizers), phenols, detergents, etc., released into water resources by direct and indirect means pollute water. Dissolved oxygen (D.O) is very essential and responsible for the existence of most of the aquatic life. D.O. in natural water is only .5 ppm. The organic pollutants run into water resources, undergo degradation by bacterial activity at the cost of precious D.O.

$$C + O_2 \longrightarrow CO_2$$

The net effect is severe shortage of DO which lead to the death of fishes and other aquatic life. In addition oil spillages cover the external surface and deprive D.O. for marine life. Sewages, food processing units and agricultural run off (like unused urea and other fertilizers) provide plant nutrients technically called eutrophication (Greek : well nourished). The excessive plant growth not only presents an unaesthetic appearance but competes in the consumption of D.O.

Pesticides like DDT are not biologically degradable and enter the life cycle through fishes and prove very harmful to entire biosphere.

Most of the currently used surfactants and detergents are biodegradable. However, the foam imparts an unaesthetic appearance. The biodegraded and hydrolysed phosphates act as nutrients and cause eutrophication by contributing for excessive growth in plants.

2. **Inorganic Pollutants :** The presence of water soluble inorganic salts, mineral acid slurries, let outs into drains and finely divided toxic metal wastes, pose three general effects namely the salinity, the acidity and the toxicity.

A large number of industries and ore processing units make use of many water soluble salts which are eventually disposed off into waste-water effluents. This contributes for the raise in salinity of water and there by changes in environment. Similarly mercury let out is converted by anaerobic bacteria into methyl mercury, a toxic compound of great concern.

Acid mine drainages have been the primary source of concern that increase acidity (fall in pH) of water reservoirs available at the mining places. At pH levels below 4.0 all vertebrates most of the invertebrates and many microorganism get killed. The lowering of pH also lead to shedding of leaves and damage of crops, corrode metal structures and monuments and make the boats unusable.

The fine metal dust, insoluble toxic metal oxides are carried from the industries and mining places through the down-stream and pose a serious concern by entering into the life cycle.

3. **Radio Active Wastes :** Cell damages and modification of DNA by ionizing x-rays and r-rays are well proven facts. With the growing nuclear power plants, nuclear reactors, under ground test explosions and use of radio-active isotopes in medical, industrial and research establishments have added a new dimension to water pollution. Water polluted by nuclear and isotopic wastes are hazardous owing to the ionizing radiations emitted by them which alter the genetic codes.

4. **Sediments :** Sediments enter water during soil erosion processes. Although, natural erosion processes may be regarded as a type of pollution, man made pollution, in fact, is playing a major role. In the natural course sediments enter water through storms and floods. Human activities like deforestation, overgrazing of pastures, strip mining, cement industries, roads and buildings, urbanization, etc., are heavily contributing to sediment formation. Sediments result in shallow lakes and rivers which contribute for easy heating of waters not suitable for aquatic life. Sediments while settling, blanket the fish nests and thereby contribute for the fall in fish population.

5. **Thermal Pollution :** Water resources close to thermal power plants are used in the generation of steam and the condensing water, at around $40^{O}C$, is allowed to flow into the reservoir. Heating of water drastically reduces dissolved oxygen content that leads to elimination of fishes and other aquatic life.

Preventive measures for Water Pollution

1. Urbanization should be discouraged.
2. Industries should be scattered.
3. Effluent treatment plants should be forced on industries and public sanitary organizations. Stringent controls should be laid out and heavy penalties be imposed on violators.
4. Public awareness should be created on the importance of protecting wells, lakes and other water resources and in keeping them away from sewages and washing places.
5. Farmers should be taught proper and controlled use of pesticides, fertilizers, etc. Minimum amount of agricultural runouts be allowed into drinking water streams.

6. In the villages hand pumps lifting water from large depths and avoiding entry of surface water should be setup.
7. Soil erosion can be avoided by undertaking a massive afforestation programme.

2.3.3 Soil Pollution

The term soil essentially means the upper layer of earth. Soil differs from air and water in its functioning as it is localized and is of motionless solid state. Any pollutant that is added to air or water is rapidly spread from one place to another and on one hand, they propagate the toxic effects from place to place and on the other they dilute the undeserved effects. Therefore, the danger of accumulation of pollutants and acute permanent effects left by them is maximum in case of **soil pollution**.

Soil pollution is effected in three different channels viz., air-borne, water borne and direct dumping of solid wastes.

Out of the above three channels the role of air-borne solid pollutants is the greatest. For example, industrial oxide gases generated by U.S industries travel all the way to Canada, get on to the soil surface in the form of acid rains. Similarly, Nuclear fallout from Chernobyl in Russia was reported to have crossed Europe continent and the Atlantic ocean and made its presence felt in the american subcontinents.

Water is a good carrier of suspended ash and soot particles through sedimentation and through rain fall.

Industrial wastes, sewage solids, domestic garbage disposal, nuclear waste disposal, fly-ash from thermal power stations is of growing concern to ecologists. Fly-ash has recently found its use in the preparation of cement and building materials but most of the other solid wastes still have no utility. Biodegradable organic wastes and sewage wastes have become breeding places for mosquitoes and insects, while non-degradable plastic carry bags, pottery pieces, building materials, syringes, etc., are spread out on vast areas. Further these solid wastes are washed by wind and rain into precious drinking water reserves and causing diseases.

Chemical pollution of the soil is caused by open mining, use of non-biodegradable pesticides like DDT, BHC, Aldrin, etc., and chemical fertilizers. They are absorbed and retained by the soil which then enters the life cycle through plants.

Preventive measures for Soil Pollution

Soil pollution may be controlled by taking the following environment friendly steps :

1. Use of plastics and other non-biodegradable materials should be limited.
2. The solid harmful wastes should be crushed in compactors, properly protected and preserved until new technology for their safe disposal is found.
3. Use of biodegradable material should be encouraged.
4. Air-borne and water borne wastes should be controlled to avoid further deterioration of soil quality.
5. Sewage wastes and agriculture wastes should be used for the production of biogas which after proper treatment becomes harmless biomanure.
6. Afforestation should be carried out to reduce soil erosion.
7. Occurrence of acid rains may be reduced by minimizing the use of fossil fuels and then any acid soils formed be naturalized by the additions of lime.

2.4 Zones of Environment

Environment consists of five zones pyrosphere, lithosphere, hydrosphere, biosphere and atmosphere (Fig. 2.1).

2.4.1 Pyrosphere

It is the inner molten part of the earth which has little impact on the environment except when it out bursts in the form of volcanic material.

2.4.2 Lithosphere

The solid crust of earth is called lithosphere. Upper part of the soil is sometimes referred to as **pedosphere**. Lithosphere comprises both minerals and organic matter.

2.4.3 Hydrosphere

The water cover, viz., lakes, rivers, seas and oceans form the hydrosphere. The aquatic plants and animals live in this zone. Living organism is found in the sea down to a depth of 4 km, although 0-100 m deep sea is rich in aquatic life and supports more animals than even land.

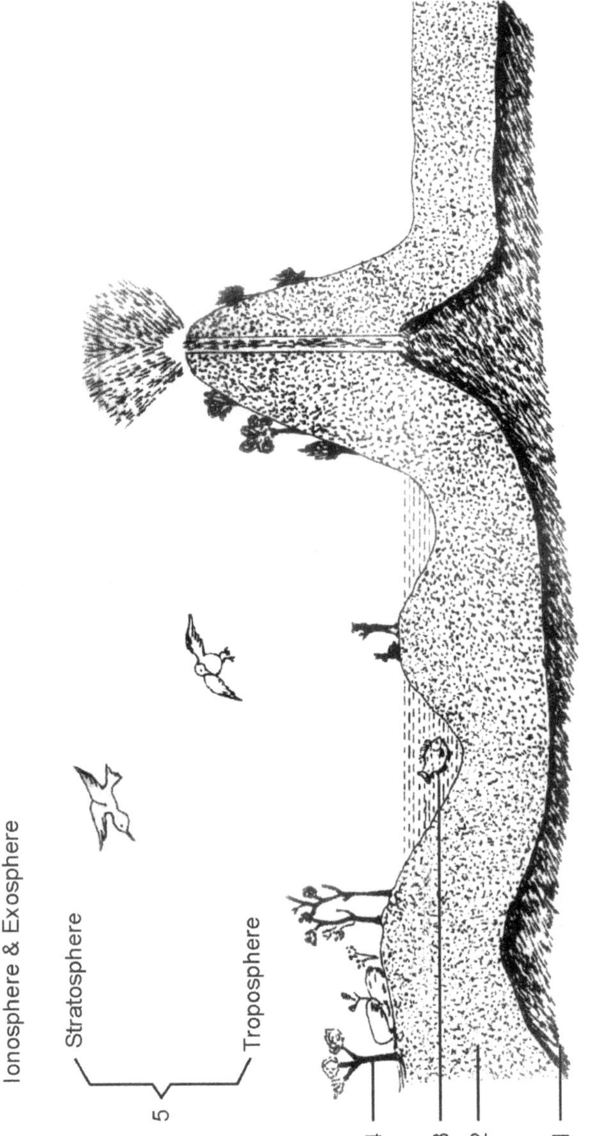

1. Pyrosphere 2. Lithosphere 3. Hydrosphere 4. Biosphere 5. Atmosphere (Troposphere + Stratosphere)

Fig. 2.1 Zones of Environment.

2.4.4 Biosphere

Biosphere is the entity of the constantly developing organic life. All the living matter viz., the plants, fishes, animals and microorganism constitute the biosphere. Vegetation (plants) absorbs the CO_2 and light and convert them into organic matter that serves as food for animals and under suitable geophysical conditions also converts into fossils fuels like coal, oil, lignite, etc. The plant kingdom is responsible for storing of energy while the animal kingdom is on the energy spending spree. Thus the biosphere is responsible for the energy circulation of the atmosphere. Man, a part of this biosphere, is the greatest exploiter of the ecosystem of the earth and solely responsible for the pollution of this planet.

Elements and compounds are distributed in the lithosphere (as different rocks and mineral deposits) and in hydrosphere (as different forms of waters) in an irregular way. These elements and compounds are transported and distributed at different places by animals, seeds, spores etc. Some plants are even responsible for the concentration of the elements. Thus cornacuspongiae contain 30% Si in comparison to 0.00005% of Si content of sea water.

2.4.5 Atmosphere

The atmosphere is further divided into **Troposphere**, **Stratosphere**, **Ionosphere** and **Exosphere**.

(A) **The Troposphere :** It is a thick cover of gases around earth that supports the biosphere. This protective cover of gases extending to a height of 8 km near poles and 18 km near equator absorb most of the cosmic rays from outer space and U.V. and ionizing radiations from the sun. This allows transmission of only near UV and visible rays required for the survival of animal kingdom. The troposphere with its water storing ability, also maintains the heat balance and biological circulation of water.

(B) **The Stratosphere :** Stratosphere lies above troposphere and extend to about 80 km from the earth. It is a low temperature zone free from dust, clouds and is rich in ozone. Within stratosphere the temperature is almost constant. Ozone readily reacts with far ultra-violet rays producing oxygen and there by relieves biosphere from the harmful effects of U.V. radiation.

Above stratosphere we have **ionosphere** and **exosphere** which are of less interest at this stage.

2.5 Classification of Pollutants

Pollutants, basing on their origin, are classified into inorganic pollutants and organic pollutants.

2.5.1 Inorganic Pollutants

In this category we find inorganic particulate matter like dust and metal oxides, gases like carbon dioxide, carbon monoxide, sulphur dioxide, chlorine, ozone, nitric oxide, etc., salts in their soluble and insoluble forms and finally as acids and oil droplets.

(A) Particulate Matter : The main levels of particulate matter varies from .10 :g m^{-3} in the village atmosphere to 2000 :g m^{-3} in the polluted urban areas. Concentration of particulate matter is especially high at places where cement and steel industries are present and at the traffic centers. Their sizes lie in the range 0.1-10 :L.

Particulate matter has both good and bad effects on the nature;

1. They are responsible for cloud formation
2. They reflect the heating infrared radiation off the earth's atmosphere and determine the heat balance
3. They provide nuclei for ice formations
4. They also absorb some of the hazardous gases, acid fumes etc., Fig. 2.2.

Sources : Smoke is a particulate matter of carbon. Around 2000 tonnes of particulate matter is thrown into atmosphere as a result of human activities. Volcanic eruptions and spraying of salt on the sea coasts are the natural source of particulate matter. Metal oxides and silicates escape into atmosphere during mining and smelting. Carbon particulates enter the atmosphere from the exhausts of transport vehicles and industrial smoke pipes. Asbestos and mica particulates are a big hazard in the mining areas and asbestos sheds. Fly ash is released into atmosphere by the thermal power stations and smelters.

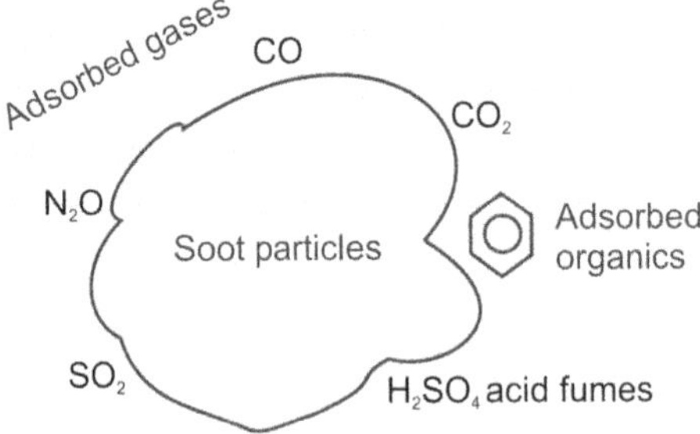

Fig. 2.2 Gases adsorbed by a soot particle.

Effects on Humans

(i) The first target of the particulate matter are the lungs. Particulate of < 3 : easily penetrate through respiratory track into lungs and cause several respiratory problems by blocking the lung capillaries.

(ii) Asbestos is a fibrous silicate. Asbestos mining workers and asbestos handling workers suffer from lung disorders like tuberculosis (T.B) and pulmonary fibrosis.

(iii) Airborne toxic metal particles like Be, Pb and other salts not only cause respiratory problems but are also carcinogenic.

(iv) Air borne soot particles adsorb the acid fumes and acidic oxides and corrode metals and historical monuments.

Control of particulate Emissions : Particulate matter is generally charged. Hence industrial particulates can be prevented from entering the atmosphere by passing the exhausts through chamber having highly charged plates. Mining workers must use masks that filter or settle the particulate materials.

(B) Carbon Monoxide (CO) : It is a colorless, tasteless and odorless gas. It is insoluble in water.

Sources : CO in general is produced due to partial combustion of carbon and carbon matter.

$$2C + O_2 \longrightarrow 2CO$$

Volcanic eruptions, natural gas emissions, germination of seeds, lightnings are the natural sources of carbon monoxide. Motor vehicles, agricultural burnings, forest fires, smelting and other industrial processes are the contribution of human activities. It is interesting to note that out of annual global emission of 350 million tonnes of CO, nature contributes only 75 tonnes and the major portion of 225 tonnes is due to ever energy craving human race.

Natural Sinks of CO : Nature has designed its own mechanism to keep under control moderate concentrations of CO. About 200 micro organism were isolated from the soil that use up CO.

Detrimental Mechanism of CO : CO is produced in large quantities in the metropolitan cities due to never stopping thick vehicular traffic and industrial exhausts. In these urban centers free unexposed soil surface is scarce. Hence we fail to exploit the natural sinks of CO mentioned above.

The biological effect of CO is due to its capability to replace O_2 from oxyhemoglobin (HbO_2).

$$CO + HbO_2 \longrightarrow HbCO + O_2$$

This renders haemoglobin incapable of transporting oxygen, a vital gas, to the cells.

Levels of 10-20 parts per million (ppm), usually present at traffic junctions are of medical concern. A concentration of 100 ppm diminishes vision and causes dizziness. Exposures to still higher levels of carbon monoxide, as in a closed garage are fatal. A person suffering from CO poisoning should be removed to a open place and treated with 100% oxygen mask.

Control of CO Pollution : Pollution due to CO may be minimized by using the natural sinks like open soil effectively. For this, concentration of industries and urbanization must be discouraged. Another important step in this direction is to develop technology to use pollution free non-fossil fuels like hydrogen or solar energy in running the automobiles.

(C) Nitrogen Oxides (NO_X) : In this category, we have to consider mainly two oxides of nitrogen. Nitric oxide NO is a colorless, odorless gas, while nitrogen dioxide NO_2 is a reddish-brown, pungent gas.

Source : They are produced by the high temperature combination of nitrogen and oxygen present in the air, as in automobile engines and industrial plants.

$$N_2 + O_2 \rightleftharpoons 2NO$$

$$2NO + O_2 \rightleftharpoons 2NO_2$$

Smog is a mixture of nitrogen oxides, hydrocarbons and other pollutants.

Sinks : Natural bacterial action is capable of removing almost all the NOx introduced by man, if distributed uniformly. But in urban atmosphere NO_x was found to be 10-100 times more than rural atmosphere, leaving the cities polluted with NO_x. Photochemical reactions also remove NO_x leaving nitric acid as end product.

$$O_2 + NO_2 \longrightarrow NO_3 + O_2$$

$$NO_3 + NO_2 \longrightarrow N_2O_5$$

$$N_2O_5 + H_2O \longrightarrow 2HNO_3$$

Impact of NO_x : The NO_x levels were found to be higher during peak traffic times and industrial hours. The yellowish clouds of smog irritate eyes and cause serious respiratory problems. These problems are particularly important in older people and patients with cardia-vascular diseases, bronchitis and asthma.

NO & NO_2 being lighter gases, easily reach stratosphere and readily react with the O_3 present there (Fig. 2.3). As O_3 provides an umbrella for the biosphere from far U.V. rays reaching earth, its depletion, thereby formation of **ozone hole**, will allow far U.V. rays reach earth without hindrance. This will in turn may kill the cells or alter the genetic code resulting in skin cancer or total disappearance of animal world.

The N_2O_5 formed contribute for acid rain formation, which corrode the metallic surfaces, buildings and historical monuments. We often read about the decoloration and corrosion of Taj Mahal due to acid fumes liberated by Mathura refineries. Further acid rains also lead to the death of forests and aquatic life.

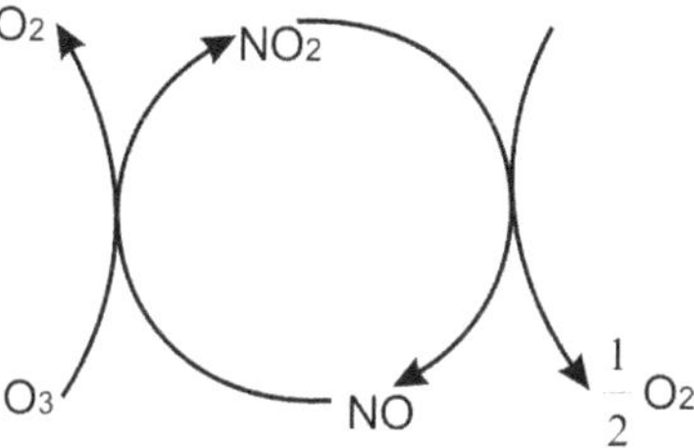

Fig. 2.3 Stratospheric reactions of nitrogen oxides.

Control of NO_x : As discussed in case of CO, urbanization and scattering of industries is an unequivocal solution. Fixing catalytic converters to vehicles to convert liberated nitrogen oxides into useful substances is another viable answer.

(D) Sulphur Dioxide (SO_2) : Sulphur dioxide is a colorless pungent gas. It is released into atmosphere by the combustion of sulphur containing coal and mineral oils. Generally SO_2 liberation is accompanied by further oxidation to give SO_3.

$$S + O_2 \rightleftharpoons SO_2$$

$$2SO_2 + O_2 \rightleftharpoons 2SO_3$$

This sulphur trioxide is highly soluble in water giving highly corrosive sulphuric acid. Similar to nitrogen oxides, sulphur oxides also contribute in destroying the ozone umbrella (Fig. 2.4).

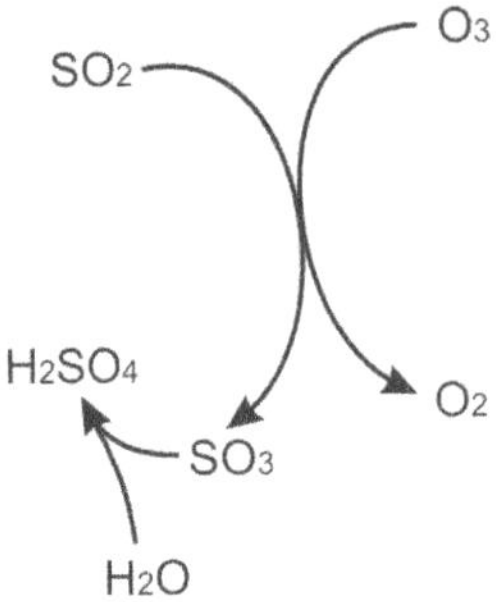

Fig. 2.4 Stratospheric reactions of sulphur oxides.

Sources of SO_2 : Volcanoes are responsible for 67% of SO_2 production but are highly limited to very few places in the world map. Though man made processes contribute only 33% of this toxic gas it is concentrated at all urban centers. Mainly SO_2 pollution is found in coal fired kilns and power stations. Unlike in the case of NO_x and CO, SO_2 pollution is not caused by motor vehicles.

Sinks of SO_2 : The sulphur dioxide and trioxide are highly soluble in water and are usually removed from air by the moisture and rains.

Impacts of SO_2 : Sulphur dioxide is responsible in thinning of ozone layer of the stratosphere.

The end product of oxidation of SO_2, i.e., SO_3, readily dissolves in water forming sulphuric acid. This acid is adsorbed on soot particles giving acid aerosol which is responsible for acceleration of decay of marble monuments and cement or line mansions.

Control of SO_2 : Sulphur oxides from industrial and power house flue gases are removed by passing through chemical scrubbers. The scrubber generally contains $CaCO_3$ which reacts with SO_2 giving unreactive $CaSO_4$.

$$CaCO_3 + 2SO_2 + O_2 \longrightarrow CaSO_4 + CO_2$$

(E) **Carbon Dioxide (CO_2) :** Carbon dioxide is transformed into sugars, proteins, fats, etc., by the process of photo-synthesis. Living organism gives off carbon dioxide in the process of respiration. Decay of leaves, wood and other organic matter also releases CO_2. So we find a perfect balance between production and consumption of CO_2 in the ecosystem.

Man, by burning fossil fuels and wood, converts them into carbon dioxide. This introduction of CO_2 into ecosystem is increasing with the increase in the civilization. Civilization has also led to felling of thick forests and their replacement by either the cultivated crops or concrete forests. While the biomass of 1 m^2 forest area can store/utilize 10-20 kg CO_2, the cultivated land stores merely 0.5 kg and concrete structures nothing. This leads to net raise of CO_2 in the atmosphere.

Sources of CO_2 : Burning of fossil fuels and wood are the main man made source of CO_2. In nature weathering of rocks, hot springs and volcanic eruption produce CO_2.

Sinks of CO_2 : The main natural sink of CO_2 is the photosynthetic routes.

Impacts of CO_2 : The i.r. radiation emitted by sun is not filtered in the stratosphere and 66% of it reaches the earths atmosphere. Under unpolluted conditions, i.e., at natural concentration of CO_2, half of this energy is reflected back into space. CO_2 strongly absorbs i.r. radiation and gets heated. It being a heavy gas can not escape into higher atmosphere and sticks to the earths atmosphere. The net result is the heating of earths surface. This phenomenon is called **Green House Effect**.

Carbon dioxide itself is not poisonous, but its formation generally uses up oxygen present in the system and animals die because of lack of enough oxygen.

(F) Toxic Metals : Many metals that are branded as toxic metals are in fact dietary trace metals required for human growth and activity. Their deficiency as well as excess (toxicity) both cause metabolic disorders and malfunctioning of glands.

1. Concentration of toxic elements at trace levels help in the growth and proper functioning of glands.
2. Concentration below the trace levels cause metabolic disorders.
3. Concentration above the trace levels or toxic levels pollute the system.

Thus metals like Hg, As, Pb are essential elements at trace levels and convert into toxic elements above trace levels. We discuss here only about toxic effects of lead and mercury.

Lead : Lead is both an occupational and an environmental hazard. Lead interferes with many enzyme catalyzed reactions in the body. For example the biosynthesis of the porphyrin component of heme is particularly sensitive to inhibition by lead.

Sources : Lead is a relatively abundant metal in nature. Silver is extracted from galena, a lead one.

The major source of air borne level is through combustion of leaded petrol. Tetraalkyl lead is added to petrol to reduce the knocking. Though about 200 :g of lead intake is excreted, still about 25 :g of lead intake is stored in bones every day. Due to chemical similarity of Pb^{2+} and Ca^{2+}, bones readily accept Pb^{2+} which on reaching soft tissues exhibit toxic effects. Lead poisoning is reversed by administering chelated calcium which on reaching bones displaces Pb^{2+} from the bones.

Lead also enters the human system through lead paints, lead batteries and lead coated pipes.

Impact of Pb : Lead poisoning damages the nerves and can cause mental retardation. Its symptoms include fatigue and poor nerve function in the extremities.

Mercury : Mercury is also a highly toxic but dietary trace element. It is used in the manufacture of thermometers, electrolytic cells and dental preparations.

Source : In nature mercury occurs as a trace component of many minerals. Cinnabar, a sulphide ore mercury, is the principal ore of Hg. Fossil fuels also contain some amount of mercury. Metallic mercury is poorly absorbed by the body but its vapors when inhaled are retained by the body. Hence mercury from broken thermometers should be immediately attended and disposed off by closing it in a corked bottle.

Impact : Exposure to mercury vapors result in vomiting, abdominal cramps and kidney damage. In children mercury poisoning causes several deformities. The biological effect of chronic mercury poisoning are irreversible.

Treatment : Mercury poisoning is treated with dimercaprol which complexes with Hg and leaves the body in the excretion.

2.5.2 Organic Pollutants

Every year thousands of synthetic chemicals are introduced into this world in the form of solvents, fibres, pesticides and medicines. Similar to toxic elements in spite of all the good they bring, these chemicals also show strong detrimental effects after a particular level. So indiscriminate usage of any organic compound should be avoided. We can categorize the organic pollutants into hydrocarbons, pesticides, medicine, carcinogens and toxic chemicals.

(A) **Hydrocarbons :** Petroleum is the starting material for most of the basic chemicals. Our industrial structure and energy requirements also prompts the usage of high quantities of natural gas and petroleum. Linear hydrocarbons like petroleum as such do not possess any long standing hazardous effects to human beings, but many aromatic compounds like benzene are carcinogenic. Petroleum being lighter than water always floats on it and its spillage covers the water resources with a thin covering. This deprives the aquatic life with required oxygen gas leading to their death.

Sources : Exhausts from motor vehicles let loose lot of unburnt hydrocarbon into air, land and water. Petroleum refineries and petrochemical industries also dump significant amount of hydrocarbons into atmosphere.

Sinks : Combustion of unwanted hydrocarbons is the best way of their disposal. Petroleum eating bacteria is still in the experimental stage.

Impacts : Hydrocarbons undergo various free radical reactions with the singlet oxygen and halogen atoms.

For example;

$$CH_4 + {}^1O_2 \longrightarrow H_3C. + .OH$$

$$CH_4 + .OH \longrightarrow H_3C. + H_2O$$

$$H_3C. + O_2 \longrightarrow H_3COO \text{ etc.}$$

When hydrocarbon vapors are inhaled, they immediately trigger the above mentioned type free radicals by combining with singlet oxygen generated in the lungs. Hence in humans hydrocarbon vapors cause irritation of mucous membrane, eye irritation, bronchial contraction and lung cancer. Plants loose their leaves due to yellowing and their stem is thickened. The aromatic hydrocarbons like benzene are proven carcinogens.

(B) **Pesticides :** The growing arable land and increasing agricultural yields necessitate the use of different pesticides (Fig. 2.5). According to the World Health Organization around 750,000 people are poisoned by pesticides every year.

$$Cl-C_6H_4-CH(CCl_3)-C_6H_4-Cl \qquad RO-\overset{O}{\overset{||}{C}}-NHR^1 \qquad CH_3-N=\overset{O}{\overset{||}{C}}$$

DDT Carbamater Methyl isocyanate (MIC)

Fig. 2.5 Structure of some pesticides.

DDT (p, p'-DICHLORO-DIPHENYL-TETRACHLORO-ETHANE)

DDT is the first pesticide introduced in 1948. The discoverer Paul Miller, a swiss chemist, was honored with a Noble Prize for this invention. It found a ready acceptance as a savior of crops and eradicator of malaria. DDT is a fat soluble pesticide and owing to its C-Cl bond (naturally occurring substances do not have C-Cl bonds) it is not biodegradable. It accumulates in the fatty tissues surrounding central nervous system. DDT can be detected in the entire living world and can also be found in significant quantities in the human body.

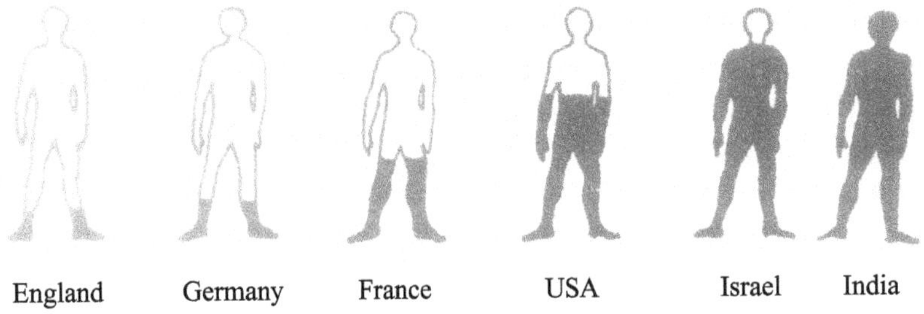

Fig. 2.6 Distribution of DDT in different countries.

The mean value of DDT accumulated in the body of population of different countries is given in Fig. 2.6. (Klausewitz et.al., 1971 U_m welt 2000, mg kg^{-1} fatty tissue). Effect of DDT on human nervous system is not recognized but it is very potent on the nervous system of insects and birds.

(C) **Methyl Isocyanate (MIC) :** This is a raw material used in the manufacture of carbamate pesticides. It has come into lime light during Bhopal gas tragedy (2nd December, 1984). It is a volatile liquid having boiling point at 43 ^{0}C and is extremely hygroscopic. It is synthesized commercially employing following scheme.

(i) Reaction of primary amine with phosgene

$$CH_3NH_3Cl + COCl_2 \longrightarrow CH_3NHCOCl + 2HCl$$

(ii) Heating of intermediate with CaO

$$2CH_3NHCOCl + 2CaO \longrightarrow \underset{MIC}{2CH_3NCO} + CaCl_2 + Ca(OH)_2$$

Impacts :

(i) MIC causes irritation to the skin, eyes and mucous membrane

(ii) It damages the fine capillaries present in the sensory organs

(iii) MIC causes chest tightness and breathing troubles like broncho-pneumonia.

(D) Carcinogens : Carcinogens are a group of chemicals that may cause cancer in humans and animals. They normally control DNA from passing commands for proper replication and continuation of genetic code. Benzene and non linear polycyclic compounds are recognized carcinogens. Structures of some of the aromatic carcinogens are given in Fig. 2.7.

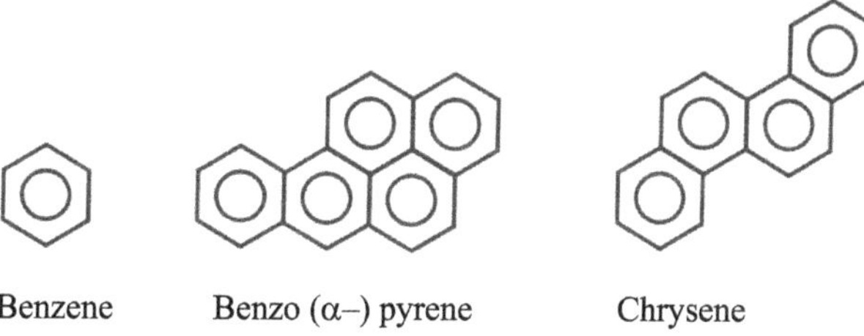

Fig. 2.7 Structures of some of the carcinogens.

Similarly unsaturated aliphatic compounds like vinyl chloride, ethylene chloride were also found to cause lung and liver cancer. Substances like benzidine, ß-naphthyl amine, 4-nitrophenyl, etc., cause bladder cancer.

(E) Hazardous Chemical Wastes : Land, water and air are equally polluted by a number of chemicals. The rivers and lakes in particular have become dumping grounds for various byproducts of paint and pharmaceutical industries and other chemical industries. Keeping their toxicity to aquatic and animal lives in view, it is the ernest duty of mankind to put controls at the disposal points.

(F) Peroxy Acetyl Nitrate (PAN) : This is a peroxide.

$$CH_3-\overset{\overset{\displaystyle O}{\|}}{C}-O-O-NO_2$$

This is equally harmful to both animals and plants.

Impacts :

(i) Young animals are easily susceptible to the toxic effects of PAN

(ii) Since peroxides produce free radicals, PAN readily interferes with radical enzymatic processes of the lung and leads to mortality by accumulation of fluids in lung.

(iii) PAN is also known to damage lung capillaries.

2.5.3 Green House Effect

The atmosphere is mainly composed of N_2 and O_2 gases and CO_2 concentrations in atmosphere is only 0.03% by volume. Sun is at a temperature of > 6000 ^{O}K and emit radiation of shorter wavelength with much energy concentrated in visible spectrum (5000 $\overset{o}{A}$). In contrast the earth is at 88 ^{O}K and peak wavelength 10100 $\overset{o}{A}$ and the emission spectrum is totally in the ir region.

Both the water vapor surrounding the earth's crust and CO_2 gas are capable of absorbing this out going radiation. Thus under normal concentration of CO_2 and H_2O the temperature of earth's surface is maintained by the energy balance of incoming and outgoing radiation.

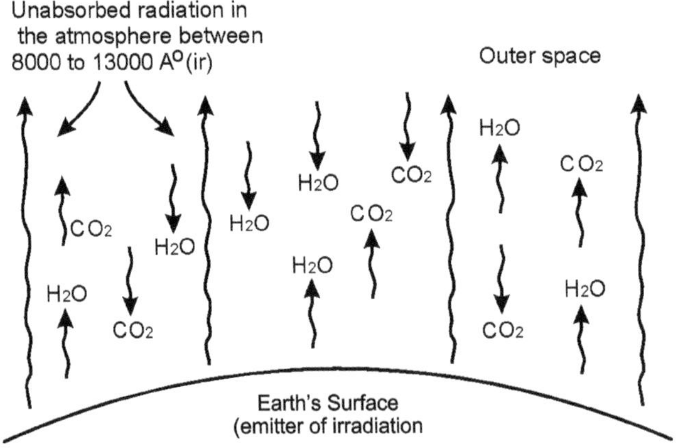

Fig. 2.8 Green House Effect.

The concentration of water vapor has not altered over years but concentration of CO_2 has dramatically increased. Between 1958 and 1988 the (CO_2) has increased from 300 ppm to 330 ppm. This influence of CO_2 in raising the temperature of earth, is comparable to glass green house. The thick CO_2 layer acts like glass panels allowing only radiation through them and

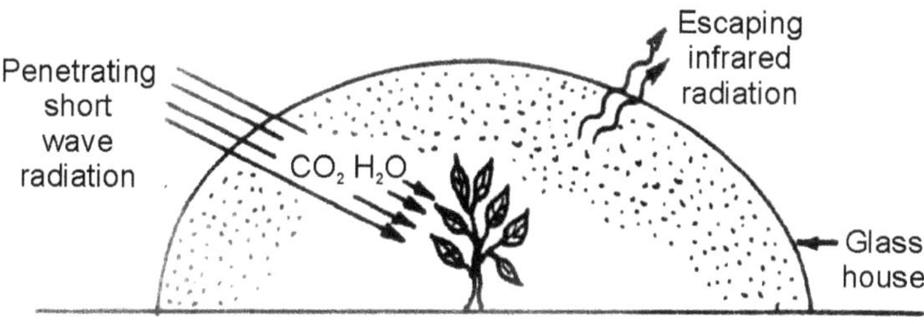

Fig. 2.9 Glass House Effect.

preventing the escape of heat outside. This phenomena could be easily realized if one enters a car standing out side, in the sun light with its windows glasses closed. The disaster lies not with any direct harm caused by heating of biosphere but because of melting of polar ice caps that control the seasons and other climatic condition, further a raise of 200 feet of water on account of this melting is enough to flood coastal cities starting from Bangkok to Boston and remove them from the world city atlas. A rise of a meter or two could permanently inundate a large fraction of world's highly fertile wetlands like Bangladesh. A 2 meter rise in the sea level is enough to submerge 20% of earth's land area. The disturbance in the seasonal conditions on account of melting of ice cap also brings many detrimental effects on the biosphere.

2.5.4 Ozone Depletion or Ozone Hole

3 or 4 billion years ago earths atmosphere consisted of ammonia, water and methane. There was little free O_2. Over the years O_2 concentration increased as a result of conversion of CO_2 into O_2 during photosynthesis. Ozone, O_3, is formed by photochemical reaction between O_2 molecules when exposed to the short uv light (< 242 mm) entering the earth's stratosphere.

$$O_2 \xrightarrow{\text{Far uv cosmic radiation}} (O) + (O) \quad \text{(atomic oxygen)}$$

$$(O) + O_2 + M \longrightarrow O_3 + M$$

where M represents a third body (N_2 or O_2) necessary to carry away the excess energy.

Opposing the above formation process is ozone removal due to photo dissociation by radiation between 200-320 nm.

$$O_3 + h\nu \xrightarrow{200-320\text{ nm}} O_2 + (O)$$

The net effect of the above reactions is the creation of a layer of ozone in the stratosphere. The concentration of O_3 in the stratosphere is maintained around 10 ppm. This ozone layer act as a protective shield against the biologically damaging short-wavelength ultraviolet radiation which can cause skin cancer, skin burns, leukemia, visual impairment and total blindness.

The peeling of ozone umbrella is accelerated by catalytic decomposition of ozone in presence of substances like NO and Cl.

$$NO + O_3 \longrightarrow NO_2 + O_2$$

$$NO_2 + O \longrightarrow NO + O_2$$

$$Cl + O_3 \longrightarrow ClO + O_2$$

$$ClO + O \longrightarrow Cl + O_2$$

Over all reaction in both the cases is $O_3 + O \longrightarrow 2O_2$.

We already discussed about the sources of NO and measures to be taken to control its concentration. The Cl atoms are also produced by unwarranted human activity. They are produced as a result of photochemical decomposition of chlorofluoro hydrocarbons (CFC), commercially called as Freons.

$$CCl_2F_2 + h\nu \longrightarrow Cl + CClF_2$$

Where $h\nu$ represents the solar radiation. Freons are commonly used as refrigerants, aerosols, foams and fire-extingushers. Each atom of chlorine was found to be capable of decomposition of 100,000 ozone molecules. So a ban on the usage of all CFC's and search for alternate materials is essential without which the mankind not only write's off its day but also that of all biosphere.

Self Test Questions

1. The killer gas of Bhopal tragedy

   (A) Sulphur dioxide (B) Carbon dioxide

   (C) Nitric oxide (D) Methyl isocyanate

2. Pollution mechanism of carbon monoxide involves

   (A) Replacement of oxygen from haemoglobin

   (B) Arrest of synthesis of porphyrin

   (C) Clotting of blood

   (D) Burning of fine capillaries

3. Major source of SO_2 is

   (A) Transportation vehicle (B) Industrial activity

   (C) Volcanoes (D) Marshy lands

4. Main sinks of CO

   (A) Soil bacterial (B) Photosynthesis

   (C) Absorption by plants (D) None of these

5. Impact of Pb toxicity is felt on

   (A) Lungs, (B) Heart,

   (C) Central nervous system, (D) Thyroid

6. DDT is soluble in

   (A) Lipids (B) Water

   (C) Proteins (D) None of these

Short and Essay Type Questions

1. What are the most common air pollutants ? Discuss them.
2. Why does DDT persist in nature ?
3. How does the organic pollutants affect oxygen content of lakes and steams ?
4. How is asbestos harmful to human body ?

5. What is the most common way that mercury enter the body ? Discuss its toxic effects.

6. What are the major sources of lead in environment ?

7. Give two sources from which the following pollutants generated (a) SO_2 (b) CO. Mention their effect on human life

8. What are pollutants ? Explain with examples.

9. How are pollutants formed ? Give suitable example.

10. Write a note on organic pollutants and their effect on human and plant life.

11. Name two common inorganic pollutants. What are their sources ?

Chapter 3

Water and Waste Water Treatment

3.1 Sources of Water

Although the problem of water supply is one of the most important of the problems that we face, it continues to draw very little attention from most people because many things are taken for granted. When the engineer is called upon to select a site for the location of a factory, he is sometimes apt to consider all other factors except that of water supply. Yet, it may be a deciding factor in the working of a factory. In some cases, the water requirement is very great. For example, about 30 million litres of water are needed per 1,000 tonnes of steel produced and about half a million litres per tonne of paper and over one and a half lakh litres per automobile. It may be noted that the largest requirement of water is for municipal supply. Besides the quantity of water, the nature and amount of impurities present in the water are of no less importance. Therefore, in selecting a proper site for a factory the following factors have to be taken in to account regarding the water supply: 1. quality and quantity of water available, 2. effect of impurities on the process and 3. quantity of water needed.

Generally, the water available contains a wide range of impurities but only a few of them may affect a process, so one is concerned with the removal, or reduction of concentration of only the few impurities and the rest may be allowed to remain as such, e.g., bacteria, if present, need not be removed from water for use in boilers. It may be noted that the water supplied for a particular industry is generally not 'consumed' in the operation in the sense of the term as electricity is consumed in a process. Almost the total quantity of water appears in one form or another and the engineer has also to deal with the disposal or re-use of waste water. The nature of the impurities in the outgoing water may be such that it is not safe or wise to let it flow out in the river stream or into the sources of water supply. Thus, connected with the water supply is the problem of the disposal of the waste or sewage. Again, in the same industry, water for several purposes is needed, e.g., steam generation, cooling, general needs of the processes and plant services, and for each purpose the specifications may be different.

Even *rain water,* which is the purest form of water found in nature, contains a considerable amount of dissolved gases and solids – both of organic and inorganic origin. Rain water either flows in to rivers or lakes and ultimately into the sea and is known as *surface water,* or it sinks underground and is available as well or spring water, when it is known as *underground water.* In general, the greater the contact that water has with the soil, or the more soluble the earthy matter with which it has come in contact, the greater is the amount of dissolved impurities. A sufficient quantity of flowing water of low hardness is available from many rivers. Even in any particular river the amount of dissolved mineral matter varies 2 to 3 times during the year. The amount of suspended and organic matter is still more variable. For cooling purposes, river water does not need any special treatment unless the amount of suspended matter is very great. Sometimes even *sea water,* in spite of its causing corrosion, can be used for cooling purposes. Sea water contains about 3.5 per cent of dissolved salts, principally sodium chloride (2.7%), and magnesium chloride and sulphate (0.6%). It is an important source of these and other substances.

Lake water has a more constant chemical analysis and the seasonal changes can be predicted quite accurately. It is softer than well water and a big lake can be taken to be a dependable source of water supply, but attention should be paid to the organic matter present. The quality of water of some of the big lakes is so good that hardly any treatment is needed for industrial uses, e.g., the water of Lake Superior, U S A contains only 2 mg/litre of turbidity and 60 mg/litre of total dissolved solids.

Usually well water is still more constant in its composition. It is clear but contains more of dissolved matter and is harder than surface water. It is more suitable for human consumption because of clarity and freedom from organic matter, harmful bacteria, unpleasant taste and odour. Deep well water is cool and is available at a more constant temperature throughout the year but contains more of dissolved minerals. Usually it is of high organic purity.

3.1.1 The Common Impurities in Water

The common impurities present in natural waters may be classified as follows.

1. *Dissolved minerals* – mostly comprise of carbonates, bicarbonates, sulphates and chlorides of calcium, magnesium, sodium and potassium.
2. *Dissolved gases* – mostly air and carbon dioxide.
3. *Suspended matter* – mostly mineral matter, giving turbid or muddy water. Organic matter may also be present.
4. *Microscopic matter* – consists mostly of plant and bacterial life giving colour, taste and odour.

In general, the removal of impurities from water of classes 1, 2 and 3 form the chief problem for industrial usage, and 3 and 4 for municipal supplies.

3.2 Temporary and Permanent Hardness

From the standpoint of an engineer the dissolved salts of calcium and magnesium are of special interest as they are responsible for some of the common and major boiler troubles and therefore will be considered in greater detail. A simple method of detecting these is by the use of soap (not detergent). A sample of water that reacts with soap to form a white scum and does not produce lather is known as *hard water. Soft water* lathers freely with soap. Calcium and magnesium salts are present mostly as bicarbonates, sulphates and also as chlorides. The difficulty in producing lather arises because of the fact that soaps are sodium and potassium salts of higher fatty acids and are soluble in water but when soap is used in water containing calcium and magnesium salts, insoluble soaps of calcium and magnesium are formed and precipitated, thus destroying the soap. A typical reaction may be illustrated as follows:

$$\underset{\substack{\text{Soap (Sodium stearate)} \\ \text{Soluble}}}{2C_{17}H_{35}COONa} + CaCl_2 \rightarrow \underset{\substack{\text{(Precipitate)} \\ \text{Insoluble}}}{(C_{17}H_{35}COO)_2\,Ca} + 2NaCl$$

Such precipitates are not formed with the newer detergents, which are sulphonates of long chain alcohols.

Calcium and magnesium salts are not the only salts that cause hardness. They are mentioned because they are most commonly present. As a general rule, all soluble salts of heavy metals cause hardness because they form precipitates with soap. The bicarbonates of calcium and magnesium are decomposed by boiling the water when carbonates are formed which are less soluble. Thus the hardness due to the bicarbonates can be reduced. The hardness due to these salts is called *temporary hardness*. As magnesium carbonate is only partially precipitated the term *cabonate hardness* is now used instead of temporary hardness. Sulphates and chlorides of calcium and magnesium are not removed by boiling and hardness due to these is called *permanent or non carbonate hardness.*

3.2.1 Determination of Hardness

In a boiler house, the degree of hardness of feed water forms a simple and important guide to its suitability for boiler feed. In a simple test, usually performed for water to be used in a low pressure boiler, 70 ml of water is taken in a 300 ml bottle. A standard soap solution is added from a burette and the bottle is shaken. The volume of soap required to produce lather stable for 5 minutes is noted and hardness is calculated. It is an empirical method but after some experience fairly accurate results can be obtained. It is estimated that if about 2 kg of common soap is added to 100 litres of water of hardness 100 mg per litre it would get precipitated by the salts present in the water. Lather would be produced only if more than this amount of soap is used.

For more accurate results the soap method has been replaced by the *complexometric method.* This depends upon the fact that when ethylenediamine tetra acetic acid (abbreviated as EDTA) or its sodium salt solution is added to a solution containing certain metals, chelated soluble complex is formed. If the pH value of the solution is made 10 and a dye, known as *Chrome Black T or Eriochrome Black T or Erio T* is added in presence of calcium and magnesium ions the solution becomes light red in colour. If EDTA is added to such a solution the calcium and magnesium ions will form complex with EDTA. When the formation of complex with free ions is completed EDTA extracts the ions from the dye and the solution turns blue. This is the end point. If certain other metallic ions are present, they interfere. This interference can be removed by adding certain salts.

The reaction taking place on the addition of Eriochrome black T to a solution containing Ca^{++} and Mg^{++} may be represented thus:

$$Ca^{++} + \text{Eriochrome black T} \rightarrow \underset{\text{light red complex}}{(Ca^{++}\ \text{chrome black T})}$$

On the addition of EDTA the free Ca^{++} and Mg^{++} form complexes with it. When they have been complexed the red complex is disrupted and a more stable complex is formed as follows:

$$\underset{\text{light red}}{(Ca^{++}\ \text{chrome black T})} + \text{EDTA} \rightarrow \underset{\text{blue}}{\text{chrome black T}} + (\text{Ca. EDTA})$$

A very useful engineering use of EDTA is in the prevention of scale formation. It is dissolved in water which is to be used in the circulatory system or for cooling, as in the radiators of automobiles. EDTA forms stable complexes with calcium and magnesium ions and prevents their deposition. It can also be used to remove some of the deposits.

The EDTA solution is also called the Versenate solution and sometimes the method is called Versenate method.

3.2.2 Degree of Hardness

At this stage the units in which hardness is usually expressed may be considered. The weights of all the substances are expressed in terms of a weight equivalent to the weight of $CaCO_3$. For example, 162 parts by weight of $Ca(HCO_3)_2$ would react with the same amount of soap as 100 parts by weight of $CaCO_3$. Therefore, the weight of $Ca(HCO_3)_2$, is multiplied by 100/162 to give weight in terms of $CaCO_3$. Similarly, weights of $CaSO_4$, $CaCl_2$, $MgSO_4$, $Mg(HCO_3)_2$ and $MgCl_2$ actually present are converted in terms of weight of $CaCO_3$ by multiplying x with

$$\frac{100}{136};\ \frac{100}{111};\ \frac{100}{120};\ \frac{100}{146};\ \text{and}\ \frac{100}{95}\ \text{respectively.}\ \left(x \times \frac{100}{\text{mol.wt of the compound}}\right)$$

The common unit of expression at present is in terms of parts of the hardness causing impurities in terms of weight of $CaCO_3$ per million parts of water. For example, if 6 kg of $MgSO_4$, are present in 100,000 kg of water, then in 1,000,000 kg of water 60 kg of it would be present. Therefore, the weight in terms of $CaCO_3$, would be $60 \times \frac{100}{120} = 50$ kg. So, the hardness of this sample of water, is 50 parts per million parts of water, which is written as 50 ppm (parts per million). In the French system it is expressed in terms of parts per 100,000 parts. In the English system, hardness is expressed in terms of grains (1/7000 lb) per gallon (10 Ibs) of water, i.e., parts per 70,000 parts. This is also called degree Clark, e.g., 14° Clark means impurities equivalent to 14 grains in terms of $CaCO_3$ are present per gallon of water or 14 parts per 70,000 parts of water. Therefore on ppm scale it is $14 \times \frac{1,000,000}{70,000} = 200$ ppm

It is becoming more common to express the results in terms of milligrams of the substance (expressed as $CaCO_3$) present in a litre of the solution (mg/l). Taking density of water as unity, milligrams per litre means 1/1000 g per 1000 ml of water (or solution), or 1/1000 g per 1000 g of water. This is the same as parts per 1,000,000 parts. Therefore, in common cases where the density of the solution is almost unity, parts per million parts is equal to milligrams per litre. Of course, for more concentrated solutions a correction has to be made. As the quantities of water (or solutions) are mostly expressed in terms of volumes, the engineer is able to make more direct calculations by using mg/l scale than by using ppm scale (although for most practical purposes both are the same).

A more recent trend is to report the various cations and anions separately in terms of their equivalents as $CaCO_3$. This has been extended to express the three forms of alkalinity, namely, bicarbonate, carbonate and caustic or hydroxide. This form of expression is now used to express even the mineral acidity (i.e., due to the sulphates, chlorides and nitrates) which might be shown as negative alkalinity. In spite of all this, free CO_2, Fe, Mn, SiO_2 and fluorides are generally expressed as these constituents. It is therefore, necessary that whenever the results are expressed in terms of equivalents of $CaCO_3$, it be clearly indicated by a statement like "expressed as p. p. m. of $CaCO_3$." It is a general practice to express the results in three significant figures; e.g., the figure 151.65 ppm is rounded off to 152 ppm.

The maximum permissible hardness of water for boilers operating at different pressures is as follows :

at 0 to 10 kg/cm^2 ……… 80 ppm

at 10 to 15 kg/cm^2 ……… 40 ppm

at 15 to 30 kg/cm^2 ……… 10 ppm

over 30 kg/cm^2 ……… 2 ppm

3.3 Water for Industries

Although the manufacturing industries need water for a variety of purposes the largest demand for water is for heat transfer. Water is the usual medium for extracting heat. Boilers, cooling towers, heat exchangers, condensers, coolers, refrigerators and air conditioning equipment illustrate the use of water for heat extraction. It is estimated that the large thermal power stations require 800 tons of water per ton of coal consumed.

Water is a *raw material* when it becomes a part of the product as in the beverage industry and some of the chemical industries. In a few cases it is a *reactant* as in the manufacture of acetylene from calcium carbide. Water being the cheapest liquid available is used widely for cleaning and washing in the homes as well as in industry. One is not conscious of its uses in *material handling,* in the pumping of materials like coal in the form of slurry for many kilometers. Harmful effluents from industries are *diluted* by large quantities of water. High pressure water jets for cleaning metal parts or the removal of bark from logs of wood illustrate the use of water for kinetic energy. Since water does not yield radio activities when exposed to neutron flux, it is used for *nuclear shielding,* or as a moderator in nuclear energy reactors.

It is obvious from the above examples that not only the quantity but also the quality of water needed for different uses vary widely. The high pressure boilers require water of very high purity for steam making necessitating very careful and elaborate treatment whereas for ordinary washing no treatment, other than simple settling for highly turbid water is needed.

As it is not possible in a small monograph to discuss the requirements of each industry separately, the treatment of water for steam making has been taken up by way of illustration showing how almost absolutely pure water can be obtained from the sea. Where the requirement is not that stringent the treatment may not be carried out to eliminate all the impurities.

3.3.1 Water for Steam Making

In water tube boilers steam is generated on the surface of the tubes which are heated by the hot gases and the flames produced by the combustion of the fuel. The steam travels through the tubes and is delivered to the boiler drums. At this stage the steam is mixed with the boiler water, foam and sludge. The problem of the production of clean-dry steam has been investigated by engineers for the last few decades. In recent times this has become a still more serious problem because boilers working at much higher pressures are being used. Therefore, all the boilers are provided with entrainment separators, i.e., devices for separating steam from the liquids and the solids. This equipment provides purification mainly by changing the direction of steam whereby the heavier particles, i.e., the liquid and the suspended matter, drop out. As a matter of fact, more heat is added to steam so that *super-heated* steam is obtained. The properties of superheated steam are almost those of a perfect gas, hence it does better work in steam engines and turbines and the steam consumption is lower. Super heated steam being at higher temperature does not have the problem due to the condensation of water in the pipelines, engines, etc., and so dry steam can be supplied over longer distances.

In water treatment complete elimination of all the impurities is not aimed at. Only those impurities which give rise to certain troubles are eliminated or kept within certain limits. In the modern high pressure boilers water purer than the distilled water used in laboratories has to be supplied. The major boiler troubles, due to the use of unsuitable water, may be classified as 1. Carry over, 2. Corrosion, 3. Scale formation, 4. Caustic embrittlement and 5. Turbine deposits. The total destruction of several large power houses has ultimately been traced to the use of unsuitable water which caused explosion of the boilers.

3.3.2 Carry Over

As steam emerges from the surface of the water, it is associated with tiny droplets of water. Such steam is called *wet steam.* The droplets may carry with them suspended matter, besides the dissolved matter present in the boiler water. Such carrying over of water by steam along with the suspended and the dissolved solids is called *carry over*. It is mainly due to priming and foaming.

The liquid contamination in the steam is expressed in percentage, by weight, of the total mixture and is called *steam quality.* If steam contains 0.3 per cent moisture steam quality is 99.7 per cent. The impurity due to solids is represented by *steam purity* and is expressed in parts per million (ppm) or milligrams per litre (mg/l) of impurity. If steam contains 1 part by weight (or 1 mg) of solid contamination per million parts by weight of steam (or per litre of water), steam purity is 1 ppm (or 1 mg/l).

Priming is such rapid or violent boiling of water occurring in the boiler that water is carried into the steam outlet in the form of a spray. It may be caused by 1. very higher water level; 2. presence of excessive foam filling the steam space; 3. high steam velocity: 4. sudden steam demand leading to sudden boiling and 5. faulty boiler design.

Control of priming is fundamentally a problem of boiler design. It is controlled by providing proper evaporation and adequate heating surfaces, uniform distribution of fire and of working load, providing anti-priming pipes or dash plates, etc. In a given boiler priming can be minimized by 1. keeping the water level as low as it can safely be kept. Even a slight lowering of the water level greatly reduces priming; 2. avoiding sudden steam demands (by suddenly taking out a quantity of steam the pressure in the boiler is reduced, thus the boiling point of water is lowered and the superheated water is suddenly converted into steam); and 3. minimizing foaming.

Foaming is the formation of small but stable bubbles above the surface of water. It becomes a serious problem when the bubbles are persistent, i.e., do not break easily. These bubbles are carried over along with steam, leading to excessive priming.

Foam formation takes place when the concentration of solids in the surface layer is different from that in the mass of the liquid. This means that the solid must alter (whether increase or decrease) markedly the surface tension. Therefore all those solids which alter markedly the surface tension of water promote foaming. Presence of fatty oil is highly undesirable as it may form soap with the alkalinity. Soaps greatly reduce surface tension of water.

Pure water has little tendency to foam but almost all the impurities cause foaming, especially the dissolved impurities like soap, which reduce the surface tension of water. It may be noted that as the steam is being generated the concentration of the dissolved impurities, especially of sodium and potassium salts is increasing. The general construction of locomotive boilers is such that these have foaming more easily than the stationary boilers have.

To a certain extent foaming can be controlled by the use of anti-foaming chemicals, some of which act by counteracting the reduction in surface tension, and some by mechanical action, like castor oil which spreads on the surface of the water. Castor oil may be used for low pressure boilers. The only satisfactory remedy for foaming is *blow down*, i.e., removal of concentrated boiler water and replacing it by fresh feed water.

Dissolved solids, suspended solids and alkalinity affect carry over. It is believed that, in general, the suspended solids are more responsible for carry over than the dissolved solids. Carry over is undesirable because it not only decreases the efficiency but also results in some practical difficulties in the successful operation of a boiler. Some of these are: 1. a considerable amount of heat is wasted by the removal of hot boiler water; 2. water reduces superheat of steam in the super heaters; 3. water takes along with it grit which impinges on the surface at high velocity, and damages the walls of the cylinders and valve seats; 4. water contains considerable amount of dissolved solids which gradually get deposited and the deposits restrict the passage in the pipelines, causing corrosion and insulation; 5. water entering the steam engine cylinder causes serious trouble because of its incompressibility and inability to escape quickly through the steam ports; 6. foaming of water makes it difficult to know the exact level of water in the gauge glass.

3.4 Corrosion

Corrosion is one of the most serious problems caused by the use of unsuitable water. Boiler tubes, drums, economizers, super heaters and condensers are the most affected parts. The corrosion problem extends even to parts which are not directly in contact with boiler water because gases like oxygen and carbon dioxide are released during heating of water. These gases have corrosive effects. Corrosion can be attributed to the following (i) dissolved oxygen; (ii) mineral acids; (iii) dissolved carbon dioxide and (iv) galvanic cell formation.

(i) ***Dissolved Oxygen***

Oxygen dissolved in water is mainly responsible for corrosion in boilers. Dissolved oxygen in the presence of moisture and at the high temperature prevailing in the boiler readily attacks iron. The higher the pressure, the greater will be the dissolve oxygen content, and the higher the temperature the lesser the oxygen content. The amount of oxygen dissolved is also influenced by the other dissolved matter present. The solubility of oxygen in pure water is greater than in water containing other dissolved matter. In a boiler, oxygen is introduced through the fresh make up water supplied to it and also through infiltration of air into the condensate system. As the water is heated the dissolved oxygen is liberated and iron is corroded, (The mechanism of corrosion has been discussed in detail in the chapter on Corrosion.)

Dissolved oxygen can be removed by :

(a) ***Mechanical deaerator*** **:** Various types of deaerators have been designed. The general principle is that in a cylindrical chamber water falls through perforated plates, or is sprayed, so that a larger surface is exposed. The deaerator is connected to a vacuum line which sucks out the gases.

(b) ***Pre-heating*** **:** Advantage is taken of the fact that the solubility of a gas decreases with rise in temperature. Air begins to be expelled at a temperature of about 65° and is almost completely removed at the boiling point. For the removal of dissolved oxygen the sample of water is sprayed in cylindrical equipment. It meets steam moving in the opposite direction. Steam heats and carries away with it the gases which have been liberated.

(c) ***Chemical treatment*** **:** The above two methods remove most of the oxygen but for a complete removal, chemical treatment is required. This is done by adding a reducing agent, sodium sulphite being commonly used, which at boiler temperature reacts rapidly as follows :

$$2Na_2SO_3 + O_2 \rightarrow 2Na_2SO_4$$

Ferrous sulphate is also used. It gives precipitate of $Fe(OH)_2$ which is oxidized to $Fe(OH)_3$.

In high pressure boilers even traces of oxygen are undesirable. As it is necessary to keep dissolved salts also to a minimum, hydrazine is added in such amounts that after reacting as under, there is a residual hydrazine of about 0.02 mg/l:

$$N_2H_4 + O_2 \rightarrow 2H_2O + N_2$$

Thus no solid is added to the boiler. Hydrazine should be handled carefully, usually its solutions are used.

(ii) *Mineral acids*

Most of the natural waters are alkaline. Waters in the mining areas are often acidic. Sometimes, in industrial areas the water may become acidic because of the discharge of acidic industrial wastes into the surface water.

Some of the inorganic salts may hydrolyse to produce acidity, the most important of such salts being magnesium chloride. Its action may be represented thus:

$$MgCl_2 + 2H_2O \rightleftharpoons Mg(OH)_2 + 2HCl$$

(As $Mg(OH)_2$ is insoluble the forward reaction goes to completion.)

$$Fe + 2HCl \rightarrow FeCl_2 + H_2$$

$$FeCl_2 + 2H_2O \rightarrow Fe(OH)_2 + 2HCl$$

As a result of these reactions even a small amount of $MgCl_2$ will cause corrosion of a large amount of iron. If the amount of the acid formed is small, it may get neutralized by the alkalinity present in the water, otherwise alkali will have to be added. It has been suggested that except in high pressure boilers, boiler water should have a pH value of about 10.5 so that the acid may not attack iron. High pressure boiler water needs separate consideration.

(iii) *Carbon dioxide*

Water contains some dissolved carbon dioxide. In case some bicarbonates are present in the boiler water, more carbon dioxide will be produced on heating because of the decomposition of the bicarbonates. Carbon dioxide coming in contact with water produces carbonic acid, H_2CO_3, which has a corrosive effect on the equipment. If carbon dioxide is released inside the boiler, it will be carried along with steam and H_2CO_3 will be formed where the steam condenses, and corrosion will take place. Local corrosion caused by carbon dioxide is called *pitting*.

Carbon dioxide in water can be removed by the addition of lime. Another method is to filter water through limestone when it not only becomes clear but carbon dioxide is also removed:

$$CaCO_3 + CO_2 + H_2O = Ca\,(HCO_3)_2$$

Of course, this reaction adds to the hardness of water.

The problem of carbon dioxide present in the steam can be tackled by the use of ammonia. It may be noted that ammonia itself has a corrosive action on copper and copper alloys; therefore its use is being replaced by amines.

(iv) ***Galvanic cells.***

Corrosion can also be due to simple galvanic cells which iron forms with some of the boiler fittings made of other materials or with impurities. This can be avoided by suspending zinc plates. Zinc is more electropositive than iron hence it is anodic to iron and gets dissolved first and iron is saved.

Galvanic couples set up cells which cause formation of corrosion products. These deposits in themselves lead to further corrosion. This type of damage can be reduced by the use of proper inhibitors. Care has to be taken to see that the surface is clean and the inhibitor film covers the surface completely. Use of organic as well as inorganic inhibitors for protection against corrosion has been discussed in the Chapter on Corrosion. It may be added that in high pressure boilers often chromates are used as inhibitors. These have to be removed from the blow down before discharge otherwise fresh water may get poisoned. The pH is reduced to < 5 and then deionized and finally the pH is increased to 6 and then discharged.

Silicates, amines, polyphosphates, tannins and lignins have been used in recirculated water systems for protection against corrosion.

3.5 Scale Formation

From the standpoint of the chemical constituents present in water, a boiler may be considered to be a huge concentrator of impurities. Water containing impurities is fed into the boiler and pure water, in the form of steam, is removed, leaving behind the impurities. In a high pressure boiler, steam equivalent to the total water capacity may be produced within 15 minutes. Boiler tubes completely blocked by deposits and fragments of scales are not very uncommon. Even the pipes feeding water in a boiler are found to have sufficient deposits.

Some of the deposits stick to the metal surface and are known as *scales*. If they are in the form of soft and muddy deposits, which can be flushed out easily, or in the form of suspensions, they are known as *sludges*.

Scale formation takes place by two distinct mechanisms. A salt like calcium bicarbonate gives off carbon dioxide and insoluble calcium carbonate is precipitated on the boiler surface giving the scale. This is the principal factor in fire tube boilers. Another important cause is the lower solubility of certain salts, notably calcium sulphate, at elevated temperatures. The data available on the solubility of various salts does not conform to the high temperature and pressure conditions of the boiler. Another complicating factor is that a number of salts are present in water which affects the solubilities of each other. Therefore, it is difficult to calculate the quantity of a salt that will crystallize out in a boiler.

Water near the hot metal surface is at a higher temperature than the main body of water. A salt like calcium sulphate crystallizes out on the metal surface because the solution there becomes super-saturated. Also, as a film of water on the metal surface is converted into steam, it leaves behind a deposit of the salts in spite of the fact that the rest of the water in the boiler is not saturated with those salts. Gradually, more and more salts are deposited giving a layer on the surface which is known as scale. The scales so formed are mostly of calcium sulphate which stick tenaciously to the metal surface and interfere seriously with the flow of heat.

The behaviour of the silicates is not well understood. Even when there are only a few milligrams of silica per litre of water it is likely to deposit silicate of calcium or magnesium or other silicates. The silicate deposits are particularly troublesome as their thermal conductivity is very poor and they stick very firmly to the metal surface making the removal of the scales very difficult.

Broadly speaking, the scales may be considered to be of three types namely, carbonate, sulphate and silicate types, whose formation has been described above. Generally, the scales found in the boilers are of mixed composition and are composed of several salts. Colloidal and suspended matter may get deposited by coagulation or may be trapped in the precipitates of other salts.

3.5.1 Disadvantages of Scale Formation

Scale formation is objectionable for the following reasons :

(a) ***Waste of fuel.*** The scales are bad conductors of heat, therefore, their effect is like that of an insulating material spread over the metal surface. This results in reduction in the rate of heat transfer with the result that a larger quantity of fuel has to be used. It has been reported that 0.25 cm thick scale would increase fuel consumption by about 2 to 3%.

(b) ***Overheating of the boiler tubes.*** A more serious effect than the waste of fuel due to overheating of the tubes is explained with the help of Fig. 3.1. The figure on the left shows a section of the boiler tube metal which is completely free from scales. In this case, in order to heat the boiler water to T_1^0 the metal surface towards the source of heat is heated to T_2^0 because of the temperature drop across the metal. The figure to the right represents a similar situation except that there has been scale formation. As the scale is an insulator, there is a greater drop in temperature. Therefore, for heating water to T_1^0 the temperature on the inner surface of the metal is T_3^0 and on the outer surface the temperature is T_4^0 which is considerably higher than the scale-free metal surface temperature of T_2^0.

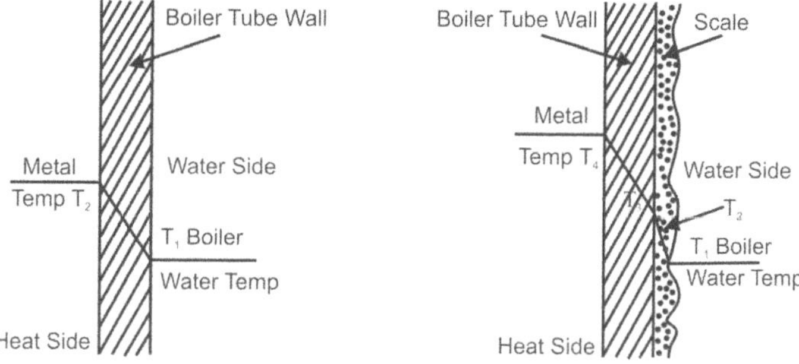

Fig. 3.1 Insulating effect of scales.

At higher temperatures the metal becomes weak and can withstand less pressure. Overheating also causes undue expansion of tubes leading to *'bagging'*. This becomes dangerous in modern high pressure boilers.

(b) ***Corrosion.*** At the higher temperature so maintained there is greater corrosion of the boiler tubes.

Thus it is seen that the disadvantages of scale formation are primarily due to the fact that scales are bad conductors of heat.

Scale formation can be prevented by the following means :

(i) ***External treatment.*** This treatment is given outside the boiler before the feed water enters it. Attempt is made to remove or reduce the amounts of that substance, mainly calcium and magnesium salts and silica, which form deposits. This requires an investment on external treatment equipment. In several cases this can be avoided, for example where the water requirement is not very exacting, the plant is small, the water available is of good quality or a combination of these reasons.

(ii) ***Internal treatment.*** In spite of the external treatment, some salts often remain in the water. Attempt is made, by adding chemicals to water entering the boiler, to convert the remaining salts into more soluble salts or such salts that the deposits will be in the form of nonstick sludge which can be easily removed. The treatment also aims to prevent corrosion not only inside the boiler but also in the pipe line and equipment handling steam in general and condensate in particular, due to the dissolved gases liberated in the boiler. In the cooling water system it prevents biological growths.

(iii) ***Blow down.*** In spite of giving external and internal treatments the concentration of impurities, especially the soluble impurities, goes on increasing as steam generation progresses and the ultimate practical solution available is to make a blow down, i.e., remove a portion of the concentrated water which contains a considerable amount of dissolved matter. Along with it some sludge is also removed. Fresh water is then fed in its place which contains a smaller amount of impurities.

The percentage of blow down is an indication of the efficiency of the water treatment. It should be less than 10%.

3.6 Caustic Embrittlement

It has been found that sometimes the boiler failure takes place because the metal plate has become brittle and inter crystalline cracks have been formed in the metal. This happens in the areas which are under stress.

The alkali metal carbonates and the bicarbonates present in the feed water give rise to caustic alkalinity at the high temperatures and pressures prevailing in the boiler. The concentration becomes greater near the rivets because of bad circulation of water and also because of leaks through which steam escapes leaving behind water with a higher concentration of caustic alkalinity.

It has been observed that those boilers which were fed with water containing sodium sulphate were free from such troubles and therefore the usual practice has been to keep the ratio of sodium sulphate to sodium hydroxide as 1: 1, 2:1 and 3:1 for boilers working at pressures up to 10, 20 and above 20 kg/sq. cm respectively. Some later authorities do not seem to favour this scheme and recommend the use of tannins, lignins, sodium nitrate or neutralization of the excess alkali or control of the pH value. The use of phosphates (discussed later in this Chapter) is now being adopted in the modern practice.

Turbine Deposits

These pose one of the most serious problems in the working of the large turbines. It would be well to remember that the turbine deposits may be traced to the following factors and remedial measures be taken :

1. The condensate should be, and is taken as, of high quality but periodically it is not of the required quality.
2. The chemical analysis throughout the steam/water circuit especially the steam portion of the system is insufficient or not of the required high standard.
3. The chemicals are either not fed properly or are not mixed properly.
4. The demineralization plant is not operated properly.

It hardly needs to be impressed on the engineers that for minimising turbine deposits these points should always be kept in mind and no slackness, for however short duration it be, should be permitted.

3.7 Methods of Boiler Water Treatment

The objective of boiler feed water treatment is firstly to avoid trouble in the boiler plant and secondly to obtain steam of sufficient purity for the steam turbine. If the steam is not of sufficient purity, there is fouling of the blades of the steam turbine which reduces the capacity of the turbines.

The actual treatment that needs to be given to boiler feed water depends to a great extent on the working pressure. As the boiler pressure goes up, the specifications for feed water become more rigid. Accordingly, a treatment has to be carefully chosen. The common treatments are described below :

3.7.1 Preliminary Treatments

Preliminary treatments may include simple treatments like sedimentation, coagulation and filtration. These will not be needed for clear water, like well water. An operation like filtration may be postponed for a later stage, say after lime-soda process. Details of these treatments are similar to those given under municipal water supply and need to be studied carefully.

In case a plant is getting some clear water, say, from a well, and some turbid water, say from a river, it is advisable not to mix them but to remove the turbidity of the river water first. Otherwise turbidity will get diluted and will be more difficult to remove. Similar principle applies for other kinds of impurities.

3.7.2 Pre-Heating

Pre-heating is heating feed water before it enters the boiler. Very often blow down water and low pressure steam is available at the plant at negligible cost and can be used for heating the feed water. Hot flue gas leaving the boiler furnace is also used for pre-heating. Water may be heated in a heat exchanger. The advantages of pre-heating are:

(a) Instead of feeding the boiler with water at that atmospheric temperature it is fed with hot water, thus saving fuel.

(b) A portion of temporary hardness may be removed. The bicarbonates are decomposed precipitating the insoluble carbonates thus :

$$Ca\ (HCO_3)_2 \rightarrow CaCO_3 + CO_2 \downarrow + H_2O$$

It may be noted that for complete removal of temporary hardness water has to be boiled for 10 to 15 minutes. Usually this does not happen in pre-heaters.

(c) Dissolved oxygen and carbon dioxide are removed. Initially dissolved gases are removed from water on heating to about 65° and are almost completely removed on boiling. As these gases have a corrosive effect their removal is advantageous.

(d) Use of hot water while giving chemical treatments, such as lime-soda treatment, greatly helps the operation. The effect of temperature is discussed along with the lime-soda process.

3.7.3 Lime-Soda Process

This process is used for feed water treatment for low and also medium pressure boilers. It is based on converting the soluble impurities into insoluble salts, which get precipitated. The precipitate is allowed to settle and is then filtered off. A calculated amount of milk of lime, generally with about 10% excess, is added to the sample. Its effects are :

(i) temporary hardness is removed by the following reactions:

$$Ca(HCO_3)_2 + Ca(OH)_2 \rightarrow 2CaCO_3 \downarrow + 2H_2O$$

$$Mg(HCO_3)_2 + Ca(OH)_2 \rightarrow CaCO_3 \downarrow + Mg\ CO_3 \downarrow + 2H_2O$$

$$Mg(HCO_3)_2 + 2Ca(OH)_2 \rightarrow 2CaCO_3 \downarrow + Mg(OH)_2 \downarrow + 2H_2O$$

$$2Na(HCO_3)_2 + Ca(OH)_2 \rightarrow CaCO_3 \downarrow + Na_2CO_3 \downarrow + 2H_2O$$

$CaCO_3$ being comparatively insoluble is readily precipitated. $MgCO_3$ formed in a similar 1:1 reaction is not so insoluble and is not precipitated. Therefore for the removal of Mg $(HCO_3)_2$ twice the amount of lime is needed that produces Mg $(OH)_2$, an insoluble magnesium salt. Excess $Ca(OH)_2$ could be removed by addition of $NaHCO_3$.

The removal of temporary hardness by the use of lime has an important advantage, namely, that of reduction in total dissolved solids. Generally speaking, a large portion of the hardness in natural water is temporary. The removal of temporary hardness by the use of lime does not introduce any soluble salt in water, as is the case with soda treatment or even zeolite process. This is a distinct advantage especially where water with low total solid content is required. As lime is inexpensive, for obtaining high quality water it may be more economical to give lime treatment to water of high carbonate hardness followed by zeolite or demineralizing treatment.

(ii) Lime also removes all permanent magnesium-hardness but is not useful to remove permanent calcium hardness.

$$MgSO_4(or\ MgCl_2)+Ca(OH)_2 \rightarrow Mg(OH)_2+CaSO_4(or\ CaCl_2)$$

An equivalent amount of permanent calcium hardness is introduced, which has to be removed by soda treatment. Therefore, in spite of almost quantitative precipitation of magnesium, there is no softening.

(iii) Lime also removes free acids, carbon dioxide, hydrogen sulphide and precipitates dissolved iron and aluminium salts as insoluble hydroxides. A large excess of lime should be avoided.

$$2HX + Ca(OH)_2 \rightarrow CaX + H_2O$$

$$CO_2 + Ca(OH)_2 \rightarrow CaCO_3\downarrow + H_2O$$

$$H_2S + Ca(OH)_2 \rightarrow CaS\downarrow + H_2O$$

$$Fe^{2+} + Ca(OH)_2 \rightarrow Fe(OH)_2\downarrow + Ca^{2+}$$

$$2Al^{3+} + 3Ca(OH)_2 \rightarrow 2Al(OH)_3\downarrow + 3Ca^{2+}$$

The lime treatment is followed by soda treatment which removes the remaining calcium permanent hardness (including Ca-hardness introduced during the removal of Mg-permanent hardness)

$$CaSO_4,(or\ CaCl_2)+Na_2CO_3 \rightarrow CaCO_3+Na_2SO_4,\ (or\ 2NaCl)$$

$$\therefore \qquad Soda = \frac{106}{100}$$

$$[Permanent(Ca^{2+} + Mg^{2+} + Fe^{2+/3+} + Al^{3+} + etc) + free\ H^+ - HCO_3^-]$$

The process may be carried on in batches or continuously. In the batch process the chemicals are added in batches in water taken in tanks, agitated for the necessary time and the precipitate allowed to settle and then filtered. Fresh water is again taken into the tank and the operation is repeated. Filtration through sand beds should be avoided as the water is alkaline and will easily dissolve silica. It may be filtered through filter press, coal, coke or even dense iron ore. Silica may also be introduced by the use of lime of inferior quality.

In the continuous process, water together with lime and soda solutions enters a smaller tank continuously at a predetermined rate, bringing the correct amount of materials together. The chemicals and the water to be treated enter a bigger tank or chamber (Fig. 3.2). The second chamber is a tall structure having a larger cross section at the top than at the bottom. As a solution moves upwards the precipitates formed as a result of the chemical reactions settle down, because towards the upper portion of the chamber the velocity is less due to the larger cross section. Clear water is removed continuously from the top and sludge from the bottom. A part of the sludge is recirculated which not only enables the use of excess chemicals left behind in the sludge but also helps in the precipitation by providing nuclei.

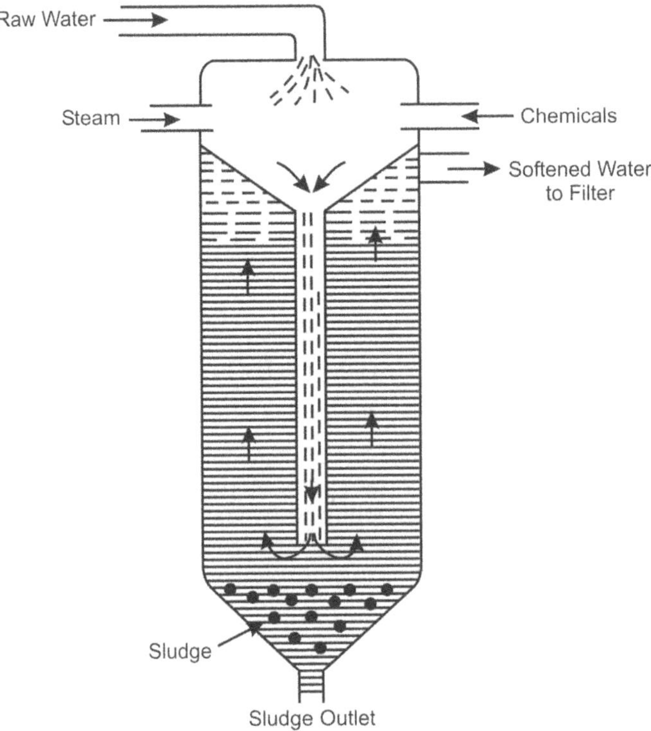

Fig. 3.2 Continuous hot lime-soda process.

When the chemicals are added to water at atmospheric temperature, the process is known *as cold lime soda process.* At this temperature the precipitates are finely divided and do not settle easily, nor can they be easily filtered. It is therefore, necessary to add coagulants like alum, aluminium sulphate, ferrous sulphate, sodium aluminate etc. These get hydrolysed and form flocculent and gelatinous precipitates which entrap the finely divided particles and settle down readily or are easily filtered off. Ferrous sulphate and sodium aluminate are costlier than alum. Ferrous sulphate helps to remove oxygen and sodium aluminate helps the removal of silica and also oil.

When this treatment is given at a higher temperature the process is known as *hot lime soda process.* Water is heated by waste steam or hot gases and a calculated amount of chemicals are added as in the cold process. The distinguishing features are the following :

(i) As the reactions proceed quickly at higher temperature, the softening reactions are completed in about 15 minutes, whereas in the cold process several hours are needed. This greatly increases the capacity of the plant.

(ii) Precipitates and sludges formed settle rapidly at higher temperatures, hence coagulants are not needed.

(iii) Filtration becomes easier because of the bigger size of the particles and the reduced viscosity of solution at the higher temperature; hence the capacity of the filter is also increased.

(iv) Residual hardness, i.e., hardness of softened water needs careful consideration. Generally, the solubility of substances is lower at lower temperatures but the solubility of calcium salts is very slightly higher. More than this, the time taken for the completion of the reactions at a lower temperature is greater. In the hot process the reactions are completed in much less time with the result that the hot process gives softer water.

In an experiment, a portion of a sample of water was treated with chemicals at 95° and another portion of the sample of the same water was treated under identical conditions except that the temperature was 10°. It was found that the residual hardness of the hot solution at the end of 10 minutes was less than that of the cold solution even after 5 hours. If the cold solution was allowed to stand for a longer period, more of the hardness would have been removed. This difference is chiefly because at higher temperatures the rates of reactions are greater, so more of the hardness is removed in less time.

Residual hardness also depends upon the excess of chemicals used. According to the Law of Mass Action the reaction is more complete and quicker if higher concentrations of the reagents are used, but large excess of the reagents should be avoided as they will appear in the softened water. Excess of chemicals required in the hot process is less than in the cold process. If precipitation is incomplete in the softening plant, salts, will be precipitated in the pipes, storage tanks and boiler.

In practice, cold lime soda process gives water with a residual hardness 50-60 mg/l and the hot lime soda process reduces the hardness to 15-30 mg/l.

In some places, water is highly alkaline and contains a considerable amount of sodium bicarbonate. The total alkalinity of this type of water is greater than the total hardness. On adding lime, calcium carbonate precipitates forming sodium carbonate.

$$2NaHCO_3 + Ca(OH)_2 \rightarrow CaCO_3 + Na_2CO_3 + 2H_2O$$

In such a case, soda may not be needed for the removal of permanent hardness. As a matter of fact, cases are known where gypsum has to be added to remove soda so produced.

3.7.4 Permutit or Zeolite Process

Permutits or zeolites are natural or synthetic complex compounds, consisting of hydrated sodium aluminium silicates approximately corresponding to the chemical formula $Na_2O.Al_2O_3.2SiO_2.6H_2O$. The artificial zeolites have greater capacities per unit weight for softening water than natural zeolites but are less durable. Zeolites are insoluble. They introduce Na^+ in place of Ca^{++} and Mg^{++}. As these exchange bases, they are also known as *base exchangers*. If Na_2Z represents permutit or zeolite, the reactions taking place may be represented thus :

$$Ca(HCO_3)_2 + Na_2Z \rightarrow 2NaHCO_3 + CaZ$$

$$MgSO_4, + Na_2Z \rightarrow Na_2SO_4 + MgZ$$

Permutits or zeolites are kept in cylinders and used as filter beds. Hard water is percolated through them at a certain specified rate. The hardness causing substances (or ions) are retained as insoluble compounds according to the reactions given above. Thus the outgoing water contains sodium salts corresponding to the calcium and magnesium salts present in the incoming hard water. After use for some time the zeolite is converted into calcium or magnesium zeolite and then it ceases to soften water. At this stage it has to be regenerated by treatment with a concentrated sodium chloride solution when the original sodium zeolite is recovered :

$$CaZ + 2NaCl \rightarrow Na_2Z + CaCl_2$$

The zeolite bed is then washed with water and is again ready for use. See Fig. 3.3.

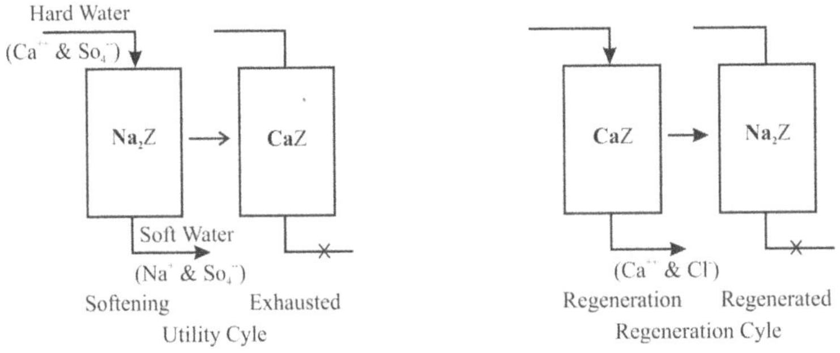

Fig. 3.3 Water softening by zeolite process.

Water having turbidity or excess of acidity or alkalinity or containing dissolved iron or manganese salts should not be fed into a zeolite softener. Turbidity will clog the pores and restrict the flow; excess of acidity or alkalinity will attack the zeolite, there fore the pH value of water to be passed through a zeolite bed should be around 7; iron or manganese salts will produce iron or manganese zeolites from which it cannot be easily regenerated. Therefore, these should be removed before passing water through the zeolite bed. Generally, hot water is not passed through a zeolite bed because zeolite tends to dissolve in hot water.

It is to be further noted that the acid radicals are not removed. The temporary hardness will give water containing $NaHCO_3$ which dissociates inside the boiler giving high alkalinity:

$$NaHCO_3 \xrightarrow{\Delta} NaOH + CO_2$$

Alkalinity causes "caustic embrittlement". Carbon dioxide goes along with steam and makes the condensed water acidic, hence corrosive. Thus waters containing a large amount of temporary hardness should not be directly treated by the zeolite process. Such water may be boiled or treated by lime to remove temporary hardness and then softened by this process.

The *advantages* of the zeolite process are several. It removes the hardness completely; the equipment is a compact unit occupying a small space; it automatically adjusts itself to different hardness and can be worked under pressure, therefore the plant can be installed in the water supply line and double pumping is avoided. Moreover, as the impurities are not precipitated, there is no danger of sludge formation at a later stage.

A combination of lime soda process and zeolite process gives better results. Zeolite process has now been superceded by deiozination process.

3.7.5 Ion-Exchange Process (Deionizers or Demineralizers)

Several organic polymeric–resins, such as polystyrene with functional groups ($-^{+}NH_3OH^-$, $-SO_3^-H^+$) on benzene rings, are now available which behave in a manner similar to the zeolites. This process differs from all other processes in the fact that by it all the ions are removed. The cations are replaced by H^+ and anions by OH^-. Water to be treated is generally passed through two cylinders containing the resins, one of which removes cations and the other anions. See Fig. 3.4.

Polystyrene resins :

$$\left[CH_2 - CH(C_6H_4SO_3^-\ H^+) \right]_n$$

Cation exchange resin

$$\left[CH_2 - CH(C_6H_4\,^{+}NH_3\ OH^-) \right]_n$$

Anion exchange resin

When the resins get used up, they are regenerated by acid and alkali treatments similar to sodium chloride solution treatment for zeolites. Suppose the first and second cylinders contain resins represented by the formulas H_2R' and $(OH)_2 R''$, the reactions may be represented thus :

$$CaSO_4 \leftrightarrow Ca^{++} + SO_4^{-}$$

(Any other hardness causing salt or ionisable substance will behave in a similar manner).

Cylinder I : $H_2R' + Ca^{++} \rightarrow 2H^+ + CaR'$

It is easy to see that the water coming out of the cation exchanger is highly acidic. This has to be kept in mind in selecting the materials handling it. Water from natural sources usually contains sufficiently large quantities of carbonates and bicarbonates. These ions like other anions can be removed by passing through the anion exchanger. CO_3^{2-} and HCO_3^- in the presence of H^+ form H_2CO_3 (or H_2O+CO_2). The load on anion exchanger is reduce by having a degasifier (which removes gases by mechanical means like blowing air or suction, as shown in Fig. 3.4, which will remove CO_2. Thus only mineral acidity will have to be removed by the anion exchanger.

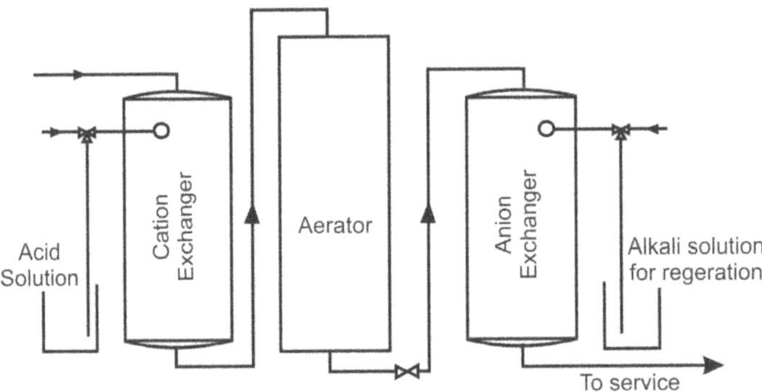

Fig. 3.4 Deionization of water.

Cylinder II : $(OH)_2R'' + SO_4^{--} \rightarrow 2OH^- + SO_4R'$

H^+ and OH^- unite to form H_2O, thus no solid is added to water. When the plant does not soften any more, the first cylinder is regenerated by passing a dilute acid solution through it and the second cylinder by passing a dilute alkali solution. The reactions during regeneration may be represented thus:

Cylinder I : $CaR' + 2HC1 \rightarrow H_2R' + CaCl_2$

Cylinder II : $SO_4R'' + NaOH \rightarrow (OH)_2R'' + Na_2SO_4$

Mixed bed deionizers are also now available. In these the sample of water is passed through only one cylinder and all the ions get removed.

This process can also be used to purify highly acidic or alkaline water. As a matter of fact, it removes all the ionizable substances and the water obtained is like distilled water. Many of the commercial plants give water with total dissolved solids as low as 3 mg/l and silica 0.1 mg/l.

The main disadvantage is that the equipment is costly and more expensive chemicals, as compared with the zeolite process, are needed for regeneration. Another important factor to be considered is the quality of water that is being softened. If it contains turbidity or a large amount of dissolved salts then the output is reduced. Turbidity must be below 10 ppm and has to be removed first by coagulation and filtration. If the water contains sufficient temporary hardness it is advisable to treat it first by lime, which removes hardness without adding dissolved salts, and then remove the remaining hardness by this process.

This process is very good for water for very high pressure boilers. For low pressure boilers, working say below 50 kg/cm^2, the equipment is rather costly and water may be treated more economically by other processes.

3.8 Feed Water Conditioning

The term 'conditioning' is a general term. It refers to a treatment given to correct undesirable tendencies or to complete softening process or to remove a difficulty created by the main process. The following are a few of such treatments.

(i) **Phosphate Treatment :** In the lime-soda process calcium is precipitated as $CaCO_3$ which is quite insoluble. However, a small amount remains in solution. This can be further removed by the addition of sodium phosphate ($Na_3 PO_4$), which precipitates the less soluble $Ca_3(PO_4)_2$; thus the hardness is further reduced. For best results, the pH value of the water should not be less than 9.5. For this purpose caustic soda is added.

1. $MgCl_2+2NaOH \rightarrow Mg(OH)_2+2NaCl$
2. $Ca(HCO_3)_2+2NaOH \rightarrow CaCO_3+Na_2CO_3,+2H_2O$
3. $3CaCO_3+2Na_3PO_4 \rightarrow Ca_3(PO_4)_2+3Na_2CO_3$

On a small scale, water almost free from hardness can be produced by adding a calculated amount of sodium hydroxide and sodium phosphate. These precipitate magnesium as hydroxide and calcium as phosphate. These reagents are costlier than lime and soda, hence for larger requirements the lime-soda treatment is preferred. To get the maximum benefit, the phosphates are added to hot water. In boilers operating at pressures up to 50 kg/cm^2 pressure about 40 mg/l of the phosphates in excess of the theoretical amount are used to complete the reactions. Excess phosphate may be 5 mg/1 or less when the pressure is 70 kg/cm^2 or higher.

Phosphate treatment also helps to control the pH value (i.e., the degree of acidity or alkalinity) of water. Trisodium phosphate, Na_3PO_4, gives alkaline reaction and can be used to increase the pH value. Sodium dihydrogen phosphate, NaH_2PO_4, is acidic and can be used to reduce the pH value.

This treatment is often given after lime-soda treatment. It has the disadvantage of causing "after precipitation" that is, gradual precipitation in the pipes, equipment, etc., (Also see internal treatment, described later in this Chapter).

(ii) Sodium Aluminate Treatment : Sodium aluminate precipitates calcium and magnesium as insoluble aluminates. Sodium aluminate hydrolyses also, giving NaOH and $Al(OH)_3$. NaOH so produced precipitates some of the magnesium as $Mg(OH)_2$. Thus on the addition of $NaAlO_2$ flocculent precipitates of $Al(OH)_3$ and $Mg(OH)_2$ are produced which enmesh finely suspended and colloidal impurities, including silica and oil droplets, which are ordinarily difficult to remove. If sodium aluminate treatment is given after lime-soda treatment no other coagulant will be necessary.

Sodium aluminate is cheaply available from bauxite refineries in the form of a thick solution and may be used as such for water softening.

(iii) Treatment for Microbiological Growth: The growth of greenish matter may be commonly observed in stagnant waters. Similar growth may take place in heat exchangers, condensers and cooling systems. They restrict the flow of water and also retard the heat transfer. Sometimes they may come out suddenly in large masses and clog the pipe or the pump. Small plant growth, which develops rapidly, is known as algae. Gelatinous matter, which sticks to surfaces and is due to micro-organism is called slime.

Micro organisms present in air and water can accumulate on heat-exchanger surface and isolate the surface from the inhibitor used to protect the surface. Bacteria produce gas which can attack stainless steel and cause pitting. A good biological control programme, which is often neglected, is needed.

The simplest way to remove them is to clean mechanically. Chlorine is very effective for the removal of both types of growth. The residual chlorine should be about 0.5 mg/l. Excess chlorine is removed by adding SO_2 or Na_2SO_3. Copper sulphate is more effective for algae than for slime. It cannot be used if pH value is above 8.5 as copper hydroxide will be precipitated. Chromates used for corrosion control help in the control of algae and slime.

Chloramine is also effective in checking these growths. Bleaching powder mixed with ammonia solution is fed in when the ammonia and chlorine combine to form chloramine and destroy the microbiological growth.

(iv) **Iron and Manganese Removal :** These are generally present in small amounts in water in the absence of dissolved oxygen. Iron and manganese give red and black precipitates, respectively, when the water containing these as dissolved impurities come in contact with air. They are highly undesirable in water for dyeing and tanning industries, and for laundries. They also give a metallic taste to water and are conveniently removed in the form of insoluble $Fe(OH)_3$ and $Mn(OH)_2$ by oxidation followed by coagulation and filtration. The oxidation may be done either by aeration i.e., bringing in contact with air (e.g., by spraying thc watcr in thc air or blowing air) or by chlorination.

(v) **Silica Removal :** About 5 to 8 mg/l of silica is present in natural waters. In some cases the amount is much more. Inside the boiler its concentration increases. Its deposits form a silicate scale which is very hard and very bad conductor and difficult to remove. One important source of silica is the sand filter. If water treated by lime-soda process is filtered though a sand filter a large amount of silica dissolves in water because of its alkalinity. Calcium and magnesium silicates act as cements and bind together other particles in to a hard scale.

In high pressure boilers the amount of silica has to be kept very low. At a pressure of about 40 kg/cm^2, 5 ppm of silica dissolves in steam and goes along with it. Its amount in steam increases to 40 ppm at 100 kg/cm^2 and is still higher as the pressure increases. Several theories of silica carry-over have been proposed. It is believed that in most cases droplets of water are carried along with steam. Water dries up in the super heaters and silica is carried forward by dry steam. It deposits on and distorts the shape of turbine nozzles and blades. Sometimes the surface becomes uneven and resistance to the flow of steam increases. In some turbines loss in efficiency of about ten per cent in five days has been reported. Catastrophic failure of the rotating parts has also been reported. Soluble silica is removed by the use of magnesium compounds during the hot lime-soda process, the sludge being recirculated. It is the sludge which absorbs silica. Magnesium oxide has been found to be very efficacious. It does not increase the dissolved solid content. Removal of silica is not a stoichiometric reaction (i.e., there are no definite combining weights). The removal proceeds by absorption from the solution.

For high pressure boilers silica is removed by strongly basic anion exchangers.

(vi) **Oil Removal :** Generally, natural waters are free from oil but small amounts are likely to come from the oils used for lubrication of the pumps. A large amount of water is obtained by condensing the exhaust steam, which may get mixed with the lubricating oil in the steam cylinder. In many cases the steam generated is used for driving the engines and exhaust steam is always available. The steam on being condensed gives water which is as good as distilled water except that it contains droplets of oil brought from the engine. In efficient plants 90 to 95% of the water requirement of a boiler is met from this source.

For the removal of oil the exhaust steam is passed through mechanical oil separators which consist of a chamber with a number of metal plates. As the steam with droplets of oil passes over the plates, droplets of oil are retained on the metal surface. Thus most of the oil is removed. The steam is then condensed for boiler use. If very high quality water is needed, sodium aluminate is added to the condensate which removes the remaining oil, as explained under sodium aluminate treatment.

According to Stromeyer, a 3 mm thick scale will raise the temperature of a plate by 150° whereas less than 0.5 mm of grease will produce a worse effect. If the oil is a vegetable or animal oil, it may get hydrolysed producing fatty acids which may have corrosive action. There is a greater possibility of the fatty acids combining with the alkalinity, producing soaps which will form sticky coatings on the metal surface. These will attract other impurities and form thicker deposits.

(vii) **Oxygen Removal :** Oxygen removal assumes greater and greater importance as the boiler pressure increases. For methods of removal see the section on "Corrosion" in this chapter.

(viii) **Boiler Compounds :** In order to check the bad effects of raw feed water, a number of substances have been used in the past, mostly without any scientific basis. A number of fraudulent compounds have been well advertised and widely used. Some of these have been found to be harmful. Therefore, one has to be cautious while using them.

The practice of using chemicals for feed water is now established on a scientific basis. Some reputed companies marketing the boiler compounds also give their compositions. If these are used by well-informed operators after taking into account the analysis of water, the water treatment becomes simpler. But such compounds are more expensive than the ingredients bought separately.

The inorganic constituents of the boiler compounds precipitate the scale forming salts and also help to control corrosion. The organic constituents are characterised by high molecular weights. These form large particles which are adsorbed on the metal or crystal surface and thus prevent formation of hard scales. The materials deposited get removed in blow down. A boiler compound used by the U. S. Navy is reported to consist of 47% anhydrous disodium phosphate, 44% soda ash and 9% starch.

It is desirable to study the analysis of water and make suitable formulations of the chemicals to be added. If such facilities are not available, boiler compounds of repute may be used.

3.9 Internal Treatment of Boiler Water

Internal treatment consists of adding chemicals to the water inside the boiler. It is mainly used as a corrective treatment, viz., to prevent scale formation, also to inhibit corrosion, steam contamination and caustic embrittlement.

After the external treatment the feed water contains only a small amount of calcium and magnesium salts. As steam generation proceeds, the concentrations of these salts increases and scales begin to form. The formation of the scales is prevented by injecting sodium phosphate directly in the boiler. This causes the residual hardness to be precipitated as calcium phosphate and magnesium hydroxide, which do not stick to the metal surface and are removed by blow down.

An ideal condition would be to convert the salts into such highly soluble ones that they are not precipitated. It is achieved to some extent by the use of sodium hexameta phosphate $(NaPO_3)_6$, which is sold under the name 'calgon'. It forms complex ions with calcium and prevent its precipitation. As it is difficult to make all the salts indefinitely soluble, it can be said that the objective of internal treatment is to precipitate them in such a physical state that scales are not formed. In such cases the physical state of the precipitate is more important than the chemical nature. Consequently the temperature, concentration pH value, reagents, etc., affect the success to a great extent. Calcium salts, if precipitated as calcium silicate, will give hard scales as they adhere to the metal surface, whereas the physical state of calcium phosphate is such that no scales are formed.

There are some cases in which the external treatment can be completely avoided and only internal treatment given. This is recommended when (i) the size of the plant is so small that installation of external treatment facilities prove expensive, (ii) the boiler is working at low or medium pressure, (iii) raw water of good quality is available and (iv) only a small amount of make up water is needed (i.e., most of the water requirement is met from water obtained by condensing exhaust steam and only a small amount of fresh water is required).

At present phosphates are widely used for internal treatment (for details see phosphate treatment). The phosphates get precipitated in the form of sludge; thus drying of the calcium salts on the metal surface and scale formation is prevented. After precipitation it is necessary to keep the precipitate as a fluid sludge which can be removed by blow-down. In low pressure boilers organic compounds like starch, lignin, tannins, etc., are added for preventing the inorganic precipitates from sticking to the metal surface. These prevent the coming together of the precipitated particles, and hence avoid their forming scales.

The above approach, of adding Na_3PO_4, which introduces caustic and phosphate in the ratio of 1:3, is known as "coordinate pH-Phosphate" method. The pH and phosphate concentrations are kept along the curve A of Fig. 3.5. Above this curve there will be free NaOH which can cause caustic embrittlement. As some PO_4^{3-} get precipitated due to Ca^{2+} and Mg^{2+} escaping prior to removal, some caustic becomes "free". To avoid this danger *"Congruent phosphate-pH control"* has been now recommended for the high pressure boilers. The ratio of caustic-to-phosphate is kept between 2.6 : 1 and 2.85 : 1. Curve B of Fig. 3.5 shows caustic to phosphate ratio of 2.6 : 1. This range is maintained by adding sodium phosphates, caustic soda, and phosphoric acid, as needed.

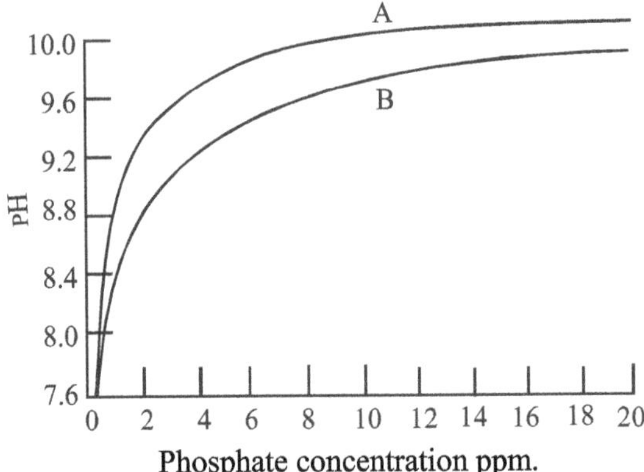

Fig. 3.5 Congruent phosphate – pH control. Exact ratio between PO_4^{3-} and OH^- is to be decided by the operating factors. For high pressure boilers, values on curves B are preferred. Above curve A there is the danger of free NaOH.

Sodium sulphite or preferably hydrazine is added to boiler water so as to remove dissolved oxygen which causes corrosion.

Chelating Agents like sodium salts of EDTA (ethylene-diamine tetra acetic acid) and NTA (nitrilo triacetic acid) are also being used, especially for treating iron. Excess of these should be avoided as they attack the metal. These produce heat-stable molecular complexes which remain in solution.

3.9.1 Blow Down

In spite of giving the external and internal treatments concentration of dissolved solids in the boiler drum and tubes increases. The low pressure boilers can have 2,500 to 5,000 ppm of solids but for the high pressure boilers it has to be 500 ppm or less so as to produce steam of acceptable purity. It therefore becomes necessary to remove a portion of water from the boiler and replace it by fresh feed water. This is a wasteful process as far as the heat aspect is concerned but is the last resort taken for the safety of the boiler. Blow down is made every few hours or made continuously.

While removing impurities that have concentrated in the boiler during steam generation blow down discharges heat that could otherwise be useful. In intermittent blow down the heat cannot be recovered. In large boilers blow down invariably is continuous. Heat recovery in such a case depends on the discharge rate and also the impurities present.

Simple calculations can be very helpful in judging the performance of the boiler and thereby showing in what direction to make efforts to improve the performance. Suppose, in a boiler the maximum total solids allowed are 1000 mg/l and it produces 50,000 kg/hour (S) of steam. The feed water has 10 mg/l of solids. To keep the concentration of solids at a constant level in the boiler, the amount removed with blow down (B) must equal the amount entering with feed water (F).

Therefore :

$$10 \text{ mg/l} \times F = 1000 \text{ mg/l} \times B$$

$$B = 10/1000 \times F = F/100$$

or $B = 1\% \times F$ i.e., 1% of the feed water is removed by making a blow down.

In case silica, suspended solids, etc. determine the blow down, calculations can be made in a similar manner.

From the above it is easy to calculate the amount of water fed in the boiler per hour.

$$F = S + B = 50{,}000 + 0.01 \times F$$

$$0.99\, F = 50{,}000$$

$$F = 50{,}000/0.99 = 50{,}505 \text{ kg/hour}.$$

These can be counterchecked from the material balance thus

$$B = F - S - 50{,}505 - 50{,}000 = 505 \text{ kg/hour}$$

As the blow down is 1% of the feed water, we have 50,505 × 0.01 = 505 kg/hour. If in actual working discrepancy is noticed, a thorough check up is required. There may be leakages of steam, water and more commonly and dangerous–leakage of cooling water in the heat exchanger, which results in the contamination of feed water.

The blow down water from a high pressure boiler is discharged into a flash tank where low pressure steam is obtained and hot water passes through heat exchanger to heat the feed water. Thus maximum heat economy is obtained (These calculations are also simple but are beyond the scope of this book).

3.10 Treating Brackish and Saline Water

The treatments discussed so far are applied to water with relatively low dissolved salt content. Due to vast regions having poor water supplies attention has been turning to water with a large amount of salt content, say, from 1,000 to 35,000 ppm. Some deep well waters, especially in the deserts, have several thousand ppm of total dissolved solids and usually water has 10,000 ppm or more of dissolved solids. Sea water has around 35,000 ppm dissolved solids. Treatment of such water directly by demineralization is very expensive.

3.10.1 Distillation

Distillation has been the common technique for getting good water from the sea. Several kinds of distillation equipment are in use. Vacuum freezing has also been adopted. Ice crystals coated with salt water are separated, washed with cold water and melted to give good water. Both these processes require a large amount of energy and the escalating cost of energy are turning attention in other directions. It may, however, be mentioned that distillation is still economically viable in the plants where a large amount of waste heat is available.

3.10.2 Membrane Processes (Electrodialysis, Reverse Osmosis)

These processes have now been developed to such an extent that some large plants are using them for treating high dissolved salt containing water prior to the common demineralization. Two of the methods are electrodialysis and reverse osmosis techniques.

In the electro dialysis or E. D. technique the dissolved mineral is driven away from the water. The basic principle is that the water is passed between paired sets of parallel membranes. An electric field is applied perpendicular to the direction of flow of water. The ions move towards the electrodes of the opposite charge. Selectively permeable membranes are chosen. The membrane near the negative electrode allows the cations to pass and the one near the positive electrode allows the anions to pass (Fig. 3.6). A large number of pairs of membranes are assembled. Plastic spacers not only separate the membranes but also direct the flow of water through a winding path across the exposed membrane surfaces. Water is sent at a pressure of about 6 kg/cm^2. This technique is favoured for water with very high dissolved salt contents, even for sea water.

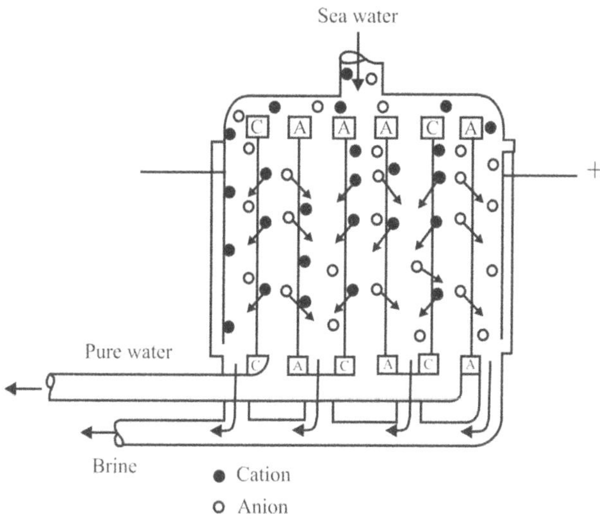

Fig. 3.6 Electrodialysis cell. In brackish water cations are removed through cation permeable membranes (C) and anions are removed through anion permeable membranes (A).

Reverse Osmosis

Diffusion is the movement of molecules from a region of higher concentration to a region of lower concentration. Osmosis is a special case of diffusion in which the molecules are water and the concentration gradient occurs across a semipermeable membrane. The semipermeable membrane allows the passage of water, but not ions (e.g., Na^+, Ca^{2+}, Cl^-) or larger molecules (e.g., glucose, urea, bacteria). Diffusion and osmosis are thermodynamically favorable and will continue until equilibrium is reached. Osmosis can be slowed, stopped, or even reversed if sufficient pressure is applied to the membrane from the 'concentrated' side of the membrane. Since reverse osmosis is thermodynamically unfavourable process, it requires a driving force to push the fluid through the membrane, and the most common force is pressure from a pump. The higher the pressure, the larger the driving force. As the concentration of the fluid being rejected increases, the driving force required to continue concentrating the fluid increases.

The reverse osmosis process is used in a wide range of applications. This includes waste water treatment, production of pure water for industries [e.g., wafer fabrication], and removal of salts from salty or sea water [desalination].

How does it work?

Let us start with the normal osmosis process. Consider a semi-porous membrane that separates equal volumes of liquid into two compartments. In one compartment is water, and the other dilute salt solution [see Fig. 3.7(a)]. Due to the effect of osmosis, water flows from side containing water to the side with the salt solution. Water flows over because it is trying to dilute the salt solution. This continues until an equilibrium state is reached.

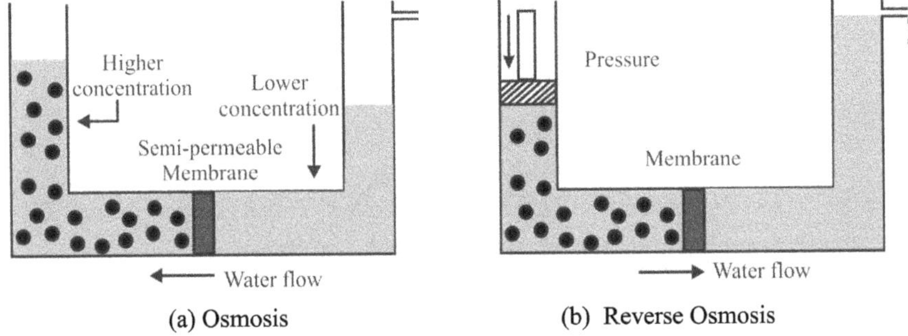

(a) Osmosis (b) Reverse Osmosis

Fig. 3.7 Reverse Osmosis Cell.

For reverse osmosis, the aim is not to dilute a salt solution, but to purify water by removing salts and other contaminants. This is done by applying a pressure to the side containing the salt solution. It causes water to flow to the other side of the membrane [see Fig. 3.7(b)]. This purifies the water, since the membrane prevents salts and other particles from passing through.

Semipermeable membrane : Many synthetic membranes that can withstand high pressures have been developed. A typical reverse osmosis membrane has 2 layers :

- a dense top layer of less than 1 μm think
- a porous sublayer of 50-150μm thick, for supporting the top layer

Reverse osmosis is also known as hyperfiltration and is the finest filtration known. Reverse osmosis is not only used to purify water and remove salts but to remove other impurities in order to improve the color, taste or properties of the fluid. Reverse osmosis is capable of rejecting bacteria, salts, sugars, proteins, particles, dyes, and other constituents that have a molecular weight of greater than 150-250 daltons. The larger the charge and the larger the particle, the more likely it will be rejected.

Further this process can be used to purify fluids such as ethanol and glycol, which will pass through the reverse osmosis membrane, while rejecting other ions and contaminants present in them.

Reverse osmosis units remove ionizable as well as non-ionizable matter, both colloidal and high molecular weight organic matter. This is a distinct advantage and sounds very attractive in dealing with the problem of colloidal silica, which is not removed by demineralization, for water for supercritical boilers. Electro dialysis units require more skilled labour who can handle the electrical and mechanical auxiliary units. Their initial cost for low dissolved solid range is slightly lower than the R.O. units. Cellulose acetate affixed to either side of perforated tube is a popular membrane. Plants giving several million litres of water are available. At the present R.O. units are being preferred. Suppliers claim efficiencies from 85 to 99%.

The E.D. and R.O. units have attained prominence during the last few years and are particularly useful when feed water has a large amount of dissolved solids. When used before demineralization the chemical requirement for demineralization is reduced and the life of the resin is increased.

In the reverse osmosis process the maintenance cost is almost entirely on the replacement of the membrane. It is reported that in a plant the life of the membrane is quite high, about three years. When the conductivity measurement indicates that excess dissolved solids are going in the water, the module can be replaced within a few minutes thus giving an almost uninterrupted supply.

In the reverse osmosis process there are certain problems connected with the fouling of the membrane and design is yet to be perfected, but the results are very encouraging and it seems that in the future R.O. technology will be used in conjunction with mixed bed demineralization for very high pressure and super-critical boilers, where the impurities are measured in parts per billion instead of parts per million.

Just to give a rough idea it may be mentioned that in one case the dissolved salt decreased in the feed water from about 4500 ppm to about 350 ppm, conductivity from about 6000 μmho to about 500 μmho and silica from 7 ppm to 2 ppm after passing through a spirally wound polyamide reverse osmosis unit. The water now becomes suitable for treatment by demineralization.

An interesting case is that of a two-unit 800 MW power station at Cadafe, Venezuela, S. America. The plant was nearing completion in 1973 but public water supply was not ready. Alternative sources considered were well and sea water. Treatment methods for water from the sea–which provides uninterrupted water supply–considered were waste–heat desalination, distillation and reverse osmosis technique. The deciding factors were capital cost, operating cost, reliability, simplicity of operation, ease with which capacity could be expanded in the future and delivery time. In that area well water was not considered a reliable source for the large quantities of water needed.

In 1978 the capital investment required for a 12 million litre per day plant was estimated to be 4.5 crore rupees for reverse osmosis technique and 3.8 crore for multistage distillation; operating costs were almost similar, the distillation being slightly costlier. In spite of higher investment costs the reverse osmosis was preferred because the energy costs are rapidly increasing and it was felt that distillation cost would increase considerably.

In the Cadafe plant the sea water is treated with chlorine and ferrous sulphate, to destroy biological growth and to coagulate, filtered twice to remove fine particles, excess chlorine removed, pH adjusted and then pumped at 60 kg/cm^2 pressure through the reverse osmosis unit which has a number of modules of the semi-permeable membranes. The primary unit reduces the total dissolved solids from 38,000 ppm of the sea water to 1,000 ppm which is treated in the secondary unit that gives water with 100 ppm of the total dissolved solids. Necessary quantity of water is taken for supply as potable water and the remaining passed through demineralizers giving high quality water for the high pressure boilers.

3.10.3 Use of Condensate

After performing the intended mechanical work the steam is condensed to give water of high purity. In the process, water picks up corrosion and erosion products which tend to deposit in boilers and super heaters and on turbine blades. Deposition of corrosion products promotes further corrosion. Leakages of condenser tubes are not uncommon. This is an important source from which considerable amount of dissolved salts are introduced. It is, therefore, absolutely necessary that chemical analysis of the condensate is made frequently and accurately (Fig. 3.8). In very high pressure and supercritical boilers it is a routine practice to filter the condensate and pass through demineralizer.

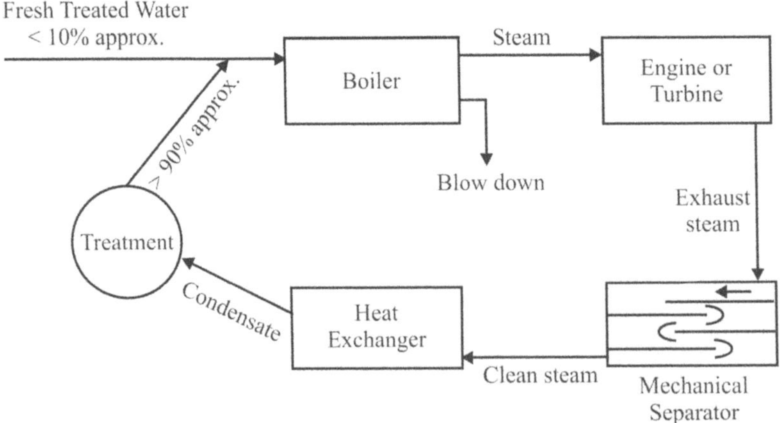

Fig. 3.8 Use of condensate.

3.11 Water for High Pressure Boilers

Although there is no universally accepted terminology, boilers working above 50 kg/cm^2 pressures are known as high pressure boilers. Above 180 kg/cm^2 steam and fluid are indistinguishable and supercritical range begins.

We have several large thermal power plants that are in operation and some more are in the planning. They generate steam at 100 kg/cm^2 pressure or at even higher pressures. Water for such boilers has to be of much superior quality to the distilled water commonly used in the laboratories. To give an example, in India large thermal power plants are having about 200 MW turbines and boiler attached to each unit generating 640 tonnes of steam per hour at a pressure of 137 kg/cm^2. The biggest unit under construction is a 500 MW turbine, with associated boiler, by the Tata Electric Co. at an estimated cost of Rs. 180 crores. If high quality water containing 1 mg/l of solids is supplied to a boiler generating 640 tonnes per hour of steam there will be almost six tonnes of solids left in the boiler per year. The working temperature and pressure being high, even thin scales have disastrous effects. Moreover, as has already been explained, silica volatilizes with steam under such conditions, and as steam expands through the turbine, silica deposits on the blades of the turbines and also on the nozzles, valves, etc. All these greatly reduce the efficiency. For all these reasons special care has to be exercised in the water treatment. Silica removal is the greatest problem for such boiler feed water. Silica in steam has to be kept below 0.02 ppm (Fig. 3.9). For this purpose silica in boiler water has to be checked. It is to be kept < 10 ppm for 60 kg/cm^2, < 3 ppm for 100 kg/cm^2, < 1 ppm for 150 kg/cm^2 and even less for still higher pressure boilers. It may be noted that more than 90% of water used in the boilers is the condensate i.e., the condensed steam after use in the turbines. Under ideal conditions the water so obtained should need no treatment, but its quality has to be checked carefully for there is the danger of contamination with the impurities most commonly due to leakage in the condensers which introduce untreated cooling water.

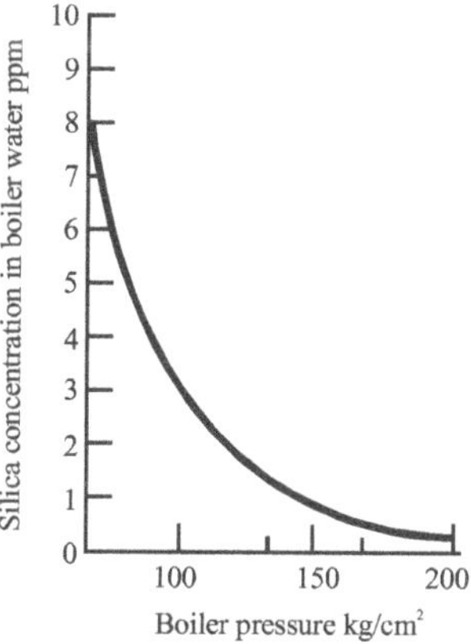

Fig. 3.9 Limits of silica in boiler water (pH about 11) as a function of boiler pressure for keeping silica in steam below 0.02 ppm.

In some countries 750 MW turbines are already in operation. The boiler for such a turbine is what is called "Once through-boiler" i.e., water is not circulated but passes only once through the heated tubes and is completely converted into steam. This means that any solid matter present will form deposits which will have disastrous effects due to the high temperature and pressure to which the tubes are exposed. A boiler for 750 MW turbines produces about 2,500 tonnes of steam per hour at a pressure of 250 kg/cm^2. The steam is superheated to 550 °C. Extreme care has to be taken for its water treatment, the impurities being measured in parts per billion parts. Even the condensate steam is passed through four polishing units of mixed bed demineraliser. There is a fifth stand-by unit. Each of these has 4.3 m^3 of cation and 1.6 m^3 of anion resin. The resins are in the form of porous beads which make them function as good filters for particulate matter and colloids.

Commonly the raw water is passed through continuous clarifiers in which coagulants, like alum, are added. If the pH value is not correct, lime is also added. It is not possible to prescribe any set pH value for a coagulant because it also depends upon the impurities present. For example, whereas the best pH value for alum in distilled water is 5.5, it may be of any value between 5.5 to 8 in natural waters. Often there is the problem of slime and algae formation in the clarifying tanks. Chlorine and, if necessary, copper sulphate can be used to deal with the organic growth. Water is then filtered through mechanical filters. Final filtration through a bed of activated charcoal is desirable to remove organic matter, colouring matter, etc. completely.

Overnight or week end shut downs expose the surfaces to air, dust, and, may be, rain with the result that the first condensate collected after the starting is highly contaminated.

The clear water is made free from all the ions by directly passing through a series of ion exchangers. Several combinations of cation, anion and mixed bed exchangers are in use. They are chosen on the basis of the types of dissolved salts present. A simple arrangement is: hydrogen cation exchanger, degasifier and finally anion exchanger, Fig. 3.10(a). Water coming out of hydrogen cation exchanger is highly acidic and contains dissolved carbon dioxide, produced from the carbonates and bicarbonates. It is cheaper to remove it by mechanical means in a degasifier than by removing in an anion exchanger. Thus the load on the anion exchanger is decreased.

It may be economical to use a weakly basic anion exchanger between the cation exchanger and degasifier as shown in Fig. 3.10(b). The weakly basic anion exchanger removes strongly ionized mineral acids but not the weakly ionized carbonic acid produced from the carbonates and bicarbonates. The weakly basic anion exchangers cost less and are cheaper to regenerate. The carbonic acid gas (CO_2) is removed in the degasifier. On passing this water through a strongly basic anion exchanger, a more complete removal of anions and of silica takes place.

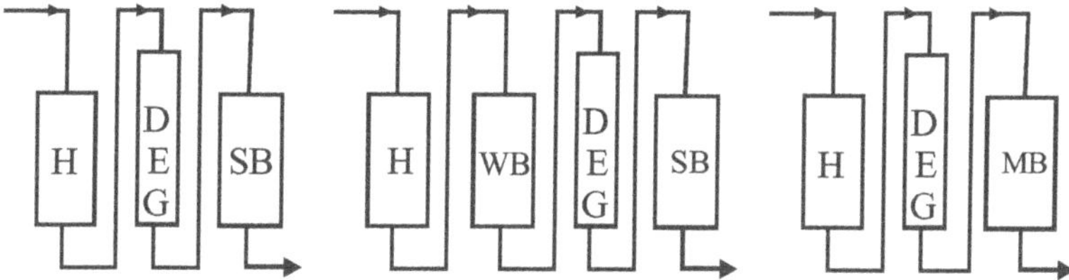

Fig. 3.10 Arrangement of exchangers. H—hydrogen exchanger; DEG—degasifier; SB—Strongly basic anion exchanger; WB—weakly basic anion exchanger; MB—mixed bed exchanger.

Mixed bed deionizers have become quite common. In these the bed consists of an intimate mixture of resins of both hydrogen cation and strongly basic anion types. As water flows through the bed, it comes into contact many times with the two types of resins alternately. The effect is like passing water through a series of several cation and anion exchangers, causing the removal of ions and silica much more completely. The effluent, i. e. the outgoing water contains less than 1 ppm of the electrolytes. For regeneration after the resins have got exhausted, water is slowly sent in the opposite direction where due to the lower specific gravity the anion resins form the upper layer and the cation resins collect in the lower portion. The two types of resins may be taken out and regenerated separately and mixed again. Exchangers are in use in which it is not necessary to take out the resins but to simultaneously regenerate the anion resins by passing caustic alkali from the top and the cation exchanger by passing acid solution from the bottom. The wastes are removed through outlets at the interface of the two beds. The beds are rinsed free from the by-products and excess regenerates and a current of air are blown upwards to effect an intimate mixing of the two types of resins. The unit is ready for use again. An important factor to be noted is that if any cation resin comes in contact with caustic soda used for the regeneration of the anion resin the cation resin picks up Na^+. It releases Na^+ and takes on NH_4^+ present in the condensate using ammonium compounds. Na^+ is harmful in super critical boilers as it deposits on turbine blades. The mixed bed exchangers cost more and are more expensive to regenerate but they are more convenient to use. They remove the electrolytes and silica much more completely.

Depending upon the quality of water available, the various units may be arranged in several different ways to give most economically water of the best quality. As an illustration, the arrangement shown in Fig. 3.10 (c) removes most of the cations in the hydrogen cation exchanger, then the degasifier removes CO_2 with the result that the effluent contains a low concentration of cations but has the sulphates, chlorides, nitrates and also silica. These are completely removed in the mixed bed exchanger. It is easy to see that this arrangement, which is only one of the several possible arrangements is cheaper and, perhaps, more efficient than a single mixed bed exchanger.

3.12 Ultra Pure Water Technology

3.12.1 Supercritical Boilers

As above about 180 kg/cm^2 pressure separation of steam from fluid is not possible, the drum type boilers cannot be used and "once-through steam generators" are used. These pose several problems. The most serious problem faced in such a system is that water entering the circuit passes right through as a supercritical fluid. This means that all the solids present in the feed-water are carried to the steam. The solids form deposits in the super-heaters and also in the turbines. The following upper limits for the impurities are prescribed:

Total dissolved solids, 50 to 500 ppb (parts per billion), Iron 10 ppb; Silica, 20 ppb; Oxygen 5 to 7 ppb; pH 9 to 9.6 (to be controlled by the use of ammonia).

Ultra-high purity water is needed for the high capacity turbines which limit steam impurities to 1 ppb and below. Even in the usual good water treatment some silica and sodium escape. Besides stricter control over demineralisers for make-up water, greater attention is being paid to condensate polishing to remove practically all dissolved solids and salts.

For lower pressure boilers weakly-acidic cation and weakly basic anion exchange resins are found to be economical. These can be regenerated more efficiently and have higher capacities than the strong electrolyte-exchange resins. Weakly basic anion-exchange resin when used ahead of the strongly basic resin, removes chlorides and sulphates and is regenerated more economically and efficiently. It removes most of the humid matter which causes fouling and poisoning of the strongly basic resin. It may be noted that weak acid and weak basic resins can be regenerated with waste regenerant discharged from the strong acid and strong basic resins respectively.

If boiler-feed water is not relatively free of suspended and colloidal matter poor utilization of the ion-exchange bed will result. Even minute amounts of suspended solids and colloids on the surface of the resin bed increase the pressure drop and cause channeling effect with the result that portions of the bed are completely bypassed and some impurities reach the boiler. Another cause of channeling is accumulation of gas bubbles around the beads causing them to float.

In condensate polishing the resins filter the condensate besides performing their main function of ion exchange. The filtration is more effective through the resin bed than through the inert media like sand, coal, etc., because the highly charged surface of the resin attracts the colloidal particles. During the back-wash, while regenerating the resins, these need to be removed.

In efficiently operated plants deep mixed resin beds have polished condensate to impurity levels as low as 0.1 ppb which approaches the limit of analytical detection. Resins ground to powder are applied on conventional tubular filters for final polishing of the condensate. These are not regenerated but are discarded after exhaustion.

Powdered resin technology has grown in importance and has become well established in the field for the production of ultra pure water.

When presence of Na^+ is not objectionable sodium cycle cation exchange resins (i.e., resins having Na^+ as the inter-changeable ion associated with the cation resin instead of H^+ as the interchangeable ion) may be used. These can treat hot condensate (70° to 15° C) causing heat economy.

For ultra-high purity water complete separation of cation and anion resins in mixed bed demineralizer is of utmost importance. If this is not achieved some cation resin beads coming in contact with caustic soda will acquire Na^+ instead of H^+. This Na^+ will be released later spoiling the purity of water. In a likewise manner Cl^- or SO_4^{2-} will be introduced from HCl or H_2SO_4 used for regeneration of anion resins. The greater the difference in the densities of the two kinds of resins the better will be the separation. It may be remembered that density difference between the anion and the cation resins decreases with age causing increased difficulty in separation. Several methods of regeneration are in operation to solve the problem. One of the simplest and latest techniques is to use inert beads having density between those of the cation and anion resins. Particle size of the three kinds of resins (cation, anion and inert resins) is also to be controlled. It is advisable to have during regeneration an inert layer of about 20 cm between cation and anion resins. The pipe for collecting the regenerants is placed in this region. Alternatively, the anion and cation resins are removed and regenerated separately. The principle of working remains the same as described earlier for demineralization.

In most plants, water enters at the top and leaves through the exit at the bottom. The regenerant also flows in the same direction. This is known as *co-current* method. In this method the resin near the exit is not fully regenerated and during working some ions are introduced in the initial stage of operation.

To overcome this difficulty the regenerants are sent in the opposite direction, i.e., in the *counter-current* direction. In this practice water to be treated enters at the top and leaves from the bottom and the regenerant enters the bottom and leaves at the top. There are some problems associated with this. To overcome them regenerant for anion resin is sent from the top and for cation resin from the bottom, simultaneously, and collected at the interface layer of the inert resin bed. This is known as *co-counter current* method. For details of the working a monograph on the subject needs to be consulted.

3.12.2 Nuclear Power Plants

In the pressurized-water reactor (PWR) the core of the nuclear reactor is surrounded by water under high pressure, say, 150 kg/cm^2, and steam is generated in a boiler or heat exchanger located adjacent to it. In the boiling-water reactor (BWR) the nuclear core is surrounded by water to produce steam directly. In both cases water treatment is similar to the high pressure boilers. The BWR exposes water to intense neutron flux and also causes activation of impurities or corrosion products. There is also a certain amount of decomposition of water to hydrogen and oxygen.

3.13 Municipal Water Supply

Perhaps the most important problem of water technology that of municipal water supply, because if the water supplied to the cities is not of the right kind, the results will be disastrous to human lives. The requirements for this purpose are very different from those of the industries. Municipalities have to supply potable water. i.e., water which is safe and good to drink. Potable water does not mean pure water. Distilled water, for example, is flat and tasteless. At the same time potability cannot be determined merely on the basis of its taste, odour and appearance. Water may have disease producing bacteria which cannot be detected by these means.

Briefly, water supplied by the municipalities should be free from disease-producing bacteria, suspended impurities and excess of dissolved salts or chemicals used for water treatment. It should be clear, colourless, odourless and pleasant to taste. A sample of clear water which has low dissolved salt content but has disease-producing bacteria can be used in a boiler, but would be unfit for drinking. On the other hand, a sample of clear water, free from bacteria, but containing a certain amount of dissolved salts may be unfit for boiler feed, yet fit to drink.

Municipalities draw their water from rivers, wells and lakes. Water so obtained contains dissolved matter which in most cases is harmless for drinking. In general, municipal water treatment is not directed towards removal of the dissolved salts present in water. Water, such as sea water which contains a large amount of dissolved matter is not commonly looked upon as a source of drinking water. But, as good quality water is not uniformly distributed on the surface of the earth, and demand for drinking water is increasing due to increase in population, a considerable amount of research is being done, especially in the U S A, on *conversion of sea water* into drinking water. Some experimental plants, each producing over half a crore litres per day of potable water from sea water, are functioning in the U S A. Two avenues of approach for obtaining potable water from sea water are distillation, and freezing out pure water as ice, leaving the salt in the mother liquor. Distillation is a well-known process but the salts present in sea water cause severe corrosion of the equipment. Freezing is more expensive but there is less corrosion and is likely to be successful in the cold climates.

The latest approach in converting sea water fit for drinking is the application of reverse osmosis technique. On account of low capital cost, low operating cost, reliability and simplicity of operation the reverse osmosis process is becoming the principal method for this purpose. Such plants have been mainly installed in the petroleum rich countries of the Middle East.

Municipalities in general do not soften the water. There are only a few municipalities in the world which partially soften the water because the water available to them is very hard.

The treatments more widely given include: 1. Sedimentation, 2. Filtration, and 3. Sterilization. See Fig. 3.11.

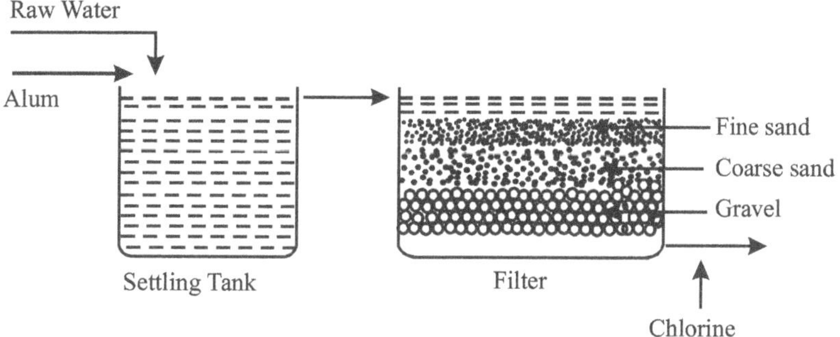

Fig. 3.11 Principle of municipal water treatment.

1. Sedimentation

Sedimentation means simple settling. It is the process by which the force of gravity acts on particles heavier than water and causes them to move downwards and settle. For this purpose water is kept in huge tanks, called sedimentation tanks (or basins) in which water moves very slowly. The basins have to be designed carefully so that the formation of eddy currents (which cause mixing) or straight channeling of water is avoided.

The time required for sedimentation depends upon the weight, size and shape of particles and the viscosity and frictional resistance of water. Because viscosity changes with temperature, the settling rate varies in different seasons. It has been calculated that the settling rate at 0° is only 45% of the rate at 30°. Consequently turbidity is more easily removed during summer than during winter. Water is kept in sedimentation tanks from 4 to 12 hours.

Finely divided suspended matter does not settle down easily. Generally, only 70% of the suspended matter settles down by plain sedimentation. For better clarification, chemical means have to be used. This is done by the addition of coagulants like alum, ferrous sulphate, ferric chloride, sodium aluminate, etc. Each of these works best within a particular pH range and has a certain function in addition to clarification. Generally about 95% of the suspended matter is removed by coagulation.

Coagulation reactions are complicated but a simplified explanation can be that the salt hydrolyses and produces a floc of the hydroxide, e.g., $Al(OH)_3$ and $Fe(OH)_3$. The flocculent precipitate so produced has an enormous surface area per unit volume. The floc removes the finely divided and colloidal matter by a variety of mechanisms, e.g., by adsorption, mechanical entanglement and electrical attraction of oppositely charged particles. Thus the smaller particles gather together to form bigger particles which quickly settle down. Some bacteria and colour are also removed. The dosage of the coagulant depends upon turbidity and may be 5 to 100 mg/l.

These years increasingly large quantities of *coagulant aids*, mostly polyelectrolytes, are being used. The polyelectrolytes are high molecular weight water soluble polymers. Some naturally occurring organic flocculants have been used. Polyacrylamides aid materially in the formation of rapidly settling flocs. Other organic polymeric materials which are strongly cationic have been used in place of aluminium or iron salts. These function through hydration. The true polyelectrolytes contain groups which undergo electrolytic dissociation giving highly charged large molecular weight ions which attract other charged particles and help coagulation. They may give anions, cations or may be non-ionic (i.e., provide both positive and negative charges in solution). Considerable research work is being done in this field but the mechanism of their activity has not been well understood.

Each coagulant gives best results in a pH range which is affected by the mineral characteristics or the water. For Example, in pure water, potash alum coagulates best at a pH of 5.5 but for most natural waters the range is 5.5 to 8.0. It should be noted that each ppm of added alum, consumes about 0.5 ppm of natural alkalinity thereby reducing the pH level. So lime or soda needs to be added. The presence of iron and manganese in water introduces additional factors affecting the choice and dosage of coagulant. It is therefore necessary that laboratory tests be made with different coagulants at varied pH levels and varied concentrations. Retention time of 2 to 6 hours is usually provided. Often coagulation can be combined with sedimentation followed by filtration. Potash alum and ferric alum are the most widely used coagulants.

2. Filtration

Filtration may be defined as 'the process of passing a liquid containing suspended matter through a suitable porous material in such a manner as to effectively remove the suspended matter from the liquid'. Generally, water is filtered after sedimentation and coagulation. In some cases where there is great shortage of space or the water is quite clear, it may be filtered directly thus saving large areas needed for the construction of sedimentation basins. Filtration is primarily a mechanical process. When water flows through a porous material many suspended particles are unable to pass through the interstices (or gaps) in the material because of their size. Thus these particles are removed and clarification takes place.

Filtration removes the suspended matter and to some extent bacteria. True colour is not removed by filtration. It can be removed by the use of coagulants.

Filters are of two types: (a) gravity filters and (b) pressure filters.

(a) *Gravity filters.* In these the water flows through the bed by gravity. These filters are of two types: (i) Slow sand filters and (ii) Rapid sand filters.

(i) *Slow sand filters* are still being used in India by many municipalities. They occupy large areas and the rate of filtration is slow. The filters are like large shallow tanks made of concrete. At the bottom there is an under-drain system through which the filtered water goes to the storage tank. Gravel is placed in the tank above the under-drain system. It consists of several layers of stone of sizes varying from 6 cm to 0.5 cm. On the gravel is placed a bed of coarse sand and above it a layer of fine sand is placed. See Fig. 3.11.

As water percolates through the filters, the suspended particles are retained on the top 5 to 10 cm layer of the fine sand. In fact, the layers of coarse sand and of gravel do not filter at all. They support the fine sand, the top layer of which retains the suspended matter. When the suspended matter has been deposited to such an extent that the rate of filtration becomes very slow, the portion of the top layer is scraped and the bed is used again. After continued use of the bed for a long time, it becomes necessary to replace the layer completely.

(ii) *Rapid sand filters* are coming into greater use primarily because their efficiency for filtration is many times that of slow sand filters, hence, they need much less space. The general arrangement of the sand bed is similar to that of slow sand filters, the distinguishing feature being the backwash system. The chief reason for a slow rate of filtration through the slow sand filter is that as the suspended matter is deposited, the gaps between the sand particles get filled up and the flow is restricted. To overcome this difficulty, the rapid sand filters are provided with backwash system. After filtering for some time the filtration is stopped and a portion of filtered water is *sent up* through the bed. This removes the suspended particles deposited in the pores and cleans the bed. When the suspended matter has been removed, the wash water supply is stopped and the filter bed which is as clean as a freshly made bed, is again realy for filtration. The cycle of filtration and washing is repeated. Thus high filter rates are obtained. The wash water is discarded. The rate of filtration may be 30 times that of the rate of filtration through slow sand filters.

(b) *Pressure filters* have essentially the same arrangement of the filter media as gravity filters. The filter material is kept in a closed cylinder and water is forced through the filter bed under pressure. Pressure filters can be installed in the water supply line, thus repumping of filtered water is avoided. As these are enclosed, they are more suitable for filtering hot water. If hot water is filtered through gravity filters, there is more loss of heat. The pressure filters occupy much less space than the other types of filters. Because of these advantages pressure filters are widely used for industrial water conditioning.

3. Sterilization and Disinfection

After filtration it becomes necessary to destroy or make inactive the residual bacteria and to prevent their further growth. This is achieved by sterilization and disinfection. Often ground water is free from disease-producing bacteria but the surface water, which is the main source of municipal water supply, is likely to be contaminated, especially because of the discharge of sewage into streams. Some epidemics, like cholera, plague and also typhoid can be controlled by taking better care of drinking water.

Chlorination is mainly used for killing disease-producing organisms. Further, it reduces (or may even remove) objectionable tastes and odours from drinking water, and oxidizes iron, manganese and hydrogen sulphide, and thereby help their removal. It also causes oxidation, decolourization and destruction of organic matter. Because of wide range of action chlorine, besides its use in water treatment, is injected into sewage and industrial waste before being discharged into streams. In such cases it causes reduction in bacteria, destruction of chemicals that may react with oxygen, such as sulphides, sulphites, ferrous ions, etc. It helps coagulation by destroying organic matter, hence the practice of pre-chlorination. In water used for food processing industries chlorine should be used cautiously. It may spoil the flavour and even corrode the metal container.

Small volumes of water can be sterilized by boiling. For larger volumes chlorine is very widely used.

Bleaching powder is convenient for use on smaller scales. Its disinfecting action is due to the chlorine made available by it. When the chlorine requirement is over about 10 kg per day, it is more economical to use chlorine gas. The reactions of chlorine and water can be represented thus.

$$Cl_2 + H_2O \rightleftharpoons HOCl + H^+ + Cl^-$$

At higher pH values hypochlorous acid ionizes thus:

$$HOCl \rightleftharpoons H^+ + OCl^-$$

Earlier it was believed that the disinfecting properties of chlorine were due to the nascent oxygen liberated, which oxidizes harmful matter. More recent work has shown that the HOCl molecules (not hypochlorite ion, OCl^-) combine with an enzyme in the cells of the organisms and thereby destroy the organisms. Therefore, chlorine is more effective as a disinfectant at lower pH values (below 6.5). A part of the chlorine added is used in combining with or oxidizing some of the impurities. The amount of chlorine added is generally such that about 0.2 mg/l of chlorine is left behind after the reactions.

If the water is exposed to the atmosphere, especially to sunlight, chlorine is dissipated quickly. In order to overcame this disadvantage, ammonia may be added along with the chlorine when chloramines like NH_2Cl and $NHCl_2$, are formed. These are not as active as chlorine but are not so easily removed, and are more desirable when water is to be distributed over long distances.

Chloramines require higher concentrations and longer contact time for comparable bacterial reduction. Chloramine tablets are used by the army for sterilizing water. Water is collected in the bottles and chloramine tablets are added followed by sodium thiosulphate tablets which remove excess, of chlorine.

Break point chlorination is a specialized method of chlorination applied to highly polluted water and for the removal of ammonia. In the beginning as chlorine is added to such samples of water there is a proportionate increase in the amount of residual chlorine as is indicated in the Fig. 3.12. Ammonia is converted to monochloramine and then to dichloramine. Further chlorination gives trichloramine or oxidizes ammonia to free nitrogen. At this stage there is a marked decline in the residual chlorine but further addition of chlorine gives a proportionate increase of free available chlorine. Because of the peculiar character of the curve obtained on such chlorination, this practice is known as "break point chlorination".

Besides its use in the removal of organic matter from highly polluted water, break point chlorination is useful in removing ammonia from boiler feed water.

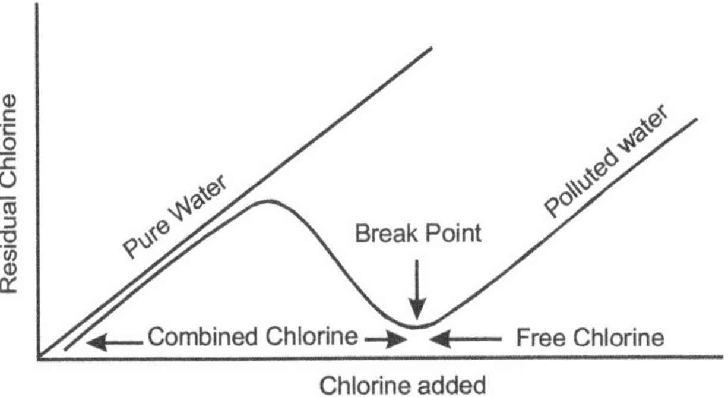

Fig. 3.12 Break point chlorination.

Disinfecting agents other than chlorine are also used. The use of ozone for disinfection has been suggested. It is expensive, hence is only used on a small scale. Bromine has been used in swimming pools. Silver iodide, in soluble form, is a good disinfectant but it is also expensive. In World War II pellets of silver iodide were used by the armed forces. Exposure of clear water to ultra violet rays also destroys the bacteria. None of these methods are adopted for disinfection on a wide scale.

In India well water is still used to a considerable extent. This can be disinfected by adding potassium permanganate solution to the well. Chlorine can also be added.

3.13.1 pH

pH is a term used to express the degree of acidity or alkalinity of a solution. It expresses the hydrogen ion concentration of the solution. In natural waters, waste waters, or solutions the engineer is not so much concerned with the total amount of the various acidic or alkaline materials but is directly concerned with their behaviour. This behaviour is responsible for the corrosion of the equipment, which is due to low pH values, and the scale or deposit forming tendency in the boilers, pipe lines, heat exchangers, condensers etc., which is more pronounced at higher pH values. In water supplies, pH control is necessary for the coagulation, disinfection and softening reactions. In sewage and waste disposal systems, pH helps to decide which particular organism will bring about decomposition most satisfactorily. In the chemical industry, adjustment of pH value enables one to have a closer control over the reactions.

Theory

According to the theory of electrolytic dissociation or ionic theory, propounded by Arrhenius in 1887, the substances which dissociate to give excess of hydrogen ions are acids and those which give excess of hydroxyl ions are alkalis. In any aqueous solution, both hydrogen and hydroxyl ions are always present. If the concentrations of these are the same the solution is neutral. If hydrogen ion concentration exceeds hydroxyl ion concentration it gives acidic reaction. Alkaline reaction is given by a solution in which hydroxyl ion concentration is greater than hydrogen ion concentration. Consequently, a measure of these concentrations is a measure of the intensity of the acid or alkaline condition of a solution.

Pure water dissociates to a very small extent as indicated below;

$$H_2O \rightleftharpoons H^+ + OH^-$$

The chemical equilibrium of the reaction can be expressed as :

$$\frac{\left[H^+\right]\left[OH^-\right]}{\left[H_2O\right]} = K$$

Since the concentration of water is very large and is almost unaffected by the slight ionization of water it can be taken as constant and the equation can be written as :

$$[H^+]\ [OH^-] = K_w$$

At 25°, 1 g of hydrogen ions is present per 10,000,000 litres of water, or hydrogen ion concentration is 10^{-7} g/l. Since pure water is neutral, this is also the concentration of hydroxyl ions. Therefore, the ionization constant for water at 25^0 is given by

$$[H^+]\,[OH^-] = 10^{-7} \times 10^{-7} = 10^{-14}.$$

On the addition of an acid to water, the H^+ concentration is increased and because the product of the ionic concentrations is a constant, the OH^- concentration is decreased, e. g. if the $[H^+]$ is increased to 10^{-5}, the $[OH^-]$ will become 10^{-9} so that the value of the constant remains unchanged :

$$10^{-5} \times 10^{-9} = 10^{-14}$$

From this it is seen that by knowing the concentration of one of the two species of ions, the concentration of the other can be known. Hence the acidic as well as the alkaline nature of a solution can be expressed in terms of concentration of either of the species of ions. For convenience H^+ concentration has been universally accepted.

Expressions of the type 10^{-5} are awkward to use in routine work. So the Fig. in the index without the sign e.g. 5 is used instead of 10^{-5} and called pH value. Exactly defined, pH *value is the logarithm of the reciprocal of hydrogen ion concentration* and can be represented by :

$$pH = log\ \frac{1}{\left(H^+\right)}$$

In simple words, pH value is a number between 0 and 14 which expresses various degrees of acidity and alkalinity, and the pH scale can be represented thus :

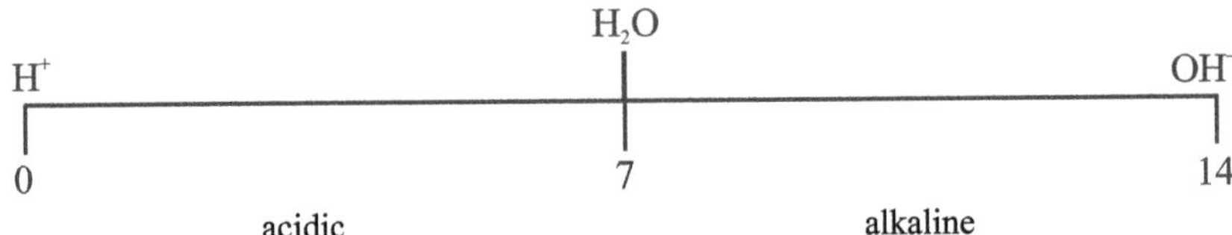

It is noteworthy that the pH scale is logarithmic. Therefore, each number on the scale represents 10 times the hydrogen ion concentration as compared with the following number. For example, pH of 4 indicates 10 times the hydrogen ion concentration than pH of 5 and 100 times than that of pH of 6 and so on.

Most natural waters have pH values, between 6.0 and 8.0 but. 7.2 to 7.8 is the more common range. Water of pH less than 7.2 is not good for the human system. For boiler feed water, pH is adjusted up to about 10.5 to prevent corrosion and also to ensure the precipitation of scale forming salts during the chemical softening reactions. At lower pH values water has corrosive action on mild steel.

3.13.2 Measurement of pH Value

1. **Colorimetric Method.** The colorimetric measurement is based on the assumption that when a definite quantity of an indicator is added to a certain fixed volume of different solutions of the same pH value, the colours of these solutions will be the same. This assumption holds good for most of the solutions which are commonly come across, but there may be interference due to some ions. The working of the colorimetric method can be easily understood with the help of an illustration from acidimetric and alkalimetric titration. If methyl orange indicator is added to a sodium carbonate solution, the solution becomes yellow. If a small amount of an acid is now added to it the colour seems to be unchanged but near the endpoint the addition of each drop of acid brings about a definite change in the colour of the solution. If only a few drops of the acid are added in excess, the amount added can be guessed by observing the colour, but with a large excess of the acid no such guess can be made. This shows that over a small range the colour can indicate the acidic or alkaline nature of the solution. In other words, colour can indicate the hydrogen ion concentration.

An indicator is a substance which gives different colours or different shades of a colour at different pH values. Therefore, by comparing the colours with the known standards, it is possible to tell the pH value. As the indication given is over a small range only, a number of indicators have to be used. Common indicators that cover the usual engineering range of pH are noted below :

Indicator	*pH range*	*Colour change*
Methyl red	4.4 to 6.2	Red to yellow
Brom thymol	6.0 to 7.6	Yellow to blue
Cresol red	7.2 to 8.8	Yellow to red
Phenolphthalein	8.2 to 10.0	Colourless to red

Standard equipments for pH value determination are available on the market. The general procedure is that a certain prescribed volume (say, 10 ml) of the sample to be tested is taken in a tube and a fixed volume (say, 0.3 ml) of the indicator is mixed with it. The colour of the sample is then compared with the colours marked on a standard disc supplied by the manufacturer. These different colours correspond to certain pH values which are read on the instrument. Different colour discs have to be used with different indicators.

2. **Electrometric Method :** pH value can be more accurately determined by the electrometric method. Compact instruments called pH meters are available on the market. The pH meter has two electrodes which are dipped in the solution whose pH value is to he determined. One of the electrodes, known as calomel electrode, is a half cell (prepared by a combination of mercury and mercurus chloride in contact with a standard potassium chloride solution) and its potential is fixed. The potential of the other electrode varies with the pH value of the solution. This second electrode is a glass electrode. Its action is not well understood. It is like a tube, made of glass of a certain composition, having in it an electrolyte of a fixed concentration. When it is dipped in a solution, a potential is developed which depends upon the relative concentrations of H^+ on the inner and the outer surfaces of the glass.

   Thus when these two electrodes (i.e., calomel electrode and glass electrode) are dipped in a solution, a potential is developed which depends upon the hydrogen ion concentration of the solution. As the potential is very small, a vacuum tube voltmeter is used. It is calibrated to read directly the pH value.

3.13.3 Alkalinity

Alkalinity in natural waters is mostly due to the presence of salts of weak acids. Carbon dioxide is formed in nature in considerable amounts and it produces bicarbonates by its action on basic materials in soils, this causes bicarbonates, to be present in large amounts, borates, silicates and phosphates present in small amounts, also contribute to alkalinity. Surface waters, which have algae, contain appreciable amounts of carbonate and hydroxide alkalinity. This happens because algae remove free as well as combined carbon dioxide from water, thereby raising the pH value. Water treated by lime soda process contains carbonates and excess hydroxide. The alkalinity of natural waters is related to its buffer capacity (i. e. capacity to receive acids without substantially lowering the pH) because the alkalinity is due to salts of weak acids.

In order of their effect on pH, the three major classes of materials may be classed as follows: (1) hydroxide, (2) carbonates and (3) bicarbonates.

Highly alkaline waters do not have a good taste. Alkalinity causes precipitates and sludges, to be deposited in the pipes and heating tubes and also makes the metal brittle.

In water-softening processes and in boiler water analysis, it is necessary to know the kinds and amounts of the various types of alkalinity present. Various formulas have been proposed to calculate these from the titration results. The formulas are mainly for the technicians who do not know chemistry, but the chemists should be able to calculate from the fundamental principles and memorizing the formulas need not be encouraged.

Any one of the following five conditions may arise in an alkaline solution:

1. *Hydroxide only.* These have pH above 10 and titration is done using phenolphthalein as indicator. In this case phenolphthalein alkalinity is equal to the hydroxide alkalinity or the total alkalinity.

2. *Carbonate only.* These have a pH of 9.5 or higher. The carbonate alkalinity is equal to the total alkalinity. The total titration, using methyl orange as indicator, is twice the titration to the phenolphthalein end point. Phenolphthalein becomes colourless at pH 8.3 when carbonate ion is converted to bicarbonate ions as follows:

$$CO_3^{--} + H^+ \rightleftharpoons HCO_3^-$$

   Further addition of acid produces carbonic acid and the reaction is complete at pH 4.5 when methyl orange turns from yellow to orange red. It can be represented as follows:

$$HCO_3^- + H^+ \rightleftharpoons H_2CO_3 \rightleftharpoons H_2O + CO_2$$

3. *Bicarbonate only.* Solutions having only bicarbonates have a pH value of 8.3 or less. Methyl orange is used as indicator.

4. *Hydroxide and carbonate.* Solutions having both hydroxide and carbonate have pH above 10. The titration from the phenolphthalein end point to the methyl orange end point indicates only half of the carbonate alkalinity. The total titration to the methyl orange end point indicates the total alkalinity. Therefore, hydroxide and carbonate alkalinities may be easily calculated as follows:

   Carbonate alkalinity = 2 × (titration from phenolphthalein to methyl orange).

   Hydroxide alkalinity = Total alkalinity - Carbonate alkalinity.

5. *Carbonate and bicarbonate.* These solutions usually have pH between 8.3 and 11. Only half the carbonate alkalinity is indicated by the titration to the phenolphthalein end point. Methyl orange end point gives total alkalinity. The alkalinities may be calculated as follows:

   Carbonate alkalinity = 2 × (titration to phenolphthalein end point)

   Bicarbonate alkalinity = total alkalinity-carbonate alkalinity.

   It is convenient to use N/50 acid solution for titration. A few drops of phenolphthalein indicator is added to the measured quantity of water and if pink colour is produced, it is titrated until just colourless. Let the number of milliliters of the acid used be P. A few drops of methyl orange are then added and titrated until the solution turns orange red. Let the total number of milliliters used for both the titrations be M.

From the foregoing discussion it is clear that M represents total alkalinity and P indicates all the hydroxides and half the carbonates. The results are summarized in the following table :

Amount and type of alkalinity			
Titration result	**Hydroxide**	**Carbonate**	**Bicarbonate**
$P = 0$	nil	nil	M
$P = \frac{1}{2} M$	nil	2 P	nil
$P = M$	M	nil	nil
$P < \frac{1}{2} M$	nil	2 P	M – 2P
$P > \frac{1}{2} M$	2 P – M	2 (M – P)	nil

Alkalinity is usually expressed in terms of calcium carbonate. 1000 ml of N/50 acid solution is equivalent to 1g of $CaCO_3$ (equivalent weight of $CaCO_3$ being 50).

$$\therefore \quad 1 \text{ ml of N/50 acid solution} = \frac{1}{1000} \text{ g or 1 mg } CaCO_3$$

From the foregoing it is easy to calculate the alkalinity quantitatively and the result can be expressed as follows:

$$\text{Alkalinity} = \frac{\text{vol. of N/50 acid} \times 1000}{\text{vol. of sample for titration}} \text{ mg/l.}$$

3.14 Waste Water Treatment

Water of certain desired standard is required for use in homes and in industry. After use, waste water has a number of impurities which it has picked up while being used. Very often it is necessary to remove the impurities or to alter their nature, otherwise the impurities will cause serious pollution of water resources. If this happens, a much more elaborate and difficult treatment will have to be given to water which is required for use. Many diseases and epidemics like the disease of the stomach, typhoid fever, etc., can be eliminated, and have successfully been eliminated in several countries by proper treatment of waste water.

The pollution and toxicity effects of the constituents from the industrial waters are of greater significance than those found in domestic waste waters. The variety and complexity of inorganic and organic components in effluent from the industries presents an increasing threat to water systems all-over the world. They include toxic metals like chromium, copper and nickel besides iron; salts of sodium, calcium and magnesium; various acids, mainly sulphuric; petroleum wastes and petroleum products like greases and oils; phenols; cyanides; toluene; ammonia, to name a few common pollutants.

The waste waters are broadly of two types : (1) Sewage which includes human, animal and household wastes. In this run off water from roofs, pavements and streets are also included. (2) Industrial wastes are the waste waters from manufacturing processes.

A number of methods are in use but they are covered by three basic processes: (1) physical, (2) chemical, and (3) biological treatments

3.14.1 Physical Treatment

The treatments by which impurities are separated from water without chemical change are known as physical treatment processes. The important methods are sedimentation, screening and filtration, multi-liquid separation, degasification, dilution, ocean discharge and underground discharge.

Sedimentation : In sedimentation particles heavier than water move downwards due to the force of gravity and settle down. This process has already been considered in some detail under Municipal Water Treatment. Waste waters contain grit – a term which includes coal particles, brick particles and gravel or sand from street washings, which are removed in grit chambers. Grit being hard will damage pumps by abrasion, clog the outlets, and cause serious difficulties in sedimentation tanks. The grit chambers are long channels in which velocity of the waste is reduced to about 20 meters per minute. The detention time depends upon the particle size and is usually between 20 seconds to 1 minute. The grit chambers are periodically cleaned by mechanical scrapers or buckets. Grit often contains a considerable amount of organic matter and this has to be quickly disposed of by incineration, i. e. heating strongly, or by burial underground, otherwise a bad smell will develop.

Water is then taken to large sedimentation tanks in which less heavy and smaller particles settle down. It is to be remembered that city sewage or industrial wastes with large amount of organic matter cannot be kept in the sedimentation tanks for a long time.

Screening and filtration : Screens and filters remove floating solids which do not settle during sedimentation. These remove suspended solids which might have escaped sedimentation. Screening as a unit process may be used singly, where factors of economy do not provide further treatments like sedimentation, filtration, etc. The waste water passed through screens with bars 2 to 5 cm apart placed at an angle of about 45° followed by screens with openings of one quarter centimeter or less. For dewatering of sludges metal gauze or cloth filters are used. In these, metal gauze or cloth may be mounted on a drum which rotates and moisture may be sucked into the drum by applying vacuum. Suspended matter is scraped from the surface of the drum. Finally, the dried sludge cake may be disposed of by incineration or can be utilized again after composting as humus. Hot, strongly acidic or strongly alkaline waters are filtered through crushed anthracite coal because of its low solubility.

Multi-Liquid Separation : Waste water from the oil refineries contains a considerable amount of oil. If it is discharged in river, lake or the sea, the oil floating on water not only looks ugly but makes bathing impossible. It prevents dissolution of oxygen by reaeration in water, thereby endangering fish and water-plant life.

In such cases, separation is done by taking advantage of the difference in specific gravities. On a small scale this is done in specially designed equipment. Centrifugal equipment is more expensive but can handle larger volumes and causes better separation. In another method water is kept under 2 to 3 atmospheric pressure and air is diffused in it. The pressure is released just before discharging into a separating tank. Fine air bubbles rise to the surface carrying droplets of oil which are then skimmed off.

Degasification : Undesirable gases like hydrogen sulphide and ammonia are conveniently removed by exposing large surfaces of the liquid to atmosphere which is deficient in the gas to be removed. In some cases it is convenient to remove them by treatment with chemical which reacts with the gases. If waste heat is available, the water can be heated to drive off the gases. Another method used is to trickle the water through stacks of wooden trays and to blow air from the bottom when the escaping air takes the gas along with it.

Dilution : Waste water from the oil industry and many chemical industries contains high concentrations of alkalies and salts. Such waters if discharged in the rivers will cause serious contamination. If uncontaminated waste water, for example, water used for cooling purposes, is available, then highly concentrated water is diluted with uncontaminated water and discharged into the stream. In case sufficient dilution water is not available, part of the waste water is diluted and the remainder is stored in storage tanks or ponds. During the rainy season the receiving stream has a large amount of water flowing and the stored up waste water is discharged with it. This is known as controlled discharge.

Ocean Discharge : If the plant is located near the ocean, waste water can be discharged into it. Before doing so, the ocean currents must be studied to ensure that the currents will not carry the suspended matter in such discharge back to the shore. This is a simple method of waste disposal but involves considerable cost in the construction of pipes carrying the waste water sewage to a safe distance in the ocean. In some large cities the pipe goes a few kilometers out in the ocean carrying sewage. The sewage of Los Angeles city, after treatment, is discharged in the Pacific Ocean at a distance of ten kilometers from the shore.

Underground Discharge : Underground discharge is adopted only where other methods are either not available or are too expensive, and has been used only at a few places. In this method the waste water is pumped into a permeable strata in the earth. Great care has to be taken to make sure that the waste will not contaminate underground sources of water. Such safe strata are usually available at the depth of about 1000 meters.

3.14.2 Chemical Treatment

Many impurities cannot be removed by physical means alone and it becomes necessary to bring about a chemical change in order to separate them. The important chemical methods employed are precipitation, coagulation, ion exchange, sequestering, oxidation, neutralization, emulsion breaking, sterilization and disinfection.

Precipitation : Chemicals are added to form insoluble compounds with the salts present in the waste water. The precipitate is either filtered off or is allowed to settle and the clear liquid is decanted. The precipitate may carry with it some suspended matter and cause further clarification.

Excess of calcium and magnesium salts are conveniently removed by the lime-soda process described under water treatment. In industrial wastes the suitable chemicals to be added depend upon the particular impurity present in the waste water. Some other aspects of precipitation are discussed under coagulation.

Coagulation. (Also see Sedimentation under Municipal Water Supply in this chapter.) Coagulation is a physico-chemical process. Use of coagulants in municipal water treatment has already been explained. Most natural waters have pH between 6 and 7.6, which is the range in which alum acts.

Each coagulant is effective in a particular pH range and this fact has to be carefully borne in mind. Ferrous sulphate can be used in a pH range 8.5 to 10 and ferric chloride needs pH of 5.5 to 10.

In industrial wastes pH adjustment may become necessary. This is done by adding lime-soda or sulphuric acid, as the case may be.

The coagulant produces a floc which is an insoluble chemical compound. The floc absorbs colloidal matter, enmeshes finely suspended solids and settles rapidly. Further, the precipitating chemical ionizes and neutralizes the electric charges of the colloidal particles causing them to coalesce to form larger particles which readily settle.

Coagulation is useful in treating industrial wastes but is not widely used in sewage treatment because this is more effectively dealt with by biological treatment. It is noteworthy that removal of organic and colloidal matter also helps to remove colour, taste and smell.

Ion exchange. Ion exchange methods are adopted to a limited extent only. Where the waste is highly acidic or alkaline, part of the acidity or alkalinity can be removed by ion exchange.

An important use of the ion exchange method is in the recovery of chromic acid from waste water of the chromium plating industry which contains about 100 to 500 ppm of Cr_2O_3. The water is passed through an anion exchanger which removes chromium as indicated below. (ROH and HR represent resins which exchange anions and cations respectively).

$$2ROH + H_2Cr_2O_7 \rightarrow R_2Cr_2O_7 + 2H_2O$$

The exchanger is regenerated with caustic soda solution producing sodium chromate:

$$R_2Cr_2O_7 + 2NaOH \rightarrow 2ROH + Na_2Cr_2O_7$$

The solution coming out is passed through a cation exchanger when chromic acid is obtained:

$$2HR + Na_2Cr_2O_7 \rightarrow 2NaR + H_2Cr_2O_7$$

The chromic acid solution so obtained is a concentrated solution and ready for use. At the same time the waste water gets rid of chromic acid which is highly corrosive. Chromium has to be removed because it is toxic to human and aquatic life. Chromium is also removed by precipitating it as chromic hydroxide.

The ion exchange method is also coming into use for other industries in which a valuable recoverable ionized product is present in the waste.

Sequestering. Some impurities may not be easily precipitated. Sequestering is the process for converting the impurities into soluble complexes. For example, calcium and magnesium ions form stable complexes with some organic compounds as well as with polyphosphates. Thus they do not precipitate soap and are rendered harmless. Sequestering is an expensive process and is adopted for removing traces of impurities, after giving other cheaper treatments.

Neutralization. Free mineral acids and alkalis must be neutralized before discharging them. This discharge must be neutral to phenolphthalein or methyl orange end point, as the case may be. In some large chemical plants acidic waste is produced in one operation and alkaline waste in another. In such cases neutralization is comparatively simple and inexpensive.

Hydrochloric and nitric acid wastes are easily neutralized by passing them through limestone beds–the resulting calcium chloride and nitrate being highly soluble, produce no particular problems. As calcium sulphate has low solubility care has to be taken in neutralizing sulphuric acid wastes, otherwise insoluble deposits will be formed.

Emulsion breaking. Waste water from the petroleum industry often contains oil in the form of an emulsion. In the course of time the emulsion may break and the oil so released will present a problem. Soda ash is added to the waste water and pH is made 9. Concentrated calcium chloride solution (2000 mg/l) is added when the emulsion breaks. In this way a considerable quantity of oil is recovered. The above-mentioned chemical methods are mostly used for industrial wastes and rarely for city sewage.

Oxidation. After the removal of a large bulk of material from the waste water or the sewage, the remaining water contains small amounts of organic waste which produce a bad smell and bad taste. These can be removed by oxidation. Oxidation may be done by spraying the diluted waste over sandy soil which gets irrigated. Some times the waste water is held in a shallow pond for several days. The impurities can also be oxidized by chlorine. Oxidation is a very important process in the disposal of sewage.

Sterilization and disinfection. (Also see under Municipal Water Supply in this chapter). Water contains bacteria, some of which may not be harmful to human beings but many are harmful. Some industrial wastes do not contain bacteria but others may contain bacteria or substances on which bacteria can develop rapidly. Therefore it is necessary to destroy the bacteria or make them inactive and to prevent their further growth. This is done by sterilization and disinfection. Chlorine is most widely used for this purpose. When the requirement is small it is used in the form of bleaching powder.

Chlorine is cheaply available for large usages in chlorine cylinders. Chloramines, formed by adding one part of ammonia to four parts of chlorine, are more persistent, i.e., their effect is not as easily lost as that of chlorine gas, but they are less powerful. All these destroy micro-organisms, tastes and odours. Sterilization and disinfection are very important treatments in handling sewage.

In this category, control of other biological growth may also be included. Algae and slime may develop rapidly in some wastes, and clog the equipment and pipelines. Chlorine destroys most of the biological growths. In some cases dilute copper sulphate solution (less than 1%) is found to be more effective.

3.14.3 Biological Treatment

Biological treatment is the most important treatment given to sewage. It is also given to industrial wastes, like the wastes from dairies, tanneries, paper and sugar industries, which contain large amounts of decomposable organic matter.

This will, therefore, be considered with special reference to sewage treatment. The sewage is first subjected to preliminary treatments like screening and sedimentation. These treatments remove large suspended or floating organic solids like wood, cloth, paper, together with some faecal matter, and excess of oil or grease. Heavy inorganic solids like sand, gravel and metallic objects settle down in grit chambers. Sometimes the equipment is so designed that the sewage comes in contact with air so that the later treatment of biological oxidation is helped.

At this stage mention of the septic tanks may be made. Septic tanks are used in isolated homes or places which are not connected to city sewerage. In the septic tank the liquid flows slowly and remains in the tank for 12 to 24 hours. The decomposition of the organic material takes place by bacteria in absence of oxygen, i.e., under anaerobic conditions producing gas. Most of the organic material decomposes and the liquid infiltrates into the ground or is led away. The Imhoff tank is an improvement on the simple septic tank and the effluent can be given further treatment.

In a large sewage treatment plant, after screening and settling in the grit chamber, the waste is led to sedimentation tanks, where some more of the heavier material settles. Generally the tank is circular with scraper arms attached. The bottom of the tank is sloping towards the centre and the scrapers move the settled solids to a sludge-hopper at the centre from which the sludge is taken out.

Coagulants are generally not added to sewage in spite of the fact that most of the solids can be removed by this method. The reason for not adding coagulants are that a large volume of sludge is produced which is difficult to dispose off.

So far about 50% of the suspended matter and 25 to 35% of decomposable organic matter gets removed. The sewage is now subjected to biological oxidation in trickling filters or by activated sludge process. The process is carried on in the presence of dissolved oxygen and is known as aerobic decomposition. Bacteria or microorganisms grow using the dissolved organic matter as their food. In this process the complex organic molecules are broken down into simpler and more stable molecules. Carbon dioxide, water, nitrates and sulphates are the usual end products.

The trickling filter is about 2 m deep with a drain at the bottom for the removal of the liquid after treatment and has air ducts for the supply of air. It is filled with broken stone, preferably cubical in shape (so that the filter does not become very compact), leaving enough space for ventilation. The sewage is sprinkled on the filter surface. The micro-organisms grow on the surface of the stones using organic material of the sewage as their food. The treated liquid is removed from the bottom. In a single unit 35 to 75% of the organic matter is removed. It may be sent through a second filter when about 90% purification is obtained.

In the activated sludge process the sewage is mixed in a tank which has certain a mount of sludge from the previous operation. This sludge, known as activated sludge, consists of organic matter inhabited by numerous bacteria and other forms of biological life. It has the property of absorbing colloidal and dissolved organic matter including ammonia. The organisms use this matter as food and convert it into insoluble matter which does not further decompose. Some bacteria decompose complex molecules into simple molecules. After mixing activated sludge, air is blown in the liquid so as to maintain an aerobic condition. Purification efficiency is 75 to 95%. A part or the sludge produced, which is rich in the organisms, is used for the next batch.

Both in trickling filters and in the activated sludge process, the nature of impurities is changed but they are not removed. The effluent (i.e., out going liquid) contains suspended solids or sludge. The effluent is kept in settling basins for 2 to 5 hours when the solids (called sludge) settle and is removed. The sludge cannot be thrown away as such because it will further decompose and give off a bad smell. It is usually kept in concrete tanks for 25 to 60 days when anaerobic (i.e., in absence of air) decomposition takes place. A considerable volume of gas, containing CO_2, CH_4, and N_2, is given off, which can be used for heating. Manure can be made from the solids remaining behind.

The effluent still contains organic matter and may be discharged into a stream of water. In case further purification is needed it is filtered through slow sand filters when further biological decomposition takes place in addition to filtration. The effluent is very good for irrigation of sandy soils and this is a valuablc usc for it.

Some of the industrial wastes, as from the paper industry, are not given such an elaborate treatment but are kept in shallow ponds, called oxidation ponds, in which air from the atmosphere is used by the bacteria and changes of similar type as described above take place.

3.15 Calculation of Water Treatment Problems

I. *Calculation of Hardness*. It has already been explained, with examples, under the Degree of Hardness that the hardness is expressed in terms of $CaCO_3$. The weights of the substances are converted in terms of weight $CaCO_3$ and then the results are expressed in units of milligrams per litre of water (or parts per million or degree Clark, etc.).

II. *Lime-Soda Requirement.* In Water Chemistry, by lime is meant $Ca(OH)_2$ and not CaO, and by soda is meant Na_2CO_3. 'The weights of the substances should be converted into weight of $CaCO_3$.

The following should be noted for lime requirement:

(i) Lime is required for $Ca(HCO_3)_2$

(ii) Lime requirement for temporary magnesium hardness is *double* that required for temporary calcium hardness.

(iii) Lime removes permanent magnesium hardness, *but it* introduces an equivalent amount of permanent calcium hardness.

As 100 parts of $CaCO_3$ are equivalent to 74 parts of $Ca(OH)_2$, the lime needed for softening = $\frac{74}{100}$ × (temporary Ca hardness + 2 × temporary Mg hardness + permanent Mg hardness).

(*Note*: the hardness is in terms of $CaCO_3$).

In case the water analysis shows the presence of acidic substances, e.g. H_2SO_4, etc., lime has to be calculated for these also. It should be noted that the result of this would be the introduction of the corresponding calcium salts, e.g. $CaSO_4$, which will cause permanent hardness and soda requirement for this will also have to be calculated.

Soda is needed for the removal of permanent calcium hardness. In the process of the removal of permanent magnesium hardness an equivalent amount of calcium hardness is introduced which also needs soda for its removal.

As 100 parts of $CaCO_3$ are equivalent to 106 parts of Na_2CO_3,

the soda requirement = $\frac{106}{100}$ × (permanent Ca hardness + permanent Mg hardness).

Of course provision has to be made for calcium hardness that may be introduced by other reactions with lime, e. g. with acids.

Example I : Calculate the amount of lime-soda needed for softening water containing the following per litre:

$Ca(HCO_3)_2$	...	162 mg;	$Mg(HCO_3)_2$	...	73 mg;
$MgCl_2$	...	95 mg;	$CaSO_4$	...	136 mg;
NaCl	...	585 mg;			

What is the temporary and total hardness of the sample ?

Solution : NaCl neither causes hardness nor has any reaction with lime-soda. The other substances on conversion in terms of $CaCO_3$ give the following results :

$Ca(HCO_3)_2$ ... $162 \times \frac{100}{162} = 100$ mg.

$Mg(HCO_3)_2$ ... $73 \times \frac{100}{146} = 50$ mg.

$MgCl_2$ ... $95 \times \frac{100}{95} = 100$ mg.

$CaSO_4$ ... $136 \times \frac{100}{136} = 100$ mg.

(i) Temporary hardness is due to the bicarbonates which are
$100 + 50 =$ **150 mg/l.**

(ii) Total hardness is the sum of all the hardness causing constituents which are $100 + 50 + 100 + 100 =$ **350 mg/1**.

(iii) Lime requirement is (for temporary Ca hardness + 2 × temporary magnesium hardness + permanent Mg hardness).

$$\therefore \text{ Wt. of lime needed } = \frac{74}{100}(100 + 2 \times 50 + 100)$$

$$= \frac{74}{100} \times 300 = \mathbf{222\ mg/l}$$

(iv) Soda is needed for permanent Ca hardness + permanent Mg hardness,

(Note : Permanent magnesium hardness appears under soda requirement also, because in the lime treatment permanent calcium hardness is introduced in place of magnesium hardness. Since the amount of calcium hardness introduced is equivalent to the hardness originally present, the weights in terms of $CaCO_3$ remain the same for calculations:

$MgCl_2 + Ca(OH)_2 = Mg\ (OH)_2 + CaCl_2$ (this $CaCl_2$ causes permanent hardness).

$$\therefore \text{ Wt, of soda needed} = \frac{106}{100} \times (100 + 100) = \frac{106}{100} \times 200$$

$$= 212 \text{ mg/l.}$$

Example II : In example I calculate the amounts of lime soda needed per 10,000 litres of water if the sample of lime is 85% pure and soda is 98% pure and 10% excess of the chemicals are to be added in order to complete the reactions quickly.

Solution : The quantities of pure lime and soda needed will remain the same but as 100 parts of the sample of lime contain only 85 parts of the pure substance the weight of the sample of lime $= 222 \times \frac{100}{85}$ mg/1.

Further, 10% excess chemical is added, i.e., 110 parts are added for 100 parts are added for 100 parts of the calculated.

$\therefore$ Actual requirement $= 222 \times \frac{100}{85} \times \frac{110}{100} = 287.3$ mg/1.

$\therefore$ 10,000 litres of water needs $287.3 \times \frac{10,000}{1,000}$

= 2873 g or 2.87 kg of lime.

Similarly soda requirement per 10,000 litres

$$= 212 \times \frac{100}{98} \times \frac{110}{100} \times \frac{10,000}{1,000} = 2380 \text{ g or } 2.38 \text{ kg.}$$

Example III : Calculate the amounts of lime-soda needed for the treatment of 1,000 litres of water containing the following:

$CaCO_3$ - 120 mg/l; $MgCO_3$ - 42 mg/l; $Mg(NO_3)_2$ - 148 mg/l;
$CaCl_2$ - 111mg/l and KCl - 50 mg/l.

Solution : $CaCO_3$ and $MgCO_3$ should be regarded as being present in the form of their bicarbonates and that their weights have been expressed in terms of $CaCO_3$ and $MgCO_3$. Permanent hardness will be caused by $Mg(NO_3)_2$. KCl will not react with lime and soda. On converting the weights in terms of the weight of $CaCO_3$ the following results are obtained :

$CaCO_3$ … 120 mg/l; $Mg(NO_3)_2$ …$148 \times \frac{100}{148} = 100$ mg/l

$MgCO_3$ … $42 \times \frac{100}{84} = 50$ mg/l; $CaCl_2$ …$111 \times \frac{100}{111} = 100$ mg/l.

Lime is required for $CaCO_3$ and $MgCO_3$ (present as bicarbonates) and for $Mg(NO_3)_2$.

$$\text{Lime requirement} = \frac{74}{100} \times (120 + 2 \times 50 + 100) = 236.8 \text{ mg/1.}$$

1000 litres of the sample needs $236.8 \times \frac{1000}{1000}$ = **236.8 g of lime**

Soda is required for $CaCl_2$ and for Mg $(NO_3)_2$ (Now present as $Ca(NO_3)_2$ as a result of reaction with $Ca(OH_2)$.

Soda required for 1000 litres of the sample

$= \frac{106}{100} \times (100 + 100) \times \frac{1000}{1000}$ = **212 g of soda.**

Example IV : A sample of polluted water on analysis gave the following results:-

Suspended matter –200 mg/l; $CaSO_4$ –250 mg/l; $MgSO_4$ –240 mg/l; $MgCl_2$ –190 mg/l; H_2SO_4 –98 mg/l and organic matter –750 mg/l. Calculate the amounts of lime soda needed per litre for its treatment if the purity of lime is 90% and that of soda is 98%. What is the hardness of the sample?

Solution : The suspended and organic matter does not need lime and soda. Sulphuric acid does not cause hardness but reacts with the alkali added for softening. Let it be assumed that lime is first added followed by soda treatment. In such a case lime will react with H_2SO_4 producing $CaSO_4$ which will cause permanent hardness for which soda will be needed.

Expressing the weights, in terms of equivalents of $CaCO_3$ the following results are obtained ;-

$$CaSO_4 \quad \ldots \quad 250 \times \frac{100}{136} = 184 \text{ mg/l}$$

$$MgSO_4 \quad \ldots \quad 240 \times \frac{100}{120} = 200 \text{ mg/l}$$

$$MgCl_2 \quad \ldots \quad 190 \times \frac{100}{95} = 200 \text{ mg/l}$$

$$H_2SO_4 \quad \ldots \quad 98 \times \frac{100}{98} = 100 \text{ mg/l}$$

Lime is needed for $MgSO_4$, $MgCl_2$ and H_2SO_4.

$$\therefore \quad \text{Weight of pure lime needed} = \frac{74}{100} \times (200+200+100) = 370 \text{ mg/1.}$$

As the sample is 90% pure, the weight of the sample of lime needed $370 \times \frac{100}{90} = 411\text{mg/1.}$

Soda is needed for $CaSO_4$ and for the calcium salts produced during the removal of $MgSO_4$, $MgCl_2$ and H_2SO_4 by lime treatment.

$\therefore$ Wt. of 98% pure soda

$$= \frac{106}{100} \times (184 + 200 + 200 + 100) \times \frac{100}{98} = 740 \text{ mg/1}$$

Hardness is produced by $CaSO_4$, $MgSO_4$ and $MgCl_2$ only

$\therefore$ Hardness = 184 +200 + 200 = 584 mg/l.

Example V : In an experiment to determine the hardness of a sample of water 25 ml of N/50 Na_2CO_3 solution was added to 100 ml of the sample of water. After the completion of the precipitation of the insoluble carbonates, the unreacted Na_2CO_3 was titrated against N/50 H_2SO_4 solution when 10 ml of the acid was required. Calculate the degree of hardness and comment on the nature of the hardness so determined.

Solution: Na_2CO_3 will precipitate the calcium hardness as $CaCO_3$. Hardness due to magnesium salts will not be precipitated because $MgCO_3$ is comparatively soluble. Therefore this method does not report the magnesium hardness.

Volume of N/50 Na_2CO_3 solution left unreacted

= volume of N/50 acid solution used for titration = 10 ml.

Volume of N/50 Na_2CO_3 solution used in precipitation

= 25 – 10 = 15 ml

1 litre of N solution = 50 g of $CaCO_3$

∴ 15 ml of N/50 Na_2CO_3 solution

$$= \frac{50}{50} \times \frac{15}{1000} = 0.015 \text{ g of } CaCO_3$$

∴ Hardness equivalent to 0.015 g $CaCO_3$ was present in 100 ml sample.

$$\text{Hardness of the sample is } 0.015 \times \frac{1000}{100} \text{ g/l}$$

$$\text{OR} \quad 0.015 \times \frac{1000}{100} \times 1000 = 150 \text{ mg/l.}$$

Example VI : The hardness of 10,000 litres of a sample of water was completely removed by passing it through a zeolite softener. The zeolite softener required 200 litres of sodium chloride solution containing 20,000 mg/l of NaCl for regeneration. Calculate the hardness of the sample of water.

Solution : As 200 litres of sodium chloride solution containing 20,000 mg/l of NaCl was needed to regenerate the zeolite through which 10,000 litres of hard water had been passed, the hardness of 10,000 litres of the sample of hard water is equivalent to 200 × 20,000 mg of NaCl.

Since 58.5 parts of NaCl $\equiv$ 50 parts of $CaCO_3$,

The hardness of 10,000 litres of the sample

$$= 200 \times 20{,}000 \times \frac{50}{58.5} \text{ mg of } CaCO_3$$

or the hardness of 1 litre of the sample

$$= 200 \times 20{,}000 \times \frac{50}{58.5} \times \frac{1}{10{,}000} \text{ mg of } CaCO_3$$

$$= 341.9 \text{ mg of } CaCO_3$$

Hardness of the sample of water is 341.9 mg/1.

Examples

1. A sample of water contains the following $Ca(HCO_3)_2$ – 162 mg/l; $Mg(HCO_3)_2$ – 146 mg/l; $CaSO_4$ – 68 mg/l; $MgC1_2$ – 95 mg/l, and KCl-745 mg/l.

   Calculate (a) temporary hardness, (b) permanent hardness and (c) amounts of lime and soda needed for the softening of 1 litre of the sample of water.

   (**Answer :** (a) 200 mg/l; (b) 150 mg/l; (c) 296 mg and 159 mg.)

2. A sample of water contains the following impurities per litre :

   $Mg(HCO_3)_2$ – 73 mg : $CaC1_2$ – 222 mg;

   $MgSO_4$ – 120 mg and $Ca(NO_3)_2$ – 162 mg.

   The sample of lime available is 74% pure and soda is 90% pure. 10% excess chemicals have to be added. Calcuiate the amounts of lime and soda required for the treatment of 5,000 litres of water.

   (**Answer :** 1100 g; 2591 g).

3. Calculate the amounts of lime and soda needed for the treatment of 10,000 litres of water containing the following:

   $Al_2(SO_4)_3$ – 34.2 mg/l; HCl – 7.3 mg/l;

   $MgCl_2$ – 9.5 mg/l; and NaCl – 58.5 mg/l.

   (Lime treatment is given first, followed by soda treatment.)

   [Hint. Al = 27. Try to solve this question by writing chemical equations also.]

   (**Answer :** 37 mg/1, 53 mg/l).

4. In an experiment to determine the hardness of a sample of water the following results were obtained;

   Volume of the sample of water to be tested = 50 ml.

   N/50 Na_2CO_3 solution added to the sample = 12 ml.

   N/50 H_2SO solution used to titrate = 7 ml.

   Calculate the hardness of water in grains per gallon and mg/l

   (**Answer :** 7 gr/gal.; 100 mg/l)

5. 100 ml of a sample of water was titrated against a N/10 acid solution using phenolphthalein as indicator when 6 ml of the acid solution was required for discharging the colour. Then on the addition of methyl orange 4 ml of the acid solution was further required for neutralization. Calculate the hydroxyl and carbonate alkalinities in terms of mg/l of $CaCO_3$.

   (**Answer :** 100; 400)

6. 100 ml of a sample of water required 5 ml of N/10 acid solution for titration while using phenolphthalein as indicator. 100 ml of the sample of water was again taken and methyl orange was used as indicator when 15 ml of the same acid solution was required for neutralization. Interpret the results of alkalinities in mg/l.

   (**Answer :** Carbonate alkalinity 500 mg/l; bicarbonate alkalinity 250 mg/l.)

7. A zeolite softener was completely exhausted and was regenerated by passing 100 litres of a sodium chloride solution containing 120 g/l of sodium chloride. How many litres of a sample of water of hardness of 500 mg/l can be softened by the softener before regenerating it again?

   (**Answer :** 20,513 litres)

Questions

1. Among chloramine and Bleaching powder, which is the powerful disinfectant ?
2. Name the indicator used in the EDTA method.
3. Why demineralization process preferred over zeolite process ?
4. Write the general formula of zeolite.
5. Differentiate between sludge and scale.

6. What is reverse osmosis ?
7. Convert 219 ppm of Mg $(HCO_3)_2$ to $CaCO_3$ equivalent.
8. Treating boiling water with soda will remove what type of hardness ?
9. What is meant by caustic embrittlement ?
10. Explain break point chlorination.
11. (A) Explain the scale and sludge formation in boilers.

    (B) What problems are caused by these?

    (C) Calculate the total hardness in ^{0}Fr and ^{0}Clarke for a sample of water obtained from a well in Khammam which showed the followed analysis:

    $MgCl_2$ = 9.5 mg/litre; $CaCl_2$ = 11.11 mg/litre

    Mg $(HCO_3)_2$ = 14.6 mg/litre; NaCl = 5.84 mg/litre

    Ca $(HCO_3)_2$ = 16.2 mg/litre; Organic impurities = 20 mg/litre
12. (A) Outline the lime-soda process for the softening of hard water.

    (B) A sample of water from Nalgonda showed the following

    NaOH = 4.0 ppm; $MgSO_4$ = 24 ppm

    $MgCl_2$ = 94 ppm ; $Ca(HCO_3)_2$ = 16.2 ppm

    $Ca(NO_3)_2$ = 16.4 ppm; KCl = 7.45 ppm

    Calculate the amount of lime and soda required (in kgs) to soften 10,000 litres of this water.
13. (A) What are boiler troubles and how they can be avoided?

    (B) Describe the ion-exchange process for softening of hard water.
14. (A) How is the softening of water carried out using the Zeolite process?

    (B) A sample of water from a well in Adilabad gave the following analysis:-

    NaCl = 5.85 mg/litre ; $MgSO_4$ = 12.0 mg/litre

    $Ca(HCO_3)_2$ = 8.1 mg/litre ; $CaSO_4$ = 13.6 mg/litre

    $Mg(HCO_3)_2$ = 14.6 mg/litre ; Organic matter = 4.0 mg/litre

    Calculate the quantities (in kgs) of lime and soda required to soften 12,000 litres of this water.

15. (A) Discuss the permutit process of softening of hard water.

(B) Calculate in kgs the amount of lime and soda required to soften 10,000 litres of water from a bore well in Medak which showed the following analysis :

KOH = 0.0056 gm/litre ; $MgSO_4$ = 0.024 gm/litre

$Ca(NO_3)_2$ = 0.0164 gm/litre ; Mg Cl_2 = 0.094 gm/litre

$Ca(HCO_3)_2$ = 0.0162 gm/litre ; KCl = 0.0745 gm/litre

16. Write a note on Boiler troubles.

17. (A) Explain the processes of scale and sludge formation in boilers.

(B) What is caustic embrittlement and how it can be prevented?

(C) Determine the temporary, permanent and total hardness in ppm units for a water sample from Kurnool which showed the following analysis:-

$Ca(CO_3)_2$ = 32.4 mg/litre ; $Mg(HCO_3)_2$ = 29.2 mg/litre

NaCl = 5.85 mg/litre ; $CaCl_2$ = 22.2 mg/litre

Mg SO_4 = 1.2 mg/litre ; Organic matter = 15.5 mg/litre

18. Discuss the hot and cold lime-soda processes for the softening of hard water.

19. (A) How is the softening of hard water carried out using Zeolite process?

(B) A sample of water from a tank in Jagtial showed the following analysis :

$CaCl_2$ = 11.1 mg/litre ; KCl = 40 mg/litre

Organic matter = 30 mg/litre ; $MgCl_2$ = 9.5 mg/litre

$Ca(HCO_3)_2$ = 16.2 mg/litre ; Mg $(HCO_3)_2$ = 14.6 mg/litre

Calculate the amount of lime and soda required in kg to soften 50,000 litres of this water.

20. Write a short note on municipal water.

21. (A) Describe the ion-exchange process for softening of hard water.

(B) One litre of water from a well in Nirmal Town, near Adilabad, gave the following analysis:-

Suspended matter = 100 mg ; $Ca(HCO_3)_2$ = 81 mg;

Mg $(HCO_3)_2$ = 146 mg ; $MgSO_4$ = 120 mg

$Ca(SO_4)$ = 136 mg ; NaCl = 58.5 mg.

Calculate the total hardness in ^{0}Fr and ^{0}Clarke of this sample of water.

22. Write a note on Sterilization of drinking water.

23. (A) Describe the lime-soda process for softening of hard water.

 (B) Sample of water from an underground reservoir in Anatapur town gave the following analyses:-

$Ca(HCO_3)$	= 81 mg/litre	;	NaCl	= 5.85 mg/litre.
Organic matter	= 25 mg/litre	;	$MgSO_4$	= 120 mg/litre.
$Mg(HCO_3)_2$	= 146 mg/litre	;	$CaSO_4$	= 136 mg/litre.

 Calculate the quantities (in kgs) of lime and soda required to soften 20, 000 litres of this water.

Fill in the Blanks

1. NaCl causes_____ hardness.
2. _____ water contains highest organic impurity.
3. Temporary hardness is caused by _____.
4. Formation of _____ causes caustic embrittlement.
5. Ca $(HCO_3)_2$ causes_____ hardness.
6. Equivalent weight of $CaCO_3$ is_____.
7. Hardness of water is measured in _____.
8. One ppm of hardness is equal to _____ ^{0}Cl
9. Dissolved oxygen in water is removed by _____.
10. Dissolved carbon dioxide in water is removed by _____.
11. Caustic embrittlement is avoided by adding _____softening agent
12. Phosphate conditioning of water is done by using _____ salt.
13. Radio active conditioning prevents formation of _____.
14. $MgCO_3$ forms_____ (scale/sludge).
15. $CaCO_3$ forms_____ (scale/sludge).
16. _____ forms hardest scale.
17. In Hot-Lime soda process the function of lime is to remove _____ type of hardness.
18. Alkalinity in water cannot be due to the simultaneous presence of _____.

19. The process of removing extra common salt from water is called as _____.

20. When P=M the alkalinity is due to the presence of _____ ions.

21. A water sample contains 136 mg of calcium sulphate per liter. The hardness of water in terms of calcium carbonate _____.

22. The hardest scale is formed from _____ ions.

23. _____ act as coagulant simultaneously as softening agent.

24. Chlorine is used in purification of drinking water for_____.

25. Removal of micro organisms in water is called as _____.

26. _____ is the indicator used in EDTA method.

27. EDTA is_____ acetic acid.

28. Sterilization of water is done to remove_____.

29. The best sterilization agent is _____.

30. Powerful disinfectant is _____.

31. Removal of microorganisms in water is called as _____.

32. Brackish water is treated by _____method.

33. O_3 can act as_____.

Chapter **4**

Thermodynamics & Thermochemistry

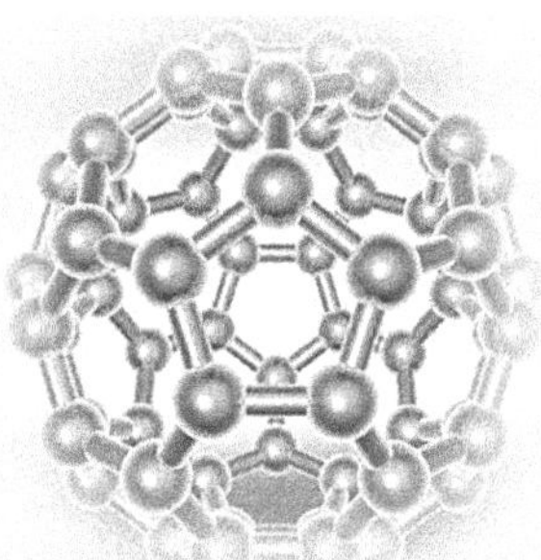

4.1 Introduction

Energy is a manifestation of nature, *prakruti*. Awareness of energy by man can be traced back to the earliest archeological evidences of prehistoric man and detailed account in the vedic scriptures. Worship of energy as *Shakti*, *Sun*, *Fire*, etc., are common features of ancient civilizations. Energy exists in two fundamental forms, kinetic energy assumed by virtue of its motion and the potential energy by virtue of its position. A moving car, a flying plane, a dropping stone, etc., speak about the kinetic energy. Commonly known energies like mechanical energy, heat energy, electrical energy, nuclear energy, chemical energy, magnetic energy, gravitational energy etc., are all simple manifestations of potential energy. The potential energy is mechanical in a stretched spring, in the attraction of the earth it is gravitational, in the attraction of unlike electrical charges it is electrical, in the attraction of atoms toward another configuration it is chemical energy, etc.

The fact that various forms of energy are manifestations of potential energy and kinetics energy is reflected by their interconvertibility. The conversion of mechanical energy into heat energy, chemical energy or electrical energy and vice versa, is a common every day experience. A few simple conversions are listed in Table 4.1.

Table 4.1 Interconversion of different forms of energy

Energy	Mode of conversion	Energy
1. Mechanical	$\underset{\text{Steam engine}}{\overset{\text{Friction}}{\rightleftharpoons}}$	Thermal
2. Mechanical	$\underset{\text{Motor}}{\overset{\text{Dynamo}}{\rightleftharpoons}}$	Electrical
3. Mechanical	$\underset{\text{Explosions}}{\overset{\text{Mixing}}{\rightleftharpoons}}$	Chemical
4. Thermal	$\underset{\text{Resistance of wire}}{\overset{\text{Steam turbine}}{\rightleftharpoons}}$	Electrical
5. Thermal	$\underset{\text{Combustion}}{\overset{\text{Heating reactants}}{\rightleftharpoons}}$	Chemical
6. Thermal	$\underset{\text{Absorption of } ir \text{ radiation}}{\overset{\text{Heating a metal red hot}}{\rightleftharpoons}}$	Light

Contd... Table 4.1

Energy	Mode of conversion	Energy
7. Electrical	Charging of battery $\rightleftharpoons$ Discharging of battery	Chemical
8. Electrical	Discharge tube $\rightleftharpoons$ Photo chemical cell	Light
9. Chemical	Chemiluminescence $\rightleftharpoons$ Photosynthesis	Light

Out of the above cited forms heat and mechanical work are the two most easily available forms of energy. The branch of science, **thermodynamics**, derives its origin from two words *thermos* meaning heat and *dynamics* meaning motion. Thermodynamics, in the classical sense, is defined as *the science which deals with the conversion of heat into motional (mechanical) work and vice versa*. With the advent of enzymes, the old definition no more stands good and thermodynamics is no more restricted to the heat and the mechanical work, but assumes a wider perspective with the definition "*The thermodynamics is an exact science that describes quantitative inter-relationship of various forms of energy accompanying the physical and chemical changes*". Thus thermodynamics deals with the relation covering the quantitative conversions of various forms of energy from one to another.

The science which deals with the properties of macroscopic particles, where the number, quantity, etc., are *well defined*, forms the subject matter of classical thermodynamics or generally referred as *thermodynamics*. The science dealing with the properties of *microscopic* particles like atoms, subatomic particles etc., where the properties are *not well defined* and we deal with only statistical average, forms the subject matter of **statistical mechanics**.

4.1.1 Scope of Thermodynamics

Thermodynamics is of great importance to chemists, physicists, technologists and engineers. The use of its principles to processes in chemistry constitutes one of its most significant achievements. All chemical processes involve energy changes of one sort or another and these changes could be defined and predicted with the help of thermodynamics. The branch of thermodynamics related to chemists and chemical engineers which deals with the processes in which only chemical energy changes are involved, is called **chemical thermodynamics**. The three fundamental generalizations, which we are going to deal later, known as the three laws of thermodynamics are very helpful in this regard.

4.1.2 Success of Thermodynamics

Before we start defining and explaining thermodynamics, we wish to show the areas of success and limitation of thermodynamics so as to fully appreciate the subject.

(i) No exceptions for the three fundamental laws of thermodynamics are reported yet. Any discrepancy between the predictions and experimental observations are attributable to errors in the latter.

(ii) Generalization of various experimental observations in physical chemistry such as Raoult's laws, Classius-Clapeyron equation, phase rule, van't Hoff equation, distribution law, law of mass action, etc., are deduced from thermodynamics.

(iii) Thermodynamical concepts are also helpful in predicting the viability of a chemical reaction and the direction of shift of a chemical equilibrium.

4.1.3 Limitation of Thermodynamics

In spite of its thumping success in resolving various chemical and physical problems and successful prediction of the changes in the physical properties few limitation were notices :

(i) Thermodynamics deals with the macroscopic or bulk properties of matter. It takes cognizance of the average of the properties of a number of molecules or atoms rather than of the individual molecule or atom. Hence it *gives no information on the molecular structure and molecular properties of the matter.*

(ii) *Thermodynamics does not consider the 'time element'* in any of its transformations. It is concerned only about the initial and final stages of a transformation but not about the time taken for the transformation. For example, thermodynamics successfully predicts that H_2 and O_2 can combine to give water, but gives no idea about how long it takes for this change. Under ordinary pressure and temperature conditions, formation of water proceeds infinitesimally slow.

(iii) Because thermodynamics is concerned about only initial and final stages, it also fails to speak about the molecular changes and intermediate steps that occur during the transformation, i.e., *it fails to speak about the molecular recourse or the mechanism* of a chemical reaction. The time of transformation and the mechanism, both form the subject matter of chemical reaction kinetics.

(iv) *It gives no information on the conditions to accelerate or inhibit the reaction.*

(v) Though mass is a form of energy as described by the Einstein's relation, $E = Mc^2$, *thermodynamics is totally silent on mass transformations* and its relation to other thermodynamic properties.

4.2 Definition of Thermodynamic Terms

Thermodynamics is better understood by considering the following terms.

(i) **System :** *A thermodynamic system is that part of the physical universe, the properties of which are under investigation and which may be real or imaginary.* Suppose if we are studying the solubility of $CuSO_4$ in water kept in a beaker, the solution in the beaker is considered as system. In simple terms *any specific part of the universe that is of interest to us may be called a system.*

(ii) **Surroundings :** The systemo is confined to a definite place in space by a boundary which separates it from the rest of the universe called the surroundings or the *surroundings are the portion of the universe from which the system is excluded.* In the above example the thermo-static bath in which the beaker is placed to maintain the required temperature and the atmosphere which maintains the external pressure and which are separated from the system by the boundary defined by the walls of the beaker is called the surroundings.

4.2.1 Classification of System

Depending upon the interaction between system and its surroundings, the systems are classified into three categories (i) Isolated system, (ii) Closed system and (iii) Open system.

(i) **Isolated system** : *A system is said to be isolated, when the boundary prevents exchange of mass or energy with the surroundings.* An isolated system produces no observable effect or disturbance in its surroundings. Hot tea poured in a thermos flask, where neither the contents nor heat is allowed to enter or leave the system, is comparable to an isolated system, Fig. 4.1(a).

(ii) **Closed system :** *A system is said to be closed when the energy exchange across the boundary with the surroundings is allowed but not the mass transfer.* Hot tea poured in a stoppered bottle exchanges its heat but not the content, is an example of closed system, Fig. 4.1(b).

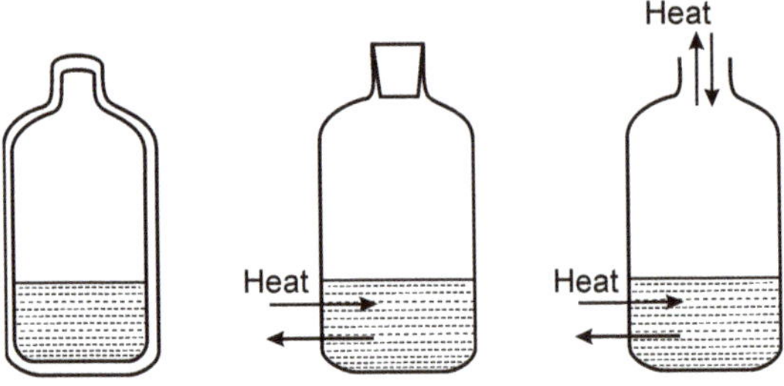

(a) Isolated system (b) Closed system (c) Open system

Fig. 4.1 Different types of systems.

(iii) **Open system :** *A system is said to be open when both energy and mass transfer is allowed across the boundary with the surroundings*. If the same hot tea, is poured in a open bottle or a cup, it exchanges its heat as well as water vapor with the surroundings, Fig. 4.1(c).

State of a system : *A system is in a definite state when all the properties are well defined and have a definite value.* The physical properties of a system may be divided into two main classes, viz., intensive properties and extensive properties. The **intensive properties** are the properties that are independent of the mass of the system like temperature, density, refractive index, pressure, etc. On the contrary **extensive properties**, like quantity of heat, volume, internal energy and entropy, depend on the mass of the system. *Two systems having same properties are said to be in the same* **thermodynamic state**. For example, for the purpose of thermodynamics the state of a mole of gas can be completely specified by giving the values of only three properties, viz., pressure, volume and temperature. Two systems having different kinds of matter are in the same thermodynamic state as long as their intensive properties are alike. Effecting a change in any one or more of these properties changes the state of the system. Since these properties are used to specify the state of a thermodynamic system, they are also called **state variables** or **state functions**. Their value depend only on the condition or state of the system and is independent of how the state is arrived at. In case of a variation of a state we are concerned only with the initial and final states of the system but not on the path used in reaching the final state.

Changes in a state : *The changes in a state are completely defined when its initial and final stages are specified.*

Path of a change : To define a path of a change in addition to specifying the initial and final changes, we have to also give information on the intermediate stages traced for reaching the final change.

4.2.2 Types of Processes

In thermodynamics we come across a number of processes. *A process is the method of operations by means of which a change in state is effected.*

(i) **Isothermal process :** A process is said to be isothermal if throughout the process the temperature is kept constant. To maintain the temperature constant an exothermic reaction transfers its excess energy to the surroundings, while an endothermic reaction taps energy from the surroundings.

(ii) **Adiabatic process :** In an adiabatic process neither heat is allowed to enter the system nor leave the system. Hence the system temperature of an adiabatic process may rise or fall from the initial state of a change.

(iii) **Isobaric process :** A constant pressure process is called an isobaric process. Boiling of water against constant atmospheric pressure is a typical example of isobaric process.

(iv) **Isochoric process :** In a path of change, if the volume remains constant at each step, then it is said to be a isochoric process. Majority of the solution phase chemical changes and solid phase transformations at constant temperature are isochoric, as practically no volume change is seen. Thus the hydrolysis of methyl acetate, at any given temperature, is an isochoric process.

(v) **Cyclic process :** If a system having undergone a change, returns back to the initial state, it is said to have traced a cycle or undergone a cyclic process.

4.2.3 Reversible and Irreversible Processes

The processes described in the preceding section could be performed in two ways. *If the process is carried out in infinitesimally slow step such that each succeeding state is only slightly different from the preceding step*, the process is called a **reversible process**. So a reversible process is a collection of infinite number of extremely small intermediate steps which are in thermodynamic equilibrium. A reversible process at any stage may be reversed to its original state by simply bringing back the initial conditions without producing any permanent effect on the surrounding or system. For example, Liquid water and ice are at equilibrium at 0 ^{O}C and 1 atm pressure. Even a minute increase in temperature, converts ice totally into water and this change simply reverts by slightly decreasing the temperature.

Consider another example where a gas is filled in a balloon. As long as a very small gas is let out or a small amount of gas is blowed in, we notice no change in the volume of the balloon and is always in a pressure-volume equilibrium. By carrying out these infinitesimally small changes it leaves no impact on the surroundings. Hence it is said to have undergone a reversible process. Instead, if the gas is allowed to escape suddenly into air (surrounding) the process cannot be reversed with ease and this sudden escape of gas disturbs the pressure balance of the surroundings. This type of change leaving a permanent impact on the surroundings is called an **irreversible process**. *An irreversible process is one which takes place so that a property of the system differ by a finite (measurable) amount from one instant to another.*

As a reversible process is infinitesimally slow, in true sense it is unattainable and requires infinitely long time periods, but near reversibility could be achieved. On the contrary all the processes in the universe like expansion of gases, flow of heat, a running motor, etc., take place in relatively short measurable period and hence are irreversible in nature.

Table 4.2 Differences between Reversible and Irreversible Processes.

Reversible process	Irreversible process
1. Occurs in infinite number of steps.	1. Occurs in finite number of steps.
2. Infinitesimally slow and cannot be completed in our life time.	2. Observable in short time periods.
3. A reversible process is only ideal and cannot be performed.	3. An irreversible process is real, natural and practical
4. The change in the properties at every step are measured.	4. The change in the properties at every step are not measurable.
5. Leaves no permanent change on the surroundings.	5. Leaves a permanent impact on the surroundings.
6. It can be made to go in both the directions of a thermo-dynamic equilibrium.	6. Unidirectional and it is difficult to reverse the process.
7. Work done in a reversible process is maximum	7. Less work is done by the irreversible process.

4.2.4 Work

An empty cart can be pushed by the application of very little force and little work is said to have been done. To move a loaded cart through the same distance we have to apply more force. Similarly to move the cart a larger distance we have to do still more work. Work is directly related to the force applied and the distance moved. Hence work is defined as the product of the distance by which a body is moved and of the component of force acting along the direction of movement. It is a scaler quantity with the units of energy.

$W = F \times l$

We talk about various types of work, viz., electrical work, gravitational work, mechanical work, etc.

(i) **Gravitational work :** Work done in moving an object against the gravitational force is called gravitational work. If an object of mass, *m*, is lifted through a height, *h*, in a gravitation field having an acceleration due to gravity, *g*, then force exerted in lifting = *mg*

Therefore,

Gravitational work = $m \times g \times h$

(ii) **Electrical Work :** The electrical work is given by the product of the quantity of current, i, and the emf across which it is flowing.

Electrical work = $E \times i$

(iii) **Pressure-Volume work :** Pressure-volume work also belongs to the above cited category of works, but because of its wider usage it is also generally referred as work. It is defined as the product of change in volume, *dV*, and the external pressure *P*, that causes this change.

Work done $*w^* = P\,dv$ .....(4.1)

(* The lower case letter for work, *w*, indicates that it is not a thermodynamic function. $*w$ indicates the small quantity of work done due to the volume change *dV*. Here *dV* is an exact differential or independent of path traced and $*w$ suggests that work depends on the path traced in doing it. A precise account of nature of path function is given in section 4.2.5).

The work expression (4.1) could be derived from the following consideration. Let a gas occupying a volume of *V* at a pressure P_{gas} and temperature *T* be placed in a cylinder and fitted with a frictionless piston, Fig. 4.2. If *P* the pressure of the surroundings is such that $P_{ext} = P_{gas} - dP$, then the gas expands till the internal and external pressure are at equilibrium. Further let *A* cm^2 be the area of the piston and *dl* be the length of displacement of the piston due to expansion of the gas under these conditions.

Work done $*w$ = External force × distance moved

but External force = External pressure × area of the piston

So $*w = P_{\text{ext}} \times A \times dl$

Substituting for P_{ext} we get $*w = (P_{gas} - dP) \times A \times dl$

As area × length = volume

$$*w = (P_{gas} - dP) \times dV = P_{gas}\, dV - dP\, dV$$

In the above expression dP and dV both being very small values their product is negligible. Hence,

$$*w = P_{gas}\, dV \text{ or } P\, dV \qquad .....(4.2)$$

For finite work

$$w = P\,\Delta\,V \qquad .....(4.3)$$

Thus work appears only at the boundary of a system and only during a change of state. Secondly work done by the system, as in the case of expansion is always positive. Therefore work of compression is assigned a negative sign.

Unit of work : When the pressure is expressed in atmospheres and the volume change in liters, then the work assumes the unit litre atmosphere (lit atm or atm dm^3). In CGS system, unit of work is erg, which is defined as the

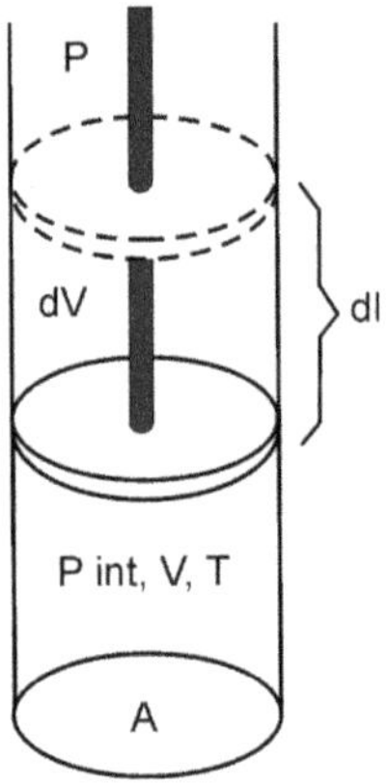

Fig. 4.2 Work of expansion.

work due to moving a body through a distance of 1 cm by the application of a force of 1 dyne. In SI system work is expressed in joule (J). The CGS and SI units are related to each other as follows :

$$1 \text{ Joule} = 10^7 \text{ ergs}$$

4.2.5 Path Functions and Inexact Differentials

We have learnt about the state functions like pressure, volume, temperature, etc, which depend only on the initial and final state of a change but not on the path traversed, in reaching the final state. Not all changes can be described in terms of initial and final states, but we also have path dependent functions like work. For example, Hyderabad and Delhi are separated by a fixed distance but the distance actually traveled depends on the mode of transportion. Thus it may be around 1500 km by air, 1600 km by train and 1700 km by bus and so on. Therefore work done depends on the mode of transort used (path traversed) for the travel. Similarly when we are pushing a heavy object by a particular distance, we encounter more resistance while doing the job on a rough flooring, but on a smooth surface it is done with relative ease. In the former case we say that we did more work than in the latter. Hence *work, a path function, depends exclusively on the path we adopt to reach the final state.* The path functions are designated by lower case (small) letters like w, q, etc.

Small change in state functions like temperature, pressure, etc., are **exact differentials** and represented as dT, dP, etc., $dT = T_{\text{final}} - T_{\text{initial}}$

On the contrary small changes in path functions are represented by **inexact differentials** as δw, δq, etc.

Large and finite changes in a state function are preceded by the symbol Δ as Δt, ΔP; and finite change in a path function is represented by removing the inexact differential Δ symbol as w.

Sign Convention of Work

Similar to expansion of a gas, work is also done during the compression process. Sign convention is adopted to differentiate these two type of works. During the expansion there is a positive change in the volume and work is said to have been done by the system on the surroundings on the system and is designated with a negative sign before the work notation. In some books totally reverse notation is also used.

Thus for an expansion $\delta w = P\,\mathrm{dV}$

for a compression $\delta w = -P\,dV$

Problem 1.

Calculate the work done during a reversible expansion of 1 mole a gas at 2 atm and 5L to 25L at the same pressure.

$$w = P\,\Delta V$$

$$= 2\,(25 - 5) = 40 \text{ L atm.}$$

Problem 2.

Calculate the work done when 2 L of a gas at 5 atm is allowed to expand into vacuum.

Work of expansion = $P_{ext}\,\Delta V$.

Since the external pressure (i.e., vacuum) = 0

$$w = 0 \times \Delta V = 0$$

So no work is done during expansion into vacuum.

4.2.6 Energy

Energy is defined as the ability to do work. Both work and energy are expressed in the same units. A detailed account of energy is already given in the introduction. The kinetic and potential energy of macroscopic particles can be estimated with ease and hence they are called macroscopic modes of energy. Any molecule at the microscopic level has atoms, subatomic particles, etc. The kinetic energy at microscopic level inturn consists of the translational (motion of molecule from one point to another), vibration (motion of any bonded atoms about a mean position) and rotational motions. The potential energy at microscopic level is determined by the attractive and repulsive forces between subatomic particles and the intermolecular forces. This microscopic energy possessed by a molecule can not be evaluated through macroscopic measurements. The total energy of a system is therefore given by

$$E_{total} = K.E + P.E. + E_{microscopic}$$

or

$$E_{total} = E_{macroscopic} + E_{microscopic}$$

4.2.7 Heat

Heat is a form of energy. It always flows from a warmer body to a colder body until both are at the same temperature, also called thermal equilibrium.

Units of Heat : Historically, heat energy is measured by the unit the calorie which is defined as the quantity of heat required to raise the temperature of 1g water from 14.5 to 15.5^{O}C. In addition to calories (cal) the other energy units are also used to define heat. The conversion factors among various units are;

$$1\ \text{cal} = 4.1833 \times 10^{7}\ \text{erg} = 4.1833\ \text{J} = 4.13 \times 10^{-2}\ \text{L atm}$$

$$1\ \text{L atm} = 1.013 \times 10^{9}\ \text{erg} = 24.2\ \text{cal.}$$

4.3 Laws of Thermodynamics

Based on several observations of the macroscopic properties three laws were postulated. These laws deal with the book keeping of energy changes and help us predict the spontaneity of a physical or chemical process.

The first law of thermodynamics serves as a basis for describing quantities of heat energy associated with chemical and physical processes.

The second law of thermodynamics deals with the direction and extent of a chemical reaction or a physical process and allows us to derive values of equilibrium constants from certain thermodynamics properties.

The third law of thermodynamics provides a starting point for the experimental evaluation of these thermodynamic properties.

4.3.1 First Law of Thermodynamics

The first law of thermodynamics is not derived from any assumption but is only a generalization of experiences. It was first stated in 1840 by Mayer and Helmholtz in Germany, Joule in England and Colding in Denmark. The first law of thermodynamics states that *whenever any process occurs, the sum of all the energy changes involved in the process must of necessity be zero*. This is simply expressing the **law of conservation of energy** according to which, *energy can be neither created nor destroyed*. This can be understood by considering the following example; when a red hot iron piece is dropped into water present in a beaker, immediately we notice a cooling of the iron piece. We can immediately frame a question, where has this heat gone. On touching the water we notice that the water is hot now. So we can infer that the heat lost by the red hot iron is trapped by the water and used in raising its temperature. Hence first law may be stated as, *whenever a quantity of one kind of energy is produced, an exactly equivalent amount of other kind must disappear.*

This statement could be substantiated by keenly observing a number of energy transformations taking around us. To run an electric pump we will be supplying energy in the form of electricity, petrol or diesel. We generate the mechanical work useful to run a car by supplying fuel. We will be able to lift a bucket only after spending the muscular energy. A sick or old man cannot lift a heavy weight because he is not having enough energy to do this job. In all these examples we notice interconvertibility of different forms of energy, viz., electrical, mechanical, thermal etc. To this list of common energies we have to also add mass as its fusion (combination) or fission (breaking) also generates energy. Einstein in 1905 proposed mass and energy inter-convertibility through mathematical expression

$$E = mc^2$$

where E is the energy produced by the destruction of mass m and c the velocity of light.

With these observations the first law could be stated as follows :

(i) The total energy of an isolated system remains constant although it undergoes a change from one form to another.

(ii) The energy of the universe is constant.

(iii) It is impossible to construct a perpetual motion machine which would produce work without consuming energy.

(iv) The total energy of a system and its surroundings taken together remain constant.

(v) The algebraic sum of all energy changes taking place in an isolated system is zero.

The microscopic energy possessed by a system is called **internal** or **intrinsic energy**. As discussed earlier this microscopic energy is a highly disorganized form of energy and has components like intermolecular forces which change continuously. So any attempt to measure all the contributions accurately and arrive at the total internal energy will prove futile. Hence in thermodynamics concentration is made only on determining the *changes* in internal energy instead of the absolute value.

Suppose a system changes from state A to state B by the absorption of a heat q, supplied by the surroundings. This heat absorbed may be totally used in doing work or in increasing the internal energy or both.

If the heat absorbed raises the internal energy by ΔE and does a work of w, then;

$$q = \Delta E + w$$

or $$\Delta E = q - w \qquad .....(4.4)$$

For an infinitesimal change, eqn. (4.4) may be written as

$$\boxed{dE = \delta q - \delta w} \qquad .....(4.5)$$

where, dE represents a small change in internal energy, δq, a small quantity of heat absorbed and $*w$ the small amount of external work done by the system.

The exact differential notation, dE, for the internal energy change suggests its path independent nature while the inexact differential notations δq and δw, suggest that they are path functions. This point could be clarified by considering (pressure-volume) work done in passing from state A to state B.

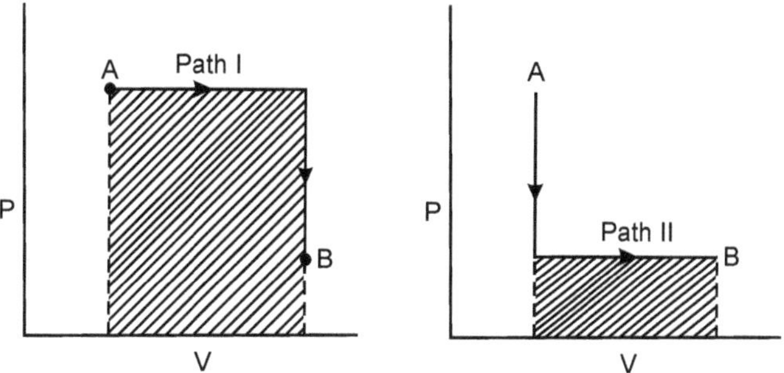

Fig. 4.3 Work is a path function.

Internal energy being a state function, the change in the internal energy, $\Delta E = E_B - E_A$, is the same for both paths. The work done in passing from A to B by any path is given by the expression $P\Delta V$ or graphically by the area under the curve in the pressure-volume diagram, Fig. 4.3. From the shaded region it is clear that the work done by path I > path II. It is important to note that though δw and δq are not exact differentials, their difference is a path independent variable, dE.

4.3.2 Heat Changes at Constant Pressure and Volume

Heat can be supplied to a system, keeping either the pressure constant or volume constant. As pressure and volume changes are generally associated with the mechanical work, we can replace w with $P\,\Delta V$ (equation 4.3) in the mathematical expression of the first law of thermodynamics (equation 4.4), so that

$$q = \Delta E + P\,\Delta V \qquad .....(4.6)$$

At constant volume : For a process occurring at constant volume, ΔV is zero and the equation (4.6) reduces to,

$$\boxed{q_V = \Delta E} \qquad .....(4.7)$$

indicating the heat supplied at constant volume is totally used in raising the internal energy of the system.

At constant Pressure : For a process occurring at constant pressure, equation (4.6) is written as

$$q_P = \Delta E + P\,\Delta V \qquad \text{.....(4.8)}$$

When this change represents a conversion of state 1 to state 2 then

$$\Delta E = E_2 - E_1, \quad \text{and} \quad \Delta V = V_2 - V_1$$

where E_2 and V_2 refer the final internal energy and volume respectively and E_1 and V_1 refer to the initial state. Substituting these limits in equation (4.8), we have,

$$q_P = E_2 - E_1 + P(V_2 - V_1)$$

or $$q_P = (E_2 + PV_2) - (E_1 + PV_1) \qquad \text{.....(4.9)}$$

Since E, P, V are thermodynamic functions which only depend on the initial and final states of matter and doesn't depend on the previous history, the term $E + PV$ may be defined as another thermodynamic function, called **enthalpy, H,** (Greek:warm).

$$\boxed{H = E + PV} \qquad \text{.....(4.10)}$$

This term *enthalpy gives a measure of heat changes* (heat absorbed or liberated) *associated with a constant pressure process*. In the present case the enthalpy of initial and final states of the change are

$$H_1 = E_1 + PV_1 \quad \text{and} \quad H_2 = E_2 + PV_2,$$

So that

$$\Delta H = H_2 - H_1 = (E_2 - E_1) + P\,(V_2 - V_1)$$

$$\boxed{\Delta H = \Delta E + P\Delta V} \qquad \text{.....(4.11)}$$

comparing equation (4.8) and (4.11), we have

$$\boxed{q_p = \Delta H}$$

4.3.3 Enthalpy

Enthalpy assumes importance as most of the physical and chemical processes are constant pressure processes. Among the physical processes, fusion, vaporization, sublimation, allotropic transformation, dissolution are some of the many important constant pressure processes, which we often encounter. Similarly most of the solid state and liquid state chemical reactions also take place at constant atmospheric pressure. Conventionally, when heat is liberated from the system into the surroundings the process is said to exothermic and indicated by a $-\Delta H$ value and the endothermicity of a reaction is suggested by the positive ΔH values. Enthalpy has the same units as heat. Hess's law of constant heat summation which we are going to discuss in thermochemistry is a special case of first law of thermodynamics. This point we will elaborate little latter.

Similar to internal energy the absolute value of enthalpy or heat content also cannot be determined but it is possible to determine only a change (ΔH) in the value. Enthalpy is also an extensive property.

4.3.4 Significance of First Law of Thermodynamics :

(i) First law of thermodynamics establishes a relation between heat and work.

(ii) It rules out creation of energy from nothing adopting any type of manipulation as a mith.

(iii) It also condemns the idea of running a perpetual motion machine, where no energy need be supplied in creating mechanical work. Work can not be created without a supply of energy.

(iv) It introduces the concept of enthalpy and internal energy and derives a relation between these two state functions

(v) It tells Hess's law of constant heat of summation as a special case of first law.

4.3.5 Limitation of First Law of Thermodynamics

Although first law of thermodynamics fully describes the relation between useful variables like heat and work, it does not give any information regarding the condition and extent of conversion of one form to other as the absolute values of internal energy and enthalpy of the initial and final states are unknown. For example, if we wish to lift water from lower level to higher level in an isolated system, according to first law of thermodynamics it is possible to do the job ceaselessly as long as the work done is supplemented by a drop in the internal energy. As a system has enormous amount of internal energy it is possible to lift water to any dizzy heights, utilizing in built internal energy and cooling of the system. It gives no information as to the condition and extent of lifting the water, which are tackled by second law of thermodynamics.

Another easily understandable example to demonstrate this fact could be derived from running. As mass has enormous amount of energy, first law expects every one to run any length of distance. But in practice we notice that every individual has different capacity, while Abebe Bikila, the greatest olympic marathon runner, can run over 42 km, there are many who cannot run even one hundredth of this distance and this is the subject matter of second law of thermodynamics.

Problem 3.

When a gas expands against a constant pressure of 1 atmosphere after absorbing 100 calories of heat, the volume of gas increases from 2 lit to 20 lit. Calculate the change in internal energy in this process.

Given : $q = 100$ cal, $P = 1$ atm, $V_1 = 2$ lit, $V_2 = 20$ lit, $\Delta E = ?$

Work done during expansion = PΔV

$$= P(V_2 - V_1)$$

$$= 1\,(20 - 2) = 18 \text{ lit atm}$$

From the first law of thermodynamics

$$q = \Delta E + w$$

or $\Delta E = q - w$

As the heat absorbed q is given in calories, we have to convert the work also into calories. The conversion factor is

$$1 \text{ lit atm} = 24.2 \text{ cal}$$

so $18 \text{ lit atm} = 18 \times 24.2 \text{ cal}$

$$w = 435.6 \text{ cal}$$

Substituting w and q in the first law equation

$$\Delta E = 100 - 435.6$$

$$= -335.6 \text{ cal}$$

4.3.6 Work Done in a Reversible, Isothermal Expansion of an Ideal Gas

It has already been derived in section 4.2.3 that the work done during an expansion caused by infinitesimal change in the volume, dV, is given by the expression (4.3).

$$\Delta w = P_{ext}\, dV \qquad \text{.....(4.12)}$$

where P_{ext} is the external pressure at a given temperature.

Any expansion carried out by such infinitesimally small steps, where the system is always in temperature and pressure equilibrium with the surrounding at every step, is called a thermodynamically reversible expansion.

The total work w done in the expansion will be the sum of these infinitesimal steps, with the external pressure decreasing steadily and corresponding volume increasing progressively. Reversible work done, w, in case of an expansion from volume V_1 to V_2 against a constant external pressure P is thus expressed mathematically as,

$$w = \int_{V_1}^{V_2} P\,dV \qquad \text{.....(4.13)}$$

In case of a reversible expansion the internal pressure of a gas at any instant and external pressure differ by infinitesimally small value and for all practical purposes external pressure (P_{ext}) may be replaced by pressure of gas.

$$w = \int_{V_1}^{V_2} P_{gas}\,dV$$

Supposing ideal behaviour for the gas and constancy of temperature (isothermal process) throughout the expansion we have

$$P_{gas} = \frac{nRT}{V}$$

Substituting for P_{gas} in the work expression (4.13)

$$w = \int_{V_1}^{V_2} \frac{nRT}{V}$$

$$w = nRT \int_{V_1}^{V_2} \frac{dV}{V}$$

$$= nRT \ln [V] \Big]_{V_1}^{V_2}$$

$$\boxed{\boldsymbol{w = nRT \ln \frac{V_2}{V_1}}} \qquad \text{.....(4.14)}$$

Changing the logarithm to base 10

$$w = 2.303\ nRT \log \frac{V_2}{V_1} \qquad \text{.....(4.15)}$$

An alternate form of equation (4.15) in terms of pressure may be obtained from the ideal gas expression at constant temperature.

$$P_1 V_1 = P_2 V_2 \quad \text{or} \quad \frac{P_1}{P_2} = \frac{V_2}{V_1}$$

$$\boxed{w = 2.303\ nRT \log \frac{P_1}{P_2}} \qquad \text{.....(4.16)}$$

Problem 4.

2 moles of an idea gas are allowed to expand from 2L to 20L at 27 ^{O}C. Calculate q, w, ΔE and ΔH.

For an isothermal reversible expansion

$$\Delta H = \Delta E = 0$$

Work done by 1 mole

$$\begin{aligned} w &= nRT \ln (V_2 / V_1) \\ &= 2 \times 2 \times 300 \ln (20/2) \\ &= 2 \times 2 \times 300 \times 2.303 \\ &= 2763.6 \text{ cal} \end{aligned}$$

From first law of thermodynamics

$$\begin{aligned} q &= \Delta E + P \Delta V \\ &= 0 + 2763.6 \\ q &= 2763.6 \text{ cal} \end{aligned}$$

Problem 5.

Calculate the isothermal and reversible work done by 10 litres of an ideal gas at 273 K and 5 atm to a final pressure of 1 atm.

Calculation of n :

22.4 litres at NTP contain 1 mole of the gas

10 litres at NTP contain ?

$$\frac{10}{22.4} \times 1 = \frac{10}{22.4} \text{ moles}$$

$$\text{No. of moles present at 5 atm} = \frac{5 \times 10}{22.4} = 2.23 \text{ mole}$$

$$\begin{aligned} w &= nRT \ln (P_1 / P_2) \\ &= 2.23 \times 8.314 \times 273 \ln (5/1) \\ w &= 8154 \text{ J} \end{aligned}$$

Problem 6.

3 moles of a monatomic gas compressed isothermally from NTP conditions to a volume of 10 litres. Find the work done.

Given : n = moles, $T_1 = 0\,^{\circ}C = 273$ K, $V_2 = 10$ L, $P_1 = 1$ atm

$$V_1 = \frac{nRT_1}{P_1} = \frac{3 \times 0.0821 \times 273}{1}$$

$$V_1 = 67.2 \text{ L}$$

$$w = nRT_1 \ln (V_2/V_1)$$

$$= 3 \times 8.314 \times 273 \ln (10/67.2)$$

$$= -12.97 \text{ kJ}$$

Work done $= -12.97$ kJ

Problem 7.

Calculate q, w, ΔH and ΔE for compressing 5 mol of an ideal gas reversibly at 27°C from 5 atm to 150 atm.

For an isothermal reversible expansion

$$dT = 0,\ \Delta E = C_v\, dT = 0 \ \ \&\ \ \Delta H = C_p\, dT = 0$$

So $\Delta H = \Delta E = 0$

$$w = nRT_1 \ln (P_1/P_2)$$

$$w = 3 \times 8.314 \times 300 \ln (5/150)$$

$$w = -25450 \text{ J}$$

Work done by the system w = Heat liberated q

Therefore Heat liberated during compression $q = -25450$ J

Problem 8.

Calculate q, w and ΔE for the conversion of 1 mole of water at 100°C and 1 atm to steam at the same pressure and temperature.

This is an isothermal, reversible, isobaric process.

Therefore, $\Delta E = \Delta H = 0$

Volume of steam at 100°C 1 atm

$$= \frac{nRT}{P} = \frac{1 \times 0.082 \times 373}{1}$$

$$= 30.65 \text{ L}$$

Assuming that the volume occupied by water is negligible;

From the first law we have,

$$q = \Delta E + P\,\Delta V$$
$$= \Delta E + 1 \times 30.65$$
$$= 30.65 \text{ L atm}$$
$$= 30.65 \times 24.218 \text{ cal}$$
$$= 742.3 \text{ cal}$$

Therefore $w = q = 742.3$ Cal

Problem 9.

Calculate the maximum work done by 5 moles of an ideal gas, when it expands from 5 atm to 2 atm at constant temperature of 27 ^{O}C.

Given : $n = 5$ moles, $T = 27 + 273 = 300\ K$, $P_1 = 5$ atm, $P_2 = 2$ atm.

The work done in an isothermal process is given by

The units of work depend on the R value that is taken.

When $R = 1.987$ cal K^{-1} mole^{-1}, w_{max} is in calories;

$R = 8.314 \times 10^7$ erg^{-1} K^{-1} mole^{-1}, w_{max} is in ergs;

$R = 8.314$ J K^{-1} mole^{-1}, w_{max} is in joules.

Substituting the given data and expressing R in J K^{-1} mole^{-1} in the maximum work expression.

$$w_{max} = 5 \times 8.314 \times 300 \ln (5/2)$$
$$= 2.303 \times 5 \times 8.314 \times 300 \log (5/2)$$
$$= 11427 \text{ J mole}^{-1} \text{ or J}$$

4.3.7 Work Done During a Phase Change

The phase changes, like, melting, vaporization, etc., are generally carried out at a constant atmospheric pressure and are constant temperature processes. Since the transformation is complete in a finite time the expressions derived for reversible conditions can not be used here.

For example during vaporization, a liquid boils at a fixed temperature when the vapor pressure P_v equal to the external atmospheric pressure. The volume change accompanied is equal to $V_{vapour} - V_{liquid}$. If the temperature is not too near the critical temperature, the volume occupied by the vapor V_v is far greater than the vapor condensed to liquid V_l.

Hence the volume change

$$\Delta V = V_v - V_l \approx V_v$$

So the maximum work for vaporization is

$$w = P_v V_v$$

At sufficiently high temperatures and low pressures the vapor obeys the ideal gas equation and under these conditions.

$$V_v = \frac{nRT}{P_v}$$

$$w = P_v \times \frac{nRT}{P_v}$$

The work done during vaporization $w = nRT$ .....(4.17)

Similar expression is obtained is case of sublimation also. In case of solid-liquid (melting) and solid-solid (allotropy) phase transformations at constant external pressure, the molar volumes (volumes occupied by 1 mole of a substance under given condition) of both phases are comparable and therefore equation (4.17) can not be used on these equilibria.

Problem 10.

1 mole of liquid water at $100^{O}C$ is evaporated into vapor at $100^{O}C$. Find the work done during evaporation.

Volume occupied by 1 mole of water vapor at 1 atm pressure and $100^{O}C$.

Therefore,

$$\Delta V = \text{Volume occupied by vapor}$$

$$V = \frac{0.0821 \times (273+100)}{1} = 0.0821 \times 373$$

$$= 30.6 \text{ L}$$

Volume occupied of 1 mole liquid of water at 100 ^{O}C is negligible when compared to the volume occupied by its vapor at the same temperature.

Therefore,

$$\Delta V = \text{Volume occupied by vapor}$$

$$W = P\,\Delta V = 1 \times 30.6 = 30.6 \text{ L atm}$$

Work done during evaporation = 30.6 L atm.

Problem 11.

Calculate the work done in the transformation of rhombic sulphur into monoclinic sulpther at their transition temperature 96 ^{O}C and at a pressure of 1 atm. Given that both exist in solid state at this temperature.

$$\text{S (rhombic)} \xrightarrow{96^{O}\text{C}} \text{S (monoclinic)}$$

Since both exist in solid state the volume change is negligible.

Therefore work done = $P\,\Delta V \approx 0$

4.3.8 Work Done in an Irreversible, Isothermal Expansion of an Ideal Gas

Let us consider gas enclosed in a cylinder filled with a frictionless piston. Suppose the external pressure is released suddenly from an initial external pressure P_1 to a final external pressure P_2 allowing the gas to expand against the new external pressure P_2 from a volume V_1 to V_2, Figure 4.4, then the work done,

$$w_{irr} = P_2 \int_{V_1}^{V_2} dV$$

$$w_{irr} = P_2\,(V_2 - V_1) \qquad \text{.....(4.18)}$$

From the combined gas law we have,

$$V_2 = \frac{nRT}{P_2} \text{ and } V_1 = \frac{nRT}{P_1}$$

$$w_{irr} = P_2 \left[\frac{nRT}{P_2} - \frac{nRT}{P_1}\right]$$

$$w_{irr} = nRT\,P_2 \left[\frac{1}{P_2} - \frac{1}{P_1}\right]$$

$$\boxed{w_{irr} = nRT\left[1 - \frac{P_2}{P_1}\right]} \qquad \text{.....(4.19)}$$

Fig. 4.4 Irreversible, Isothermal expansion of an ideal gas against a constant external pressure.

4.3.9 Comparison of w_{rev} and w_{irr}

The work done in an isothermal reversible expansion and irreversible expansion of an ideal gas are given as,

$$w_{\text{rev}} = nRT \ln \frac{P_1}{P_2} \qquad \ldots\ldots(4.20)$$

$$w_{\text{irr}} = nRT \left[1 - \frac{P_2}{P_1}\right] \qquad \ldots\ldots(4.21)$$

Rearranging equation (4.20)

$$w_{\text{rev}} = nRT \ln \left[1 - \left(1 - \frac{P_2}{P_1}\right)\right] \qquad \ldots\ldots(4.22)$$

But $\ln(1 - X) = -X - \frac{X^2}{2} - \frac{X^3}{3} \; \ldots\ldots \approx -X$

when X is a positive simple fraction.

Therefore equation (4.22) becomes

$$w_{\text{rev}} = nRT \left\{-\left(1 - \frac{P_1}{P_2}\right)\right\} = nRT \left(\frac{P_1}{P_2} - 1\right) \qquad \ldots.(4.23)$$

The difference of revesible and irreversible works is equal to

$$w_{\text{rev}} - w_{\text{irr}} = nRT \left(\frac{P_1}{P_2} - 1\right) - nRT \left(1 - \frac{P_2}{P_1}\right)$$

$$w_{\text{rev}} - w_{\text{irr}} = nRT \left(\frac{P_1}{P_2} - 1 - 1 + \frac{P_2}{P_1}\right)$$

$$w_{rev} - w_{irr} = nRT\left(\frac{P_1^2 + P_2^2 - 2P_1P_2}{P_1P_2}\right)$$

$$w_{rev} - w_{irr} = nRT\,\frac{(P_1 - P_2)^2}{P_1P_2} \qquad \text{.....(4.24)}$$

In the equation (4.24) P_1, P_2, R, T and $(P_1 - P_2)^2$ are all positive values.

Therefore, $w_{rev} - w_{irr} = +ve$

Which means $\boxed{w_{rev} > w_{irr}}$

Thus, it may be concluded that the work derived in an isothermal *reversible process is greater than that derived in an irreversible process.*

4.4 Heat Capacities at Constant Pressure and Constant Volume

Heat capacity of a system is the quantity of heat required to raise the temperature of the system by one degree. When the weight of the material under consideration is 1 g. the heat capacity is then called the **specific heat of that substance**. When the quantity under consideration in 1 mole, the heat capacity is then called the **molar heat capacity**. Generally from chemists point of view the latter definition is more appropriate.

The specific heat is generally expressed in cal deg^{-1} g^{-1} while the molar heat capacity has units cal deg^{-1} mol^{-1}. In place of calories other heat units could also be used in expressing the heat capacity.

While raising the temperature of a system, we can either keep the pressure constant or volume constant but not both constant. Hence to define heat capacity it is essential to define whether the pressure is kept constant or the volume. When the volume is maintained constant while raising the temperature, consequently raising the pressure, it gives the heat capacity at constant volume. Similarly when the pressure is maintained constant while raising the temperature, consequently raising the volume it gives the **heat capacity at constant pressure**. These are designated by the symbols C_v and C_p respectively. The heat capacity at constant volume is given by

$$C_v = \frac{\text{Heat given at constant volume}}{\text{Change in temperature}} = \frac{dq_v}{dT}$$

Heat capacity at constant pressure is given by

$$C_p = \frac{\text{Heat given at constant Pressure}}{\text{Change in temperature}} = \frac{dq_p}{dT}$$

According to equation (4.7), the heat q_V absorbed at constant volume is equal to the change in the internal energy dE_V under the same conditions.

Therefore $C_V = \dfrac{dE_V}{dT}$

The above expression is generally written in partial derivative form as

$$C_V = \left(\frac{\partial E}{\partial T}\right)_V$$

Thus heat capacity at constant volume may be defined as the rate of increase in internal energy with increase in temperature at constant volume.

For an ideal gas internal energy is independent of change of volume ($dE_V = dE$). Therefore for an ideal gas C_V is expressed as

$$C_V \text{ (ideal gas)} = \frac{\partial E}{\partial T} \quad \text{or} \quad \boxed{dE = C_v\,dT} \qquad \text{.....(4.25)}$$

In a similar manner from the heat, q_P, absorbed at constant pressure is equal to the change in enthalpy dH_p under the same conditions.

$$C_P = \frac{dH_P}{dT} \quad \text{or} \quad \left(\frac{\partial E}{\partial T}\right)_P$$

Thus heat capacity at constant pressure may be defined as the rate of increase in enthalpy with the increase in temperature at constant pressure.

For an ideal gas H is independent of change in pressure ($dH_p = dH$). Therefore equation (4.25) for an ideal gas is expressed as

$$C_P \text{ (ideal gas)} = \frac{dH}{dT} \quad \text{or} \quad \boxed{dH = C_P\,dT} \qquad \text{.....(4.26)}$$

4.4.1 Determination of the Value $C_p - C_v$ for an Ideal Gas

From the first law of thermodynamics equation (4.5) we have

$$dE = \delta q - \delta w$$

Considering 1 mole of an ideal gas undergoing only reversible pressure-volume work we have by equation (4.13)

$$\delta w = P\,dV$$

where P is the pressure of the gas and by equation (4.25)

$dE = C_V\,dT$ where C_V is molar heat capacity.

Upon differentiating the equation of state from ideal gas, $PV = RT$, at constant pressure (where $\delta q = \delta q_P$), we have

$$P\,dV = R\,dT$$

Substituting these values in the first law expression

$$C_V\,dT = \delta q - P\,dV \qquad .....(4.27)$$

Substituting the value of $P\,dV$ in equation (4.27)

$$C_V\,dT = \delta q_P - R\,dT$$

Rearranging above eqn.

$$\frac{dp_P}{dT} = C_V + R \qquad .....(4.28)$$

But the heat supplied at constant pressure, δq_P is equal to the enthalpy change, dH, therefore, eqn. (4.28) becomes.

$$\frac{dH}{dT} = C_V + R \qquad .....(4.29)$$

As expression on the left hand side is equal to C_P, we have,

$$C_P = C_V + R$$

or $$\boxed{C_P - C_V = R} \qquad .....(4.30)$$

The difference between the molar heat capacities of an ideal gas at constant volume and at constant pressure should be equal to R.

4.4.2 Determination of the Value C_P/C_V for an Ideal Gas

From the first law of thermodynamics eq. (4.12) we have

$$dE = \delta q - \delta w$$

At constant volume, $\Delta V = 0$, no pressure volume work is possible and all the energy supplied should be utilized only to raise the internal energy of the system. It is reasonable to assume that an increase in temperature of a gas mainly raises the microscopic kinetic energy, because any change in electronic forces (which form potential energy) between subatomic particles needs much larger energy. The main constituent of kinetic energy is the translational energy possessed by virtue of movement of the molecules from one place to another. In particular monoatomic gases do not possess vibrational and rotational modes of motion and have exclusively the translational motion. The kinetic energy of 1 mole at any temperature is given by the expression $\frac{3}{2}RT$. If the temperature is raised from T to $T + 1$ (i.e., by 1^0),

Increase in kinetic energy per degree = $\frac{3}{2}R(T+1) - \frac{3}{2}RT$

So $\quad dE = \frac{3}{2}R$ .....(4.31)

But we known at constant volume $dE = C_v \, dT$ and for a raise of 1^0 temperature.

$$\Delta E = C_v \qquad .....(4.32)$$

Comparing eqn. (4.31) and (4.32) we have for one mole of an ideal gas.

$$C_v = \frac{3}{2} R \qquad .....(4.33)$$

Substituting the value of C_v obtained in the equation (4.30),

$$C_p - \frac{3}{2} R = R$$

or $\quad C_p = \frac{3}{2} R + R = \frac{5}{2} R$ .....(4.34)

Hence the ratio of heat capacities C_p and C_v, represented by the symbol γ is thus equal to

$$\gamma = \frac{C_p}{C_v} = \frac{(5/2)R}{(3/2)R}$$

So $\quad \gamma = \frac{5}{3} = 1.667$ .....(4.35)

Thus it can be stated that for an ideal gas possessing only translational energy, the ratio of the two heat capacities should be equal to 1.667. This ratio is essentially temperature independent. This constancy was found to be strictly obeyed by monatomic gases, which maintain ideal behavior over a wide temperature range and have insignificant inter-atomic forces.

Diatomic gases like N_2, O_2, H_2, etc., have both translational and vibrational components in their kinetic enrgy. Hence for these gases the ratio

$\frac{C_p}{C_v}$ was found to be equal to 1.44

Problem 12.

Calculate the heat required to raise 36g of H_2O vapor at 110^0C to 220^0C (i) at constant volume (ii) at constant pressure. Given C_v of water in this temperature range is 6 cal K^{-1} mol^{-1}.

Moles of $H_2O = 36/18 = 2$

We know $C_p = C_v + R$

$$= 6 + 2 = 8 \text{ cal K}^{-1} \text{ mol}^{-1}$$

$$T_1 = 110 + 273 = 383 \text{ K}, \quad T_2 = 220 + 273 = 493 \text{ K}$$

$$\Delta E = q_v \quad = n\, C_v\,(T_2 - T_1) = 2 \times 6\,(493 - 383) = 1220 \text{ cal}$$

$$\Delta H = q_p \quad = n\, C_p\,(T_2 - T_1) = 2 \times 8\,(493 - 383) = 1760 \text{ cal}$$

4.4.3 Adiabatic Changes

As already described adiabatic change (Greek : not passing through) is a thermodynamically isolated process where no heat enters or leaves the system. As the system is thermally insulated, the energy needed for an adiabatic expansion is supplied within the system met by a drop in temperature. A decrease in temperature amount to a fall in internal energy. On the contrary an adiabatic compression is accompanied by a rise in temperature. This can be explained by the mathematical formulation of first law, equation (4.5). In a reversible adiabatic change the heat absorbed $\delta q = 0$ and the equation (4.5) reduces to;

$$dE = -\delta w \qquad .....(4.36)$$

As the work done during an expansion is positive, the change in internal energy turns out to be a negative value and the reverse is true for adiabatic compression.

From the equations (4.3) and (4.25) we know that

$$\delta w = P\, dV \quad \text{and} \quad dE = C_v\, dT$$

Substituting these values in equation (4.36) we have,

$$C_v\, dT = -P\, dV \qquad .....(4.37)$$

For 1 mole of an ideal gas $P = RT/V$, substituting this pressure expression in equation (4.37) and rearranging we get

$$C_v \frac{dT}{T} = -R\frac{dV}{V} \qquad .....(4.38)$$

If C_v is assumed independent of temperature and integration of the above expression gives

$$C_v \int_{T_1}^{T_2} \frac{dT}{T} = -R \int_{V_1}^{V_2} \frac{dV}{V}$$

Where V_1 and V_2 are initial and final volumes and T_1 and T_2 are the corresponding temperatures of an adiabatic change.

The result of these integrations is

$$C_v \ln \frac{T_2}{T_1} = \frac{P_2 V_2}{P_1 V_1}$$

Substituting for T_2/T_1 in equations (4.38)

$$C_v \ln \frac{P_2 V_2}{P_1 V_1} = -R \ln \frac{V_2}{V_1}$$

$$C_v \left[\ln \frac{P_2}{P_1} + \ln \frac{V_2}{V_1} \right] = -R \ln \frac{V_2}{V_1}$$

Bringing the $\ln \frac{V_2}{V_1}$ terms on one side

$$(C_v + R) \ln \frac{V_2}{V_1} = -C_v \ln \frac{P_2}{P_1}$$

But we know from equation (4.30), $C_v + R = C_p$

or $$C_p \ln \frac{V_2}{V_1} = -C_v \ln \frac{P_2}{P_1}$$

$$\frac{C_p}{C_v} \ln \frac{V_1}{V_2} = \ln \frac{P_2}{P_1}$$

rearranging above equation

or $$\ln \frac{P_2}{P_1} = \ln \left(\frac{V_1}{V_2} \right)^{C_p/C_v}$$

$$\frac{P_2}{P_1} = \left(\frac{V_1}{V_2} \right)^{\gamma}$$

or $$\boxed{P_1 V_1^{\gamma} = P_2 V_2^{\gamma}} \qquad \text{.....(4.39)}$$

This is the **Pressure-Volume relation of an adiabatic change.**

Volume-Temperature relation of an adiabatic change :

For an ideal gas

$$\frac{P_2}{P_1} = \frac{T_2 V_1}{T_1 V_2}$$

Substituting for P_2/P_1 in the equation (4.39)

$$\frac{V_1 T_2}{V_2 T_1} = \left(\frac{V_1}{V_2}\right)^{\gamma}$$

or $$\frac{T_1}{T_2} = \left(\frac{V_1}{V_2}\right)^{\gamma}\left(\frac{V_2}{V_1}\right) = \left(\frac{V_1}{V_2}\right)^{\gamma-1}$$

or $$\boxed{T_1 V_1^{\gamma-1} = T_2 V_2^{\gamma-1}} \qquad .....(4.40)$$

Pressure-Temperature relation of an adiabatic change :

From the ideal gas equation

$$\frac{V_1}{V_2} = \frac{T_1 P_2}{T_2 P_1}$$

Substituting the value of V_1/V_2 in eqn. (4.39)

$$\frac{P_2}{P_1} = \left[\frac{T_1 P_2}{T_2 P_1}\right]^{\gamma}$$

or $$\left(\frac{P_2}{P_1}\right)\left(\frac{P_1}{P_2}\right)^{\gamma} = \left(\frac{T_1}{T_2}\right)^{\gamma}$$

or $$\left(\frac{P_2}{P_1}\right)^{1-\gamma} = \left(\frac{T_1}{T_2}\right)^{\gamma}$$

or $$\boxed{P_1^{1-\gamma}\, T_1^{\gamma} = P_2^{1-\gamma}\, T_2^{\gamma}} \qquad .....(4.41)$$

4.4.3 Work done in an adiabatic reversible expansion of an Ideal Gas

Work done δw, due to a reversible process is given by the expression

$$\delta w = P\,dV$$

But in an adiabatic process, from equation (4.36)

$$dE = C_v\,dT = -\,\delta w$$

So $$\delta w = P\,dV = -\,C_v\,dT$$

An adiabatic expansion is accompanished by a change in volume and in temperature, from V_1 to V_2 and T_1 to T_2 respectively, on integrating we get,

$$w = \int_{V_1}^{V_2} P\,dV$$

$$= -C_V \int_{T_2}^{T_1} dT$$

Considering only the latter value

$$w = -C_V (T_2 - T_1) \quad .....(4.42)$$

$$\boxed{w = C_V (T_1 - T_2)}$$

Alternatively considering the former expression

$$w = \int_{V_1}^{V_2} P\,dV$$

For an adiabatic change $PV\gamma$ = constant

or $$P = \frac{\text{Constant}}{V^\gamma}$$

Designating the constant by k, we have

$$w = \int_{V_1}^{V_2} \frac{k}{V^\gamma} dV$$

$$w = \frac{k}{1-\gamma} \left[V^{1-\gamma}\right]_{V_1}^{V_2}$$

$$w = \frac{k}{1-\gamma} \left[V_2^{1-\gamma} - V_1^{1-\gamma}\right]$$

$$w = \frac{k}{1-\gamma} \left[\frac{k}{V_2^{\gamma-1}} - \frac{k}{V_1^{\gamma-1}}\right]$$

But $\quad k = P_1 V_1^\gamma = P_2 V_2^\gamma$

$$w = \frac{k}{1-\gamma} \left[\frac{P_2 V_2^\gamma}{V_2^{\gamma-1}} - \frac{P_1 V_1^\gamma}{V_1^{\gamma-1}}\right]$$

$$w = \frac{k}{1-\gamma} \left[P_2 V_2 - P_1 V_1\right] \quad .....(4.43)$$

Since $P_1 V_1 = RT_1$ and $P_2 V_2 = RT_2$ equation (4.42) may be written as

$$w = \frac{1}{1-\gamma}(RT_2 - RT_1)$$

$$w = \frac{R}{1-\gamma}(T_2 - T_1) \quad .....(4.44)$$

Therefore, adiabatic work, $w_{adiabatic}$ is equal to

$$\boxed{w_{adiabatic} = -C_v(T_2 - T_1) = \frac{R}{1-\gamma}(T_2 - T_1)} \quad .....(4.45)$$

4.4.4 Comparison of Isothermal and Adiabatic Processes

(i) For an isothermal reversible process, the pressure volume product is constant.

$PV =$ constant

In case of an adiabatic reversible process

$PV^{\gamma} =$ constant

(ii) In an adiabatic process $P \, \alpha \, \frac{1}{V^{\gamma}}$ while $P \, \alpha \, \frac{1}{V}$ is true for an isothermal process.

Hence for the same volume change, the pressure drop in adiabatic process is greater than in an isothermal process. This is illustrated in the pressure *vs* volume diagram where the adiabatic *P-V* curve has a larger slope than the isothermal *P-V* curve, Fig. 4.5.

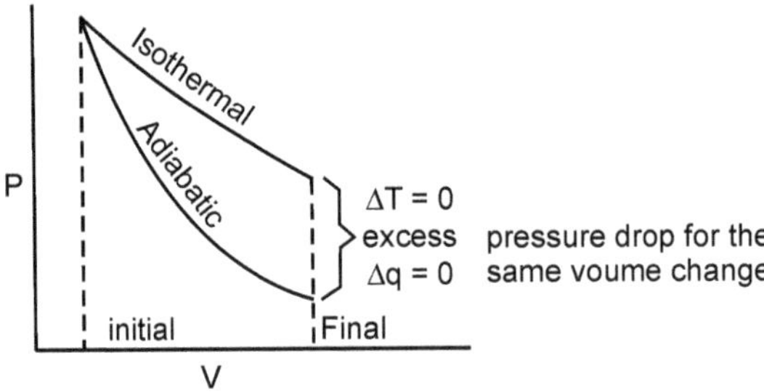

Fig. 4.5 The Pressure-Volume curves for adiabatic and isothermal reversible changes.

(iv) Since work done is equal to the area under the pressure-volume curve, it is clear from the curves in the Fig. 4.5 that more work is derived from an isothermal expansion than by an adiabatic expansion for the same volume change.

Problem 14.

Calculate q, w, ΔH, and ΔE for an adiabatic reversible expansion of 2 mol of a monatomic gas from 1L and 27° to 5L.

For a monatomic gas $\gamma = 1.66$

For an adiabatic reversible change $T_1 V_1^{\gamma} = T_2 V_2^{\gamma}$

Given $T_1 = 27^{\text{O}}\text{C} = 300$ K

$T_2 = ?$ $\quad V_1 = 1$ L $\quad V_2 = 5$ L

On substituting in the temperature – volume expression

$$300\,(1)^{1.66} = T_2\,(5)^{1.66}$$

$$T_2 = \frac{300}{(5)^{1.66}} = 20.74 \text{ K}$$

Work done in an adiabatic reversible process

$$w = \frac{nR(T_2 - T_1)}{1-\gamma}$$

$$w = \frac{2\times 8.314(20.74-300)}{1-1.66} = \frac{2\times 8.314(-279.26)}{(-0.66)}$$

$$= 7035.7 \text{ J}$$

For an adiabatic process $w = -\Delta E$

$$\Delta E = -\,7035.7 \text{ J}$$

$$\Delta E = \int_{T_1}^{T_2} nC_p\,\mathrm{d}T \;= nC_{\mathrm{p}}\,(T_2 - T_1)$$

$$\Delta E = n \times (5/2)\,R\,(T_2 - T_1)$$

$$= 2 \times (5/2) \times 8.314\,(20.74 - 300) = -11609 \text{ J.}$$

Problem 15.

Find the adiabatic and reversible work done for the expansion of an ideal gas at 50 L, 5 atm and 300 K to a pressure of 1 atm.

Given : $n = ? P_1 = 5$ atm, $\quad V_1 = 50$ L $\quad T_1 = 300$, $\quad P_2 = 1$ atm

$$n = \frac{5\times50}{300\times0.0821} = 10.15 \text{ moles}$$

$$\frac{T_2}{T_1} = \left(\frac{P_1}{P_2}\right)^{\frac{1-\gamma}{\gamma}}$$

$$\frac{T_2}{300} = \left(\frac{5}{1}\right)^{-0.44/1.44}$$

$$T_2 = 300 \times (5)^{-0.44/1.44}$$

$$T_2 = 300 \times 0.61 = 183.5$$

$$w = \frac{nR(T_2 - T_1)}{1-\gamma}$$

$$= \frac{10.15\times8.314(183.5-300)}{1-1.44}$$

$$= 22.343 \text{ kJ}$$

Problem 16.

A quantity of air at 25°C was allowed to expand reversibly and adiabatic from 100 atm to 50 atm. Taking C_v = 5 cal K^{-1} mol^{-1}, calculate the final temperature of the air.

Given : $T_1 = 25^{\mathrm{O}}\text{C} = 25 + 273 = 298$ K

$C_v = 5$ cal, $P_1 = 100$ atm, $P_2 = 50$ atm, $T_2 = ?$

$C_p - C_v = R$

$C_p - 5 = 2$

$C_p = 7$

Therefore,

$$\gamma = \frac{C_p}{C_v} = \frac{7}{5} \text{ and } (1-\gamma) = 1 - \frac{7}{5} = -\frac{2}{5}$$

$$\frac{(1-\gamma)}{\gamma} = \frac{(-2/5)}{(7/5)} = -\frac{2}{7} = -0.286$$

$$\frac{T_2}{T_1} = \left(\frac{P_1}{P_2}\right)^{\frac{1-\gamma}{\gamma}}$$

$$\frac{T_2}{298} = \left(\frac{100}{50}\right)^{-0.286}$$

$$T_2 = 244.5 \text{ K}$$

Final temperature of air is 244.5 K

Problem 17.

5 mole of oxygen at 27°C and 5 L was allowed to expand adiabatic and reversibly to a volume of 15 l. Calculate the final pressure, temperature and the work done.

Given : $n = 5$, $\gamma = 1.44$, $T_1 = 27\ ^{\text{O}}\text{C}$, $T_2 = ?$

$V_1 = 5$ L, $V_2 = 15$ L, $P_2 = ?$

$$\frac{T_2}{T_1} = \left(\frac{V_1}{V_2}\right)^{\gamma-1}$$

$$\frac{T_2}{300} = \left(\frac{5}{15}\right)^{1.44-1}$$

$$T_2 = 0.61 \times 300 \quad = 184.2$$

$$P_2 = \frac{nRT_2}{V_2} = \frac{5 \times 0.0821 \times 184.2}{15} = 5 \text{ atm}$$

$$w = \frac{nRT(T_2 - T_1)}{1-\gamma}$$

$$= \frac{5 \times 8.314(184.2 - 300)}{1 - 1.44} = 10.94 \text{ kJ}$$

Work done $= 10.94$ kJ

Problem 18.

5 moles of nitrogen at 27°C was allowed to expand reversibly and adiabatically from 200 atm to 50 atm. Taking $C_V = 5$ cal K^{-1} mol^{-1}, calculate the final temperature and volume.

Given : $n = 5,\ V_2 = ?$ $\quad P_1 = 200$ atm, $\quad P_2 = 50$ atm

$T_1 = 300$ K, $\quad T_2 = ?$ $\quad C_v = 5$ cal K^{-1} mol^{-1}

$C_p = C_v + R$

$C_p = 2 + 5 = 7$

Therefore,

$$\gamma = C_p/C_v \qquad = 7/5 \qquad = 1.44$$

$$1 - \gamma = 1 - 1.44 \quad = -0.44$$

For an adiabatic process

$$\frac{T_2}{T_1} = \left(\frac{P_1}{P_2}\right)^{\frac{1-\gamma}{\gamma}}$$

$$\frac{T_2}{300} = \left(\frac{200}{50}\right)^{-0.44/1.44}$$

$$T_2 = 300 \times (4)^{-0.3}$$

$$T_2 = 196.4 \text{ K}$$

$$V_2 = \frac{nRT_2}{P_2} = \frac{5 \times 0.0821 \times 196.4}{50} = 1.6 \text{ L}$$

Final volume= 1.6 L

Final temperature = 196.4 K

4.5 Joule-Thomson Effect

Joule-Thomson effect is the result of a collaborative work of the two British scientists Joule and Thomson from 1852-1862. They noticed that when a compressed gas like CO_2 at low temperature is allowed to expand suddenly into vacuum or low pressure region it generally gets cooled and in some cases it even leads to liquification of the gas. This sudden expansion into vacuum is also called **throttle expansion**. In a throttle expansion the system is not left with enough time gap to equilibrate with the surroundings and hence is comparable to an adiabatic (where no heat is allowed to enter or leave the system) process.

4.5.1 Joule-Thomson experiment

To mimic the throttle or adiabatic expansion conditions Joule and Thomson employed a long tube, thermally insulated from outside, Fig. 4.6.

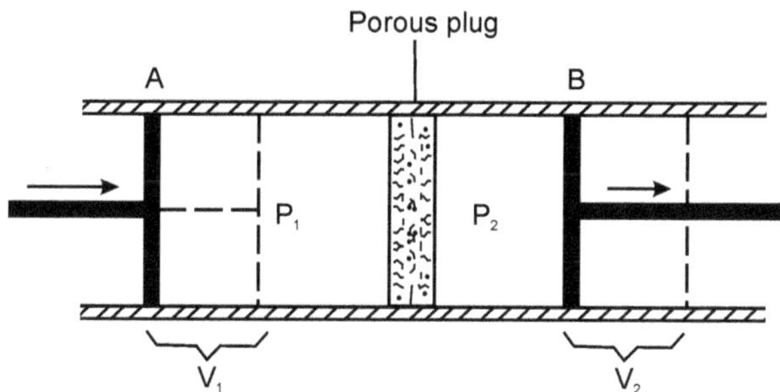

Fig. 4.6 Expansion of gas through porous plug (Joule-Thomson experiment)

In the middle of the tube a porous plug (made of cotton or wool) was placed to permit the slow passage of gas from a high pressure region to low pressure region. Two pistons A and B were fixed on either side of the porous plug. The pressure on the left and right side compartments of the porous plug were kept at pressure P_1 and P_2 respectively such that $P_1 > P_2$. Both the compartments were also fitted with temperature and pressure measuring devices to detect any minor variations of these parameters. Now piston A is forced inside by a volume V_1 with a simultaneous pulling out of piston B such that the pressure P_1 and P_2 of these compartments respectively, are unaltered. As $P_1 > P_2$, to keep the pressure constant on both the sides of the porous plug, there is a overall expansion of the volume of the gas. The temperature changes across the plug were recorded for different volume changes at different initial temperature and pressure conditions.

Experimental Observations

(i) All the gases show a change in temperature across the porous plug

(ii) The change in temperature is directly proportional to the difference in pressure on both the sides of porous plug or in other words the total volume change of the gas.

(iii) At sufficiently low initial temperature (before throttle expansion), all gases experience a fall in temperature in moving from left side (high pressure) compartment to right side (low pressure) compartment.

(iv) At moderate initial temperature conditions all the easily liquefiable gases (eg., methane, CO_2, N_2, etc.) suffer a fall in temperature and gases like H_2, He etc., record a rise in temperature.

(v) Every gas at a fixed initial pressure has an initial temperature above which the gas shows heating effect and below this temperature it shows cooling effect. This temperature is called the **inversion temperature**.

4.5.2 Definition and Theoretical Explanation

Basing on the above observations the Joule-Thomson effect may be defined as *a phenomena where a change in temperature is produced, when a gas expands adiabatically across a porous plug from a higher pressure zone to a lower pressure zone.*

The drop in temperature across a porous plug is the result of an adiabatic expansion, where, in over-coming the Van der Waals forces of attraction existing between the molecules, adiabatic work done is compensated by a fall in internal energy $(dE = -\delta w)$ or thermal energy.

4.5.3 Mathematical formulation of Joule-Thomson Effect

In the Joule-Thomson experiment adiabatic conditions are maintained.

So $\quad q = 0$

Substituting this value is the first law expression

$$\Delta E = 0 - w$$

or $$\Delta E = -w$$

or $$E_2 - E_1 = -w \quad .....(4.46)$$

In this experiment in the left side compartment work is done on the system in compressing the gas at a constant pressure P_1 by a volume V_1 and in the right compartment work is done by the system in the expansion of the gas at a constant pressure P_2 by a volume V_2.

As the pressure-volume work is given by the product of pressure and volume change;

work done on the system in the left compartment $= -P_1 V_1$

(The negative sign indicates that the work is done by the surroundings on the system).

work done by the system in the right compartment $= P_2 V_2$

total work done $= (-P_1 V_1) + (P_2 V_2)$

$= P_2 V_2 - P_1 V_1$

Substituting this work expression is equation (4.46)

$$E_2 - E_1 = -(P_2 V_2 - P_1 V_1)$$

or $$E_2 + P_2 V_2 = E_1 + P_1 V_1$$

or $$H_2 = H_1$$

or $$\Delta H = H_2 - H_1 = 0$$

Thus, it can be said that throttle expansion is an **isoenthalpic process** or the enthalpy across the porous plug remains the same during an adiabatic expansion.

Because the temperature change in the Joule-Thomson experiment depends on the initial temperature and initial pressure of the gas and is isoenthalpic, Joule and Thomson defined a constant μ known as **Joule-Thomson coefficient**,

$$\mu = \left(\frac{\partial T}{\partial P}\right)_{\mathrm{H}} \qquad .....(4.47)$$

The Joule-Thomson Coefficient μ may be defined as the *temperature change per unit pressure drop across porous plug under conditions of constant enthalpy.*

4.5.4 Thermodynamic Postulation of Joule-Thomson Coefficient

From a knowledge of calculus, the enthalpy change may be expressed as a complete differential with temperature and pressure as variables as given below;

$$dH = \left(\frac{\partial H}{\partial P}\right)_{\mathrm{T}} dP + \left(\frac{\partial H}{\partial T}\right)_{\mathrm{P}} dT$$

But $\left(\frac{\partial H}{\partial T}\right)_{\mathrm{P}}$ the enthalpy change with change in temperature at constant pressure condition, is equal to C_{p}.

So $$dH = \left(\frac{\partial H}{\partial P}\right)_{\mathrm{T}} dP + C_{\mathrm{p}}\, dT$$

Joule-Thomson effect is an isoenthalpic ($dH = O$) process, hence

$$0 = \left(\frac{\partial H}{\partial P}\right)_{\mathrm{T,H}} dP + C_{\mathrm{p}}\, dT$$

The subscript H represents that the operation is carried out at constant enthalpy. Rearranging the above expression,

$$\left(\frac{\partial H}{\partial P}\right)_{\mathrm{H}} = -\frac{1}{C_{\mathrm{p}}}\left(\frac{\partial H}{\partial T}\right)_{\mathrm{T,H}} \qquad .....(4.48)$$

substituting $(\partial P/\partial T)_{\mathrm{P}}$ by the Joule-Thomson coefficient μ (eq. 4.47)

$$\mu = -\frac{1}{C_p}\left(\frac{\partial H}{\partial T}\right)_{T,H} \qquad \text{.....(4.49)}$$

At constant enthalpy, eqn. (4.48) may be written as

$$dT = -\frac{1}{C_p}\left(\frac{\partial H}{\partial T}\right)_{T} \times dT \qquad \text{.....(4.50)}$$

From the expression (4.50) it is clear that under isoenthalpic conditions (assuming constancy of C_p over the temperature range under consideration) :

1. The change in temperature is directly proportional to the drop in the pressure
2. Under low initial temperature conditions since variation of enthalpy with pressure at constant temperature is positive, the right hand side, of the expression (4.50) becomes a negative value resulting in a drop in temperature.
3. At moderate initial temperature, for gases like H_2 & He, $(\partial H/\partial P)_T$ has a negative value making the overall right hand side of the expression (4.50) a positive quantity, resulting in a rise in temperature of the system.
4. When the enthalpy change with change in pressure, $(\partial H/\partial P)_T$ at a constant initial temperature is zero, the right hand side expression vanishes and we call that temperature as inversion temperature.

   Thus the mathematical formulation explains all the experimental observations.

4.5.5 Application of Joule-Thomson Effect

Liquefaction of gases employing Joule-Thomson effect is a widely used industrial process. Pictet (1877) used this technic for the first time in solidifying carbon dioxide and for the liquefaction of oxygen. Hampson (1897) & Linde (1895) developed this into a full fledged technique for the liquefaction of various gases. In this a gas at an initial pressure of about 250 atm is allowed to steam through a throttle and the gas so cooled is employed in cooling the second stage incoming gas. After repeating this process of compression and expansion and cooling the incoming gas, the temperature falls enough to liquefy the gas. The fact that inert gases like He do not liquefy at ordinary temperature where O_2, CO_2 and N_2 liquefy is taken advantage in separating and concentrating them from air.

Problem 19.

The Joule-Thomson coefficient for μ for N_2 follows the equation $\mu = 0.0140 - 2.5 \times 10^{-4}$ P. Calculate approximate cooling observed as the gas undergoes expansion from 0 to 100 bar.

According to Joule-Thomson principle

$$\mu = \left(\frac{\partial T}{\partial P}\right) \quad \text{or} \quad \int_{T_1}^{T_2} dT = \int_{P_1}^{P_2} \mu \, dP$$

$$\Delta T = \int_{P_1}^{P_2} \left(0.0140 - 2.5 \times 10^{-4} P\right) dP$$

Integrating

$$\Delta T = \left[0.014\text{P}\right]_{100}^{1} - 2.5 \times 10^{-4} \times \frac{P^2}{2}$$

$$= 0.014\,(1 - 100) - 2.5 \times 10^{-4}\,[1^2 - 100^2]$$

$$= -1.4 + 1.25 \times 10^{-4} \times 100^4$$

$$= -1.4 + 1.25 = -0.15\ ^{\text{O}}\text{C}$$

Extent of cooling = $-0.15\ ^{\text{O}}$C

4.6 Thermochemistry

Thermochemistry is a branch of thermodynamics dealing with the heat changes accompanying chemical reactions. All the reactions are accompanied by heat changes, although extremely small in some reactions.

In our day to day life we often notice heating or cooling of the reaction medium due to chemical reaction. On burning a candle or match stick enormous amount of heat is liberated. Similarly quick lime (CaO, burnt shells, etc.) when put in water, also liberates heat giving the lime used for mortar and white washing. Cooking of food is perhaps the best known chemical reaction that needs absorption of heat from the surroundings. Similarly dissolution of sugar in water is a process where the system is significantly cooled. If heat is liberated in the reaction, the process is said to be **exothermic** and if heat is absorbed in a reaction, it is said to be **endothermic process**. Heat can be absorbed or liberated during a reaction, accompanied either by a change in volume, designated by Q_v or by a change in pressure, designated by Q_p. Normally, reactions in gas phase or reactions where one of the product or reactant is a gas are accompanied by a change in the volume. In liquid and solid phase reactions, the volume changes are insignificant and are carried out at constant pressure. As majority of the reactions take place at atmospheric pressure, the heat change q_p, equal to the enthalpy difference between the products and reactants, ΔH, assumes significance and is said to be **heat of reaction**. It is defined as the difference as enthalpies of the reaction products and of the reactants, present in a definite physical state at constant temperature. The definite physical state of each substance is specified by the letters *s*, *l*, *g* in the parenthesis after the formula of the substance. For example when 1 mole (12 g) of carbon is burnt in oxygen to produce carbon dioxide gas at 25°C and 1 atm pressure, the system generates a heat of 94.05 k cal of heat and transfer it to the surroundings. This is represented in the form of a thermochemical equation,

$$C(s) + O_2(g) \longrightarrow CO_2(g) + 94.05 \text{ Kcal}$$

The initial and final states of this reaction may be represented as

Physical variable	Initial	Final
Temperature	25°C	25°C
Pressure	1 atm	1 atm
Substance	1 mole of C (s)	1 mole of CO_2 (g)
	1 mole of O_2 (g)	

The liberation of heat in this reaction indicates a decrease in enthalpy as we proceed from reactants to products.

$$\Delta H = H_{\text{final}} - H_{\text{initial}}$$

At 25°C and 1 atm pressure

$$\Delta H = H_{CO_2(g)} - H_{C(s)} - H_{O_2(g)}$$

This fall in enthalpy is indicated by a negative sign

Therefore, $\Delta H = -94.05$ kcal

So the above thermochemical equation is also written as

$$C(s) + O_2(g) \longrightarrow CO_2(g), \quad \Delta H = -94.05 \text{ kcal}$$

It should be born in mind that a negative enthalpy change always indicates an exothermic process, while a positive enthalpy change always describes an endothermic reaction.

Thus the thermochemical equation

$$H_2(g) + I_2(s) \longrightarrow 2HI(g), \quad \Delta H = 12.4 \text{ kcal}$$

suggests that it is an endothermic reaction. The heat of reaction, ΔH, always refers to a 100% completed reaction and allowance must be made for any incomplete reaction. The exothermic and endothermic process may be represented by Fig. 4.7.

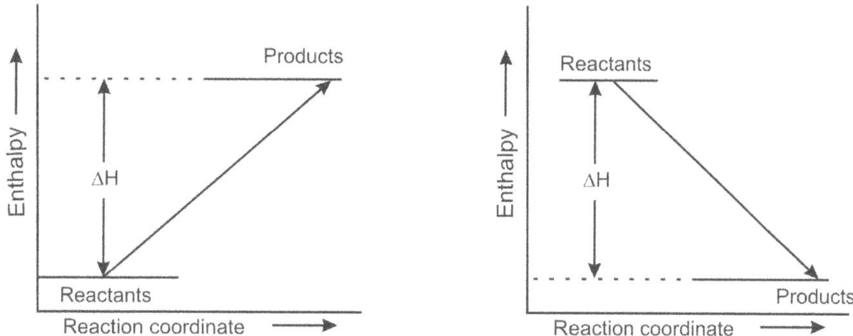

Fig. 4.7 Exothermic and Endothermic processes.

Substances present in aqueous phase are indicated by *aq.* written in parenthesis as shown in the following precipitation reaction.

$$AgNO_3(aq) + NaCl(aq) \longrightarrow NaNO_3(aq) + AgCl(s), \ \Delta H = -30.4 \text{ kcal}$$

4.6.1 Factors Affecting Heat of Reaction

Heat of reaction depends on the physical state of the reactants and products and temperature at which conversion is effected.

(a) **Physical state of reactants and products :** Carbon is present mainly in two allotropic forms, viz., graphite and diamond. Their oxidation gives different heats of reaction.

$$C\text{ (diamond)} + O_2\text{ (g)} \longrightarrow CO_2\text{ (g)}, \quad \Delta H = -94.5 \text{ kcal}$$

$$C\text{ (graphite)} + O_2\text{ (g)} \longrightarrow CO_2\text{ (g)}, \quad \Delta H = -94.05 \text{ kcal}$$

(b) **Temperature :** The heat of reaction depends on the temperature at which the reaction is carried out. This variation of enthalpy change of a reaction with temperature changes is given by Kirchoff's equation.

(c) **Heats of dilution or solution :** The heats of reaction values depend on the concentration of reactants. The anhydrous reactants give different heats of reaction compared to the same taken in solution.This is due to the extra enthalpy changes due to dilution.

$$H_2S\text{ (g)} + I_2\text{ (g)} \longrightarrow 2HI\text{ (g)} + S\text{ (s)}, \quad \Delta H = -17.18 \text{ kcal}$$

$$H_2S\text{ (g)} + \underset{\textit{Saturated solution}}{I_2\text{ (aq)}} \longrightarrow 2HI\text{ (aq)} + S\text{ (s)}, \quad \Delta H = -21.86 \text{ kcal}$$

4.6.2 Heat of Reaction at Constant Pressure and Constant Volume

The heat changes carried out at constant pressure and at constant volume are given by ΔH and ΔE respectively. The eqn. (4.11) gives the relation between these two values

$$\Delta H = \Delta E + P\,\Delta V$$

where $P\Delta V$ gives the external work done due to volume change affected during the chemical reaction.

If n_R is the total number of moles of the reactants and n_P the total number of moles of product formed after complete reaction, then $n_p - n_R = \Delta n$ is the change in number of moles.

If V is the volume occupied by 1 mole of the reactant at a given pressure and temperature and all the reactants and products are assumed to be gases, $V\Delta n$ gives the change in the volume, ΔV, as a result of reaction.

Substituting the value of ΔV

$$\boxed{\Delta H = \Delta E + PV\,\Delta n} \qquad .....(4.51)$$

But for an ideal gas $PV = RT$

Hence $$\boxed{\Delta H = \Delta E + RT\,\Delta n} \qquad .....(4.52)$$

From a knowledge of ΔH and Δn, the heat of reaction at constant volume could be calculated.

For reaction in which all the reactants and products are completely in the solid and/or liquid phase, for example;

$$\text{S (Rhombic)} \longrightarrow \text{S (Monoclinic)}$$

$$\text{NaOH (aq)} + \text{HCl (aq)} \longrightarrow \text{NaCl (aq)} + H_2O$$

$$\text{Na (s)} + \tfrac{1}{2}Br_2\,(l) \longrightarrow \text{NaBr (s)}$$

The change in the volume is negligible and we can take $\Delta H = \Delta E$

Similarly, in case of reactions involving *same number of gaseous molecules*, on both the sides of the thermochemical equation;

$$\text{C (s)} + O_2\,\text{(g)} \longrightarrow CO_2\,\text{(g)}$$

$$H_2\,\text{(g)} + I_2\,\text{(g)} \longrightarrow \text{2HI (g)}$$

also the heat of reaction at constant pressure (ΔH) and constant volume (ΔE) are equal.

4.6.3 Heat of Formation

Heat of formation of a compound is defined as the change in enthalpy when one mole of the substance is formed from its elements. Elements exist in different forms depending on temperature and pressure. Thus, at room temperature and 1 atm pressure carbon is a solid, mercury is a liquid and oxygen is a liquid. The state of aggregation at normal temperature (25^{0}C) and pressure (1 atm) is referred as standard state. When the elements are present in their standard states of aggregation the heat of formation is referred as standard heat of formation, ΔH_f.

For the reaction.

$$C\,(s) + O_2\,(g) \longrightarrow CO_2\,(g)$$

as 1 mole of CO_2 is formed from its elements the heat of reaction and heat of formation are the same.

But is case of formation of HCl from its elements the thermochemical equation may be written as

$$H_2\,(g) + Cl_2\,(g) \longrightarrow 2HCl\,(g), \quad \Delta H = -44.12 \text{ kcal}$$

as 2 moles of HCl are formed, the heat of formation of HCl is half the heat of the reaction; i.e., $\Delta H_f = -44.12/2 = 22.06$ kcal. The standard heats of formation of some compounds are given in Table 4.3.

In the Table 4.3 we notice that though majority of the compounds are formed with the liberation of heat there are some compounds which are formed by the absorption of heat from the surroundings.

Table 4.3 Standard heats of formation

Thermochemical equation			Standard heat of formation in kcal
$H_2\,(g) + \frac{1}{2}O_2\,(g)$	$\longrightarrow$	$H_2O\,(l)$	$\Delta H_f^0 = -68.32$
$\frac{1}{2}H_2\,(g) + \frac{1}{2}Cl_2\,(g)$	$\longrightarrow$	$HCl\,(g)$	$\Delta H_f^0 = -22.06$
$\frac{1}{2}H_2\,(g) + \frac{1}{2}Br_2\,(l)$	$\longrightarrow$	$HBr\,(g)$	$\Delta H_f^0 = -8.66$
$\frac{1}{2}H_2\,(g) + \frac{1}{2}I_2\,(s)$	$\longrightarrow$	$HI(g)$	$\Delta H_f^0 = 6.20$
$\frac{1}{2}N_2\,(g) + \frac{1}{2}O_2\,(g)$	$\longrightarrow$	$NO(g)$	$\Delta H_f^0 = 21.6$
$\frac{1}{2}N_2\,(g) + O_2\,(g)$	$\longrightarrow$	$NO_2\,(g)$	$\Delta H_f^0 = 8.09$
$C\,(s) + \frac{1}{2}O_2\,(g)$	$\longrightarrow$	$CO\,(g)$	$\Delta H_f^0 = -26.4$
$C\,(s) + O_2\,(g)$	$\longrightarrow$	$CO_2\,(g)$	$\Delta H_f^0 = -94.05$
$C\,(s) + 2H_2\,(g)$	$\longrightarrow$	$CH_4\,(g)$	$\Delta H_f^0 = -17.89$
$2C\,(s) + 3H_2\,(g)$	$\longrightarrow$	$C_2H_6\,(g)$	$\Delta H_f^0 = -20.24$
$S\,(s) + O_2\,(g)$	$\longrightarrow$	$SO_2\,(g)$	$\Delta H_f^0 = -70.96$
$2C\,(s) + H_2\,(g)$	$\longrightarrow$	$C_2H_2\,(g)$	$\Delta H_f^0 = 54.19$

As the heat of formation is the difference between the heat content (enthalpy) of the compound and its elements, a knowledge of enthalpies of the elements will help us in calculating the absolute enthalpy of the compound formed. In the following equation.

$$C\ (s) + \tfrac{1}{2} O_2\ (g) \longrightarrow CO\ (g), \qquad \Delta H_f^0 - 26.4\ \text{kcal.}$$

$$C\ (s) + O_2\ (g) \longrightarrow CO_2\ (g), \qquad \Delta H_f^0 = -94.05\ \text{kcal.}$$

The first equation suggests that carbon monoxide, enthalpy is less by 26.4 kcal than its elements, carbon and oxygen, while the second equation suggests that carbon dioxide is having an enthalpy less by 94.05 kcal than oxygen and carbon. *Arbitrarily, the enthalpies of all the elements in their standard states are taken to be zero at all temperature.* Adopting this convention we can plot an enthalpy diagram, Figure 4.8. From this we find that the enthalpy of a compound is equal to the standard heat of formation ΔH_f^0. This arbitrary fixing of enthalpies of elements at a zero level in no way violates our concept of heat of reaction or heat of formation as these values

$$\begin{array}{llcll} CH_4\ (g) & +\ 2O_2\ (g) & \longrightarrow & CO_2\ (g) & +\ 2H_2O\ (l) \\ -17.9 & \quad 0.0 & & -94.1 & \quad 2\ (-68.3) \end{array}$$

are always a difference in the enthalpies of products and reactants, on the contrary it allows us to predict the heat of reaction of other chemical changes. Considering complete oxidation of methane, and writing the enthalpies of all the reactants and product (obtained from above convention) we have

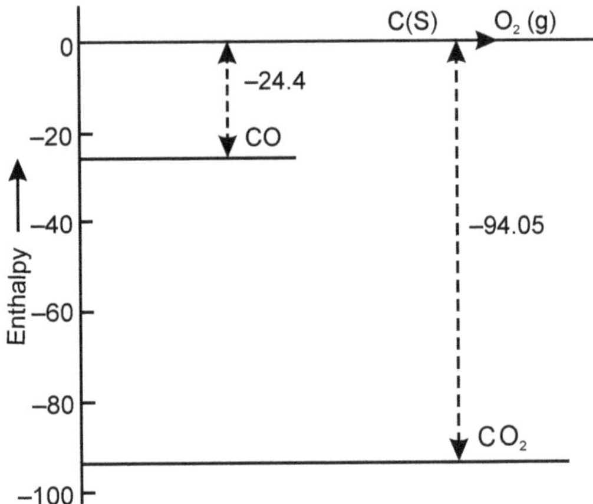

Fig. 4.8 Enthalpies of CO and CO_2.

Enthalpy of the products $= -94.1 + 2\,(-68.3)$

$= -230.7$ kcal

Enthalpy of the reactants $= -17.9$

Enthalpy change of the reaction $= (-230.7) - (-17.9)$

$= -212.8$ kcal

The heat of reaction values obtained this way were found to be accurate by 0.01 kcal. A knowledge of heat of reaction and the heat of formation of all but one reaction constituent allows us to calculate the heat of formation of that left out constituent. Thus in the above example it is possible to calculate the heat of formation of methane, if we know the heat of reaction and heat of formation of CO_2 (g) and H_2O (*l*).

4.6.4 Heat of Combustion

Heat of combustion is a particular case of heat of reaction. *It is defined as the heat evolved when 1 mole of a compound undergoes complete combustion at a given temperature.* As the final products of combustion of organic matter are always CO_2 and H_2O we can write their thermochemical equation conveniently. For example;

$$\underset{\text{Methane}}{CH_4} + 2O_2\,(g) \longrightarrow CO_2 + 2H_2O\,(l),\ \Delta H = -22.8 \text{ kcal}$$

$$\underset{\text{Graphite}}{C\,(s)} + O_2\,(g) \longrightarrow CO_2\,(g),\ \Delta H = -94.05 \text{ kcal}$$

$$\underset{\text{Benzene}}{C_6H_6\,(l)} + 7\tfrac{1}{2}\,O_2(g) \longrightarrow 6CO_2\,(g) + 3H_2O(l),\ \Delta H = -780.98 \text{ kcal}$$

If the substance involved, oxygen and the products of oxidation (CO_2 & H_2O) are all in their respective standard states, the change in enthalpy, ΔH_c^0 is said to be standard heat of combustion. As combustion of any organic substance invariably accompanied by the evolution of heat, heat of combustion is always an exothermic process.

The heat of combustion of hydrocarbons is industrially very important and is called the calorific value of that fuel. Pricing of any fuel is done on the basis of its calorific value.

The human body gets energy by the combustion (oxidation) of organic food material consumed by us. Here calorific values of food material are fixed depending an ease of digestibility by the human system. Thus carbohydrates, (rice, potatoes etc) which are readily degraded and absorbed are given a high calorific value compared to cellulose material (e.g., green vegetables, etc.) which are hard to be degraded by the human enzymes.

Heats of combustion are determined by burning the substance in the presence of excess of oxygen in a bomb calorimeter. Calculation of heat of combustion helps us in arriving at the heat of formation of different organic substances which cannot be determined by direct experimentation.

For example; synthesis of benzene by the direct combustion of carbon and hydrogen is not possible in an one step process. Instead to find the heat of formation of benzene, it is subjected to combustion and heat of combustion is determined.

$$\underset{x}{C_6H_6\,(l)} + \underset{0}{(15/2)O_2\,(g)} \longrightarrow \underset{6(-94.05)}{6CO_2\,(g)} + \underset{3(68.32)}{3H_2O\,(l)},\ \Delta H = -780.98$$

As the heat of formation of CO_2 and H_2O are –94.05 and –68.32 kcal respectively, by adopting the convention that the enthalpy of all elements in their standard states is zero, we have

Enthalpy of product

$$= 6\,(-94.05) + 3\,(-68.32)$$

$$= -769.26 \text{ kcal}$$

Enthalpy of reactants

$$= x + 0 \quad = x$$

Heat of combustion $=$ Heat of reaction

$$-780.98 = (-769.26) - (x)$$

So $x = (-769.26) + (780.98)$

Heat of formation of benzene = 11.72 kcal.

4.6.5 Heat of Neutralization

This is also a particular case of heat of reaction involving acid base neutralization. It is defined *as the heat change accompanying the neutralization of a dilute aqueous solution of one gram-equivalent of an acid (or a base) by a dilute aqueous solution of one gram-equivalent of a base (or an acid).*

The heat of neutralization of a strong acid by a strong base was always found to be equal to –13.7 kcal. This can be understood as the strong acids and bases are 100% ionized in the aqueous solution and the basic process involved is the combination of H^+ and OH^- ions to give one mole of H_2O.

For example;

$$NaOH\,(aq) \longrightarrow Na^+\,(aq) + OH\,(aq) \quad 100\%\ \text{ionized}$$

$$HCl\,(aq) \longrightarrow H^+\,(aq) + Cl^-\,(aq) \quad 100\%\ \text{ionized}$$

Neutralization is simply

$$Na^+(aq) + OH^-(aq) + H^+(aq) + Cl^-(aq) \longrightarrow Na^+(aq) + Cl^-(aq) + H_2O\,(l), \quad \Delta H = -13.7 \text{ kcal}$$

or $$H^+(aq) + OH(aq) \longrightarrow H_2O\,(l), \ \Delta H = -13.7 \text{ kcal}$$

In case either acid or base or both being weak, the heat of neutralization was found drop and less heat was found to be liberated. This drop in the enthalpy change is due to utilization of a fraction of heat liberated for the ionization of the weak acid or base present in the system.

For example; in the neutralization of HCl (aq) by NH_4OH (aq), before neutralization HCl being a strong acid is present in a fully ionized form while NH_4OH being a weak base is only partially ionized.

$$H^+(aq) + Cl^-(aq) + NH_4OH\,(aq) \longrightarrow NH_4Cl\,(aq) + H_2O, \ \Delta H = -12.3 \text{ kcal}$$

The drop in the heat of neutralization from –13.7 to –12.3, i.e., by 1.4 kcal, is utilized for dissociation of NH_4OH to NH_4^+ and OH^- ions.

$$NH_4OH\,(aq) \longrightarrow NH_4^+(aq) + OH^-(aq), \quad \Delta H = 1.4 \text{ kcal}$$

The heat used for the dissociation or ionization of one mole of a compound is called the **heat of dissociation** or **ionization**.

4.6.6 Heat of Solution

When a solute is dissolved in a solvent, we notice an increase or drop in temperature.

Thus, on dissolving sugar in water we notice cooling of the solution, while dissolving HCl in water generates heat in the system. This heat change due to dissociation is a concentration dependent factor, as shown in Fig. 4.9.

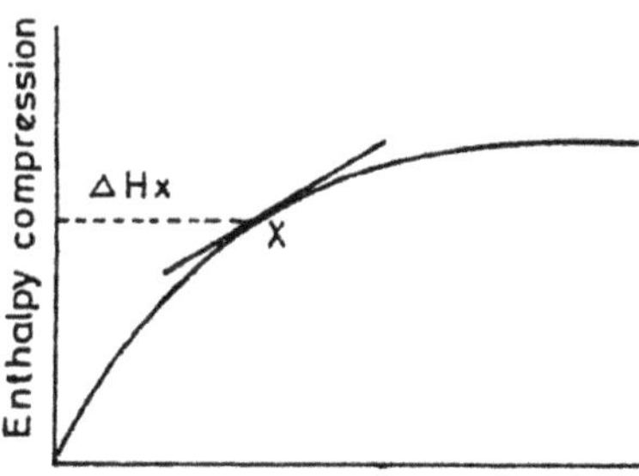

Fig. 4.9 Heat of solution.

The enthalpy change per mole of solute when it dissolves to form a solution of particular concentration is called the ***integral of solution.*** Initially for concentrated solution the enthalpy change is larger and it decreases on subsequent dilution. The following data demonstrates this fact further. The heats of formation in kcal of crystalline $ZnCl_2$ in various amounts of water at 25 ^{O}C are;

$$ZnCl_2\,(s) + 0\,H_20 = -99.40$$

$$ZnCl_2\,(s) + 10\,H_20 = -108.98$$

$$ZnCl_2\,(s) + 15\,H_20 = -109.88$$

$$ZnCl_2\,(s) + 25\,H_20 = -110.00$$

$$ZnCl_2\,(s) + 50\,H_20 = -112.55$$

$$ZnCl_2\,(s) + 100\,H_20 = -114.00$$

$$ZnCl_2\,(s) + 200\,H_20 = -114.86$$

$$ZnCl_2\,(s) + 800\,H_20 = -115.43$$

$$ZnCl_2\,(s) + \infty\,H_20 = -116.48$$

This addition 1 mole of $ZnCl_2$ to first ten moles of H_2O brought an enthalpy change of (–108.98) – (99.40) = –9.58 kcal, while near infinte dilution an aquation from 200 moles of H_2O to 800 moles of H_2O brought an enthalpy change of only 0.57 kcal. The **heat of solution** of a substance in a particular solvent is defined as *the change in enthalpy of the system when 1 mole of the substance is dissolved in such a large volume of the solvent that further addition of solvent produces no heat change.* In case of $ZnCl_2$ this value is,

$$ZnCl_2 \longrightarrow ZnCl_2\,(aq),\ \Delta H = -116.48\ \text{kcal}$$

On the same lines, the **differential heat of solution** is defined as *the enthalpy change per mole of solute when an infinitesimal amount of solute is added to a solution of definite concentration.* This is given by the slope of the tangent at any concentration x on the curve projecting the variation of enthalpy with concentration, Fig. 4.9.

For defining the **heat of dilution**, one has to define the initial and final concentration of the solute in the solvent. *It is defined as the enthalpy change when a solution containing 1 mole of the solute is diluted from one concentration to another.*

4.7 Effect of Temperature on Heat of Reaction - Kirchoff's Equation

The enthalpy changes of both physical and chemical processes are temperature dependent. An exact relation to describe the variation of enthalpy changes with temperature was given in 1858 by Kirchoff and is known as **Kirchoff's equation** or **Kirchoff's law**. It states that *for a small difference in temperature the molar heat capacity is practically constant, although variation in molar heat capacity is significant for larger temperature variations.*

4.7.1 Deduction of Kirchoff's equation

Consider a chemical reaction where reactant A converts to products B with a heat of reaction of ΔH at any given temperature.

$$\underset{\text{(Reactant)}}{A} \longrightarrow \underset{\text{(Product)}}{B} \qquad \Delta H$$

Suppose H_A is the enthalpy of the reactant and H_B that of product at the described temperature and pressure. As heat of reaction is equal to difference in enthalpy of products and reactants, we have;

$$\Delta H = H_B - H_A \qquad .....(4.53)$$

In order to determine the variation of heat of reaction with temperature at a given pressure, this expression is differentiated with respect to temperature as shown below.

$$\left(\frac{\partial \Delta H}{\partial T}\right)_p = \left(\frac{\partial H_B}{\partial T}\right)_p - \left(\frac{\partial H_A}{\partial T}\right)_p \qquad .....(4.54)$$

The quantity $(\partial H / \partial T)_P$ is equal to C_p, the heat capacity at constant pressure.

$$\left(\frac{\partial \Delta H}{\partial T}\right)_p = C_{p(B)} - C_{p(A)} = \Delta C_p \qquad .....(4.55)$$

This equation is referred to as Kirchoff equation. This may be stated as *the rate of variation of heat of reaction with temperature, at constant pressure, is equal to the increase in the heat of capacity accompanying the reaction.*

The practical applicability of Kirchoff equation can be realized if it is integrated between the temperature limits T_1 and T_2, where the heat of reaction changes from ΔH_1 to ΔH_2.

$$\int_{\Delta H_1}^{\Delta H_2} d\Delta H = \int_{T_1}^{T_2} \Delta C_p \, dT \qquad .....(4.56)$$

Here molar heat capacity at constant pressure of the products and reactants also varies with temperature. To simplify the situation, it is assumed that although individual heat capacities vary with temperature, ΔC_p the difference is constant and temperature independent over a limited range. Then the equation (4.56) may be written as

$$\int_{\Delta H_1}^{\Delta H_2} d\Delta H = \Delta C_p \int_{T_1}^{T_2} dT$$

On integrating and applying the limits we have

$$\Delta H_2 - \Delta H_1 = \Delta C_p (T_2 - T_1) \quad .....(4.57)$$

ΔC_p in the equation may be preferably taken as equal to $\overline{C}_{p(B)} - \overline{C}_{p(A)}$, where $\overline{C}_p$ gives the mean heat capacity of the substance for the required temperature limits.

Equation 4.57 may also be written in the form

$$\frac{\Delta H_2 - \Delta H_1}{T_2 - T_1} = \Delta C_p$$

This relation is called Kirchoff's equation at constant pressure.

From a knowledge of change in molar heat capacity and heat of reaction at any temperature it is possible to predict heat of reactions at any given temperature.

4.7.2 Kirchoff's equation at constant volume

From the first law we know that the heat of reaction and internal energy are related through the expression $\Delta H = \Delta E + P\Delta V$.

At constant volume $\Delta V = 0$ $\quad \therefore \ \Delta H = \Delta E$

Therefore for a reaction carried out at constant volume the Kirchoff's equation takes the form

$$\frac{\Delta E_2 - \Delta E_1}{T_2 - T_1} = \Delta C_v$$

Thus the *Kirchoff's equation at constant volume may be stated as the rate of variations of heat of reaction with temperature, at constant volume, is equal to increase in the heat capacity accompanying the reaction.*

4.8 Thermochemical Laws

The calculation of heat of reaction, heat of formation, heat of combustion and heat of neutralization revolve around the first law of thermodynamics; which states that it is impossible to derive energy from nowhere and what we experience is only a transformation of energy from one form into another. Two laws are stated basing on this fundamental law.

4.8.1 The Lavoisier-Laplace Law

This was put forward by A. L. Lavoisier and P. S. Laplace in 1780. It states that *the heat absorbed (or evolved) in a chemical reaction is reemitted (or absorbed) when the reaction is reversed.* Thus if for the formation of CO_2 gas 94.05 kcal are liberated.

$$C(s) + O_2(g) \longrightarrow CO_2(g), \qquad \Delta H = -94.05 \text{ kcal}$$

then the decomposition, of CO_2 takes place by the absorption of 94.05 kcal.

$$CO_2(g) \longrightarrow C(s) + O_2(g), \qquad \Delta H = 94.05 \text{ kcal}$$

This law was stated much before the fist law was enunciated and its importance can be realized by considering its failure. A failure of this law gives us a chemical reaction, where energy can be minted from nowhere by repeated reversal of the reaction, which is obviously against the first law of thermodynamics.

4.8.2 Hess Law of Constant Heat Summation

G. S. Hess in 1840 enunciated this law and its states that *the amount of heat evolved or absorbed in a chemical change is always constant and is independent of path adopted.*

or

The resultant heat change in a chemical reaction carried out either at constant pressure or constant volume is the same whether it takes place in one or several stages.

Hess law is a direct consequence of first law of thermodynamics. The heat changes of a chemical reaction at constant pressure and constant volume are given by ΔH and ΔE respectively. Enthalpy H, and internal energy E are thermodynamic functions and hence their changes are path independent and depend only on the initial and final states of the system. This implies that the ΔH and ΔE values are same irrespective of the path and number of stages adopted in tracing a particular path.

Consider the conversion of Reactant A to product B. Let Q calories be liberated in the direct conversion of A to B. B could also be reached by a three step process, via intermediates, L and M or by a two step process via intermediate P. So the total heat in a one step, two step and three step processes are

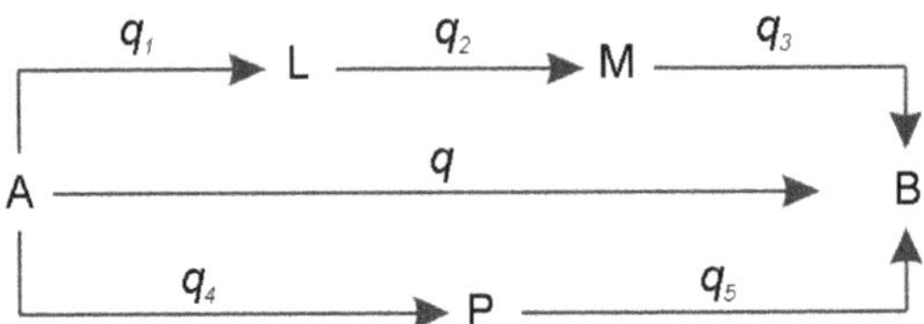

In a one step process A-B total heat liberated $= q$

In two Step process A-P-B total heat liberated $= q_4 + q_5$

In three step process A-L-M-B total heat liberated $= q_1 + q_2 + q_3$

According to Hess's law of constant heat summation

$$q = q_1 + q_2 + q_3 = q_4 + q_5$$

A failure of Hess's law, similar to Lavoisier-Laplace's law, leads to an energy minting process, which is against the first law of thermodynamics.

This fact could be seen in the following example;

(i) Single stage combustion of carbon to give CO_2

$C(s) + O_2(g) \longrightarrow CO_2(g)$, $\Delta H = -94.05$ kcal

(ii) Two stage combustion of carbon to give CO_2

$C(s) + \frac{1}{2} O_2(g) \longrightarrow CO(g)$, $\Delta H = -26.4$ kcal

$CO + \frac{1}{2} O_2(g) \longrightarrow CO_2(g)$, $\Delta H = -67.65$ kcal

On adding $C(s) + O_2(g) \longrightarrow CO_2(g)$, $\Delta H = -94.05$ kcal

Thus heat evolved in both the processes is virtually same.

4.8.3 Application of Hess's Law

Hess's law treats the thermodynamical equations similar to algebraic equations, where they can be added, subtracted, multiplied and divided.

These manipulations allow one to calculate the heat of reaction or formation of reactions which are very slow or cannot carried out in a direct one step, but theoretically feasible. Following are few applications that fully demonstrate the importance of Hess law of constant heat summation.

(i) Calculation of heat of formation of organic compounds : Thermodynamically it is possible to synthesize organic compound like ethyl alcohol, benzene, sugar, etc., by suitable combination of elements, viz., carbon, hydrogen and oxygen. But practically many complications are involved in these syntheses and their heats of formation values are evaluated from other thermochemical data. One such example, the heat of formation of benzene from the heat of combustion values was described in 4.6.4. Here method for calculation of heat of formation of ethanol is described. Three thermochemical equation are considered for this purpose

$$C_2H_5OH(l) + 3O_2(g) \longrightarrow 2CO_2(g) + 3H_2O\ (l), \quad \Delta H^o = -326.7 \text{ kcal} \qquad \text{(i)}$$

$$C(s) + O_2(g) \longrightarrow CO_2(g),\ \Delta H^o = -94.05 \text{ kcal} \qquad \text{(ii)}$$

$$H_2(g) + \tfrac{1}{2}\,O_2(g) \longrightarrow H_2O(l),\ \Delta H^o = -68.32 \text{ kcal} \qquad \text{(iii)}$$

Multiplying, equation (ii) by 2 and equation (iii) by 3 and adding we have

$$\text{(ii)} \times 2,\quad 2C(s) + 2O_2(g) \longrightarrow 2CO_2(g) + 188.1 \text{ kcal}$$

$$\text{(iii)} \times 3,\quad 3H_2(g) + \frac{3}{2}\,O_2(g) \longrightarrow 3H_2O(l) + 204.96 \text{ kcal}$$

$$\text{(iv)}\quad 2C(s) + 3H_2(g) + \frac{7}{2}\,O_2\,(g) \longrightarrow 2CO_2(g) + 3H_2O(l) + 392.06 \text{ kcal}$$

$$\text{(i)}\quad C_2H_5OH(l) + 3O_2(g) \longrightarrow 2CO_2(g) + 3H_2O(l) + 326.72 \text{ kcal}$$

Subtracting (iv) - (i)

$$-C_2H_5OH + 2C(s) + 3H_2(g) + 1/2\ O_2 \longrightarrow + 65.34 \text{ kcal}$$

on rearranging

$$2C(s) + 3H_2(g) + \tfrac{1}{2}\,O_2 \longrightarrow C_2H_5OH \quad \Delta H^o = -65.34 \text{ kcal}$$

Thus heat of formation of ethanol, $\Delta H^o = -65.34$ kcal

(ii) Calculation of heat of combustion : Some of the heats of combustion values, like partial combustion of carbon to give carbon monoxide without any contamination of fully oxidized product CO_2, are very difficult to obtain experimentally. These values are obtained using Hess law. To demonstrate this consider the following thermochemical equation.

(i) $C(s) + O_2(g) \longrightarrow CO_2(g) \quad \Delta H = -94.05$ kcal

(ii) $CO(g) + \frac{1}{2} O_2(g) \longrightarrow CO_2(g) \quad \Delta H = -67.71$ kcal

subtracting (i) - (ii)

$C(s) - CO(g) + \frac{1}{2} O_2(g) \longrightarrow 0 \quad \Delta H = -26.34$ kcal

Rearranging the above equation

$C(s) + 1/2\ O_2(g) \longrightarrow CO(g) \quad \Delta H = -26.34$ kcal

(iii) Calculation of heat of slow reaction : Allotropic transformations like transition of sulphur from monoclinic form to rhombic form are slow processes and generally are accompanied by small amounts of heat changes. These values are difficult to determine experimentally and are obtained from Hess law. For this consider the following thermochemical equations.

(i) S (rhombic) + $O_2(g) \longrightarrow SO_2(g)$, $\Delta H^o = -71.1$ kcal

(ii) S (monoclinic) + $O_2(g) \longrightarrow SO_2(g)$, $\Delta H^o = -71.7$ kcal

subtracting (ii)-(i)

S (Monoclinic) – S (rhombic) $\longrightarrow 0 \ \Delta H^o = -0.6$ kcal

on rearranging

S (Monoclinic) $\longrightarrow$ S (rhombic) $\quad \Delta H^o = -0.6$ kcal

Thus transformation of 1 mole of monoclinic sulphur to rhombic sulphur liberates a heat of 0.6 kcal.

(iv) Calculation of lattice energies : The heat of formation of ionic compounds depend on various factors, viz., the ionization potential, electron affinity, heat of vaporization, heat of dissociation of the molecules and the lattice energy of the compound. Born-Haber cycle is a thermodynamical cycle showing the inter relation of these quantities. The formation of NaF from its elements is expressed by the following thermochemical equations.

$Na(s) \longrightarrow Na(g),$		$\Delta H_{sublimation}$
$Na(g) \longrightarrow Na^+(g) + e^-$		$I.P._{Na}$
$\frac{1}{2} F_2(g) \longrightarrow F(g)$		$\frac{1}{2} \Delta H_{diss\ of\ F_2}$
$F(g) + e^- \longrightarrow F^-(g)$		E.A.
$Na^+(g) + F^-(g) \longrightarrow NaF(s)$		U
$Na(s) + \frac{1}{2} F_2(g) \longrightarrow NaF(s)$		$\Delta H_f^o(NaF)$

where

$$\Delta H_f^o{}_{(NaF)} = \Delta H_{sublimation} + I.P_{Na} + \frac{1}{2} \Delta H_{diss\ of\ F_2} - E.A. + U$$

Out of the above thermochemical quantities generally the lattice energy U is difficult to determine experimentally and this value is calculated from Hess law. In the **Born-Haber** cycle given above the energies are

$$\Delta H_f^o{}_{(NaF)} = -136\ \text{kcal},\ \Delta H_{sublimation} = 26\ \text{kcal},\ I.P_{Na} = 118\ \text{kcal}$$

$$\frac{1}{2} \Delta H_{diss\ of\ F_2} = 18\ \text{kcal and E.A.} = -83\ \text{kcal}$$

According to Hess law

$$-136 = 26 + 118 + 18 - 83 + U$$

So lattice energy $U = -215$ kcal

4.9 Bond Energy

For the purpose of calculation of heat of reaction, heat of formation, heat of combustion, etc., the thermochemical equations can be added, subtracted and multiplied similar to simple algebraic equations.

In defining the heat of formation, say of HBr, the combining elements are considered in their standard states of aggregation.

$$\frac{1}{2} H_2(g) + \frac{1}{2} Br_2(l) \longrightarrow HBr(g)$$

This in reality involves the following steps

(i) Vaporization of liquid bromine to gaseous bromine

$$\frac{1}{2} Br_2(l) \longrightarrow \frac{1}{2} Br(g)$$

(ii) Conversion of bromine molecule to atomic bromine

$$\frac{1}{2} Br(g) \longrightarrow Br(g)$$

(iii) Conversion of gaseous molecular hydrogen to atomic hydrogen

$$\frac{1}{2} H_2(g) \longrightarrow H(g)$$

(iv) H-Br bond formation by the combination of atomic hydrogen and bromine.

$$H(g) + Br(g) \longrightarrow HBr\,(g)$$

The fourth step actually corresponds to H-Br bond formation. The **bond formation energy, bond energy** or **bond enthalpy** is defined as *the heat evolved when the free atoms in the gaseous state combine to give one mole of molecular product also in the gaseous state.*

It is convenient to start with the molecule and study its dissociation into gaseous atoms. The energy liberated is called **bond dissociation** energy and is defined as *the average amount of energy required to break a particular bond in 1 mole of molecules producing free atoms or radicals in gaseous state.* The bond energy is equal in magnitude with the bond dissociation energy but opposite in sign.

The dissociation of methane involves fragmentation into different radicals accompanied with the liberation of different energies.

$$CH_4(g) \longrightarrow CH_3(g) + H(g), \ \Delta H = -102 \text{ kcal}$$

$$CH_3(g) \longrightarrow CH_2(g) + H(g), \ \Delta H = -105 \text{ kcal}$$

$$CH_2(g) \longrightarrow CH(g) + H(g), \ \Delta H = -81.3 \text{ kcal}$$

Overall equation $CH_4(g) \longrightarrow C(g) + 4H(g), \ \Delta H = -396.3$ k cal

Average bond dissociation energy $\Delta H_D = \dfrac{-396.3}{4} = -99.07$ k cal (ave)

So bond energy of C-H bond = 99.07 k cal mol^{-1}

From a knowledge of heat of atomization (heat liberated in atomizing a mole of gaseous molecule) and heat of formation is sufficient to evaluate the bond energies of different bonds.

Evaluation of O-H bond energy : Water has two O-H bonds and the O-H bond energy is calculated from the following thermochemical equations.

(i) $H_2(g) \longrightarrow 2H(g)$, $\Delta H = -104.2$ kcal

(ii) $O_2(g) \longrightarrow 2O(g)$, $\Delta H = -118.4$ kcal

(iii) $H_2(g) + ½\, O_2(g) \longrightarrow H_2O$ $\Delta H = 57.8$ kcal

(iii)-(i)-$\frac{1}{2}$(ii) $2H(g) + O(g) \longrightarrow H_2O(g)$, $\Delta H = 221.2$ kcal

The enthalpy change of 221.2 kcal corresponds to formation of two OH bonds. So average O-H bond energy = 221.2/2 = 110.6 kcal

Evaluation of C-C bond energy : For this consider ethane molecule which has six C-H bonds and one C-C bond. The thermochemical equation for the formation of ethane is

(i) $2C(s) + 3H_2(g) \longrightarrow C_2H_6(g)$ $\Delta H = -20.24$ kcal

(ii) $C(s) \longrightarrow C(g)$ $\Delta H = 172$ kcal

(iii) $H_2(g) \longrightarrow 2H(g)$ $\Delta H = 104.2$ kcal

(i) -2x(ii)-3x(iii)

$$2C(g) + 6H(g) \longrightarrow C_2H_6(g) \quad \Delta H = -674.8 \text{ kcal}$$

Total bond energy of $C_2H_6 = 6 \times C-H + 1 \times C-C$

$$-674.8 = 6 \times (-99) + \Delta H_{C-C}$$

$$\Delta H_{C-C} = -674.8 + 594$$

∴ C-C bond dissociation energy = –80.8 kcal

So C-C bond energy = 80.8 kcal

Evaluation of C = C bond energy : Consider ethylene for this purpose. It has four C-H bonds and one C = C bond.

(i) $2C(s) + 2H_2(g) \longrightarrow C_2H_4(g)$, $\Delta H = 12.5$ kcal

(ii) $C(s) \longrightarrow C(g)$, $\Delta H = 172$ kcal

(iii) $H_2(g) \longrightarrow 2H(g)$, $\Delta H = 104.2$ kcal

(i)-2x(ii)-2x(iii)

$$2C(g) + 4H(g) \longrightarrow C_2H_4(g) \quad \Delta H = 12.5 - 344 - 208.4$$

$$\Delta H = -539.9 \text{ kcal}$$

Total bond energy of C_2H_4 $= 4 \times C-H + 1 \times C=C$

$= 4 \times (-99) + \Delta H_{C=C}$

$\Delta H_{C=C} = -539.9 + 396$

C=C bond dissociation energy = –142.1 kcal

The mean bond energies of some important chemical bonds are given in Table 4.4.

Table 4.4 Bond energies of some important chemical bonds

Bond	Energy		Bond	Energy	
	kJ mol^{-1}	kcal mol^{-1}		kJ mol^{-1}	kcal mol^{-1}
H-H	436.4	104.2	Cl-Cl	242.7	57.9
C-C	347	82.8	Br-Br	192.5	45.9
C=C	620	148.0	I-I	151.0	36
C/C	812	193.8	C-Cl	327	78
C-H	414	98.8	C-Br	226	54
C-N	276	65.9	C-I	193	46
C-O	351	83.8	N-H	393	93.8
C=O	781	186.4	H-Cl	431.9	103.1
O=O	498.7	119	H-Br	361.1	87.4
O-O	142	33.9	H-I	298.3	71.2
N-N	193	46.1	O-H	460	109.8
N/N	941.4	224.7	C–N	276	65.9

Bond energies are highly useful in the estimation of heat of reaction or formation which are difficult to determine experimentally. Thus the heat of reaction of oxidation of methanol to formaldehyde may be estimated as follows;

$$CH_3OH(g) + \tfrac{1}{2} O_2(g) \longrightarrow HCHO(g) + H_2O(g)$$

The bond energies are as follows

Reactant		**Products**		
$3 \times$ C-H	$3 \times 98 = 294$,	$2 \times$ C-H	2×98	$= 196$
$1 \times$ C-O	$1 \times 80 = 80$	$1 \times$ C=O	1×173	$= 173$
$1 \times$ OH	$1 \times 110 = 110$			
CH_3CHO	484 kcal	HCHO		369 kcal
		$2 \times$ OH	$= 2 \times 110$	
$\tfrac{1}{2}O_2(g)$	118/2 =59 kcal	H_2O	= 220 kcal	

Total bond enthalpy of reactants = 484 + 59 = 543 kcal

Total bond enthalpy of products = 369 + 220 = 589 kcal

Heat of reaction ΔH = Bond energy of reactants − Bond energy of products

= 543 − 589 = −46 kcal

In the calculation involving the estimation of enthalpies of a substance or heat of reaction, proper allowance has to made for structural effects such as resonance for compounds like benzene, carboxylic acids, carbon dioxide etc.

Problem 20.

If the heat of combustion of benzene at 27 ^{O}C is 754.346 kcal mol^{-1}. Find the internal energy change accompanying the reaction. Under the given temperature conditions we know that benzene and water exist as liquids and CO_2 and O_2 as gases.

$$C26H_6\ (l) + 7\tfrac{1}{2}\ O_2\ (g) \longrightarrow 5CO_2 + 3H_2O + 754.346\ \text{kcal}$$

Workdone during the process $w = P\,\Delta V = \Delta nRT$

$$\Delta n = 6 - \frac{15}{2} = -\frac{3}{2}$$

$$w = -\frac{3}{2} \times 2 \times 300$$

$$w = -900\ \text{cal}$$

Heat reaction $\Delta H = -754.346$ kcal $= -754346$ cal

From first law of thermodynamics

$$\Delta H = \Delta E + w$$

$$\Delta E = \Delta H - w$$

$$= -754346 - (-900)$$

$$= -754346 + 900$$

$$\Delta E = -753446\ \text{cal}$$

Internal energy change accompanying the reaction = −753.446 kcal

Problem 21.

Heat of formation of ammonia at 27 ^{O}C and 1 atm is 21.9 kcal. Calculate the heat of formation at constant volume.

$$N_2\ (g) + 3H_2\ (g) \longrightarrow 2NH_3\ (g) - 21.9\ \text{kcal}$$

$$\Delta n = 2 - 4 = -2$$

$$\Delta H = \Delta E + \Delta n\, RT$$

$$-21900 = \Delta E + (-2)\, 2 \times 300$$

$$\Delta E = -21900 + 1200 = -20700$$

Heat of formation at constant volume = –20.8 kcal

Problem 22.

Heat of combustion of rhombic sulphur was –296.83 kJ, while that of monoclinic form is –297.16 kJ. Predict heat of reaction for transformation of rhombic sulphur to monoclinic sulphur.

Given :

$$\text{S (rhombic)} + O_2\ (g) \longrightarrow SO_2\ (g) + 296.83\ \text{kJ} \qquad \text{I}$$

$$\text{S (monoclinic)} + O_2\ (g) \longrightarrow SO_2\ (g) + 297.16\ \text{kJ} \qquad \text{II}$$

$$\text{S (rhombic)} \longrightarrow \text{S (monoclinic)} + (296.83 - 297.1) \qquad \text{I-II}$$

$$\text{S (rhombic)} \longrightarrow \text{S (monoclinic)} - 0.33\ \text{kJ} \qquad \text{III}$$

Heat of reaction of transformation = 0.33 kJ

Problem 23.

Pure calcium carbonate reacts with dilute carbonic acid forming corresponding bicarbonate salt.

$$CaCO_3\ (s) + H_2CO_3\ (aq) \longrightarrow Ca(HCO_3)_2\ (aq)$$

If the heat of foramtion of $CaCO_3$, H_2CO_3, Ca^{2+} and HCO_3^- are –1207, –700, –543 and –692 kJ mol^{-1} respectively, calculate the heat of reaction of the above reaction.

$$Ca + C + 1\tfrac{1}{2}\, O_2 \longrightarrow CaCO_3 + 1207 \qquad \text{I}$$

$$H_2 + C + 1\tfrac{1}{2}\, O_2 \longrightarrow H_2CO_3 + 700 \qquad \text{II}$$

$$Ca \longrightarrow Ca^{2+} + 2e^- + 543 \qquad \text{III}$$

$$\tfrac{1}{2}H_2 + C + 1\tfrac{1}{2}O_2 + e^- \longrightarrow HCO_3^- + 692 \qquad \text{IV}$$

(III – 2 × IV) – (I + II)

$$CaCO_3 + H_2CO_3\ (aq) + Ca + 2e^- + H_2 + 3O_2 + 2e^-$$

$$\longrightarrow Ca + 2e^- + 3O_2 + H_2 + Ca^{2+} + 2HCO_3^- + 2e^-$$

$$CaCO_3 + H_2CO_3 \longrightarrow Ca(HCO_3)_2 + \Delta H$$

$$\Delta H = (-543 - 2 \times 692) - (-1207 - 700)$$

$$= -20\ \text{kJ mol}^{-1}$$

Problem 24.0

71.7 kcal of heat was absorbed when SO_2 was formed from its elements in their standards states at 27 ^{O}C and 1 atm. Calculate the heat of formation at constant volume.

$$S(s) + O_2(g) \longrightarrow SO_2(g) - 71.7 \text{ kcal}$$

$$\Delta n = 1 - 1 = 0$$

$$\Delta H = \Delta E + P\,\Delta V$$

$$\Delta H = \Delta E + \Delta n\,RT$$

$$71.7 = \Delta E + 0 \times RT$$

Therefore; $\Delta E = 71.7$ kcal

Heat of formation at constant volume $q_v = \Delta E = 71.7$ kcal

Problem 25.

Using the data given in the above problem find the contribution of cyclohexane ring towards enthalpy change. ΔH_f of cyclohexane is –936.9 kcal.

ΔH (combustion) of linear hexane = -995 kcal

ΔH (Combustion) of cyclohexane = -936.9 kcal

n Hexane $\longrightarrow$ $6CO_2 + 7H_2O + 995$ kcal

Cyclohexane $\longrightarrow$ $6CO_2 + 7HCO + 936.9$ kcal

n-hexane $\longrightarrow$ cyclohexane + (995-936.9) kcal

n hexane $\longrightarrow$ cyclohexane + 58.1 kcal mol^{-1}

Therefore, contribution of cyclohexane ring towards heat of formation of cyclohexane = 58.1 kcal mol^{-1} .

Problem 26.

If heat of solution of $AlCl_3$ is -76.85 kcal mol^{-1}, calculate the heat evolved when 267 g of aluminium chloride dissolved in excess of water.

Molecular weight of $AlCl_3$ = $24.98 + 35.5 \times 3$ = 131.48

No. of moles of $AlCl_3$ dissolved = 267/131.48 = 2

Heat evolved = 2×76.85 = 153.7 kcal mol^{-1}.

Problem 27.

Calculate the bond energy of C=C from the following data.

$2C\ (s) + 3H_2\ (g) \longrightarrow C_2H_4\ (g)\ \Delta H = 12.5$ kcal I

$2H_2\ (g) \longrightarrow 4H\ (g)\quad \Delta H = 4 \times 52.1$ kcal II

$2C\ (s) \longrightarrow 2C\ (g)\quad \Delta H = 2 \text{ x } 170.9$ kcal III

$C\text{-}H \longrightarrow C + H\quad \Delta H = 99$ kcal IV

II + III – I gives

$2C\ (g) + 4H\ (g) \longrightarrow C_2H_4\ (g)$

$\Delta H_r = (4 \times 52.1 + 2 \times 170.9) - (12.5)$

ΔH_f of $C_2H_4 = 537.7$ kcal mol^{-1}

IV × 4 = 4 × 99 = 396 kcal

Energy present in 4 C-H bonds = 396 kcal

Therefore,

C = C bond energy = 537.7 – 396 = 141.7 kcal

Problem 28.

Find the C-H bond energy from the following data:

$C\ (s) + 2H_2\ (g) \longrightarrow CH_4\ (g) + 17.89$ kcal mol^{-1} I

$C\ (s) \longrightarrow C\ (g) - 171.7$ kcal mol^{-1} II

$H_2\ (g) \longrightarrow 2H\ (g) - 104.18$ kcal mol^{-1} III

(II + 2 × III) – I gives

$C\ (s) + 2H_2 - C\ (s) - 2H_2 \longrightarrow C\ (g) + 4H\ (g) - CH_4\ (g)$

$CH_4\ (g) \longrightarrow C\ (g) + 4H\ (g)$

$\Delta H = (171.7 + 2{\times}104.18) - (-17.89) = 397.95$

CH_4 is equivalent to 4 C-H bonds

Therefore, C-H bond energy = 397.95/4 = 99.5 kcal mole^{-1}

Problem 29.

Calculate the lattice energy from the following data

$Na\ (s) + \frac{1}{2}F_2\ (g) \longrightarrow NaF\ (s)\quad \Delta H_1 = -136$ kcal mol^{-1} I

$Na\ (s) \longrightarrow Na\ (g)\quad \Delta H_2 = 26$ kcal mol^{-1} II

$Na\ (g) \longrightarrow Na^+\ (g) + e^-\quad \Delta H_3 = 118$ kcal mol^{-1} III

$\frac{1}{2}F_2\ (g) \longrightarrow F\ (g)\quad \Delta H_4 = 18$ kcal mol^{-1} IV

$F\ (g) \longrightarrow F\ (g)\quad \Delta H_5 = -83$ kcal mol V

Lattice energy is equal to the enthalpy of reaction between Na (g) and F (g) to give NaF (s).

$$Na^+ (g) + F (g) \longrightarrow NaF (s) \qquad \Delta H_2 = ?$$

I – II + III – IV + V gives

$$Na^+ (g) + F (g) \longrightarrow NaF(s)$$

So $\quad \Delta H_L = \Delta H_1 + \Delta H_2 + \Delta H_3 + \Delta H_4 + \Delta H_5$

$$= -136 + 26 + 118 + 18 - 83$$

$$= -57 \text{ kcal mol}^{-1}$$

Lattice energy of NaF is equal to –57 kcal mol^{-1}

Problem 30.

The heat of sublimation of iodine is 6096 cal mol^{-1} at 200 ^{O}C. If the molar heat capacities of solid iodine and iodine vapor are 13.97 and 7.87 cal mol^{-1} respectively, what will be the heat of submlimation at 250 ^{O}C.

The reactionis $I_2 (s) \rightleftharpoons I_2 (g)$

$$\begin{matrix}\text{Mean molar heat} \\ \text{capacity } \Delta C_p\end{matrix} = \begin{bmatrix}\text{Molar heat} \\ \text{capacity of products}\end{bmatrix} - \begin{bmatrix}\text{Molar heat capacity} \\ \text{of reactants}\end{bmatrix}$$

$$\Delta C_p = 7.87 - 13.97$$

$$= -6.10 \text{ cal mol}^{-1}$$

i.e. ΔH decreases by 6.1 calories for every 1 degree rise. According to Kirchoff's equation.

$$\left(\frac{\partial \Delta H}{\partial \Delta T}\right)_v = \Delta C_p$$

$$\Delta H_2 - \Delta H_1 = \Delta C_p (250 - 200)$$

So $\quad \Delta H_{250^0} = \Delta H + \Delta Cp (250 - 200)$

$$= 6096 + (-6.1) (50)$$

$$\Delta H_{250^0} = 5791 \text{ cal mol}^{-1}$$

Heat sublimation at 250 ^{O}C = 5791 cal mol^{-1}.

Problem 31.

The molar heats of CO, O_2 and CO_2 are 6.93, 6.95, 9.34 cal mol^{-1}. If the heat of combustion of CO at 20 ^{O}C is 67950 cal mol^{-1}, what is the heat of combustion at 300 ^{O}C and at the same pressure.

$$CO + \tfrac{1}{2}O_2 \rightleftharpoons CO_2$$

$$\Delta C_p = C_p \text{ (products)} - C_p \text{ (reactants)}$$
$$= (C_p) - (C_p) - \tfrac{1}{2}(C_p)$$
$$= 9.34 - 6.93 - \tfrac{1}{2}\, 6.95$$
$$= -1.065.$$

i.e. ΔH decreases with increases in temperature by 1.065 cal for every degree rise per mol

$$\Delta H_{573}{}^{0} = \Delta H + C_p \;(573 - 273)$$
$$= 67950 + (-1.06)\,(+280)$$
$$= 67950 - 296.8$$
$$= 67653.2 \text{ cal mol}^{-1}$$

Heat of reaction at 573 K = 67653 cal mol^{-1}

Problem 32.

Calculate ΔH of the reaction

$$CH_4\,(g) + Cl_2(g) \longrightarrow CH_3Cl(g) + HCl(g)$$

Given bond energy values C–H = 98 kcal, Cl – Cl = 57 kcal, C – Cl = 78 kcal and H – Cl = 102 kcal

Heat of reaction = Enthalpy of products – Enthalpy of reactants

$$\Delta H = \left(H_{CH_3Cl} + H_{HCl}\right) - \left(H_{CH_4} + H_{Cl_2}\right)$$
$$= [3 \times 98 + 78 + 102] - [4 \times 98 + 57]$$
$$= 474 - 4119$$
$$= 25 \text{ kcal}$$

Problem 33.

Standard heat of formation of HCl gas at 25 ^{O}C is –92.3 kJ mol^{-1}. What would be the value at 60 ^{O}C. The average C_p values in this temperature range for $H_2(g)$, $Cl_2(g)$ and HCl(g) are 28.54, 32.27 and 28.50 J mol^{-1} K^{-1} respectively.

ΔC_p = (Mean molar heat capacity of reactants) – (Mean molar heat capacity of products)

For the formation of HCl

$$\frac{1}{2} H_2(g) + \frac{1}{2} Cl_2(g) \longrightarrow HCl\,(g)$$

So $\Delta C_p = 28.5 - \frac{1}{2}(28.54) - \frac{1}{2}(32.27)$

$$= -1.905 \text{ J mol}^{-1} \text{ K}^{-1}$$

From the Kirchoff's equation we have

$$\frac{\Delta H_2 - \Delta H_1}{T_2 - T_1} = \Delta C_p$$

or $\Delta H_2 - \Delta H_1 = (T_2 - T_1)\,\Delta C_p$

From the data given we have

$$\Delta H_{25\,^\circ C} = -92.3 \text{ kJ mol}^{-1},\ \Delta H_{60} = ?,$$

$$T_1 = 273+25\,^{O}C = 298K,\quad T_2 = 60 + 273 = 333 \text{ K}$$

Substituting these values is Kirchoff's equation

$$\Delta H_{60\,^\circ C} - (-92.3) \quad = -1.905\,(333\text{-}298)$$

$$\Delta H_{60\,^\circ C} = (-92.3) - 1.905 \times 35$$

$$= -158.98 \text{ kJ mol}^{-1}$$

Problem 34.

When 1 mole of aqueous HCN is neutralized according to the reaction

$$HCN(aq) + OH^-(aq) \longrightarrow H_2O + CN^-(aq)$$

2.9 kcal is evolved at constant pressure and 298 K. Calculate the heat of dissociation of HCN. Given that the heat of dissociation of water is –13.7 kcal.

The dissociation equlibria of HCN is given by equation.

$$HCN\,(aq) \longrightarrow H^+(aq) + CN^-(aq) \quad \Delta H_n = -2.9 \text{ kcal}$$

The neutralization reaction may be considered to proceed in two steps

I $\quad HCN(aq) \longrightarrow H^+(aq) + CN^-(aq)$

II $\quad H^+(aq) + OH^-(aq) \longrightarrow H_2O \qquad \Delta H_{II} = 13.7 \text{ kcal}$

It is given that the heat of dissociation of water is –13.7.

Therefore step II is accompanied by the absorption of 13.7 kcal of heat (i.e., $\Delta H_{II} = 13.7$)

But $\quad \Delta H_I + \Delta H_{II} = \Delta H_n = -2.9$

$$\Delta H_I + (+13.7) = -2.9$$

$$\Delta H_I = -13.7 - 2.9 = -16.6 \text{ kcal mol}^{-1}$$

ΔH of dissociation of HCN = -16.6 kcal mol^{-1}

4.10 Introduction to the Second Law of Thermodynamics

The first law of thermodynamics is successful in achieving three ends;

(i) It rules out the possibility of designing a perpetual motor

(ii) It extends the scope of law of conservation of energy for mass related functions like heat and work

(iii) It systematizes thermochemistry.

It imposes no restriction on the heat-work relationships as to the source of heat or on the direction of its flow, as long as the total energy of the system and surroundings remain constant. Thus it is possible to use ice as a heat source in heating water at room temperature with the same ease as using water at 50°C as a heat source. But our experience negates this contention and suggests that always heat flows readily from a higher temperature to lower temperature. Similarly in the looks of first law of thermodynamics, there is no surprise if water flows from a lower to a higher level, without any external aid, but the converse is the reality. A process which proceeds on its own accord, without the help of any external aid is called a **spontaneous process**. The idea of spontaneity is well demonstrated in the following physical and chemical processes.

(i) A water fall runs down the hill, but never up, spontaneously.

(ii) A sugar cube spontaneously dissolves in a cup of coffee but dissolved sugar does not reappear in its original form.

(iii) A lighted incense stick kept in a corner of a room, spreads its aroma throughout the room spontaneously.

(iv) Water freezes readily below $0^{O}C$ and ice melts spontaneously above $0^{O}C$, or, in other words, to stop ice from melting requires an additional effort.

(v) An inflated tire or balloon is spontaneously deflated. The reverse process, that is inflation of a tube is not a spontaneous process and is achieved by an external aid like hand pump or motor pump.

(vi) Heat flow from a hotter object to colder body but the reverse never happens spontaneously.

(vii) A piece of zinc spontaneously reacts with acid liberating hydrogen gas and forming a zinc salt. However, zinc does not reappear when hydrogen gas is pumped into a solution of zinc salt.

(ix) Iron exposed to water and oxygen gets rusted spontaneously, but a rusted-iron piece does not spontaneously change back to iron.

As far as chemists are concerned a reaction that does occur under a given set of conditions like pressure, temperature, concentration, etc., is called a spontaneous reaction. A spontaneous reaction could be an extremely fast process like neutralization of an acid by a base to give salt and water, or it could be unobservably slow process like formation of water by the combination of gaseous hydrogen and oxygen. The slow spontaneous process proceeds at a measurable rate with the mercury touching the arc temperature.

It is also very common to think that only exothermic reactions are spontaneous as they are down hill energy processes.

Thus we find many spontaneous exothermic processes like

$$CaO\ (s) + H_2O \longrightarrow Ca\ (OH)_2\ (aq)$$

$$CH_4\ (g) + 2O_2(g) \longrightarrow CO_2(g) + 2H_2O(l),\ \Delta H_{298}{}^{o} = -212.8\ \text{kcal}$$

But experience tells us that melting of ice above $0^{O}C$ or dissolution of sugar in water, even through are endothermic, proceed spontaneously. Similarly ammonium nitrate ionizes in water spontaneously, although it is an endothermic process.

$$NH_4NO_3\ (s) + H_2O\ (l) \longrightarrow NH_4^{+}\ (aq) + NO_3^{-}\ (aq),\ \Delta H_{298}{}^{o} = 108\ \text{kcal}$$

$$H_2\ (g) + I_2\ (s) \longrightarrow 2HI \quad \Delta H_{298}{}^{o} = 12.4\ \text{kcal}$$

These observations suggest that it is possible for an exothermic reaction to be nonspontaneous, just as it as possible for an endothermic reaction to be spontaneous. This also further suggests that consideration of energy changes is not enough to predict spontaneity or non-spontaneity of a reaction/process.

This necessitate the hunt for another thermodynamic quantity, other than enthalpy, that determines the spontaneity of a process and a new law that lays the conditions on the spontaneity of a reaction.

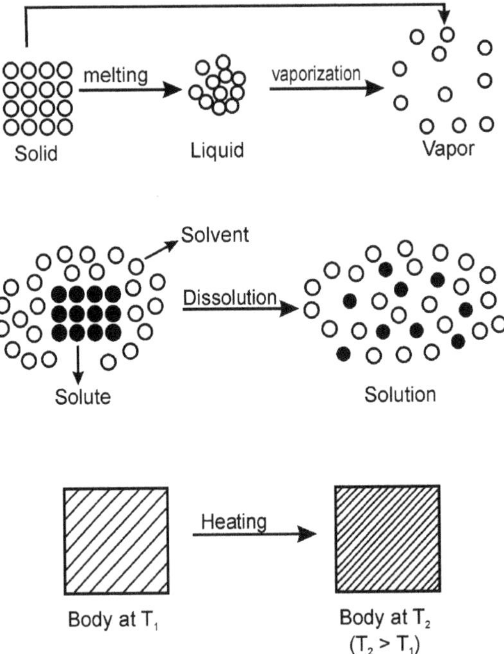

Fig. 4.10 The spontaneous processes lead to increase in entropy of the system

4.10.1 Entropy

One common feature of spontaneous processes either it is melting of ice or evaporation of water or dissolution of the salt or escape of gas from a high pressure region to low pressure region is the loss of orderliness. Ice is a crystalline form where water molecules are arranged in a particular pattern and its melting gives liquid water where the regular pattern of solid ice vanishes and the water molecule start moving in a space defined by the container. This randomness further increases as we heat water to boiling where they evaporate giving water vapor, Fig. 4.10. The new thermodynamic function called **entropy** is defined as *a measure of the randomness or disorder of a system. The greater the disorder, the greater is the entropy.* Conversely, the more-ordered system has a smaller entropy. Both the enthalpy and the entropy together predict the spontaneity of a process. The following points give a comparative account of entropy and enthalpy.

1. Entropy and enthalpy, both are thermodynamic state functions. They do not depend on the path or previous history but depend only on the initial and final stages.

$$\Delta S = S_{\text{final}} - S_{\text{initial}}$$

2. Absolute energy or enthalpy values can not be determined but it is possible to determine the absolute entropy of a system. The entropy values for 1 mole of substance at 1 atm and 25°C are called **standard entropies (S^o)**.

3. Entropies of both elements and compounds are positive (that is $S^o > 0$). By contrast the standard enthalpy of formation, ΔH_f^o, for elements in their standard states is zero, while the compounds have both positive and negative heats of formation.

4. The molar enthalpy changes have dimensions of energy and are expressed in kcal mol^{-1} or kJ mol^{-1}. Entropy values have dimension energy per Kelvin or cal K^{-1} or Jk^{-1} but *standard molar entropies* have the dimension of energy per Kelvin per mole or JK^{-1} mol^{-1} and this unit is also called entropy unit represented by the symbol e_u.

The following standard entropy values inform more about the significance of entropy.

Substance	S^{O} J K^{-1} mol^{-1}	Substance	JK^{-1} mol^{-1}
H_2O (l)	69.9	I_2 (s)	116.7
H_2O(g)	130.7	I_2 (g)	260.6
P (white)	44.0	CO (g)	197.9
P (red)	29.3	CO_2 (g)	213.7

Water vapor has more entropy than liquid water. Similarly, iodine vapor has greater entropy than the solid iodine. The entropies of a given substance in different phases follows the order;

Solid < Liquid < Vapor

Within the solid phase the white allotropic form of phosphorous has higher entropy than red phosphorous form suggesting that the red phosphorous is more closely packed when compared to white phosphorous.

Among CO and CO_2 gas molecules, CO_2 is relatively more stable than CO as it has a greater entropy value.

4.10.2 The Second Law of Thermodynamics

Second law of thermodynamics mainly deals with the relations connecting the entropy and spontaneity of a chemical reaction. All the spontaneous processes are unidirectional or totally irreversible. As already discussed, for any spontaneous process

$$S_{final} > S_{initial} \quad \text{or} \quad \Delta S > O.$$

Hence an irreversible or spontaneous processes are always accompanied by an increasing entropy.

The universe is made up of the system and the surroundings. Entropy change of the universe is the sum of entropy changes of the system and its surroundings.

$$\Delta S_{\text{universe}} = \Delta S_{\text{system}} + \Delta S_{\text{surroundings}}$$

Reversibility is only an ideality as it occurs infinitesimally slow. As almost all the natural processes occur at a finite rate they are irreversible or spontaneous. So ΔS of universe is always positive.

Spontaneous Process :

$$\Delta S_{\text{universe}} = \Delta S_{\text{system}} + \Delta S_{\text{surroundings}} > 0 \quad \text{.....(4.58)}$$

Hence the second law, of thermodynamics may be stated as

(i) *The entropy of universe is ever increasing.*

(ii) *All natural or spontaneous processes, i.e., processes taking place without any external aid, are thermodynamically irreversible.*

We can also define second law in terms of heat or enthalpy and the spontaneity of a process. Conduction of heat along a rod from a hotter end to colder end is a spontaneous process and is unidirectional. After some time the hotter and colder ends disappear and the rod attains a uniform temperature, where heat is allowed to flow in either direction. Such a state is termed as equilibrium. This observations may be consolidated into another statement of second law and is given that;

All the naturally occurring processes always tend to change spontaneously in a direction which will lead to equilibrium.

or

At equilibrium the change in entropy of the universe is zero.

At equilibrium :

$$\Delta S_{\text{universe}} = \Delta S_{\text{system}} + \Delta S_{\text{surroundings}} = 0$$

In the above example heat flow from a colder end to the hotter end, a non-spontaneous process, is possible with the external aid. But experience suggests that it is impossible to achieve complete reversal. This is because ;

Heat cannot be completely converted into an equivalent amount of work without causing other changes in some part of the system or surroundings.

This forms another statement of **second law of thermodynamics**. The difficulty in the conversion of heat into 100% work, without leaving some change in some part of the system or surroundings may also be expressed by another statement that ;

There is no machine that works with 100% efficiency.

It is impossible to take heat from a reservoir convert it totally into work without transferring a part of it to a cold sink. These definitions of second law will be discussed in detail later under Carnot cycle. Even considering a theoretical case like the reversible expansion of a gas where heat energy supplied is totally converted into work, we notice, at the end of the process, a change in the volume of the gas when compared to the start. Any attempt to retain the work without effecting a change on the system, like coming back to original volume, is a futile effort. Hence it can be stated that;

Heat cannot be converted to work, even under ideal conditions, without leaving permanent changes either in the system or its surroundings.

4.10.3 Entropy changes in reversible processes

We already learnt from the second law of thermodynamics, entropy change of the universe is the sum of entropy change in the system and its surroundings. When the system gains heat, the surroundings lose the same amount of heat and vice versa. The consequence of gaining heat by a system is two folds;

(i) The entropy of the system increases and

(ii) The temperature of the system may rise or the heat may be utilized in increasing the internal energy or in doing work by the system.

In an isothermal reversible expansion of a gas, q_{rev}, the heat absorbed is totally used in raising the entropy of the gas at constant temperature T.

Then entropy change of the system is given by;

$$\Delta S_{sys} = \frac{q_{rev}}{T} \quad \text{or} \quad T\Delta S = q_{rev} \qquad .....(4.59)$$

or for an infinitesimal change;

$$dS = \frac{dq_{rev}}{T}$$

Since the expansion is reversible, the surroundings lose heat qrev, and then the entropy change of surroundings is expressed as;

$$\Delta S_{surr} = \frac{-q_{rev}}{T}$$

4.10.4 Entropy change of the universe for a reversible operation

$$\Delta S_{universe} = \Delta S_{system} + \Delta S_{surroundings}$$

or $$\Delta S_{univ} = \frac{q_{rev}}{T} - \frac{q_{rev}}{T}$$

So entropy change of the universe for a reversible operation is zero.

4.10.5 Entropy change of the universe for an irreversible process

Consider an isothermal, irreversible expansion of a gas and let q_{rev} be the amount of heat absorbed by the system in this operation.

From the first law of thermodynamics, the irreversible heat change is equal to

$$q_{rev} = \Delta E + w_{irrev}$$

If the same process is repeated reversibly we can write the reversible heat change as equal to

$$q_{rev} = \Delta E + w_{rev}$$

we know that work done is a reversible process maximum and is greater than an irreversible work.

$$w_{rev} > w_{irrev}$$

So we have

$$q_{rev} > q_{irrev}$$

On dividing both sides with T,

$$\frac{q_{rev}}{T} > \frac{q_{irrev}}{T}$$

we already know $\frac{q_{rev}}{T} = \Delta S_{sys}$

Therefore, $\Delta S_{sys} > \frac{q_{irrev}}{T}$ or $T\Delta S > q_{irrev}$

Here it should be noted, that *enthalpy being a state function, the enthalpy change of the system is a constant value and always corresponds to the reversible change.* The value of q_{irr} can have a minimum value of zero and maximum value of q_{rev} depending upon the nature of operation.

However, when we consider the surroundings, the nature of operation carried out in the system have no bearing on the surroundings and its entropy change exclusively depends on the amount of heat given to the system. As q_{irrev} is the heat lost by the surroundings at temperature, T,

So $$\Delta S_{surr} = \frac{q_{irrev}}{T}$$

The total entropy change of the universe

$$\Delta S_{universe} = \Delta S_{sys} + \Delta S_{surr}$$

$$= \Delta S_{sys} - \frac{q_{irrev}}{T}$$

Since $\frac{q_{rev}}{T} > \frac{q_{irrev}}{T}$, $\Delta S_{sys} > \frac{q_{irrev}}{T}$

or Entropy change of the universe $\Delta S > 0$

This was the conclusion behind, the statement of second law that the entropy of the universe is ever increasing.

4.10.6 Entropy changes for an Ideal Gas

Considering that an ideal gas is doing only pressure-volume work, a heat, q_{rev}, supplied reversibly to the system is utilized in changing the internal energy and in doing pressure-volume work.

$$\delta q_{rev} = dE + PdV$$

or $$\delta q_{rev} = C_v\, dT + PdV$$

For 1 mole of ideal gas $P = RT/V$

Therefore, $\delta q_{rev} = C_v dT + \frac{RT}{V}\, dV$

But the entropy change $dS = \frac{\delta q_{rev}}{T}$

Therefore, $dS = C_v \frac{dT}{T} + R\frac{dV}{V}$

Assuming C_V is independent of temperature in an operation involving a raise in temperature from T_1 to T_2 accompanied by a volume change from V_1 to V_2, we have;

$$\int_{S_1}^{S_2} dS = C_V \int_{T_1}^{T_2} \frac{dT}{T} + R \int_{V_1}^{V_2} \frac{dV}{V}$$

$$\Delta S = S_2 - S_1 = C_V \ln \frac{T_2}{T_1} + R \ln \frac{V_2}{V_1} \quad \text{.....(4.60)}$$

From the equation of state for an ideal gas we have

$$\frac{P_1 V_1}{T_1} = \frac{P_2 V_2}{T_2} \quad \text{or} \quad \frac{V_2}{V_1} = \frac{P_1 T_2}{P_2 T_1}$$

Substituting the value for V_2/V_1 in the equation (4.60)

$$\boxed{\Delta S = C_V \ln \frac{T_2}{T_1} + R \ln \frac{T_2}{T_1} - R \ln \frac{P_2}{P_1}} \quad \text{.....(4.61)}$$

or $$\Delta S = (C_V + R) \ln \frac{T_2}{T_1} - R \ln \frac{P_2}{P_1}$$

But we already know $C_p = C_V + R$

$$\Delta S = C_p \ln \frac{T_2}{T_1} - \text{R} \ln \frac{P_2}{P_1} \quad \text{.....(4.62)}$$

From the equation (4.61) and (4.62), for an isothermal process the entropy change ΔS_T is equal to

$$\boxed{\Delta S_T = R \ln \frac{V_2}{V_1} = -\text{R} \ln \frac{P_2}{P_1}} \quad \text{.....(4.63)}$$

Similarly for an isobaric process (constant pressure process) from equation (4.62), we have

$$\boxed{\Delta S_P = C_p \ln \frac{T_2}{T_1}} \quad \text{.....(4.64)}$$

For an isochoric process (constant volume process) from equation (4.61), we have

$$\boxed{\Delta S_V = C_V \ln \frac{T_2}{T_1}} \quad \text{.....(4.65)}$$

4.10.7 Entropy changes accompanying phase change

The melting, vaporization, sublimation and solid-solid transformations and their converse processes (i.e., freezing, condensation, etc.) all accompanied by phase changes. For example,

$$\text{Ice (solid)} \underset{\text{Freezing}}{\overset{\text{Melting}}{\rightleftharpoons}} \text{Water (liquid)}$$

$$\text{Water (liquid)} \underset{\text{Condensation}}{\overset{\text{Vaporization}}{\rightleftharpoons}} \text{Vapor (gas)}$$

$$\text{Carbon (diamond)} \underset{\text{Solid-Solid tranformation}}{\rightleftharpoons} \text{Carbon (graphite)}$$

During a phase change, the temperature of transition remains constant and all the heat is either evolved or absorbed from the surroundings is utilized in bringing a change in the latent heat. Thus during evaporation of water at 1 atm pressure the liquid boils at 100 ^{O}C, and what ever heat is supplied is simply utilized in raising the latent heat of vaporization. During condensation once again the temperature remains constant at 100 ^{O}C, but heat is given to the surroundings.

In general if ΔH_L represents the latent heat of a physical transition, the entropy change ΔS_L is given by

$$\boxed{\Delta S_L = \frac{\Delta H_L}{T}} \qquad \text{.....(4.66)}$$

Since evaporation is accompanied by a transition of more organized liquid phase to less organized vapor phase the entropy of the system increases and ΔS_L is positive. For the converse process (condensation) we have a decrease in entropy or ΔS_L is negative.

4.10.8 Entropy change in an Isothermal Process

As derived already, for an ideal gas the entropy change, equation (4.63), of an isothermal process is given by

$$\Delta S_T = \text{R ln } \frac{V_2}{V_1} \quad = -\text{R ln } \frac{P_2}{P_1}$$

4.10.9 Entropy change in an Adiabatic Process

In an adiabatic process heat is neither allowed to enter nor leave the system and q_{rev} always remains zero. It follows, therefore, from equation (4.59) a reversible adiabatic processes is associated with no entropy change and is referred to as isentropic process.

Problem 35.

What is the change in the entropy in evaporating 56 g of benzene at its boiling point (80.1 ^{O}C) into vapor at the same temperature. Given that the heat of evaporation of benzene is 7.35 kcal mol^{-1}.

$$\Delta S = \frac{-q_{rev}}{T} = \frac{\Delta H}{T}$$

Given $\Delta H = 7350$ cal mol^{-1} and T = 273 + 80.1

$$= \frac{7350}{353.1} = 20.82 \text{ cal K}^{-1} \text{ mol}^{-1}$$

Therefore change in entropy of evaporation of benzene at its boiling point 20.82 eu.

Problem 36.

Calculate the entropy change in melting 5g of ice at 0 ^{O}C. Given that the molar heat of fusion of ice is 1440 cal.

Given : $\Delta H_f = 1440$ cal mol^{-1}, n = 5/18 = 0.28 moles

$T = 0 + 273 = 273\ K$

Change in entropy

$$\Delta S = \frac{\Delta H}{T} = \frac{1440}{273} = 5.27 \text{ cal K}^{-1} \text{ mol}^{-1}$$

Entropy change for 0.28 moles

$$= 0.28 \times 5.27$$

$$= 1.47 \text{ cal K}^{-1}.$$

Problem 37.

Heat of sublimation of iodine is 6946 cal mol^{-1} at 200°C. Calculate the entropy change accompanied with sublimation of 10 g of iodine.

Given : ΔHs = 6946 cal mol^{-1}, $T = 200 + 273 = 473\ K$

Change in entropy

$$\Delta S = \frac{\Delta H}{T} = \frac{6946}{473} = 14.7 \text{ cal K}^{-1} \text{ mol}^{-1}$$

Entropy change for 10 g

$$= \frac{10}{254} \times 14.7 = 0.58 \text{ cal K}^{-1}$$

Problem 38.

A monatomic gas with $C_V = 5$ cal mol^{-1} is subjected to expansion from a volume of 5 L and 300 K to 25 L and 400 K. Calculate the entropy change of the expansion.

$$\Delta S = C_V \ln \frac{T_2}{T_1} - R \ln \frac{V_2}{V_1}$$

$$= 5 \ln \frac{400}{300} + 8.314 \ln \frac{25}{5}$$

$$= 1.44 + 13.4$$

$$= 14.8 \text{ J K}^{-1} \text{ mol}^{-1}$$

4.10.10 Carnot cycle and maximum efficiency of a heat engine

An important consequence of the second law of thermodynamics was deduced by a french engineer, Sodi Carnot. In 1820 in his study of efficiencies of heat engines, Lord Kelvin actually recognized the importance of Carnot's work, which led him to introduce the celebrated absolute temperature scale. The basic principle of heat engine is that heat (Q_2) is absorbed by the working substance of the engine from a source at a high temperature (T_2). Part of this heat is utilized in doing work and the rest (Q_1) is rejected to the sink at a lower temperature (T_1).

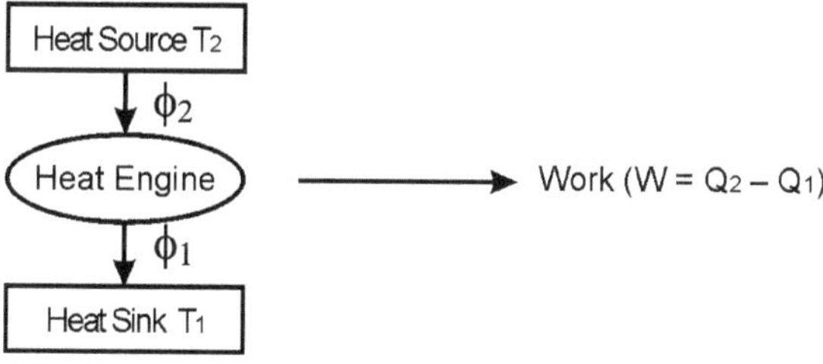

To derive continuous work we must have a cyclic operation. A cycle or cyclic operation is a succession of changes, as a result of which the system returns its original state.

Carnot designed a theoretical heat engine consisting of a cycle of four steps to prove that the efficiency of a heat engine depends not on the magnitude of the temperature of heat source or sink but on their difference. The efficiency of a heat engine is defined by the ratio w/Q_2. If all the heat that is absorbed is converted into work, i.e., when Q_1 = O; then the heat engine is said to have efficiency of unity or 100% efficient.

$$\text{Efficiency} \quad \phi = \frac{Q_2 - Q_1}{Q_2} = \frac{W}{Q_2} \qquad .....(4.67)$$

The Carnot cycle has four reversible operations.

Step I : Isothermal expansion

The gas in a piston or heat engine absorbs a heat of Q_2 from the heat source and expands at a constant temperature T_2, equal to that of the heat source. The entropy change, ΔS_1 of this process is Q_2/T_2.

Step II : Adiabatic expansion :

The isothermal expansion is followed by an adiabatic expansion. As no heat transaction is allowed in the adiabatic process, the adiabatic expansion results in a drop of the temperature from T_2 to T_1 where T_1 is the temperature of the heat sink. Since all the adiabatic processes are isentropic, ΔS_2 of the step is equal to zero.

Step III : Isothermal compression :

In the isothermal compression, the temperature is kept constant at T_1, the temperature of the heat sink, and a heat of Q_1 liberated during the compression is transferred to the sink. The entropy change of this step ΔS_3 is $-Q_1/T_1$.

Step IV : Adiabatic compression :

The adiabatic compression is carried to complete the cycle and return the gas to its original temperature T_2. This process being an adiabatic process, the ΔS_4 is zero.

All the four steps for a pressure-volume work executed by a gas sealed in a piston and cylinder set-up is represented in Fig. 4.11.

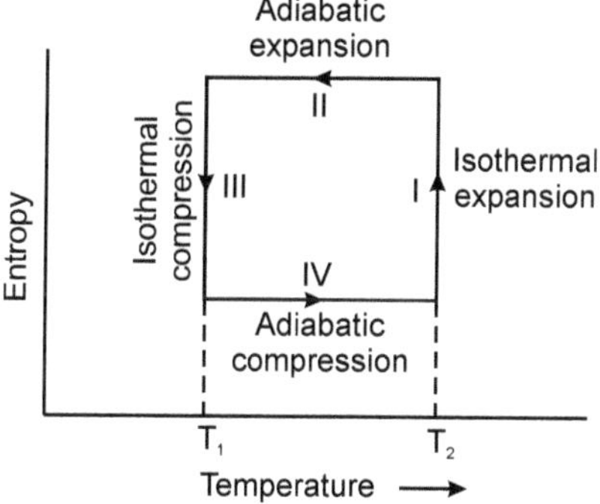

Fig. 4.11 Representation of expansion and compression processes.

The corresponding carnot cycle is shown diagrammatically in Fig. 4.12.

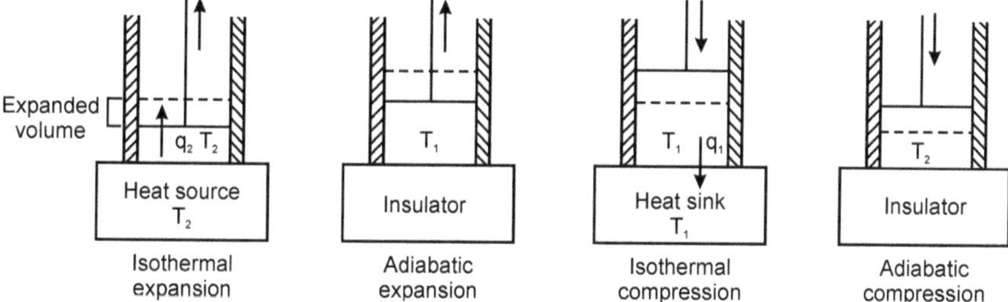

Fig. 4.12 The Carnot cycle

Entropy being a state function, the total entropy change of the carnot cycle in zero.

i.e., $\Delta S = \Delta S_1 + \Delta S_2 + \Delta S_3 + \Delta S_4 = 0$

or $$\frac{Q_2}{T_2} - \frac{Q_1}{T_1} = 0$$

or $$\frac{Q_2}{T_2} = \frac{Q_1}{T_1}$$

or $$\frac{T_1}{T_2} = \frac{Q_1}{Q_2}$$

Subtracting from 1

or $$1 - \frac{Q_1}{Q_2} = 1 - \frac{T_1}{T_2}$$

or $$\frac{Q_2 - Q_1}{Q_2} = \frac{T_2 - T_1}{T_2}$$

Hence efficiency of a reversible heat engine (ϕ) given by expression (4.67) may be written as;

$$\phi = \frac{Q_2 - Q_1}{Q_2} = \frac{T_2 - T_1}{T_2}$$

This suggests that greater the difference in temperature between source and sink, greater is the efficiency of an engine. To achieve an efficiency of unity or in constructing an engine with 100% efficiency where all the heat taken from the source is totally converted into work without rejecting any in the sink we require to have the sink at absolute zero or a source at infinite temperature. But neither of these conditions are physically realizable, so the efficiency of even a hypothetical reversible engine is less than unity. This is in the tune with the statements of second law of the thermodynamics and may be stated as;

No machine has yet been made that has an efficiency of unity or 100 percent.

Problem 39.

A heat engine operating between heat source at 600 K and sink at 300 K absorbs 4 kJ from the source. Calculate the efficiency of heat engine, heat rejected to the sink and work done per cycle.

Given $T_1 = 300\text{K}$ $\quad T_2 = 600\text{ K}$

$Q_1 = ?$ $\quad Q_2 = 4\text{KJ}$

$w = ?$ $\quad \phi = ?$

Efficiency of a heat engine

$$\phi = \frac{T_2 - T_1}{T_2} = \frac{600 - 300}{600} = 0.5$$

But $\quad \phi = w/Q_2$

$$w = \phi \times Q_2 = 0.5 \times 4\text{ k J} = 2\text{ kJ}$$

Heat rejected

$$= 4 - 2 = 2\text{ kJ}$$

Problem 40.

Calculate the efficiency of a steam engine operating between 105 ^{O}C and 55 ^{O}C. If the temperature of the steam boiler is raised to 135 ^{O}C what will be the rise in the efficiency of the engine.

Heat engine is operating between 105 and 55 ^{O}C

i.e., $T_1 = 55 + 273 = 328$ K $\quad T_2 = 105 + 273 = 378$ K

$$\text{efficiency } \phi = \frac{378 - 328}{378} = 0.132 \text{ or } 13.2\%$$

When the temperature of the heat reservoir is raised

$$T_1 = 328 \text{ K} \quad T_2 = 135 + 273 = 408 \text{ K}$$

$$\phi = \frac{408 - 328}{408} = 0.196 \text{ or } 19.6\%$$

Problem 41.

A carnot engine absorbs 375 kJ at 250 ^{O}C and converts 225 kJ into work. Calculate (i) the efficiency of the engine, (ii) the temperature of the sink.

Given : $T_2 = 250 + 273 = 523K$, $T_1 = ?$, $Q_2 = 375$, $Q_1 = 225$ kJ

Efficiency of heat engine

$$\phi = \frac{Q_2 - Q_1}{Q_2} = \frac{T_2 - T_1}{T_2}$$

$$\phi = \frac{375 - 225}{375} = 0.4 \text{ or } 40\%$$

$$0.4 = \frac{523 - T_1}{523}$$

$$T_1 = 523 - 0.4 \times 523 = 313.8 \text{ K}$$

Efficiency $= 0.4$, Temperature of the sink = 313.8 K

Problem 42

A heat engine works with an efficiency of 60%. If it does a work equivalent to 225 kJ, calculate the heat absorbed from the reservoir and heat rejected to the sink.

Given : $\phi = 60\%$ or 0.6 and $w = 225$ kJ

$$\phi = w/Q_2$$

So $\quad 0.6 = 225/Q_2$

Heat absorbed $Q_2 = \frac{225}{0.6} = 375$ kJ

Heat rejected $Q_1 = Q_2 - w = 375 - 225 = 150$ kJ

4.11 Gibbs Free Energy

The second law puts the conditions for a chemical reaction, or any other process for that matter, to be spontaneous the entropy of the universe has to increase or mathematically $\Delta S_{univ} > 0$. In order to determine ΔS_{univ} we require to calculate ΔS_{syst} and ΔS_{surr}. Because *we are mainly concerned with what happens in the system rather than entire universe and because the calculation of* ΔS_{surr} is often difficult. Hence it is desirable to have another thermodynamic function that will help us determine the spontaneity only by the consideration of the system.

In addition to entropy, enthalpy is another quantity that helps in the determination of the direction of a reaction. Together enthalpy and entropy enable us to define a new thermodynamics function called **Gibbs free energy** or simply **free energy** (G) as follows;

$$\boxed{G = H - TS} \quad \text{.....(4.68)}$$

This new thermodynamic function G has the units of energy as both H and TS have the units of energy. Like E, H and S, G is also a thermodynamics function and therefore independent of the previous history or path traced and depends only on the initial and final states.

4.11.1 Variation of free energy with temperature and pressure

According to definition of free energy

$$G = H - TS$$

Substituting the value of enthalpy, $H = E + PV$

$$G = E + PV - TS$$

Differentiating the above equation

$$\mathrm{d}G = \mathrm{d}E + P\mathrm{d}V + V\mathrm{d}P - T\mathrm{d}S - S\mathrm{d}T$$

But we know from first law of thermodynamics

$$\mathrm{d}H = \mathrm{d}E + P\mathrm{d}V$$

and from second law of thermodynamics

$$\mathrm{d}H = T\,\mathrm{d}S$$

Therefore,

$$dG = dH + VdP - dH - S\,dT$$

$$\boxed{dG = VdP - SdT} \qquad .....(4.69)$$

This equation is called **total differential equation of free energy**.

At Constant pressure equation (4.69) becomes,

$$(dG)_p = -(SdT)_p$$

or $$\left(\frac{\delta G}{\delta T}\right)_p = -S \qquad .....(4.70)$$

At Constant temperature equation (4.69) becomes,

$$(dG)_T = (VdP)_T$$

or $$\left(\frac{\delta G}{\delta P}\right)_T = V \qquad .....(4.71)$$

For an ideal gas $V = \dfrac{nRT}{P}$

Equation (4.71) takes the form

$$(dG)_T = \frac{nRT}{P}(dP)_T$$

Integrating the above equation for a pressure change from P_1 to P_2

$$\int_{G_1}^{G_2}(dG)_T = \int_{P_1}^{P_2}\frac{nRT}{P}(dP)_T$$

$$\boxed{G_2 - G_1 = \Delta G = nRT\,ln\,\frac{P_2}{P_1}} \qquad .....(4.72)$$

For an isothermal process

$$\frac{P_2}{P_1} = \frac{V_1}{V_2}$$

Therefore equation (4.72) in terms of volumes may be written as,

$$\boxed{\Delta G = nRT\,ln\,\frac{V_1}{V_2}} \qquad .....(4.73)$$

4.11.2 Free Energy change and criteria for spontaneity

Expansion of a gas from a higher pressure (P_1) to lower pressure (P_2) is always spontaneous. Substituting this condition in equation 4.72, we have

$$\Delta G = RT \ln \frac{P_2}{P_1} = -\text{ve}$$

At constant temperature a spontaneous process is always accompanied by decrease in free energy. Similarly extending the argument for cases $P_1 = P_2$ and $P_1 < P_2$. We can draw the following conclusions for constant temperature processes :

If $\Delta G < 0$ (or ΔG is negative) the reaction/system is spontaneous.

If $\Delta G > 0$ (or ΔG is positive) the reaction/system is nonspontaneous.

If $\Delta G = 0$ (or no change G) the reaction/system is at equilibrium.

Let us also assume that

Let G_1, H_1 and S_1 correspond to an initial state of a process at temperature T; after carrying out an operation under constant temperature conditions, let G_2, H_2 and S_2 correspond the final state; then from the equation (4.68) we have;

$$G_1 = H_1 - TS_1 \qquad \text{(i)}$$

$$G_2 = H_2 - TS_2 \qquad \text{(ii)}$$

Subtracting equation (i) from equation (ii);

$$(G_2 - G_1) = (H_2 - H_1) - T(S_2 - S_1) \qquad .....(4.70)$$

or

$$\boxed{\Delta G = \Delta H - T\Delta S} \qquad .....(4.71)$$

Thus spontaneity of an isothermal process is safely decided, just from a collective knowledge of variations in enthalpy and entropy of a reaction system without knowing what is happening in the surroundings.

There are three permutations by which ΔG could become negative, there by make the reaction spontaneous.

1. If ΔH is a negative and ΔS is positive, then ΔG is invariably negative regardless of temperature. Hence an exothermic reaction with a positive ΔS is spontaneous at all temperatures.

   e.g : Any combustion reaction.
2. If both ΔH and ΔS are negative, then ΔG will be negative only when $T\Delta S < \Delta H$. This conditions is met at sufficiency low temperature. So an exothermic reaction with a negative ΔS could become spontaneous, at low temperature such that $T\Delta S < \Delta H$.

3. If both ΔH and ΔS are positive, then ΔG will be negative only when $T\Delta S > \Delta H$. This conditions is normally met at high temperatures. Thus an endothermic reaction with positive ΔS could become spontaneous at relatively high temperature fulfilling the conditions $T\Delta S > \Delta H$.
4. If ΔH is positive while ΔS is negative, ΔG is positive regardless of temperature and there is no way that it could be made spontaneous. So an endothermic reaction, with a negative ΔS always sounds a thermodynamically improbable propositions and only its reverse reaction is expected to occur.

4.11.3 Helmholtz Energy

This is a new thermodynamics function connecting the internal energy and entropy and is defined as

$$\boxed{A = E - TS} \quad .....(4.72)$$

The letter A, assigned to **Helmholtz free energy** is derived from the german word Arbeit meaning work. Hence Helmholtz free energy is also called **work function**. The Helmholtz free energy A has also the same units of energy.

At constant temperature, addressing a similar argument as for Gibbs free energy, we can write

$$\Delta A = \Delta E - T\Delta S \quad .....(4.73)$$

But we already know $T\Delta S = q_{\text{rev}}$

So $$\Delta A = \Delta E - q_{\text{rev}} \quad .....(4.74)$$

From the first law of thermodynamics we know $\Delta E - q_{\text{rev}}$ is equal to the maximum work done on the system, i.e., $-w_{\text{max}}$. We then can rewrite equation (4.74) as

$$\Delta A = -w_{\text{max}} \quad .....(4.75)$$

A decrease in the Helmholtz free energy, thus gives the maximum work obtainable from a system.

4.11.4 Free energy and Work

At constant temperature the change in free energy is given by

$$\Delta G = \Delta H - T\Delta S \quad .....(4.76)$$

At constant pressure we know

$$\Delta H = \Delta E + P\Delta V$$

Substituting this value of ΔH is equation (4.76) we get

$$\Delta G = \Delta E + P\Delta V - T\Delta S$$

or $$\Delta G = (\Delta E - T\Delta S) + P\Delta V$$

The expression in the parenthesis is equal to ΔA

So $$\Delta G = \Delta A + P\Delta V$$

If ΔA is replaced by its equivalence $-w_{max}$, we have at constant temperature and pressure.

$$\Delta G = -w_{max} + P\Delta V$$

or $$\Delta G = w_{max} - P\Delta V \qquad .....(4.77)$$

Hence a decrease in free energy gives the maximum work at constant pressure and constant temperature conditions other than pressure - volume work and is referred to as ***net work.***

Problem 43.

Calculate the free energy change accompanied by an isothermal compression of 2 moles of an ideal gas at 300 K from 10 atm to 20 atm.

$$(\Delta G)_T = nRT \ln \frac{P_2}{P_1}$$

$$= 2 \times 8.314 \times 300 \times 2.303 \log \frac{20}{10}$$

$$= 2 \times 8.314 \times 300 \times 2.303 \times 0.3010$$

$$= 3458 \text{ J}$$

The positive ΔG value indicates that this process is nonspontaneous.

Problem 44.

Calculate the free energy change for an isothermal expansion of 3 moles of an ideal gas at 400 K from 2 lit to 20 lit.

$$(\Delta G)_T = nRT \ln \frac{V_1}{V_2}$$

$$= 3 \times 8.314 \times 400 \times 2.303 \log \frac{2}{20}$$

$$= 3 \times 8.314 \times 400 \times 2.303 \times (-1.0)$$

$$= -22.98 \text{ kJ}$$

The negative ΔG value indicates that the expansion is spontaneous.

4.11.5 Free energy and phase transition

The phase transition involves an equilibrium at the transition temperature between the phases under consideration. We know that for a system at equilibrium, at constant temperature ΔG =0. Substituting this value in eqn. (4.71) we have;

$$0 = \Delta H - T\Delta S$$

or $$\Delta S = \frac{\Delta H}{T}$$

This expression is same as the one discussed under phase transition and entropy changes.

4.11.6 Free energy and chemical reaction

The change in free energy is made out of two terms; an energy term and an enthalpy term as shown in equation (4.71).

$$\Delta G = \underset{\text{energy term}}{\Delta H} - \underset{\text{enthalpy term}}{T\Delta S}$$

All chemical reactions (or systems) achieve stability by acquiring a minimum energy. From discussion on spontaneity we also understand that all systems go for maximum freedom or randomness or maximum entropy. In a chemical reaction the opposing tendencies enthalpy and entropy strike a balance and free energy turns out to be a measure balance between these factors.

4.11.7 The Claperyon Equation

Melting of a solid, boiling of a liquid, sublimation of a solid or transition between two forms of a solid, all involve phase transformations and they take place at constant temperature. At this temperature both the phases of any transformation are at equilibrium.

The free energy change of any isothermal converion A $\rightarrow$ B is equal to zero.

$$\Delta G = 0 \quad \text{or} \quad G_A = G_B$$

If the temperature and pressure are altered in infinitesimally small steps, the system remains in equilibrium under new conditions and the changes in free energies of reactants and products are also same.

$$\underset{dG_A}{A} \longrightarrow \underset{dG_B}{B}$$

But $dG = VdP - S_A dT$ and $dG_B = V_B dP - S_B dT$

or $V_A dP - S_A dT = V_B\, dP - S_B dT$

$$\frac{dP}{dT} = \frac{S_B - S_A}{V_B - V_A} = \frac{\Delta S}{\Delta V}$$

For any phase change

$$\Delta S = \frac{\text{Latent heat}}{\text{Temperature}} = \frac{\Delta H}{T}$$

$$\frac{dP}{dT} = \frac{\Delta H}{T\Delta V} \qquad \text{.....(4.78)}$$

This equation is called Claperyon equation. The change in volume ΔV is the difference in the molar volumes of both the phases.

(i) **For solid-liquid (fusion) equilibria :**

$$\frac{dP}{dT} = \frac{\Delta H}{T\Delta V} = \frac{\Delta H_f}{T(V_l - V_s)}$$

(ii) **Equilibrium between two crystalline forms :**

Let α and β be any two crystalline forms (eg. Monoclinic and rhombic forms of sulphur)

$$\alpha \rightleftharpoons \beta$$

$$\frac{dP}{dT} = \frac{\Delta H_f}{T(V_\beta - V_\alpha)}$$

(iii) **Liquid – vapor equilibira :**

Liquid $\rightleftharpoons$ Vapor

$$\frac{dP}{dT} = \frac{\Delta H_f}{T(V_v - V_l)}$$

4.11.8 The Clausius – Claperyon equation

Clausius simplied the Claperyon equation for specific equilibria like sublimation and vaporization where molar volume of one phase is far greater than the other.

Solid $\rightleftharpoons$ Vapor $\quad V_{vapor} >> V_{solid}$ and $V_{vapor} - V_{solid} \approx V_{vapor}$

Liquid $\rightleftharpoons$ Vapor $\quad V_{vapor} >> V_{liquid}$ and $V_{vapor} - V_{liquid} \approx V_{vapor}$

Substituting the approximation in Claperyon equation.

$$\frac{dP}{dT} = \frac{\Delta H_v}{TV_v}$$

But for any ideal gas $PV = RT$ or $V = \frac{RT}{P}$ substituting for the volume of vapor.

$$\frac{dP}{dT} = \frac{\Delta H_v.P}{T.RT}$$

or

$$\frac{dP}{dT} = \frac{\Delta H_v}{RT^2}$$

$$\frac{d\ln P}{dT} = \frac{\Delta H_v}{RT^2} \qquad .....(4.79)$$

Although the Clausius-Claperyon equation is approximate for it neglects the volume of the liquid/solid and supposes ideal behaviour of the vapor it has the advantages of great simplicity.

Integrating equation 4.79 we have,

$$\ln\frac{P_2}{P_1} = \frac{-\Delta H_v}{R}\left(\frac{1}{T_2} - \frac{1}{T_1}\right)$$

$$\ln\frac{P_2}{P_1} = \frac{\Delta H_v}{R}\left(\frac{T_2 - T_1}{T_1\,T_2}\right) \qquad .....(4.80)$$

Here ΔH_v is assumed to remain constant within the temperature and pressure changes of phase equilibria.

The validity of Clasius-Claperyon equation can be tested by either directly substituting the data in the expression 4.80 or from a straight line obtained by plotting ln P vs $1/T$.

4.11.9 Gibbs-Helmholtz equation

Gibbs-Helmholtz equation relates the rate of change of free energy with the corresponding energy term under isothermal conditions. There are two forms of Gibbs-Helmholtz equation.

(i) Obtained by relating the free energy (G) with the enthalpy (H) of the process.

(ii) Obtained by relating the work function (A) with the internal energy (E) of the process.

(i) Gibbs-Helmholtz equation interms of free energy and enthalpy changes

We know that the free energy change

$$dG = VdP - SdT$$

At any given pressure

$$dG_p = -SdT \text{ or } \left(\frac{\partial G}{\partial T}\right)_p = -S$$

Substituting this entropy value in the general free energy expression $G = H - TS$ we have

$$G = H + T\left(\frac{\partial G}{\partial T}\right)_p$$

Dividing through out with T.

$$\frac{G}{T} = \frac{H}{T} + \left(\frac{\partial G}{\partial T}\right)_p$$

For any small change in free energy ΔG the corresponding enthalpy change is ΔH. Therefore,

$$\frac{\Delta G}{T} = \frac{\Delta H}{T} + \left(\frac{\partial \Delta G}{\partial T}\right)_p$$

or $$\boxed{\Delta G = \Delta H + \Delta T\left(\frac{\partial \Delta G}{\partial T}\right)_p} \quad .....(4.81)$$

Gibbs-Helmholtz equation is applicable to constant pressure processes like chemical reactions carriedout at atmospheric pressure. An easily comprihendable form could be obtained in the following way.

Dividing equation 4.81 throughout by T^2.

$$\frac{\Delta G}{T^2} = \frac{\Delta H}{T^2} + \frac{1}{T}\left(\frac{\partial \Delta G}{\partial T}\right)_p$$

or $$\frac{1}{T}\left(\frac{\partial \Delta G}{\partial T}\right)_p - \frac{\Delta G}{T^2} = \frac{-\Delta H}{T^2}$$

or $$\left(\frac{\partial \Delta G}{\partial T}\right)_p = \frac{-\Delta H}{T^2} \quad .....(4.82)$$

Therefore, from a knowledge of free energy change values at any two temperatures it is possible to calculate the enthalpy change of a reaction.

(ii) Gibbs-Helmholtz equation interms of changes in internal energy

For isochoric reactions like Gibbs-Helmholtz equation (4.82) suggested above is not suitable. For this analogous expressions interms of work function and internal energy were derived. Starting with the expression for work function $A = E - TS$ and proceeding in a similar way we get

$$\left(\frac{\partial \Delta A}{\partial T}\right)_V = \frac{-\Delta E}{T^2} \qquad .....(4.83)$$

Application of Gibbs-Helmholtz equation

Any chemical reaction is accompanied by liberation or absorption of heat (a change in enthalpy). A direct measurement of small changes in enthalpy values should be done only with abundant caution because of absence of 100% thermal insulating material. This could be achieved indirectly, by measuring ΔG values from corresponding *emf* values at any two temperatures and substituting the same in the Gibbs-Helmholtz equation.

Decrease in free energy $\Delta G = -nF\varepsilon$

Where ε is emf of the cell, n = no. of electrons transferred and F is the Faraday (96500 coulombs).

Substituting the value of ΔG in equation

$$-nF\varepsilon = \Delta H + T\left(\frac{\partial(-nF\varepsilon)}{\partial T}\right)_P$$

$$-nF\varepsilon = \Delta H - TnF\left(\frac{\partial \varepsilon}{\partial T}\right)_P$$

$$\Delta H = TnF\left(\frac{\partial \varepsilon}{\partial T}\right)_P - nF\varepsilon \qquad .....(4.84)$$

4.11.10 Van't Hoff Isotherm

Van't Hoff isotherm gives a relation between the equilibrium constant and free energy change of a chemical reaction. For this consider a general reaction

$$a\text{A} + b\text{B} + \ldots\ldots\ldots \longrightarrow l\text{L} + m\text{M} + \ldots\ldots\ldots$$

The change in free energy is given by the expression

$$dG = VdP - SdT$$

At constant temperature, $dT = 0$

$$\therefore \quad dG_T = VdP$$

or for 1 mole of a gas V = RT/P

$$\int dG = RT \int \frac{dP}{P}$$

$$G = RT \ln P + G^O$$

Where G^O is the *integration constant* and called **standard free energy.**

For a moles of A the equation (4.84) takes the form

$$aG_A = aG^O{}_A + a\, RT \ln P_A$$

Similarly for b moles of B

$$bG_B = bG^O{}_B + b\, RT \ln P_B$$

Free energy change of the reaction

$$\Delta G = \text{Free energy of products} - \text{Free energy of reactants}$$

$$\Delta G = (lG_L + mG_M + \ldots..) \quad - (aG_A + bG_B + \ldots..)$$

$$\Delta G = [(lG^O{}_L + mG^O{}_M + \ldots..) - (aG^O{}_A + bG^O{}_B + \ldots..)]$$

$$+ [(l\, RT \ln P_L + m\, RT \ln P_M + \ldots..)$$

$$- (a\, RT \ln P_A + b\, RT \ln P_B + \ldots..)]$$

$$\Delta G = \Delta G^O{}_L + RT \ln \frac{P_L^l \cdot P_M^m \ldots\ldots}{P_A^a \cdot P_B^b \ldots\ldots}$$

The ratio $\frac{P_L^l \cdot P_M^m \ldots\ldots}{P_A^a \cdot P_B^b \ldots\ldots}$ corresponds the partial pressures of reactants and is called **reaction quotient, Q**

$$\Delta G = \Delta G^O + RT \ln Q \qquad \ldots..(4.85)$$

When the reaction attains equilibrium, we know

$$\Delta G = 0$$

and $Q = K_p$ where K_p is the quilibrium constant.

Therefore $\boxed{\Delta G^O = RT \ln K_p}$ .....(4.86)

Substituting the ΔG^O value in equation 4.85 may be written as

$$\Delta G = -RT \ln K_p - RT \ln Q \quad .....(4.87)$$

This relation is called **Van't Hoff isotherm** and *it gives a relation between the free energy change, the equilibrium constant and the partial pressures of reactants and products at any given temperature.*

4.11.11 Van't Hoff Isochore or Van't Hoff Equation

The effect of temperature on the equilibrium constant is rationalized by Van't Hoff equation. This is achieved by comparing the Van't Hoff isotherm and the Gibbs Helmholtz equation.

Under equilibrium conditions

The Gibbs Helmholtz equation is written as

$$\left[\frac{\partial\left(\Delta G^0 / T\right)}{\partial T}\right] = \frac{-\Delta H^0}{T^2}$$

The Van't Hoff equation is written as

$$\frac{\Delta H^0}{T} = -R \ln K_p$$

Therefore $$\left[\frac{\partial\left(-R \ln K_p\right)}{\partial T}\right]_p = \frac{-\Delta H^0}{T^2}$$

Since R is a temperature independent term

$$R\left[\frac{\partial\left(\ln K_p\right)}{\partial T}\right]_p = \frac{\Delta H^0}{T^2}$$

$$\left[\frac{\partial\left(\ln K_p\right)}{\partial T}\right]_p = \frac{\Delta H^0}{RT^2} \qquad \text{....(4.88)}$$

The utility of this Van't Hoff equation or Van't Hoff isochore could be realized by taking its integral form

$$ln\ K_p = \frac{-\Delta H^0}{RT} + \text{integral constant} \qquad \text{.....(4.89)}$$

This suggests that a plot of ln K_p vs $1/T$ should give a straight line.

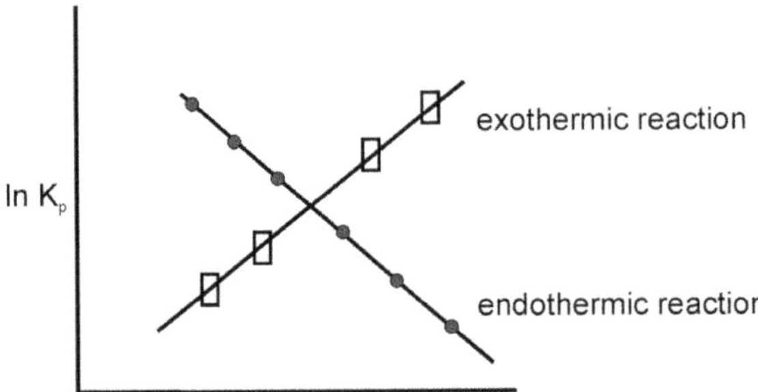

This graphical form of Van't Hoff equation explains the Lechatelier principle according to which the exothermic reactions are favored by a fall in temeperature and endothermic reaction are favored by a rise in temperature.

Integrating equation 4.89 between any two temperatures T_1 & T_2 at which equilibrium constants are $(K_p)_1$ and $(K_p)_2$, we get

$$\ln \frac{\left(K_p\right)_1}{\left(K_p\right)_2} = \frac{-\Delta H^0}{R}\left(\frac{1}{T_1} - \frac{1}{T_2}\right) \qquad \text{.....(4.90)}$$

Thus, from a knowledge of equilibrium constants at any two temperatures, we can calculate the change in enthalpy of the reaction.

Applications of Van't Hoff equation

1. Standard heat of reaction could be calculated from a knowledge of equilibrium constants at any two temperatures.
2. From a knowledge of $\Delta H^0_{Reactions}$ and equilibrium constant at any temperature allows one to estimate the equilibrium constant at any other temperature.

Numerical Problems

First Law of Thermodynamics

1. Calculate the work done when 2 moles of an ideal gas at 10 atm expands isothermally and reversibly at 27 ^{O}C to a pressure of 1 atm.
2. Calculate the work done by 2 moles of an ideal gas when it expands isothermally and reversibly from 4 to 5 litres at 27 ^{O}C.
3. Calculate the work done by 5 moles of an ideal gas in expanding reversibly and isothermally from a volume of 5 lit to 50 lit at 300 K.
4. Calculate the work done in calories, when 3 moles of an ideal gas at 27 ^{O}C expands isothermally and reversibly from a volume of 10 litres to 50 litres.
5. Two moles of an ideal gas at 300 K, expands isothermally and reversibly from 5 litres to 20 litres. Calcuate the work done in this process.
6. n-Hexane boils at 68.74 ^{O}C and heat of vaporization is 6896 cal mol^{-1}. The vaporization process is isothermal and reversible. Find the change in entropy.
7. 2 moles of a diatomic gas expands adiabatially and reversibly from a volume of 2L at 20 atm to 10 L. Calculate the final pressure and work done.
8. 3 moles of a monatomic gas was compressed adiabatically from NTP conditions to a volume of 10 litres. Find the work done.
9. When a gas expands against a constant pressure of 1 atmosphere after absorbing 100 calories of heat, the volume of the gas increases from 2 L to 5 L. Calculate the change in internal energy of this process.
10. The variation of μ for air at 0 ^{O}C in the pressure range 1-20 bar is expressed by the equation $\mu = 0.264 - (8.95 \times 10^{-4})\,P$.

    Calculate the cooling accompanied by expansion from 20 atm to 1 atm at this temperature.

Thermochemistry

1. The heat of combustion of C_2H_4 at 17 ^{O}C at constant volume –332190 cals. What is the heat of combustion at constant pressure.

2. Calculate heat of formation of acetic acid, given the following

Heat of combustion	***Compound***
–208.34 kcals	acetic acid
–94.05 kcals	carbon
–68.32 kcals	hydrogen.

3. The heat of combustion of methane is –215 kcal mol^{-1}. The heat of formation of CO_2 (g) and H_2O (l) are –96.4 and –68 kcal $mole^{-1}$ respectively. Calculate the heat of formation of methane.

4. The heat of combustion of benzene, carbon and hydrogen are – 754.3, –954.4 and –68.4 kcal respectively. Calculate the heat of formation of benzene.

5. If the standard heat of formation of HCl gas at 25 ^{O}C is –92.3 kJ mol^{-1}, what would be its value at 60 ^{O}C. The average C_p values in this temperature range for H_2 (g), Cl_2 (g) and HCl (g) are 28.54, 32.27 and 28.50 J mol^{-1} deg^{-1} respectively.

6. Heat of combustion of n-Hexane is –995 kcal mol^{-1}. If the heat of combustion of carbon and hydrogen are –94.05 and –68.32 kcal mol^{-1}, calculate the heat of formation of n-hexane.

7. The molar heat capacity at constant volume for n-hexane in the temperature range 30 ^{O}C --- 75 ^{O}C is 92.45 cal K^{-1} mol^{-1}. Calculate the heat required to raise 86 g of the sample from 35 to 65 ^{O}C; (i) at constant volume (ii) at constant pressure.

8. When 1 mole of aqueous NH_4OH is neutralized according to the reaction

$$NH_4OH(aq) + H^+(aq) \longrightarrow NH_4(q) + H_2O$$

liberating 51.6 kJ at 298 K, calculate the heat of ionization of NH_4OH. Given heat of dissociation of water is 577.32 kJ.

9. Calculate the heat of reaction for the following reaction.

$$H_2 + \underset{H}{\overset{H}{\diagdown}}C = C\underset{H}{\overset{H}{\diagup}} \longrightarrow H_3C - CH_3$$

The C-H, H-H, C-C and C = C bond energies are 99, 124.2, 83.1, 141.7 kcal mol^{-1} respectively.

10. Calculate the bond dissociation energy of O-H bond at 25 ^{0}C with the help of following equations

H_2 (g) ⟶ 2H (g) $\Delta H = 104.2$ kcal I

O_2 (g) ⟶ 2O (g) $\Delta H = 118.0$ kcal II

½H_2 (g)+ ½O_2 (g) ⟶ OH (g) $\Delta H = 10.1$ kcal III

11. The bond dissociation energies of Br_2, H_2, C-H and C-Br are 46, 124.2, 99 and 66 kcal mol^{-1} respectively. Calculate the heat of formation of bromomethane.

12. Compute the heat of reaction for the formation of ethyl alcohol from ethylene and water.

$CH_2 = CH_2 + H_2O \longrightarrow CH_3CH_2OH$

13. Molar heats of combustion of C_2H_2(g), C(s) and H_2(g) are 310.62 kcal, 94.05 kcal and 68.32 kcal respectively. Calculate the standard heat of formation of C_2H_2(g).

14. On complete combustion of ethanol

C_2H_5OH (l) + $3O_2$ (g) ⟶ $2CO_2$ (g) + $3H_2O$ (l)

326.7 kcal is liberated. Calculate ΔE of the reaction at room temperature.

15. Calculate the heat of formation of ethylene from the following data.

H_2 (g) + ½O_2 (g) ⟶ H_2O (l), $\Delta H = -68$ kcal

H_2 (g) + ½ O_2 (g) ⟶ H_2O (l), $\Delta H = -68$ kcal

H_2 (g) + ½ O_2 (g) ⟶ H_2O (l), $\Delta H = -68$ kcal

16. For the reaction

$4C_2H_2Cl$ (g) + $13O_2$ ⟶ $2Cl_2$ (g) + $8CO_2$ (g) + $10H_2O$ (g)

ΔH at 298 K = 1229.6 kcal. Calculate ΔE of the reaction at the same temperature.

17. When 1 mole of liquid benzene is burnt completely with oxygen 783.4 kcal of heat are liberated at 18 ^{0}C. Calculate the heat of formation at constant volume at the same temperature.

Second Law of Thermodynamics

18. Calculate the entropy change for melting of SO_2 at –76 ^{O}C, given heat of fusion is 1769 cal mol^{-1}

19. Calculate ΔS for evaporation of diethyl ether at 35 ^{O}C and 1 atm, given $\Delta H_{vap} = 27.2$ kJ mol^{-1}.

20. What is the entropy change if 2 moles of a gas at 25 ^{O}C and 2 L is allowed to expand to 150 ^{O}C and 20 L. Given molar heat capacity at constant volume of the gas in this temperature range is $C_v = 5$ cal mol^{-1} K^{-1}.

21. Evaporation of 1 mole of n-hexane at its boiling point, 35 ^{O}C, is effected by absorption of 6500 cal mol^{-1}. Calculate the entropy change accompanying the process. Express your result in both CGS and SI unit systems.

22. Calculate the entropy change in expanding one mole of a gas at 5 L and 300 K to 25 L under isothermal conditions.

23. Calculate the efficiency of a heat engine operating between 400 K and 27 ^{O}C.

24. Calculate the efficiency of a heat engine which operates between the temp 25 ^{O}C and 120 ^{O}C. How can you increase the efficiency of the engine.

25. Calculate the efficiency and heat rejected by a Carnot's engine operating between 400 ^{O}C and 150 ^{O}C and absorbing 225 kJ at 400 ^{O}C.

26. Find the work done by a steam engine operating between 383 K and 303 K and absorbing 840 cal.

27. A Carnot cycle operates between the temperature limits 200 K and 300 K and absorbs 300 kJ in each cycle. Calculate the (i) efficiency of heat engine (ii) work done (iii) heat rejected.

28. A heat engine is capable of converting 30% of the heat it absorbs. If it absorbs 800 J per cycle, with the reservoir at 500 K, find the (i) efficiency of heat engine, (ii) Heat rejected.

Self-Test Questions

1. The cooling effect produced by an expression of compressed gas through an orifice is known as
   (A) Linde's method of liquefaction (B) Isothermal reversible expansion
   (C) Joule-Thomson effect (D) Claudes method of liquefaction
2. The first law of thermodynamics can be mathematically expressed as
   (A) $C_p - C_v = R$, (B) $\Delta H = \Delta G + T\Delta S$,
   (C) $A = E - TS$, (D) $\Delta E = \mathrm{q} - \mathrm{w}$
3. The temperature at which Joule-Thomson coefficient is zero for a given gas is known as;
   (A) Transition temperature (B) Critical temperature
   (C) Inversion temperature (D) Critical solution temperature
4. Which one of the following statements is true for an adiabatic expansion of a gas;
   (A) $\Delta E = q$, (B) $\Delta E = 0$, (C) $-\Delta E = w$, (D) $w = q$
5. Maximum work is derivable in a
   (A) Reversible isothermal expansion of a gas from volume V_1 to V_2
   (B) Reversible, adiabatic expansion of a gas from volume V_1 to V_2
   (C) Sudden expansion of a gas from volume V_1 to V_2
   (D) Irreversible expansion of a gas from V_1 to V_2
6. Which one of the following process is reversible in the thermodynamics sense
   (A) Mixing of two gases by diffusion
   (B) Dissolution of a solute in a solvent
   (C) Expansion of a gas into vacuum
   (D) Evaporation of water at 373 K and 1 atm pressure
7. Which one of the following changes does entropy decrease
   (A) Evaporation of water, (B) Solidification of liquid
   (C) Sublimation of a solid (D) Expansion of a gas

8. Free energy change for a reaction is zero. The reaction is therefore

   (A) A spontaneous reaction,

   (B) An Irreversible reaction

   (C) A reversible reaction in a state of equilibrium

   (D) A reaction accompanied by phase transfer

9. For the reaction $H_2(g) + I_2(g) \longrightarrow 2HI(g)$

   (A) $\Delta E < \Delta H$, (B) $\Delta E = \Delta H$, (C) $\Delta E > \Delta H$, (D) $\Delta E = w$

10. The heat of formation of ammonia is –46 J mol^{-1} the heat of reaction for the reaction $2NH_3(g) \longrightarrow N_2(g) + 3H_2(g)$ is

    (A) –46 kJ, (B) –138 kJ, (C) +46 kJ, (D) +92 kJ

11. kcal heat is absorbed when

    (A) 1 dm^3 water is heated by 1°C at a pressure of 1 atm

    (B) 1 L of water is heated at its boiling point

    (C) 1 cc of water is heated from 14.5 – 15.5 °C

    (D) 1 L of water is heated from 14.5 – 15.5 °C

12. A system is having a ΔE of 50 J liberates 100 J of heat during a process, then

    (A) 150 J of work is done on the system

    (B) 150 J of work is done by the system

    (C) 50 J of work is done on the system

    (D) 50 J of work is done by the system

13. The change in free energy of a system is equal to

    (A) The maximum work obtainable from the system

    (B) The pressure-volume work

    (C) The entropy change of a system

    (D) The net work obtainable from the system

14. For any chemical reaction irrespective of the operation carried out the heat of reaction is equal to

    (A) $\Delta E - w$, (B) $q_p - w$, (C) $\Delta E + w$ (D) $q_v - w$

15. A thermodynamic function is one

(A) which depends on the path traced

(B) which is independent of the path traced

(C) which depends on the initial and final temperatures of the system.

(D) Which depends only on the final pressure of the system.

16. For a spontaneous process

(A) $\Delta H > 0$, (B) $\Delta S > 0$, (C) $\Delta S < 0$, (D) $\Delta G > 0$

17. Second law of thermodynamics states that

(A) Energy can neither be created nor destroyed

(B) For any action there is an equal and opposite reaction

(C) Entropy of the universe remains always the same

(D) Entropy of the universe always increases

18. In an adiabatic reversible process

(A) $q = 0$, (B) $\Delta A = 0$, (C) $\Delta G = 0$, (D) $\Delta E = 0$

19. The ΔG of a reaction at any instant is a measure of

(A) The heat transferred from system to surroundings

(B) The heat transferred from surroundings to the system

(C) The disorder of a reaction

(D) The direction in which the net reaction occurs in the systems

20. A process is spontaneous when

(A) The entropy of the system increases

(B) The entropy of the surroundings increases

(C) The entropy of both the system & surroundings increases

(D) The entropy of the universe increases

21. The efficiency of a heat engine operating between $100^{o}C$ and $0^{o}C$ is

(A) 100%, (B) 0%, (C) 27%, (D) 53%

22. Heat of neutralization of HCl and NaOH is

(A) > 13.7, (B) –13.7, (C) < 13.7, (D) None of these

23. When 2 lit of a compressed gas expands suddenly into a vacuum space of 20 lit, the work done

(A) positive (B) 0 (C) Negative (D) none of these

24. Change in work function gives
    (A) Maximum work done by the system
    (B) Maximum work done by the surroundings
    (C) Maximum entropy change of a system
    (D) None of the above

Short and Essay Type Questions

1. (a) State and explain first law of thermodynamics.
   (b) Derive an expression for the maximum work done during isothermal reversible expansions of an ideal gas.
2. What is entropy? What are its units? Does the entropy change when ice melts into water? Explain your answer.
3. With the help of Carnot's cycle, derive an expression for the efficiency of a reversible engine working between the temperatures T_1 and T_2.
4. Write a short notes on "Joule-Thompson" effect.
5. Explain the terms (a) free entropy and (b) maximum work function.
6. Derive an expression for the work done in isothermal, reversible expansion of an ideal gas.
7. The heat of neutralization of all strong acids - strong bases is 13.7 kcal mole^{-1}. Explain.
8. Explain the statement "The entropy of the universe tends always towards a maximum".
9. State first law of thermodynamics in its various forms.
10. (a) State the first law of thermodynamics.
    (b) Derive the expression PV= constant.
11. Explain the terms (a) Entropy and (b) Free energy.
29. Explain Hess's law with a suitable example.
30. Explain the terms state function. Which of the following are state functions (a) free energy (b) enthalpy (c) heat absorbed (d) work of expansion.

31. Give the relations between heat absorbed, work done and change in internal energy of a system. Explain why the internal energy is called a state function and heat absorbed is not a state function. Under what conditions does the heat absorbed have a definite value ? derive the equation $C_p - C_v = R$

32. Define heat of formation and heat of combustion. State Hess's law.

33. State Hess's law of constant heat summation and explain how it can be applied in the indirect determination of heat of chemical reaction, mention its applications.

34. Define second law of thermodynamics.

35. Give the equation relating ΔH and ΔE.

36. Define the terms "heat of formation", "heat of combustion" "heat of neutralization" and "integral heat of solution".

37. Which would have the higher entropy in the gas phase at the same temperature, Argon or ethane ? Explain.

38. What are isothermal and adiabatic process? Explain.

39. What is cyclic process? Describe in detail the Carnot's reversible cycle for getting maximum efficiency and derive an expression for it.

40. Derive an expression for relating heat capacity at constant volume (C_v) and heat capacity at constant pressure (C_p).

41. Discuss the variation of heat of reaction with temperature, and derive Kirchoff's equation.

42. Explain briefly the significance of entropy.

43. Show that internal energy of a system is a state function.

27. (a) Derive the expression for the maximum work done in an isothermal reversible expansion of a gas.

    (b) Explain the terms in equation $\Delta H = \Delta E + P\Delta V$. Under what conditions is $\Delta H = \Delta E$.

28. How heats of formation of a compound is determined from heat of combustion data? Explain your answer with suitable example.

29. Compare the reversible works done in isothermal and adiabatic processes. Which process gives maximum work and why?
30. Distinguish between :
    (i) System and surroundings
    (ii) Isothermal and adiabatic process
    (iii) State function and path function
    (iv) Heat of formation and heat of reaction
    (v) Gibb's free energy and Helmholtz free energy
31. Heat of formation of CO can not be determined directly because of unavoidable, simultaneous formation of CO_2. Suggest a method by which the value can be determined.
32. Explain how free energy hel-ps in predicting the spontaneity of a chemical reaction.

Chapter 5

Chemical Fuels

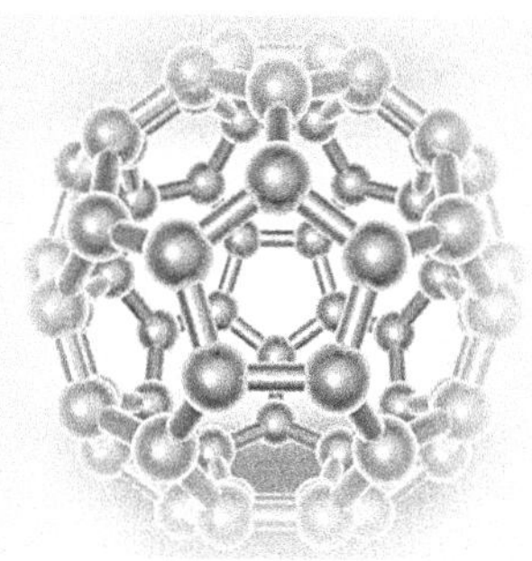

5.1 Introduction

A chemical fuel is any substance used to produce heat and power by combustion. Only *chemical* fuels will be discussed in this chapter. *Nuclear* fuels are now being used in several large plants for the generation of electrical power and propulsion of ships and will be considered in a subsequent chapter. In these cases the energy liberated is by fission reactions. By common usage the term fuel also implies that it can be stored and transported.

Fuels are primarily used for heating but they have other very important functions. For example, in an internal combustion engine petrol is not used for heating the engine. In several metallurgical operations, as in a blast furnace, coke not only supplies heat but also acts as a reducing agent.

5.1.1 A Survey of Fuel Resources

The fuel resources of the world have been estimated differently by different authorities. It gives a better picture of the problem if we consider for how many years a particular kind of fuel is likely to last rather than to give estimates of its known reserves. To give a general idea, it may the said that the world's gas supplies are likely to last fifteen years, oil eighty years, coal five hundred years, and nuclear fission materials two thousand years. Materials for fusion reaction seem to be unlimited. India is very well placed in its thorium deposits in which she holds world monopoly. By breeder reaction thorium gives an isotope of uranium which can easily undergo fission.

India has sufficiently large deposits of coal but these are small when compared with some other developed countries, and are perhaps insignificant when calculated on the per capita basis. Hence much foresight in planning and prudent use of coal is necessary. Our best grade coal, which comprises only 13% of the total reserve, contains about 17% ash. This in itself is too high an ash content when compared to coal ash of a big producer like England (which is about 5%, for comparable quality). 70% of our coal is of rather low quality containing as high as 35% ash which virtually means that one third of the coal is ash. To meet the challenge several coal washeries have been established which make several fractions of coal, the better grades being used for metallurgical purposes. Production of gas from such inferior coal needs to be given consideration.

In the early part of the 1960's the oil position of India seemed to be quite bright. Major oil fields were found between 1952 and 1964 in Naharkatiya, Cambay, Ankaleswar, Kalol, Rudrasagar and Lakwa. Some of these are drying up. Also due to the very heavy rise in oil prices the economy of the country has been upset besides the uncertainty of supply during any war or crisis. India has high technical skill in oil exploration due to which her experts are helping other countries, like Iraq, to find new oil fields. Oil exploration under the sea near the shore has received much impetus and has yielded rich dividends as evidenced by the production from the Bombay High.

When people think of petroleum, their attention naturally goes to petrol. It may be mentioned here that petrol is about one-eighth of the total petroleum products. Diesel oil, furnace oil, lubricants, etc. and numerous other materials for the manufacture of which petroleum is the basic raw material also have their importance.

The fertiliser industry is another large consumer of naphtha. It is the same fraction which gives petrol for the internal combustion engines. It is said that one part by weight of fixed nitrogen in the form of a fertiliser gives ten parts of the cereals. Hence the fertiliser industry has to be given priority. But how does naphtha come into the fertiliser scene? Ammonia is the basic material for the fertilisers for the manufacture of which atmospheric nitrogen is used and hydrogen is produced by catalytic decomposition of the hydrocarbons thus:

(1) $CH_4 = C + 2H_2$

(2) $CH_4 + H_2O = CO + 3H_2$

(3) $CO + H_2O = CO_2 + H_2$

(CO_2 being removed by washing with water under pressure).

In the coal-based fertiliser plants the hydrogen and nitrogen will be produced by passing steam and air through hot coal. This has the advantage of utilizing poor grade coal and the saving of naphtha.

A very large number of useful products are made from petroleum fractions. Some of the major consumers, to name a few out of a large number, are plastics, rubber, organics, synthetic fibres, pharmaceuticals. Our earlier Five Year Plans were more oil-oriented than coal-oriented, as is clear from the Dieselisation of the railways, oil based fertiliser plants, encouragement given to small consumers to have Diesel engines, and oil-fired furnaces. This approach has undergone radical rethinking. As petroleum has become a material of strategic significance, several countries are planning to have stocks equivalent to '90 days' use. Our country should take immediate steps to do this. Serious attention is being given to the production of synthetic petrol also to basic raw materials for petro-chemicals utilising low grade coal.

A sizeable amount of our electric power comes from hydroelectric power plants which depend upon rainfall. Paradoxically, when there is less rainfall, with concomitant less electricity available more water pumps need to be operated for irrigation.

It is absolutely necessary to economise and efficiently use the fossil fuels, but that is not enough. To burn away coal and petroleum products, which are essential raw materials for many chemicals is dangerous. These must be preserved for certain vital chemical industries. All this forces one's attention to nuclear power stations. One major apprehension against nuclear fuels is the safety factor. The present safety record of the nuclear power plants is good. An expert has suggested that the plants may be built underground at suitable impermeable geological formations so that in case of accidental wreckage, the radio-active wastes may not spread. According to him, the underground construction will add only a small fraction to the total cost. The safety and security of handling and transporting a fuel like plutonium will remain. According to the famous Nobel Laureate, Glenn T. Seaborg, even the large scale use of nuclear energy for power production will increase the radioactivity of the environment by less than one percent of the natural background radioactivity. He further says that this is less dangerous to man than billions of tonnes of carbon dioxide, millions of tonnes of sulphur dioxide and large quantities of other harmful matter emitted by plants using coal and petroleum products. When nuclear fusion plants are established, radioactive wastes will be much less. Processes have been suggested for the disposal of the radioactive wastes mainly by converting them to solids and burying them safely. It is easier to exercise control over a few large nuclear power plants producing electricity for the entire country than to control the activities of a large number of thermal power plants scattered all over the country.

5.2 Classification

Fuels are classified according to whether they occur in nature (primary fuels) or are prepared (secondary fuels) and according to whether they are in the solid, liquid or gaseous state. The classification is summarized in Table 5.1. They are classified according to their occurrence so that it is known whether a material is available in nature and can be used directly or it has to be specially prepared. Each physical state has its advantages and disadvantages over the other states, hence it is convenient to classify according to the physical state.

Table 5.1

State	Natural (or primary)	Prepared (or secondary)
Solid	Wood, Peat Lignite, Coal	Charcoal Coke
Liquid	Petroleum	Petroleum distilates, Alcohols, Tar
Gaseous	Natural gas	Coal gas, Coke oven gas, Producer gas, Water gas, Blast furnace gas

5.3 Wood

In India wood is used to a considerable extent as a domestic fuel. Its use as an engineering fuel is restricted. On a moisture-free basis, wood contains about 50% C, 6% H_2, 43% O_2 and less than 1% ash. Freshly cut wood contains 25 to 40% moisture and air-dried wood contains about 15% moisture. The calorific value of common air-dried wood is about 3,500 cal/g, but in some cases it may be as high as 5,000 cal/g.

A reference to bagasse, as a fuel, may be made here. Bagasse is the residue left after extracting the juice from sugar cane. It is an important fuel in sugar mills. Its calorific value on 15% moisture basis is 3,800 cal/g. Attempts are being made in India to make paper on a large scale out of bagasse instead of using it as a fuel in the sugar mills.

A recent study shows that wood has surpassed nuclear power as an energy source. Brazil, which is the eleventh largest steel producing country of the world, meets 40% of its energy needs of the steel industry by burning wood. Of course, Brazil is a vast country with thick forests and low population density. Sweden, a leading country in paper and pulp industry, meets 60% of the energy needs of this industry by burning wood wastes. Several countries are experimenting with mixing methanol (produced from wood) which is pollution free, with petrol.

A warning must be given against indiscriminate use of wood. Cutting down of forests has serious environmental effects. Although attention has been directed to wood due to shortage of petroleum and also the increasing price, methods of increasing wealth by planting fast growing trees are being explored. As a matter of fact there is already a need in many countries, like India, to have more forests.

5.3.1 Wood Charcoal

Wood charcoal is obtained by the destructive distillation (heating out of contact with air) of wood. Water and various kinds of volatile matter are expelled. If the process is carried out carefully the original form and structure of wood is left. Charcoal consists of about 80% C, 15% O_2 and N_2, 2% H_2 and 3% ash. Its calorific value is high being 8,000 cal/g. Usually the weight of charcoal obtained does not exceed 20% of the weight of wood. Its volume is 50-75% of the original volume. The physical properties largely depend upon the temperature at which the destructive distillation was carried out and on the rate of heating.

Because of its purity and low ash content, charcoal was the most important metallurgical fuel. Its chief defect is its poor mechanical strength which results in its being crushed to powder in the operation. The powdered charcoal can easily be swept away in a current of gases and also it may prevent the proper flow of gases in the furnace. Consequently it has been largely replaced by coke. Because of its very large surface area, its main industrial use is in the adsorption of gases, colouring matter, etc., in the various purification processes of the chemical industry.

5.4 Coal

5.4.1 Introduction

Coal is a complex substance which originated several million years ago from the various plants that existed profusely on the surface of the earth. The plants were caught up in swamps, buried underground and slowly changed due to heat, pressure and complex biochemical and chemical processes. Now it is found as beds or layers in between clay or sediments which were deposited in course of time. Coal is an important fuel. The fact that on proper processing it yields gas, oil, fertilizers and chemicals increases its usefulness. As coal is easy to mine and handle, its use will continue even against severe competition from mineral oil, hydroelectric power and nuclear fuels.

5.4.2 Origin

Coal has been defined as 'a stratified rock consisting of the more or less altered remains of plants, together with associated mineral matter', The process which changed plant into coal has taken millions of years, and has been accomplished by bacteria, heat and pressure.

Geologists have put forward two theories of coal formation. Millions of years ago when the temperature was moderate and rain-fall was heavy, vegetation was thick and grew to far greater size than it does at present. This was so, especially in low-lying wet land. The first theory, known an *'in situ'* theory, contends that the coal formation took place where the vegetation grew and accumulated. The second theory, known as *'drift'* or *'transportation'* theory, holds that the trees, etc., were uprooted and the material drifted or was transported by rivers to lakes and estuaries and deposited there. Thus huge piles of wood collected and got buried underground. There are arguments for and against each of these theories but it is now accepted that in some cases the first theory and in other cases the second theory is more applicable. For example, the great purity of many coal seams can be better explained by the *'in situ'* theory and great thickness of coal seams appears more reasonable on the basis of the *'drift'* theory (About 20 m depth of vegetation is needed to give 1m depth of coal seam). The Gondwana coals of India show more evidence in favour of the drift theory.

As the plants accumulated under water, some silt also settled, along with them. Thus, at a later stage some mineral matter was mixed with the coal as it formed. The initial transformation of wood is believed to be due to bacterial action. In the earlier stages aerobic bacteria caused some oxidation but mostly anaerobic bacteria brought about the degradation of organic matter and the removal of oxygen. The bacterial action produced acidity, which if accumulated, prevented further action and coal formation could not proceed. In some situations the soil was alkaline which removed the acidity and the bacterial action continued to take place bringing about further degradation of matter. This explains why coal is found associated with certain kinds of rocks. There is still some difference of opinion among the authorities regarding the exact mechanism of the bacterial action. But many now accept that the bacterial action took place only in the early stages and is responsible for the breaking up of organic matter into simpler molecules. The degraded matter, as time elapsed, gave rise to peat. Beyond this stage heat, pressure and time became the chief factors responsible for further conversion. Coal formation cannot be fully explained on the basis of heat and pressure theory, even though they are high at the depths at which coal is found. The process is extremely slow and hence coal formation must have been completed only in the course of millions of years. It is interesting to note that in most cases, although there are some exceptions, with the increase in vertical depth, the carton content increases and volatile matter decreases. Anthracite and graphite are formed under extreme conditions.

This process of coal formation may be complete or may be stopped at any stage, thus giving rise to material of varying maturity. This is expressed by the term 'rank' of coal.

The stage in the formation of coal from vegetation can best be summarised thus :

Plant debris $\rightarrow$ peat $\rightarrow$ lignite $\rightarrow$ brown coal $\rightarrow$ sub-bituminous coal $\rightarrow$ bituminous coal $\rightarrow$ semi-anthracite coal $\rightarrow$ anthracite coal $\rightarrow$ graphite.

This series is known as *'peat to anthracite'* series. Each member is more *'mature'* or is of higher *'rank'* than the one which precedes it in the series. Thus anthracite coal is of higher 'rank' than semianthracite coal. With the progress of coal forming reactions, moisture and oxygen were removed and the percentage of carbon increased. Heating value increased from 4,500 cal/g for lignite to 8,600 cal/g for bituminous coal. There is a slight fall in the heating value of anthracite coal because of the removal of hydrogen.

5.4.3 Coal Classification

Coals have been classified in several ways. The most common method of classification is on the basis of rank. From the origin of coal it is clear that wood, after a long interval of time under certain conditions, was converted into coal. Rank represents the extent to which this process of coal formation has proceeded or, in simple words, maturity or age of coal. The important ranks are described in the following pages.

5.4.4 Peat

In the process of the formation of coal from wood, the first stage is the formation of peat. Strictly, it is not regarded as coal. Peat is mostly dug by manual labour, not much machinery is employed and it is air-dried. It is uneconomical to obtain peat lying buried deep under the earth's surface. One principal difficulty in its use is that peat deposits are associated with large amounts of water. There may be as much as 90% water and only 10% solid. To make it an economical fuel, its moisture content should be reduced to 30% or less. For this purpose it is dried in the sun for about one to two months. Its calorific value is only slightly higher than that of wood. Large quantities of peat are used as domestic fuel in Europe. In some parts of Russia, due to the shortage of coal, a considerable amount of peat is also used for power generation. It is not one of the important fuels used in India. Small amounts of peat deposits are found in Nilgiri Hills (Tamil Nadu) and Sunderban (W. Bengal).

5.4.5 Lignites and Brown Coal

In lignites the transformation of wood to coal has taken place to a greater extent than in peat. Their properties are intermediate between peat and coal. Lignites are found near the earth's surface and usually in thick seams, sometimes as much as 30 m in thickness. These are classed into two groups.

The true lignites are those which contain high proportions of relatively unchanged woody matter, have a high percentage of moisture, say, up to 60% and have low calorific value. The other form is commonly known as brown coal. Brown coals contain much less moisture, around 20% and have a higher calorific value.

Lignites are usually amorphous and break easily, they are therefore difficult to transport. They burn with long smoky flames.

Large amounts of lignites are consumed in Germany and Russia. In German plants large quantities, about 2,000 tonnes per day, are continuously dried by steam or hot gases obtained by the combustion of a portion of the lignite. The moisture content is thereby reduced to about 15% and then the product is made into briquettes by pressing. On the basis of a moisture content of 15%, the calorific value is about 4,500 cal/g.

Lignite is found scattered in several parts of India but the most important deposits are at Neyveli, about 200 km south of Madras. These are spread over an area of about 250 sq. km. The deposits in this area are estimated to be about 200 million tonnes and 5.9 million tonnes of lignite was mined in 1981. These deposits are at a depth of 55 m below the ground level and the average thickness is about 16 m. One of the main difficulties faced in the working of the project has been that water exists in the mines under high pressure which can completely flood the mines. This difficulty has been overcome by drilling a number of artesian wells and constantly pumping out the water.

The Neyveli Lignite Project is a complex project and plays an important role in the economy of the area which lacks fuel deposits. A 250 MW thermal power station is operational at Neyveli. The project manufactures 1.5 million tonnes of urea, to be used as fertilizer and 3.8 million tonnes of carbonized briquettes for use in the homes and the industries.

Out of the total estimated lignite deposits of 2,100 million tonnes in India, 1,900 mt are in Tamil Nadu alone. Smaller amounts of lignite are also found in Kutch, Rajasthan and Kashmir. The Kashmir lignite is of low calorific value (about 200 cal/g).

5.4.6 Sub-bituminous Coals

These are black in colour and are more homogeneous and smooth in appearance. The moisture and volatile matter contents are high, and on exposure to air, these slowly crumble into smaller pieces. These are non-caking coals. The calorific value of sub-bituminous coals on a mineral free basis, is 7,000 cal/g or a little higher. Carbon content varies from 75 to 83% and oxygen content from 13 to 20%.

5.4.7 Bituminous Coals

The term bituminous coal covers coals with a wide range of properties. Therefore the term is used after prefixing caking, non-caking, low volatile, medium volatile, high volatile, etc. (e.g., caking bituminous coal). The carbon content varies from 78 to 90% and volatile matter from 20 to 45%. Coals of this class are used in the largest quantities in industries for making metallurgical coke, coal gas and for domestic heating. The calorific value, on a mineral free basis goes up to about 9,000 cal/g.

Bituminous coals are black, comparatively hard and brittle and burn easily with a yellow smoky flame. The moisture content is below 6 percent. The bituminous coals are of two broad types. The 'caking coals' are rich in hydrocarbons. On heating, these swell and fuse together forming larger lumps. More air therefore has to be supplied for their combustion. The 'non-caking coals' are usually called 'steam coals'. These do not fuse and are largely used for industrial purposes and are especially suitable for steam raising.

5.4.8 Anthracite Coals

These are of the highest 'rank' and in them the coal formation process has proceeded up to the maximum extent. The quantity of these coals available is very limited and hence they are expensive. Small amounts of anthracite are found in Darjeeling and Jammu. The carbon content is 90 to 98% and volatile matter is less than 8%. Therefore they ignite with difficulty. Anthracite coals are hard and brittle and produce very little dust. These are black but can be handled without soiling the fingers. They are non-caking and burn without smoke and with a short non-luminous flame giving intense local heating.

5.4.9 Indian Coals

In the late 18th and early 19th century, exposures of coal were noticed in different parts of India. In 1774 the first colliery of India was opened in Raniganj coalfield and in 1775 the first consignment of 100 tonnes of coal was received in Calcutta. Systematic exploration of coalfields was started 1845.

The coal deposits in seams more than 1.2 m thick and occurring down to depth of 600 m (which is the present working limit) are estimated to be about 1,20,000 million tonnes. Out of this only about 4,000 million tonnes are Tertiary coal and the remainder is Gondwana coals. The coking coals are in short supply, being only about 2,000 million tonnes. Deposits of inferior variety coals are very vast. The quantity of coking coal is insufficient for the requirements of Indian metallurgical industries, therefore the inferior coals are being washed to remove ash and coking coal is being suitably blended with weakly coking coal.

Most of the Indian coal is what is known as Gondwana coal. 'Gondwana' is a geological term which refers to a certain rock system which is about 200 million years old. Gondwana coal is a laminated bituminous coal with alternate dull and bright layers. In India, anthracite coal is found to a negligible extent. Coal deposits of the Tertiary era (60 million years old) are found in Assam, Rajasthan and Jammu. There are nearly 850 collieries are working.

Coalfields of Bihar and Bengal produce about 80% whereas the coalfields of Madhya Pradesh 10%, Orissa 5% and the rest of the country produce about 5% of the coal. This shows the uneven geographical distribution of coal and the consequent distribution problems. The important coal-fields of Bihar and Bengal are Barjora, Bokaro, Giridih, Karanpura, Jharia and Raniganj. The last two named are the biggest and best coalfields of the country.

The coalfields have seams or layers of coal beds from which coal is mined. Most of the coalfields have less than half a dozen seams but Raniganj, Jharia and Karanpura fields have more than twenty seams. Most of the coal seams are less than 10 meters thick. Some of the thickest coal seams of India are the Upper Seam No.1 (45 m) of the Talchir coalfield, the Argada Seam (47.5 m) of the south Karanpura and the Jhingurda seam (152 m) of Singrauli which is the thickest seam of the world.

Most of the Indian coals have a high ash content. This is a serious disadvantage. Coal suitable for metallurgical industry is not at all sufficient for the needs of the country and is mostly available from some of the seams of Jharia, Raniganj and Bokaro coalfields.

The Gondwana coal, which contributes about 98% of our coal, has been classed into the following two broad groups

Raniganj coal has greater moisture content (about 5%) and more volatile matter (30-38%) and therefore burns easily with a long flame. It has lower calorific value, the ash does not form clinkers, and it is unsuitable for making coke but is good for making gas.

Jharia or Barakar coal has less than 2% moisture and the volatile matter content is about 28%; consequently it burns with a shorter flame, and is more suitable for steam raising in locomotive boilers (because less combustion space is needed). Its calorific value is higher and ash clinkers at higher temperature. It is suitable for making coke.

Singrauli coalfield situated in M. P. on the border of U. P. is becoming the largest coalfield of the country. An estimate puts its reserves at 9,000 million tonnes. It is mined mostly by opencast mining, which is a great advantage. In this coalfield Jhingurda and Gorbi projects are already producing large quantities of coal. Bina and Jayant mines have also been developed. Bina has a reserve of 130 m. t. It has placed U. P. on the coal map of India.

Coal from Singrauli coalfield has high ash content and most of it cannot be transported economically to long distances. Therefore, several super thermal power plants have been built in this area so that coal has not to be transported to a long distance and electricity is distributed over large parts of the country.

The Tertiary coal is much younger than Gondwana coal and is found in comparatively small amounts in widely separated regions like Assam, Jammu and Kashmir, Rajasthan and Tamil Nadu. The seams are quite thin, being 0.5 to 3 m thick. The coal being younger in age is mostly of lower rank like the lignite. Due to the geological upheavals, some coalfields of Jammu and Kashmir and Assam are of higher ranks. These have low moisture (< 3%) and ash (2 to 10%) contents. The Tertiary coal has higher sulphur content–Assam coals have up to 7% sulphur–making them unsuitable for metallurgical purposes.

5.4.10 Grading of Coal

The basis of grading is different from classification of coal. Classification is, generally, based on the origin of coal or the rank, which indicates how far coal formation has proceeded. The objective of grading is fixing its price for marketing. In India up to June 1975 low moisture coal was graded on the basis of ash content and high moisture coal on the basis of ash and moisture content. Coal containing more than 2% moisture is regarded as high moisture coal. This system did not take into account the calorific value of coal which was assumed to be practically the same in a particular grade. Actually this has been found to vary by 10% or so even in a particular grade. As coal is primarily used for heating purposes, a new grading system has been introduced from July 1, 1975. In this system *useful heat value* of coal has been incorporated in the basic system which was in force till that time. This new system is given in Table 5.2.

Table 5.2

Grading Criteria	Grade	Useful heat K cal/kg
NON-COKING COALS		
(a) *High moisture*:		
Ash + Moisture percentage		
(i) Below 17.5	Selected A	6135
(ii) 17.5 to below 19	Selected B	6135-5930
(iii) 19 to below 24	Grade I	5930-5240
(iv) 24 to below 28	Grade II	5240-4600
(b) *Low moisture:*		
Ash percentage		
(v) Below 15	Selected A	6620
(vi) 15 to below 17	Selected B	6620-6350
(vii) 17 to below 20	Grade I	6350-5940
(viii) 20 to below 24	Grade II	5940-5340
(ix) 24 to below 28	Grade III A	5340-4710
(x) 28 to below 35	Grade III B	4710-3800
COKING COAL (Bengal & Bihar)		
Ash percentage		
Below 13	Grade A	7110
13 to below 14	Grade B	7110-7010
14 to below 15	Grade C	7010-6880
15 to below 16	Grade D	6880-6760
16 to below 17	Grade E	6760-6605
17 to below 18	Grade F	6605-6480
18 to below 19	Grade G	6480-6350
19 to below 20	Grade H	6350-6190
20 to below 24	Grade HH	6190-5560
OUTLYING FIELDS (M, P., Orissa, Maharashtra; Gujarat)		
Ash + Moisture percentage		
Below 19	Selected Grade	5960
19 to below 24	Grade I	5960-5170
21 to below 28	Grade II	5170-4570
28 to below 35	Grade III	4570-3770

(Coals of Assam and Andhra are not graded but Andhra coal may be grouped under "Outlying fields". For Assam coals also useful heat value is calculated as in other cases)

Out of the gross calorific value of coal, as determined in a bomb calorimeter a certain amount, which depends on the ash content of coal, is not available for useful purposes. As the percentage of ash increases the amount of heat not available increases. Accordingly certain deductions, as summarized in Table 5.3, are to be made from the gross calorific value for different percentages of ash to obtain the calculated useful calorific value.

Table 5.3 Deductions to be made from gross calorific value to obtain the useful calorific value of coal.

Ash %	Deduction K cal/kg	Ash %	Deduction K cal/kg
2	20	30	1030
5	65	35	1200
10	145	40	1370
15	260	45	1540
20	480	50	1710
25	790		

In order to preserve and make proper use of coal resources of the country, the Coal Controller allots coal of certain grades to the consumers. The consumer is not allowed to use coal of a grade higher than that fixed for him by the Coal Controller. The consumers are classified according to the industry to which they belong and are further subdivided on the basis of the use to which the coal is finally put. For example, the railways are allotted Selected A non-coking steam coal for mail and express trains, Selected B non-coking steam coal for passenger and through goods trains, and so on. The boilers using pulverized coal are allotted Grade III A and III B. Such restrictions are imposed so that superior coal is not used for purposes for which inferior coal can be used. Thus superior coal, which is in short supply, is reserved for specialized use.

With a view not to exhaust the limited reserves of coking coal a limited amount of coke is being imported.

5.5 Analysis of Coal

The composition of coal may vary widely from one mine to another, hence it is important to analyse and be able to interpret correctly the results from the point of view of fixation of price for each grade and also for proper utilisation of coal. The results are often expressed in terms of the following three bases:

(i) basis of fuel as used or as received.

(ii) moisture free basis, and

(iii) moisture and ash-free basis, or on the basis of combustible matter.

The analysis of solid and liquid fuels is expressed on a percentage weight basis and of gases on percentage weight or volume basis. Gas analysis should be taken to be on a volume basis unless specifically stated that it is on a weight basis. Of course while stating the calorific value of a gas on the volume basis the temperature and pressure of the gas needs to be stated.

Coal analysis is of two types :

(i) Proximate analysis and (ii) ultimate analysis. All analysis of high moisture (more than 2% moisture) coal is to be carried out on 72 mesh sample after equilibrating for 48 hours at $40 \pm 2°C$ and relative humidity $60 \pm 2\%$.

5.5.1 Proximate Analysis

Proximate analysis is the simplest form of analysis of coal and gives information which is useful in the practical utilisation of coal. It consists in the determination of moisture, ash, volatile matter and fixed carbon. As a matter of fact, proximate analysis alone with sulphur content and calorific value gives enough information regarding coal for the requirements of most of the industries.

Moisture is determined by heating 5 g of finely powdered coal at 105° for an hour. The loss in weight is reported as moisture on a percentage basis. The moisture present in a fuel evaporates in the furnace and for this purpose some heat is needed and consequently the efficiency of the fuel is reduced. Moreover during transportation, freight has to be paid on the increased weight due to the presence of moisture in coal. Therefore the presence of moisture is undesirable.

Ash determination can be made on the same sample after determining moisture. The sample is burnt at $700\text{-}750^o$ in a muffle furnace until a constant weight is obtained. Ash content is calculated from the weight of the residue. Ash is highly undesirable for several reasons. It not only reduces the heating value of the fuel but also creates cleaning and disposal problems. In metallurgical usages it is still more undesirable as it adds to the impurities. Freight charges have to be paid for its transportation even though indirectly. Indian coals, in general, have a high ash content.

For determining *volatile matter* content about I g sample is taken in a platinum crucible with a properly fitting cover. It is then heated at $950 \pm 20^{\circ}$ for exactly 7 minutes. The percent loss in weight minus the percent moisture is reported as the percent volatile matter. High volatile matter content means that a larger portion of the fuel will be distilled and burned as a gas or vapour. For engineering purposes this means that a larger combustion space will be required for complete combustion. For example, in a locomotive boiler, use of coal rich in volatile matter is not recommended because there is a shortage of space and a large combustion space cannot be provided, consequently much of the volatile matter will escape unburnt.

Fixed carbon is reported as the difference between 100 and the sum of the percentages of moisture, ash and volatile matter of the fuel. This helps in designing the furnace and the shape of the fire box because it is the fixed carbon that burns in the solid state.

5.5.2 Ultimate Analysis

Ultimate analysis includes ash, carbon, hydrogen, sulphur, nitrogen and oxygen. For good comparison and accuracy, the calculations should be on an ash free basis but very often ash is also included in the analysis though it is not necessary to include it. The ultimate analysis is required for establishing a detailed and accurate heat balance for the equipment which the fuel is used, otherwise for most engineering purposes proximate analysis, determination of calorific value, and sometimes sulphur content are sufficient. The various elements, are estimated by methods similar to those adopted for the common organic analysis and only a brief outline is given here.

Carbon and *hydrogen* determination is made on a 0.2 g sample. It is necessary, therefore, to obtain a very representative sample. The sample is burnt in a current of oxygen in the combustion apparatus commonly used in the organic analysis. As coal may contain sulphur and also chlorine, it is necessary to use a silver gauze heated to about 800° to retain the oxidation products of these elements. Carbon and hydrogen are converted into carbon dioxide and water which are absorbed in weighed potassium hydroxide and calcium chloride tubes and the percentages of the elements are calculated from the results.

Nitrogen is determined by the Kjeldahl method. About 1 g of powdered coal is heated with concentrated sulphuric acid along with potassium sulphate and mercuric sulphate. After the solution has become clear it is treated with caustic alkali and the liberated ammonia is distilled over and absorbed in a known volume of standard acid solution. From this the weight of ammonia liberated is obtained and then the percentage of nitrogen in the coal is calculated.

Sulphur is often determined on the washings obtained from a sample of coal used in a bomb calorimeter for the determination of calorific value.

However, it can also be determined separately. The washings are treated with barium chloride solution when barium sulphate is precipitated and the percentage of sulphur in the coal is then calculated.

Ash determination is carried on as in proximate analysis. The sum of the percentage of carbon, hydrogen, nitrogen, sulphur and ash is subtracted from 100 and recorded as the percentage of *oxygen*.

5.5.3 Significance of the Constituents

Carbon and hydrogen add directly to the heating value of a fuel and are certainly most desirable. The amount of carbon depends on the type of coal and increases with rank from lignite to anthracite. Hydrogen is mostly associated with the volatile matter and hence it affects the use to which the fuel is put.

Nitrogen is found in small amounts only, say 1%, in coal and does not contribute any useful property.

Sulphur, when present as sulphide, contributes to the heating value but it has several undesirable effects. It is highly undesirable in coal which is to be used for making coke for the iron industry as it is transferred to the metal and adversely affect the properties of iron and steel. Even otherwise the oxidation products, especially when moisture is present, have corrosive action on the equipment and also cause atmospheric pollution.

Oxygen content decreases as the coal becomes more mature. It is not desirable as it reduces the calorific value. As the oxygen content increases, the capacity of coal to hold moisture increases and the caking power decreases.

Phosphorus is undesirable in metallurgical coal because it is transferred to the metal and badly affects the properties of the metal.

Moisture reduces the calorific value. Some of the practical aspects of moisture in coal have been discussed in the chapter on Combustion. It promotes corrosion of storage equipment.

Volatile matter in coal affects the furnace design. Higher the volatile matter content the larger is the combustion space needed. Coal with lower volatile matter content burns with a shorter flame. Volatile matter content is of special significance in coal gas manufacture and in carbonization plants, specially when by-product recovery is the main objective.

Ash adds to the handling and disposal costs. It prevents complete combustion of carbon. It may be intimately mixed within the mass of coal (called fixed or inherent ash) or may be present in definite layers in the coal (called free or extraneous ash). The inherent ash cannot be removed by washing whereas extraneous ash can be washed. Indian coal is rich in ash.

The ash found after burning coal is in the form of oxides. It consists mostly of silica and alumina and also oxides of iron, magnesium, calcium, etc. Its composition is of great importance in metallurgical operations because it affects the slag and metal composition and consequently is an important consideration in selecting the flux.

When coal is to be used in a boiler, the fusion temperature of the ash is of great significance. If the constituents of the ash fuse at the working temperature, clinkers (i.e., lumps of ash) begin to form on the grates and the primary air supply is reduced. Coal particles may also get embedded in the fused ash and thereby cause loss. Fused particles of the ash may stick to the tubes of the boiler and reduce heat transfer. In common boilers, it is desirable that the fusion temperature of the ash be high. In the U. S. A. ash with fusion temperature above 1430° is known as refractory and below 1200° as fusible ash.

Ash with a low melting temperature forms molten slag which is absorbed in the pores of the refractory lining of the boiler furnace. The coefficients of expansion and contraction of the ash and the refractory are significantly different with the result that the life of the refractory is reduced and spalling, (i.e., breaking into pieces) of refractory results.

5.6 Calorific Value

As the most important use of a fuel is in heating, the calorific value determination is of importance. The *gross* or *higher* calorific value of a fuel is the amount of heat, expressed in calories per gram, which is liberated when the fuel is completely burnt in an enclosed vessel at constant volume in the presence or excess of oxygen, the products of combustion having been cooled to the room temperature. The heat evolved due to the formation of nitric and sulphuric acids during combustion are subtracted from the heat evolved. In such a case, water vapour produced by the combustion of hydrogen and evaporation of moisture will condense and will also give heat. This is what happens in calorific value determination in the laboratory but in actual use water vapour is not condensed; it escapes along with hot combustion gases, and a lesser amount of heat is available. The *net* or *lower* calorific value is the gross calorific value minus latent heat of the water vapour. In the net calorific value it is assumed that the products have cooled to room temperature (15°) but the water vapour has not condensed.

Dulong proposed a formula for the calculation of the calorific value from the chemical composition. Several modifications to it have been proposed, one of the formulae is:

$$\text{Gross C. V.} = \frac{1}{100}\left(8{,}080\ \text{C} + 34{,}500(\text{H} - \text{O}/8) + 2240\ \text{S}\right)\ \text{cal/g.}$$

(C, H, O and S are percentages of these elements in coal.)

In the formula oxygen is assumed to be in combination with hydrogen and only surplus hydrogen is available for giving heat. Calorific values can easily be converted from one unit to another as follows :

$$1 \text{ cal/g} = 1 \text{ K cal/kg} = 1 \text{ G.H.U/lb} = 1.8 \text{ B.Th.U/lb.}$$

5.6.1 The Bomb Calorimeter

Several manufacturers have put on the market bomb calorimeters differing slightly in design. The simple sketch, given in the Fig. 5.1, shows only the bomb of a calorimeter used for determining the calorific values of solid and liquid fuels.

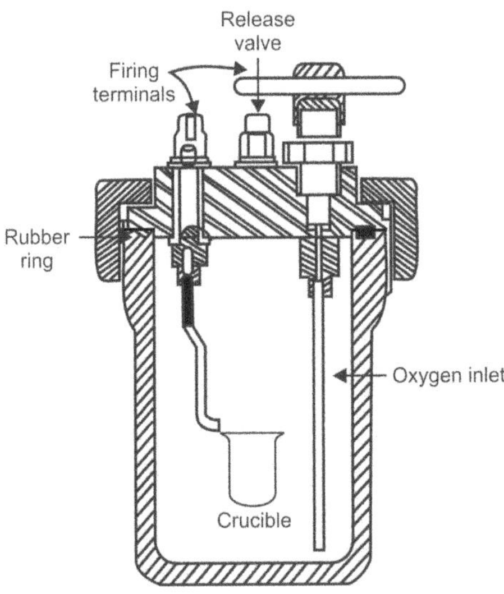

Fig. 5.1 Bomb for determining calorific values of solid and liquid fuels.

The calorimeter has a carefully made, strong cylindrical bomb in which the combustion takes place. It is made of stabilized austenitic steel which is resistant to corrosion and can withstand a pressure of at least 50 atmospheres. The bomb has a cap which is screwed to the body making a perfect gas-tight seal. Two stainless steel electrodes are fitted to the cap. One of these is a tube through which oxygen is introduced. A small ring is attached to this electrode on which the crucible containing the fuel is supported. A valve for releasing the pressure is also provided. The capacity of the bomb is about 300 ml. The bomb is placed in a calorimeter which is surrounded by an outer jacket to prevent heat losses. A stirrer is provided for stirring water in the calorimeter.

Procedure. About 1 g of the fuel is weighed into the crucible and a weighed piece of firing wire is stretched between the electrodes in such a manner that it is in close contact with the fuel so that it may be ignited. Often the powdered coal is made into a small pellet with the fuse wire inserted in it, weighed and placed in the crucible. 2 ml of water is poured in the bomb so that the products of combustion of sulphur and nitrogen may be absorbed in it. The cap is screwed down on the bomb and oxygen is filled in up to a pressure of about 25 atmospheres. The bomb is then placed in a weighed amount of water taken in the calorimeter. Electrical connections are made, stirring is started and temperature reading is taken with a thermometer reading to 0.01^o. When the thermometer shows a steady temperature, the electrical contact is made to fire the fuel and temperature readings are continued for 5 minutes after the maximum temperature is attained. The water is stirred during the experiment. The bomb is then removed and allowed to stand so that the acid mist may settle down. The pressure is slowly released and the contents of the bomb are carefully washed, the washings are collected and analysed for sulphuric and nitric acids formed. As sulphur and nitrogen do not give sulphuric acid and nitric acid in actual practice, correction needs to be made for the heat given out in their formation. Heat from the fuse wire has also to be taken into account and cooling corrections have to be made. The calorific value is then calculated from the following formula:-

$$\text{Calorific value} = \frac{(T+t) \times (W+w) - \text{corrections}}{\text{Wt. of fuel in g}} \text{ cal/g}$$

Where T = observed rise in temp; t = radiation correction;

W = weight of water; w = water equivalent of calorimeter.

Correction includes heats of formation of nitric and sulphuric acids and heat given out by the fuse wire while igniting the fuel.

5.6.2 Calorific Intensity or Theoretical Flame Temperature

This is defined as the temperature that would be attained if the fuel burned completely in the theoretical amount of air and the heat produced was totally utilised in heating the products of combustion. This temperature is never attained in actual combustion, but it represents the maximum temperature to which a material could be heated under ideal conditions by burning a particular fuel. This is important in judging the maximum possible efficiency of an internal combustion engine (Carnot cycle efficiency), the reactions in metallurgical furnaces, and rate of heating in any operation. The calorific intensity is calculated by dividing the heat of combustion of fuel by the heat capacity of the products of combustion.

The temperature actually attained in the furnace depends upon several factors like the weight of air supplied, heat losses and the specific heat of the products. It has been estimated that if bituminous coal is burned in 25% excess air, theoretically, the temperature will be 2200° but if 100% excess air is supplied the temperature will be only 1500°. For obtaining good heat transfer and for generating high pressure steam it is necessary to obtain high temperature in the furnace. The temperature can be increased by heating the air supplied.

5.6.3 Pulverized Coal

The rate of combustion of solid fuels is slow because of the difficulty of contact between the fuel and oxygen. The combustion rate can be increased by increasing the rate at which air is forced in but this means a supply of excess air resulting in waste of heat. The combustion rate can also be increased by finely powdering or pulverizing the fuel so that air and fuel easily come in close contact. Generally high volatile coal is more satisfactory for making pulverized coal. The volatile matter is quickly liberated and burnt, and thus helps the fixed carbon to burn. Pulverized fuel has some of the advantages of liquid or gaseous fuels.

Coal is dried and powdered finely. It can then be transported by screw conveyers or by a stream of air. Thus the chief advantages of pulverized coal are :

1. A lower percentage of excess air is necessary for obtaining complete combustion; thus there is less waste of heat, higher temperatures are obtained, and thermal efficiency is higher.
2. The rate of combustion can be changed as easily as in the case of a liquid or gaseous fuel. Heating of the furnace can be started or stopped as and when needed; thus fuel is saved.
3. Low grade coal with high ash content can be used. This is important from the point of view of Indian coals because of the high ash content of many coal deposits.
4. Labour charges are lower.

The main disadvantages of pulverized coal are the cost of pulverizing and as the ash is in a finely divided state, about 75% of it goes along with the flue gases and is discharged on the surrounding area. These disadvantages have been overcome to a considerable extent by developed methods of pulverizing and the use of dust catchers for the removal of ash from the combustion gases. On the whole the advantages are greater than the disadvantages.

Coal with low melting ash is not very satisfactory for using in the powdered state because the fused ash would be deposited on the boiler tubes and interferes with heat transfer. The properties and behaviour of ash have already been discussed.

5.6.4 Selection of Coal

Broadly speaking the use of coal can be divided into the following categories, on the basis of uses:-

(i) power production,

(ii) direct or indirect heating,

(iii) manufacture of other fuels, mostly manufacture of coke and coal gas.

For different uses the factors to be taken into consideration vary to some extent. Factors for coking are considered under the Section on coke. For the selection of coal for other uses the following are the main factors.

(i) *Calorific value* should be high so that a small quantity of coal can give a large quantity of heat, thereby reducing handling and storage costs also.

(ii) *Ash* content should be low. Ash content not only means added handling, storage and disposal costs but reduction also in heating. Each percent of ash means a heat loss of 1.5%. For most purposes the ash should not have a low melting point.

(iii) *Moisture* should be as low as possible because it is paid for at the same rate as coal. It also reduces heating value.

(iv) *Size* should be as uniform as possible as it facilitates handling and regulation of combustion. Smaller particles mixed with larger lumps interfere with the movement of the gases.

(v) *Caking* properties are undesirable except for making coke. For most other purposes non or weakly caking coals are preferred.

(vi) *Quality* of coal received in various installments should be as uniform as possible so that its behaviour while burning does not change. Frequent changes in quality make it necessary to reconsider the problems of combustion space, behaviour of ash, air supply, etc.

5.6.5 Cleaning of Coal

On mining, coal of various qualities is obtained and is separated into several types and sizes for marketing. Indian coals contain a considerable amount of ash. Coking coal, i.e. coal suitable for making coke, is in short supply. Importance of coal cleaning is clear form the fact each percent rise in ash content decreases the output of a blast furnace by 3 to 4%.

As coals are being used up; mines of poorer quality coal are left behind. Further, due to use of highly mechanised mining methods, coal of inferior quality gets mixed up. Indian coking coals contain 26 to 28% ash whereas the metallurgical industry requires coking coal with not more than 15 to 17% ash. On making coke, because of the removal of volatile matter the ash percentage in creases. Due to large reserves of coal of very poor quality, other large consumers like the thermal power plants, railways and cement factories have started considering the use of washed coal. Thus need for coal washing i. e. removal of ash from raw coal by physical, physico-chemical or mechanical treatment, has increased.

In washing coal, advantage is taken of the difference in specific gravity between clean coal and mineral matter. The average specific gravity of good coal is about 1.5, of average coal is 1.65 and of waste coal with 25 to 50%, combustible matter is about 2. There are several ways by which this separation can be done. One method is to treat coal of proper size in a Baum Jig in which pulsation is created in water by air under pressure. The whole mass moves forward. Coal with less ash content forms the upper layer and moves out through the coal outlet. Coal of inferior quality, with a large amount of minerals, forms the lower layer and is removed separately.

The gravity separation method is more common. An aqueous solution of calcium chloride or a suspension of a solid, such as sand, magnetite, baryta, etc., is made in water. The specific gravity is carefully adjusted. Coal is agitated with the solution so that the clean coal floats and dirt sinks. The inferior coals which have a large amount of finely dispersed ash are treated in cyclone washers. The coal is crushed to size below 1 cm and subjected to strong centrifugal force in the cyclone washers. Froth flotation process has also been adopted. In all these cases coal has finally to be washed with water. Special care has to be taken in removing unnecessary water, otherwise much of the advantage of cleaning the coal will be lost. In an experiment with coal having about 15% ash, use of liquid of sp. gr. 1.5 gave a yield of 82% of coal with an ash content of 4.5%.

In the coal washery of Hindustan Steel Ltd., at Dugda, Hazaribagh Dist., coal as received is screened, and coal above 7.5 cm in size is crushed to less than 7.5 cm size. The coal is then fed to a Baum Jig Washer having a suspension of sand in water (sp. gr. 1.7 to 1.8). The coal floats and the heavier impurities settle and are rejected. The clean coal is separated into three different sizes, viz. 7.5 to 2.5 cm, 2.5 to 0.6 cm and below 0.6 cm. Coal below 0.6 cm in size is not washed because it has low ash content. The other two sizes are treated in a liquid of sp. gr. 1.45 to 1.55. An idea of the final results can be had from the following annual rated yield:

(a) 1.44 million tonnes of clean (17% ash);

(b) 3.70 million tonnes of middling (17 to 35% ash),

(c) 0.26 million tonnes of rejects (approximately 50% ash).

5.6.6 Storage of Coal

Coal should be stored for the minimum period necessary because, besides locking up of the capital and deterioration in quality, there is the danger of spontaneous combustion leading to outbreak of fire. Generally, the more mature the coal is the less are the chances of its spontaneous combustion. Accordingly, unless considerable precautions are taken, it is risky to store dried lignite. Storage of anthracite coal may be considered to be quite safe.

The primary cause of the self-heating of coal is oxidation. Oxidation takes place very easily just after mining because of the presence of hydrocarbons on the surface. It is advisable to keep coal for about six weeks before transporting it to the consumer. The oxidation process is exothermic and if heat is not removed the temperature rises with the consequent increase in the rate of reaction and a point comes when the coal begins to burn. This process is known as *spontaneous combustion* and is helped by the following factors :

1. *Fineness of the coal.* The finer the coal, the larger is the surface area exposed on which oxidation can take place. If fine coal is mixed with larger pieces of coal, the gaps are filled up and cooling air does not circulate. This is quite risky.
2. *Sulphides.* Sulphides are easily oxidized giving heat. The oxidation products occupy more volume, therefore the coal is broken into smaller pieces. This increases the surface area, thereby promoting further oxidation. Previously the presence of sulphides was considered to be an important factor in promoting spontaneous combustion but now it is not considered to be so important.

3. *Temperature.* At higher temperatures the reactions proceed faster. Immediate attention should be paid if the temperature has gone up to, say, 90°. Coal should not be stored against walls which get heated like the walls of the boiler house.

The best way of storing coal is to store it under water or in airtight bins so as to exclude oxygen. In many cases this is found to be difficult. An alternative solution to the problem is to have free air flow throughout the stack so that heat is carried away. Coal may be stored in an open space in piles not higher than 2 meters. Care should be taken not to mix fine coal with bigger lumps. It has been suggested that sized hard coal can be piled up to 7 meters and anthracite coal up to 10 meters without undue risk.

If a hot spot or fire is located in a coal pile, the portion may be isolated and coal spread out. Water should not be used as it might quench the fire but it will promote conditions which will cause heating within a short period.

5.6.7 Coking Coals

Coals which on heating soften, become plastic and fuse together giving large coherent masses are known as *caking* coals. If the residue, i. e. the coke, is porous and strong it can be used on an industrial scale; then the original coal is known as *coking* coal. All coking coals are caking but all caking coals are not coking coals. Because the largest industrial use of coke is for metallurgical purposes, it is implied, by general usage, that coke should be suitable for metallurgical purposes. Coals which have little or no fusing action are called *free-burning coals.*

In India coke is manufactured mainly for the iron and steel industry. Iron ore reserves of India are estimated to be about 21,000 million tonnes whereas coking coal estimates are for 2,000 million tonnes and are totally insufficient for the needs of the country. Therefore, low grade coking coal is washed and ash content reduced to below 17% and mixed with weakly coking coal.

Accordingly, coking coal should have as low ash (< 17%), sulphur and phosphorus contents as possible. It is noteworthy that coke in most cases will come in direct contact with the metal; hence its impurities are likely to be transferred to the metal, especially sulphur and phosphorus which badly affect the properties of the metal. Ash will have to be removed as slag for which purpose more flux will need to be used; thus the output capacity of the furnace will be further reduced. Besides this, the ash also has to be heated to a high temperature and there is wastage of heat. The best results are said to be obtained with coals having 83 to 90% carbon content.

5.7 Coke

5.7.1 Properties of Metallurgical Coke

Coke is the residue left on heating bituminous coal out of contact with air. Its largest industrial use is in the metallurgical industry, especially in the blast furnace. For such purposes the important properties are the following :

1. *Purity*. Ash, sulphur and phosphorus contents should be as low as possible. Coke is an important source from which sulphur is transferred to pig iron in the blast furnace. Ash has to be removed as slag. It is estimated that 2% of coke is required for the removal of 1% of ash.
2. *Porosity*. Coke should be porous so that oxygen can easily come in contact with the carbon of the coke and the rate of combustion is high.
3. *Strength*. During handling and also while dropping into the furnace if coke breaks into fine particles, the flow of gases will be retarded. In a blast furnace coke rubs against other hard materials, therefore it should be strong and resistant to abrasion.
4. *Combustibility*. Obviously coke should burn quickly. This property depends upon several factors, e.g. quality of coal used, coking temperature, coking time, etc.

Coal does not possess as much porosity and strength as coke. If coal is used in a blast furnace, it will be crushed under the load. By coking much of the sulphur is removed but the ash content in creases. Because of these reasons coke is preferred to coal for a blast furnace. Due to the expulsion of volatile matter, coke burns with a short flame, and is unsuitable for a reverberatory furnace.

5.7.2 Carbonization Process

Carbonization means heating or distilling coal out of contact with air. This produces a solid residue, called coke, and liquid and gaseous products of distillation. This process is also known as coking. Carbonization is carried out mostly for making metallurgical coke. In countries which have poor petroleum resources gas making is also an important objective of carbonization. Carbonization is also used to make smokeless fuel for domestic purposes. Ammonia liquor, crude benzene, and crude tar are the other by-products of carbonization.

On heating out of contact with air when the temperature of bituminous coal reaches a little over 100°, water and absorbed gases are given off. The decomposition of coal beings at about 200°. First oxygen is removed in the form of carbon monoxide and carbon dioxide, the hydrogen sulphide and olefin gases are liberated followed by the appearance of oily products. Around 350°, the active decomposition of coal begins and the evolution of plenty of gas and tarry vapours can be seen. This is also accompanied by the softening of coal, and the tarry viscous materials produced gradually dissolve the remaining mass. This process is completed when the temperature reaches, 450°. At this stage the mass is in a plastic state and the escaping gases bubble through it and thus the mass solidifies into a porous structure which is a characteristic of coke.

Coke is a bad conductor of heat. Therefore the movement of the plastic zone slowly occurs from the heated surface to the interior. It is well marked and proceeds at the rate of 1 to 2 cm per hour. Thus if a 40 cm thick bed is heated from both sides the coking time required will be about 15 hours. It is reported that there may be a temperature drop of 200° across a centimeter thick layer of coke. Most of the gases while escaping pass through the porous coke, and due to the higher temperature of the coke further decomposition occurs giving various organic compounds.

It is because of this slow rate of heating that coke ovens are built only 40 to 45 cm wide, 3 to 6 m high and about 10 to 12 m long so that there may be proper heating and some sizeable quantity of coke is produced in each oven.

The quality of coke is influenced by various factors, the main factors being the quality of the coal carbonized, the rate and the temperature of carbonization.

Because of the slow rate of carbonization and of heat conduction of caking coals it becomes necessary to suitably blend non-caking coal with caking coal. Non-caking coals neither soften on heating nor give coherent coke but have higher rate of carbonization and heat transfer. Therefore a mixture of two kinds combines the advantages of both besides helping to overcome the shortage of coking coals.

Coal carbonization may be carried out at low-temperature (i.e. 500° to 700°) or at high temperature (900° to 1100°). Low temperature coke contains 5 to 15% volatile matter and is week for metallurgical purposes. As it ignites easily, it is more satisfactory as domestic fuel. Another objective of low-temperature carbonization is to get high calorific value gas. High-temperature carbonization gives coke which is strong and can withstand load and abrasion in a metallurgical furnace.

5.8 Manufacture of Coke

5.8.1 Beehive Coke Ovens

Especially where the consumption of coke is not very large, it is made in beehive coke ovens. The shape of the oven is like a dome. An oven for 5 tonne coke using about 8 t of coal is about 2.5 meters high and 4 meters in diameter. It is made of brick work and has an opening at the top for charging the coal. A door is provided at the side through which the coke is removed. A number of such ovens are operated in series so that the heat of each oven helps to heat the adjoining ovens. While the oven is still hot from a previous operation, the coal is charged. The side door is then closed leaving a small opening through which a limited amount of air enters for the combustion of volatile matter within the dome. The heat of the oven liberates the volatile matter which burns within the dome, thereby further heating the charge. The evolved heat is sufficient to distil the volatile matter and also to fuse the coal. The time taken for completion of the entire operation is about 3 days. The coke is cooled by spraying water and then taken out through the side door.

On account of the very large capital investment in the by product coke oven plants, there seems to be a revival of interest in beehive plants.

5.8.2 By-product Coke Oven Practice

A modern coke oven is a long, narrow, rectangular chamber made of silica bricks, about 10 to 12 meters long, 3 to 6 meters high and 40 to 45 cm wide. It tapers slightly (about 4 to 10 cm) towards the discharge end so that coke can be easily pushed out. Mechanically operated cast iron doors lined with fire bricks are provided at both ends. The oven is provided at the top with an ascension pipe connected to a collecting main. There are 3 or 4 openings at the top, as shown in Fig. 5.2, through which blended coal of the proper size (0.5 to 0.3 cm) is fed. Every oven is separated from its neighbouring one by a vertical flue in which gases burn. Thus each oven is heated from both the sides. The ovens are built in a block, consisting of ovens with alternating flues for heating. Such a block is known as a coke oven battery and has 20 to 100 ovens. Each oven holds about 20 tonnes of coal and the yield is about 70%. The time taken for coking varies from 12 to 20 hours and the temperature is 1100° or more. The various operations involved in manufacture of coke have been schematically shown in the Fig. 5.3.

Underneath the battery there is a regenerative system which consists of chambers filled with a chequer work of firebricks. The hot gas leaving the flues heats a set of chambers while air and fuel gas pass through other chambers which have already been heated. After some time the directions are reversed and the waste hot gas heats other chambers through which the air and fuel gas were passing. Air and fuel gas pass through the heated chambers. Thus the sensible heat of the waste gas is recovered and instead of the fuel gas burning in air at atmospheric temperature hot fuel gas burns in hot air, thereby producing a higher temperature (1300° or more) and increasing the efficiency of the process.

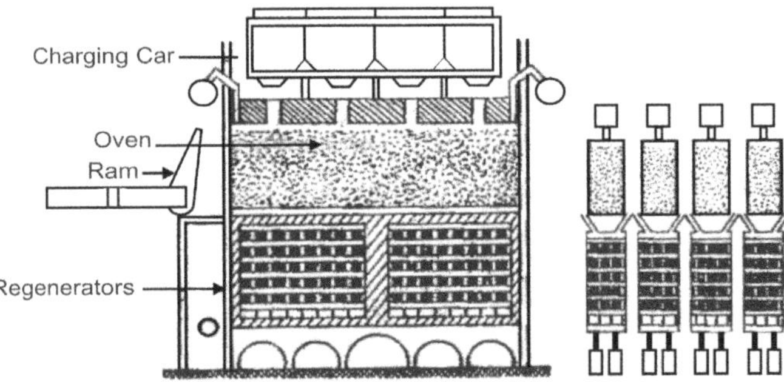

Fig. 5.2 Coke oven.

Theoretically about 40% of the coke oven gas generated is enough to heat the ovens and 60% is available for other uses. The gas is of high calorific value and can be transported to distant places and sold or used in steel-making operations. Therefore in integrated steel plants only a small portion of the coke oven gas mixed with blast furnace gas is used for heating the ovens.

After carbonization both the doors of the oven are opened and red hot coke is pushed out by a pusher or a ram, into a coke car standing at the other end. The car takes the coke to a quenching station where a spray of water cools it. Excess water is allowed to evaporate and the coke screened into different sizes for various uses, such as blast furnace, foundry and domestic requirements.

In coke oven plants in India coal containing, on as average, about 18% ash is used. It yields about 75% coke containing about 25% ash. The yield of coke suitable for use in a blast furnace is about 63% on the weight of coal charged in the coke oven.

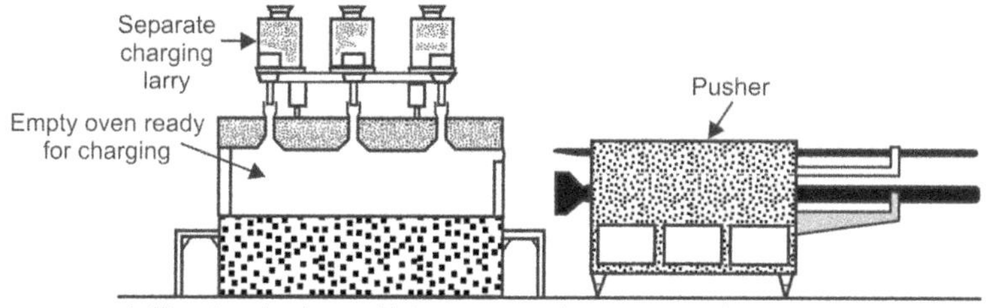

A. The charging larry, with hoppers containing measured amounts of coal, is in position over charging holes from which covers have been removed. The pusher has been moved into position

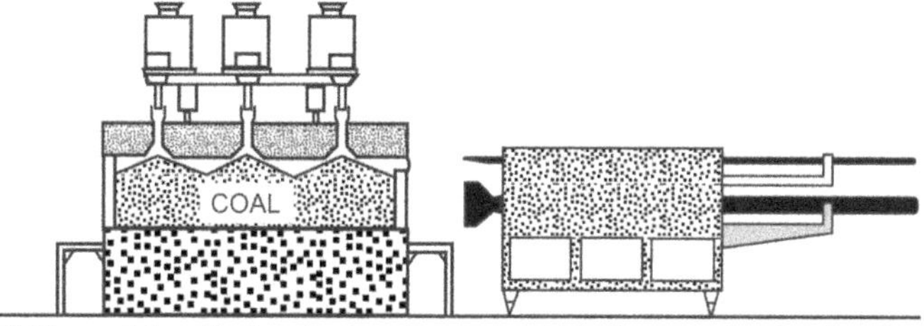

B. The coal, from the larry hoppers has dropped into the oven chamber, forming peaked piles.

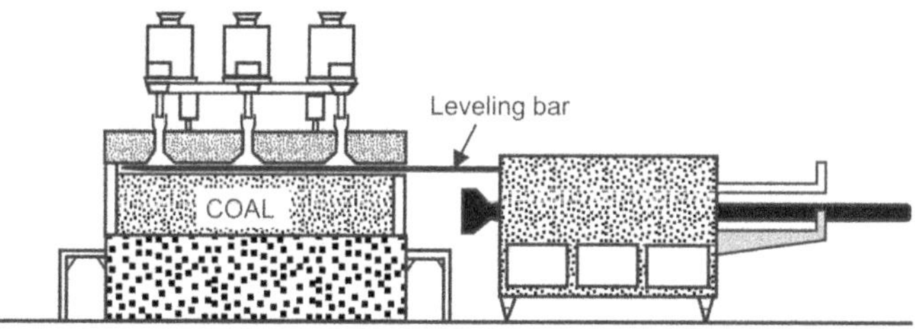

C. The leveling door at the top of the oven door on the pusher side has been opened, and the leveling bar on the pusher has been moved back and forth across the peaked coal piles to level them. The bar next is withdrawn from the oven, the leveling door and charging holes are closed, and the coking operation begins.

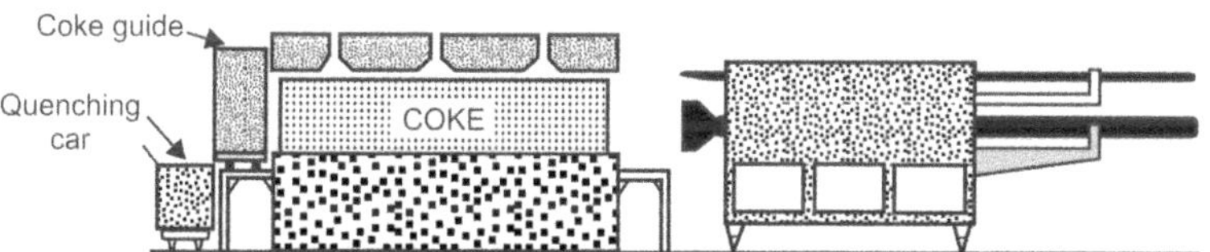

D. Coking of the coal originally charged into the oven has been completed (in about 18 hours) and the oven is ready to be "pushed". The oven doors are removed from each end, and the pusher, coke guide and quenching car are moved into position.

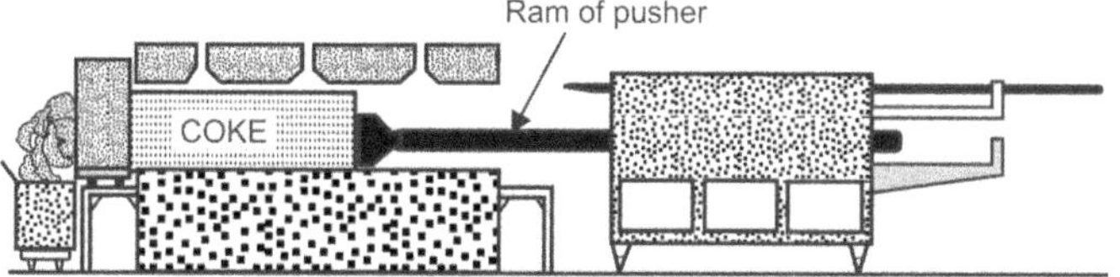

E. The ram of the pusher advances to push the incandescent coke out of the oven, through the coke guide and into the quenching car.

Fig. 5.3 Schematic representation of the sequence of operations involved in charging, leveling and pushing in one coke cycle of a by-product coke oven.

5.9 Liquid Fules

5.9.1 Sources

The main source of liquid fuels is petroleum which on distillation gives several fractions. The lower boiling fractions find use in internal combustion engines and the higher boiling fractions in Diesel engines and in oil-fired furnaces. Alcohols, tar and colloidal fuels are used to a lesser extent.

5.9.2 Advantages

The use of liquid fuels has increased in recent times because of the following advantages :

1. Transportation, storage and handling are easier. Liquid fuels can be transported by pipes to distant places and in a plant the storage may be at one place and it may be piped to points of consumption. For this reason liquid fuels can be stored in a place not otherwise useable. This is of special advantages in ships where storage tanks of even irregular shapes can be made and otherwise useless space utilised. Labour charges are also less.

2. For the same storage space the heating value of a liquid fuel is 1.5 to 2 times more than that of a solid fuel. Difference in storage space needed by liquid and gaseous fuels is great. Liquid fuels take much less storage space than is needed for the storage of gaseous fuels. This is a distinct advantage of liquid fuel over gaseous fuels.

3. Combustion can be started immediately. The rate of combustion can be changed readily and stopped as soon as heating is not needed. This results not only in better control over the process but also in the economy of fuel.

4. Liquid fuels have several other advantages, for instance, cleanliness, absence of ash, no danger from spontaneous combustion, less excess air needed for combustion, facility in refueling, etc. They also require less furnace space for combustion.

The chief disadvantages of many of the liquid fuels are that they are highly inflammable and are volatile. Of course some of the liquid fuels, like fuel oils, are not so inflammable and volatile.

5.10 Petroleum

5.10.1 Origin

The theories of origin of petroleum have been more controversial than that of coal. It is now accepted that it originated from organic matter like the bodies of fish or other aquatic and plant life. They fell to the bottom of a very calm, shallow sea where they got covered by a layer of silt. Under such conditions, the organic matter which was settling continuously decayed under almost anaerobic (i. e., in absence of air) conditions. The steady accumulation of silt above the organic matter led to an increase in temperature and rise in pressure. As with coal, bacterial action played an important role. The oil produced moved to porous sandstone and limestone rocks. For petroleum deposits to form it is necessary that there be an overlaying rock which is impervious so that the petroleum cannot migrate upwards, and also that escape in other directions be blocked by various rock formations. There are several types of rock formations for oil trap. It is easy to see that in the same deposit one could strike oil, or gas or even salt water. Often there is an accumulation of gas above the oil and salt water underneath. Any natural gas or oil moved to the surface would be evaporated and lost, leaving the high boiling fraction as a natural deposit of pitch.

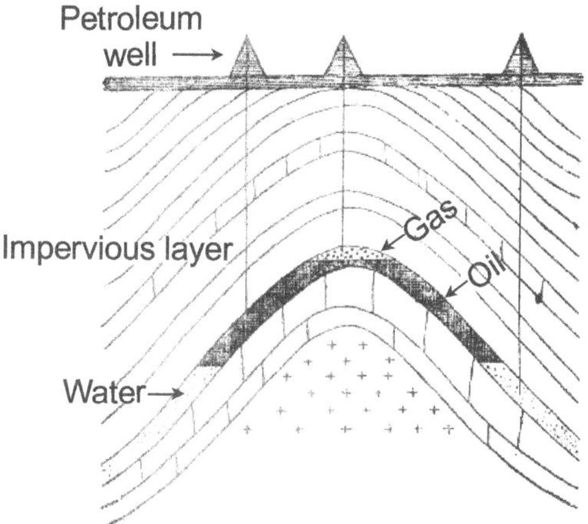

Fig. 5.4 Oil, gas and water in a petroleum reservoir.

Petroleum is brought to the surface by drilling wells. There is a possibility that the well may strike the water layer, Fig. 5.4, and no oil be brought to the surface by that particular well, or that it may strike oil or gas. Initially the gas pressure may be enough to force some of the oil to the surface but later it has to be pumped. The oil so obtained is known as crude oil.

5.10.2 Indian Resources of Petroleum and Petrochemicals

The history of petroleum in India begins with its accidental discovery in Digboi, Assam, in 1889-90. The Digboi oilfield has so far produced over 9 million tonnes of the crude oil for which about a thousand wells have had to be drilled. A few years after getting independence, extensive oil exploration began which changed the picture of the petroleum industry in India faster than could have been imagined in pre-independence days. Due to our heavy dependence on imported crude oil, the country's economy received a serious jolt in the latter part of 1973 when supplies from the Middle East were interrupted and prices were increased several times. Upto August 1977 under the auspices of Oil and Natural Gas Commission 1,468 wells were drilled. Of these 781 were found oil-bearing and 95 gas producing.

In 1981 the exploration cost, which includes cost of the seismic and other surveys and test drilling was about Rs. 8 per barrel of petroleum produced whereas the international price is more than, Rs. 200 per barrel (as compared to Rs. 15 in 1973).

After independence, the Government of India revised the petroleum policy completely. Instead of importing the more expensive refined products, it started importing crude petroleum for distillation and refineries set. With the discovery of newer oilfields, mostly in Gujarat and Assam and the construction of long pipelines for carrying crude oil and distillates to different parts of the country, the number of refineries set up in the private and public sectors and their capacities have expanded. The chief oil-producing areas are Ankleshwar, Nawagam and Kalol in Gujarat and Rudrasagar and Lakwa in Assam. Crude oil production was 0.45 million tonnes in 1960, which increased to 3 m. t. in 1965, 7.2 m. t. in 1973 and 12.8 m. t. in 1979 with import of about 14 m. t. In 1980-81 India imported 16.7 million tonnes of crude oil and about 8 m. t. of petroleum products at an estimated cost of Rs. 5,600 crores.

Shortage of petroleum led man to search for oil under the sea. In 1974 oil and gas were found in the Arabian sea, about 160 km from Bombay. This opened a new chapter and high hopes in Indian economy. Wide scale under-water search has been made but so far the main source is what is known as the *Bombay-High,* under the high and rough sea. The first exploratory well was drilled with the help of Sagar-Samrat. The Bombay-High-oilfield covers an area of about 1500 sq. km. The crude oil is brought on the process platform and most of the water is separated. Initially oil was transported by tankers but since June 1979 both oil and gas are brought by pipelines to the oil terminal at Uran on the main land of our country (Fig. 5.5). Oil with less than 5% water is pumped under high pressures through the trunk pipe lines to which are connected other pipe lines. Similarly gas is pumped under high pressure after dehydration. Laying of pipelines under the sea is a very difficult work. There is no hard and steady ground to rest the pipeline on and also there are sea currents which tend to wash it off. Moreover, sea water is highly corrosive. The under-sea technology, a very challenging technology, is still in the developing stage. From 1982, when production is reached the maximum production the Bombay High oil field is expected to produce 1,60,000 to 2,00,000 barrels per day of oil for the next 15 years. (20,000 barrels per day means 1 million tonnes per year). The off-shore oil production of 5 million tonnes in 1980 is to be increased to 12 million tonnes. The oil is associated with gas, which is called associated gas, as it is a co-product with oil. Its production is expected to reach 3 to 4 million m^3 per day. It is to be noted that production of associated gas is not under independent control as it is produced along with the oil. The Bassein oil field produces gas by itself and is known as non-associated gas and its production can be increased or decreased independently, according to requirement. Typical composition of these gases is given in Table 5.4.

Table 5.4 : Typical Composition of gas from Bombay-High.

Composition Hydrocarbons	**Associated gas % by vol.**	**Non-Associated gas % by vol.**
C_1	69.1	78.0
C_2	13.1	6.7
C_3	9.3	4.0
C_4 & higher	5.3	2.0
Others (O_2, N_2, CO_2 etc.)	3.2	9.3

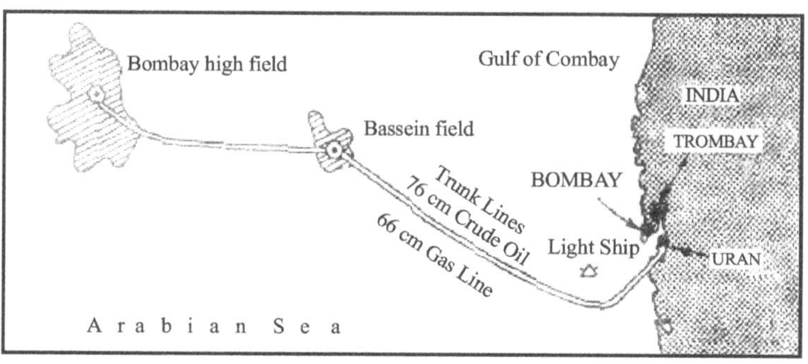

Fig. 5.5 Location of offshore oil fields.

As petroleum obtained from the Bombay-High is of low sulphur content, not much hydro-treatment of sulphur removal is needed. It gives oil with cyclic naphtha and high aromatic content. It is being processed at Koyali (Gujarat) and Mathura (U. P.) besides at the refineries around Bombay. The large quantities of the resultant benzene, toluene and xylene will help in the development of organic chemical industries.

As a very large quantity of gas became available unexpectedly there were no facilities to utilize it in any way other than burning in the thermal power plants as an interim measure. It is a wasteful process but there was no other way. The gas production from the Offshore sources is expected to be 20 million cubic meters per day for the next 30 years. Ten fertilizer plants are being built which will use methane to produce about 13,500 tonnes of ammonia and 22,500 tonnes of urea per day. Thus fertilizer production of the country will be doubled.

C_2 and C_3 fractions of the gas on cracking yield ethylene and propylene which are used for the manufacture of various synthetic organic chemicals like low and high density polyethylene, polyvinyl chlorides (PVC), ethylene glycol (to be used by the aromatic plants for making polyester fibers), polystyrene, acetone, etc. C_4 and the higher fractions are liquified and transported to the various bottling centres for supply in cylinders as cooking gas for domestic use.

The cheapest mode of transporting oil and gas is by taking them through pipelines. Several large networks of pipelines have been built in India. The Gauhati product pipeline, constructed at a cost of about Rs. 7 crores, is designed to transport petroleum products from Gauhati Refinery. Rudrasager-Gauhati-Barauni pipeline system transports crude oil from Rudrasagar oil fields to Gauhati and Barauni refineries. It brings oil from Assam through 1200 km of difficult terrain crossing many rivers. The Barauni refinery pipeline system is very long and has cost Rs. 26 crores. It is in two sections, Barauni-Kanpur and Barauni-Haldia. Barauni to Kanpur is a product line, i.e. finished products are pumped from Barauni (Bihar) to Kanpur. At Haldia port crude oil will be brought by tankers from abroad and pumped to Barauni. Gujarat pipeline system, costing about Rs. 7 crores handles gas, crude oil and refined products. Cambay-Dhuvaran, Ankleshwar-Baroda and Ankleshwar-Uttaran lines are for gas; Ankleshwar-Koyali line is for crude oil and Koyali-Ahmedabad line is a product line. Not going into details, it may be mentioned that, in 1974, there were 1990 km of pipelines for carrying crude oil with diameters ranging from 10 cm to 40 cm, 1581 km of pipeline (diameter 10 cm to 30 cm) for carrying finished products from the refineries to the marketing centres and 205 km for carrying gas (diameter 20 to 40 cm). A 1,256 km long pipeline system is under construction from Salaya in the Gulf of Kutch to Koyali and Mathura Refineries via Virumgam (Gujarat) to transport imported crude, and that obtained from the Bombay-High. The crude will be received at an off-shore terminal which is the first of its kind to be set up in the country.

In the public sector there are nine refineries. The oldest refinery is at Noonmati near Gauhati, Assam, and has an annual capacity for handling 0.88 million tonnes of crude oil. The second refinery at Barauni in Bihar has a capacity of 3 million tonnes. The third refinery constructed is at Koyali, near Baroda (Vadodara) in Gujarat with an annual capacity of 7.3 million tonnes of crude oil. The capacity of Cochin Refinery is 3.3 million tonnes. The refinery at Madras (3 mt) is specializing in the production of lubricants. Other refineries in the public sector are at Bombay, Trombay, Vishakhapatnam and Haldia. The Assam Oil Company, in the private sector operates a refinery at Digboi. One of the biggest refineries (6 mt) of the country has started production near Mathura. This site has been chosen because it is in the interior of the country and is safe from defence point of view. Also it is well connected by rail transport with different parts of the country so that the products can be cheaply marketed. Crude oil will be brought to it from Gulf of Kutch by 1200 km long pipeline.

The Petrochemical Complex of the Indian Petrochemical Corporation Limited, near Vadodara, Gujarat, was inaugurated on the 30th March 1979. It has 14 individual production units constructed at a cost of Rs. 400 crores. It is helping the production of petrochemicals, thermoplastics, elastomers, synthetic fibres and other organic intermediates for drugs, dyestuffs, and paints. The IPCL was established in 1969. The Aromatics Complex for the production of xylenes, and dimethyl terephthalate (DMT) was commissioned in 1973. Orthoxylene is used for the manufacture of phthalic anhydride which is the chief raw material for plasticizers used with polyvinyl chloride (PVC) resin. DMT is the raw material for polyester fibre and films used in textile and industrial fabrics, electrical industry and manufacture of X-ray films. The production of xylenes was about 60,000 tonnes and of DMT, 24,000 tonnes in 1978.

The construction of the Olefins Complex was started in 1974. In 1978 it produced about 2,00,000 t of ethylene, 70,000 t of propylene and 31,000 t of butadiene. Its downstream units produce a number of chemicals used in the plastics, rubber, synthetic fibre and allied industries.

It is estimated that aromatic and olefin units will ultimately yield about 3,00,000 tonnes of products which, after conversion to the final products will be sold for about Rs. 3,000 crores per annum. Some of the widely used products are polypropylene (30,000 t), low density polyethylene (80,000 t) and polybutadiene rubber (20,000 t). Details of products and their preparations are beyond the scope of this book but it is sufficient to give a glimpse into its importance.

5.10.3 Classification of Crude Petroleum

Petroleum consists of a mixture of gaseous, liquid and solid hydrocarbons together with small amounts of nitrogen and sulphur compounds. The hydrocarbons may be of the paraffinic, naphthenic (cyclo-paraffinic), olefinic or aromatic series. According to the nature of the main constituents present, the crudes are classified into three groups:

1. *Paraffin-base* petroleum mainly consists of hydrocarbons of the paraffin series. On distillation solid paraffin wax is left behind.
2. *Asphalt base* petroleum is rich in non-paraffinic hydrocarbons, like aromatic and naphthenic hydrocarbons, and on distillation asphalt or bitumen is left behind.
3. *Mixed-base has* a composition in between the above two types and on distillation some wax appears in the distillates and the residue mostly consists of asphalt.

From the standpoint of lubrication engineering, it is noteworthy that a lubricant from paraffin-based crude is of lower specific gravity, and change in its viscosity with change in temperature is less than that of a lubricant from the naphthenic series. At lower temperatures more of waxy solid separates from the paraffin-base lubricant. The petroleum base also affects the performance in the internal combustion engines, as will be presently explained.

5.10.4 Petroleum Refining

In general, crude petroleum is not marketed. Crude petroleum is distilled and refined to give several fractions which are more valuable. A landmark in the distillation process was the use of the bubble-cap distillation column in 1920. Catalytic conversions have been increasingly dominating the petroleum industry since 1940.

The crude oil is obtained in the form of a water-in-oil emulsion and the emulsion is broken up by adding de-emulsifying agents or by electrical treatment. Care is taken to remove sodium and magnesium chlorides as these corrode the refining equipment and deposit scales in the heating pipes.

The emulsion-free crude is heated in pipes in a furnace heated by burning gas or oil. The vapours are passed through a tall cylindrical fractionating column which is provided with a number of carefully designed fractionating plates, see Fig. 5.6. As the vapours move upwards, the higher boiling fractions condense and fall back and only lower boiling vapours move to higher plates where further condensation takes place and some more liquid falls back. As the freshly condensed liquid comes to the lower plates, it gets heated and vapourized and again condenses on higher plates. This process of condensation and vapourization takes place many times causing separation of hydrocarbons according to their boiling points. Thus the higher boiling fractions condense towards the lower part of the column. The lower the boiling point the higher is the position of the plate on which a fraction condenses. Pipes are led off from different points taking out mixture of vapours and liquids of the desired boiling points which are condensed separately and different Fractions are obtained. This gives *straight-run* petrol, kerosene (or white oil) and gas oil fractions. Operations from drilling to the final use are shown in Fig. 5.7.

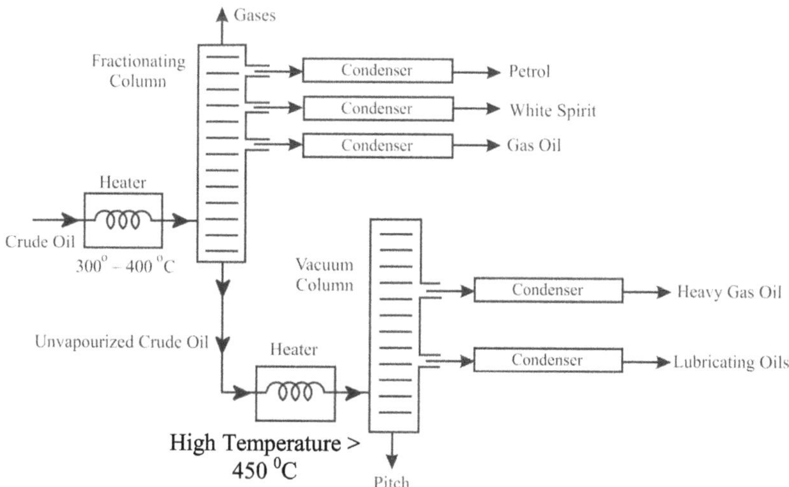

Fig. 5.6 Petroleum distillation.

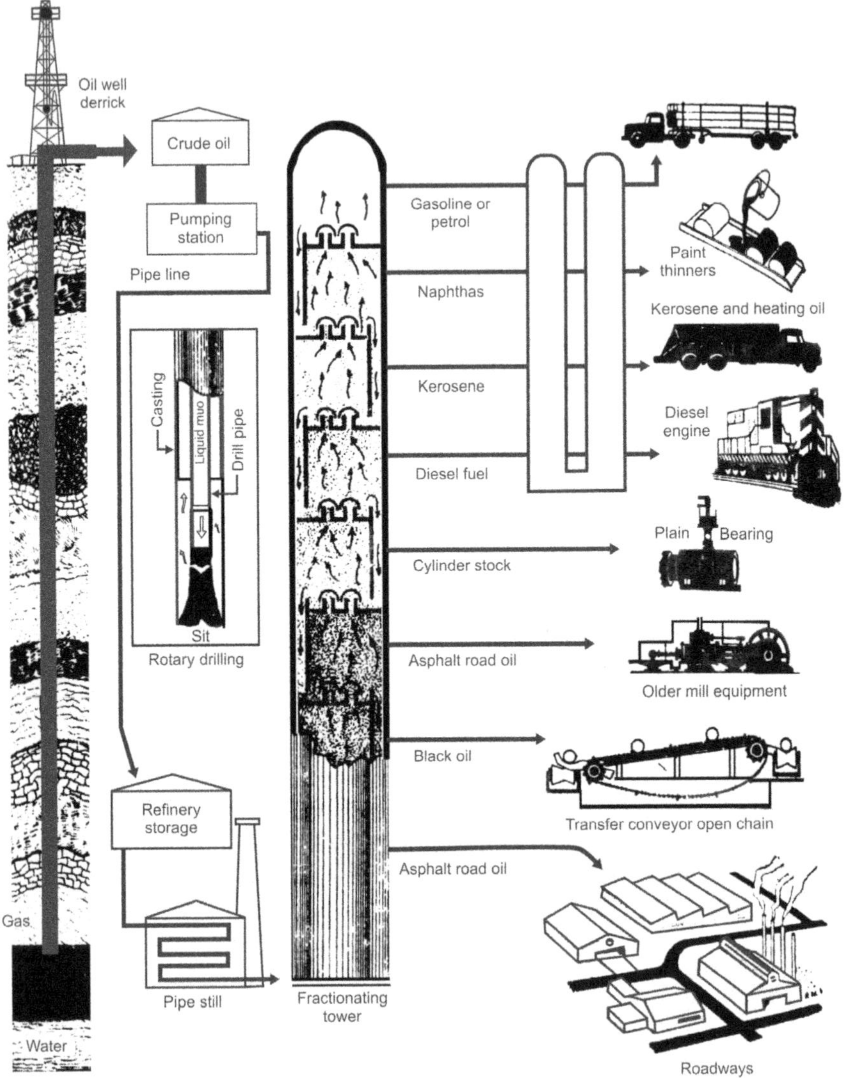

Fig. 5.7 Flow-sheet showing pumping and refining of petroleum and applications of different fractions,
(*Courtesy, U.S. Steel Corporation*)

If the entire crude is distilled at atmospheric pressure, as described above, considerable thermal decomposition of the higher boiling hydrocarbons would take place. About half of the unvapourized crude oil is removed from the bottom and distilled under reduced pressure in a similar fractionating column and heavy gas, oil and lubricating oil fractions are separated. The residue gives pitch.

Of all the fractions collected petrol has the largest demand, but the yield, by the above-described method, is only 20% of the crude. The result is that there is surplus of heavier fractions. Moreover, the straight-run petrol is not of sufficiently high quality for direct use in internal combustion engines. It has to be properly blended. To overcome these difficulties, the heavier fractions are subjected to cracking whereby the larger molecules are broken down to smaller molecules giving lighter fractions which are more useful. Two methods of cracking in use are the thermal and the catalytic cracking.

5.10.5 Thermal Cracking

In thermal cracking the heavy oil is subjected to high temperature when the molecules break down to give smaller molecules of the paraffin and olefin series and some hydrogen, but some molecules polymerize to give molecules larger than those originally present. The process can be carried out in vapour phase or in liquid phase. In the vapour phase, higher temperature and lower pressure are employed and the time required is less than in the liquid phase. The heavy oil is heated under pressure (vapour phase, 600° and up to 23 kg/cm^2; liquid phase, 450° and up to 100 kg/cm^2) when cracking takes place. It is then led to a chamber maintained at a lower pressure where the lighter fraction vapourizes and the vapours are taken to a fractionating column and different fractions are separated. The unconverted oil is returned to the heating chamber and processed again. Even under ideal condition only 50- 60 litres of pertol are obtained from 100 litres of the heavy oil. The petrol contains some unsaturated molecules which are undesirable in aviation petrol. The petrol obtained from vapour phase cracking has better anti-knock properties but poorer stability than petrol from liquid phase cracking.

5.10.6 Catalytic Cracking

In catalytic cracking reactions similar to those in thermal cracking are brought about in the presence of catalyst. The result is that lower temperature and much lower pressure is necessary and the process is under better control. It produces some aromatic and branch-chained hydrocarbons. The amounts of unsaturated hydrocarbons produced and the sulphur content are lower than in thermal cracking. The petrol produced has better anti-knock characteristics than the petrol from the thermal cracking.

The process is almost similar to thermal cracking. The heated oil is brought in contact with the catalyst. There are several methods of making the catalyst bed. Essentially the catalyst consists of certain almunium silicates. These are either naturally occurring clays or are synthetically made. After use for sometime deposits of tarry matter are formed which poison the catalyst. It is regenerated by burning off the deposits. The oil is then led to a flash chamber where the product vapourizes due to the lower pressure (Fig 5.8). Vapours are led to a fractionating column and different fractions are separated in the usual manner.

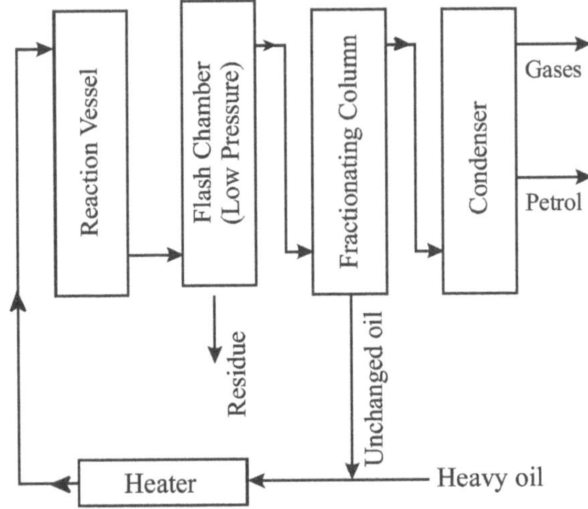

Fig. 5.8 Cracking of heavy oil.

5.10.7 Synthesis of Petrol

A large quantity of gaseous hydrocarbons, saturated as well as unsaturated are produced by primary distillation and cracking processes. These can be reformed by various processes to give higher molecular weight portions suitable for being used as fuel for internal combustion engines. There are several processes which are quite complex in nature and are beyond the scope of the present volume, and will only be described briefly.

An important variant of catalytic cracking is known as *hydrocracking.* Here the cracking is done in the presence of hydrogen and a suitable catalyst under very high pressure (200 kg/cm^2) and at high temperature (550°). Hydrogen combines with the unsaturated products and ultimately gives better petrol.

Another technique adopted is *dehydrogenation* i.e. removal of hydrogen, thus creating more unsaturation, followed by polymerization to give the desired kind of hydrocarbons. The refinery gases consist of saturated as well as unsaturated hydrocarbons. The saturated ones are less reactive and have to be converted to unsaturated ones by dehydrogenation and then all the hydrocarbons are polymerized. If necessary, after polymerization the product can be hydrogenated to remove residual unsaturation. In *aromatization* the straight-chain hydrocarbons are converted into aromatic hydrocarbons which are more useful for blending. Isomerization is used to improve anti-knock properties.

Synthetic petrol from coal. Much work has been done in Germany for obtaining petrol and other useful substances from coal by passing steam through a hot bed of coal. For converting coal to synthesis gas (syngas) or liquid hydrocarbons five major items to be controlled are pressure, temperature, steam, hydrogen and catalyst. The most important of the processes is the Fischer-Tropsch process in which first water gas ($CO + H_2$) is made from low grade coal. Water gas is mixed with more hydrogen and passed over a suitable catalyst whose main constituents are cobalt, nickel and iron. By regulating the composition of the gas mixture, the temperature and the catalyst, a variety of products can be produced at atmospheric pressure. When the same gas mixture is passed over a catalyst containing 85% thoria and 15% alumina at 400° and at 300 to 600 kg/cm^2 pressure, the oil produced contains high proportions of isobutane and is used as aviation petrol. By adjusting the proportion of the reactants and the working conditions products having mixtures of different constituents are obtained. The reactions may be represented as follows :

(i) $nCO + 2nH_2 \rightarrow nH_2O + C_nH_{2n}$

(Synthesis gas) (Olefins)

(ii) $C_nH_{2n} + H_2 \rightarrow C_nH_{2n+2}$

(Paraffins)

Another approach to the problem has been direct hydrogenation of coal, credit for which goes to Bergius. Powdered coal of low ash content is made into a paste by mixing with heavy oil. The paste, mixed with hydrogen, is passed over a tungsten or molybdenum oxide catalyst at about 450° at 250 kg/cm^2 pressure when hydrogen combines with the carbon of coal to produce hydrocarbons which are separated into various fractions. The heavy oil produced is used to make paste with a fresh batch of coal powder.

The Fischer-Tropsch process can use a low grade coal, which is abundant in India, but requires a much higher pressure for conversion. The advantage of the Bergius process is the lower pressure employed for carrying out the operation, but it requires high grade coal, which is in short supply. In the world market, synthesis of petrol from coal has not been economically competitive. One large coal-based synthesis gas (syngas) and liquid hydrocarbon plant has been in operation from the midfifties in South Africa near Johannesburg. Another plant, three time the size of the first plant, has gone into production. Coal-based projects require much more capital than petroleum based. For example, it is estimated that a coal-based ethylene plant required three times the investment as compared to the naptha based of comparable size. In the Fischer-Tropsch process about two-thirds of the coal is wasted in one form or the other. Due to the repeated increase in the petroleum prices after 1973, the interest in such processes has been revived and various countries including India, are considering setting up synthetic petrol plants. Due to abnormal increase in the petroleum prices and by improving the technology, it is hoped that in the future the process may become economical. At any rate, it has become necessary to pay attention in this direction with a view to reducing dependence on foreign countries.

There is no commercial synthetic petrol plant in India. At the Central Fuel Research Institute, Dhanbad, one 450 litres/day pilot plant has been working for the vapour phase hydrogenation of suitable fraction of coal tar as well as coal hydrogenation products. A 500 kg/day slurry phase coal hydrogenation pilot plant has been sanctioned for installation at CFRI, Dhanbad.

5.10.8 Refining of Petrol

The petrol obtained by straight distillation or by synthesis contains some unsaturated hydrocarbons which get oxidized and polymerized causing gum and sludge formation on storing. Also, there are some sulphur compounds which adversely affect tetra ethyl lead which needs to be added for obtaining better anti-knock properties.

The petrol is treated with concentrated sulphuric acid which removes these constituents. It is then treated with sodium hydroxide to remove the acid and finally washed with water. Some compounds, known as inhibitors, are also added which retard the oxidation reaction and thus improve storing quality.

After refining, the products are blended to give petrol that meets the various specifications like octane number, volatility, heating value, etc.

5.10.9 Petrol or Gasoline

Straight run petrol consists of a mixture of paraffinic, naphthenic and aromatic hydrocarbons, with 5 to 8 carbon atoms, in varying proportions. Cracked and reformed petrol is a more complex mixture. Its specific gravity is about 0.73 and the boiling range is 40° to 200°. The most important use for petrol is in internal combustion engines. The following factors should be considered in evaluating the quality of petrol for the purpose :

1. In spark-ignition engines petrol has to be partly vapourized on mixing with air. If it does not vapourize, there is difficulty in starting the engine. On the other hand too high a volatility will cause vapour formation in the fuel line thereby reducing fuel supply and stopping the engine. This is known as vapour-locking.
2. In the engine it should not form gum or cause corrosion. It should be stable in storage.

3. It should have proper 'anti-knock' properties which are explained in detail in the following paragraphs.

As calorific value is almost directly related to specific gravity a fuel of higher specific gravity is preferred where space is limited as in aeroplanes.

5.10.10 Knocking

Although theoretically the power output of an internal combustion engine increases with increase in the compression ratio, in actual practice it falls rapidly after a certain compression ratio. If the fuel-air mixture is compressed beyond this point two difficulties arise. Firstly, because of the high temperature due to the higher compression ratio, the fuel ignites even before the regular spark occurrence. This is one of the causes of pre-ignition. The second difficulty arises if after the start of the combustion and the flame proceeding at a particular rate, the last portion of the fuel burns instantaneously. This instantaneous combustion causes a shock wave to be set up. This is known as 'knocking', 'pinking', 'detonation'. It depends on the characteristics of the fuel, besides design factors.

5.10.11 Octane Number

The most common way of expressing the knocking characteristics of a fuel is by octane number. It has been found that n-heptane knocks very badly and its anti-knock value has been arbitrarily taken as 0 and *iso-octane* gives very little knocking and its anti-knock value has been taken as 100. The octane number is numerically the percentage by volume of *iso-octone* in a mixture, of *iso-octane* and n-heptane which just matches the knocking characteristics of the fuel. Thus, if a sample of petrol gives as much knocking as a mixture of 80 parts of *iso-octane* and 20 parts of n-heptane its octane number is 80. Now-a-days fuels with octane numbers above 100 are available and the value is determined by extrapolation.

The maximum permissible power increases with increase in the octane number. Based on experimental results graphs have been drawn to show this relationship. Another unit, known as performance number, is now being used to enable direct deduction of the power output that can be expected of a fuel.

5.10.12 Chemical Structure and Knocking

A detailed relationship between the knocking tendency of a fuel and the chemical structure has been worked out. This is important because the maximum power that can be developed is related to the anti-knock property of the fuel. In general, the anti-knocking property increases with increase in the compactness of the molecule, cyclic structure and double bonds. Thus in the paraffin series butane has an octane number of about 90 whereas heptane has 0. Short side chains increase antiknock properties. Cycloparaffins have better anti-knock properties but aromatic hydrocarbons have still better properties. Olefins of the same chain length as the paraffins have better properties.

It is because of these reasons that cracked petrol gives better engine performance than straight run petrol. The olefins, and the diolefins in particular, do not have very good keeping qualities. They are more easily oxidized and tend to form objectionable gums. This tendency, however, can be checked by the addition of inhibitors. The aromatic compounds are more susceptible to pre-ignition.

5.10.13 Anti-knock Agents

From the foregoing discussion the necessity of improving anti-knock properties is obvious. At the same time it is clear that it is not always easy to alter the chemical structure so that its performance is improved. To improve anti-knock properties some organo-metallic compounds are added of which tetra ethyl lead, (TEL), $Pb(C_2H_5)_4$, is most widely used. TEL is a colourless oily liquid of sp. gr. 1.62 and B. P. 200°. It has a sweet odour but is extremely poisonous and a fuel containing it should be handled with great care lest it enter the body through a cut or otherwise. In motor spirit about half ml and in aviation fuels one ml or a little more of TEL are used per litre of petrol. Because of its poisonous nature petrol containing TEL is always coloured.

It has already been explained that knocking is due to the spontaneous ignition of the compressed mixture of petrol vapour and air. In the cylinder tetra ethyl lead decomposes giving a fog of lead oxide particles. This fog influences combustion reactions of petrol and air mixture and prevents knocking. On burning, TEL deposits lead oxide and metallic lead on the spark plug and on the cylinder walls and piston head (necessitating more frequent cleaning). This results in ignition at these points. To remove this difficulty ethylene dibromide is also added so that lead is removed as volatile bromide along with the exhaust gases.

Tetra ethyl lead is more effective on alkanes than on aromatic compounds. Sulphur compounds reduce its effectiveness.

5.11 Diesel Fuel

The working of a Diesel engine is different from that of a petrol engine, consequently different properties are required for the fuel. In a Diesel engine air is highly compressed, to say 40 kg/cm^2, whereby its temperature rises to 500° or more. At this stage the oil is injected in the from of a spray. Droplets of the oil get vapourized and heated to the temperature at which spontaneous combustion takes place causing a sudden rise in pressure to about 70 kg/cm^2.

While the pressure rises, the injection of the fuel continues and the pressure begins to fall at the end of the injection of the fuel. In high speed engines the combustion process takes about 500th part of a second. If the oil is such that combustion does not begin quickly, the entire amount of oil gets injected before combustion takes place. Then all of a sudden total quantity of the oil burns and causes a sudden and great rise in pressure making a thumping sound, this is known as *Diesel knock*. Diesel knock depends upon the engine design as well as on the properties of the oil. If an oil has 'ignition lag', i. e. spontaneous combustion takes place after an interval, there will be knocking. In other words, the greater the ignition-lag the higher is the Diesel knock. Paraffins give less knock than aromatic hydrocarbons.

The Diesel oil is a higher boiling petroleum faction than petrol. Its distillation range depends upon the engine in which it is to be used. Fractions distilling from 180 to 400° are in use. For the sake of safety, the flash point of Diesel oil should not be below 65°. (It is noteworthy that flash point has nothing to do with self-ignition.) The *chief advantage* of using a Diesel engine is that its thermal efficiency is greater than that of the engines employing other types of fuel. It has been reported in a comparative study that in ordinary locomotive steam engines 6-8 percent of the heat of combustion of coal was converted into power, 16-30 percent was the conversion in steam turbines, 22-28 percent in petrol engines, whereas in Diesel engines 32-38 percent of heat energy of the fuel oil was converted into power. Other advantages of the Diesel engine are: (i) the engine can be started easily and full power can be developed quickly. This is not possible in a steam engine. (ii) Storage of fuel oil is easier and it occupies less space with the result that a Diesel locomotive runs a much longer distance without stopping to take fuel. (iii) Water is not consumed–only cooling water is needed which is recirculated and used. (iv) On the long runs, the working of a Diesel engine has proved to be more economical than the steam engines. Diesel oil certainly has all the advantages explained under 'Advantages of Liquid Fuels'. Because of these reasons Diesel engines are being used by various railways all over the world.

5.11.1 Cetane Number

The knocking characteristic described above is expressed in terms of cetane number just as for a petrol engine it is expressed in terms of octane number. Cetane ($C_{16}H_{34}$) has a very short ignition lag and its number is taken as 100 and that of *a*-methyl naphthalene, $C_{11}H_{10}$, which has a very long lag, as zero. The cetane number is defined as 'the percentage by volume of cetane in a mixture of cetane and *a*-methyl naphthalene which is equivalent in ignition quality to the fuel under test'. (Also see aniline point).

Oils for high speed, medium speed and low speed Diesel engines should have cetane number not less than 45, 35 and 25 respectively.

Knocking in a Diesel engine is primarily due to the oil not igniting in time. The cetane number can be improved by adding substances like ethyl nitrate, acetone peroxide and iso-amyl nitrate, which help in starting ignition. These substances are known as "dopes", and are added in small amounts (1 to 1.5%). The dopes do not improve starting from a cold engine and also are not so effective on low cetane number fuels. For these reasons the use of a dope for Diesel oil is not as common as the use of tetra ethyl lead is for petrol.

Knocking in a petrol engine and in a Diesel engine are due to two different reasons, as has already been explained in detail. In a petrol engine it is due to sudden spontaneous combustion of the last part of the mixture, and in a Diesel engine it is due to delay in the spontaneous combustion of the first part of the oil. The result is that an oil of high octane number has low cetane number and vice versa. Therefore, a petroleum crude which gives petrol of high octane number gives Diesel oil of low cetane number.

5.12 Other Liquid Fuels

5.12.1 Kerosene

Kerosene is the fraction distilling between 140° and 280° and has been used in our country for illuminating as well as for heating purposes. In recent times its use as aviation fuel in jet engines has opened a new avenue for it. The reasons for its use for aviation purposes are chiefly its high calorific value per unit weight and high combustion temperature which enables the engine to develop more power.

5.12.2 Fuel Oils

The term 'fuel oil' is used to designate higher molecular weight petroleum fractions. Their calorific values are around 10,000 cal/g. These are mostly used for steam raising, heating furnaces, etc. Their use as Diesel fuel has already been discussed.

For the Same calorific value the theoretical amount of air needed for the combustion of a fuel oil is almost the same as for a solid fuel but for actual combustion of liquid fuel in a furnace much less excess air is needed. According to Haslam and Russell, the excess air needed for different fuels is as follows:

Oil	... 15-25%	excess air
Pulverized fuel	... 15-20%	" "
Coal (fired on chain grates)	... 55-70%	" "
Coal (hand firing)	... 55-150%	" "

For this reason and because of the ease of combustion, the heat liberated per unit combustion space is much greater for fuel oils. This constitutes an advantage, specially when shortage of space is problem as in the case of ships.

The most important property to be considered for a particular burner is the viscosity of the oil. The viscosity varies from 7,000 seconds in Redwood Viscometer no. 1 down to 40 seconds at 40°. In the furnace, the oil is injected through a burner, generally with the help of steam or air. If oil of higher viscosity is used a sufficient quantity will not be injected through the burner. The oil is heated to reduce its viscosity and then injected through an atomiser which forms it into a spray. This helps rapid and complete combustion. The temperature inside the furnace should be high and the combustion should be complete before the oil droplets come in contact with cold surfaces.

In order to prevent fire hazards, it is essential that the flash point of the oil should be above 65°. The oil should certainly be free from water, sediment, etc.

5.12.3 Colloidal Fuels

Colloidal Fuels are suspensions of finely powdered coal in fuel oils stabilized by the addition of an emulsifier like sodium stearate. A colloidal fuel can be used just like an ordinary fuel oil in the same equipment. A colloidal fuel having 30-40% of coal in an oil has a calorific value of about 9,500 cal/g.

5.12.4 Tar

Tar obtained from coke ovens is used as a fuel, along with gaseous fuels in open hearth furnaces. It has to be heated and injected along with air or steam. Its calorific value is about 9,000 cal/g. Its use has reduced fuel costs in steel plants.

5.12.5 Benzol

Benzol is another product of coal carbonization which is used as a liquid fuel. Its distillation range is 70 to 150° and contains about 70% benzene, 18% toluene and 6% xylenes. Its calorific value is 10,200 cal/g and its octane number is 90. Mixed with petrol it is used as a fuel for internal combustion engines.

5.12.6 Power Alcohol

Ethyl alcohol, besides synthesis from hydrocarbons, is manufactured on a large scale by fermenting carbohydrates. In India its most important source is fermentation of molasses which is the residue left after crystallizing sugar from cane juice. The dilute solution of alcohol obtained by fermentation is concentrated by repeated distillation. The concentration of alcohol cannot be increased by distillation above 97.6% as it forms a constant boiling mixture with water which has a lower boiling point than alcohol. The properties of alcohol can be greatly improved, especially for use in internal combustion engines, by removing traces of moisture.

For removing the remaining quantity of water, alcohol is usually distilled with benzene. When benzene passes over with a portion of alcohol and water, it leaves behind absolute (or power) alcohol. Power alcohol can also be obtained by distilling alcohol in the presence of a dehydrating agent which keeps back the moisture.

Power alcohol has a lower calorific value (7,000 cal/g) than the petroleum products but has a high octane number (90). Its compression ratio is also higher. Power alcohol mixed with petrol has found favour in those countries which have a shortage of petrol. Power alcohol was mixed with petrol during the Second World War. If the design of the internal combustion engine is slightly modified, its mixture with petrol should give satisfactory results. At the present, serious thought needs to be given to this as India can produce cheaply large quantities of power alcohol from molasses but its large scale use as fuel seems to be doubtful because it is needed for the chemical industries. An expert opinion is that it takes 50% more energy to produce alcohol than is contained in it when used as a fuel. If this is true, it does not make sense in using it as a fuel. However, Brazil, which has vast stretches of unused land has successfully launched a programme for producing power alcohol from the bio-mass.

5.13 Gaseous Fuels

5.13.1 Advantages

The use of gaseous fuels is on the increase because of the following advantages :

1. Combustion is under better control. Gases burn instantaneously and the rate of burning can be changed very easily. Gases are miscible with air, therefore, the excess air needed is much less than that needed for liquid or solid fuels. Oxidizing or reducing atmosphere can be easily maintained. All these contribute to fuel economy and efficiency of furnace operation.
2. Heat from hot flue gases can be recovered and gaseous fuel can be pre-heated in regenerative furnaces, as is done in steel making by the open hearth process in glass manufacture. This gives economy of heat and at the same time a higher temperature is produced.
3. Gaseous fuels may be prepared at a central place and distributed over a wide area through pipelines. In some cases solid fuels unsuitable for direct use can be converted into gaseous fuels and valuable by-products recovered.

4. Neither smoke nor ash is produced. Thus gaseous fuels are very clean.
5. If the supply is obtained from a public supply there is no problem of storage.

The principal disadvantage of gaseous fuels is that they occupy large volumes, therefore, storage is difficult.

5.13.2 Coal Gas and Coke-oven Gas

Coal gas and coke-oven gas may be considered together because both are obtained by the destructive distillation of coal and their composition and calorific values are similar. Coal gas is made in horizontal or vertical retorts by heating-coal rich in volatile mater and gas is the main product. Coke-oven gas is made in coke-ovens by heating coking coal. Metallurgical coke is the main product and gas is the by-product. In India coke-oven gas is made to a greater extent than coal gas and is used in steel factories. In European countries coal gas is widely used as 'city gas' for lighting and heating purposes. The average calorific value is around 5,000 kcal/m^3. Coal gas has a slightly higher calorific value than coke-oven gas. It has about 50% hydrogen, 30% hydrocarbons (mainly methane), 5% carbon monoxide and smaller amounts of other gases.

The changes taking place on heating coal have been described in detail in this chapter under 'Coke'. It only needs to be added that the temperature of carbonization is governed by the end products needed. A very low temperature (300-400°) gives a higher yield of volatile products, like benzene. As the temperature is increased the yield of gaseous products increases and the yield of liquid products decreases.

The volatile products obtained from the retorts of coke-ovens are drawn off through ascension pipes connected to a collecting main which runs horizontally along the entire row of ovens or retorts. Here the volatile matter gets slightly cooled and some of the tar and water is condensed. The tar is led off to a tar well. The hot gas passes up the condensers where further condensation takes place and the liquid separates into tar and ammonia liquor layers. The gas is then washed by a spray or water or sulphuric acid to remove the remaining ammonia. The cooled gas is finally washed with oil, and benzol is recovered. The various portions so obtained are given elaborate treatment for by product recovery.

The gas has hydrogen cyanide, hydrogen sulphide and carbon disulphide as impurities. It is passed through purifiers which consist of a series of chambers having several shelves loosely packed with ferric hydroxide which retains hydrogen sulphide.

$$Fe_2O_3.\, xH_2O + 3H_2S = Fe_2S_3 + (x + 3)\, H_2O$$

Hydrogen cyanide also gets removed as ferric cyanide. Iron sulphide is exposed to air when sulphur separates and is recovered. The presence of carbon disulphide is undesirable because on oxidation, it forms corrosive oxides of sulphur. This is removed by passing the gas over a heated nickel catalyst when some of the hydrogen of the gas reduces carbon disulphide to hydrogen sulphide and carbon. The gas is stored in huge gas holders floating on water.

5.13.3 Producer Gas

Producer gas is essentially a mixture of carbon monoxide and nitrogen but often steam is blown along with air, thereby giving some hydrogen also. A typical producer giving the reactions is illustrated in the Fig. 5.9. The producer consists of an air-tight cylindrical furnace made of steel plates lined with fireclay bricks. There is water-seal at the base which helps the removal of ash. Air, or a mixture of air and steam, is blown through a pipe at the base. At the top there is an inlet for coal and also an outlet for the gas generated.

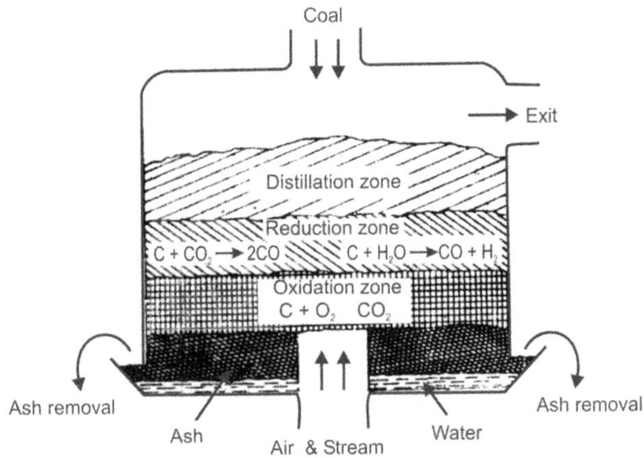

Fig. 5.9 Producer gas plant.

Coke or even inferior coal of high ash content is generally used for making producer gas. Bituminous coal gives gas of higher calorific value. The producer is filled with the fuel and the fuel is ignited. Air is forced in through the hot fuel and in the lower zone, known as tie oxidation zone, carbon burns to give carbon dioxide. The reaction is strongly exothermic.

$$2C + O_2 = 2CO + 52.8 \text{ kcal}$$

$$2CO + O_2 = 2CO_2 + 135.3 \text{ kcal}$$

(The positive sign indicates exothermic reaction and the negative sign indicates endothermic reaction).

Carbon dioxide passes upwards through the bed of coal and is reduced to carbon monoxide. The reaction is endothermic.

$$CO_2 + C = 2CO - 41.2 \text{ kcal}$$

In the uppermost part of the fuel bed there is fresh coal which is at a much lower temperature and no reaction takes place but the coal gets pre-heated and volatile matter distils. Gas goes out through the outlet and ash is removed through the water-seal. In efficient plants the producer gas so obtained contains about 33% carbon monoxide, 64% nitrogen and the rest is carbon dioxide and hydrogen.

In working the producer in the manner described above, some serious difficulties are faced. The combustion of carbon in air heats the lower part with the result that the ash forms clinkers which are difficult to remove. The firebars, on which the coal rests, and the refractory lining of the plant get heated to a high temperature with the result that their lives are reduced. In addition, the gas coming out is at a high temperature and if it is transported to distant parts of the factory much of its sensible heat is lost.

To reduce these difficulties some steam, about 0.35 kg per kg of coal, is blown in along with the air. In the reduction zone, steam reacts with carbon, producing carbon monoxide and hydrogen;

$$C + H_2O = CO + H_2 - 41.9 \text{ kcal}$$

This reaction is endothermic and helps to keep the temperature lower. The calorific value of producer gas is about 900 kcal/m^3. Because of this low value and also as it comes out at a high temperature it is generally not transported to distant places but is used on the spot where it is generated. Producer gas is widely used in steel plants where the producers are built just below the open hearth furnaces. Manufacture of producer gas makes possible the utilization of lower grade coal.

5.13.4 Water Gas or Blue Gas

This consists essentially of a mixture of carbon monoxide (40%) and hydrogen (50%) along with some carbon dioxide and nitrogen. It is manufactured in a manner similar to the manufacture of producer gas. The change in the method is that first only air is passed through the coal bed when carbon burns to carbon dioxide and the entire bed gets heated to 1000° or more. The outgoing gas, which is a mixture of carbon dioxide and nitrogen is not used as a fuel but is used to preheat the incoming air and steam. Steam is then blown through red hot coal when carbon monoxide and hydrogen are produced and are collected. This reaction is endothermic, therefore the bed cools down. Steam supply is stopped and air is again blown in to raise the temperature. Water gas is also called 'blue gas' because both carbon monoxide and hydrogen burn with a blue flame. It is of little value for lighting purposes. Its calorific value is about 2,700 kcal/m^3.

The calorific value of water gas being low, it is also not suitable for supplying to distant places. The calorific value can be increased by mixing oil gas. This is done by first heating a checker work of firebricks by burning a portion of the gas. Then oil is sprayed on this heated checker work when it gets vapourized and partly cracked and the water gas is blown through it. Thus *carburetted water gas* is obtained which contains about 35% each of carbon monoxide and hydrogen, 20% of various lower hydrocarbon gases and 10% of carbon dioxide and nitrogen. Its calorific value is about 4,500 kcal/m^3.

5.13.5 Blast Furnace Gas

The reactions taking place in a blast furnace are such that gas similar in composition to producer gas is obtained. It contains about 25% carbon monoxide and has a calorific value of 800 kcal/m^3. The quantity of gas produced is very large but its calorific value is so low that it cannot be used for many purposes nor can be supplied to distant places.

The gas contains a considerable quantity of dust and is passed through dust catchers where the heavier particles drop down. It is then washed with a spray of water and passed through electrical precipitators to remove the remaining particles. Gas purification is important because the dust is deposited in the pipelines, burners and furnaces and thus reduces efficiency.

In the past a quantity of the gas was used for gas-driven blower engines for the blast furnace but this use is on the decline. The present tendency in the iron and steel industry is to have integrated steel plants. 25% of the gas is used up in heating the blast air, and the balance finds important and large scale use for steam raising in the power plant and heating the coke ovens, open-hearth furnaces and reheating furnaces. The use of blast furnace gas for heating coke ovens makes available a larger quantity of coke-oven gas, which has a higher calorific value, for other purposes. This has resulted in savings on fuel cost in steel plants.

5.13.6 Natural Gas

Almost every petroleum well gives out some gaseous product consisting mostly of methane along with some ethane, propane and other hydrocarbons. In addition, there are large deposits of the gas in gas wells without any liquid petroleum being present. The gas is given the name "natural" gas. In the U.S.A. 4,00,000 million m^3 of the gas is produced and distributed through about 0.8 million km of pipelines.

Natural gas is 'wet' as it contains hydrocarbons which could condense in the pipeline. These are removed by compression and cooling or by absorption in liquid oil. The 'dry' gas so obtained contains over 80% methane, the balance being mostly ethane. Its calorific value is about 10,000 kcal/m^3.

As the facility of pipelines is lacking in many place, easily condensable hydrocarbons like propane, isobutane and butane are liquified at room temperature by compression. The liquified petroleum gases (known as L. P. G.) are sold in cylinders for use as domestic fuel.

The use of natural gas supplied through pipelines and in gas cylinders is increasing rapidly in India.

5.14 Determination of Calorific Value

5.14.1 Junker's calorimeter

Figure 5.10 illustrates a gas calorimeter originally designed by Junker. It consists essentially of an inner combustion chamber in which the gas burns. The chamber is surrounded by a water jacket through which a number of tubes, called flues, pass. The gas on combustion heats the water jacket and then passes down the flues in the water jacket, thus complete transfer of the heat of combustion to water is ensured.

Water enters the calorimeter through a constant head and its temperature is noted by a thermometer. The gas whose calorific value is to be determined passes through a gas meter, which records its volume, and burns in a burner, as shown in the figure. The hot products of combustion move upward in the chamber and then downward through the flues provided in the water jacket and escape through the exit. A thermometer at the exit indicates its temperature which should be as near the room temperature as possible so that the transfer of heat to water is complete. Water, moving in a direction opposite to that of the gases in the flues, goes out from the upper portion. Its temperature and volume are recorded. Water formed by the condensation of steam is also collected.

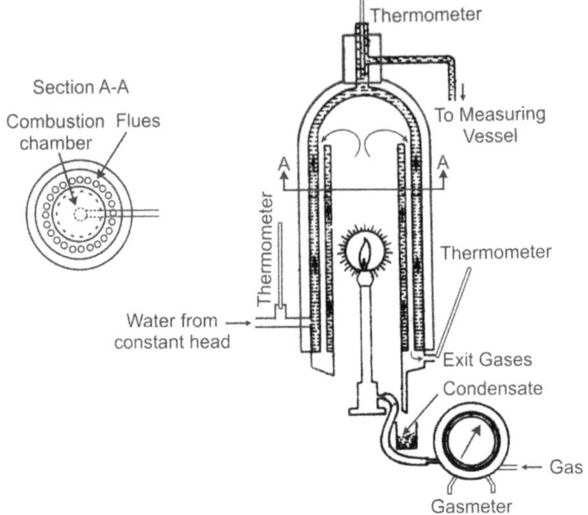

Fig. 5.10 Junker's gas calorimeter.

The volume of gas burnt during the experiment is converted to volume at 15° and 760 mm pressure. Knowing the weight of the water flowing and the rise in its temperature the heat given out by the gas is known. The gross calorific value is then obtained by dividing the heat given out by the corrected volume of the gas. For obtaining the net calorific value, the condensed water is weighed and the weight multiplied by the latent heat of vaporization of water. This is subtracted from the total heat and the result is then divided by the corrected volume of the gas.

5.15 The Non-Conventional Fuels

At the present several methods of non-conventional energy sources are being tapped. In this Section, on fuels, only those which involve direct and simple chemical reactions will be considered.

Why do we hear these days so much about the non-conventional fuels? The reason is that the existing energy resources are inadequate for several reasons, the principal ones being:

1. Wood, grown in course of years, and fossil fuels produced in millions of years get used up in decades. Their resources are getting exhausted at a very much faster rate than they are produced.

2. Hydro-electricity is cheap to produce but besides construction of dams being expensive, there are not many sites at which the dams can be built. Moreover, hydro-electricity depends upon rainfall. Ironically in an agricultural country like India, where more electricity is needed for operating pumps when there is a shortage of rainfall, generation of hydro-electricity also falls.

3. The technology for nuclear fuels based on fission reactions is well established but there is the problem of the disposal of the radio-active wastes. Fusion reactions are certainly promising for the future but expert opinion is divided on how long it will take to develop them.

The problem though known to the technologists for many years has been receiving attention and support from the Governments from about 1970 due to recurring conflicts in several oil producing countries. To tackle the problem, attention has to be paid to (i) increasing domestic energy production from both depletable as well as non-depletable sources, (ii) reducing energy consumption, (iii) increasing efficiencies of the various fuel consuming devices and (iv) increasing energy conservation (e. g. by better insulation).

5.15.1 Hydrogen as Energy Source

Attention of the applied chemists goes, naturally, to hydrogen as an energy source on account of the following reasons:

1. As it is produced from water its source is practically inexhaustible.

2. On combustion in air there is no pollution except for small amount of NO_2 produced, but if hydrogen is burnt in oxygen there is no danger of even this.

3. The mass-energy density (i.e., energy obtained from a unit mass) of hydrogen is much higher than for other fuels, though on a volume basis it is very much lower.

4. It can burn over a wide range of fuel-air ratios, i.e., its flammability limits are very wide. For example, 4% to 75% by volume of hydrogen can burn in air as compared to 5% to 15% for methane. This is a great advantage in the internal combustion engines.

5. High calorific value and clean burning qualities are distinct advantages for domestic and industrial heating.

6. As on combustion only water is produced, the hydrogen fired furnaces are more compact and the chimney stacks are not necessary.
7. Heat recovery from the burnt gases is more complete.
8. If hydrogen and oxygen are simultaneously manufactured, combustion can be in oxygen alone.
9. Hot plates (similar to electrical hot plates) of porous ceramics with fine dispersion of platinum (catalyst) on which hydrogen gas is directed have been developed. These are very convenient to use.

It may be added that hydrogen in itself is not a primary energy source because a large amount of energy is required to produce it. Hydrogen has been suggested as *the energy carrier* of the future. It can serve as a medium through which energy produced by a primary source, like the nuclear or solar energy, can be stored and utilized.

This leads one to consider, in outline, the **methods of manufacture** of hydrogen which may be grouped as under:

1. *Electrolysis* : Several types of electrolysers have been developed. It has to be remembered that the cost of electricity is prohibitive. Unless the cost of electricity comes down the electrolytic methods will not be economical. It has been suggested that off-peak hour electricity (i.e., when it is not required for other uses in large amounts) may be used to generate hydrogen. The cost can also be brought down by the sale of oxygen and heavy water which are valuable by-products.
2. *Thermo-chemical decomposition* : Several methods are in use utilizing a combination of heat and chemicals for decomposition of water. A serious disadvantage of such methods is that they require some fuel. An expert report says that these are cheaper than electrolytic methods even if the present fuel costs are doubled.
3. *Photolysis*: Photolysis of water by ultra violet rays being explored as a future possibility.

5.15.2 Storage of Hydrogen

Besides the cost of production a very major draw back in the use of hydrogen as a fuel is that of its storage. Under ambient conditions it occupies very large a volume.

Liquid hydrogen occupies much less volume but liquifaction in itself requires considerable energy, thus this does not seem to be a solution for energy saving. Moreover, liquid hydrogen requires very heavy containers which add to the transportation costs.s

A more practical approach to the storage of hydrogen is in the form of metal hydrides. The formation of hydrides has to be an easily reversible reaction. Considerable research has gone into the study of hydrides derived from the rare earths. For this purpose several Fe-Ti and Mg alloys are quite promising. The Fe-Ti alloys are more efficient but due to their high weights they are recommended for stationary appliances where weight is not a limiting factor. Magnesium based alloys are preferred for mobile applications.

Formation of hydrides is exo-thermic reactions and heat must be removed to allow the reactions to proceed to completion.

5.15.3 Gas from Organic or City Wastes

The gober-gas plants have become so common that they hardly need a detailed description here. The cow dung or human excreta is collected in a pit and allowed to putrify under anaerobic conditions when in the absence of oxygen the bacteria decompose much of the organic matter yielding gas. The gas can be used as a domestic fuel. In large plants it can be used to raise steam which drives a turbine to produce electricity.

The municipalities and the corporations have the problem of disposal of solid wastes. They are used as landfills. The gas formed gives off foul smell in the neighbourhood and causes pollution. If it is collected it can be utilised as a fuel and at the same time help to solve the problem of pollution.

The process is quite simple in principle. Perforated pipes are sunk in a well bored in a sanitary landfill and a vacuum is created at the bottom as a result of which gas formed due to anaerobic decomposition of the organic waste is sucked. It is used to produce electricity in the usual manner.

One well is enough for one acre of landfill. It produces 100 KW of electricity, i.e., ten acres of landfills are required for each megawatt of electricity. The cost of a well in 1985 was about Rs. 10 lakhs, inclusive of installation of generator, compressor, pumps and pipes. It is reported that the cost of the generation of electricity is 10 paise per unit as compared to nearly

50 paise through thermal power plants. A simple landfill is calculated to last for about ten years. After this the refuse needs to be taken out and used as a valuable manure. The emptied landfills are available for dumping the garbage and the cycle can be repeated.

To give a specific example, it is estimated that if the entire refuse of Delhi is taken to landfills which are then used for generating electricity in the manner described above about 60 MW of electricity a day will be produced. The peak consumption of electricity in Delhi in 1985 was 850 MW per day.

Chapter 6

Combustion

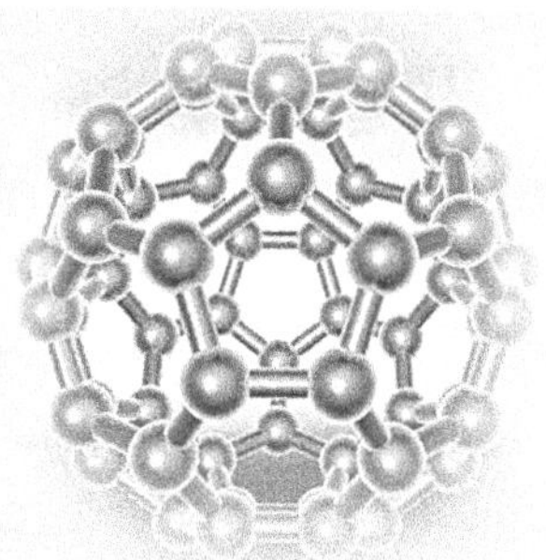

6.1 Introduction

Combustion is defined as a chemical reaction which is accompanied by heat and light. In a narrow sense it is the union of an element or a compound with oxygen. For practical purposes, it is used to designate only those cases in which the development of heat is rapid and is accompanied by a considerable rise in temperature. In the common fuels it involves the burning of carbon and hydrogen in air and also to a much smaller extent of sulphur. For proper combustion, the substance must be brought to its *kindling* or *ignition temperature*, which is defined as the temperature at which it ignites and continues to burn without further addition of heat from outside. The ignition temperature depends upon several factors; the presence of a catalyst, for instance, lowers the temperature.

6.1.1 Spontaneous Combustion

Some materials like coal, especially powdered coal, oily rags or wastes, are known to burn by themselves. If such materials are stored in such a manner that they are almost insulated and the loss of heat from them is slow, then heat produced by their slow oxidation gradually raises the temperature. As the temperature begins to raise the rate of oxidation increases with a further rise in temperature. The result is that the ignition temperature is soon reached and the material begins to burn. This is an example of spontaneous combustion and must be carefully guarded against in storage of such materials, as it is highly risky.

6.1.2 Some Practical Aspects of Combustion

When a fuel like bituminous coal is fired into a furnace, its volatile constituents are vapourized. Moreover, if the furnace is already not at a sufficiently high temperature, the volatile gases (which mostly consist of hydrocarbons) will not be heated to their ignition temperatures and thereby escape unburned, causing wastage. Besides the proper temperature, there should be a sufficient supply of air for combustion of the gases to take place. The amount of air that will pass through the grates also depends upon the size of the coal fired. Smaller pieces, like coal dust, restrict the flow.

Theoretically, the presence of *moisture* in coal is undesirable as it causes waste of heat, but in practice a small quantity of it is advantageous. Moisture may be present in the coal itself or may be added by moistening the coal before use. Water coming in contact with red hot carbon vapourizes and then reacts with carbon giving carbon monoxide and hydrogen. This reaction is endothermic and lowers the temperature. Thus, moistening of coal helps to keep the temperature of the fire bars low and also prevents the formation of clinkers. If sufficient air and combustion space are available the gases produced will burn to give heat and form water again, which generally escapes as steam. There is some loss of heat in the form of the latent heat of steam, but the other advantages outweigh this disadvantage. It is estimated that if 10% moisture is present, loss of heat due to it will be about 1% of the total heat of the fuel but if moisture was not present about 10% of heat might be wasted, due to loss in the ash-pit and incomplete combustion. The process of the addition of moisture to coal just before it is fed into the furnace is known as *coal tempering*. Optimum free moisture content is 7 to 9% when coal has minimum density. For this reason air flow increases, thereby combustion is facilitated. Free moisture also reduces the escape of the fines (i.e., fine particles of coal). Moisture may be added by sprinkling water on coal or, better still, by admitting steam through the bottom of the coal hopper.

In practice, the efficiency of combustion is judged by the *percentage* of CO_2 in the flue gas. As air contains 21% by volume of O_2 and 1 volume of O_2 gives rise to 1 volume of CO_2, there can be up to 21 % of CO_2 in the flue gases by the combustion of carbon. Naturally, attempts are made to get as high a percentage of CO_2 in the flue gas as possible. If there is excess of air, there will be loss of heat as some heat is carried away as sensible heat. Hence excess air is usually reduced. This practice, carried on without proper consideration, may lead to lower efficiency. By cutting down excess air, the combustion may be incomplete, resulting in the formation of CO. In this reaction only 2,416 calories of heat per gram of carbon is produced as against

8,100 calories when a gram of carbon burns to give CO_2. Hence the percentage of CO is a more important consideration, as greater saving can be made by eliminating CO in flue gases than trying to prevent some loss of sensible heat in excess air. In actual practice, some CO is present even when there is a small amount of excess air.

The optimum excess air varies from furnace to furnace as it depends upon its design and the fuel burnt. The optimum excess is that which has the minimum of heat, both potential (e.g., in the form of CO which could burn to give more heat) and sensible heat going in the flue gases. In coal fired boilers usually 50 to 100% excess air is supplied.

6.1.3 Calculations

Simple combustion calculations can be studied in three categories :

(A) Burning of solids or liquids.

(B) Burning of gases.

(C) Calculation of air supplied.

When the composition of the fuel is known, the theoretical amount of air needed for its combustion and the products of combustion can be easily calculated with the help of a simple chemical equation. For this purpose the composition of air is taken to be:

	By weight	By volume
Oxygen	23%	21%
Nitrogen	77%	79%

This means that when 1 kg of O_2 is npeeded for a reaction

$$\frac{1\times100}{23}=4.35\text{ kg} \text{ of air is to be supplied.}$$

In the same manner for

$$1\text{ m}^3\text{ of O}_2,\ \frac{1\times100}{21}=4.76\text{ m}^3\text{ of air is to be supplied.}$$

This gives the amount of N_2 that will go along with the products of combustion.

Hydrogen burns to give water which is supposed to be in a liquid state and the volume, being comparatively very small, is neglected. The volumes of solids are also neglected.

If the analysis of fuel shows O_2, it is supposed to be in chemical combination; therefore, its amount is to be subtracted from the calculated total requirement.

Some useful conversions are :

(i) Gram molecule of gases occupy 22.4 litres at N. T. P.

(ii) Pound molecule of gases occupy 359 cu. ft. at N. T. P.

(iii) 1 lb. of air occupies 12.39 cu. ft. at N. T. P.

Since air contains 23% by weight of oxygen, the weight of air needed is the weight of oxygen divided by 0.23. Let C, H, O and S represent the weights of the elements in kg per kg of fuel, the weight of air theoretically needed is :

$$W = 11.6\,C + 34.8\,(H - O/8) + 4.35\,S$$

(A) Burning of Solid or Liquid Fuels : In such cases the analysis of solid or liquid fuel is given and amount of O_2 needed for combustion is to be calculated. From the amount of O_2 other requirements can easily be found. The combustion of each element is calculated separately.

(a) *Combustion of Carbon.*

(i) $2C + O_2 = 2CO$

In terms of weights $2 \times 12 + 2 \times 16 = 2 \times 28$

or 24 parts of C need 32 parts of O_2 for combustion to give 56 parts of CO,

or

1 g of C needs $\frac{32}{24} = 1.333$ g of O_2 giving 2.333 g of CO.

(ii) Complete combustion of C to CO_2, can be calculated as follows

	Carbon		Air	Flue gases
	C	+	$O_2 + (N_2)$	$= CO_2 + (N)$
Wt. in g or Ib.	12	+	32 + (107)	= 44 + (107)
	1	+	2.667 + (8.93)	= 3.667 + (8.93)
	1	+	11.6	= 12.6
Volume in liters at 0° and 760 mm.	12g	+	22.4 + (84.3)	= 22.4 + (84.3)
	1g	+	1.87 + (7.0)	= 1.87 + (7.0)
Volume in cu.ft. at 0° and 760 mm.	12 lbs	+	359.0 + (1350)	= 359.0 + (1350)
	1 lb	+	30.0 + (112.5)	= 30.0 + (112.5)
			142.5 cu. ft.	142.5 cu. ft.

(b) *Combustion of Hydrogen.*

$$2H_2 + O_2 = 2H_2O;\ 4 + 32 = 36$$

or 1 g of H_2 needs 8 g of O_2 to produce 9 g of water.

making use of Avogadro's hypothesis;

2 vols. of H_2 need 1 vol. of O_2 to produce water (which occupies negligible volume).

(c) *Combustion of Sulphur.*

$$8 + O_2 = SO_2;\ 32 + 32 = 64$$

or 1 g of S needs 1 g of O_2 to produce 2 g of SO_2.

Example 1 : Calculate the weight and volume of air needed for the combustion of 1 kg of carbon. Give the composition of the combustion products.

Solution :

$$C + O_2 = CO_2;\ 12 + 32 = 44$$

$\therefore$ Wt. of O_2 needed for 1 kg of C $= 1 \times \frac{32}{12} = 2.667$ kg.

As 23 parts by wt. of O_2 is present in 100 parts of air, the wt. of air needed

$$2.667 \times \frac{100}{23} = 11.6 \text{ kg}$$

Calculation of volume of air needed:

As 1 gram mole of a gas occupies 22.4 litres at N. T. P.

32 g of O_2 occupies 22.4 litres at N. T. P.

$\therefore$ 32 of O_2 is Present in $22.4 \times \frac{100}{21}$ litres of air at N. T. P.

$\therefore$ Volume of air at N. T. P. needed for the combustion of 1000 g of $C = \frac{2667}{32} \times \frac{22.4 \times 100}{21} = 8890$ litres.

As 1 m^3 = 1000 litres, the volume (at N, T. P.) of air needed for the combustion of 1 kg of carbon = 8.89 m^3

Wt. of N_2 present in 11.6 kg of air = 11.6 – 2.667 = 8.933 kg

Wt. of CO_2 produced in the reaction $= \frac{1 \times 44}{12} = 3.667$ kg,

$\therefore$ 1 kg of C needs 11.6 kg of air for combustion and the products contain 8.933 kg of N_2 and 3.667 kg of CO_2 or 70.9% N_2 and 29.1% CO_2 by weight are present in the flue gases.

Volumetric composition :

1 g mole of a gas occupies 22.4 litres at N, T. P.

$\therefore$ Volume of $N_2 = \frac{22.4 \times 8933}{28}$ = 7145 litres.

and volume of $CO_2 = \frac{22.4 \times 3667}{44}$ = 1866 litres.

Total volume = 9011 litres.

$\therefore$ Volume of $N_2 = \frac{7145 \times 100}{9011}$ = 79.3%

and volume of $CO_2 = \frac{1866 \times 100}{9011}$ = 20.7%

This principle can be used to solve actual problems as illustrated.

Example II : The percentage composition of a sample of bituminous coal, by weight was found to be C = 76%; H_2 = 5.2%; O_2 = 12.8%; N_2 = 2.7%; S = 1.2%; Ash = 2.1%. Calculate the minimum (a) weight and (b) volume at N. T, P. of air necessary for complete combustion of I kg of coal and percentage composition by weight of dry products,

Solution : The calculations maybe made in a tabular form:

Wt. of element per kg of fuel (in kg)	Wt. of O_2 needed (in kg)	Wt. of dry products of combustion (in kg)
C = 0.76	$0.76 \times \frac{32}{12} = 2.026$	$0.76 \times \frac{44}{12} = 2.786\ (CO_2)$
$H_2 = 0.052$	$0.052 \times \frac{16}{2} = 0.416$	
$O_2 = 0.128$	-	-
$N_2 = 0.027$	-	$0.027\ (N_2)$
S = 0.012	$0.012 \times \frac{32}{32} = 0.012$	$0.012 \times \frac{64}{32} = 0.024(SO_2)$
Ash = 0.021	-	-
Total = 1.000	2.454	

Wt. of oxygen to be supplied = Wt. of oxygen needed-Wt. of oxygen present =2.454 - 0.128 = 2.326 kg,

$\therefore$ Wt. of air needed $= 2.326 \times \frac{100}{23} = 10.113$ kg.

$\therefore$ Wt. of nitrogen supplied with air = wt. of air - wt. of oxygen

= 10,113 – 2,326

= 7.787 kg.

The term "products of combustion" refers to the gaseous products of combustion (i.e., not the ash, which has not taken part in combustion).

$\therefore$ products in the combustion gases, excluding water, are

CO_2 = 2.786 kg; SO_2 = 0.024 kg; N_2 = 0.027 kg (from coal) + 7.787 kg (from air supplied) =7.814 kg. Total = 10.624 kg.

$\therefore$ Percentage composition by wt.:

$$CO_2 = \frac{2.786 \times 100}{10.624} = 26.22\% \qquad N_2 = \frac{2.786 \times 100}{10.624} = 73.55\%$$

$$SO_2 = \frac{0.024 \times 100}{10.624} = 0.23\%$$

Example III : The percentage composition of a sample of anthracite coal is C, 90; H_2, 3.5; O_2, 3.0; N_2, 1.0; S, 0.5; the remainder being ash. Estimate the minimum weight of air required for the combustion of 1 kg of this fuel, and the composition of the dry products of combustion, by volume, if 50% excess air is supplied.

Solution :

Wt. of element kg	Wt. of O_2 needed kg	Wt. of dry product kg
C = 0.9	$0.9 \times \frac{32}{12} = 2.40$	$0.9 \times \frac{44}{12} = 3.30\ (CO_2)$
H_2 = 0.035	$0.035 \times 8 = 0.28$	...
O_2 = 0.03	...	...
N_2 = 0.01	...	0.01 (N2)
S = 0.005	$0.005 \times \frac{32}{32} = 0.005$	$0.005 \times \frac{64}{32} = 0.01\ (SO_2)$
Total	2.685	

Wt. of O_2 to be supplied = Wt. of O_2 needed - Wt. of O_2 present

= 2.685 - 0.03 = 2.655 kg,

Wt, of air needed for combustion $= 2.655 \times \frac{100}{23} = 11.54$ kg.

As 50% excess air is supplied, the wt. of excess air supplied

$$= 11.54 \times \frac{50}{100} = 5.77 \text{ kg.}$$

$\therefore$ Wt. of air supplied = 11.54 + 5.77 = 17.31 kg.

Wt. of present in air $\frac{17.31 \times 77}{100}$ = 13.33 kg.

Wt. of N_2 from the fuel = 0.01 kg,

$\therefore$ total wt. of N_2 = 13.34 kg.

Wt. of free O_2 (due to excess air) $= \frac{5.77 \times 23}{100} = 1.33$ kg.

$\therefore$ The Wt. of dry products in flue gases are: CO_2, 3.30 kg;

N_2, 13.34 kg; O_2, 1.33 kg and SO_2, 0.01 kg.

Molecular wt. of these gases are CO_2 – 44; N_2 – 28; O_2 – 32 and SO_2 – 64 respectively.

Let 1 kg mole of a gas occupy V m^3

∴ The volumes of the gases are:

$$CO_2 = \frac{3.30 \times V}{44} = 0.0750 \text{ V m}^3; \quad O_2 = \frac{1.33 \times V}{32} = 0.0416 \text{ V m}^3$$

$$N_2 = \frac{1.334 \times V}{28} = 0.4764 \text{ V m}^3; \quad SO_2 = \frac{0.01 \times V}{64} = 0.00015 \text{ V m}^3$$

∴ total $= 0.5932 \text{ V m}^3$

∴ Percentage by volume of the gases is:

$$CO_2 = \frac{0.0750V \times 100}{0.5932V} = 12.64\% \qquad O_2 = \frac{0.0416V \times 100}{0.5932V} = 7.01\%$$

$$N_2 = \frac{0.4764V \times 100}{0.5932V} = 80.32\% \qquad SO_2 = \frac{0.00015 \times 100}{0.5932V} = 0.03\%$$

Example IV: A sample of petrol was found to contain 15.2% H_2 and 84.8% C by weight. Calculate (i) the weight of air needed for the combustion of 1 kg of the fuel; (ii) the volumetric composition of the products of combustion if 15% excess air is supplied,

Solution :

I kg of petrol contains 0.152 kg of H_2 and 0.848 kg of C

Wt. of O_2 needed for 0.152 kg of $H_2 = 0.152 \times 8 = 1.216$ kg

Wt. of O_2 needed for 0.848 kg of C $= 0.848 \times \frac{32}{12} = 2.261$ kg

Total wt. of O_2 needed for combustion $= 3.477$ kg

∴ Wt. of air needed $= 3.477 \times \frac{100}{23} = 15.12$ kg

Excess air supplied $= 15.12 \times \frac{15}{100} = 2.268$ kg,

Wt. of O_2 in excess air $= 2.268 \times \frac{23}{100} = 0.522$ kg,

Total air supplied $= 15.12 + 2.268 = 17.388$ kg.

∴ Wt. of N_2 in flue gases $= 17.388 \times \frac{77}{100} = 13.39$ kg.

Wt. of CO_2 produced per kg of fuel $= 0.848 \times \frac{44}{12} = 3.109$ kg.

∴ The flue gases from the combustion of 1 kg of petrol contain 3109 g of CO_2; 522 g of O_2 and 13390 g of N_2

Let 1g mole of a gas occupy V litres.

∴ The relative volumes of the gases are:

$CO_2 = \frac{3109\ V}{44} = 70.7$ V litres $O_2 = \frac{522\ V}{32} = 16.3$ V litres

$N_2 = \frac{13390\ V}{28} = 565.2$ V litres

Total vol. $= 565.2$ V litres

∴ Percentage of CO_2 by vol. $= \frac{70.7V \times 100}{565.2\ V} = 12.5\%$

Percentage of O_2 by vol. $= \frac{16.3V \times 100}{565.2\ V} = 2.9\%$

Percentage of N_2 by vol. $= \frac{478.2V \times 100}{565.2\ V} = 84.6\%$

(B) Burning of gases : These calculations are comparatively easy and their principles are well worked out in junior classes. The following example illustrates how to solve the problems:

Example V: The volume analysis of a producer gas is: H_2 – 14%, CH_4 – 2%, CO – 22%, CO_2 – 5%, O_2 – 2%, N_2 – 55%. Find the air required for the perfect combustion of 1 m^3 of the gas. If 40% excess air is supplied, find the volume analysis of the dry products.

Solution : The Combustion reactions are :

1. $2H_2 + O_2 = 2H_2O$
   2Vol. 1 Vol. (nil)
2. $CH_4 + 2O_2 = CO_2 + 2H_2O$
   1 Vol. 2 Vol. 1 Vol. (nil)
3. $2CO + O_2 = 2CO_2$
   2Vol. 1Vol. 2Vol.

Applying these reactions to the combustion of 100 m^3 of the gas:-

Gas	Vol. in m^3	Vol. of O_2 needed	Vol. of dry CO_2	products N.
H_2	14	7	-	-
CH_4	2	4	2	-
CO	22	11	22	-
CO_2	5	--	5	-
O_2	2	[-2]	-	-
N_2	55	-	-	55
Total	100	(22-2 = 20)	29	55

$\therefore$ For the perfect combustion of 100 m^3 of the gas, 20 m^3 of O_2 is needed and the dry products contain 29 m^3 of CO_2 and 55 m^3 of N_2.

$$\therefore \text{ Vol. of air needed per } 100 \text{ m}^3 = 20 \times \frac{100}{21} = 95.24 \text{ m}^3$$

$$\therefore \text{ Vol. of air needed per m}^3 = 0.95 \text{ m}^3$$

$$\text{Vol. of N}_2 \text{ in the air supplied} = 95.24 - 20 = 75.24\text{-m}^3$$

$$\text{Vol. of excess air supplied} = 95.24 \times \frac{40}{100} = 38.1 \text{ m}^3$$

$$\text{Vol. of O}_2 \text{ in excess air} = 38.1 \frac{21}{100} = 8.0 \text{ m}^3$$

$$\therefore \text{ Vol. of N}_2 \text{ in excess air} = 38.1 - 8.0 = 30.1 \text{ m}^3.$$

Dry combustion products of 100 m^3 of gases contain 29 m^3 of CO_2 + 55 m^3 of N_2 (as calculated in the table) + 75.24 m^3 of N_2 (from air supplied for perfect combustion) + 30.1 m^3 of N_2 (from excess air supplied) + 8.0 m^3 of O_2 (from excess air)

$$= 29.0 \text{ m}^3 \text{ CO}_2 + 160.34 \text{ m}^3 \text{ N}_2 + 8.0 \text{ m}^3 \text{ O}_2 = 197.34 \text{ m}^3.$$

$$\therefore \quad CO_2 = \frac{29.0 \times 100}{197.34} = 14.7\% \quad O_2 = \frac{8 \times 100}{197.34} = 4.1\%$$

$$N_2 = \frac{160.34 \times 100}{197.34} = 81.2\%$$

(C) **Calculation of Air Supplied :** It has already been stated that if theoretical minimum air is supplied in a furnace, the combustion reactions will not be complete and there will be considerable wastage. In coal furnaces 50 to 100% excess air is generally supplied. In practice it is difficult to measure the air supplied. For proper control and economy it is also necessary to know the actual state of combustion, i.e., whether or not complete combustion is taking place, how much of CO is being formed, how much of free oxygen goes in the flue gas, etc. All this information can be had from the flue gas and fuel analyses, as will be shown in the examples that follow.

6.2 Flue Gas Analysis by Orsat Apparatus

The principle of analysis is that the gas to be analysed is taken in a burette which is connected to several pipettes containing suitable solutions. The gas is forced through these pipettes one after another. Thus certain constituents get absorbed and the volumes of the unabsorbed gases are found, from which the volume of the particular constituent is known.

The most widely used apparatus for the purpose is the Orsat apparatus, Fig. 6.1. It consists of a capillary glass tube manifold connected at one end to a gas burette *B* and to a three way stop cock *d* at the other end. The manifold has three side tubes connected to three stop cocks *a*, *b* and *c*, through which are connected three absorption pipettes, p_1, p_2 and p_3, each of which has two bulbs as shown in the figure. Just below the cocks there are marks to show the levels of the solutions. The gas burette is surrounded by a water jacket so that the temperature of the gas remains unchanged during the experiment. The lower end of the burette is connected by a rubber tube to an aspirator bottle *L*, filled with water. Three-way stop cock d is connected to the flue gas source through a U-tube containing glass wood for retaining smoke particles. The working of the three-way cock should be well studied before use because one position connects it to the flue gas source, the second to the atmosphere and the third shuts off the main apparatus.

The pipettes are usually filled with glass tubes so that the surface area of contact between the gas and the solution is increased. The three pipettes p_1, p_2 and p_3 have a solution for the absorption of CO_2, O_2 and CO respectively. p_1 has potassium hydroxide solution (200 g in 500 ml of water). It absorbs only CO_2. p_2 has a solution of pyrogallic acid (25 g) and potassium hydroxide (200 g) in water (500 ml). It can absorb CO_2 and O_2. p_3 contains cuprous chloride (100 g) solution in liquor ammonia (125 ml) and water (75 ml). It can absorb CO_2, O_2 and CO. Hence it is necessary that the flue gas is passed first through pipette *Pi* when CO_2 is removed, then through p_2 when only O_2 will be removed (as CO_2 has already been removed in p_1) and finally through p_3 when only CO will be removed. The whole apparatus is readily available in compact wooden cabinets.

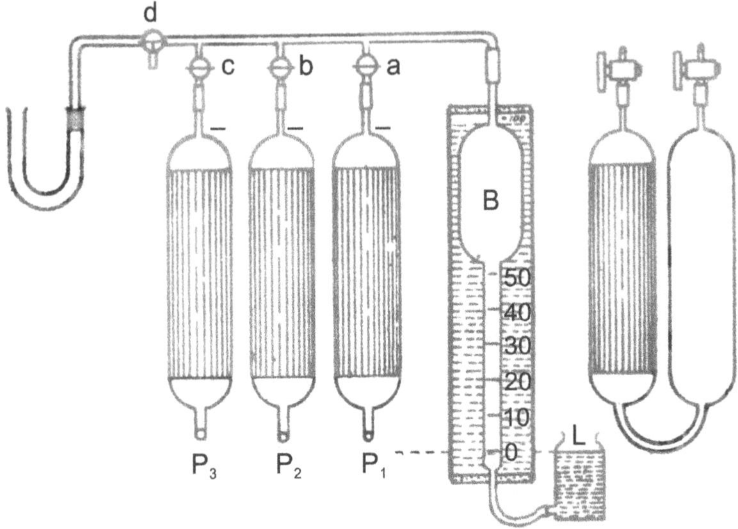

Fig. 6.1 Orsat apparatus.

The stoppers are greased and the solutions taken as described. Cocks *a*, *b* and *c* are closed. Cock *d* is opened to the atmosphere and aspirator *L* is raised till the burette is filled with water but water does not enter the capillary tube. Cock d is closed, cock a is opened and the leveling bottle *L* is lowered. Water from the burette runs into *L,* thereby sucking out gas from the apparatus which is now completely closed except for an opening through pipette p_1, the back limb of which is open to the atmosphere. The result is that air enters the back limb and the solution is pushed to the front limb of the pipette. Thus the level of the solution in the pipette is raised up to the fixed mark and cock a is closed. In a similar manner the levels of the solutions are brought to fixed marks in pipettes p_2 and p_3.

The apparatus is now ready for taking in the gas for analysis. Cock *d* is opened to the atmosphere and *L* is raised till the burette is filled with water. Cock d is now connected to the gas supply and *L* is lowered drawing in the gas to be analysed in the burette. This sample is mixed with the air which was present in the apparatus. Cock *d* is opened to the atmosphere and the gas expelled by raising aspirator bottle L. This process of taking in and expelling the gas is repeated twice. Finally the burette is completely filled with the gas, cock *d* is opened to the atmosphere and bottle *L* is very carefully raised till the level of water in it is the same as in the burette which stands at zero mark. Now the burette has 100 ml gas at atmospheric pressure. Cock d is closed and a is opened and L raised to drive the gas into p_1. Bottle *L* is lowered to take the gas back in the burette. The gas is again sent into pipette p_1. This process is repeated several times to ensure full absorption of CO_2 in KOH solution. Bottle L is lowered till the solution in the pipette stands at the fixed mark and cock a is closed. The levels of water in the bottle L and burette B are equalized and the volume noted. To be certain that all CO_2 has been absorbed the process is repeated with the same sample till the two volume readings are the same. The decrease in volume gives the volume of CO_2 in 100 ml gas sample. The volumes of O_2 and CO are determined in a similar manner by passing the remaining gas through pipettes p_2 and p_3 respectively.

The Orsat apparatus is simple and easy to work. It gives valuable information for regulating combustion in the furnace. Presence of free CO shows incomplete combustion and air supply has to be increased. If there is a large amount of free O_2 it shows a large excess of air. Both CO (except in small amounts) and excess O_2 should not be present together. Presence of CO and O_2 together indicates non-uniform combustion i. e., excess air is being sent in, and in certain part of the furnace there is insufficient supply of air. If sulphur is present in the fuel, SO_2 will be produced which will be dissolved in water over which gas is collected and will not be recorded by this method. The results help to make important calculations as shown in the following examples.

In smaller plants the facility of ultimate analysis is not available but Orsat apparatus, which is not expensive, is available. With some practice and due care one should be able to get good results which can indicate the approximate combustion state. This is satisfactory for most purposes. More accurate results are, however, obtained by detailed analysis and calculations. The following example illustrates the method of calculation.

Example VI. A sample of flue gas has the following analysis CO_2, 12.5%; O_2, 7.1%; N_2, 80.4%. Calculate the excess air supplied.

Solution. On the assumption that there is no nitrogen in the fuel, oxygen and nitrogen must have come from air.

100 m^3 of flue gas has 80.4 m^3 of N_2

$\therefore$ Vol. of air supplied per 100 m^3 of flue gas

$$= 80.4 \times \frac{100}{79} = 101.8 \text{ m}^3$$

$$\therefore \text{ Vol. of } O_2 \text{ in air actually supplied} = 101.8 \times \frac{21}{100} = 21.4 \text{ ml}$$

Vol. of excess O_2 present in 100 m^3 of flue gas 7.1 m^3

$$\therefore \text{ Vol. of } O_2 \text{ used in combustion} = 21.4 - 7.1 = 14.3 \text{ m}^3$$

$$\therefore \text{ Percentage of excess air} = \frac{7.1}{14.3} \times 100 = 49.7\%.$$

Example VII. The analysis of the coal in a boiler trial was:

C – 88%, H_2 – 3.6%, O_2 – 4.8%, other matters 3.6% and the flue gas analysis by volume was CO_2 – 10.9%, CO – 1%, O_2 – 7.1%, N_2 – 81%. Find the proportion of carbon burned to CO, and the air required per kg of fuel for the combustion as it actually occurred and also the weight of flue gas per kg of fuel burned.

Solution. For converting the volumetric composition of flue gas into composition by weight, the percentages are to be multiplied by the molecular weights of the respective gases. This gives their relative weights, which are:

CO_2, 10.9 × 44; CO, 1 × 28; O_2, 7.1 × 32; N_2, 81 × 28.

$$\text{Wt. of C in } CO_2 \text{ present} = 10.9 \times 44 \times \frac{12}{44} = 10.9 \times 12 \text{ parts}$$

$$\text{Wt. of C in CO present} = 1 \times 28 \times \frac{12}{28} = 1 \times 12 \text{ parts}$$

∴ Total wt. of C = (lO.9 +1) × 12 parts

∴ 11.9 × 12 parts by weight of C is associated with 81 × 28 parts of N_2 by weight.

Wt. of C in 1 kg of the coal = 0.88 kg.

∴ In the flue gases 0.88 kg. of C is associated with

$$0.08 \times \frac{81 \times 28}{11.9 \times 12} \text{kg of N2}$$

∴ Wt. of air supplied per kg of coal

$$\frac{0.88 \times 28}{11.9 \times 12} \times \frac{100}{77} = 18.15 \text{ kg.}$$

$$\text{Percentage of C burning to CO} = \frac{\text{Wt. of C in CO}}{\text{Wt. of C in } CO_2 \text{ and Co}} \times 100$$

$$= \frac{1 \times 12}{10.9 \times 12 + 1 \times 12} \times 100 = 8.4\%$$

Weight of flue gas per kg of fuel burnt can be calculated thus-

11.9 × 12 parts by Wt. of C is present in flue gas whose wt. is 10.9 × 44 (CO_2) + 1 × 28 (CO) + 7.1 × 32 (O_2) + 81 × 28 (N_2) parts. or 142.8 parts by wt. of C is present in 3003.8 parts by wt. of flue gas.

∴ 1 kg of fuel contains 0.88 kg of C

∴ the wt, of dry flue gas from it

$$= \frac{0.88 \times 3003.8}{142.8} = 18.51 \text{ kg.}$$

As the analysis of the flue gas does not include moisture, this is the weight of dry gases. Generally, in problems, the weight of dry flue gas is asked, but in this particular problem 'dry' flue gas has not been mentioned. Therefore the weight of moisture needs to be added. The moisture produced per kg of fuel is obtained from the weight of hydrogen.

Wt. of H_2 per kg of fuel = 0.036 kg

$\therefore$ Wt. of water = $0.036 \times 9 = 0.324$ kg.

$\therefore$ Wt. of flue gas per kg of fuel, including moisture

$$18.51 + 0.32 = 18.83 \text{ kg.}$$

Example VIII. In a boiler trial the dry coal as burned contained 84% of carbon and 3% of free hydrogen. The flue gas analysis gave 11.5% of CO_2, 8.4% of O_2 and 80.1% of N_2. Calculate per kg of dry fuel, the weight of necessary air and the weight of excess air.

Solution. Multiplying the volumetric results of analysis by the respective molecular weights of the gases the following relative weights are obtained:

$$CO_2 = 11.5 \times 44;\ O_2 = 8.4 \times 32;\ N_2 = 80.1 \times 28$$

Relative wt. of C in 11.5×44 parts of CO_2 is

$11.5 \times 44 \times \frac{12}{44} = 11.5 \times 12$ parts which is associated with 80.1×28 parts of N_2.

1 kg of fuel contains 0.84 kg of C, and the wt. of N_2 associated with it

$$= 0.84 \times \frac{80.1 \times 28}{11.5 \times 12} \text{ kg.}$$

$$\therefore \text{Wt. of air} = \frac{0.84 \times 80.1 \times 28}{11.5 \times 12} \times \frac{100}{77} = 17.73 \text{ kg per kg of fuel.}$$

Wt. of necessary air can be found by calculating the wt. of O_2 needed for C and H_2 in 1 kg of fuel.

Wt, of C in 1 kg of fuel = 0.84 kg.

$\therefore$ Wt. of O_2 needed for it = $0.84 \times \frac{32}{12} = 2.24$ kg.

Wt. of H_2 in 1 kg fuel = 0.03 kg.

Wt. of O_2 needed for it = $0.03 \times \frac{16}{2} = 0.24$ kg.

$\therefore$ Total wt. of O_2 needed per kg. of fuel = 2.48 kg,

$\therefore$ Wt. of air needed = $2.48 \times \frac{100}{23} = 10.78$ kg.

Wt. of excess air supplied.

= Wt. of air supplied – wt. of air needed

= 17.73 – 10.78 = 6.95 kg.

Alternative method for calculating weight of excess air:

N_2 associated with 8.4% excess $O_2 = 8.4 \times \frac{79}{21}$ parts by volume

Ratio of excess N_2 to total $N_2 = \frac{8.4 \times 79}{21} \times \frac{1}{80.1}$

$\therefore$ Excess air $= 17.73 \times \frac{8.4 \times 79}{21} \times \frac{1}{80.1} = 7.00$ kg.

$\therefore$ Wt. of air used = wt. of air supplied - excess air

= 17.73- 7.00 = 10.73 kg.

The following is much simpler method of calculating the percentage of excess air but if nitrogen and sulphur are present in the fuel the results are slightly incorrect, In the above example, 100 m^3 of flue gas has 80.1 m^3 of N^2

$\therefore$ Vol of air actually supplied per 100 m^3 of flue gas

$$= 80.1 \times \frac{100}{79} \text{ m}^3$$

Vol. of O_2 in the air supplied $= 80.1 \times \frac{100}{79} \times \frac{21}{100} = 21.3 \text{ m}^3$

Vol. of excess O_2 present in 100 m^3 of flue gas = 8.4 m^3

$\therefore$ Vol. of O_2 used in combustion = 21.3 - 8.4 = 12.9 m^3

$\therefore$ Percentage of excess air $= \frac{8.4}{12.9} \times 100 = 65.1\%$

(Excess air on the basis of previous method $= \frac{7}{10.73} \times 100 = 65.2\%$)

Note : Whatever method of calculation is applied the results must be same provided the question has also been framed correctly. Often the setter of the question does not do detailed and full calculations and simply alters the figures here and there e.g., one may lazily make the percentages of CO_2 as 12.0% and O_2 9% and N_2, 79% in the above question. This is wrong because by altering percentage of CO_2 the other percentages get altered according to the laws of combustion and not arbitrarily, and answers by different correct methods would be different. If this happens the obvious conclusion is that the question is wrong. The teachers framing the question have to be careful.

Examples

1. A sample of coal contains 87% carbon, 2% hydrogen, 1% oxygen, 1% sulphur and 9% ash. Calculate the theoretical weight and volume (at N. T. P.) of air required for complete combustion of I kg of the sample of coal.

   (**Answer :** 10.78 kg; 8.27 m^3).

2. A boiler is fired with coal having the following percentage composition by weight: C, 74.0; H_2, 6.8; O_2, 13.2; N_2, 2.1; S, 1.0 and ash, 2.9.

   Calculate the minimum quantity of air required for the complete combustion of 1kg coal and the percentage composition by weight of the dry products of combustion if 20% excess air is supplied.

   (**Answer :** air, 10.4 kg; CO_2, 21.11; O_2, 3.74; N_2, 74.99; SO_2, 0.16)

3. A sample of coal has 80% C; 5% H_2; 0.5% Sand 14.5% ash. Calculate the theoretical quantity of air necessary for the combustion of 1 kg of coal. Find the analysis of the flue gas by weight if 20 kg of air are supplied per kg of coal.

   (**Answer :** air, 11.04 kg; N_2, 73.83%; CO_2, 14.07%; O_2, 9.89%; H_2O, 2.16% and SO_2, 0.05%).

   Note: This question does not say, "dry flue gas."

4. A sample of dry coal has the following composition: C, 84%; H_2, 4%; O_2, 6%; S, 1% and the remainder ash. If 50% excess air is supplied estimate the percentage composition by volume of the dry products of combustion.

   (**Answer :** CO_2, 12.51; O_2, 7.04; N_2, 80.40; SO_2, 0.05).

5. A sample of coal was found to have the following percentage composition:

   C, 81%; H_2, 5%; S, 0.5%; N_2, 3% and ash, 10.5%. Calculate the percentage composition by volume of the flue gases, when air, 30% in excess by weight is supplied for combustion.

   (**Answer :** CO_2. 13.74%; N_2, 81.33%; O_2, 4.90% and SO_2, 0.03%).

6. The fuel oil used in a Diesel engine has 85.8% C; 12.5%. H_2; 1.5% Sand and 0.2% O_2. Analysis of the products of combustion by Orsat apparatus shows 8% CO_2 and 10% O_2 by volume. Calculate the weight of air supplied per kg of the oil.

   (**Answer :** 26.7 kg.)

7. During a boiler trial, the average analysis of the flue gas gave the following results; CO_2, 10.75%; CO, 1.16%; O_2, 6.8%; N_2, 81.29%. The chemical analysis of the coal fired gave: C, 88.1%; H_2, 3.5%; O_2, 4.8%; ash, 3.6%. Determine the number of kg of air supplied to the boiler furnace per kg of coal. Calculate also the weight of flue gas per kg of coal.

   (Answer : air, 18.23 kg; dry flue gas, 18.49 kg. (with moisture 18.805 kg).

8. The flue gas analysis by volume in a boiler trial was CO_2 10.5%; CO, 1%; O_2; O_2, 8%; N_2 80.5%: and the coal analysis, as burned, was C, 82%; H_2, 4.2%; O_2, 48%; other matter, 9%. Calculate the minimum weight of air needed for complete combustion and the weight of excess air supplied per kg of coal.

   (Answer : air, 10.76 kg; excess air 6.64 kg).

9. The percentage composition of a sample of a producer gas, by volume is as follows: H_2, 16; CH_4, 1; CO, 22; CO_2, 6; O_2, 1; N_2, 54. Find the minimum volume of air necessary for the complete combustion of 100 m^3 of this gas and the composition of the dry products of combustion by volume.

   (Answer : air, 95.23 m^3, CO_2, 18.3%; N_2, 81.7%)

10. A producer gas has the following composition by volume; CH_4, 3.5%; CO, 25%; H_2, 1O.4%; CO_2, 10.8%; N_2, 50.3%. Calculate the theoretical minimum quantity of air required per m^3 of the gas. If 20% excess air is supplied, find the percentage composition of the dry products of combustion.

    (Answer : air, 1.175 m^3 ; CO_2, 19.07%; N_2, 78.53%; O_2, 2.40%)

11. A gas used in an internal combustion engine had the following composition by volume; H_2, 45%; CH_4, 36%, CO, 15%; N_2, 4%. Find the volume of air required for the combustion of 1 m^3 of the gas.

    (Answer : 4.86 m^3.)

12. In a boiler trial the following analysis of the flue gas was obtained, by volume: CO_2, 13%; O_2, 6%. Find the kg of air per kg of coal if the carbon content of coal was 85%.

    (Answer : 16 kg.)

13. A sample of coal was found to contain the following: C, 81%; H_2, 4%; O_2, 2%; N_2, 1%; the remaining being ash. Estimate the quantity of minimum air required for complete combustion of 1 kg of the sample. Find the composition of dry flue gas by volume if 40% excess air is supplied.

    (Answer: 10.7 kg; CO_2, 13.23%; N_2, 80.77%; O_2, 6%)

Chapter 7

Nuclear Fuels

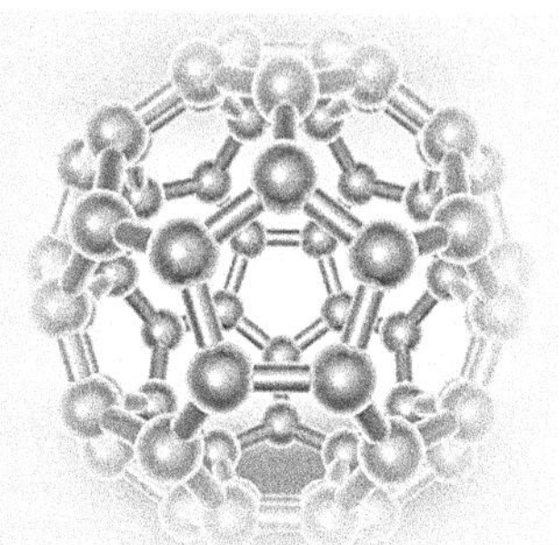

7.1 Nuclear Energy

The term *'atomic energy'* so commonly used really refers to *'nuclear energy'*. In the reactions involved, energy stored in the nucleus of the atom is released. The use of the term 'atomic energy' began because the energy is released by the splitting of the atom or more correctly the splitting of the nucleus of the atom (or by fusion of nuclei of atoms). It may be added here that the nature and origin of nuclear forces has not yet been fully understood. In the chemical fuels, the energy is available because of the rearrangement of the valence electrons of the atoms and the resulting recombination of the atoms forming new compounds. These new compounds have less energy stored in them and are, therefore, more stable. In the chemical reactions involved, the energy released is the difference in the energy of the reactants and of the reaction products. Nuclear fuels release energy by an entirely different mechanism. In the process, different elements are produced and some of the binding energy of the nucleus is released. Moreover, the nuclear reactions are tens of thousands of times faster than the chemical reactions. Therefore a *very large amount of energy is released in less than a millionth of a second.* This causes a terrific explosion.

The fission of 1 kg of uranium yields the same amount of heat energy as is released by the combustion of 3,000 tonnes of coal. If the fusion reactions can be made possible under suitable conditions, the deuterium present in all natural water to the extent of 0.015 percent will represent an almost inexhaustible source of energy. It has been calculated that the energy equivalent of the deuterium in one litre of ordinary water is the same as that obtained from the combustion of 300 litres of petrol. Thus more than 50×10^{20} litres of water present in the oceans could supply the world's power requirements of several million years, if the use of deuterium can be made to produce energy by fusion reactions. One of the most important aspects of nuclear fusion as a source of energy is that the basic fuel material is water.

7.1.1 Nucleus of the Atom

The nucleus of the atom consists of particles collectively called *nucleons.* These are mainly the protons and the neutrons. The mass of the atom as well as of the smaller particles is represented on a scale taking the mass of C atom as 12. On this scale, the mass of the neutron is 1.00867, of the proton

1.00728 and of the electron is 0.00055. The proton has a positive charge and the electron has a negative charge. Neutrons are electrically neutral. The *atomic weight* of an element takes into account the mass of all the particles making the atom. The *atomic number* is the number of protons in the nucleus. The chemical properties of the elements mainly depend on the protons in an atom of the element. Different elements have different atomic numbers. The total number of nucleons (i.e. the number of protons plus the neutrons) is the *mass number* of the element.

The atoms of an element may have different numbers of neutrons. These atoms, therefore, have different mass numbers and are called *isotopes.* Hence, isotopes may be defined as elements with the same atomic number but different mass numbers. All elements have isotopes. In a discussion on nuclear reactions it is necessary to identify the isotopes. This is done by using the mass number as the superscript e.g. ^{235}U, ^{238}U, etc. A subscripts of the left is used to indicate the atomic number e.g., $^{238}_{92}U$, $^{4}_{2}He$, etc.

A different element is formed if a proton is added to the nucleus of an element. If neutrons are added to the nucleus, the element remains the same, but an isotope of the element is produced, i.e., chemical properties of the element remain unchanged, but its fission properties change. Some of the elements or their isotopes (more correctly their nuclei) break down by themselves. These are known as *radio-active* elements or isotopes. In this process of breaking down they give rise to different elements and emit electromagnetic waves or charged particles or both.

Emission of the particles takes place in several ways and is due to the instability of the nucleus. Beta particles or high velocity electrons are emitted by the conversion of neutrons into protons $^{1}_{0}n \rightarrow ^{1}_{1}H + ^{0}_{-1}e + \gamma$. If a proton is converted into a neutron, a positron (positive beta particle) is emitted. $^{1}_{1}H \rightarrow ^{1}_{0}n + ^{0}_{1}e + \gamma$. A proton may also be converted into a neutron by capturing an electron from the orbit of the atom. In addition to these processes, there may be the emission of alpha particles, $^{4}_{2}$ He, which are fast moving helium nuclei. $^{x}_{y}Z \rightarrow ^{x-4}_{y-2}Z + ^{4}_{2}He$ (Z stands for a nucleus). These processes excite the nucleus with the result that electromagnetic waves are usually emitted. These have shorter wavelengths than the X-rays and are known as gamma rays.

The conventional symbols of these particles are: electron ${}^{0}_{-1}e$ positron ${}^{0}_{+1}e$, neutron ${}^{1}_{0}n$ and proton ${}^{1}_{1}H$.

As a result of the aforesaid processes the nuclei undergo decay. The rate of decay of a radio-active element at any instant is proportional to the number of its atoms present at that instant. The kinetics of the decay strictly follows the first order law as:

$$-\frac{dN}{d\lambda} = \lambda N \qquad .....(7.1)$$

where λ is the proportionality constant and is known as the decay constant, and N refers to the number of atoms at any time t.

The integral forms of the eq. (5.1) most commonly used are:

$$N = N_0 e^{-\lambda t} \qquad .....(7.2)$$

$$\text{or} \quad \log_e \frac{N}{N_0} = -\lambda t \qquad .....(7.3)$$

$$\text{or} \quad \log \frac{N_0}{N} = \frac{\lambda t}{0.693} \qquad .....(7.4)$$

N_0 refers to the number of atoms of the parent element at zero time.

The quantity which is very frequently used to characterize a radioactive element is its *half life* period. This is defined as the time taken by a given amount of a radio-active element to decay exactly to its half of the original value and is expressed mathematically in terms of decay constant as:

$$t_{\frac{1}{2}} = \frac{0.693}{\lambda} \qquad .....(7.5)$$

Sometimes *mean life* $\bar{\tau}$ is also used which is given by the relation;

$$\bar{\tau} = \frac{1}{\lambda} \qquad .....(7.6)$$

7.1.2 Binding Energy and Mass Defect

The diameter of the nucleus of an atom is very small, even when compared to the diameter of the atom. All the protons of an atom are located in the small space of the nucleus. The protons repel each other. The question naturally arises as to how the protons are held together in the nucleus. They are held together by *binding energy*.

As the atoms are made up of electrons, protons and neutrons, the atomic mass of an element should be the sum of the masses of these fundamental particles. Actually the atomic mass of an element is always less than the sum of the masses of electrons, protons and neutrons present in the atom. For example, consider an element whose mass number (total number of protons and neutrons) is A and its atomic number is Z, hence the number of neutrons is given by (A -Z). Let atomic masses of electron, proton and neutron be m_e m_p and m_n respectively. Atomic mass of element (M') is given by the expression:

$$M' = \text{mass of electrons} + \text{mass of protons} + \text{mass of neutrons},$$

$$= Z.m_e + Z.m_p + (A\text{-}Z)\, m_n$$

$$= Z\,(m_e + m_p) + (A\text{-}Z)\, m_n \quad \text{.....(7.7)}$$

But $m_e + m_p = m_H$ .....(7.8)

where m_H is the atomic mass of hydrogen containing one electron and one proton. From eq. (5.7) and (5.8) we have,

$$M' = Z\, m_H + (A\text{ -}Z)\, m_n \quad \text{.....(7.9)}$$

Suppose, the actual atomic mass of the element under consideration is given by M. Then,

$$\text{Mass defect} = M' - M = \Delta M \quad \text{.....(7.10)}$$

$$\Delta M = Zm_H + (A - Z)\, m_n - M \quad \text{.....(7.11)}$$

Thus the mass defect (ΔM) is the loss of mass in the formation of nucleus from its constituents. This loss of mass represents the amount of energy which would be released if an atom of mass number A were synthesized from its constituents. This energy is known as the binding energy and is responsible for holding the nucleons in the nucleus. A quantitative relationship between mass defect (ΔM) and binding energy (E) is given by Einstein's equation given below:

$$E = \Delta Mc^2 \quad \text{.....(7.12)}$$

where c is velocity of light (2.9979×10^{10} cm/sec). The velocity of light being very high indicates that the binding energy of a nucleus is also very high. It can also be concluded from eq. (5.12) that the higher the value of mass defect the higher will be the binding energy and hence higher is the stability of the nucleons.

Mass defect and binding energy can be further illustrated by taking the formation of a helium atom which contains 2 electrons, 2 protons and 2 neutrons. From eq. (5.9) M' is given by:

$$M' = 2m_H + 2m_n \qquad .....(7.13)$$

On ^{12}C scale $m_H = 1.007825$ and $m_n = 1.008665$.

Substituting these values in eq. (5.13) we have,

$$M' = 2 \times 1.007825 + 2 \times 1.008665$$

$$M' = 4.03298 \text{ amu (atomic mass unit)}$$

The actual mass of helium atom (M) on the same scale is 4.00260 amu and hence mass defect ΔM is given by

$$\Delta M = M' - M = 4.03298 - 4.00260 = 0.03038 \text{ amu}$$

Making use of Einstein equation and the conversion table given below, the binding energy of helium is given by

$$E = 0.03038 \times 931.9 = 28.30 \text{ MeV}$$

It is more convenient to express the binding energy in terms of specific binding energy or binding energy per nucleon. For example, in the case of a helium atom containing four nucleons specific binding energy is given by 28.30/4 = 7.075 MeV.

Conversion Table :

I g mass $= 9 \times 10^{20}$ ergs $= 56.16 \times 10^{25}$ MeV

$= 2.12 \times 10^{18}$ calorie $= 8.53 \times 10^{10}$ B. t. u.

$= 2.5 \times 10^{7}$ KWH

For most calculations in nuclear reactions the unit of mass is what is called atomic mass unit or amu. It is 1/12 of the weight of one ^{12}C atom.

According to Avogadro's number, 1 g atom (i. e. 12g) of ^{12}C has 6.02×10^{23} atoms of ^{12}C and by definition the mass of 1 atom of ^{12}C is 12 amu.

$\therefore \quad 6.02 \times 10^{23} \times 12$ g

or $\quad 6.02 \times 10^{23}$ amu = 1g = 9×10^{20} ergs = 56.16×10^{25} MeV

or $\quad$ 1 amu = 931.481 MeV

(Electron volt is the energy a unit electric charge gains when it falls through a potential of one volt. Million electron volts is abbreviated as MeV)

The significance of the mass defect can be better understood by considering a concrete example. The mass defect of helium has already been calculated as 0.03038 amu, which is equivalent to 28.3 million electron volts of energy. This means that 0.03038 amu of mass has been converted into energy in the formation of an atom of helium. Therefore, 28.3 MeV of energy is required for splitting an atom of helium. In other words, if the helium atom is synthesized from the elementary particles, this amount of energy will bc released.

7.2 Fission and Fusion

As the mass number increases the binding energy per nucleon or specific binding energy increases rapidly and reaches a maximum when the mass number is about 56 (i.e., for iron). See Fig. 7.1. Since the greater the specific binding energy, the greater is the stability of the nucleus, the nuclei of the atoms of mass numbers around 53 are the stablest. The specific binding energy of the heavier atoms is slightly less. Since there is the tendency to attain the state of maximum stability, the heavier atoms tend to break down giving more stable atoms (mass numbers 56 to 140) and release energy which can be calculated from the mass defect. This process is known as *fission.* The nuclear fission of an atom takes place only if energy equal to its activation energy is supplied. This has been practicable only in the cases of ^{233}U, ^{235}U and ^{239}Pu. These atoms readily attain activation energy when neutrons of appropriate velocity strike the nuclei.

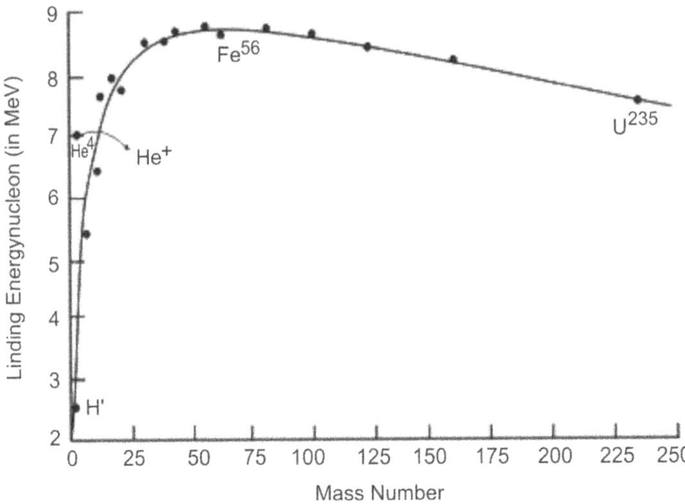

Fig. 7.1 Binding energy per nucleon as a function of its number.

If binding energy per nucleon is plotted versus mass number the curve for binding energy per nucleon rises rapidly as the mass number increases to about 56. To put it in other words, the mass defect of the lighter elements is small and it rises rapidly as the mass number increases. Therefore, if the lighter atoms join together (or fuse together) to form heavier atoms, a considerable amount of mass will be converted into energy. This process is known as *nuclear fusion.* Thus lighter atoms release energy by fusion whereas heavier atoms release energy by fission and give rise to elements whose nuclei are more stable. The fusion reactions can start only when the atoms are heated to very nigh temperatures. The high temperature required is produced by fission reaction. This seems to be the basis of hydrogen bomb manufacture.

A mixture of oxygen and hydrogen forms water only if certain activation energy is supplied, say by heating or by an electric spark. In a similar manner, fission reactions start only if certain activation energy is supplied. Thus if a slow-moving, or thermal neutron, is absorbed by the nucleus of an ^{235}U atom, the energy is increased by 6.6 MeV. This is greater than the required activation energy and fission reaction starts. If a thermal neutron raises the energy of an ^{238}U atom to 5.3 MeV only and fission reaction does not start because the energy is less than the required activation energy. If a fast neutron is absorbed by the nucleus of an ^{238}U atom, the activation energy is reached and fission reaction starts. The fast neutrons are more difficult to capture and a large number of them are wasted. Therefore, ^{238}U is not used as fuel in a reactor. Uranium occurring in nature contains 0.7% of ^{235}U and 99.3% of ^{238}U, which does not produce a chain reaction (0.005% of ^{234}U is also present). Therefore, natural uranium is enriched so that the proportion of ^{235}U is increased.

This discussion shows the necessity for slowing down the neutrons. The neutrons on being released from a fission reaction have a velocity of about 10,000 km per second. They are slowed down with the help of a material, known as a *moderator,* to a velocity of about 2 km per second. The moderator has a low atomic mass so that the fast neutrons on passing through it loose energy by collision with the atoms. The neutrons then have small energies nearly in equilibrium with the average kinetic energy of the moderator molecules (which are also light weight particles). The material to be used as a moderator must also have a low cross-section for the absorption of neutrons and a high scattering cross-section so that, when neutrons collide with its atoms, they are not lost. Heavy water, (D_2O), and graphite are commonly used as moderators. Beryllium atoms are small, do not absorb neutrons, and they can withstand high temperature. All these properties make it a very good moderator but it is expensive and toxic. Only heavy water can be used with natural uranium oxide fuel and carbon only can be used as a moderator with natural uranium metal fuel. The moderating ratios (which is a measure of moderating property) of various substances are: H_2O – 72; D_2O – 21,000; Beryllium – 129; Boron – 0.0009; Carbon – 170 and He – 83. Ordinary water is not used as a moderator because it absorbs neutrons. Water can only be used as a moderator if the fuel is enriched by increasing its ^{238}U content. The neutrons with low energy or slow moving neutrons are known as *thermal neutrons.* The thermal neutrons are more useful neutrons in the nuclear reactions because they spend more time in the neighbourhood of each nucleus and are more likely to be captured in a collision to cause further fission.

7.2.1 Probability of Reaction

On directing a beam of neutrons on a matter it is found that only some of them initiate a nuclear reaction. Only Some of the neutrons are captured by the nucleus, some are scattered, and the remainder pass through the matter unchanged. The probability of interaction of a neutron with the target nucleus is expressed as 'cross-section'. Cross-section depends upon the target nucleus and the energy of the neutrons. The cross-sections of different elements and of different isotopes of an element are different. A large cross-section indicates that the probability of interaction between a neutron and the target nucleus is high.

Cross-section is usually denoted by σ and has the dimensions of an area. The cross-section for scattering is denoted by σ_s and for absorption by σ_a. The cross-section for total interaction of a neutron and the target nucleus is denoted by σ_T and is the sum of σ_s and σ_a.

The unit for σ is the 'barn' and is equal to 10^{-24} cm^2. (The diameter of a nucleus is about 10^{-12} cm.)

The probability of absorption of a neutron increases as its velocity is slowed down. On absorption it may either be captured or may lead to a fission reaction.

7.3 Nuclear Fission

Nuclear fission is brought about by bombarding the nucleus of an atom by high energy particles. Of the various particles, neutrons are most widely used because they, being neutral, are not repelled when they come near the nucleus. In order to have a self-sustaining reaction or a chain reaction it is necessary that more neutrons be produced as a result of the fission reaction than are captured during the reaction. For example, fission of each ^{235}U nucleus on an average produces 2.5 neutrons. About 200 isotopes of about 30 elements, of mass numbers 70 to 160, have been isolated in the fission products of ^{235}U. A typical reaction can be represented thus:

$$^{235}_{92}U + ^{1}_{0}n \rightarrow ^{145}_{56}Ba + ^{88}_{36}Kr + 3\,^{1}_{0}n + 200\text{ MeV}$$

The actual reactions taking place is quite complicated and there are several intermediary stages before the final products are obtained. For example, the products in the above equation are radioactive and give further intense radiations (That is why one reads in the newspapers of radio-active dust from nuclear tests.) It is to be noted that the atomic as well as mass numbers on both the sides of the equation balance, but the mass of the products is slightly less than the mass of the reactants. It is this mass which is converted into energy as depicted by Einstein's equation.

The exact calculations of the energy released are beyond the scope of this book. An approximate calculation may help to understand the principles already enunciated. Let the equation given above be taken as the basis. The approximate binding energy per nucleon of ^{235}U is 7.6; that of ^{145}Ba is 8.3 and of ^{88}Kr is 8.7. Therefore the total binding energy of an atom of $^{235}U = 7.6 \times 235 = 1786$ MeV. The sum total of the binding energies of ^{145}Ba and $^{88}Kr = 8.3 \times 145 + 8.7 \times 88 = 1969$ MeV. In this simplified calculation, energies of the neutrons have not been taken into account. Thus it is seen that fission of one atom of ^{235}U releases 1969 -1786 = 183 MeV of energy, since ^{235}g of ^{235}U has 6.02×10^{23} atoms the energy released by 1 g of ^{235}U

$$= \frac{173 \times 6.02 \times 10^{23}}{235} \text{ Mev} = 1.68 \times 10^{10} \text{ cal per g of 235 U}$$

Since the above-mentioned reaction takes place in about 10^{-12} seconds, the energy released by one g of ^{235}U in one second is about 1.68×10^{22} cal. This is a terrific rate of energy release.

As the number of neutrons liberated in the above reactions is greater than the number of neutrons consumed, new nuclei are split, which liberate a still larger number of neutrons, and so on. Thus a divergent chain starts. See Fig. 7.2.

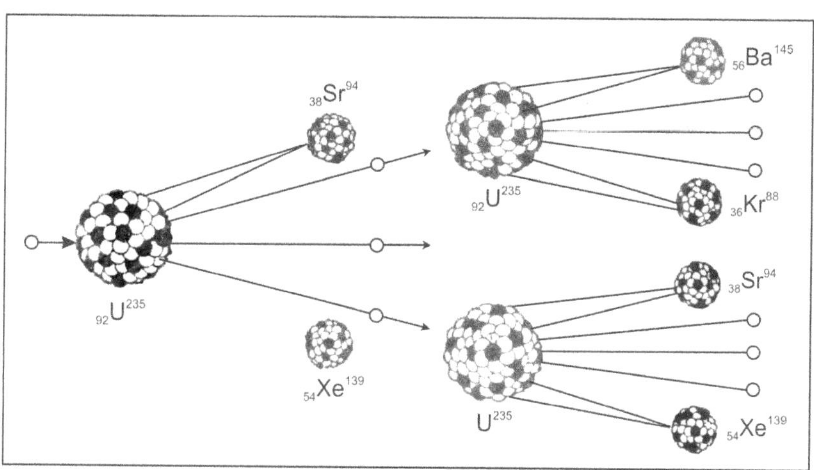

Fig. 7.2 Fission of uranium and the chain reaction. In this diagram, the small white circles represent neutrons and the black circles represent protons.

(From G. B. King and W. E. Caldwell, *College Chemistry*)

Courtesy, American Book Co., New York.

Natural uranium contains about 99.3% of ^{238}U isotope which reacts with neutrons giving out electromagnetic radiations known as gamma rays. The new isotope, ^{239}U, is radioactive and a negative charge is emitted from the nucleus. Thus the positive charge of the nucleus increases by one and a new element, neptunium, is produced. Neptunium is also radio-active and a negative charge is emitted with the result that the positive charge of the nucleus again increases by one and the element plutonium is produced. Fission of plutonium takes place in a manner similar to that of uranium (235). The reactions can be expressed as follows:-

$$^{238}_{92}U + ^{1}_{0}n \rightarrow ^{239}_{92}U + \gamma$$

$$^{239}_{92}U \rightarrow ^{239}_{93}Np + ^{0}_{-1}e$$

$$^{239}_{93}Np \rightarrow ^{239}_{94}Pu + ^{0}_{-1}e$$

Plutonium, used in India's nuclear explosion, on 18 May 1974, is supposed to have been produced in the country by such a reaction. Neutrons for the fission of plutonium do not need to be slowed down much and the whole structure can be much smaller. Such a reactor is called a "fast" reactor since it operates with fast neutrons.

Large deposits of thorium are found in Kerala and Tamil Nadu. ^{232}Th nucleus captures a neutron and gives out gamma rays and a radio-active isotope is formed. The nucleus of this isotope emits negative charges and palladium and uranium successively are synthesized. Like ^{235}U and ^{289}Pu, ^{233}Uranium is also fissile. The reactions can be depicted thus:

$$^{232}_{90}Th + ^{1}_{0}n \rightarrow ^{233}_{90}Th + \gamma$$

$$^{233}_{90}Th \rightarrow ^{233}_{91}Pa + ^{0}_{-1}e$$

$$^{233}_{91}Pa \rightarrow ^{233}_{92}U + ^{0}_{-1}e$$

The reactions mentioned above, (i.e. producing ^{239}Pu and ^{233}U) called *breeder reactions* are important because they produce fissionable material from non-fertile material. Fissionable materials are not available in nature in large quantities, hence it is important to produce these from more abundant materials. Thorium is a very important raw material for the production of artificial isotope ^{233}U which is fissionable by thermal neutrons.

The importance of the fast breeders can be seen from the fact that one tonne of uranium in an ordinary nuclear reactor will produce 480 million units of electricity, whereas one tonne of uranium converted into plutonium and used in a fast breeder reactor will produce 18,000 million units of electricity.

Thermal neutrons cause fission of ^{233}U, ^{235}U and ^{239}Pu. Of these three nuclides ^{239}Pu has been used largely for making weapons. ^{233}U has been produced in small amounts and has found use for experimental purposes. ^{235}U is being used for industrial purposes.

As a reactor fuel, uranium is used in three forms, uranium metal or its alloys; uranium compounds made compact by sintering ceramic fuels; and uranium compounds dispersed in a metal matrix (cermets).

7.3.1 Controlled Fission

For using nuclear reactions for power production it is necessary to control the fission. Only 0.7% of natural uranium is ^{235}U, the balance being almost wholly ^{238}U. The fast moving neutrons which are very few in numbers are mostly captured by ^{238}U with almost no fission reaction taking place. Further, if only a small amount of uranium is taken, most of the neutrons may escape from the surface of the metal. Therefore, it is necessary that there should be a certain minimum mass of the metal for producing a self sustaining reaction. This mass is known as the *critical mass.* The reactor is also surrounded by a reflecting medium, so that the neutron loss is minimized. The reactor is said to be critical when the number of the thermal neutrons in it remains constant. If the neutrons are being lost, the reactor becomes sub-critical and chain reaction will not be sustained. If the neutron density is increasing it becomes supercritical and the rate of energy release will increase.

It has already been explained that by slowing the neutrons the possibility of their capture increases. The moderators like graphite and heavy water reduce the energy of the fast neutrons, enabling them to be captured by ^{235}U instead of ^{233}U.

Since on an average 2.5 neutrons are released for each split of an ^{235}U nucleus, it is necessary that the neutron population be adjusted. For maintaining a constant and uniform power supply only one of the neutrons should split another nucleus. Therefore, 1.5 neutrons per split are available for non-fission reactions. Of the non-fission reactions the important ones are those which convert ^{238}U and ^{232}Th into fissionable materials as explained in the previous section.

The neutron population of the reactor is controlled by inserting rods made of a material which can absorb neutrons. Boron steel and cadmium have high absorption cross-section. The neutron absorption reaction can be represented as follows:

$$^{10}_{6}B + ^{1}_{0}n \rightarrow ^{7}_{3}Li + ^{4}_{2}He$$

Hence control rods of boron-steel or cadmium are commonly used for controlling neutron population. If the power supply is to be increased the rods are partially withdrawn, thereby making more neutrons available for starting more fission reactions. By pushing the control rods further inside the reactor, more neutrons are absorbed and the power output is diminished. Since the nuclear reactions take place in 10^{-14} to 10^{-12} seconds no man can adjust the depth of the control rod in the core. For this purpose electronic and electro-mechanical devices are used.

As the fission reactions proceed, the fission products accumulate. These elements can absorb neutrons, thus tending to slow down the useful reactions. It has been found that when about 1 per cent of the fuel has been consumed the fission products begin to slow down the fission reactions effectively and have to be removed. Therefore, the fuel has to be periodically taken out, the fission products, sometimes called 'ashes', are separated from it and the purified fuel is used again.

As a result of fission reactions, some matter is converted into energy. The heat energy can be used for steam production in a manner similar to steam production by the combustion of chemical fuels.

7.3.2 Power Reactors

A reactor is an assembly in which the chain reaction is kept going steadily at a controlled rate so that heat is produced continuously. A power reactor plant essentially consists of a reactor in which heat energy is produced by nuclear fission and arrangement for transferring heat from the reactor core to a heat exchanger in which water is heated to produce steam under pressure. The high pressure steam is sent to a turbine which drives a generator thereby producing electric power.

Power reactors are of several types, the essential difference in these being the type and arrangement of fuel, moderator, etc. and the heat transfer unit.

A typical reactor has a huge sphere made of steel. Fuel is kept in the sphere. The fuel is either natural uranium or uranium enriched with ^{235}U isotope. Uranium rods are sealed in alloy cans so that they may not get oxidized. The canning material should be inert towards coolant. There must be a good thermal contact between the fuel and cladding material and both of these must be as resistant as possible to radiation damage. To ensure good heat transfer the cans are provided with fins (or projections). These cans are surrounded by a moderator like graphite-blocks or heavy water. This arrangement of the fuel and moderator is called the reactor core. It is provided with pipes or channels carrying the cooling material. The core is surrounded by graphite layers which reflect the neutrons back in to the core. Control rods, made of boron steel or cadmium are also provided in the core. These rods can be inserted in the core to varying depths, thus the rate of reaction can be controlled. See Fig. 7.3.

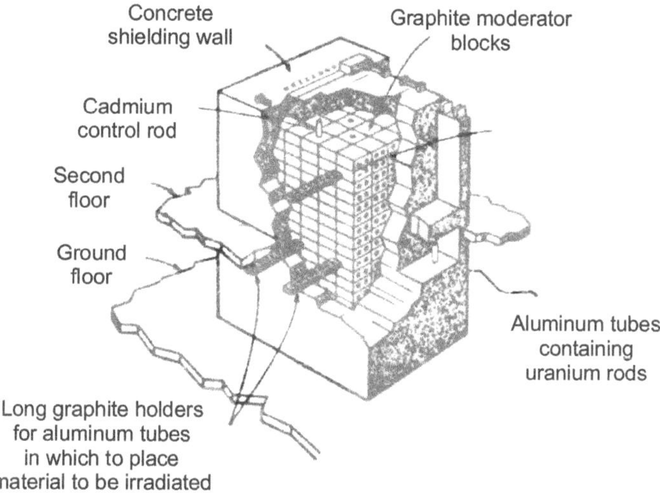

Fig. 7.3 General plan of a nuclear reactor using graphite as a moderator.
(C. H. Sorum, fundamentals of general chemistry, courtesy, prentice-hall, Inc, Englewood cliffs)

The reactors which use natural uranium are more bulky than those which use synthetic fuel. The reactors which produce more fissionable material (by reactions already explained in the earlier section) than is consumed are known as breeder reactors and their principal objective is the production of fissionable material and power production is a secondary objective. The reactors are shielded by heavy concrete containing iron ore so that radiation may not affect the workmen outside.

Gases, like carbon dioxide and hydrogen, water and liquid metals are used for carrying away heat from the reactor proper. The gas is circulated at a high rate and under pressure. Water is a very good and cheap cooling agent, but it has to be kept under pressure. Liquid sodium metal does not need to be kept under pressure but it is difficult to handle and also attacks materials of construction. Air and carbon dioxide are easy to handle but they absorb neutrons and acquire undue radioactivity.

Water-cooled reactors are more common. The core is immersed in water under pressure. The pressure may be as high as 150 kg/cm^2. The hot water under pressure is taken to a heat exchanger or boiler. See figure 5.4. Saturated steam at a pressure of about 40 kg/cm^2 is generated in the steam generator. The steam drives a turbine which drives a generator. The exhaust steam is condensed and sent back to the steam generator.

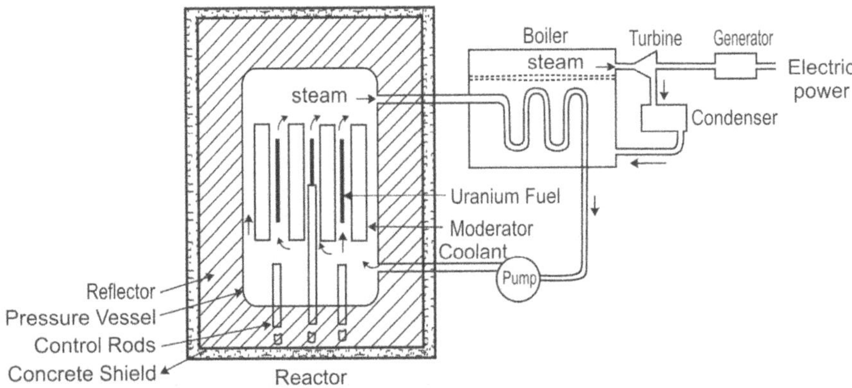

Fig. 7.4 Flow sheet diagram of a power reactor plant.

A simplification of the above arrangement is made by boiling the water in the reactor and directly generating high pressure steam which is sent to the turbine. In this case, the turbine and the condenser are subjected to radiations and have to be properly shielded.

In the arrangements discussed above, the coolant is circulated through the reactor under high pressure which necessitates the use of pipes capable of withstanding high pressure. This difficulty is removed by using molten sodium, or an alloy of sodium and potassium, as the cooling medium. The added advantage of liquid metal coolant is that the reactor can be operated at higher temperatures without needing high pressure equipment. Besides this, is the fact that the metal is a better conductor of heat with the result that high thermal efficiency is obtained and the equipment does not have to work under pressure. Liquid metals are preferred as coolants, but they have to be handled with care. They also give problems for construction materials which can withstand corrosion and high temperature.

Reference has already been made to breeder reactions. In a breeder reactor the fuel instead of being surrounded by a moderator is surrounded by a blanket of thorium or uranium which reacts with the neutrons liberated and is converted into fissile material. It is periodically removed and the fissile material is separated. Such a unit is more compact. Heat is transferred from the core and used for power generation. For fast breeder reactors solid solutions of CO_2-PuO_2 and UC-PuC are the favoured fuel materials.

One serious drawback of fission reactors is the production of radio active waste. The fission products are in a highly excited state and their nuclear particles rearrange giving off energy in the form of photons, electrons, neutrons and alpha particles. The gamma rays are particularly dangerous as they have an extremely damaging effect on living tissues. The half-life period, that is the time taken for the isotope to radiate half its surplus energy, varies from fractions of a second to millions of years. The isotopes with short half-lives can be stored in a safe place until the radiation is given off. Disposal of the others poses a serious problem. Several suggestions have been made but work is in progress to find still better solutions for disposal.

7.4 Fusion Reactions

The fission reaction poses two problems. One is the shortage of the fissible materials and the other is the disposal of radioactive wastes. Fusion reactions use deuterium, 0.015 percent of it being present in water (Tritium, ${}_{1}^{3}H$ has to be manufactured as plutonium is manufactured). Sea water contains deuterium in plenty. For every 6,700 hydrogen atoms one atom of deuterium is present. Deuterium contained in one m^3 of sea water can theoretically produce about 8×10^3 Joules of power. According to this, theoretically, deuterium contained in 4 km^3 of sea water can produce energy equivalent to the coal produced annually in India. These do not produce radioactive waste. Moreover, weight for weight, much more energy is released. It is believed that the heat and light given out by the sun and the starts is due to fusion reaction in which hydrogen gas present at several hundred million degrees is being converted into helium. As very high temperature is required, the reaction is also known as thermo nuclear reaction. This may happen in one of the following ways :

Reaction	Mass Loss	Energy Release per g He
${}^{2}_{1}H + {}^{2}_{1}H \rightarrow {}^{4}_{2}He$	0.026 amu	1.39×10^{8} kcal
${}^{2}_{1}H + {}^{3}_{1}H \rightarrow {}^{4}_{2}He + {}^{1}_{0}n$	0.019 amu	1.03×10^{8} kcal
$4{}^{2}_{1}H + {}^{4}_{2}He + 2{}^{0}_{1}e$	0.029 amu	1.55×10^{8} kcal

It is the last reaction which is believed to be the predominant reaction taking place in the sun and the stars whereas the first reaction is the likely reaction in the hydrogen bomb. The last reaction does not proceed in one step, but represents the summation of a series of reactions. The energy release shown in the last column is not the net energy available. Sufficient energy is needed in bringing the charged particles close enough together for them to react.

The main problem is how to produce the high temperature required to start the reaction. In the hydrogen bomb this is obtained by a fission reaction (or an atom bomb which acts as a trigger). For peaceful utilization of the energy even if the temperature is produced the problem of the material comes up because no material can withstand the temperature. The problem of the material has been solved by keeping the hot gas (which is in the form of plasma i. e. a gas of positively charged nuclei and of electrons) within a magnetic field that it does not come in contact with any material. One method that has been suggested is to shoot deuterons (i. e. the nuclei of deuterium) at suitable targets when some of them react to produce very high temperature. This hot plasma is kept within a magnetic field for sufficient time for the reaction to proceed. It has been possible to keep the plasma at the required temperature for a few hundredths of a second. For a plant to be successful it has to be kept for at least a second. The use of laser beams to achieve the required conditions is also being tried.

It is reported that the scientists at Princeton University have succeeded in attaining a temperature of 60 million degrees which is quite close to 100 million degree temperature required for the sustained release of fusion energy. It may be added that even the design for such a thermo-nuclear fusion power plant has not yet been made.

Some scientists have suggested that it may be possible to bypass the inefficient way of steam generation and that the stream of charged particles emerging from a fusion reaction may directly give electricity.

7.5 Radiation and Thermal Pollution Hazards

All the reactors have been designed to provide sufficient shielding so that the radiation level outside the shield is well within the accepted levels. The primary cooling circuit and heat exchangers are also shielded since the primary coolant becomes highly radioactive through exposure to reactor neutrons. The escape routes of activity through pump or pipe or valve failure are carefully controlled.

The thermal pollution is another environmental hazard. That part of a reactor's thermal power which is not converted to electrical energy is eventually released as heat, often to a pond of water. If this heat is not appropriately dispersed this may lead to ecologically harmful temperature rises. Cooling towers are now being used. It must be mentioned here that the thermal pollution problem in nuclear plants is greater than in the fossil fuel power plants as the conversion of thermal into electrical energy in the latter is much more efficient. However, the over-all hazards are less than, in many other industries, perhaps much less than in many chemical industries.

7.5.1 Safety in Nuclear Power Plants

Contrary to some popular misconceptions a reactor cannot explode like an atom bomb as the conditions required for a major explosion are most unlikely to be achieved in a nuclear reactor. It has been estimated that for a major explosion to take place, about 50 g of fissile material will have to be fissioned before the critical mass can blow itself apart within the order of a microsecond. The neutron multiplication is far too slow in a reactor and as such possibility of a major explosion is almost out of question.

Nuclear Power Plants have maintained a fairly clean record as regards safety. Control systems are provided for adequate redundancy. Safety reactor shut down systems as well as emergency core cooling systems are provided with reliable power supply. Some of the Reactor designs provide a built in safety system which stops the chain reaction or gives negative effect when the temperature in the reactor core shows an increasing tendency. Elaborate procedures for operation and maintenance of the plants are formulated and operating personnel are thoroughly trained in the various aspects of radiological protection. In addition to the above, the Health Physics Units attached to the power plants monitors and regulates the working of the personnel to safe-guard them from radiation exposure. Regular monitoring on the environment around the plants is done for the radio activity in the air, drinking water sources, soil, marine life, milk and animals to safeguard the public in case objectionable levels are observed. Radiological exposure limits have been specified by International bodies which are mandatory for all power plants. The population growth and agriculture around the plants up to a radial distance of 50 kilometers is controlled. In addition to the above, emergency evacuation plans are worked out to be implemented in co-ordination with the local civic authorities in case of accidents to minimise the radiation exposure to the public.

However some accidents in the nuclear power plants have been reported. Most of them have been of minor magnitude. Two major accidents need special mention. The first is that which took place in the Three Mile Island plant in U.S.A. In it the leakage of the radioactive material could be successfully restricted to the containment and reactor buildings, with the result that the radio activity in the neighbourhood did not raise to dangerous level. No loss of life was reported.

The second, and a much more serious accident, took place at the end of April, 1986 in the Chernobyl plant in the southern part of the U. S. S. R., and the first report came on account of increased radio activity noticed in Scandinavian countries. It is said that the plant did not have containment and there were escape routes for the fission products which got released into the atmosphere. The intensity of fire is reported to be so high that it could not be controlled for days and the reactor got completely damaged. Some deaths have also been reported with excessive exposure to radio activity to many more persons.

The disasters in nuclear power plants involve spread of radio activity having long term and unknown effects and naturally attract much public attention. Although the small thermal power plants continuously give out harmful emissions of CO_2 and SO_2 which one hardly notices, but the cumulative effects of such plants scattered allover the world due to ever increasing number of large thermal power plants using coals of poorer and poorer quality cannot be completely overlooked.

7.6 Sources of Nuclear Fuels

Uranium is certainly the most important element used as fuel for nuclear energy. It is more abundant than gold and silver. The amount of uranium in the earth's crust is about the same as that of a common metal like tin. It is widely distributed, but rich deposits have not been found in India.

Uraninite and pitchblende are the principal ores, U_3O_8, of uranium. The ore is first concentrated by physical means and then is leached by sulphuric acid. It is precipitated from the solution as sodium uranyl vandate which is reduced to give UO_2. It is dissolved in nitric acid, impurities are separated, and more concentrated UO_2 is precipitated. The oxide is then reduced by hydrogen.

It has already been explained that ^{235}U isotope is more important for nuclear energy than ^{235}U. The separation of isotopes cannot be done by chemical means. The metal is converted into fluoride which volatilizes at 57°C. The fluorides of the two isotopes have slightly different densities and are separated by diffusion. The diffusion has to be repeated many times to obtain any significant concentration of ^{235}U.

Next to uranium, thorium is the most important material for nuclear energy. In the earth's crust, it is three to four times as abundant as uranium but rich deposits of it are not so common. Moreover, its minerals are not easily attacked by chemicals. This makes its extraction difficult. Thorium is mostly extracted from the *monazite* sand of which large deposits are found in India. Important deposits of monazite are also found in Brazil. Rock monazite is believed to have been disintegrated by the weathering action. Later, in the course of thousands of years, sea waves and tides eroded the material, resulting in concentrated deposits of the mineral as beach sand which contains 1 to 10% monazite. Monazite contains 1 to 10% ThO_2 and 0.1 to 1% U_3O_8. It also contains rare earths like cerium and lanthanum.

The monazite-bearing sand collected from beaches contains quartz as the chief gangue. It is subjected to ore-dressing operations like gravity separation, froth flotation and magnetic separation. The coarser deposits are pulverized. The raw material is then treated either by the acid or alkali process.

In the acid process the concentrated monazite sand is heated to about 200° with concentrated sulphuric acid. The thick liquid so obtained is diluted with water and allowed to stand so that the sand settles. The solution from the top is filtered and ammonia is added to bring its pH value to 1.0 when thorium precipitates as hydrated oxide. Rare earths also begin to precipitate at this pH value but most of them precipitate at pH value 2.3. Uranium precipitates when the pH value of the solution is about 6.0. The thorium fraction is dissolved in nitric acid and is treated with an organic solvent which extracts thorium salt from the solution.

In the alkali process the monazite sand is heated with concentrated sodium hydroxide solution and filtered. The residue is dissolved in concentrated hydrochloric acid. Thorium and uranium are then precipitated from the acid solution by the addition of alkali. Final purification is done by extraction with an organic solvent.

Thorium oxide obtained by either of the processes is heated with carbon in an atmosphere of chlorine and the chloride is obtained thus:

$$ThO_2 + C + 2Cl_2 = ThCl_4 + CO_2$$

The chloride on treatment with calcium or sodium gives the metal:

$$ThCl_4 + 4Na \rightarrow Th + 4NaCI$$

Thorium is also obtained by electrolysis of a fused mixture of thorium and sodium (or potassium) chlorides (or fluorides). Thorium melts at a high temperature (1845°). The metal obtained by the above mentioned processes is in powder form and is made into blocks by the combined effect of pressure and temperature.

Besides being used for nuclear power production, thorium finds several other uses.

7.6.1 Reprocessing of Fuels

The necessity to reprocess the spent fuel is due to ecological as well as economic considerations. The ecological concern is the radio-toxicity of plutonium and fission products. The cladding of the nuclear fuel cannot remain intact and stable for about a thousand years required for the fission product activity to decay, whereas plutonium would have to be kept under seal for a still longer period. The β and γ activities of spent fuel are of the order of 10^6 to 10^7 ci/t which is to be brought down to the level of natural uranium (0.67 ci/t). The economic reason for the reprocessing stems from the reusability of uranium and plutonium as nuclear fuels. The recovered uranium can be used to save about 13% of the separative work needed for isotope enrichment.

Reprocessing of the spent fuel and the storage of the resultant fission products are an important part of the nuclear fuel cycle. The nuclear fuel cycle gives a large number of elements including the fissile and fertile materials. The principal separation technique used in the reprocessing is solvent extraction and particularly the PUREX (Plutonium Uranium Recovery and Extraction) process. The main emphasis is laid on the minimization of nuclear waste, which is achieved by (i) recycling the process chemicals (solvents, acid etc.) (ii) reducing the consumptive process chemicals and (iii) using chemicals which can be decomposed into gaseous form.

The major process steps in nuclear fuel reprocessing in aqueous media are reduction and oxidation steps. These steps are used either to destroy process chemicals or to alter the valency of plutonium, uranium or neptunium. In the PUREX process the solvent used is a 20 to 30% solution of tri-n-butylphosphate (TBP) in kerosene or n-dodecane. TBP is the most suitable solvent because of its high radiation stability. The broad stages involved in the reprocessing of the spent fuels are as follows:

(i) The spent fuel is digested with hot NaOH if the cladding material is aluminium. This removes aluminium in the form of $NaAlO_2$. If the cladding material is other than aluminium, the fuel is dissolved in hot HNO_3 (leaving cladding material undissolved).

(ii) For the subsequent separation of plutonium from uranium, advantage is taken of the very low partition coefficient of Pu (III) for extraction into organic solvents. Plutonium is reduced to Pu (III) by using reducing agents such as Fe (II), SO_2 or NH_2-NH_2 whereas, uranium remains in the +6 state. The organic layer after solvent extraction contains uranium. Plutonium goes to aqueous layer.

(iii) For the separation of the fission products and purification of uranium and plutonium, additional solvent extraction steps are required followed by other techniques such as ion-exchange.

7.6.2 Possibilities of Nuclear Power

The sources of chemical fuels are quite limited, and the rate at which the fuels are being consumed is ever-increasing. Nuclear reactions provide an alternative source for power generation. Electric power is generated by turbines. The difference in the various power plants arises from the methods used to turn the turbines. In a hydro-electric project, running water turns the turbines. In a thermal power plant, a chemical fuel, such as coal, is used to heat water and generate steam which does this work. In a nuclear power station, the fuel is uranium, which supplies heat by fission reaction to produce the steam which turns the turbine blades, which in its turn drives the generator to produce electricity in the conventional way.

Thus the only saving that is achieved in a nuclear power plant is in the cost of the fuel consumed. Although it is likely to supply cheaper power, the cost of most finished industrial products will not be affected much because the cost of fuel forms only a fraction of their cost of production.

One of the achievements of nuclear power has been the construction of small nuclear reactors which are used to drive ocean going vessels. Submarines have been built which have gone round the world without coming to the surface for refueling.

The nuclear power plants also call for special requirements. The operation of a nuclear plant requires special skill, which, of course, is acquired as they become more common. The reactor has to be carefully shielded to prevent radiation affecting the workmen. The plants, at the present, are heavy and cannot be used in aircraft or automobiles.

Advantages and Ddisadvantages of Nuclear Power Plants

Advantages :

- Essentially no greenhouse gas emissions
- Does not produce air pollutants such as carbon monoxide, sulfur dioxide, mercury, nitrogen oxides or particulates
- The quantity of waste produced is small
- Small number of accidents
- Low fuel costs
- Large fuel reserves
- Ease of transport and stockpiling of fuel
- Future designs may be small and modular (SSTAR, etc.)

Disadvantages :

- Nuclear waste produced is dangerous for thousands of years
- Consequences of an accident might be disastrous
- Risk of nuclear proliferation associated with some designs
- High capital costs
- In the past long construction periods, imposing large finance costs and delaying return on investment
- High maintenance costs
- High cost of decommissioning plants

7.7 Indian Resources

The nuclear power stations require high initial capital outlay but have low operation costs as compared to the thermal power plants. In India, coal is found mostly in the eastern region. The western and the southern regions have poor energy resources. Nuclear power plants can help to meet the energy requirements of such regions.

Raw materials: The Atomic Minerals Division of the Department of Atomic Energy, Government of India, surveys, prospects and develops the atomic minerals in India. Industrial production of these is handled by the Indian Rare Earths Ltd. and the Uranium Corporation of India. Indian reserves of uranium and thorium are estimated to be 73,000 tonnes of U_3O_8 in grades ranging from 0.015 to 0.07% of U_3O_8 and 3,78,000 tonnes (ThO_2).

The Singbhum district of Bihar has perhaps the highest potential for uranium deposits in the country. A uranium mill located at Jaduguda close to the mine site, is extracting uranium from the ore by acid leaching. The mill processes 1,000 tonnes of ore (0.06% U_3O_8) per day. Narwapahar, in the same region, has the largest uranium deposit. There are several other localities in the neighbourhood, like Bhatin, which also have deposits and are being handled. Incidentally, these deposits are associated with valuable metals like molybdenum and nickel which are in short supply in India. Apart from Bihar, uranium deposits have been discovered near Udaipur in Rajasthan, Kulu district in Himanchal Pradesh, Dehradun District in U. P. and at several other places in the country. New mines and mills are being planned at Narwapahar and Turamdih (East) together with a full scale recovery plant at Mosabani for recovery of uranium from copper tailings.

It is believed that India has the largest thorium reserves of the world. Monazite, which is the chief mineral, is widely distributed in the granites, meta-sediments, and sedimentary rocks of Peninsular India, as well as along many river and stream courses. The rivers transport it to the sea and deposit it along with other alluvial matter. Some of it might have been collected by the disintegration of monazite rock by weathering near the sea shore itself. In the course of thousands of years the sea waves and tides eroded and washed the various materials and redeposited monazite mineral in somewhat concentrated form, on the beaches of Kerala and South Madras coasts. Significant reserves of monazite have also been found in the Ranchi district, Bihar and the adjacent plains of Purulia district in West Bengal and in the Chatrapur coast of the Ganjam District of Orissa.

The Indian Rare Earths Ltd., separates monazites from the beach sand in a plant at Manavalakurichi and the monazite is treated in another plant situated at Alwaye, Kerala State. The solution is processed at this plant for the separation of the rare earths. The solid residue is sent to the Thorium Plant at Trombay, near Bombay, where thorium nitrate and uranium fluoride are produced. Most of the thorium nitrate is consumed by the gas mantle industry. Uranium fluoride is further processed at the Uranium Metal Plant, Trombay, and nuclear grade uranium metal and uranium oxide is obtained.

Nuclear Fuel Complex, Hyderabad, started producing in 1972 valuable nuclear materials like fuel elements for power reactors (natural UO_2), enriched uranium oxide fuel elements (for Tarapur station), nuclear grade UO_2, zircalloy cladding and other components. Uranium Metal Plant, Trombay, has produced nuclear grade uranium metal since 1959. About 10 kg per year of plutonium is produced in 50 MWe (MWe, is megawatt electrical, refers to power produced as electricity) research reactor in Trombay. Another plutonium plant is under trial production (To make an atomic device the minimum amount or the "critical size" needed is about 8 kg of plutonium). Since 1955 thorium nitrate and crude uranium fluoride are in production at the Thorium Plant, Trombay.

Fabricated fuel requirements for a 235 MWe pressurised heavy water type reactor (like RAPP, MAPP, etc.) are 61 tonnes of uranium in the first year and 33 tonnes of uranium per year subsequently and those for 500 MWe reactor 115 tonnes and 61 tonnes respectively. Zircaloy requirements are 19 tonnes and 43 tonnes for the 235 MWe and 500 MWe PHWRs, respectively.

Research Reactors : As a first step in the nuclear research programme, APSARA, a swimming pool reactor, capable of operating at a power level of 1 MW was designed and constructed entirely by Indian effort, except for the enriched uranium fuel elements (140 g of ^{235}U). This reactor attained criticality in 1956 and has been in operation since then. It is primarily used for producing radio-isotopes and for investigating fission reactions.

For undertaking a materials testing programme and large scale production of radio-isotopes experimental investigations of solid state physics and reaction engineering and also for training personnel in reactor technology, a 40 MW research reactor now known as CIRUS was then constructed with Canadian collaboration. This reactor became critical in 1960 and uses 55 kg of natural uranium metal as fuel, heavy water as moderator and ordinary water as coolant.

To enable further investigations on new reactor concepts, a zero energy reactor ZERLINA, was built. It attained criticality in 1961. The reactor also uses 45 kg of natural uranium as fuel and heavy water as moderator. It is normally operated at power levels not exceeding 100 watts.

In order to understand the physics of fast neutron multiplying systems, a zero energy experimental fast reactor 'PURNIMA' has been constructed by scientists and engineers of Bhabha Atomic Research Centre. After carrying out the necessary modifications in the reactor, the advanced Purnima-II reactor was made critical on May 10, 1984. This is the first reactor in the world to utilise ^{233}U as fuel. The experiments carried out with this reactor will help the detailed designing of the Pulsed Fast Reactor being set up at the Indira Gandhi Centre for Advance Research, (the earlier name was Reactor Research Centre) at Kalpakkam near Madras,

For the production of radio-isotopes, and developing the technology related to the power reactors, a high flux reactor DHRUA (the earlier name R-5) of 100 MW thermal power was made critical on August 8, 1985 at Trombay.

In view of the fact that the long term future of the nuclear power lies with fast reactors, a Fast Breeder Test Reactor (FBTR) has been successfully commissioned at Kalpakkam. This reactor has been mostly setup with the indigenous efforts. This is a liquid sodium metal coolant reactor and the mixed carbide fuel used has been developed by Indian scientists and engineers. In addition to providing necessary experience in design, construction and operation of fast reactors, this reactor will serve as a useful irradiation facility and also produce about 15 MW of electricity. To commission the Fast Breeder Reactor, a wholly owned enterprise of Government of India, Bharatiya Nabikiya Vidyut Nigam limited (BHAVINI) was incorporated on 22nd October 2003. BHAVINI is currently constructing a 500 MWe Fast Breeder Reactor (FBR) at Kalpakkam with a cost of Rs 3492 crores. The FBR will generate power by recycling plutonium and depleted uranium recovered from the spent fuel of the pressurized Heavy Water Reactor. The FBR technology would thus allow the nuclear power generation capacity to grow to 3.5 lakhs MWe without needing additional uranium.

7.7.1 Power Programme

India's programme for nuclear power development envisages the construction of natural uranium fueled reactors in its first phase to be followed in the second phase by fast breeders fueled with plutonium produced in the first phase reactors. This will eventually lead to a self-sustaining thorium-uranium nuclear fuel cycle.

India's first nuclear power station, Tarapur Atomic Power Station (TAPS) has been built at Tarapur (96 Km north of Bombay) in Maharashtra State, at a cost of Rs. 65.2 crores. This part of the country has large power requirements but sources of power supply are inadequate. Regular supply of 420 MW of electric power began from July 1969. The U. S. A. has supplied much of the equipments and has also provided 40 tonnes of enriched uranium oxide which has been fabricated into fuel elements needed as the initial fuel. This fuel remains intact for 2 years of operation. After that one fifth of it needs to be replaced by new fuel. Partial replacement is required each succeeding nine months. Later, 22 tonnes of fuel will be required every year (A thermal power plant of the same size would require about a million tonnes of superior grade coal every year). The plant has two reactors of boiling water type and uses enriched uranium fuel. Cooling water for the plant is drawn from the Arabian Sea and fresh water for making steam is drawn from a reservoir several miles away.

India's second nuclear power station Rajasthan Atomic Power Station (RAPS) is located at Rana Pratap Sagar on Chambal River at Rawatchatta, Kotah, Rajasthan. It has two reactors of 220 MW (electrical) each. The first reactor attained criticality in 1972 while the second reactor was made critical in 1980. Both units are now supplying electrical power to the Rajasthan grid. These reactors, unlike TAPS reactors, use natural uranium (UO_2 in zircaloy cladding) as fuel and heavy water as moderator as well as coolant. The total cost of the power station is about Rs. 150 crores. India's third Atomic Power Station and the first fully indigenously designed and constructed 235 MW electrical first reactor using natural uranium as fuel and heavy water as moderator as well as coolant at Kalpakkam was connected to the grid in July, 1983. The second 235 MW reactor of this station was made critical on August 12, 1985 and has already been connected to the grid. The MAPP reactors will use natural uranium as fuel and heavy water as moderator as well as coolant as in RAPS reactors.

India's Nuclear Power Programme currently has fourteen reactors in operation that includes 2 Boiling Water Reactors (BWR) and 12 Pressurised Heavy Water Reactors (PHWR). The Total Nuclear Power Generation Capacity of these reactors is 2770 MWe.

Operating Nuclear Power Reactors.

Location	Type / Capacity
Tarapur	BWR / 2 × 160 MWe
Rajasthan	PHWR / 1 × 100, 1 × 200 MWe and 2 × 220 Mwe
Kalpakkam	PHWR / Unit 1-170 MWe, Unit 2-220 MWe
Narora	PHWR / 2 × 220 MWe
Kakrapara	PHWR / 2 × 220 MWe
Kaiga	2 PHWR units of 220 MWe

There are eight power reactors under construction at present of which two PHWRs of 540 MWe capacity each are at Tarapur in Maharashtra, four 220 Mwe capacity power reactors two each at Kaiga (Karnataka) and Rawatbhata (Rajasthan) and two VVERs of 1000 MWe capacity each at Kudankulam(Tamilnadu).

Nuclear Power Reactors Under Construction.

Location	Type / Capacity	Expected Criticality Date
Tarapur (3 & 4)	PHWR / 2 × 540 MWe	▪ Unit 3- July 2006 ▪ Unit 4- October 2005 (Attained Criticality on 06-March-2005)
Kaiga (3 & 4)	PHWR / 2 × 220 MWe	▪ Unit 3- December 2006 ▪ Unit 4- June 2007
Rajasthan (5 & 6)	PHWR / 2 × 220 MWe	▪ Unit 5- May 2007 ▪ Unit 6- November 2007
Kudankulam Tamilnadu	VVER / 2 × 1000 Mwe	▪ The project is being implemented under the Inter-Governmental Agreement (IGA) signed between India and a Supplement to the IGA signed between India and Russian Federation. ▪ Unit – 1–2007 ▪ Unit – 2–2008

Nuclear Power Corporation Ltd. (NPCIL), a public sector company which owns, constructs and operates these nuclear power plants in India, is spearheading the nuclear power programme in India and plans to put up a total installed nuclear power capacity of 20,000 MWe by the year 2020.

7.7.2 Other Uses

Besides power generation, the nuclear reactors have other peaceful uses. The reactors are also used to produce radioactive isotopes of various elements. These isotopes have the same chemical behaviour as other isotopes. For example, radioactive carbon or sulphur can be used in exactly the same way as normal carbon or sulphur. Minute quantities of these can easily be traced due to their radioactivity. Such isotopes are incorporated in the material of a machine or in a material whose movement is to be traced or given to a human being and the movement of radioactivity is traced showing the passage of the element. Radioactive isotopes are being put to an increasing number of medicinal purposes. The isotopes have proved to be of immense use for the treatment of cancer in the various parts of the body. Radio isotopes are used for the sterilisation of medical instruments, preservation of food products, and are used for producing various improved varieties of food grains. The use of isotopes for the radiography of welds and materials to locate defects under surface is of great importance to the industry.

The nuclear test conducted by India in May 1974 was directed towards peaceful use of the nuclear energy. Five such uses mentioned are building the harbours, building water reservoirs, the construction of dams, copper mining and stimulating oil wells that are in the process of drying up.

India has thus become the sixth country in the world, and the only developing country with an independent capability starting from the mining of uranium, fabrication of fuel, design, construction, commissioning and operation of nuclear power plants, waste management and reprocessing of spent fuel. The latest manufacturing capabilities have been developed in the country as part of the indigenization effort to manufacture large and complicated reactor components like calandria, end shields, steam generators and steam turbines, heaters, pumps etc. for the conventional system.

Chapter **8**

Explosives and Rocket Fuels

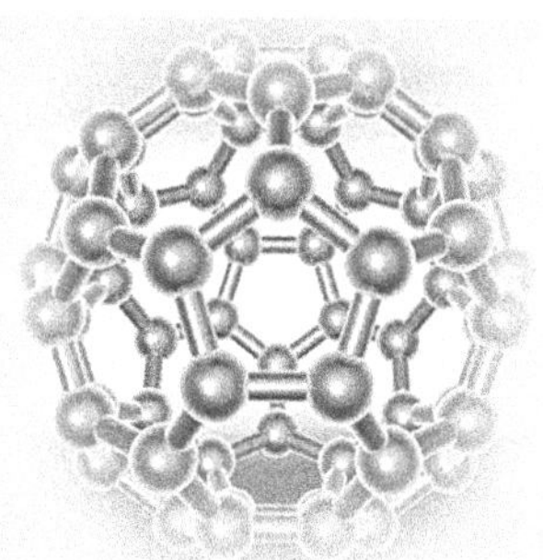

To a certain extent explosives and rocket fuels are related materials. In both the cases, chemical reactions take place rapidly releasing a large amount of energy. In fact many materials were first used as explosives and later were used as rocket fuels. Due to this reason, explosives have been considered in this chapter before rocket fuels.

8.1 Explosives

8.1.1 Introduction

Explosives are compounds or mixtures which, on receiving thermal or mechanical shock, are rapidly converted into products of greatly increased volume. This is caused by the formation of gaseous products and the generation of a great deal of heat. Since the gases occupy a much larger volume, and the volume further increases because of the high temperature produced, extremely high pressure is exerted on the surroundings giving disruptive or propelling effect.

Most people associate explosives with their military uses. Explosives have enabled man to perform such engineering feats as would not have been possible without their aid. Large engineering constructions, like the various dams which have been built in a short span of time, would have taken decades or even hundreds of years and thousands of labourers, reminding one of medieval ages in which the great pyramids were built in Egypt, or huge and magnificent temples constructed by the Incas in South America. Much more common examples of the use of explosives are in the blasting of the ores of iron and other metals, coal mining, quarrying limestone, construction of tunnels and mountain roads and numerous similar heavy works. Less powerful explosives are increasingly being used for such odd jobs such as digging holes for tree plantations, making ditches for draining water and even for rivetting and welding. In this section the industrial explosives will be considered and military explosives will be referred to only in passing.

In India there are three large plants manufacturing industrial explosives with an installed capacity of 58,500 tonnes per annum. Of this 36,000 tonnes is for nitroglycerine based and the balance for slurry explosives. Units with additional capacity of 62,5000 tonnes (15,000 t nitroglycerine and 47,500 t slurry type) are being set up. About 50,000 t of industrial explosives were manufactured in 1981.

8.1.2 Classification

Explosive materials may be solid (e.g., trinitrotoluene, TNT) liquid (e.g., nitroglycerine) or gaseous (e.g., mixture of oxygen and acetylene). They are usually divided into three broad groups, low explosives, primary explosives and high explosives.

Low explosives, also called *propellants*, do not explode suddenly but burn. The chemical reactions take place comparatively slowly and burning proceeds from the surface inwards in layers at a rate of about 0.2 m per second. The evolved gases disperse readily, without building high pressures, and they can be controlled.

Primary explosives or initiating explosives are commonly known as *detonators*. These are highly sensitive explosives which explode on receiving a slight shock or fire. They are used in small amounts in cartridges and shells to start the explosion of the main explosive which is comparatively less sensitive. They should be handled with the utmost care.

High explosives have higher energy content than the primary explosives but are quite stable and insensitive to mechanical shock and fire. A small amount of a primary explosive is placed in contact with the high explosive so as to give it an explosive shock. This starts the rapid chemical reaction in the body of the high explosive which proceeds at the rate of 1,500 to 10,000 m per second. The large volume of the evolved gases, especially under the influence of the high temperature produced, has a shattering effect. The high explosives may be subdivided into military high explosives, and blasting or industrial explosives. The former consists of pure substances, and in the latter the pure explosive is mixed with non-explosive materials as in the dynamites.

8.1.3 Requirements

An explosive must fulfill the following basic requirements :

1. The explosive must be stable under ordinary conditions of storage. If it undergoes a rapid decomposition at 40°, it is considered thermally unstable and unsafe. If rapid decomposition does not take place up to 95°, it is considered very stable. It is noteworthy that the character of reaction on heating depends upon the rate of heating. In general, the higher the rate of heating the more violent is the chemical reaction.

2. The chemical reaction must be highly exothermic. Ammonium nitrate gives 346 cal/g whereas TNT gives 925 cal/g and nitroglycerine produces 1,486 cal/g of heat. In general, the energy released is of the order of 1000 cal/g.

3. The rate of reaction must be very rapid so that the heat of reaction accumulates faster than it is dissipated and the temperature rises sharply.

4. The products of reaction must be gaseous, preferably having small molecular weights so that the volume change is high. Under the influence of a sharp rise in temperature the gases expand rapidly. For example, 1 g of nitroglycerine yields about 40 ml of gases when cold, which occupy about 750 ml at the temperature of explosion. In general, the volume of the hot gases is about 10 to 15 times as much as at the room temperature (Water, which is a liquid at the room temperature, is in a gaseous state at the moment of explosion).

8.1.4 Properties and Tests

Since the explosives are required to do different kinds of work it is necessary to standardize, control and evaluate their properties so that the work is performed satisfactorily. In the military explosives this is all the more necessary so that the target may be hit accurately. Most of the tests are empirical and are useful in comparing one explosive with another.

Strength and velocity of detonation. These two properties determine the power or *brisance* of an explosive and are responsible for its shattering effect. Strength depends upon the gas pressure developed from a given weight or volume of the explosive. The velocity of the detonation wave decides the rate at which the effective strength is developed. A soft rock is blasted satisfactorily by an explosive of lower strength and lower velocity. A hard rock requires higher strength and higher velocity of detonation.

Brisance is usually tested by *Trauzl lead block* test in which a given weight of the explosive is exploded in a standard lead cylinder and the increase in its volume forms the basis of the test. Brisance is also evaluated by exploding the explosive in a closed cylinder in contact with standard sand. The amount of sand crushed is determined by the use of standard sieves and indicates the shattering power of the explosive.

Sensitivity to mechanical impact is of special importance to primary or initiating explosives. The minimum height from which a standard weight must be dropped to detonate the explosive is a measure of sensitivity.

The *heat* of explosion (determined calorimetrically) and the volume of the gaseous products are very important and indicate the work capacity of the explosive. This is different from the power of the explosive which is governed by the rate at which the heat energy is liberated.

It is extremely difficult to directly correlate the various properties of an explosive with its behaviour when it is used in a particular way. The different explosives do not differ very much from each other in the properties indicated above, but their actual performance varies considerably in different situations. Therefore, a number of practical field tests have been developed, especially for military explosives.

It is noticed that most of the explosives contain nitrogen in one form or another. The reason is that the strength of the bond linking the nitrogen atom to the adjoining atom is low with the result that the compound is relatively unstable and is sensitive to shock. In most compounds nitrogen is linked to oxygen which moves from nitrogen to carbon and hydrogen forming carbon monoxide, carbon dioxide and water. Nitrogen itself is liberated as a gas. In this process on an average, about 1000 cal/g of heat is liberated which helps the gases to expand and cause the destruction.

8.2 Manufacture and Properties of useful Explosives

8.2.1 Primary Explosives

As has already been explained, these are high explosives which are sensitive to heat, impact and friction and therefore explode easily. They have lower shattering power than the secondary explosives. Because they explode so easily and have small shattering effect, they are not used as the main explosive. The secondary explosives do not explode as readily, therefore, explosion is initiated in the primary explosive, which causes explosion in the secondary explosive. The most important primary or initiating explosives are lead azide, PbN_6, and mercury fulminate, $Hg(CNO)_2$.

Lead azide is very popular for military use because of the low cost, stability in storage, and excellent initiating action. It cannot initiate explosion in the less sensitive secondary explosive like cast TNT. As it reacts with brass, the caps of the containers loaded with it are made of aluminium. It is manufactured cheaply by the action of ammonia on sodium giving sodium amide which, with nitrous oxide, gives sodium azide. A soluble lead salt on treatment with sodium azide produces lead azide.

Mercury fulminate is more expensive and more sensitive than lead azide. It is slightly toxic and is used to a lesser extent than lead azide. For its manufacture, mercury is dissolved in excess of nitric acid and the solution is added to ethyl alcohol. The solution begins to boil spontaneously and mercury fulminate is precipitated.

DiazodinitroPhenol (DDNP) is widely used in commercial blasting caps. It is quite sensitive and has high brisance and therefore can initiate explosion in less sensitive explosive. DDNP and lead azide have replaced mercury fulminate for most uses.

Tetracene may also be mentioned as it is used in explosive rivets. Its initiating efficiency is low and it is not used by the military. It is easy to ignite, has high heat of explosion and produces a large volume of the gases.

8.2.2 Low Explosives or Propellants

Black powder and smokeless powder are the important low explosives.

Black powder or gun powder is cheap and is still used to a considerable extent in spite of a number of new explosives being available. It is an intimate mixture of potassium nitrate, charcoal and sulphur (75:15:10). The powdered mixture is moistened and pressed into a cake and then broken up into grains. A small spark is sufficient to explode the whole mass. Although it has some desirable properties (e.g., non-shattering action) it is not good for coal mines because it produces a flame which can ignite gas and coal dust mixture and it also produces carbon monoxide, which is harmful to human beings. It is used for blasting where there is good ventilation, in fireworks and for igniting smokeless powders, etc.

Smokeless powder is nitrocellulose or more correctly cellulose nitrate. It is prepared by treating cellulose with nitric and sulphuric acids (The nitrocellulose plastics are not nitrated as fully as nitrocellulose explosives are nitrated). Nitrocellulose is dissolved in a mixture of alcohol and ether and the solvent is evaporated when a jelly-like solid is left behind. As it may react with traces of acid liberated during storage, a stabilizer (e.g., diphenylamine) is added to it. It is then pressed into cylindrical rods of smokeless powder.

It is called smokeless powder because on ignition it produces carbon monoxide, carbon dioxide, nitrogen and water vapour and almost no smoke.

8.2.3 Secondary High Explosives

These are quite insensitive and explode by detonation of a primary explosive. They release a large amount of energy required for various military purposes, demolition work and for certain recent methods of metal working operations. Some of them can be cast to give the required shape, some are pressed and some are available even as a plastic sheet–which makes them very useful for metal working operations. They are often classed into four groups: (1) Single compound explosives, (2) binary explosives, (3) plastic explosives and (4) dynamites.

8.2.4 Single Compound Explosives

Single Compound Explosives as the name suggests, contain only one chemical compound. Except ammonium nitrate, all other explosives are organic compounds–nitrates, nitro–compounds and nitramines of aliphatic, aromatic and heterocyclic series. The early explosives were obtained by nitration of glycerol and cellulose. As the coal tar industry developed, the nitro-compounds, like TNT came into prominence. The manufacture of nitramines started recently with the development of the synthetic organic chemicals industry.

Ammonium nitrate is cheap, has low brisance value, is quite insensitive, very stable and non-toxic. It cannot be used in contact with alloys of copper as it forms tetraminocupric nitrate which detonates easily. It is dangerous to store near any inflammable material. It is a high explosive though about half as powerful as TNT and is mostly used along with another explosive in making binary explosives. Its use is increasing.

Trinitrotoluene, TNT, is perhaps the best known of the high explosives and is widely used for firing shells. It is melted (m. p. 80.8^{o}) and poured into the containers. Its great importance is because: (1) it is not sensitive and therefore is safe to handle during manufacture, storage and transportation, (2) it does not absorb moisture, (3) it does not react with metals to form unstable compounds and (4) it has high brisance value. For all these reasons TNT is widely used for military as well as industrial purposes.

Pentaerythritol tetranitrate, PETN, is a standard, extremely powerful and sensitive military explosive. *Cylonite* or cyclotrimethylenetrinitramine, usually called RDX, came into prominence during World War II as a military as well as an industrial explosive. It is highly powerful. It is more sensitive but less toxic than TNT. *Picric acid*, trinitrophenol, was an important high explosive but has been largely replaced by TNT. *Tetryl* is a nitroaryl nitromine and its use as a single base explosive has been discontinued but it is used in binary explosives.

8.2.5 Binary Explosives

These consist of mixtures of various explosives, particularly of TNT with other explosives. For certain applications, they are superior to single compound explosives and are more convenient to make. TNT is an important ingredient of these explosives because it has a low melting temperature. Generally they are heated to liquid state and poured into the container and allowed to solidify. The widely used combinations are: *amatol* (ammonium nitrate and TNT, 50% of each), *pentolite* (PETN and TNT, 50% of each), *tetryol* (70% tetryl and 30% TNT), *torpex* (40% RDX, 40% TNT and 20% aluminium powder) and *tritonal* (80% TNT and 20% aluminium flakes). Each of these combinations has some special effect, hence is more suited for a particular use. Their chief advantages are higher energy output and lower cost of making the ammunition because they can be melted and poured.

8.2.6 Plastic Explosives

A combination of explosives which is in a plastic state and can be press-loaded, hand-moulded and made into various shapes without serious risk, is known as a plastic explosive. Plastic explosives are usually available in bulk, and are largely needed for military use, and in the form of plastic sheets which are mainly used for industrial applications. A simple combination consists, of a high explosive, which gives the destructive power, mixed with wax or oil, which makes it less sensitive and plastic. It may be made into a putty-like mass or a sheet.

On account of their engineering applications, special mention may be made of flexible plastic sheet explosives which are commercially available. These contain a high explosive like PETN with other materials, and are made into tough and flexible sheets of various sizes and thickness. The sheet can be cut into the desired shape and can be easily fixed to a metal part by an adhesive. Explosion is started by special caps. On explosion, due to the sudden impact, the metal can be made to take the desired shape under suitable conditions of metal forming.

8.2.7 Dynamites

Alfred Nobel, the founder of the Nobel prizes, gave the name dynamite to a solid mixture of nitroglycerine and kieselguhr (which is the mineral remains of unicellular algae). In a simple form it can be made by absorbing nitroglycerine in wood flour mixed with sodium nitrate. The mixture is encased in a cardboard cylinder to form sticks. By suitable compounding the dynamite can be used for blasting hard rocks or breaking soft minerals like coal into large lumps. It seems that detonation of dynamite is related to the porous, granular structure, because it is noticed that it often fails when highly compressed (as happens in wet bore holes when high water pressure acts on the dynamite cartridge).

Nitroglycerine is a high explosive and is used almost wholly for dynamite manufacturing. Nitroglycerine is manufactured by the action of strong nitric acid on high purity glycerine in the presence of strong sulphuric acid. The mixture is agitated and kept cold by circulating cold brine through closed coils. The temperature is not allowed to exceed 15°. The reaction may be represented as follows :

$$
\begin{array}{lcl}
CH_2OH & & CH_2ONO_2 \\
| & & | \\
CHOH + 3HNO_3 + (H_2SO_4) & \longrightarrow & CHONO_2 + 3H_2O + (H_2SO_4) \\
| & & | \\
CH_2OH & & CH_2ONO_2
\end{array}
$$

It is thus seen that the compound is glyceryl trinitrate and the term nitroglycerine is a misnomer. Nitration is complete in about an hour and the mixture is run into a separating tank in which nitroglycerine rises to the top. It is separated and washed, first with water and then with dilute sodium carbonate solution. The operation is highly risky and must not be attempted by unskilled persons.

Nitroglycerine is a viscous white to yellowish liquid resembling glycerine. Its specific gravity is 1.6; it freezes at 13° and is very sensitive to shocks.

By varying the constituents, several types of dynamites are made. The important ones are the following:

Guhr dynamite is the name given to original dynamites made by absorbing nitroglycerine in porous material like charcoal or kieselguhr. This makes nitroglycerine safer to handle. It is dry and is granular in structure and relatively insensitive to shock. It is not used in the U. S. A. but is still used in Europe.

Straight dynamites of various strengths are on the market. A typical 40% dynamite used by the U. S. Bureau of Mines contains 40% nitroglycerine, 44% sodium nitrate, 15% wood pulp and 1% calcium carbonate (to act as an antacid). As nitroglycerine has relatively high freezing temperature, the dynamite is unsatisfactory for low temperature use. For low temperature use, nitroglycerine is partly replaced by nitration products of other related alcohols like glycol. Water resistance of straight dynamite is not good and should be fired within 24 hours if used under water.

In *ammonia dynamites* nitroglycerine and sodium nitrate are partly replaced by ammonium nitrate. This makes the dynamite cheaper. It has a great heaving (i.e., lifting or moving) ability and is less shattering.

Gelatin dynamites contain a small amount (usually $< 6\%$) of nitrocellulose. Nitrocellulose is dissolved in nitroglycerine when a jelly-like mass is produced. It is mixed with other ingredients in the usual manner. The gelatin dynamites are highly water resistant and can be used for under-water operations. Another advantage of these over straight dynamites is that they are safer because the jelly-like mass has fewer tendencies to exude than the liquid nitroglycerine.

Ammonia gelatin dynamites have been developed to combine the advantages of ammonia dynamites with gelatin dynamites. They are similar to gelatin dynamites but a part of nitroglycerine is replaced by ammonium nitrate.

Military dynamites contain high explosive like TNT, RDX with binders. They do not contain nitroglycerine. They neither freeze in cold climate nor exude in hot weather. Their transporation, storage and handling are safer than those of the straight dynamites.

Permissible or permitted explosives have been developed for use in coal mines where there is the danger of violent explosion due to the frequent presence of methane and coal dust mixed with air. It is found that a hot and long lasting flame is more likely to cause explosion in such a surrounding than a cool flame of a short duration. The explosives for such uses contain about 80% ammonium nitrate and a small percentage of nitroglycerine as sensitizing agent soaked in a fibrous material. The black powder is certainly a better explosive in the sense that it produces large lumps of coal and not much fine dust but it is risky in the presence of methane and coal dust mixed with air.

The percentage *marking* on dynamites is for comparing strength on a weight for weight basis with that of straight dynamite with the percentage of nitroglycerine. For example, 60% ammonia dynamite was found to contain 20% nitroglycerine, 75% ammonium nitrate and 5% combustible material. It is marked 60% because its strength is comparable to that of straight dynamite which contains 60% nitroglycerine.

8.2.8 Some Recent Uses of Explosives

During the last few years explosive working of metals is being adopted on an increasingly large scare, especially in the aircraft and missiles industry, to solve difficult metal forming problems.

Generally the explosive is separated by air, water or oil from the work piece, which may be a metal plate or a metal sheet. Very high pressure, say of the order of a few thousand kg per cm^2, may be exerted on the work piece by the explosion and a variety of shapes of varying sizes may be produced.

Such operations have a number of advantages over the traditional methods of metal working. Some of these are:

(l) Capital investment of press is eliminated; (2) only female die is needed and a punch is not required; (3) the explosive pressure is uniformly distributed over the entire surface, thereby local stresses are absent. Details of the latest applications are worth the careful study (see bibliography) of the mechanical engineers (ASME, Detroit, has several publications on Advanced High Energy Rate Forming).

8.3 Rocket Fuels

8.3.1 Introduction

The term rocket is used to designate a wide variety of missiles, vehicles, thrust devices, etc. The fundamental principle governing the forward motion of the rockets was given three centuries ago by Newton in his third law of motion when he stated that for every action there is an equal and opposite reaction. The working of this law is noticed even by a small child when he releases an inflated rubber balloon and as air escapes from the small opening, the balloon moves forward against the force of gravity. The child has then "fired a rocket" just as the great scientists and engineers fire mighty rockets to the outer space.

The rocket engine or motor is unique in that both a fuel and an oxidizing agent are carried by it. Fuel and oxidizing agent are collectively referred to as the *propellant.* The oxidizing agent and the fuel component may be present in the same molecule, or the propellant may be a mixture. As the fuel does not depend upon atmospheric oxygen for combustion and the thrust produced is independent of the medium through which the rocket travels, the rocket engines are the only type of mechanism available for travel in the upper atmosphere (i. e. above about 30,000 meters) and the interplanetary space. The rockets are, however, being used for military purposes also for throwing explosives against targets thousands of miles away.

8.3.2 Principles of Rocket Propulsion

The rocket moves forward in accordance with Newton's third law of motion, viz. for every action there is an equal and opposite reaction. The thrust is not due to the jet of gases pushing against the surrounding air but is due to the 'equal and opposite reaction'. The propellant reacts chemically and produces gaseous products in the combustion chamber. The gases are at a high pressure and high temperature, of the order of 300 kg/cm^2 and 3,000°, and expand and emerge through a nozzle at supersonic velocity producing a thrust for propulsion.

The *propulsive force or thrust* is due to the momentum of the gas and can be calculated thus:

$$F = \frac{m}{g} \times v + (P_e - P_a)\, A_e$$

Where F= thrust (in lbs), m = mass flow rate of the propellant gases (in lbs/sec), g = the acceleration of gravity (in ft/sec), v = effective exhaust velocity (in ft/sec), P_e = exit gas pressure at the nozzle (in lbs/ft^2), P_a = ambient pressure (in Ibs/ft^2), A_e = Nozzle exit area (ft^2)

The equation shows that the rocket engine performance increases with increase in exhaust velocity. Since P_a decreases with altitude, and becomes zero in perfect vacuum, the performance increases with altitude. The exhaust velocity increases with increase in adiabatic (i. e. without heat change) flame temperature and with decrease in molecular weight of the combustion gases. The chemical fuels contain carbon, hydrogen, oxygen and nitrogen and give, in order of increasing molecular weights, H_2, H_2O, CO, N_2, O_2 and CO_2, Thus it is seen that lower molecular weight gases are produced by fuels rich in hydrogen and poor in carbon. For such materials the maximum temperature attainable is about 3500° as at higher temperatures highly endothermic reactions take place giving rise to atomic hydrogen, oxygen and nitrogen and the free hydroxyl group.

Another important term commonly used is *Specific impulse* (or *specific thrust* for liquid propellants). The term refers to thrust per pound per second of fuel burnt. In the simplest form specific impulse is Isp = F/m sec., where m is rate of propellant consumption (in lbs/sec) to produce the total thrust F (in Ibs). Detailed calculations show that specific impulse increases as the chamber pressure and temperature increases and the molecular weight of the exhaust gases decreases. In these relationships, the liquid propellants containing hydrogen and powerful oxidizing agents such as liquid oxygen, fluorine and ozone, show advantage over the solid propellants.

The *velocity* of a rocket, V, is approximately given by the following expression:

$$V = v \ \text{loge} \frac{Mi}{Ms}$$

where, v = effective exhaust velocity, Mi = initial mass of rocket, Mf = final mass of rocket

As the rocket moves, the mass of the propellant decreases, and the final mass of the rocket when all the propellant has been used up is less than the initial mass. In the expression given above the effects of air drag and gravity, which reduce the velocity have not been included.

In the design of an efficient rocket engine an attempt is made to keep the structural mass to a minimum and the mass of propellant to a maximum. In doing so several complicating factors have to be carefully considered. For example, the maximum theoretical specific impulse is for a propellant system of hydrogen and ozone. As the density of hydrogen is very low even under high pressures, the total mass of the complete rocket becomes high and a rocket using other propellants with lower specific impulse proves to be more efficient. Therefore, rockets are conveniently compared on the basis of their *over-all impulse ratio*, i.e., the ratio of the total impulse to total initial mass of the rocket.

In order to obtain very high velocities for rocket vehicles, as required for interplanetary flights, a series of rockets, one on top of the other, are built and fired successively and the earlier ones dropout. The advantage of such an arrangement is that each successive stage is smaller than the previous one and starts firing at a higher initial velocity. Theoretically the larger the number of such stages, the greater will be the final velocity but each additional stage means duplicating the entire mechanism with added costs. This puts a limit on the number of stages to be provided.

Most of the rocket engines burn chemical propellants. The engine has a combustion chamber. In the case of a liquid propellant it is injected in the chamber. There is a complicated arrangement of valves, regulators, igniters, etc. The propellant, either solid or liquid, burns in the combustion chamber and provides hot gas under high pressure, which escapes in a jet through a nozzle to the rear. This gives thrust for the rocket vehicle to move forward.

It has been reported that commencing at the launch, in the first 10 seconds the Saturn-5 moon rocket consumed 15 tonnes of fuel each second, producing a thrust of 3750 tonnes. More than 95% of the weight of the Staurn-5 rocket was made up of liquid fuels. The Apollo spacecraft was placed on top of the rocket. The rocket was detached successively and was not meant to return intact to the earth.

The above discussion helps one to understand the requirements and the working of the rocket fuel.

Although the nuclear reactor, plasma jet or hydrogen heaters have been used, most rockets at present use chemical propellants for providing the necessary power. The chemical propellants are classified into solid and liquid categories.

8.3.3 Solid Propellants

Depending upon their physical structure, the solid propellants are divided into two main groups: *homogeneous* and *composite* (or heterogeneous). Solid propellant mixtures in which the propellant or mixture of propellants are intimately mixed in a colloidal state are known as homogeneous solid propellants. Nitrocellulose and nitroglycerine combination is an example of this type. Composite propellants, like gun-powder or other combinations, in which an oxidizing agent is dispersed in a rubber-like fuel mass, have distinctly separate phases. Historically speaking, gun-powder was the only propellant used from the early 14th century to the late 19th century. In rocket engines it gives a specific impulse of about 60 sec., flame temperature of 800° to 1650° and the volume of the gases is about 400 times the volume of the charge.

Nitrocellulose, popularly known as smokeless powder or guncotton, replaced gunpowder. It is known as a *single base propellant.* Further improvement came by the discovery of the colloiding and solvent action of nitroglycerine (a powerful explosive in itself) on nitrocellulose. It is a powerful propellant and the name *ballistite* is given to such powerful propellants. The propellants which contain two materials, such as nitroglycerine and nitrocellulose, are called *double-base propellants.* The name *cordite* was given to a propellant using a plasticizer such as petroleum jelly. A typical composition is 65% nitrocellulose, 30% nitroglycerine and 5% jelly.

For smooth burning of a double-base propellant the operating chamber pressure of about 30 kg/cm^2 is required. The improvement over the earlier propellants can be seen from the fact that the flame temperature is about 2,750°, heat evolved is 850 to 1100 cal/g and volume of gas is about 1,500 times the original propellant volume.

The main constituents of a double-base propellant are nitrocellulose (50 to 55%) and nitroglycerine (40 to 45%). Smaller amounts of diethylphthalate (up to 5%) are also added, which along with nitroglycerine acts as solvent and plasticizer. They make the mixture into a homogeneous plastic mass which can be worked smoothly and is strong. As nitrocellulose decomposes in course of time, a stabilizer, like diphenylamine or ethylcentrallite is added (up to 1%). For military purposes, a flash suppressor (so that it may not be easily seen) like potassium nitrate (or sulphate) is added (about 1%).

The organic ingredients are mixed to form a paste and dried. Then the water soluble inorganic salts are added and the whole mass is made homogeneous by rolling in special mills and obtained in the form of flakes. The flakes are extruded in a press to give the required shape to the final grain.

8.3.4 Composite Propellants

Gun powder is the oldest composite propellant. During World War II 'Galcit series' of propellants, with about 75% potassium perchlorate and 25% asphalt oil, were developed. Compositions of some other typical propellants of this type are; 80% ammonium perchlorate and 20% resin binder; 80% ammonium nitrate, 18% resin binder with 2% catalyst; 46% each of ammonium picrate and sodium nitrate with 8% plastic resin binder.

It will be noticed from the above data that oxidizing agents, commonly perchlorates and nitrates, have to be used in composite propellants. Obviously, the most important criterion in the selection of an oxidizing agent is the percentage of available oxygen. It should not be hygroscopic, should be stable in contact with the fuel at the atmospheric temperature, and should not form corrosive products. Generally it is desired that smoke is not formed. Metallic ingredients form smoky exhaust gases. For example, ammonium perchlorate leaves no solid residue but the products contain hydrogen chloride and water which form a corrosive and toxic fog. Potassium perchlorate leaves white particles of potassium chloride.

The composite propellants contain 50% to 90% solid oxidizer in the final product. The solid constituents of suitable particle size are combined with the blended liquid constituents. The whole mass is mixed at suitable temperature and pressure. The product now obtained is in a dough-like mass or a thick liquid and is cast into a rocket motor or into moulds.

One of the most important properties of the propellant, whether double-base or composite, is its burning characteristics. It should burn in layers parallel to the burning surface. The burning rate is the rate, expressed in cm per second, at which the burning surface recedes. Before putting a product into actual use, the effect of temperature and pressure of the gases produced on the burning rate needs to be studied. Once burning starts, the temperature and pressure on the remaining propellant increase greatly and may alter the burning characteristics. Irregular combustion is very dangerous and may give an unpredictable and erratic course to the projectile. To some extent the burning characteristics are controlled by adjusting the grain shape and size of the propellant.

The propellants should be mechanically strong and should be able to withstand high pressures developed during firing. They are essentially high energy, unstable substances and it should be ensured that the inner layers will not suddenly explode under the influence of the high pressures.

The combustion chamber is made or metal and expands at a rate different from that of the propellant. In the design, allowance should be made for this difference.

During storage, the propellant may be exposed to different conditions of temperature and humidity. Sometimes it shrinks during storage causing the formation of cracks which increase the surface area thereby increasing the combustion rate to a dangerous level. Hygroscopic materials may absorb moisture and retard combustion. All these factors have to be carefully considered in the actual formulation of the propellant.

8.3.5 Liquid Propellants

Liquid propellants have several advantages over solid propellants. They have higher specific impulse and are more versatile. Their engines can be checked and calibrated more precisely. The chief disadvantage is that the engine is more delicate and less suitable for rough handing than the engine for solid propellants.

Liquid propellants are classified into two groups: mono-propellants and hi-propellants. A mono-propellant may be a single compound, like ethyl nitrate, or it may be a solution of a fuel and an oxidizing agent, like alcohol and hydrogen peroxide.

Bi-propellants consist of two liquids, one a fuel and the other an oxidizer. They ignite spontaneously on contact and have to be handled, stored and injected separately into the combustion chamber. They are capable of developing more energy than the monopropellants and are commonly used for high energy rockets. On the other hand, mono-propellants are more convenient to use as they require a single storage tank and only one fuel injection and control system.

8.3.6 Mono-propellants

Very few available substances are satisfactory as monopropellants. The reason is that requirements of a mono-propellant are conflicting. The substance must be safe to store and at the same time it should easily ignite and burn smoothly.

Of the common monopropellants, a mixture of 21.4% methanol and 78.6% hydrogen peroxide (87% conc.) gives the maximum specific thrust. Next comes nitromethane. Pure hydrogen peroxide has also been used but it is not easy to store and handle because of its reactivity. Metal oxides catalyse its decomposition. Therefore the storage tanks are made of special materials and must be clean. Hydrogen peroxide ignites in contact with organic materials. One has to be careful in dealing with concentrated hydrogen peroxide as its decomposition is self-catalytic and may become violent. Other important mono-propellants are ethylene oxide, hydrazine and propyl nitrate. Each of these has certain advantages and certain disadvantages.

8.3.7 Bi-propellants

Bi-propellants are more widely used. The common *oxidizers* are liquid oxygen (LOX). hydrogen peroxide, ozone, fuming nitric acid and liquid fluorine. Liquid oxygen has to be stored under pressure in insulated containers made of special materials. It is safe, non-toxic and a good oxidizing agent. Ozone is a more powerful oxidizing agent but is toxic and may explode at high concentrations. Fuming nitric acid is insensitive to shock but attacks most metals and organic materials. White fuming nitric acid (WFNA) contains up to 5% water. Red fuming nitric acid (RFNA) contains up to 22% dissolved nitrogen peroxide.

Liquid fluorine is very reactive, corrosive, toxic and volatile. It is very good oxidizer but is difficult to store and handle.

Fuels. Liquid hydrogen with 100% ozone gives the maximum specific impulse but it has a very low boiling point and low specific gravity. Hydrazine, N_2H_4, (B. P. 1.7^o) is also a very good fuel but is expensive. 75% ethyl alcohol is good and readily available. Addition of water to alcohol reduces flame temperature but the molecular weight of combustion gases is also reduced which, to some extent, compensates for reduction in performance. Kerosene and aniline are the other important fuels. The specific impulse in all these cases (except hydrogen and ozone combination) is higher with liquid fluorine as oxidizer; next in order is liquid oxygen.

The Saturn rocket, which is perhaps the most powerful rocket in use, is reported to use kerosene oil and liquid hydrogen as fuel and liquid oxygen as oxidizer.

Of all the *properties* desirable in a liquid propellant, specific thrust and velocity considerations are the most important. These are obtained by high heat of reaction, smooth and rapid burning are next in importance. High specific gravity and low vapour pressure help to decrease the size and mass of the storage tanks. As long burning and large engines require cooling, at least one of the propellants must be satisfactory for use as a coolant by circulating it around the combustion chamber. This means it should have high specific heat, high thermal conductivity and high thermal stability. For facility in pumping, low viscosity is desired. It should be safe to store and should have low corrosivity and toxicity.

Although on the whole liquid bi-propellants are more satisfactory, the choice between a solid and liquid propellant depends upon the nature of launching and flight requirements and on the missile system. Liquid propellants give higher specific impulse but the arrangements for handling, storage and injection are more complicated. Solid propellants are easy and safer to handle, transport and store but have lower specific impulse. To combine the advantages of both, some of the large rockets use multi-stage systems employing solid and liquid propellants in different stages. Each stage has a separate engine and gets detached after the fuel has burnt thereby reducing the mass of the rocket that needs to be propelled further.

Questions

1. Differentiate between proximate and ultimate analysis of coal, how is ultimate analysis carried out in the laboratory?

2. Calculate the gross and net calorific value of a sample of coal having the following composition

   C = 80%, H = 7%, O = 3%, S = 3.5%, N = 2.1% and Ash = 4.4%.

3. Explain the following

   (A) Octane and cetane numbers

   (B) Wood fuel must have low ash content

4. What is calorific value? How is it determined by calorimeter?

5. Calculate minimum weight of air required for complete combustion of 1 kg of fuel containing C = 90%, H = 3.5%, O = 3.0%, S = 0.5%, water = 1.0%, N = 0.5% and the rest is ash.

6. What is meant by "fixed carbon" in coal? What is its importance?

7. What is the significance of flue gas analysis?

8. Differentiate low and high temperature carbonization.

9. Discuss refining of petroleum and its important functions.

10. Define the term GCV and NCV of a fuel.

11. Discuss flue gas analysis by Orsat's apparatus

12. What is cracking? Discuss one method of catalytic cracking

13. Explain knocking in internal combustion engine and prevention of knocking

14. What is carbonization of coal? Explain its purpose

15. Mention advantages of catalytic cracking over thermal cracking

16. Why is coke and not coal, is used as a fuel in metallurgical processes?

17. Discuss the fractionation of naturally occurring petroleum, indicating the names, boiling ranges and uses of the various fractions.

18. What are the advantages of gaseous fuels over solid and liquid fuels?

19. The determination of the calorific value of a coal sample gave the following data:

    Wt of coal sample = 0.9 g, Water equivalent of the calorimeter = 440 g, Weight of water = 2560 g, Rise in temperature = 2.42 °C. Calculate the gross and net calorific values if the coal contains 6% hydrogen and assume latent heat of steam is 587 cal/g

20. Differentiate between chemical and nuclear fuels.

21. Explain, why the octane number of cracked gasoline is higher than that of the straight-run gasoline?

22. Why calorific value of producer gas is less than water gas?

23. What are the functions of the following in a nuclear reaction

    (i) Heavy water (ii) Reflector

    (iii) Molten alloy of Na or K (iv) Biological shield

24. "Nuclear fission is preferred to nuclear fusion as a source of nuclear energy" – comment.

25. Explain mass defect and nuclear binding energy

26. What are the advantages of gaseous fuels over other forms of fuels?

27. How is gasoline manufactured by fluid-bed catalytic cracking method?

28. (A) What do you understand by "Cracking"? Discuss the cracking processes with their applications.

    (B) 2.5 gms of mesitylene was burnt in a bomb calorimeter. The temperature of 860 gms of water placed in the calorimeter rose from 21.5 °C to 24.0 °C. If the water equivalent of the bomb calorimeter is 140, calculate the gross calorific value of mesitylene as fuel.

29. Write an account of the different types of gaseous fuels and their domestic/industrial uses.

30. (A) What is the composition of coal and how is it formed?

    (B) Discuss the proximate and ultimate analysis of coal.

    (C) How is metallurgical coke manufactured? What are its chief uses?

31. (A) Discuss the analysis of flue gases by Orsat apparatus.

    (B) What is knocking and what is its significance in petrol engines?

32. (a) Draw the diagram of the Junker's gas calorimeter and explain its working.

    (b) What is Octane number and what is its importance?

    (c) 0.8 gm of Cresol was burnt in Oxygen in a bomb calorimeter when the temperature increased from 22 ^{o} to 26 ^oC. The quantity of water in the calorimeter was found to be 1380 gms and the water equivalent of the calorimeter was 120 gms. Calculate the Gross Calorific Value of the fuel.

33. Calculate the volume of air needed for complete combustion of 1000 m^3 of a fuel gas having the following composition by volume: $CH_4 = 6\%$, $CO = 8\%$, $H_2 = 34\%$, $N_2 = 48\%$

34. (a) What is crude petroleum? How is it refined?

    (b) What are the various factors obtained from it, what are their boiling ranges and what are their uses?

35. What are the various types of coals? Write briefly on proximate and ultimate analysis of colas.

36. (a) What is petroleum? How is it refined and what are the various fractions obtained with their boiling ranges? What are their uses?

    (b) Determine the volume of air needed for complete combustion of 2000 cubic metres of a fuel gas having the following composition by volume –

    $CO = 10\%$; $CH_4 = 3\%$; $H_2 = 32\%$; $N_2 = 53\%$

Objective Questions

1. The temperature used for making high grade metallurgical coke by high temperature carbonation is .....

   (A) 900-1200 ^oC (B) 500 – 700 ^oC

   (C) 800-900 ^oC (D) 120-1350 ^oC

2. Cetane number of diesel fuel is improved by adding .....

   (A) Tetraethyl lead (B) methyl tertiary butyl ether

   (C) Ethyl nitrite (D) methanol

3. Analysis of nitrogen in coal is carried out by

   (A) Bomb calorimeter (B) Boys gas calorimeter

   (C) Proximate analysis (D) Ultimate analysis

4. LPG consists of

   (A) Liquefied methanol ethane

   (B) Vapour of petrol

   (C) Liquefied butane and isobutene

   (D) Compressed natural gas

5. _____ is obtained by passing a mixture of steam over hot coke

   (A) Producer gas (B) Water gas

   (C) Coal gas (D) Natural gas

6. Main constituent of natural gas is _____

   (A) Hydrogen (B) Carbon monoxide

   (C) Methane (D) Ethane

7. _____ is a good fuel for a jet engine

   (A) Petrol (B) Diesel oil

   (C) Kerosene (D) Alcohol

8. Stellar energy is due to _____ reaction

   (A) Fusion (nuclear) (B) Fission (nuclear)

   (C) Both fusion and fission (D) Electrochemical

9. The gaseous fuel which is also used as a source of hydrogen is _____

   (A) Producer gas (B) Water gas

   (C) Coal gas (D) Natural gas

10. The metal used for absorbing neutrons in a nuclear reactor is _____

    (A) Cadmium (B) Radium

    (C) Uranium (D) Zinc

11. Which of the following is a flue gas analyzer

    (A) Bomb calorimeter (B) Junker's calorimeter

    (C) Orsat apparatus (D) Boy's calorimeter

12. Producer gas is a mixture of

(A) $CO_2 + H_2$ (B) $CH_4 + H_2$

(C) $CO + N_2$ (D) $CO + CH_4$

13. In a nuclear reactor Boron steel is used as a

(A) Coolant (B) neutron absorbing material

(C) Moderator (D) Reflector

14. For improving anti-knock property of petrol it is mixed with –

(A) Lead bromide (B) lead tetraethyl

(C) allyl bromide (D) tetraethyl lead + ethyl bromide

Chapter 9

Lubricants

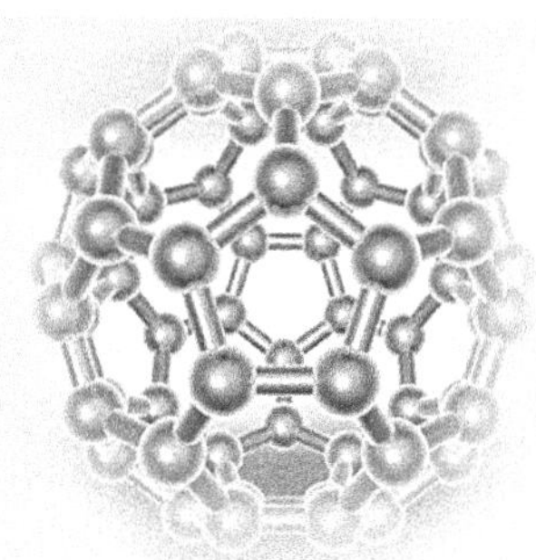

9.1 Principles of Lubrication

9.1.1 Introduction

When one solid surface is moved over another solid surface, there is a frictional resistance to motion. Even if the two surfaces were perfectly clean and smooth, this frictional resistance, which is primarily due to molecular attraction, would be great. These forces are much weaker than the atomic bonds and have been referred to in Chapter I under 'Secondary Forces of Attraction'. These are known as *van der Waals'* forces. Experiments have been made with clean and smooth surfaces made of different materials such as glass, copper, steel and various alloys. It has been found that on pressing the surfaces with a sliding motion, they stick so strongly that the lower surface may be lifted and even held in the vertical position without separating. Even the plane surfaces with which one deals in actual practice are not perfectly plane and smooth. These have minute projections which interlock each other, and also the surfaces are not perfectly clean. There may be even on the so-called 'clean' surface, some adsorbed moisture, dirt etc. Such impurities tend to reduce the friction. Thus, in practice, when one solid is moved over another, the projecting particlcs arc torn off causing wcar of the surface and a loss of energy which appears as heat. In general, both these effects are undesirable. Friction is desirable in cases like brakes, clutches and heat generation in friction-welding. Surface wear necessitates the replacement of the parts. As the equipment gets heated up; it is damaged, and even results in *seizure* i.e., welding of surfaces.

Any substance introduced between two moving surfaces with a view to reducing the frictional resistance is known as a lubricant. A thin film of lubricant, such as oil, is bound to the metal surface by the van der Waals' forces. In the fatty oils the glycerol end of the molecule is firmly held at the surface and the hydrocarbon chain sticks up normal to the metal surface. Thus, these oils have more oiliness and reduce metal-to-metal contact even under high load. Mineral oils are less polar but their behaviour is similar. Most of the lubricants commonly used are liquids, generally oils, but they may be solids or even gases. Graphite is a common solid lubricant. Air acts as a lubricant in high speed spindles. Quantitatively petroleum supplies the largest amount of material for making lubricants, but lubricants may be drawn from widely different sources, e.g., in cream separators milk acts as a lubricant; in certain very low temperature equipment, liquid air may be the lubricant.

Besides reducing energy losses due to friction and reducing wear, a lubricant may perform several other functions, such as providing a seal around the piston ring in an internal combustion engine so that the high pressure gases do not leak past the piston in the combustion chamber. In many instances, the lubricant helps in the dissipation of heat and, therefore, acts as a coolant. Yet another valuable function is that it prevents the entry of dirt and moisture between the moving parts.

9.1.2 Metallic Friction

The most important function of a lubricant is to reduce frictional resistance. The friction phenomena have been studied for many years. There are the classical works of Amontous (1699), Coulomb (1785), Towers (1885), Reynolds (1886), Erving (1892) and many others. Later, valuable contributions were made by Blok, Hardy, Bowden and Tabor. Such studies have led to the development of specialized branches of study, like Physics and Chemistry of Surfaces and Tribology (which deals with the motion of surfaces).

The well-known laws of friction are:

1. Frictional force is proportional to the normal load; and
2. Frictional force is independent of the apparent area of contact. (The real area of contact is often less than one ten-thousandth of the apparent area.)

   To these laws a third law can be added:

3. Frictional force is nearly independent of the sliding velocity.

These laws are fairly well obeyed over a wide range of experimental conditions. Very hard materials like diamond and very soft materials like Teflon show deviations from the first law. Surfaces with a very good finish and polish exhibit deviations from the second law. Deviations from the third law are quite common. The force required to start sliding is greater than to keep the surfaces in motion. The former is known as static friction and the latter as kinetic friction. The kinetic friction remains constant over changes in speeds which are commonly come across. The coefficient of friction almost doubles over speed range of a million times.

According to the modern adhesion theory, developed chiefly by Bowden and Tabor at Cambridge, even the best polished flat surface of a metal on examination under a high magnification (e.g., with an electron microscope) shows the surface to be like a hilly area. Peaks and valleys of different height, and depth, are observed. This is known as surface roughness and the highest peaks are called surface asperities (Fig. 9.1).

When two flat surfaces are placed over one another, contact is established between the asperities i.e., the peaks of the upper surface rest over the peaks of the lower surface (Fig. 9.2(a)).

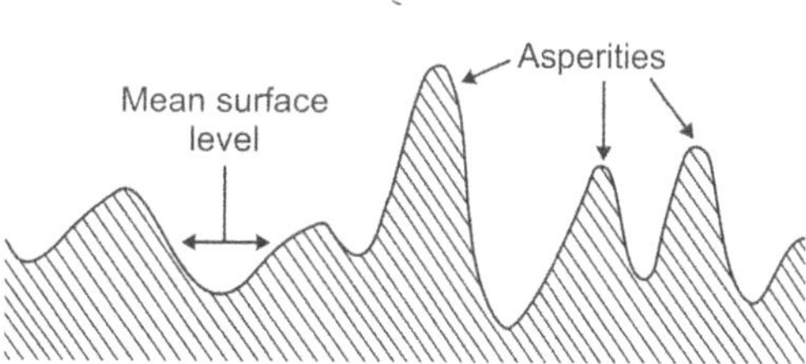

Fig. 9.1 Surface roughness.

Thus the real area of contact is much smaller than the apparent area of contact. A load of a few kilograms per square centimeter develops very high pressures (of the order of a few tonnes per sq. cm) at the real area of contact. Under such high loads the asperities are crushed and deform plastically (i.e., permanently) until the area of contact is large enough to support the load. This real area of contact is independent of the size and geometry of the surface but is proportional to the applied load (Fig. 9.2(b)). If the real area of contact = A; applied load = W and mean pressure =P_m, then $A = W / P_m$.

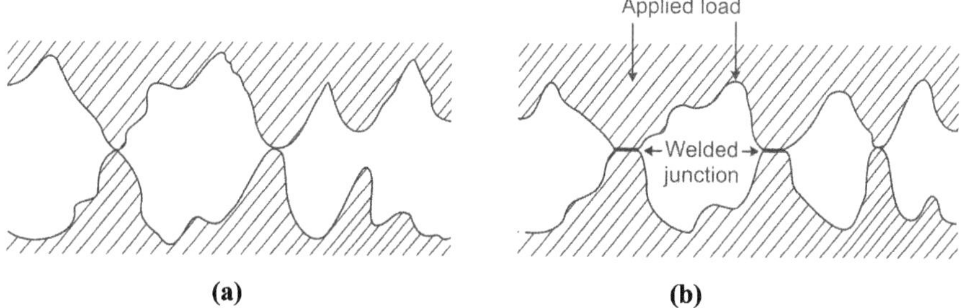

(a) (b)

Fig. 9.2 Contact is established between the asperities (a) which get flattened under load (b).

When the surface of a metal is made to slide over another surface, there is resistance to motion. This resistance is known as frictional resistance. At the points of contact cold welding of asperities takes place. Any attempt to slide the surfaces involves the shearing of the metal-to-metal junctions.

The mean shear stress S_m at which sliding takes place depends on the strength of adhesion of the junction. Hence the force required to cause motion, $F = A \times S_m$ = friction force.

$$\text{Coefficient of friction} = \mu = \frac{F}{W} = \frac{A \times S_m}{A \times P_m} = \frac{S_m}{P_m}$$

Thus it is seen that coefficient of friction is independent of the apparent area of contact. In this process of shear, the peaks of the softer metal get broken more easily than the peaks of the harder metal. This fact is also supported by experimental evidence. This accounts for the surface wear and is diagrammatically illustrated in figure 7.3.

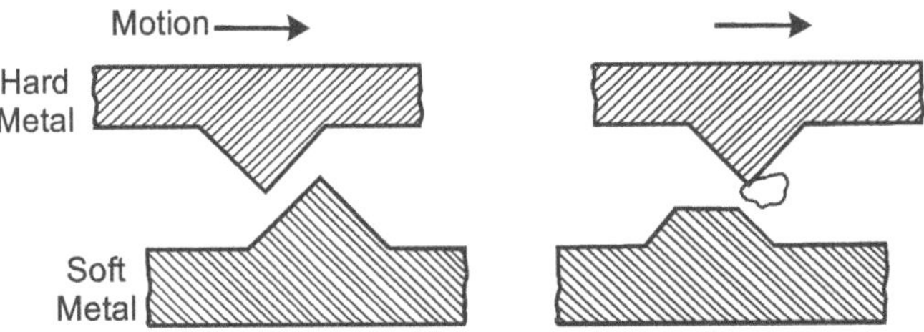

Fig. 9.3 Surface wear is due to shearing of asperities.

From the foregoing it is clear that metallic friction is not purely a surface effect. When solids slide over one another, deformation and distortion take place to a relatively greater depth below the sliding surfaces. Frictional force is, therefore, the sum of ploughing force P and shearing force S. Therefore, $F = P + S$.

Ploughing is due to hard metal asperities ploughing or cutting grooves in the softer metal. This becomes a significant factor when a hard metal slides over a relatively soft metal. In other cases its effect is not significant.

As it is the softer metal whose particles are broken off, the mean shear stress S_m is approximately equal to the shear strength S of the softer metal. In the same way since plastic deformation of the softer metal takes place more readily, the factor P_m can be assumed to be the yield pressure P of the softer metal. Therefore, the coefficient of friction can be expressed as follows:

$$\mu = \frac{S_m}{P_m} = \frac{\text{shear strength of softer metal}}{\text{yield pressure of softer metal}} = \frac{S}{P}$$

Shear strength and yield pressure of a metal vary together and their ratio is approximately the same for all metals. This explains the experimental observation that the coefficients of friction of different clean metals do not vary greatly. Since small changes in temperature affect both shear strength and yield pressure of a metal to the same extent, the coefficient of friction of clean metals does change with small changes in temperature.

9.1.3 Friction of Non-metals

The non-metals can be classed in two broad groups: (a) brittle materials, and (b) plastics.

The frictional characteristics of metals are primarily due to the fact that under load plastic deformation and welding take place. The resistance to motion is due to the shearing of the welds. Such a mechanism is exhibited by brittle non-metals like glass and common salt. These break easily. However, when they are subjected, static load some plastic deformation similar to that in metals takes place. Under such circumstances the nature of friction is similar to that met within the metals. A noteworthy point is that the metals have higher surface reactivity and energy than non-metallic surfaces. This fact modifies the frictional behaviour of metals to a great extent.

Some of the plastics like polyvinyl chloride, polystyrene and nylon are visco-elastic; therefore, the real area of contact is not directly proportional to the load. In most cases, as the load increases, the increase in the real area of contact is less rapid than the increase in load, consequently the frictional force increases less rapidly than the load. However, the frictional behaviour of even widely varying classes of non-metal is quite uniform.

9.2 Types of Lubrication

Science of lubrication deals with the problems met within the movement of solid surfaces in contact with each other. Lubrication is broadly classified as hydrodynamic or fluid film lubrication, boundary or thin film lubrication and extreme pressure lubrication. In the case of a shaft running at a fair speed in a well-lubricated bearing, with not too high a load; a film of the liquid lubricant covers the shaft as well as the bearing surfaces. In such a case the two metal surfaces do not come in contact with each other and the coefficients of friction for unlubricated surfaces range from 0.5 to 1.5 and for well-lubricated surfaces from 0.001 to 0.03. These figures show how great an advantage can be achieved because of reduction in friction and also reduction in wear and heating. Figure 7.4 shows formation of a fluid film of lubricant in a journal bearing with shaft under vertical load. When the journal is at rest, Fig. 7.4 (A), almost all the lubricant is squeezed out of the line of contact between the journal and bearing. The clearance between journal and bearing is full of the lubricating oil and its layers adhere to the journal and bearing surfaces. As the journal begins to rotate on the application of a torque to it, it tends to roll up the bearing surface. Rotation of the journal drags the oil between the journal and bearing.

Besides other factors, the quantity of the oil so dragged depends upon its viscosity. As the speed of rotation increases, more and more oil is pumped between the journal and bearing. The pressure exerted by the oil further increases as the clearance between the journal and bearing decreases (due to the rolling action of the journal on the bearing surface). This causes the formation of a wedge-shaped oil film between the journal and bearing surfaces which is essential for hydrodynamic lubrication. A stage comes when the oil pressure exceeds the load and the journal floats, and the journal and bearing surfaces separate completely with an oil film between them as shown in Figure 7.4 (B).

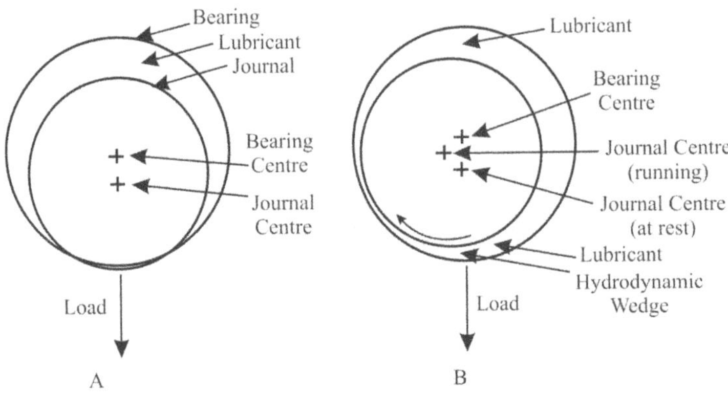

Fig. 9.4 Fluid film formation in a bearing (A) at rest (B) running.

If an oil of low viscosity is used, the bearing force, due to the rotation of the journal, will not be sufficient to drag the oil between the journal and the bearing. Therefore, for hydrodynamic lubrication viscosity is an important consideration. Right clearance space between the moving surfaces, filled with fluid of correct viscosity and such a design of the equipment that a wedge-shaped film is formed between the moving surfaces are the essentials for hydrodynamic lubrication. Theoretically in such a system, friction is independent of the load but depends on the viscosity and the thickness of the film of the lubricant, and relative velocity and area of the moving surfaces. Only the intermediate planes of the lubricant shear over one another. This type of lubrication is known as *hydrodynamic* or *thick film* or *fluid-film lubrication*. In hydrodynamic lubrication, the oil film is sufficiently thick to cover the irregularities of the surfaces and the metal surfaces do not come in contact with each other. Therefore, there is no wear and the resistance to movement is only due to the resistance between the particles of the lubricant moving over each other. This internal resistance of the lubricant is known as its *viscosity*. A liquid which flows easily has low viscosity and the one which flows sluggishly has high viscosity. Therefore, in hydrodynamic lubrication, the lubricant chosen should have the minimum viscosity under working conditions, and yet should remain in place and separate the surfaces.

The second kind of lubrication takes place when a continuous fluid film cannot be maintained. This happens (i) When a shafting comes into action from rest or (ii) the speed is very low or (iii) the load is very high or (iv) the viscosity of the oil is too low. In such cases it is necessary that a layer of oil be adsorbed by physical, chemical or both physical and chemical forces on the metal surfaces. The property which is responsible for this kind of adsorption is known as *oiliness*. An oil which has more oiliness than another oil of the same viscosity, gives a lower coefficient of friction at low speeds or high loads. Under such conditions when the fluid film cannot be maintained but the lubricant is absorbed on the metal surfaces, the metal surfaces come very close to each other, yet they do not touch each other and the load is carried by layers of the lubricant which have been adsorbed on the metal surface or which chemically reacts with the metal surface forming a thin film of metal soap which acts as lubricant. This kind of lubrication is known as *boundary* or *thin film lubrication.* The co-efficient of friction is usually between 0.05 and 0.15. It does not depend so much on the viscosity of the lubricant as it depends on other physical and chemical properties of the lubricant and the nature of the underlying surface. Vegetable and animal oils have greater oiliness than the mineral oils. Therefore vegetable oils or some additives are added to the mineral oils to improve their properties. Such oils are called *blended oils* or *compounded oils.*

The fatty oils, soaps and additives added for providing boundary lubrication does not have high melting points e.g., m. p. of stearic acid is 70° and of its soap i.e. sodium stearate is 280°. Further, the organic compounds decompose at high temperatures. In many sliding processes where load and speed are high, the local temperatures attained are much higher with the result that the boundary lubricants fail. For such extreme conditions certain compounds are added to mineral oils. These are called extreme pressure additives. They have active chlorine (as in chlorinated esters), sulphur (as in sulphurized oils) and phosphorus (as in tricresyl phosphate) group. At high temperatures the active groups react with the metal forming films of metallic chlorides, sulphides and possibly phosphides. These compounds have much higher melting points (e.g., m. p. of iron chloride is 650°, of iron sulphide is 1100° and of iron phosphide is 1290°) and low shear strengths. Thus they provide lubrication under extreme pressures and temperatures. This type of lubrication is called *extreme pressure lubrication.* In a sense it is really extreme temperature lubrication.

9.3 Testing of Lubricants

The suitability of a lubricating oil for a particular use can be decided finally by its actual behaviour in the machine. Laboratory tests help to classify the lubricants and to specify them when obtaining a fresh supply. Comparison and evaluation of different products available in the market is possible by correct specifications and interpretations of the tests.

Correct assessment of the lubricant saves risk of destruction of expensive machinery. It is necessary that standard equipment and procedure, say, as laid down by Indian Standards Institute, be followed so that the results are comparable and accurate

The important tests for lubricants are the following:-

1. Viscosity, 2. Flash and the fire points, 3. Cold tests – cloud and pour points, 4. Emulsification, 5. Volatility, 6. Gravity, 7. Colour and Fluorescence, 8. Carbon residue, 9. Oxidation stability, 10. Aniline point, 11. Neutralization number, 12. Saponification number and 13. Mechanical tests.

9.3.1 Viscosity

Viscosity is the most important single property of a lubricating oil. It may be defined as 'a measure of the internal resistance to motion of a fluid by reason of the forces of cohesion between molecules'. In simpler words, viscosity is resistance to flow or 'thickness' or 'body' of the oil. The formation of a fluid lubricating film, the viscous frictional losses in the bearings and heat generation under particular conditions of load, bearing speed, and lubricant supply, depend on viscosity and to some extent on the oiliness of the lubricant. 'Therefore it is necessary that the lubricant selected has viscosity sufficient to maintain a lubricating film under the working conditions but does not have excessive viscosity which would cause undue friction.

Absolute viscosity may be defined as the force required to move a plane surface of one square centimeter in area at a speed of one cm per second over another fixed parallel plane separated by one cm thick layer of the liquid. Poise is the unit of viscosity and is equal to one dyne per second per square centimeter. This unit being very large, viscosity is often expressed in terms of centipoise (0.01 poise). The viscosity of water at 20° is about 1 centipoise.

In commercial viscometers, a fixed volume of the liquid is allowed to flow through a standard capillary tube and the time of flow is noted. There are three makes of commercial viscometers in wide use. The Redwood viscometer is used in the Common wealth countries, Saybolt in the U.S.A. and Engler in Europe. The results are sometimes expressed in terms of time taken by a given quantity of the oil to flow through the particular instrument, e.g., 'viscosity is 100 Redwood seconds at 20°'.

The instruments are of standard dimensions; therefore, viscosity in terms of poise can be calculated from the time taken by the oil to flow.

It is noteworthy that in most cases the viscosity of oil decreases with rise in temperature. Therefore viscosity at the working temperature needs to be considered. The change in viscosity with change in temperature is expressed by the *viscosity index*. The viscosity index is an empirical number indicating the rate of change in viscosity of an oil within a given temperature range. Low viscosity index means a large change in viscosity with change in temperature. A high viscosity index shows a relatively small change in viscosity with change in temperature. The naphthenic oils have higher rates of change of viscosity with temperature than the paraffinic oils. An oil of high viscosity index does not become very thick when cooled or very thin when heated. Therefore in equipment if the change in temperature in service is large, as in hydraulic system in aircraft the viscosity index of the oil should be high.

It may be noted that viscosities of gases *increase* with rise in temperature. The viscosity of a liquid decreases as temperature rises because intermolecular attraction is reduced due to expansion. In the case of a gas, as the temperature rises, the velocity of the molecules is increased with the result that the collisions of the molecules is increased, thereby the collisions of the molecules become more frequent, therefore there is increased resistance to motion.

The viscosity of an oil may be significantly higher at high pressures; hence this change also needs to be taken into account. The viscosity of a gas also increases with increase in pressure.

Redwood Viscometer : The Redwood viscometer is made in two sizes, No. 1 and No. 2. Viscometer No. 1 is commonly used for determining viscosities of lubricating oils and is shown in the Fig. 7.5. Viscometer No. 2 is similar to it but the jet for the outflow of the oil is of larger diameter, and therefore it is more satisfactory for oils of higher viscosities, e.g., the fuel oils. The bottom of the oil cylinder is also heated by the liquid in viscometer No. 2.

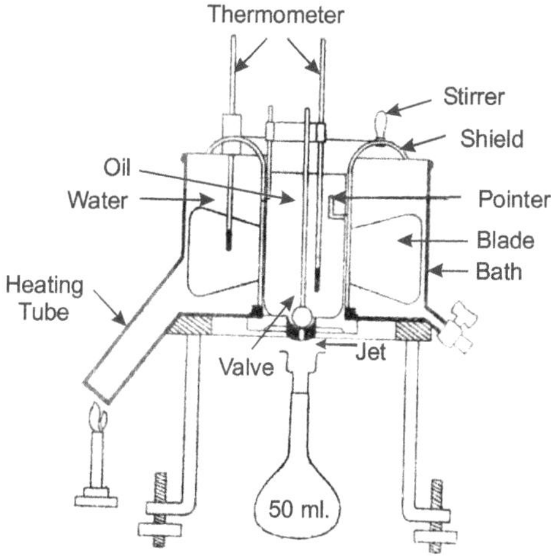

Fig. 9.5 Redwood viscometer.

The viscometer may be heated by gas, as shown in the figure, or electrically heated. The oil is taken in the inner cylinder which is of silver-plated brass, 90 mm in height and 46.5 mm in diameter. At the bottom of the cylinder a jet made of agate is fixed. It is 10 mm long and its bore is 1.62 mm in diameter. It is opened or closed by a valve which is a small silver-plated brass ball fixed to a stout wire. There is a pointer which indicates the level up to which the oil is to be taken. A thermometer indicates the oil temperature. Outside the oil cylinder is a stirrer carrying blades for stirring water for maintaining the desired temperature uniformly. It has a shield which prevents water splashing into the oil cylinder. A thermometer indicates the water temperature.

The oil cylinder is cleaned and the valve is closed. Oil, free from suspensions or emulsions, is taken in the cylinder up to the pointer level. Water is taken in the bath and the temperature is carefully regulated. A 50 ml measuring flask is kept just below the jet. When the oil is at the desired temperature and the water is as nearly as possible at the same temperature, the ball valve is lifted and suspended from the thermometer bracket. The time taken for 50 ml of the oil to collect is noted and the ball valve is closed to prevent overflow of the oil. The result may be expressed in Redwood seconds or converted into absolute units at the particular temperature. The Saybolt Universal viscometer is worked in a similar manner.

Viscosity is the most important single property of lubricating oils. Under conditions of hydrodynamic lubrication, the bearing friction, heat generation, the rate of flow of the lubricant under given conditions of load and speed, and the design of the bearing are determined by the viscosity of the lubricant at the operating temperature.

There is no simple method of deciding the optimum viscosity for a particular use. Experience is the guiding factor. In some cases much higher viscosity oils have been used because the gear boxes are difficult to make tight, thus leakages are minimised. High viscosity oils reduce noise and help in making a good fit.

9.3.2 Flash and Fire Points

The flash point is the lowest temperature at which the oil gives off vapours that will ignite when a flame is passed over the surface of the oil. The fire point of an oil is the lowest temperature at which it ignites and continues to burn for at least five seconds.

The two tests are carried out in the same apparatus on the same sample. The apparatus used may be the 'open' or 'closed' cup. In the open cup apparatus the oil is heated with its upper surface exposed to the atmosphere. The closed cup gives more reproducible results. The Pensky Marten's closed tester is commonly used and is shown in Fig. 7.6. It consists of an oil cup about 5 cm in diameter and 5.5 cm deep. The level up to which oil is to be filled in is marked inside the cup. The cup is supported by its flange over a heating vessel in such a manner that there is clearance between the cup and the heating vessel. The oil is dried because the presence of water makes it difficult to determine the flash point. After taking oil in the cup to the level indicated, it is closed by a cover. Through the centre of the cover passes a stirrer which stirs the oil as well as the vapours as shown in the Fig. 7.6. The cover has four openings of standard dimensions. Through one of these passes a thermometer. Through the second opening shown in the Figure, passes the test flame and the other two openings are for the admission of air. The shutter opens the openings for the test flame and air, and dips the burner and brings the flame over the oil Surface. As the test flame gets extinguished it is automatically lighted by the pilot burner.

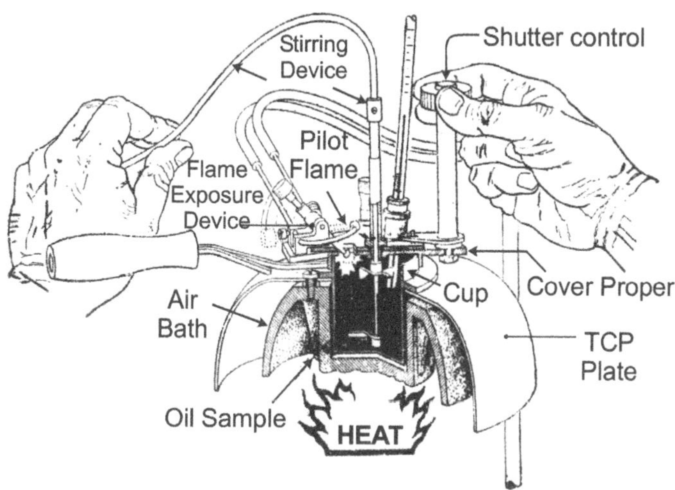

Fig. 9.6 A section through Pensky Marten's flash point apparatus.
(Lubrication Engineering Manual, Courtesy, U.S. Steel Corporation)

Heat is applied to raise the oil temperature at the rate of about 5° per minute. The stirrer is worked between tests at the rate of about 1 to 2 revolutions per second. The flame is applied to the oil every 1° up to 105° and afterwards at increments of 3°. The temperature at which a distinct flash is seen inside the cup is noted as the flash point.

The heating of the oil is continued and the test flame applied as before. When the oil ignites and continues to burn for at least 5 seconds, the temperature reading is recorded as the fire point. In most cases the fire points are 5° to 40° higher than the flash points.

The flash point and fire point do not indicate directly the lubricating value of the oil but these are important when the oil is exposed to high-temperature service. Thus these tests provide safeguards against risks of decomposition and fire during storage and use. Flash and fire points are now being omitted from many specifications although some insurance companies insist on it from the point of view of safety during storage.

Formerly an oil of high flash point was supposed to offer more resistance to spontaneous combustion and explosion in the lubrication of air compressors. Flash point does not indicate tendency to undergo spontaneous combustion. Spontaneous ignition temperature is the temperature at which ignition can take place in contact with a hot body. It is reported that a hot bearing disintegrated and a portion of it fell into the crank case of large Diesel engine with the result that violent explosion took place.

9.3.3 Cloud Point and Pour Point

When an oil is cooled at a specified rate, the temperature at which it becomes cloudy or hazy in appearance is called the cloud point of the oil. This is due to the separation of tiny crystals of wax. The temperature at which the oil just ceases to pour or flow is called the pour point.

The oil under test is taken up to a fixed level in a glass test jar of standard dimensions. The jar is kept in a jacket which is surrounded by a freezing mixture. A thermometer is introduced in the oil. The test jar is withdrawn from the jacket with every degree of fall in the oil temperature and examined. It is replaced in the jacket within three seconds. The temperature at which cloudiness is noticed is taken as the cloud point.

Cooling is continued and the test jar is withdrawn with every 3^{0} fall in temperature and tilted to observe the flow of oil. The temperature at which it does not flow even when kept horizontal for five seconds is taken as the pour point.

The cloud point is not so important for lubricating oils as it is for fuel oils which have to pass through unheated filters of fine mesh. For example, a jet plane may be exposed to 60^{0} and if solids separate from the fuel oil the carburetor of jets may be blocked up. The pour point is of greater significance for lubricating oils exposed to low temperature. It indicates the temperature below which the oil cannot be poured out from its container or the equipment cannot be lubricated by it by gravity flow.

9.3.4 Emulsification

There are several tests to indicate the emulsification tendencies of the oils and a particular test is chosen according to the conditions met with by an oil in actual use. In a test, oil and water are stirred and the time taken for the emulsion to break is noted. In a more widely used test, steam is passed into the oil under specified conditions, and the time in minutes taken for the separation of the oil from the emulsion is recorded as the steam emulsion number.

The time taken by the oil for separation from the emulsion depends upon its refining and viscosity. Good refinement and lower viscosities are conducive to quicker separation of the oil from the emulsion.

The emulsification tendency is undesirable in most applications of the oil. Formation of a little emulsion in a circulating oiling system may not be harmful but sticky emulsions and sludges interfere with oil circulation and are certainly undesirable. The formation of stable emulsions with water may be desirable in the case of certain steam-cylinder oils and water-soluble cutting lubricants. The general principle is that formation of emulsion is desirable when an oil is to be used in the presence of water. In the case of steam turbines the oil should separate quickly from the emulsion so that it can be recirculated for use.

9.3.5 Volatility

The volatility test has not been standardized. Volatility may be found by exposing a weighed quantity of the lubricant to a current of hot air, under specified conditions and determining the loss in weight. Its application is limited to those conditions in which the oil is exposed to a temperature at which a fraction of it volatalises leaving a residual oil which may have different lubricating properties.

9.3.6 Gravity

Besides specific gravity, A.P.I. (American Petroleum Institute) gravity is in use. A.P.I. gravity is an arbitrary scale in which the specific gravity of water is taken as 10. Liquids lighter than water are assigned values higher than 10 and the heavier liquids, values lower than 10. Naphthene base oils have lower A P I gravity than those of paraffin base oils.

Gravity need not be mentioned in the specifications of lubricating oils because it gives little indication of the quality or usefulness of the lubricant. It is used for inter-conversion of weight and volume measurements.

The terms 'density', 'specific gravity' and 'gravity' are often loosely used. Density is the mass of unit volume of a substance under specified conditions. It has a unit e.g. grams per milli litre, pounds per cubic foot, etc. Specific gravity is the ratio of the weight of a substance to the weight of an equal volume of water. Gravity is a reading of an arbitrary scale on a hydrometer. e.g., degree Baume, A. P. I. gravity, etc.

9.3.7 Colour and Fluorescence

Colour is determined by the usual colorimetric method of comparison of the colour of the oil with that of the standard colour discs. Colour indicates the degree of refinement. A pale colour does not indicate lower viscosity, as many think. Distillates are lighter in colour than the residual oils. A pale-coloured lubricant is desirable in such applications as in the textile industry where staining of fabrics has to be avoided. In other cases it is unnecessary to insist on colour requirements.

The mineral oils show fluorescence in reflected light. Fixed oils do not show fluorescence; hence fluorescence helps to detect adulteration.

9.3.8 Carbon Residue

The carbon residue test determines the percentage of carbon residue left in an oil after subjecting it to an extreme heating condition out of contact with air. In the process, evaporation and pyrolysis of the oil take place. The Conradson method has been widely used but recent recommendations are in favour of the Ramsbottom method. The Conradson apparatus is shown in Fig. 7.7. In the Conradson method, a weighed quantity of the lubricant is taken in a crucible which is placed in another iron crucible having a cover with a small opening. These are then placed in a third crucible, and covered with a hood and strongly heated at a specified rate and the residue is cooled and weighed. The result is expressed as percentage of the residue on the original weight taken.

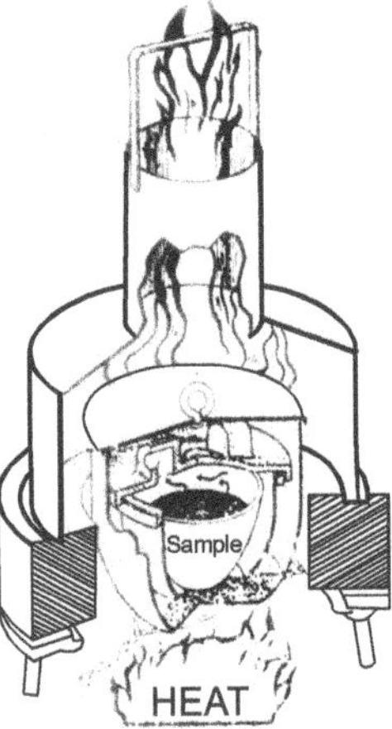

Fig. 9.7 Conradson method for carbon residue determinations.
(Lubrication Engineers' Manual, Courtesy, U.S. Steel Corporation)

In the Ramsbottom method about 4 g of the oil are accurately weighed in a coking bulb made of hard glass, 57 mm in height and 25 mm in diameter. It is then placed for 20 minutes in a furnace which has been previously heated to 550°. The bulb is then removed from the furnace, cooled in a desiccator and weighed, and the result expressed in percentage.

Efforts have been made to correlate the results obtained by the two methods but not much success has been achieved. The Ramsbottom method gives more reproducible results.

The carbon residue test indicates the carbon-forming tendency of lubricating oils in internal combustion engines, especially as regard piston-crown deposits around the ring, and on valve seals in the combustion chambers. Under very good conditions some indication of carbon deposits in the cylinder may be obtained. It has been suggested that carbon residue varies directly with the viscosity. In good lubricating oils, stabilizing additives are added to the oils to enable them to resist chemical decomposition at the high temperature developed in internal combustion engines. It is noteworthy that the black deposits in the engine do not entirely consist of carbon, though it is the main constituent. Dust from the air may also be present. The carbon residue test is important for lubricants used for internal combustion engines and air compressors.

9.3.9 Oxidation Stability

Lubricating oils tend to oxidize in contact with air, slowly at room temperature but rapidly as the temperature increases above 100°. The reaction is further accelerated in the presence of metal particles which act as catalysts. These conditions are common in internal combustion engines, steam turbines, and other machinery which have oil circulating systems. Oxidation products are undesirable. Some of these are insoluble, gummy and sticky materials which clog oil holes, filters, pipe lines and other parts of the oiling systems. The soluble products are acidic and corrode the surface of the bearings and also form varnish like deposits. Therefore, resistance to oxidation is a desirable property of a lubricating oil. Oxidation inhibitors are added to oils to improve their resistance to oxidation. The inhibitors get oxidized first and the oil is protected.

A common test is the Sligh oxidation test. In this test a 10 g sample of the oil is taken in a special flask and air is displaced by oxygen. The flask is kept in an oil bath maintained at 200°. After 2½ hours the flask is removed, cooled, and the contents are diluted with petroleum naphtha and allowed to stand for an hour. The precipitate is filtered, washed, dried and weighed. The weight is expressed as percentage of the weight of the sample taken to give Sligh oxidation number.

9.3.10 Aniline point

The aniline point is defined as 'the minimum equilibrium solution temperature for equal volumes of aniline and sample'. For its determination equal volumes of the sample and aniline are taken in a tube and mechanically mixed. The mixture is heated until homogeneous and then cooled at a controlled rate. The temperature at which the two phases separate is recorded as the aniline point.

The aniline point of an oil gives an indication of the possible deterioration of rubber sealing, packing, etc. in contact with the oil. The aromatic hydrocarbons have a tendency to dissolve natural rubber and certain types of synthetic rubber. Therefore, in such cases a low aromatic content in the lubricant is desirable. As the aniline point is a measure of 'paraffinicity' of an oil, the higher the aniline point the lower is the percentage of aromatic hydrocarbons and the higher is the content of paraffins. Since paraffins give easy starting and a minimum of knock in a Diesel engine, a high aniline point usually indicates a high cetane number (See Diesel fuels).

9.3.11 Neutralization Number

The neutralization number refers to the determination of acidic or basic constituents. Determination of acidic constituents is more common and is referred to as acid value (or number). It is the number of milligrams of potassium hydroxide required to neutralize the free acid in one gram of the oil.

About 10 g of the oil is weighed in a conical flask and 50 ml of neutral alcohol (or alcohol toluene mixture) is added and boiled for a minute. The acidity is titrated against N/l0 standard potassium hydroxide solution using phenolphthalein as the indicator. The results are expressed in terms of the number of milligrams of potassium hydroxide required per gram of the sample. Basic constituents are determined in a similar manner by titrating against acid solution using the appropriate indicator.

Generally, mineral acids are not present in the lubricants unless refined in a faulty manner. Organic acids are not as corrosive as the mineral acids and in good samples the acid value should be very low, say, less than 0.1. Increase in acid value should be watched because it indicates oxidation of the oil which will lead to gum and sludge formation, besides corrosion. For certain uses, storage of the oil in galvanised iron tanks is discouraged because the free acidity produced attacks the zinc forming zinc soaps.

9.3.12 Saponification Number

The saponification number of an oil is the number of milligrams of potassium hydroxide required to saponify the fatty material present in 1 g of the sample.

About 5 g of the sample is weighed in a flask and 50 ml of semi-normal potassium hydroxide solution in alcohol is added to it and boiled for an hour under a condenser. The unreacted potassium hydroxide is titrated back and the result calculated.

Mineral oils do not form soaps but vegetable and animal oils do. Thus this test helps to determine approximately the proportions of these in a mixture.

9.3.13 Mechanical Testing

Physical and chemical tests do not give a completely satisfactory basis upon which to select a lubricant for a particular purpose. Several machines are in use for determining frictional resistance. Recently, there has been the problem of lubrication under conditions of very high pressure. To judge the suitability of a lubricant under extreme conditions several machines have been designed which determine the load-carrying capacity of a lubricant, torque, friction and wear. One such machine is a Shell Four Ball Extreme Pressure Lubricant Tester which is shown in the Fig. 7.8.

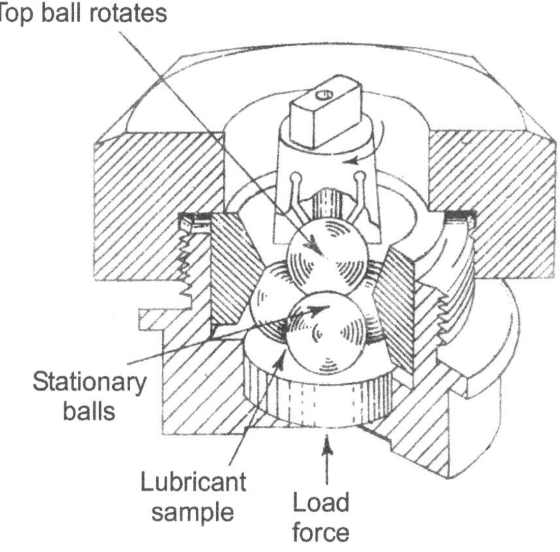

Fig. 9.8 Shell four ball extreme pressure lubricant tester.
(adopted from Lubrication Engineer's Manual, Courtesy, U.S. Steel Corporation)

The Shell Four Ball machine has four steel balls which are immersed in the lubricant under test. The lower three balls remain stationary while the top ball is rotated. Any desired load can be applied. If the lubricant is satisfactory under the test conditions the balls after the test come out undamaged. The load is gradually increased and the balls withdrawn and examined at specified intervals. In the earlier stages formation of scars on the balls is observed. When the load is further increased, at a certain value the lubricant fails completely and so much heat develops that the balls weld together. The test enables the determination of the load that can be safely carried by a lubricant. Similarly torque, friction and wear test can be performed. The results can be approximately correlated to conditions met within actual use. Comparison between two lubricants can also be made.

9.4 Classification of Lubricants

On the basis of consistency, lubricants may be classified as liquid, semi-solid and solid lubricants.

9.4.1 Liquid Lubricants

Liquid lubricants are used to a greater extent than all the semi-solid and solid lubricants put together. The important groups of liquid lubricants are the following:

9.4.2 Petroleum Oils

These provide more lubricants than all other sources because of their abundance, stability under severe service conditions and lower cost. From the petroleum crude, the lighter fractions are distilled off at atmospheric pressure as described under 'Petroleum Refining'. The residue is distilled under reduced pressure and, if necessary, steam distillation may also be adopted. Distillates of various boiling ranges and properties are collected separately. The fractions so obtained are not quite satisfactory for direct use because of the presence of three kinds of constituents. The solid waxes crystallize out at a lower temperature and interfere with the flow of the lubricant. Certain constituents get easily oxidized under working conditions and cause sludge formation. Some constituents, mainly asphalt, break down at higher temperatures causing carbon deposition and sludge formation. All these need to be removed before the lubricant can be expected to give satisfactory service.

In the older methods of refining, the oil was chilled and the crystallized wax was filtered off. The oil was then given sulphuric acid treatment to remove the asphalt and the unstable constituents. These methods being slower and less efficient are being supplemented and/or replaced by solvent refining.

Essentially, solvent refining consists of brining a suitable solvent and the oil into intimate contact with each other. The solvent chosen should be such that it is immiscible with the oil but the undesirable constituents should be more soluble in it than in the oil. The solvents commonly used are furfural, dichloro-ethyl ether, nitrobenzene, mixtures of liquid propane and cresol, sulphur dioxide and benzene, etc. Each of the various solvents has some advantages over the other and the treatment processes differ slightly from one another. The oil and the solvent meet in a counter current manner, i.e., the fresh solvent comes in contact with the oil which has already come in contact with the solvent and has a minimum of impurities. The oil portion, free from impurities, contains some solvent which is removed by distillation and the refined oil is left behind. The solvent portion is also distilled and the solvent is recovered and used again. The residue from solvent distillation contains naphthenic, asphaltic and resinous substances and is used as a source of these or as a fuel oil.

The cost of solvent refining is lower than that of acid refining. It removes asphaltic materials more satisfactorily, thereby reducing the formation of carbon deposits. The yield of refined oil is higher and it shows less change in viscosity with temperature due to the removal of naphthenic compounds. On the other hand, the solvent refined oil is less resistant to oxidation due to the removal of natural inhibitors and has lower oiliness, It, therefore, becomes necessary to add substances to improve the properties.

Paraffinic oils have lower specific gravity, less change in viscosity with temperature and a lesser tendency to oxidation than the naphthenic oils.

9.4.3 Fixed Oils

The oils obtained from vegetable and animal sources consist of glycerides (esters with glycerol) of higher fatty acids. They decompose on heating and do not distil, hence are called fixed oils. These were the principle sources of lubricants before the advent of petroleum oils. They are still used, but to a comparatively smaller extent, for compounding petroleum oil and for making greases. The fixed oils have higher oiliness due to which they offer lower friction and have a higher load-carrying capacity than the petroleum oils. They are, however, more expensive, hydrolyse in presence of water and have a greater tendency to oxidation and decomposition at elevated temperatures than the petroleum oils. Only oils of the non-drying group, i.e., the oil which has fatty acids with a small amount of unsaturation, are used for lubrication.

About 5% of oils like castor oil, sperm oil, lard, etc or their fatty acids are added to petroleum oils to improve their properties. Such mixtures are known as *compounded oils* and are used as steam cylinder oils, marine engine oils, etc. Such oils are now used to a much smaller extent than formerly. A quantity of the fixed oils is converted into calcium, sodium and lithium soaps and used in the preparation of greases and cutting emulsions.

9.4.4 Addition Agents

The requirements of some of the modern machinery are such that petroleum oils by themselves fail to lubricate them satisfactorily. Addition of fixed oils improves them but even this is inadequate. This technique of adding other substances for improving the properties has led to extensive research which is ever developing newer substances called addition agents or *additives*. Specific agents or combinations have been developed which improve one or more of the desired properties. Some of the typical properties that are improved by incorporating addition agents are following:

Extreme pressure: Under extreme pressure a thick film of the oil is difficult to maintain and the oil needs to have higher oiliness. Besides trying to improve oiliness directly, additives are used which are absorbed on the metal surfaces or chemically react with the metal, and give a film which is easily sheared. This enables the oils to be retained in place and carry the load. Some additives react at higher temperatures forming low melting point alloys and give anti-weld properties. Important additives of this group are metallic (e.g., lead, zinc and molybdenum) compounds and chlorine, sulphur or phosphorus compounds.

Viscosity Index. Certain high molecular weight compounds improve the viscosity- temperature properties of the oil.

Higher viscosity index means less change of viscosity with change in temperature. Generally the V. I. for naphthenic stocks vary from 30 to 60, and for paraffinic it varies from 90 to 120. By suitable additives it can be increased to about 150. There is improvement by refining to a high degree but a cheaper method is by the use of additives.

Polymeric compounds like polymathylacrylates, polyacetates, polybutenes and polymerized alkylstyrenes are generally used as the additives. The base stock has a molecular weight of about 600 and the additives used have molecular weights of about 20,000. The additives used are not completely soluble in the oil at the lower temperatures and tend to be compact as show in Fig. 7.9a.

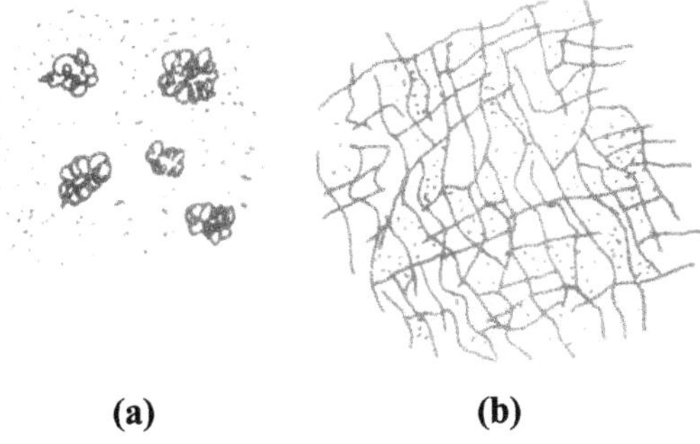

(a) (b)

Fig. 9.9 Effect of additives on the control of viscosity.

As the temperature is increased the additive begins to dissolve which means their molecules gradually surround the oil molecules as shown in Fig. 7.9b. This makes it difficult to shear oil particles. Thus resistance to shear between the layers of oil increases. In other words viscosity increases due to the use of the additives but as the viscosity of oil decreases as the temperature increases these two effects tend to cancel each other and the net result is that there is less change in viscosity with change in temperature.

Oxidation. Under certain conditions, as in internal combustion engines, turbines, etc,. oxidation of the oil is a serious problem. Aromatic, phenolic or amino compounds retard oxidation of the oil by getting themselves oxidized. These substances are called antioxidants or inhibitors.

Corrosion. Corrosion may be due to water or due to oxidation products of the oil. Certain compounds of phosphorus, sulphur and sulphur-chlorine protect the metal surface by preventing contact between the metal surface and the corrosive substances.

Pour point. Certain condensation products of chlorinated wax with naphthalene or phenol prevent the separation of wax from oil, thus the pour point is depressed.

Deposit formation. In internal combustion engines, deposits are formed due to imperfect combustion. These are dispersed and cleaned by additives which have detergent properties.

The outline given above shows how the properties of petroleum oils can be improved. As a matter of fact, most of the lubricating oils on the market have one or more of the additives. The terms *'compounded oils'* and *'blended oils'* have been used for oils in which other substances, mainly fatty materials, are added. The term is not so widely used now but can be used for oils having the additives.

9.4.5 Synthetic Lubricants

In many machines, operating conditions are so severe that petroleum lubricants fail to serve the purpose. For example, in die casting, furnace equipment, hot rolling mills, etc. the temperatures and the pressures to which the lubricant is exposed are very high and the petroleum and fatty oils will decompose under these conditions. In addition, the lubricant should be non-inflammable. The oil for the bearings of a high altitude jet plane may have to be pumped at -50°. When the plane is landing or taking off, the same oil may get heated to as high a temperature as 250°. Under both conditions the oil has to maintain satisfactory properties. To meet such exacting requirements viscous fluids have been synthesized from various organic and inorganic materials and used as synthetic lubricants. These are quite expensive and in some cases may be 100 times as expensive as petroleum lubricants.

The more widely used synthetic lubricants are the silicones, phosphates and esters. The distinguishing features of these may briefly be stated as follows.

1. *Inflammability.* The silicones, fully chlorinated and fluorinated hydrocarbons and some phosphates are non inflammable.
2. *Flash point.* The esters, though inflammable, have relatively high flash points.
3. *Thermal stability.* The silicones and fluorocarbons do not decompose on exposure to high temperatures.
4. *Viscosity.* Many of the synthetics have a high viscosity index, i.e., the change in viscosity with temperature is small.

The important hurdle in the wide use of the synthetic lubricants is the high cost. Often they need to be mixed with petroleum products and some of them do not mix very easily. Some of them are toxic and some are corrosive. Their solvent effect on the packing and sealing material should also be studied before using them.

Because of all these reasons the use of synthetic lubricants is quite limited and is restricted to specialized purposes. For most of the common purposes, however, synthetic lubricants are not needed, as petroleum oils with additives give good service.

An idea of the mechanism of synthetic lubricants can be had from a brief consideration of silicone fluids. These have a chain of silicon atoms in which these atoms are joined to each other through oxygen atoms. It is this structure that makes them stable at high temperatures and resistant to oxidation. Various organic groups are attached to the silicon atoms satisfying the remaining bonds. The nature and size of these groups control physical properties like viscosity, viscosity index, volatility, flash and fire points, oiliness, etc. A typical structure, shown below, is that of polymethyl siloxane:

$$\begin{array}{ccccccc} & CH_3 & & CH_3 & & CH_3 & \\ & | & & | & & | & \\ CH_3- & Si & -O- & Si & -O- & Si & -\ldots\ldots\ldots \\ & | & & | & & | & \\ & CH_3 & & CH_3 & & CH_3 & \end{array}$$

Silicones are especially noted for high viscosity index and also for very high oxidation resistance and very low volatility. They do not decompose in the presence of moisture. Silicones have two main disadvantages. Firstly, under extremely severe conditions they may break down and produce silica (SiO_2) which has serious abrasive action on the rubbing surfaces. Secondly, they are poor lubricants under boundary lubrication conditions.

9.4.6 Semi-Solid Lubricants

There are situations in which a liquid lubricant is unsatisfactory for various reasons and a semi-solid lubricant has to be used. Primarily these are the situations in which the oil does not remain in place because of bad seal, high load, slow speed or intermittent operation. There are cases where dripping of oil is undesirable as in a ceiling fan; the bearing also needs to be sealed against entry of dust, dirt, grit or moisture. In all such cases, a semi-solid lubricant, called a grease, is used. Grease has been defined as 'a combination of a petroleum product and a soap or a mixture of soaps, suitable for certain types of lubrications'. For incorporating special properties other ingredients may be added. Thus mostly it is a petroleum oil made semi-solid by mixing with it a sodium, calcium, lithium or aluminium soap. The term grease is also used for other semi-solid lubricants prepared by adding solids such as graphite, mica, talc, chalk, etc. to mineral oils. Certain combinations of waxes, fats and petroleum residues are also called grease. Just as for oils, additives may be added to improve the properties of greases. Extreme pressure additives or solid lubricants may be added for services under high loads and/or high temperatures.

Greases do not require as much attention as the liquid lubricants. In most cases a small quantity of grease serves the purpose satisfactorily for a long time. For example, in a truck the wheel bearings need to be greased only once in 150,000 km of running. As dripping and spattering is almost completely eliminated the working is clean and products (especially food and textile) are not spoilt. Further, the accessories like pumps and filters needed for fluid lubrication are eliminated when grease is used. Because of heavier consistency, the frictional resistance and wear caused by greases is greater than that caused by oils. On storage they have a tendency to separate into oil and soap.

In a grease, soap has several functions. It certainly is the thickening agent. It sticks to the metal surfaces more firmly. The nature of the soap decides the temperature up to which the grease can be used, its consistency and its resistance to water and to oxidation.

The lubrication property of the grease is mostly decided by the oil which has to be chosen according to the use to which the grease is put. For example, for a grease to be used at high pressure and low speed, the oil used should be of higher viscosity than for low pressure and high speed use.

From the foregoing, it is clear that while selecting a grease for a particular application, the properties of both the oil and the soap need to be considered.

Before considering some examples of greases, mention may be made of two important tests performed on greases. The stillness of a grease is determined by a *penetration test.* In the test, a cone of standard weight and size is dropped into the sample of grease at a particular temperature. The depth in tenths of millimetres to which the cone penetrates is known as penetration number. The apparatus used is shown in Fig. 7.10. The test does not show the lubricating property of the grease but indicates the ease with which it can be pumped in greasing equipment. The temperature to which it can be used is expressed as its *dropping point*, this is the temperature at which it passes from the semi-solid to the liquid state. The apparatus used is shown in Figure 7.11 and is self-explanatory. The sample is taken in a metal cup which has an opening in the bottom. As the temperature is raised the sample passes from a semi-solid to a fluid state. The temperature at which its first drop falls is called its dropping point.

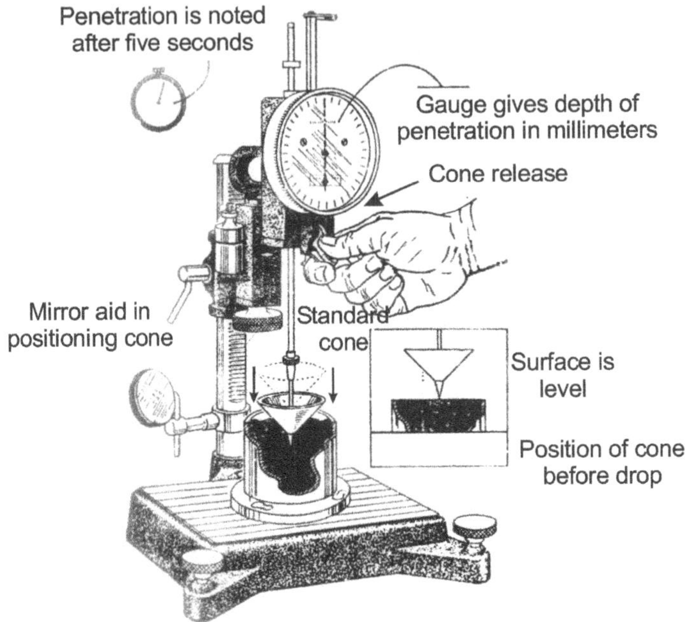

Fig. 9.10 Penetration test for determining consistency of grease.
(*Lubrication Engineer's Manual, Courtesy, U.S. Steel Corporation*)

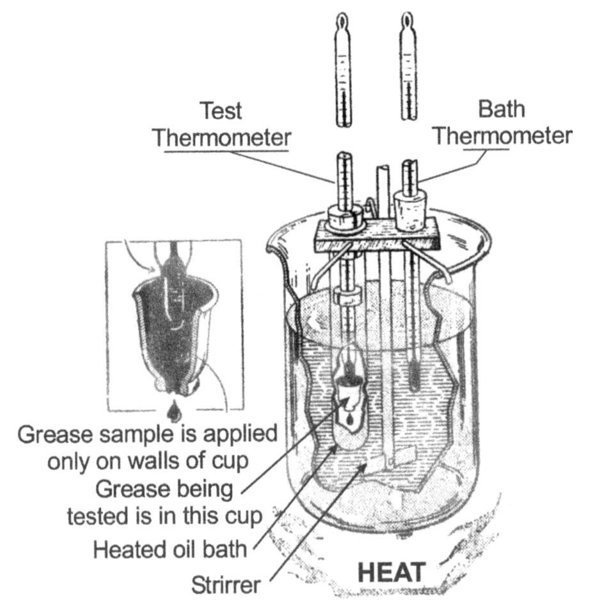

Fig. 9.11 Dropping point test.
(*Lubrication Engineer's Manual, Courtesy, U.S. Steel Corporation*)

The greases are named after the soap used in their manufacture, for example, calcium base, soda base grease, etc.

Calcium base greases are prepared by adding the required amount of calcium hydroxide solution to an oil, like tallow, heated and stirred till soap formation is completed. Hot petroleum oil is then added and mixed. Oil of the correct viscosity is chosen to give grease for the particular use. The calcium base greases are the cheapest and most commonly used. They are also known as *cup greases.*

Calcium soap being insoluble in water, these greases are water-resistant. They are unsatisfactory for use above 80° because water begins to be expelled causing the separation of oil and soap.

Soda base greases are not resistant to water because the soap is soluble in water. They hold water more firmly consequently can be used up to 175°. They are suitable for use as ball and roller bearing greases where the lubricant gets heated due to friction. They are prepared in a manner similar to that of calcium greases except that caustic soda is used instead of calcium hydroxide.

Lithium base greases are comparatively new in the field and combine the good qualities of both calcium and soda base greases. These can resist the action of water as well as of heat. Various additives are added to lithium base greases to make them suitable for difficult conditions like extreme pressure, oxidizing conditions, etc.

Various other greases are made to suit different needs, *Mixed base*, i.e., calcium and soda base greases have to some extent the properties of both the types. Railway carriage *axle greases* are made by using rosin soap, made by saponifying rosin with lime, with or without oil. Fillers like talc and mica are also added. These greases are cheaper than others and are satisfactory for less delicate equipment working under heavy load and at low speed.

9.4.7 Solid Lubricants

The use of solid lubricants becomes necessary where it is not possible to maintain a fluid film, or the presence of a fluid or semi-fluid is undesirable (as in commutator brushes for electric generators and motors), or the temperature or load is too high even for a semi-solid lubricant. The crystal lattices of such materials consist of layers. The bonds between the atoms within the layers are great but between the adjacent layers it is small. The result is that the force required to shear the crystals in a direction parallel to the layers is low. Thus the layers slide over one another easily. Usually some organic substances are mixed with solid lubricants so that they may stick firmly to the metal surfaces.

Graphite is the most widely used solid lubricant. Natural graphite is in the form of flakes. Artificial graphite, produced from coke or anthracite coal, is amorphous. Graphite is more satisfactory on uneven surfaces as it fills the low spots and makes the surface more even and then graphite particles slide over each other while the equipment is in motion. Graphite is also used as dispersion with the help of tannins, in oil or in water when it is called *oildag* or *aquadag* respectively. In internal combustion engines oildag has been found to be useful as it forms a film between the piston rings and the cylinder and gives a tight fit, thus increasing compression. Aquadag is useful where a lubricant free from oil is needed. It is used in air compressors, general machine shop work, lathes and for machinery used in the foodstuff industry. Graphite is also incorporated in greases.

Graphite is stable in air up to 550° but above this temperature it begins to oxidize. In the absence of air it can be used up to very much higher temperatures.

Molybdenum disulphide is a new entrant in the field. It has a very low co-efficient of friction and its fine powder may be sprinkled on surfaces sliding at high velocities. It is stable in air up to 400°. It is also used in greases. Molybdenum disulphide has applications in space, vacuum and aero space. Also, it has good radiation stability and therefore, can be used in nuclear reactors.

Boron nitride can be used up to 800°.

A number of other substances, like talc, mica, etc., are used as solid lubricants. The use of talc power on carom board is a common example of the use of solid lubricant.

9.5 Selection of Lubricants

A lubricant, for a particular application, has certainly to be chosen on the basis of service requirements and its own properties, but during use certain properties may change which need to be watched carefully. The properties of a good lubricant should not change under service conditions. It should have proper viscosity under the working conditions. This is to be watched particularly if there is a temperature change during service. At certain temperature the lubricant may cause certain problems. At lower temperatures wax may separate. At higher temperatures a portion of it may volatalize leaving behind a residual oil which has different lubricating properties. If the lubricant is exposed to a high temperature, its flash and fire points and volatility become factors in the selection. Oxidation may cause formation of deposits as well as give rise to corrosive products. Reactions with water need to be further watched if the lubricant comes in contact with water or steam as in steam turbines.

A careful study of the various tests and their correct interpretation is necessary for the selection of lubricants. Unfortunately tests which have no valid significance for a particular work are often specified. Low carbon residue and high oxidation stability have been specified, e.g., for simple shafting lubrication, where high temperature and oxidation never occur. Demulsificaion has been specified where water or steam do not come in contact with the lubricants. Such mistakes must be avoided. The guiding principles for selecting a lubricant may be illustrated by taking a few concrete examples.

9.5.1 Cutting Tools

Cutting fluids are needed for tools used in the machine shop for cutting, sawing, threading, planing, turning etc. In these cases the cutting fluid, cool or lubricate or serve to cool as well as to lubricate. In these machining operations there is close contact between the tool and the work piece and the friction is very high. This generates a large amount of heat and the tool gets overheated and may loose its temper. Therefore, besides lubrication, cooling is also required. To give satisfactory service, the cutting fluid should have good lubricating property, low viscosity (to enable the lubricant to fill in the cracks formed on the work piece), high thermal conductivity, chemical stability and absence of corrosive action on the metals.

In certain situations, e.g., rough grindings and turning, only cooling may suffice. Because of its cooling capacity and cheapness water may be used in such cases in spite of its tendency to cause rusting. Water is generally used in high speed operations as in such cases lubrication is not so effective and cooling is primarily needed.

When fine work has to be done, a stream of a fixed oil, like sperm oils, rape oil or lard, is directed over the job. Compounded oils, prepared by mixing fixed and petroleum oils, are also used for the purpose. These act more as lubricants and less as heat-carrying agents. Oils are mostly used when the speed is low and pressure is high, as in broaching.

The use of water or oil alone may be considered to be two extremes. Water is predominantly a cooling agent and oil a lubricating agent. In most cases both cooling and lubrication are almost equally needed. For such situations cutting emulsions are used. These are prepared by making an emulsion of petroleum oil and water with the help of emulsifying agents like soap or, better still, sulphonated oils (say, prepared by the action of sulphuric acid on castor oil which has a large amount of an ester of a hydroxy acid).

An important property that is required in a cutting emulsion is that it should not cause rusting of the metals. A cutting emulsion serves two purposes. It lubricates to reduce frictional heat and serves as a medium for the transfer of the heat produced. Thus it reduces power consumption and tool wear.

9.5.2 Internal Combustion Engines

In an internal combustion engine the situation is quite different from that met with while cutting with tools but in both cases high temperature develops. In the engine, combustion of the fuel takes place and its latent energy is converted to mechanical energy. In the process the engine gets heated. The heat is mostly removed by cooling systems but the lubricating oil also is needed to remove some heat. It must lubricate surfaces at 150 to 200° and should not be swept away by gases at 1000 to 2000°.

The lubricating oil is subjected to high temperatures, therefore, besides low variation in viscosity with temperature (i.e., high viscosity index), it should have high thermal stability. The oil should neither evaporate easily nor decompose at the operating temperatures of the engine. The lubricating oil should also help to remove a part of the heat. The lubricant has to perform yet another function. A portion of the fuel and the lubricant decompose and form carbon and lacquer deposits which restrict the free movement of the piston. The lubricant should disperse these and clean up the engine.

Accordingly, additives for giving high viscosity index, oxidation stability and detergency are added to petroleum oils for making them suitable for use as lubricating oils for internal combustion engines. A passing reference may be made to the cleaning of the engine by a cheap thin oil, like kerosene oil. This removes the solid deposits by flow of oil but the sludge remains. A better method is to use a solvent which quickly cleans the lubricated parts. Tetrahydro naphthalene has been patented for such a purpose.

9.5.3 Steam Engine Cylinders

Lubrication of the steam cylinder is done by injecting the lubricating oil drop by drop, into the steam line. The oil is atomised by steam and is spread by it on the surface of cylinder walls, pistons, valves and valve seats. Therefore, a new quality, metal-wetting quality, is essential for these oils. The oil should have comparatively high viscosity (20 to 55 centipoise at 100°). As some condensation of steam takes place, it is necessary that the oil should emulsify with water (It is noteworthy that this property will cause difficulty in the use of exhaust steam because it will be difficult to remove oil from steam and condensed water). It should not have a tendency to gumming in the presence of steam at high temperatures. Oil and water emulsions do not lubricate as efficiently as oils, but in a steam engine cylinder formation of emulsion is considered desirable.

The steam engine cylinder oil generally used is a steam refined paraffin base oil which is compounded by the addition of fixed oils and additives.

9.5.4 Steam Turbines

Oils for lubricating steam turbines have to face some of the most difficult conditions. The oil is exposed to high temperatures and oxidizing conditions. Oxidation causes rapid deterioration of lubricating properties. There are chances of leakage of steam and cooling water, which may form emulsions and sludges, restricting proper circulation. There may be foaming due to entanglement of air bubbles. The oil is also required to carry away the heat so that the bearing temperature does not exceed, say, 80°. Under all these exacting conditions, the oil has to be used continuously for a long time after reconditioning it by settling, centrifuging and filtering.

The turbine oil should, therefore, have high oxidation resistance and good demulsification, i.e., easy separation of oil from emulsion. For getting satisfactory properties, it becomes necessary to compound the oil by the addition of additives for giving oxidation stability, rust protection, anti-foaming and good water separation.

9.5.5 Gears

There are several types of gears operating under some what different conditions but all of them are subjected to extreme pressures. In general, the lubricant should stick well to the gear teeth, should not be removed by centrifugal force, should have a high load-carrying capacity and high resistance to oxidation. In some cases it is exposed to water, heat, dirt, etc., and has to work under boundary lubrication conditions.

The extreme pressure additives which are not removed by centrifugal force are used for these lubricants.

Questions

1. Explain the theories of lubrication. Write about the applications of lubricants in industry.
2. Describe the important properties of lubricants.
3. What are lubricants and what are their important properties?
4. Define lubricants. Describe their important characteristics.

5. "The analysis of trace metals in oils can be indication of extensive wear". Explain.

6. Describe the mechanism of extreme pressure lubrication.

7. What is the importance of viscosity and viscosity index of a lubricant?

8. What are the factors that govern the choice of lubricant?

9. What is the significance of carbon residue in the selection of a lubricant?

10. Explain how Graphite and Molybdenum disulphide act as lubricants.

11. What are solid lubricants? Explain one of them in detail.

12. The boiling point of an oil is 325 ^{0}C. Using Troutons Rule ($\Delta H_{vap} / T_{bp} = 88kJ$) and the ratio $\Delta H_{vap} / E_{vis} = 5$, determine E_{vis} for the lubricant

13. What lubricants would you use for operating in a space craft intended for a journey of 6 months

14. Distinguish between hydrodynamic lubrication, boundary lubrication and extreme pressure lubrication.

15. How can a viscous lubricant be converted into a grease.

16. What are synthetic lubricants? How are they superior to petroleum based lubricants?

17. Explain viscosity index and aniline point as applied to a lubricating oil.

18. What is meant by oiliness of a lubricant? How can this be improved?

19. Explain the following properties of lubricants giving their significance.

    (i) viscosity & viscosity index (ii) flash & fire points

    (iii) cloud & pour points (iv) saponification number

    (v) Neutralisation number (vi) carbon residue

20. Explain different additives employed for improvement of lubricants

Objective Questions

1. Machines operating under high temperatures and loads are lubricated by ....

   (A) grease (B) mineral oil

   (C) solid lubricant (D) synthetic oil

2. Molybdenum disulphide is a ....

   (A) solid, fuel (B) liquid, lubricant

   (C) solid, lubricant (D) semisolid, lubricant

3. Mineral oils are

   (A) suitable at low T (B) unstable

   (C) thickened in use (D) oxidation resistant

4. Vegetable oils are

   (A) suitable at high T (B) stable

   (C) thickened in use (D) oxidation resistant

5. Oiliness is less for

   (A) Vegetable oils (B) Mineral oils

   (C) Greases (D) Waxes

6. Greases are not suitable for

   (A) engine parts (B) gear boxes

   (C) watches (D) instruments with bearings

7. Lubricants are aimed to reduce

   (A) abrasion (B) corrosion

   (C) wearing (D) all

8. Lubricant used in machines working at low temperature should have

   (A) high pour point (B) low pour point

   (C) high flash point (D) low flash point

9. Graphite dispersed in oil is called

   (A) dispersion lubricant (B) grease

   (C) oildag (D) graoil

10. Generally for liquid lubricants

    (A) flash and fire points are identical

    (B) flash point > fire point

    (C) flash point < fire point

    (D) flash point > pour point

11. Silicones are known for

    (A) high viscosity index, high oxidation resistance and low volatility

    (B) high viscosity index, high oxidation resistance and high volatility

    (C) high viscosity index, low oxidation resistance and low volatility

    (D) low viscosity index, high oxidation resistance and low volatility

Chapter 10

Electrochemistry

Conductance, Transport Number, EMF

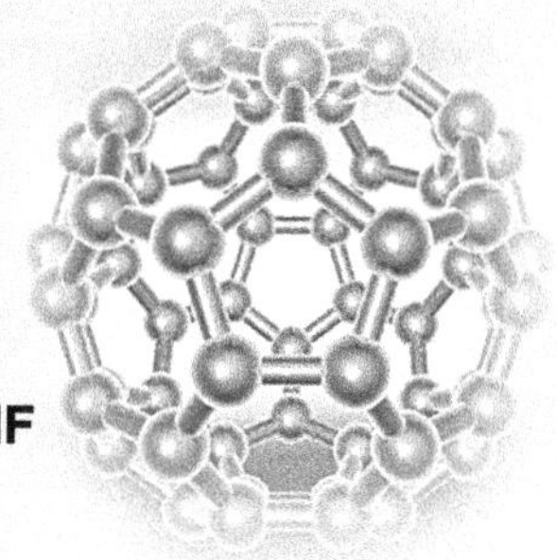

10.1 Introduction

All matter is electrical in nature. An atom the basic particle is made up of subatomic particles, viz., electrons, protons, neutrons, etc., where the positively charged nuclei is surrounded by the revolving cloud of electrons. Any chemical change may be considered as rearrangement of atoms and their electron clouds. Thus the formation of HI by the combination H_2 and I_2 may be represented as

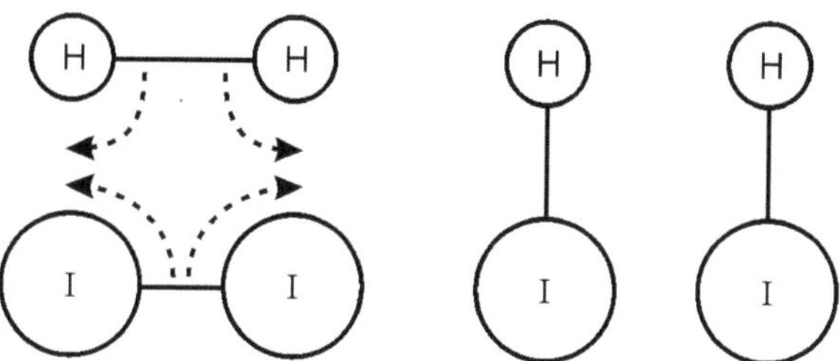

Here the atoms rearrange themselves by changing partners with simultaneous rearrangement of covalent bonds. This suggests that a chemical change and an electrical change are two faces of the same coin.

Electrochemistry is the branch of physical chemistry that is concerned with the inter-relationship of electricity and chemistry. It has a two pronged coverage, firstly, with the nature and properties of substances containing ions and charged particles and, secondly, with the relationship between electric currents and chemical reaction. The former is sometimes called **ionics** and the latter **electrodies**.

10.2 Electrical Conduction

The substances of this universe can fall into four different categories depending upon their capability of flow of electrons, viz., electrical conductors, insulators, semiconductor and super-conductors.

A. Electrical Conductors (or Simply Conductors)

With few exceptions like mica, generally all the good thermal conductors are good electrical conductors. They allow electrons to pass through them with minimum resistance, and show no preference of direction of flow. Some examples of this class are metals, metal sulphides, acids, bases, ionizable aqueous and nonaqueous salt solutions and fused salts. The electrical conduction of a solid depends on the spacing and state of occupancy of the energy bands. Their valence and conduction bands are placed closely and side by side.

B. Insulators or Non-Conductors

Substance of this category do not allow the electricity to pass through them. Some examples of this class are pure water, majority of the substances of plant origin (Wood, rubber, paper, etc.), many synthetic organic compounds (benzene, CCl_4, ether, etc.), mica and all non-metals except carbon. In an insulator the gap between the valence band and conduction band is considerably large making the flow of electrons difficult.

C. Semiconductors

Elements of 4A group, especially Si and Ge have properties that are intermediate between those of metals and nonmetals and therefore are called semiconductors. Their conducting properties are considerably enhanced by the addition of certain impurities, called doping elements. Addition of impurities like phosphorus (P) that have more valence electrons than Si generally called donor impurities give ***n*-type semiconductors**. Similarly acceptor impurities like boron (B) that have lesser number of valence electrons produce ***p*-type semiconductors**. In addition to conducting properties these impurities also restrict the electron movement in the crystal.

The most valuable property derived out of doping and unidirectional flow of electrons is replacement of cumbersome valve versions with small semiconductor pieces of thickness of a pencil eraser called **"Chips"**. This led to invention of solid state devices like calculators, mini computers, pocket radios, etc. Further, variation of conducting property of a semiconductor with temperature is exploited by making temperature measuring and controlling devices.

D. Superconductors

Electrical conductivity decreases with decrease in temperature. Lead was found to conduct electricity with zero resistance below 7.3 *K*. In 1987, physicists found that certain mixtures of copper, barium and rare earth metal oxides exhibit this zero resistance to as high as 130 *K*. The property of certain metals, alloys and compounds by virtue of which they conduct electricity with zero or negligible resistance is known as **super conductivity**. The characteristics temperature at which a metal becomes superconducting is called transition temperature T_c.

Because of zero resistance, super conductors attain the capacity to store infinite amount of current. This allows transfer of power from the place of generation to place of utility without laying few hundred away kilometer lines and without any loss of power during transmission. Further superconductivity is expected to create very high field strength electromagnets that can be used for many purposes like lifting super heavy weights.

10.2.1 Types of Electrical Conductors

In this chapter our concern is mainly on various types of electrical conductors and the physical and chemical phenomena that accompany on account of passage of electricity through them. The electrical conductors are of three type.

(i) **Metallic or Electronic Conductors :** This class of conductors undergo no chemical change when current is passed through them. However, the passage of electricity through these conductors could accompany physical changes such as heating, generation of light, change of color, etc.

Mechanism : In electronic conductors flow of electricity will be in the form of a stream of electrons and this flow is akin to flow of a liquid through a pipe. Metals have relatively low ionization potential and electrons in the outer shell or valence shell are loosely bound to the kernel. Thus a metal is conceived as a kind of electron fluid in which metallic ions are regularly arranged and the electrons have complete freedom of movement in the vacant valence orbitals.

When electrons are injected into a metal strip by connecting it to an electron source like a battery, an equal number of them leave the metal strip to join the electron acceptor or the sink or the positive pole. Thus at any given instant the electron population in a metal strip remains constant. The quantity of charge or electron flow in coulombs per second is defined as an ampere. Since each electron bears a charge of 1.6×10^{-19} coulombs, one ampere of current means $1 / 1.6 \times 10^{-19}$. 6×10^{18} electrons in one second. Mathematically the charge (Q) in coulombs is a product of current (I) in amperes and time (t) in seconds

$$Q = I \times t \quad .....(10.1)$$

The electron flow primarily depends on the arrangement of metal ions. During their motion the electrons experience a resistance due to collisions with metal ions vibrating about a fixed mean position. An increase in temperature increases the frequency of vibration, thereby, the flow of electrons decrease. This was demonstrated by Davy by means of a beautiful experiment. For this a 4-5 inch fine platinum wire was connected in a voltaic circuit. This heats the whole wire to red hot. Now heating any part of the wire with a spirit lamp cools instantly the rest of the wire due to poor flow of electrons. On the contrary cooling any part of the wire led to heating of the rest of wire as it favors the flow of electrons.

(ii) **Electrolytic Conductors :** Substances like salt melts and salt solutions conduct electricity through them. However, unlike metallic conductors this class of conductors undergo chemical change during the passage of current through them. For example, when current is passed through molten NaCl, sodium metal is deposited at the negative electrode while chlorine gas escapes at the positive electrode. Substances whose melts or aqueous solutions allow the passage of current undergoing a simultaneous chemical transformation are called **electrolytes** and the conduction as **electrolytic conductance**. On the other hand, a substance that do not conduct electricity either in molten state or in solution phase are called **non-electrolytes**. Thus urea in its molten and in solution phase behaves as a bad-conductor of electricity.

Electrolyte molecules are made up of oppositely charged species called ions. In the solid state they are firmly held in a fixed lattice, characteristic of the substance and can not have any kinetic motion. But in fused state or in solution, the ions have independent existence and are free to move. Unlike metals electrolytes do not have any free electrons. Electricity is conducted only by virtue of movement of electrically charged ion. For example, in case of sodium chloride, the solid state is defined by face centered cubic lattice and the sodium and chloride ions do not have translational freedom. In fused state and solution state these ions are free to move towards oppositely charged electrodes.

Fused State Equilibrium :

$$\underset{\text{solid}}{NaCl} \rightleftharpoons \underset{\text{ionized}}{Na^+\ Cl^-}$$

Solution State Equilibrium :

Salt and ions are solvated (aquated)

$$NaCl\,(aq) \rightleftharpoons Na^+_{(aq)} + Cl^-_{(aq)}$$

The negatively charged ions, called anions, move towards **anode** or positive electrode. The electrode connected to negative terminal of the battery is called **cathode** or negative electrode and the cations move towards it. This assembly of cathode and anode in an electrolytic medium is referred as **cell**. Electricity or flow of electrons from a battery enters the cell through cathode and leaves through anode. Regardless of the kind of cell or battery, the cathode is defined as an electrode at which reduction occurs i.e., the species around it gain electrons and the anode is defined as an electrode at which oxidation occurs i.e., the species around it lose electrons. In the fused sodium chloride, the sodium ions are reduced at cathode to liberate sodium metal and at anode the chloride ions are oxidized to chlorine gas. Thus the mechanism of electrolytic conduction may be summarized as follows.

1. Electrolytic conduction mainly occurs through movement of cations and anions, and therefore depends on their concentration.
2. Electrolytic conduction is always accompanied by electrochemical changes.
3. When there are no secondary reactions due to solvent or any other foreign species, there is a progressive decrease in conduction. This fall is attributed to the fall in the concentration of electrolyte on account of deposition or development of non-conducting atmosphere around electrode.
4. A rise in temperature decreases the viscosity of the medium and increases the kinetic motion of the ion. Both these factors act cumulatively and the electrolytic conduction invariably increases with increase in temperature.
5. The electrodes may participate in the electrochemical secondary reactions, in which case they are corroded and sometimes may even contribute for the rise in electrolytic conduction.

A comparison between electrolytic and metallic conduction is given in Table 10.1.

Table 10.1 A comparative account of electrolytic and metallic conduction.

Metallic Conduction	Electrolytic Conduction
1. Metallic conduction is a result of flow of electrons in the valence orbitals.	1. Electrolytic conduction is a result of movement of ions, which themselves carry the charge.
2. There is no chemical change during conduction of electricity	2. It is always accompanied by some electrochemical change.
3. Metallic conduction is unchanged with time under same temperature	3. Electrolytic conduction decreases with time under same temperature conditions, provided there are no secondary reactions.
4. Electrodes are unaffected during conduction.	4. Electrodes may be affected during conduction.
5. Metallic conduction decreases with a rise in temperature	5. Electrolytic conduction increases with increase in temperature

10.2.2 Electrolysis

The process of decomposition of electrolyte, in its molten state or solution phase by passing electricity is known as electrolysis. Some typical changes caused by a flow of electrons through two of the electrolytes under different conditions is discussed below.

A. Electrolysis of Molten Sodium Chloride

Sodium chloride melts at 801 ^{O}C to give a clear, colorless liquid melt. When two inert electrodes like graphite are dipped into this melt and connected to a source of direct current the following changes take place.

(i) A pale green gas, identified as chlorine, is liberated at the anode.

(ii) Silvery white metallic sodium is formed at cathode. The metal remains liquid because its melting point is only 97.8 ^{O}C and floats owing to its low density.

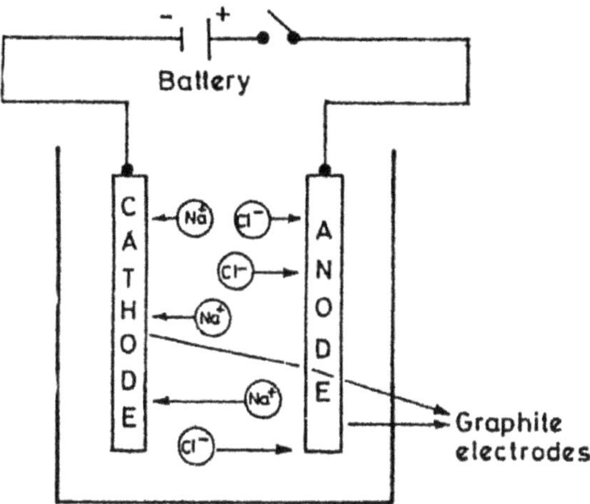

Fig. 10.1 Electrolysis of Molten NaCl.

B. Electrolysis of Aqueous Sodium Chloride

When a solution of sodium chloride is subjected to electrolysis using unattackable electrodes, we notice,

(a) Liberation of hydrogen gas at the cathode

(b) Liberation of chlorine gas at the anode

(c) The overall solution becoming basic due to formation of NaOH.

The electrolysis of aqueous solutions of electrolytes is some what more complex because of the ability of water to be oxidized as well as reduced.

The oxidation reaction (at anode) for water is,

$$2H_2O \longrightarrow O_2\uparrow + 4H^+ + 4e^-$$

The reduction reaction (at cathode) for water is,

$$2H_2O + 2e^- \longrightarrow H_2\uparrow + 2OH^-$$

In acidic solution, water is reduced through the reaction,

$$2H^+ + 2e^- \longrightarrow H_2\uparrow$$

Therefore, in the electrolysis of aqueous NaCl, the following anodic reactions are possible.

$$2Cl^- \longrightarrow Cl_2 + 2e^- \qquad \text{I}$$

$$2H_2O \longrightarrow O_2 + 4H^+ + 4e^- \qquad \text{II}$$

Therefore, in the electrolysis of aqueous NaCl, the following anodic reactions are possible.

$$2Cl^- \longrightarrow Cl_2 + 2e^- \qquad \text{I}$$

$$2H_2O \longrightarrow O_2 + 4H^+ + 4e^- \qquad \text{II}$$

Similarly at cathode the possible reactions are

$$Na^+ + e^- \longrightarrow Na \qquad \text{III}$$

$$2H_2O + 2e^- \longrightarrow H_2\uparrow + 2OH^- \qquad \text{IV}$$

Out of these four possible reactions, experimentally reaction (I) & (II) were found take place.

$$2Cl^- \longrightarrow Cl_28 + 2e^- \qquad \text{(at Anode)}$$

$$2H_2O + 2e^- \longrightarrow H_2\uparrow + 2OH^- \qquad \text{(at Cathode)}$$

$$2Cl^- + 2H_2O \longrightarrow Cl_2\uparrow + H_2\uparrow + 2OH^- \quad \text{(Net cell reaction)}$$

$$+ 2Na^+ \qquad\qquad + 2Na^+ \quad \text{(Spectator ions)}$$

$$2NaCl + 2H_2O \longrightarrow Cl_2\uparrow + H_2\uparrow + 2NaOH \quad \text{(Cell reaction)}$$

The results suggest that H_2O has a greater tendency to be reduced than Na^+ and similarly Cl^- has a greater tendency to be oxidized than H_2O.

C. Electrolysis of Aqueous $CuSO_4$

In the electrolysis of $CuSO_4$ there are two possible oxidation and reduction reactions.

Oxidation (Anode) :

$$2SO_4^{2-} \longrightarrow S_2O_8^{2-} + 2e^- \qquad \text{V}$$

$$H_2O \longrightarrow O_2\uparrow + 4H^+ + 4e^- \qquad \text{VI}$$

Reduction (Cathode) :

$$Cu^{2+} + 2e^- \longrightarrow Cu \qquad \text{VII}$$

$$2H_2O + 2e^- \longrightarrow H_2\uparrow + 2OH^- \qquad \text{VIII}$$

Liberation of oxygen gas at anode and deposition of copper metal on the cathode suggest the preference of steps (VI) and (VII) in the electrolysis over others. Therefore we could write the electrolysis of aqueous $CuSO_4$ as,

$$2H_2O \longrightarrow O_2 + 4H^+ + 4e^- \quad \text{(at Anode)}$$
$$2Cu^{2+} + 4e^- \longrightarrow 2Cu \quad \text{(at Cathode)}$$

$$2H_2O + 2Cu^{2+} \longrightarrow O_2 + 2Cu + 4H^+ \quad \text{(Net cell reaction)}$$
$$2SO_4^{2-} \qquad 2SO_4^{2-} \quad \text{(Spectator ion)}$$

$$2CuSO_4 + 2H_2O \longrightarrow 2Cu + 2H_2SO_4 + O_2\uparrow \quad \text{(cell reaction)}$$

10.2.3 Quantitative Aspects of Electrolysis

Faraday's Laws : Michael Faraday (1832) made a detailed studies of electrolysis process and deduced quantitative relationships between electricity used and the extent of chemical reactions. His results can be expressed in the form of two laws of electrolysis.

Faraday's First Law : The mass of substance generated at an electrode is proportional to the amount of electricity passed during electrolysis.

In other words the amount of electrolyte deposited during electrolysis is directly proportional to the number of moles of electrons passed. For example, in the reduction of silver ion to silver metal,

$$Ag^+(aq) + e^- \longrightarrow Ag(s)$$

1 mole Ag is deposited by 1 mole of electrons used for reduction. A better way of this kind of expression is obtained by considering the charge carried by Avogadro number electrons. Since each electron carries a charge of 1.6×10^{-19} coulombs, the total charge carried by 1 mole of electrons is

$$1 \text{ mole } e^- \longleftrightarrow 6.022 \times 10^{23} \times 1.6024 \times 10^{-19} \longleftrightarrow 96494 \text{ C}$$

This charge carried by 1 mole of electrons is known as a **faraday** (F) and is generally expressed up to three significant figures.

$$1F = 96500 \text{ C}$$

A coulomb is thus defined as the amount of electrical charge that passes a given point in an electrical circuit when a current of 1 ampere flows for 1 second.

Therefore,

$$1 \text{ coulomb} = 1 \text{ ampere} \times 1 \text{ second}$$

or $\quad 1 \text{ C} = 1\text{A.s}$

During electrolysis, if I amperes of current is passed for t seconds the total current passed in coulombs is equal to It.

$$\underset{\text{(coulomb)}}{Q} = \underset{\text{(amperes)}}{I} \times \underset{\text{(second)}}{t}$$

But according to Faraday law, the mass of substance generated (W) is directly proportional to the charge (Q)

$$W \propto Q$$

or

$$\boxed{W = zQ = z\,It} \quad .....(10.2)$$

where z is a proportionality constant called **electrochemical equivalent**.

When $I = 1$ A and $t = 1$ s

$$W = z$$

Thus z, the electrochemical equivalent can be defined as the amount of substance deposited by 1 ampere of current when passed for 1 second.

or The amount of substance deposited by 1 coulomb.

Because 1g equivalent of an ion is discharged by 1 Faraday (96500 coulombs) of electricity the electrochemical equivalent may be mathematically expressed as

$$z = \frac{\textit{Equivalent weight of ion or substance}}{96500} \quad .....(10.3)$$

Experimental verification involves passing a constant current (I amps) for a fixed amount of time (t seconds) and noting the fall in concentration of electrolyte or the amount of an ion deposited on one of the electrode. Doubling the current passed or duration, simply doubles the amount deposited, there by proving Faraday's first law. In other words a plot between W and Q should give a straight line to prove first law.

The first law has been shown to apply very accurately to molten electrolytes as well as solutions of electrolytes and is independent of temperature, pressure and the nature of the non-interfering solvent or electrode.

Faraday's Second Law

When the same quantity of electricity is passed through different electrolytes, the masses of substances generated at respective electrodes are directly proportional to their chemical equivalents.

The second law can be verified by passing the same quantity of current for a fixed amount of time through a series of electrolytic cells containing different electrolytes, say, dilute H_2SO_4, $AgNO_3$ and $CuSO_4$, Fig. 10.2.

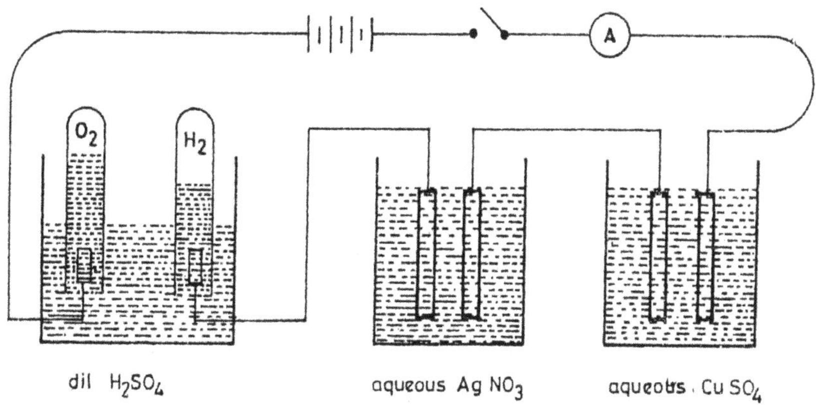

Fig. 10.2 Voltameters in a series circuit.

The amounts of substances deposited on cathode of these electrolytic cells, often called as **voltameters** and **coulometers**, on passing 1 coulomb of electricity were

Hydrogen	)))	0.00001006 g
Copper	)))	0.0003179 g
Silver	)))	0.001078 g

Because the amounts deposited by fixed current are proportional to the equivalent weight of the corresponding substances, the ratio of these masses should correspond to the equivalent weights.

$$\frac{\text{Weight of copper deposited}}{\text{Weight of hydrogen evolved}} = \frac{\text{Equivalent weight of Cu}}{\text{Equivalent weight of } H_2} \quad \text{.....(10.4)}$$

$$\frac{0.0003179}{0.00001006} = \frac{31.79}{1.006}$$

Similarly,

$$\frac{\text{Equivalent weight of Ag}}{\text{Equivalent of weight of } H_2}$$

$$= \frac{0.001078}{0.00001006} = \frac{107.8}{1.0006}$$

Thus for a series of electrolytes, the amount deposited by the same current is directly proportional to their equivalent weights.

Problem 1.

What is the electrochemical equivalent of copper, when 0.3950 g of copper is deposited by a current of 0.5 amperes in 40 minutes.

Quantity of current passed = $0.5 \times 40 \times 60$

= 1200 coulombs

Copper deposited by the current = 0.3950 g

So the amount of copper deposited per coulomb

$$= \frac{0.3950}{1200} = 0.0003292 \text{ g}$$

Therefore,

Electrochemical equivalent of copper $= z = 0.0003292$ g

Problem 2.

What volume of O_2 would be liberated at NTP from an aqueous solution of NaOH by a current of 3 amperes passed for 3 hours.

$$Q = It = 3 \times 3 \times 60 \times 60 = 32400 \text{ coulombs}$$

In this electrolysis the evolution of oxygen at the anode is given by the equation;

$$2H_2O \longrightarrow O_2 + 4H^+ + 4e^-$$

Therefore 1 mole of oxygen is liberated by passing 4 faradays. (4×96500 coulombs) of charge. We also know that at NTP each mole of oxygen occupies 22,400 cm^3.

Hence the volume of O_2 liberated

$$= 32400 \times \frac{1}{4 \times 96500} \times 22400$$

$$= 1880.2 \text{ cm}^3$$

Problem 3.

What is the current strength, when 0.25 g of silver is deposited on silver electrode in a silver coulometer, when electricity is passed for 30 minutes. What would be the gain in weight of copper cathode in copper coulometer placed in the same circuit.

96500 coulombs of current is required to deposit 1g equivalent (i.e. 107.87 g) of silver.

So the quantity of current required to deposit 0.25g of silver is

$$= \frac{0.25}{107.87} \times 96500 \quad = 224 \text{ C}$$

But $\quad Q = I \times t \quad$ or $\quad I = \frac{Q}{t}$

Current strength $I = \frac{224}{30 \times 60} = 0.124$ A

We also know,

$$\frac{\text{Equivalent weight of Cu}}{\text{Equivalent weight of Ag}} = \frac{\text{Mass of Cu}}{\text{Mass of Ag}}$$

$$\frac{31.77}{107.87} = \frac{\text{Mass of Cu}}{0.25}$$

Mass of $\text{Cu} = \frac{31.77}{107.87} \times 0.25 = 0.074$ g

Gain in weight of copper anode = 0.074 g

Problem 4.

How long will it take to deposit 1 g of Cu from a solution of $CuSO_4$, with a constant current of 1000 A, if the only electrode reaction is

$$Cu^{2+} + 2e^- \longrightarrow Cu$$

To deposit 1 atomic weight of Cu, i.e, 63.54 g, we need 2 faraday of current. Therefore 31.77 g is deposited by 1 faraday or 96500 C.

Amount of charge needed to deposit 1g of Cu is $\frac{1}{31.77} \times 96500$ C

But Faraday's first law we have $Q = It$

$$\frac{1}{31.77} \times 96500 = 1000 \times t$$

Time needed for deposition $t = 3036$ s

Problem 5.

A current of 15 millie amperes is passed through silver nitrate for 5 hours. Find the number of grams of silver deposited in the coulometer (At. wt. of Ag 107.88)

$$\begin{aligned}\text{Amount of charge (C)} &= \text{Amount of current (A)} \times \text{Time (s)} \\ &= (15 \times 10^{-3}) \times (5 \times 60 \times 60) \\ &= 15 \times 10^{-3} \times 18000 \\ &= 270 \text{ C}\end{aligned}$$

1 Faraday or 96487 coulombs of current deposits 107.88 g of Ag

1 coulomb of current deposits $\frac{107.88}{96487} = 0.00118$ g of Ag

Therefore,

270 coulombs of current deposits $0.001118 \times 270 = 0.3$ g of Ag.

Problem 6.

Find the volume of chlorine generated at 273 K and 3 atm from another coulometer arranged in series with the above coulometer.

When some current is passed through two coulometers in series, same equivalents are discharged.

$$\frac{\text{Weight of chlorine}}{\text{Weight of silver}} = \frac{\text{Equivalent weigth of chlorine}}{\text{Equivalent weight of silver}}$$

$$= \frac{35.5}{107.88}$$

$$W = \frac{35.5}{107.88} \times 0.3 = 0.0987 \text{ g}$$

We know 35.5 × 2 (= 71 g) of chlorine at NTP occupies a volume of 22.4 L. 0.0987 g of chlorine at NTP occupies a volume of ?

$$\frac{0.0987}{71} \times 22.4 = 0.03 \text{ L}$$

So at STP ($P_1 = 1$ atm, $T_1 = 273$ K) the gas occupies a volume $V_1 = 0.03$L.

But we want volume V_2 at $P_2 = 3$ atm, $T_2 = 273$ K.

$$\frac{P_1V_1}{T_1} = \frac{P_2V_2}{T_2} \quad \text{or} \quad V_2 = \frac{P_1V_1}{T_1} \times \frac{T_2}{P_2}$$

$$V_2 = \frac{1 \times 0.03}{273} \times \frac{273}{3} = 0.01 \text{ L}$$

Volume of Cl_2 evolved = 0.01 L

10.3 Electrical Resistance and Ohms' Law

We already discussed that the metallic conduction increases with fall in temperature. With the approach of absolute zero the resistance will be almost equal to zero and therefore *all substances behave like perfect conductors at absolute zero.* G. S. Ohm in 1827 found that resistance (R) is related to the current passed (I) and the potential difference across the conductor (E)

$$I = \frac{E}{R} \qquad \text{.....(10.5)}$$

Ohm's law may thus be defined that the current strength flowing through a conductor at uniform temperature is directly proportional to the potential difference applied across the conductor and inversely proportional to the resistance of the material.

When current is measured in amperes and potential difference in volts the resistance offered is a fixed value called ohm (represented by symbol omega 'Ω')

$$\text{Ohm} = \frac{\text{Volt}}{\text{Ampere}}$$

Thus, Ohm is the resistance offered by a body when a current of 1 ampere flows across it under an application of a potential difference of 1 volt. Ohm's law is strictly obeyed by all metallic conductors. Electrolytic conductors obey the law, when the current passed is small enough not to bring any significant concentration changes of the electrolyte.

In addition Ohm also found that the resistance of a conductor of uniform cross section is directly proportional to its length and inversely proportional to the area of cross section.

$$R\ \alpha\ \frac{1}{a} \quad \text{or} \quad R = \rho\ \frac{1}{a} \qquad .....(10.6)$$

where ρ (rho) is a proportionality constant called **specific resistance** or **resistivity**. Specific resistance is nothing but the resistance offered by a material of unit length and unit area of cross section. The units of specific resistance are

$$\rho = \text{Ohm} \times \frac{\text{cm}^2}{\text{cm}} = \text{Ohm cm}$$

10.3.1 Conductance and Specific Conductance

Resistance and specific resistance are commonly used for metallic conductors, where the atoms are static. In case of electrolytic solutions the electricity is virtually conducted by constantly moving ions. So it was thought more meaningful to define another quantity called **conductance** (this term should not be equated to conductivity, as done in older literature, as the term conductivity is reserved by IUPAC to represent specific conductance introduced below). It gives the ease with which electricity flows through an electrolytic conductor. The conductance of an electrolytic solution is defined as the reciprocal of its resistance.

$$\textbf{Conductance } (L) = \frac{1}{R}\ \textbf{Ohm}^{-1} \textbf{ or mho or } \Omega^{-1} \textbf{ or Siemens (S)} \qquad .....(10.7)$$

The conventional unit of conductance is ohm^{-1} or mhos or Ω^{-1}. In SI system the same unit is called Siemens and is represented by symbol S.

The reciprocal of specific resistance is defined as **conductivity** or **electrolyte conductivity** or **specific conductance**. Of these three terms only conductivity and electrolytic conductivity are recognised by IUPAC and therefore specific conductance and specific resistance are no longer in use. This quantity, designated by κ (kappa), may be defined as the conductance of unit volume of the solution. In other words electrolytic conductivity is the conductivity of a solution taken in between electrodes of unit area of cross section and separated by a unit distance, Fig. 10.3.

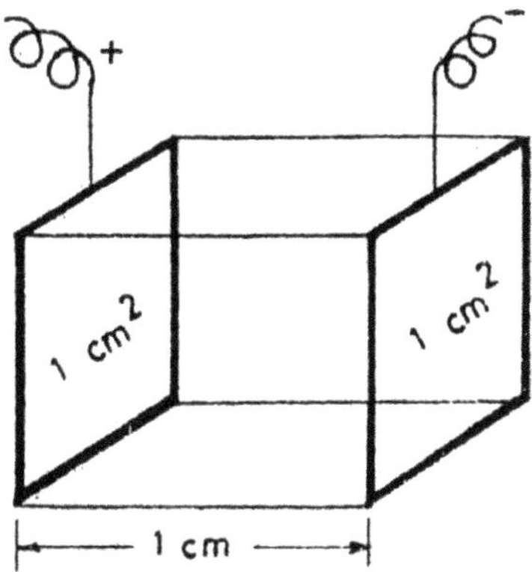

Fig. 10.3 Electrolyte conductivity.

In CGS system the conductivity is expressed in S cm^{-1} or mho cm^{-1}

$$\text{Conductivity } \kappa = \frac{1}{\rho} = \frac{1}{\text{Ohm cm}} = \text{mho cm}^{-1} = \text{S cm}^{-1} \qquad .....(10.8)$$

In SI system, meter being the fundamental unit of distance, the conductivity is expressed as S m^{-1}. The conversion factor of both CGS & SI systems is,

$$1 \text{ S cm}^{-1} = 100 \text{ S m}^{-1} \quad \text{or} \quad 1 \text{ S m}^{-1} = 0.01 \text{ S cm}^{-1}$$

The electrolyte conductivity is dependent upon the concentration of electrolyte. The conductivity increases with increase in the concentration of electrolyte at a given temperature, Table 10.2. A relation between conductivity and conductance could be obtained starting from our definition represented by eqn. (10.8).

$$\text{Conductivity } \kappa = \frac{1}{\rho}$$

$$\text{But from eq. (10.6) } \frac{1}{\rho} = \frac{1}{R} \times \frac{l}{a}$$

i.e., $$\kappa = \frac{1}{R} \times \frac{l}{a}$$

Substitute L for $1/R$ (from eqn. 10.7)

$$\kappa = L\,\frac{1}{a} \qquad \text{.....(10.9)}$$

$$L = \kappa\,\frac{a}{l}$$

Because the concentration of all pure solids is taken as unity, the electrolytic conductivity of electrolytes remains constant at a constant temperature.

10.3.2 Equivalent Conductivity

Since conductance and electrolyte conductivity are concentration dependent terms, to allow a comparison between different electrolytes, F. W. Kohlrausch (1876) suggested the use of **equivalent conductivity/ equivalent conductance**, represented by Λ_v (Greek : Capital Lambda).

It is defined as the *conductance of all the ions produced by the dissociation of one gram equivalent of an electrolyte, dissolved in a certain volume, V, of the solvent at a constant temperature.*

Mathematically it is expressed as;

$$\mathbf{\Lambda_v = 1000\,\frac{\kappa}{C}}\ \text{ohm}^{-1}\ \text{cm}^2\ \text{equivalent}^{-1} \qquad \text{.....(10.10)}$$

where C is the concentration of the solution in gram equivalents per litre. Since V is the volume of solution containing 1 equivalent of electrolyte

$$V = \frac{1000}{C}$$

Therefore eqn. (10.10) becomes

$$\mathbf{\Lambda_v = \kappa V}\ \text{Ohm}^{-1}\ \text{cm}^2\ \text{eq}^{-1} \qquad \text{.....(10.11)}$$

The equivalent conductivity of a solution can thus be readily obtained from a knowledge of concentration (or volume of solution containing one equivalent of electrolyte) and electrolyte conductivity. Eq. (10.11) also suggests that equivalent conductivity of a solution of unit volume containing one gram equivalent of a substance is equal to the electrolyte conductivity.

In SI system equivalent conductance is expressed as;

$$\Lambda_V = \frac{1000\ \kappa\left(S\ m^{-1}\right)}{C\left(eq\,m^{-3}\right)}\ S\ m^2\ eq^{-1}$$

Conversion of SI units into CGS units and vice-versa is done using the conversion factor ;

$$1\ S\ m^2\ eq^{-1} = 10^4\ S\ cm^2\ eq^{-1}$$

10.3.3 Molar Conductivity

In recent times it has become a practice to express the concentration in moles per litre rather than equivalents per litre. Keeping in pace with this new practice the term **molar conductivity/molar conductance** was introduced. This quantity is also sometimes referred to as **molecular conductivity**. Molar conductivity is defined as the conductance of all the ions produced by the dissociation of one gram molecular weight of an electrolyte dissolved in a certain volume, V of the solvent at constant temperature. This quantity is generally denoted by the symbols : and Λ_m. Similar to equivalent conductance, molar conductance is given by the expressions,

$$\mu = 1000\ \frac{\kappa}{C}$$

and $: = \kappa V$ .....(10.12)

where V is the volume of the liquid containing 1 mole of the electrolyte.

The units of : in CGS system are $Ohm^{-1}\ cm^2\ mol^{-1}$ or $S\ cm^2\ mol^{-1}$ and in SI system it is $S\ m^2\ mol^{-1}$. The conversion factors between these systems is

$$1\ S\ m^2\ mol^{-1} = 10^4\ S\ cm^2\ mol^{-1}$$

10.3.4 Relation between Molar Conductivity and Equivalent Conductivity

The relationship between molar conductivity and equivalent conductivity can be obtained by comparing eqn. (10.11) and (10.12).

$$\frac{\Lambda_v}{\mu} = \frac{\dfrac{1000\kappa}{C\left(\text{in equivalents per litre}\right)}}{\dfrac{1000\kappa}{C\left(\text{in moles per litre}\right)}}$$

$$\frac{\Lambda_v}{\mu} = \frac{C(\text{in moles})}{C(\text{in equivalents})}$$

But $\quad C(\text{mol/L}) = \frac{W}{M},\ C(\text{eq/L}) = \frac{W}{E}$

and E = $\frac{W}{z}$ where z is the total charge carried by cations (or anions) liberated out of dissociation of the electrolyte.

$$\therefore \quad C(\text{eq/L}) = \frac{W}{(m/z)} = Z.\ C(\text{mol/L})$$

Therefore

$$\frac{\Lambda_v}{\mu} = \frac{C(\text{mol/L})}{z\ C(\text{mol/L})} = \frac{1}{z} \qquad .....(10.13)$$

This can be well demonstrated by considering the following examples.

For $KCl \longrightarrow K^+ + Cl^-$ $\quad z = 1$ Therefore Λ_v = :

For $ZnCl_2 \longrightarrow Zn^{2+} + 2Cl^-$ $\quad z = 2$ Therefore $2\Lambda_v$ = :

For $Na_2HPO_4 \longrightarrow 2Na^+ + HPO_4^{2-}$ $\quad z = 2$ Therefore $2\Lambda_v$ = :

For $AlK(SO_4)_2 \longrightarrow Al^{3+} + K^+ + 2SO_4^{2-}$ $\quad z = 4$ Therefore $4\Lambda_v$ = :

Problem 7.

The equivalent conductivity of 0.1 N barium chloride was found to be 105.2 S cm^2 eq^{-1} at room temperature. What would be its molar conductivity at the same temperature.

Barium chloride dissociated through following equation,

$$BaCl_2 \longrightarrow B^{2+} + 2Cl^-$$

The total charge carried by individual ions z = 2

We know from eqn. (10.13) $\frac{\Lambda_v}{\mu} = \frac{1}{z}$

given Λ_v = 105.2 S cm^2 eq^{-1} & z = 2 (sum of change on cation/anion)

$$\frac{105.2}{\mu} = \frac{1}{2}$$

or Molar conductivity : $= 105.2 \times 2$

$= 210.45$ cm^2 mol^{-1}

10.3.5 Measurement of Conductivity of Electrolytes

Experimental determination of conductivity of electrolytic solutions is important in covering both the basic and applied aspects of electrochemistry. The basic aspects such as nature and quantification of intermediate ionic attractions, ion-ion and ion-solvent pairings, etc., can be done through conductometric measurements. On the applied front the study of corrosion and passivity occupies the top priority. In addition, these measurements are widely used for testing the purity of water obtained from an ion-exchanger to determine quantities such as solubility of sparingly soluble salts, dissociation constants of weak electrolytes, ionic product of water, to compare the strengths of acids and bases and to carry out conductometric titrations of pure and mixture of electrolytes.

Basically determination of conductance, *L*, involves estimation of resistance, *R*, of the electrolytic solution. This is usually done by means of **Wheatstone bridge circuit** which involves a comparison of unknown resistance with a standard resistance. The Wheatstone bridge circuit is shown in Fig. 10.4.

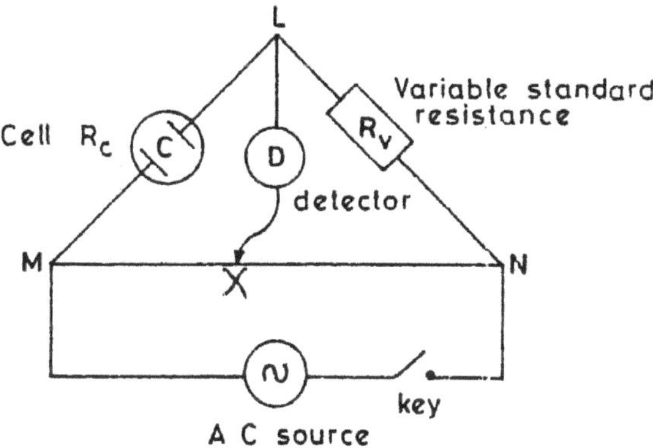

Fig. 10.4 Wheatstone bridge circuit.

The electrolyte of known concentration is taken in a container called **conductivity cell**. It consists of two platinum plates of area of cross section *a* cm^2 and separated by a distance 1 cm. These plates are generally coated with platinum black to decrease the polarization effects. This forms one arm of the circuit. The other arm of the circuit is fitted with a variable standard resistance. These two arms are attached to both ends of a meter bridge. A source of alternating current is also attached to both extremes of the meter bridge. The preference of alternating current over direct current is for the following reasons.

(i) Direct current changes the concentration of electrolyte due to electrolysis.

(ii) Gases liberated at the electrodes undergo polarization and thereby affect the conductance.

(iii) By employing alternating current the direction of current is reversed few thousand times and pulses produced in opposite direction mutually cancel the polarization effects.

The current balance detector, D, fixed between R_c and R_v is generally a head piece or magic eye for less accurate measurements and a cathode ray oscillator for precession measurements.

Now the sliding contact, often called jockey, is moved over the long and uniform meter bridge wire MN, the point of least current through the detector, X, indicated by least buzzing noise of a head phone or smallest signal of an oscilloscope is noted. Under these conditions Wheatstone bridge principle suggests that the ratio of the resistances in the meter bridge arms, i.e., MX to NX is equalto the ratio of LM to LN.

$$\frac{\text{MX}}{\text{NX}} = \frac{\text{LM}}{\text{LN}}$$

But the resistance in LM is that of conductivity cell, i.e., R_c, and the resistance in arm LN corresponds the variable resistance, R_v.

Therefore,

$$\frac{\text{MX}}{\text{NX}} = \frac{R_c}{R_v}$$

Since R_v is known and MX and NX are determined through experiment, the resistance of the cell R_c can be calculated. Inverse of R_c gives the conductance of the experimental solution.

To calculate the electrolyte conductivity of the solution we take help of the eqn. (10.9).

$$\kappa = L\,\frac{l}{a} \qquad .....(10.14)$$

A knowledge of l and a allow us to calculate the electrolytic conductivity of the solution. But in the normally available cells the electrodes are far from parallel to obtain an exact value for l and far from perfect geometrical shape to obtain an exact value for a. So, instead of attempting to find the values of l and a of the conductivity cell, it is a matter of great simplification to use the literature electrolytic conductivity values of standard electrolytes like KCl

Table 10.2. With these values one can determine its conductivity and use these two values to find the ratio *l/a*. This ratio is a characteristic of a conductivity cell and is known as cell constant, X. The eqn. (10.14) may thus be written as

or $$\boxed{\kappa = L \times X} \quad .....(10.15)$$

Electrolyte conductivity = Conductance × Cell constant.

10.3.5 Determination of Cell Constant

In finding the cell constant, first and foremost step is to remove organic and oily impurities sticking to the walls of the cell and electrodes by treating with dilute chromic acid. It is then thoroughly rinsed with conductivity water. The electrodes of the cell are then dipped in a standard KCl solution, say N/10 KCl, thermostated to required temperature and its conductance is measured. The electrolytic conductivity of N/10 KCl, as determined accurately by Kohlrausch and verified by other scientists at 25 ^{o}C is equal to 0.012856 S cm^{-1}. If our experimental temperature is also 25 ^{o}C, then cell constant is obtained from the following expression.

$$\text{Cell constant} = \frac{0.012856}{\text{Observed conductance}}$$

The cell constant so obtained can be used in finding the electrolytic conductivity of experimental electrolyte solution.

Various conductivity cells are designed to meet the requirements of different conductance measurements. For solutions of high conductance measurements the electrodes are widely spaced, Fig. 10.5(a), while for low conducting liquids they are closely spaced Fig. 10.5(b).

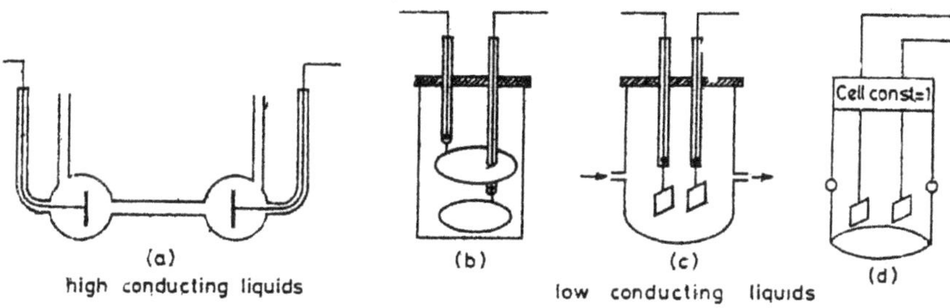

Fig. 10.5 Types of conductance cells.

Fig. 10.5(c) gives the arrangement of a conductivity cell used for on line conductance measurements and Fig. 10.5(d) is a commercial dipping type cell.

For liquids having large conductance values normal distilled water is satisfactory. For low conducting substances, the electrolyte is dissolved in a special grade water called conductivity water. It is prepared in an all quartz still. First the water is refluxed, for sufficiently long time, with alkaline permanganate to oxidize the volatile organic impurities. Then it is distilled and collected into an air tight quartz container. Purity of the water can be established by comparing the conductivity with the literature value.

Conductivity of water after single distillation in glass at 25 ^{O}C

$$= 2.0 \times 10^{-6}\ \text{S cm}^{-1}$$

Conductivity of triple distilled water at 25 ^{O}C

$$= 0.8 \times 10^{-6}\ \text{to}\ 1 \times 10^{-6}\ \text{S cm}^{-1}$$

Conductivity of conductivity grade water at 25 ^{O}C

$$= 0.056 \times 10^{-6}\ \text{S cm}^{-1}$$

Conductivity of purest water available at 25 ^{O}C

$$= 0.038 \times 10^{-6}\ \text{S cm}^{-1}$$

If the measured conductivity is low, the real conductivity of an electrolytic solution is obtained by subtracting conductivity contribution (due to dissociation) of water used for making the solution.

Problem 8.

The resistance of a N/50 KCl solution at 25 ^{O}C present in a conductivity cell is 350 S cm^{-1}. Calculate its cell constant. Given the electrolytic conductivity of N/50 KCl at 25 ^{O}C is 0.002765 S cm^{-1}.

$$\text{Cell constant } (X) = \frac{\text{Electrolytic conductance } (\kappa)}{\text{Observed conductance at the same temp.} (L)}$$

Since resistance (R) is inverse of conductance (L),

$$X = \kappa \times R$$

$$= 0.002765 \times 350$$

$$X = 0.9678\ \text{S cm}^{-1}$$

Problem 9.

The resistance of an 0.1 N solution of an electrolyte is 40 ohm. If the distance between the electrodes is 1.2 cm and area of cross section is 2.4 cm^2, calculate the equivalent conductivity.

Given $l = 1.2$ cm, $a = 2.4\ cm^2$, $C = 0.1$ N, $R = 40$ ohm

$$\text{Cell constant} = \frac{1}{a} = \frac{1.2}{2.4} = 0.5\ cm^{-1}$$

$$\text{Observed conductivity} = \frac{1}{40} = 0.025\ S$$

$$\text{Electrolyte conductivity } \kappa = L \times X$$

$$= 0.5 \times 0.025$$

$$= 0.0125\ S\ cm^{-1}$$

Equivalent conductivity

$$\Lambda = \frac{1000\kappa}{C} = \frac{1000 \times 0.0125}{0.1}$$

$$= 125\ S\ cm^2\ eq^{-1}.$$

10.3.6 Effect of Dilution on Conductance

We already learnt that conductance depends on three factors, viz.,

(i) on the number of ions,

(ii) on the charge of the ions

(iii) on the mobility (mass) of ions.

Addition of an electrolyte, till saturation limit, increases the population of free ions and it is reasonable to expect a simultaneous increase of conductance. But the number of ions of an electrolyte are decided by the ionic character of a bond. Further, with the increase in concentration of a solute, the ions get closer amounting to an increase in the electrostatic forces of attraction and viscosity of the medium. These factors tend to reduce the conductance of the solution. A cumulative effect of all the above factors is :

(i) The conductance increases with increase in the concentration of the electrolyte to a certain maximum level and,

(ii) Any further increase will only result in a fall in the conductance value.

This trend is well demonstrated in case of highly soluble electrolytic solutions of KOH and acetic acid, Fig. 10.6. However, in majority of the cases, saturation limit precedes the segment representing a fall in the conductance. Therefore we often notice only an increase in conductance with the increase in concentration followed by a simple plateau representing the arrival of saturation limit. Because electrolytic conductivity is defined as the conductance of a unit cube of solution, it adopts the trends followed by conductance.

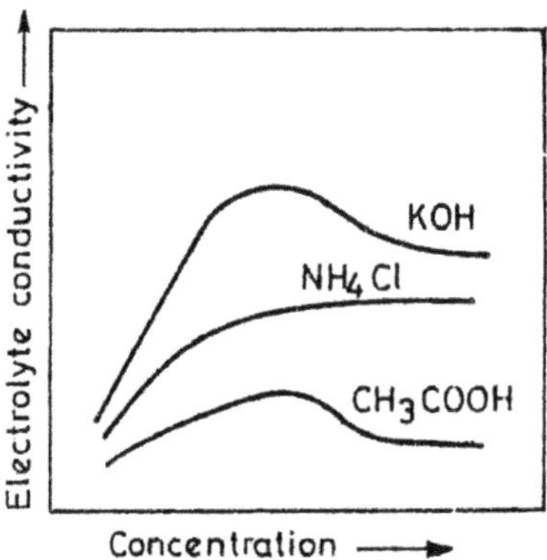

Fig. 10.6 Variation of conductivities of electrolytes with concentration.

On the other hand equivalent conductivity and molar conductivity are inversely proportional to the concentration and therefore, increase with increase in dilution (decrease in concentration), Fig. 10.7.

Equivalent conductivity

$$\Lambda_v = 1000\,\frac{\kappa}{C} = \kappa V$$

Molar conductivity

$$: = 1000\,\frac{\kappa}{C} = \kappa V$$

The trends of decrease in electrolytic conductivity and increase in equivalent and molar conductivity values with dilution is shown in Table 10.2.

Table 10.2 Electrolytic and molar conductivities of aqueous solutions of KCl at different concentrations (25 °C).

Concentration C mol L^{-1}	Volume containing 1 mole, V mL	Electrolytic conductivity κ S cm^{-1}	Molar* conductivity : S cm^2 mol^{-1}
1.0	1,000	0.11173	111.73
0.1	10,000	0.012856	128.56
0.05	20,000	0.00667	133.4
0.02	50,000	0.002766	138.3
0.01	1,00,000	0.001413	141.3
0.005	2,00,000	0.0007175	143.5
0.001	10,00,000	0.0001469	146.9
0.0005	20,00,000	0.0000739	147.8

* KCl being uni-univalent electrolyte its equivalent and molar conductances are same.

A comparative information on the variation of equivalent conductivities with concentration of various electrolytes is given in Table 10.3. A close observation of data given in Table 10.2 & 10.3 and the curves (in Fig. 10.7) obtained basing on the data, prompt us to draw the following conclusions.

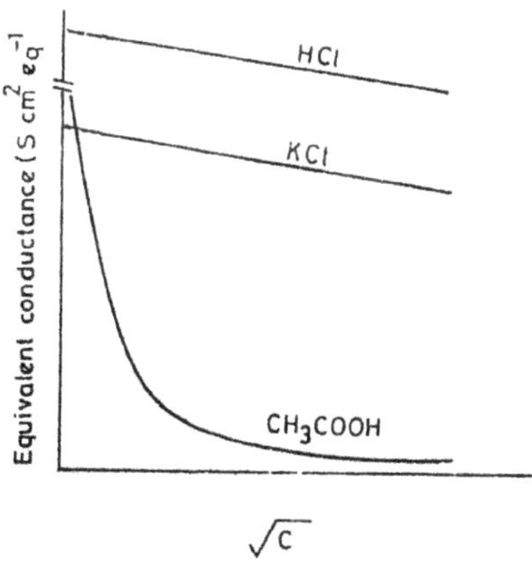

Fig. 10.7 Variation of equivalent conductivity of HCl, KCl and CH_3COOH with square root of concentration.

(i) The electrolytic conductivity increases with increase in concentration of electrolyte.

(ii) After reaching a particular concentration level, the conductance of some electrolytes may marginally drop and ultimately reach a constant value.

(iii) In case of less soluble or sparingly soluble salts the conductance may reach the constancy without showing any fall in its value.

(iv) Equivalent conductivity invariably falls with the increase in the concentration or increase with increase in dilution.

(v) A gradual, linear change in Λ or : with square root of concentration is found in case of strong electrolytes like KCl and HCl, known for their complete dissociation in aqueous media.

(vi) Electrolytes like acetic acid, known for their low ionization, show insignificant change in conductance at higher concentrations but abruptly rise at higher dilutions (lower concentrations).

(vii) The maximum, value of equivalent conductivity obtained by the extrapolation to zero concentration or infinite dilution of an electrolyte is a characteristic value and is called equivalent conductivity at infinite dilution, Λ_4.

Table 10.3 Variation of equivalent conductivity (S cm^2 eq^{-1}) with concentration of various electrolytes at 25 ^{O}C.

Concentration	**Strong electrolytes**					**Weak Electrolyte**
C mol L^{-1}	**HCl**	**KCl**	**NaOH**	**$AgNO_3$**	**½$BaCl_2$**	**CH_3COOH**
0.10	391.3	128.6	221	109.1	105.2	5.2
0.05	399.1	133.4	227	115.2	111.5	7.4
0.02	407.2	138.3	233	121.4	119.1	11.6
0.01	412.0	141.3	237	124.8	123.9	16.2
0.005	415.8	143.5	240	127.2	128.0	22.8
0.001	421.4	146.9	245	130.5	134.3	48.6
0.0005	429.16	149.86	248.11	133.36	139.98	390.71

10.3.7 Classification of Electrolytes

On the basis of curves obtained in case of variation of equivalent conductivity with square root of concentration the electrolytes are grouped into two categories.

A. Strong Electrolytes

Strong electrolytes are the substances whose equivalent conductivity do not change appreciably with the changes in concentration. The equivalent conductances at any dilution are of the same order as the equivalent conductance at infinite dilution. The solutions of mineral acids, bases and many of their simple salts fall in this category of compounds.

B. Weak Electrolytes

Weak electrolytes are the substances which possess low and constant equivalent conductivity values at high concentrations, and these values increase very rapidly in extremely dilute solutions. The carboxylic acids, amines and their salts fall in this category of compounds.

The linear variation of equivalent conductivity with $\sqrt{C}$ of different strong electrolytes, led to the formulation of an empirical relation called **Kohlrausch equation,**

$$\boxed{\Lambda_v = \Lambda_4 - b\sqrt{C}} \qquad .....(10.16)$$

where b is a constant.

Kohlrausch equation suggests that for strong electrolytes the intercept of Λ_v vs $\sqrt{C}$ plots is equal to Λ_4. Weak electrolytes do not obey Kohlrausch relation and therefore this method of calculation of Λ_4 is not suitable for weak electrolytes. This is achieved through indirect calculation based on **Kohlrausch's law of independent migration of ions** discussed later.

Problem 10.

The resistance of 0.2 N solution of an electrolyte was found to be 250 ohms at 25 ^{O}C. Calculate the equivalent conductivity if the cell constant is 0.75 cm^{-1}.

Observed conductance

$$L = \frac{1}{R} = \frac{1}{250}$$

but Electrolytic conductivity

$$\kappa = X \times L$$

$$= 0.75 - (1/250)$$

$$= 0.003 \text{ mho cm}^{-1}$$

Equivalent conductivity of 0.2 N electrolyte

$$\Lambda_V = \frac{1000\kappa}{C} = \frac{1000 \times 0.003}{0.2}$$

$$\Lambda_V = 15 \text{ mho cm}^2 \text{ eq}^{-1}$$

Problem 11.

The equalent conductivity of HCl of concentrations 0.1, 0.01 and 0.001 N solutions are 391.3, 412.0 and 421.4 S cm^2 $equiv^{-1}$ respectively. Calculate Kohlrausch constant and Λ_4. (HCl.)

According to Kohlrausch law $\Lambda_v = \Lambda_4 - b\sqrt{C}$

Substituting the given data, we have;

$$391.3 = \Lambda_4 - b\sqrt{0.1} \qquad \text{I}$$

$$412.0 = \Lambda_4 - b\sqrt{0.01} \qquad \text{II}$$

$$421.4 = \Lambda_4 - b\sqrt{0.001} \qquad \text{III}$$

II-I gives,

$$(412 - 391.3) = 0 - b\left[\sqrt{0.01} - \sqrt{0.1}\right]$$

$$20.7 = -b\,(0.1 - 0.316)$$

$$b = \frac{20.7}{0.216} = 95.8$$

Substituting the value of b is equation II

$$412 = \Lambda_4 - (95.80)\sqrt{0.01}$$

$$\Lambda_4 = 412 + 95.8\,(0.1)$$

$$\Lambda_4 = 421.6$$

Similar result can be obtained by taking any other pair of equations given above.

10.4 Migration of Ions

The concept of conductivity supported our contention that the dissolved electrolyte dissociates to give ions carrying different charges. These ions can have independent existence in the solutions with opposite charges, to maintain electrical neutrality. When electricity is passed, hither to freely moving ions, move towards the oppositely charged electrodes with different velocities. This phenomenon is called **migration of ions**. The independent existence of ions, their different mobilities and their migration can be demonstrated by the following experiment designed by Noyes. To give a clear visual impression he chose two colored ions, viz., blue colored Cu^{2+} and yellowish red colored dichromate ion. 5 grams of agar-agar powder is taken in 100 mL of water and boiled with copper dichromate prepared by mixing equimolar proportion of copper sulphate and potassium dichromate. This hot solution is taken in a U-tube and allowed to settle, Fig. 10.8. After cooling, this sets to a dark green jelly. A small amount of charcoal dust is scattered on the both ends of the jelly to mark the initial position of the copper dichromate jelly. It is then sandwiched between equal quantities of potassium nitrate jelly. Finally, a solution of potassium nitrate in pure water is added in the two limbs of the U-tube and platinum electrodes are inserted into the solutions.

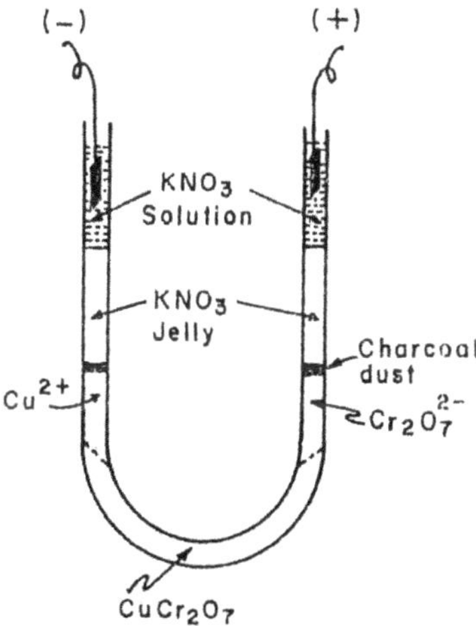

Fig. 10.8 Noyes experiment.

On passing current the blue colored copper ion rise into the potassium nitrate jelly beneath the cathode (negative electrode). Similarly the yellowish dichromate ions move into the anodic arm.

The movement of both the colored ions in opposite direction is indicated by the formation of blue and yellow colored bands in the two limbs. The different thicknesses of these bands further suggests that the Cu^{2+} and $Cr_2O_7^{2-}$ have different velocities.

10.4.1 Relative Speeds of Ions and Amounts Liberated at the Electrodes during Electrolysis

The above cited Noyes experiment of colored ions clearly demonstrated that the ions move with different velocities. But Faraday's second law of electrolysis states that the masses of substances generated at several electrodes in series are proportional to their chemical equivalents, i.e., equal number of ions are discharged at cathode and anode. This paradox can be understood by considering a hypothetical electrolytic cell with two electrodes. Let the electrolyte dissociate to give equal number of cations and anions represented by the + and – signs respectively. To follow the changes

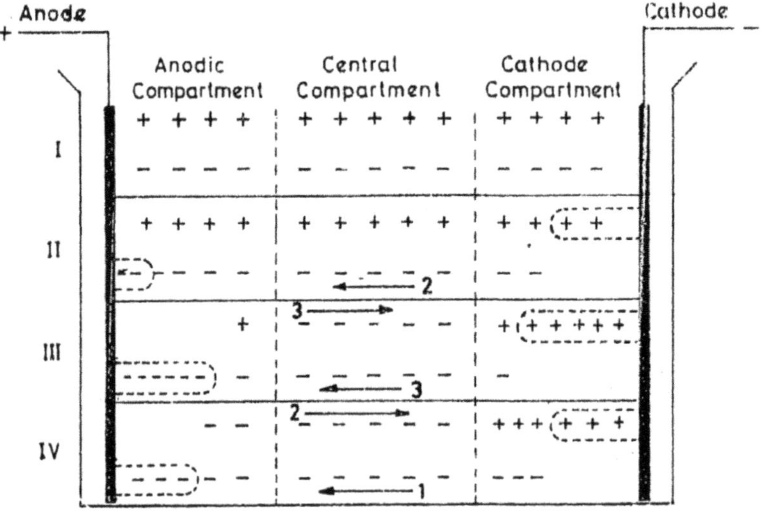

Fig. 10.9 The migration of ions.

occurring around cathode and anode, it is desirable to split the cell into three compartments using two porous diaphragms ∀ and ß as shown in Fig. 10.9. The compartment holding the anode and cathode are called anodic compartment and cathodic compartment respectively and the portion lying between these compartments is called central compartment. The porous diaphragm prevents any mixing of the solutions, but allows the ions to pass through only under the influence of electric field.

Let the electrolytic solution be represented by 13 pairs of ions, out of which 4 pairs each are present in anodic and cathodic compartment and 5 pairs in the central compartment.

The condition of solution at the commencement of electrolysis is shown in Fig. 10.9.

Now, let us make some assumptions and try to record the changes taking place.

A. The Anions (–) only can move and cations do not move

Let us assume that anions only are mobile and any two of them cross the diaphragm and move from right to left in a given time. The condition attained is given in Fig. 10.9, II. Though the cations have not moved, we find that two unpaired ions are discharged at each electrode. Further we also notice that the concentration of electrolyte in cathode compartment has fallen by two units while the concentrations in the anodic and central compartments remains the same.

B. The anions and cations move with the same speed

Let us consider movement of three ions of each type in a given time. This situation is shown in Fig. 10.9, III. The number of discharged cations and anions is once again same, i.e., 6 ions at each electrode. However, the concentration of cathodic and anodic compartments have fallen by 3 units each and the concentration of central compartment remains unchanged.

C. The speed of cations is twice the speed of anions

Let us suppose that in a given time when two cations have passed the diaphragm from left to right only one anion passes from cathode compartment to anode compartment, Fig. 10.9 IV. The total number of discharged ions is 3 each. The anodic compartment concentration has fallen by two units and cathodic compartment has fallen only by one unit. As usual the concentration of central compartment is unchanged.

The observations of the above three situations are summarized below :

(i) Ions are always discharged in equivalent amounts at the respective electrodes, irrespective of the relative speeds of individual ions.

(ii) Fall in the concentration of any compartment (around any electrode) is directly proportional to the speed of the ion moving away from it.

So

Fall in conc. around cathode α **Speed of anion**

&

Fall in conc. around anode α **Speed of cation** .....(10.17)

Thus, we have.,

$$\frac{\textbf{Fall in conc. around cathode}}{\textbf{Fall in conc. around anode}} = \frac{\textbf{Speed of anion}}{\textbf{Speed of cation}} = \frac{u_-}{u_+} \quad .....(10.18)$$

While deriving the above relation, it is presumed that the discharged ions do not react with the electrode material, i.e., the electrodes are unattackable. However, often we encounter situations where anode material passes into the solution due to attack by anions. Electrolysis of silver nitrate using silver electrodes is a typical example of such a case. Supposing that, only nitrate ions are mobile and we employed unattackable platinum electrodes, the state of affairs would have been as described by Fig. 10.9, II. But with attackable silver electrodes, however, we have the condition illustrated by Fig. 10.10.

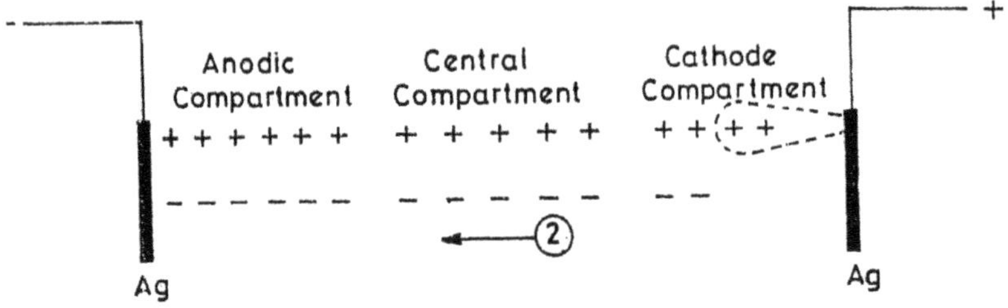

Fig. 10.10 Migration of ions in cells with attackable electrodes.
(The bold sign + marks represent the Ag^+ ions entered the solution by the dissolution of silver anode)

In the cathodic compartment, as expected, there is a fall in silver nitrate concentration by two units. In the anodic compartment, we expect to have a discharge of two anions on the anode, without any change in the concentration of the electrolyte (refer Fig. 10.9, II). But the experimental results show an increase in the concentration of $AgNO_3$ by two molecules. Further the total number of Ag+ deposited on the cathode is equal to the total rise in Ag^+ concentration in the anodic compartment. Exhaustive studies using attackable electrodes gave way to the following conclusions.

(i) The rise in concentration of electrolyte around one electrode is equal to fall in concentration around counter electrode.

(ii) Cations deposited on cathode is equal to the sum of the ions lost at both the electrodes.

(iii) The rise in concentration around anode correspond to the difference in speeds of both the ions.

10.5 Transport Numbers

The current is carried by the anions and cations during electrolysis. Faraday's first law of electrolysis states that the amount of substance deposited or the number of ions discharged, is proportional to the electricity passed through the solution. Therefore it can be said that the total electricity passed through an electrolytic solution is directly proportional to the sum of speeds of the ions. In other words the amount or fraction of current carried by each ion is proportional to corresponding speeds. The fraction of total current carried by an ion is called its **transport number** or **transference number** or **Hittorf number**. If t_+ and t_- designate the transport numbers of cation and anion respectively, then

$$t_+ = \frac{\textbf{Current carried by cation}}{\textbf{Total current passed through the solution}} \quad .....(10.19)$$

$$t_- = \frac{\textbf{Current carried by anion}}{\textbf{Total current passed through the solution}} \quad .....(10.20)$$

Since current carried by every individual ion is directly proportional to the speed, we have

$$t_+ = \frac{u_+}{u_+ + u_-} \quad \text{and} \quad t_- = \frac{u_-}{u_+ + u_-} \quad .....(10.21)$$

where u_+ and u_- indicate the velocities of cations and anions respectively. The eq. (10.21) clearly suggest that faster the speed of an ion, relative to other ions in the same solution, the greater will be the fraction of current carried by it. In calculating the transport number of an ion it is not required to have absolute ionic velocities and it is sufficient if we know the relative ionic velocities.

10.5.1 Important relations of Transport Numbers

A. Sum of transport numbers of anions and cations

From eqn. (10.21), we have

$$t_+ + t_- = \frac{u_+}{u_+ + u_-} + \frac{u_-}{u_+ + u_-} = \frac{u_+ + u_-}{u_+ + u_-} = 1 \qquad .....(10.22)$$

Therefore, sum of transport numbers of ions of an electrolyte is **unity**. This statement is true only when the current is totally carried by ions and not used in performing any other function like heating the solution etc. The expression (10.21) also allows one to calculate the transport number of an ion given the value of transport number of counter ion.

B. Transport number and speed ratio of ions

From equation (10.21), we have

$$t_+ = \frac{u_+}{u_+ + u_-}$$

Dividing numerator and denominator of right hand expression by u_+,

$$t_+ = \frac{(u_+/u_-)}{(u_+ + u_-)/u_+} = \frac{1}{1+\gamma} \qquad .....(10.23)$$

where $\gamma = (u_-/u_+)$ or speed ratio of anion to cation

$$t_- = (1 - t_+) = 1 - \frac{1}{1+\gamma} = \frac{\gamma}{1+\gamma}$$

10.5.2 Determination of transport number by Hittorf Method

PRINCIPLE : Adding 1 on both sides of equation (10.18), we have

$$\frac{\text{Fall in conc. around cathode}}{\text{Fall in conc. around anode}} + 1 = \frac{u_-}{u_+} + 1$$

$$\frac{\text{Fall in conc. around cathode} + \text{Fall in conc. around anode}}{\text{Fall in conc. around anode}} = \frac{u_+ + u_-}{u^+}$$

or $$\frac{\text{Total fall in conc. of the cell}}{\text{Fall in conc. around anode}} = \frac{u_+ + u_-}{u_+} = \frac{1}{t_+}$$

Therefore,

$$t_+ = \frac{\text{Fall in conc. around anode}}{\text{Total fall in conc. of the cell}} \quad .....(10.24)$$

or $$t_+ = \frac{\text{Equivalent migrating from anode}}{\text{Equivalents of current passed}} \quad .....(10.25)$$

and $$t_- = \frac{\text{Equivalent migrating from cathode compartment}}{\text{Equivalents of current passed}} \quad .....(10.26)$$

These expressions provide basis for the determination of transport numbers by Hittorf's method proposed in 1853.

The equivalents migrating from either anode compartment or cathode compartment is determined using conventional analytical techniques. Equivalents of current passed is usually estimated employing a silver or copper **coulometer/voltameter** connected in series. For this electricity is passed through $AgNO_3$ or $CuSO_4$ solution using chemically inert platinum electrodes. The amount of metal deposited directly gives the equivalents of current passed. Determination of transport number of one ion gives the transport number of other ion.

Method : As shown in Fig. 10.11 the apparatus consists of a U-tube joined on both sides with two vertical tubes fitted with bulb shaped glass vessel at the bottom. All the three tubes are provided with stop cocks at their bottom. The junctions of U-tube and vertical tubes are fitted with two stop cocks. This arrangement is illustrated in Fig. 10.11. In this set up the two vertical tubes act as anodic and cathodic compartments and the middle U-tube as the central compartment. The procedure involves following steps.

(i) Hittorf apparatus is filled with the experimental solution of known concentration, the transport numbers of whose ions is of interest.

(ii) Suitable electrodes A and C are inserted in the vertical compartments and connected in series with a coulometer and a source of current.

(iii) A current of 10-20 milli amperes is passed for 2-3 hours and the stop cocks S_1 and S_2 are closed.

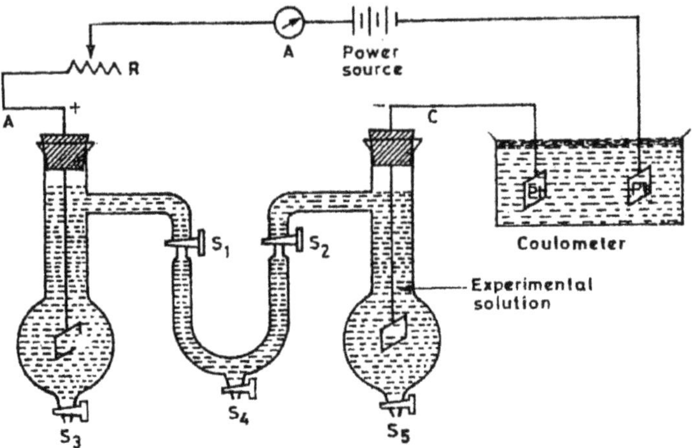

Fig. 10.11 Hittorf's apparatus.

(iv) Samples of Experimental liquids from all the three compartments are collected by opening the stop cocks S_3, S_4 and S_5 and analyzed.

(v) The metal deposited in the coulometer, e.g., silver in the silver coulometer, is estimated and total current passed during electrolysis is established.

(vi) Loss of material in a given volume of liquid in the cathode and anode compartments is determined.

Calculations : For a clear understanding and appreciation we take a specific example of silver nitrate as electrolyte and silver coulometer in series. It is possible to carry out the electrolysis using either unattackable electrodes like platinum electrodes or attackable silver electrodes.

A. Electrolysis using Unattackable Electrodes

In this electrolysis Ag^+ migrate from anode and get deposited in the cathodic compartment, while NO_3^- that migrates into anodic compartment do not contribute for any change of concentration of electrolyte. As discussed already the net effect will be a fall in the concentration of $AgNO_3$ in both cathodic and anodic compartments. The concentration of central compartment is unchanged. The amount of Ag deposited in the silver coulometer gives the amount of charge passed through the circuit.

Suppose *a* grams of anodic solution collected after electrolysis contains *b* grams of $AgNO_3$.

Weight of water in the collected solution = $(a - b)$ grams

So $(a - b)$g of water **finally** contains b g of $AgNO_3$

or $b/170.8$ g eq of $AgNO_3$

or $b/107.8$ g.eq of Ag.

Let this quantity be represented by N_f g eq of Ag

Similarly **before electrolysis** let c g of solution contain d g of $AgNO_3$.

Therefore, weight of water = $(c - d)$ g

So $(c - d)$ g of water contains d g of $AgNO_3$

or $d/170.8$ g eq of $AgNO_3$

or $d/107.8$ g eq of Ag

Therefore;

$$(a - b) \text{ g of water contain } \frac{d}{107.8} \times \frac{(a-b)}{(c-d)} \text{ g eq of Ag}$$

Let this quantity be designated by N_i g eq of Ag.

Fall in the concentration of Ag as a result of electrolysis

$$= (N_i - N_f) \text{ g eq of Ag}$$

Further let us also suppose that W g of Ag is deposited in the silver **coulometer**

$$= \frac{W}{107.87} \text{ g eq of Ag}$$

Let this value be designated by N_c g eq of Ag

From eqn. 10.24, we know that;

$$t_+ = \frac{\text{Fall in conc. of Ag in anodic compartment}}{\text{Amount of Ag deposited in coulometer}}$$

$$\boxed{t_+ = \frac{(N_i - N_f)}{N_c}} \quad .....(10.27)$$

Therefore,

$$\boxed{t_- = 1 - \frac{N_i - N_f}{N_c} = \frac{N_c - N_i + N_f}{N_c}} \quad .....(10.28)$$

(ii) Electrolysis using Attackable Electrodes

When a silver anode is used, the excess NO_3^- that entered the anodic compartment during electrolysis attack the anode. Therefore, instead of a fall in $AgNO_3$ concentration the anodic compartment registers a raise in its concentration. Carrying out similar calculations, as in the case of platinum electrodes, we can find N_i and N_f values, where $N_f > N_i$. Increase in the concentration around anode = $(N_f - N_i)$ g eq of Ag.

Let N_c be the amount of Ag deposited in the silver coulometer.

If there was no migration of Ag^+ ions from anodic to cathodic compartment the actual raise in $AgNO_3$ concentration would have been equal to the NO_3^- entered the anodic compartment or current passed (N_c equivalents of Ag).

Since $N_c > (N_f - N_i)$

Fall in concentration of anodic compartment = $N_c - (N_f - N_i)$

Therefore;

$$t_{Ag+} = \frac{N_c - (N_f - N_i)}{N_c} = \frac{N_c - N_f + N_i}{N_c} \quad .....(10.29)$$

$$1 - t_{ag+} = 1 - (N_C - N_F + N_i) = \frac{N_f + N_i}{N_c} \quad .....(10.30)$$

Precautions :

(i) Low currents and low conversions are recommended. Large currents may heat the electrolytic solution and may subsequently set-up convection currents. One of the prime conditions of this experiment is the 100% utilization of charge passed through the solution.

(ii) Concentration of central compartment should be unchanged after the electrolysis. Any change in concentration indicate a mixing of electrolytic solutions.

(iii) When copper coulometer is used in place of silver coulometer, the silver equivalent of charge is obtained by 107.87 multiplying the copper deposited with $\frac{107.87}{31.77}$.

10.5.3 Factors influencing Transport Number

(i) **Presence of other electrolytic impurities :** All the cations and anions in the solution influence the transport number.

(ii) **Concentration of electrolytes :** Due to changes in intermolecular forces, transport numbers of cations and anions are affected by changes in concentration, Table 10.4. An increase in transport number of one ion is accompanied by decrease in the transport number of counter ion. *In an infinitely dilute solution transport numbers reach a limiting value.*

Table 10.4 Variation of Transport number K^+ ions of KCl with concentration and temperature

Temperature ^oC	Concentration (M)				
	0.01	**0.02**	**0.05**	**0.100**	**0.200**
0	0.493	0.493	0.493	0.492	0.491
18	0.496	0.496	0.496	0.495	0.495
30	0.498	0.498	0.498	0.497	0.496

(iii) **Temperature :** The influence of temperature, though small, is more pronounced than the influences of concentration on transport numbers, Table 10.4.

(iv) **Complexation : Abnormal transport numbers** are found in electrolytes where the ionic state differs with the concentration. For example electrolytes like ZnI_2, CdI_2, etc., dissociate in water giving various ions depending on the concentration. Taking CdI_2 as a specific case we have;

(a) *In very dilute solutions :* In very dilute solutions CdI_2 dissociates completely giving Cd^{2+} and $2I^-$.

$$CdI_2 \rightleftharpoons Cd^{2+} + 2I^-$$

(b) *In moderately dilute solutions :* In moderately dilute solutions we find the following equilibria.

$$CdI_2 \rightleftharpoons Cd^{2+} + 2I^-$$

$$CdI_2 \rightleftharpoons CdI^+ + I^-$$

Thus part of the CdI_2 instead of dissociating into Cd^{2+}, leaves twice as heavy and uni-positive CdI^+ ions in the solution. Under these conditions the transport number of cation falls but still will be on the positive side.

(c) *In concentrated solutions :* With the increase in concentration of CdI_2 in the solution, the simple ions Cd^{2+} and I^- combine to give CdI_3^- and CdI_4^{2-} complex ions as indicated by the following equilibria.

$$CdI_2 \rightleftharpoons Cd^{2+} + 2I^-$$

$$CdI_2 \rightleftharpoons CdI^+ + I^-$$

$$CdI_2 + I^- \rightleftharpoons CdI_3^-$$

$$CdI_2 + 2I^- \rightleftharpoons CdI_4^{2-}$$

Under these conditions positively charged cadmium constituents, viz., Cd^{2+} and CdI^+, move only towards cathode and get deposited in the cathodic compartment, while the negatively charged cadmium constituents, viz., CdI_3^- and CdI_4^{2-} migrate to anode and discharge there. Due to migration of more and more cadmium ions towards anode as a result of increase in CdI_2 concentration, the transport numbers of cadmium ion constituents gradually fall to zero (Indicating migration of equal number of cadmium constituents in both the direction) and then assume a negative transport number value, Table 10.5.

Table 10.5 Transport numbers of Cadmium ion constituent in aqueous solution of cadmium iodide at 18°C.

[CdI_2] g. eq.L^{-1}	0.01	0.02	0.05	0.1	0.2	0.5	1.0
$t_{Cd^{2+}}$	0.444	0.443	0.396	0.296	0.127	–0.003	–0.42

Problem 12.

0.01 N HCl is taken in a Hittorf's apparatus and a current of fixed rating is passed. At the end of electrolysis concentration of acid in the anodic compartment was found to be 0.0083 N while the Ag coulometer connected in series deposited 0.2225 g of Ag. Calculate the transport number of H^+ and Cl^- ions.

Initial concentration of H^+ = 0.01 equivalents per liter

Final concentration of anodic compartment = 0.0083 equivalents/L

Fall in the concentration of the anodic solution = 0.0017 eq L^{-1}

Amount of Ag deposited in the Ag coulometer = 0.2225 g

$$\text{Equivalents of Ag deposited} = \frac{0.2225}{107.87} = 0.0021 \text{ equivalents}$$

$$t_+ = \frac{\text{Fall in conc. of H}^+ \text{ in anodic solution}}{\text{Amount of Ag deposited in coulometer}} = \frac{0.0017}{0.0021} = 0.81$$

Transport number of $H^+ = 0.81$

Transport number of $Cl^- = 1 - 0.81 = 0.19$

Problem 13.

The speed ratio of La^{3+} to Cl^- in $LaCl_3$ at infinite dilution is 0.9115. Calculate the transference number of La^{3+} and Cl^- at infinite dilution.

$$\text{Speed ratio } \gamma = \frac{\text{Speed of La}^{3+}}{\text{Speed of Cl}^-} = 0.9115$$

$$\text{Transport number of cation } t_+ = \frac{\gamma}{1+\gamma}$$

Substituting the value of γ in the above formula

$$t_{La^{3+}} = \frac{0.9115}{1+0.9115} = 0.4768$$

But $\quad t_{La^{3+}} + t_{Cl^-} = 1$

$$t_{Cl^-} = 1 - 0.4768 = 0.5232$$

Transport number of La^{3+} and Cl^- at infinite dilution are 0.4768 and 0.5232 respectively.

Problem 14.

During electrolysis of 0.00186 molal aqueous NaOH between platinum electrodes, 215 coulombs charge was carried by solute particles. After electrolysis, if 100 mL alkali in the cathodic compartment contains 0.00935 g per 100 mL of the solution, determine the transport number of Na^+ and OH^-.

Initial concentration of NaOH

$$= 0.00186 \quad = 0.00186 \text{ eq per 1000g of solvent.}$$

No. of equivalent of current passed

$$= \frac{215}{96500} \quad = 2.228 \times 10^{-3} \text{ equiv.}$$

If there is no migration of OH^- from cathode compartment, the incrase in weight would have been equal to 2.228×10^{-3} equiv.

Concentration of NaOH after electrolysis $= 0.00935$ g per 100 mL

or $= 0.0935$ g per 1000 mL

or $= \frac{0.0935}{40}$ g eq per 1000g

$= 0.00234$g eq per 1000g

Increase in conc. of NaOH in cathodic compart

$= 0.00234 - 0.00186 = 4.8 \times 10^{-4}$

No. of g eq lost from cathodic compartment $= 2.238 \times 10^{-3} - 4.8 \times 10^{-4}$

$= 1.748 \times 10^{-3}$

No. of g eq lost from cathodic compartment

$$t_- = t_{OH^-} = \frac{\text{No. of g eq lost from cathodic compartment}}{\text{No. of equivalents deposited in the coulometer}}$$

$$= \frac{1.748 \times 10^{-3}}{2.228 \times 10^{-3}} = 0.785$$

Therefore,

$$t_{Na^+} = 1 - 0.785 \quad = 0.215$$

Problem 15.

0.0536 molal aqueous $AgNO_3$ was taken in a Hittorf cell and electrolysed using silver electrodes by passing 150 mA for 40 min. If 50 g of anodic solution contained 300.2 mg of Ag, what is the transference number of Ag^+.

Before electrolysis concentration of $AgNO_3 = 0.0536$ M

$= 0.0536$ eq in 1000 g

No. of equivalents of current passed

$$= \frac{(150 \times 10^{-3}\,\text{A})(40 \times 60\,\text{s})}{96500\,\text{F}}$$

$= 3.731 \times 10^{-3}$ equiv.

If there is no migration of Ag^+, the rise in concentration of $AgNO_3$ in anode compartment would have been = 3.73×10^{-3} equiv.

After electrolysis 50 g of anodic solution contains

$$= 300.2 \text{ mg of Ag}$$

$$= \frac{300.2 \times 10^{-3} \times 1000}{50}$$

So 1000 g of anodic solutions contains $\frac{300.2}{50} \times \frac{1}{108}$ equiv.

$$= 5.560 \times 10^{-2} \text{ equiv.}$$

Increase in the concentration of Ag

$$= 5.56 \times 10^{-2} - 0.0536$$

$$= 1.992 \times 10^{-3} \text{ equiv.}$$

Loss in the weight of anodic solution

$$= 3.731 \times 10^{-3} - 1.992 \times 10^{-3}$$

$$= 1.7384 \times 10^{-3} \text{ equiv.}$$

$$t_+ = \frac{\text{Loss in anodic compartment}}{\text{Equivalents of current passed}} = \frac{1.7384 \times 10^{-3}}{3.731 \times 10^{-3}} = 0.466$$

Transport number of Ag^+ = 0.466

Problem 16.

A 1.25 % solution of $CuSO_4$ was electrolyzed between copper electrodes. At the end of electrolysis 50 g of the anodic solution was found to contain 0.843 g of $CuSO_4$ and silver coulometer placed in series was found to have 0.405 g of Ag. Calculate transference number of participating ions. Atomic weights of Cu, S,O and Ag are 63.5, 32, 16 and 108 respectively.

After electrolysis :

Weight of anodic solution = 50 g

Weight of $CuSO_4$ present = 0.843 g

Weight of water present N_f = 50 – 0.843 = 49.157 g

Before electrolysis :

Weight of anodic solution = 100 g

Weight of $CuSO_4$ present = 1.25 g

Weight of water present = 100 – 1.25 = 98.75 g

98.75 g of water contains 1.25 g of $CuSO_4$

49.157g of water contains ?

$$N_i = \frac{49.157}{98.75} \times 1.25 = 0.622 \text{ g}$$

Equivalent weights of $CuSO_4$

$$= \frac{63.5+32+64}{2} = 79.5$$

Increase in concentration of $CuSO_4$ around anode ($N_f - N_i$)

$$= 0.843 - 0.622 = 0.221 \text{ g}$$

$$= \frac{0.221}{79.5} = 2.78 \times 10^{-3} \text{ equiv.}$$

Equivalents of Ag deposited (N_c)

$$= \frac{0.405}{108} = 3.75 \times 10^{-3} \text{ equiv.}$$

Fall in concentration of anodic compartment due to migration of Cu^{2+}:

Increase in weight of Cu^{2+} in the absence of Cu^{2+} migration (N_c)

$$= 3.75 \times 10^{-3} \text{ equiv.}$$

Increase actually observed ($N_f - N_i$) = 2.78×10^{-3} equiv.

Fall in concentration of anodic compartment = $N_c - (N_f - N_i)$

$$= 3.75 \times 10^{-3} - 2.78 \times 10^{-3}$$

$$= 0.97 \times 10^{-3}$$

$$t_+ = \frac{\text{Fall in conc. of } Cu^{2+} \text{ in anodic compartment}}{\text{Equivalents of current passed}} = \frac{N_c\left(N_f - N_i\right)}{N_c}$$

$$= \frac{0.97 \times 10^{-3}}{3.75 \times 10^{-3}} = 0.259$$

Transference number of Cu^{2+} $= 0.259$

Transference number of SO_4^{2-} $= 1 - 0.259 = 0.741$

Problem 17.

A solution of $CuSO_4$ of strength 0.25 g of copper in 46 g of solution is electrolyzed by passing 200 mA for 30 min. If the transport number of Cu^{2+} is 0.275, calculate the final concentration of Cu^{2+} present in the anode compartnent, (i) When electrolyzed between Pt electrodes; (ii) When electrolyzed between Cu electrodess.

(i) When electrolyzed between Pt electrodes (Unattackable)

46 g of $CuSO_4$ solution initially contains 0.25 g Cu

or 45.75 g of water initially contains 0.25 g of Cu

N_i = Equivalents of Cu^{2+} present in 45.75 g of

$$N_i = \frac{W}{\text{eq. wt of Cu}} = 0.25/31.75 = 0.0079 \text{ eq. of Cu}$$

N_c = Equivalents of current passed

$$= \frac{(200\times10^{-3}\text{ A})\times(30\times60\text{ s})}{96500}$$

$$= 3.73 \times 10^{-3} \text{ equiv.}$$

But $$t_+ = \frac{\text{Loss in the weight of anode solution}}{\text{Equivalents of current passed}} = \frac{N_c - (N_i - N_f)}{N_c}$$

$$0.275 = \frac{\text{Loss in the weight of anode solution}}{3.73\times10^{-3}}$$

Loss weight of anode solution $= 0.275 \times 3.73 \times 10^{-3}$

$= 1.026 \times 10^{-3}$ equiv

$$N_c - (N_i - N_f) = 1.026 \times 10^{-3}$$

So,

Expt. increase in equiv of Cu $= (N_i - N_f) = 3.73 \times 10^{-3} - 1.026 \times 10^{-3}$

$$N_i - N_f = 2.704 \times 10^{-3} \text{ equiv}$$

$$\begin{aligned} N_f &= N_i - 2.704 \times 10^{-3} \\ &= 7.9 \times 10^{-3} + 2.704 \times 10^{-3} \text{ equivalents} \\ &= 5.196 \times 10^{-3} \text{ equivalents} \end{aligned}$$

Concentration of Cu^{2+} left in the anode compartment after electrolysis

$$= 5.196 \times 10^{-3} \text{ equiv.}$$

(ii) When electrolyzed between Cu electrodes (Attackable)

Fall in concentration of Cu^{2+} in anode solution $N_c - (N_f - N_i)$

$$t_+ = \frac{\text{Fall in conc. of } Cu^{2+} \text{ in anode solution}}{\text{Equivalents of current passed}} = \frac{N_c - (N_f - N_i)}{N_c}$$

$$0.275 = \frac{3.73 \times 10^{-3} - (N_f - 7.9 \times 10^{-3})}{3.73 \times 10^{-3}}$$

$$0.275 \times 3.73 \times 10^{-3} = 3.73 \times 10^{-3} - (N_f - 7.9 \times 10^{-3})$$

$$N_f - 7.9 \times 10^{-3} = 3.73 \times 10^{-3} - 1.026 \times 10^{-3}$$

$$N_f = 7.9 \times 10^{-3} + 3.73 \times 10^{-3} - 1.026 \times 10^{-3}$$

$$N_f = 10.604 \times 10^{-3}$$

Concentration of Cu^{2+} left in the anode compartment after electrolysis

$$= 10.604 \times 10^{-3}$$

10.6 Kohlrausch Law of Independent Migration of Ions

In a state of infinite dilution all electrolytes ionize or dissociate cent percent, and the ions can be expected to be wide apart not to have any electrostatic or any other type of attraction or repulsion.

Under these conditions, we achieve an ideal situation where all the charge is expected to be carried by the ions independent of each other. During our discussion on the variation of equivalent conductivity with square root of concentration we noticed that equivalent conductivity increases with dilution to reach a limiting value characteristic of the electrolyte and named it as equivalent conductivity at infinite dilution, Λ_4. In 1875 Kohlrausch made a series of measurements, involving electrolytes with common cations and common anions and calculated their differences. He obtained a constant difference in Λ_4 values of an ion pair, irrespective of the nature of the common counter ion employed, as exemplified by following results obtained at 25 ^{O}C.

Difference between Λ_4 values of some pairs of electrolytes having K^+ and Na^+ cations and common anions :

KCl	149.85	KBr	151.64	KOH	272.10	KNO_3	144.96
NaCl	126.45	NaBr	128.24	NaOH	248.70	$NaNO_3$	121.56
Diff	23.40		23.40		23.40		23.40

Difference between Λ_4 values of some pairs of electrolytes having OH^- and Cl^- ions and common cations :

NaOH	248.70	NH_4OH	272.15	LiOH	237.28	½$Ba(OH)_2$	262.23
NaCl	126.45	NH_4Cl	149.9	LiCl	115.03	½$BaCl_2$	139.98
Diff	122.25		122.25		122.25		122.25

Basing on similar observations, Kohlrausch advanced a hypothesis called **Law of independent migration of ions**. This law states that *at infinite dilution each ion makes a definite contribution towards the equivalent conductivity of an electrolyte, irrespective of nature of the co-ion with which it is associated in the solution.* Therefore equivalent conductivity at infinite dilution is made up of two independent terms, called **ion conductivities**, characteristic of each ionic constituent of an electrolyte in solution. We can express the same conclusion mathematically as

$$\boxed{\Lambda_4 = \lambda_+ + \lambda_-} \qquad .....(10.31)$$

where λ_+ and λ_- are the ion conductances of cation and anion respectively. The ion conductance of any ion is a fixed characteristic value in a given solvent. (In the absence of any specific mention, these values are always for aqueous solutions), and are dependent on temperature. The ion conductivity values of some common ions at 25 ^{O}C is presented in Table 10.6.

Table 10.6 Ion conductivities at infinite dilution at 25°C in S cm^2 equiv.

Cation	λ_+	Anion	λ_-
H^+	349.81	OH^-	197.6
Li^+	38.68	Cl^-	76.35
Na^+	50.10	Br^-	78.14
K^+	73.50	I^-	76.84
Ag^+	61.92	NO_3^-	71.46
NH_4^+	73.55	ClO_4^-	67.36
½Mg^{2+}	53.05	IO_4^-	54.38
½Ca^{2+}	59.50	HCO_3^-	44.50
½Sr^{2+}	59.46	½SO_4^{2-}	80.02
½B_a^{2+}	63.63	Acetate$^-$	40.90
½Cu^{2+}	54.00	Benzoate$^-$	32.30
½Zn^{2+}	53.00	½Oxalate^{2-}	24.00

10.6.1 Ionic Conductivity and Ionic Mobilities

Equivalent conductivity of any electrolyte at any dilution is directly proportional to the charge carried by the ions and their velocities. The conductivity is thus given by the products of charge and velocity of individual ions. At infinite dilution the ionization is complete and solution containing one equivalent of various electrolytes contain equivalent number of ions. Hence at infinite dilution total charge carried by all ions is same in every case. Because the total charge is constant at infinite dilution, the Λ_4 must depend exclusively on ionic velocities.

Defining the **ionic velocity or mobility** as the speed with which a charged particle at infinite dilution moves under a potential gradient of one volt per cm, we have,

$$\lambda_+ \propto u_+ \qquad \lambda_- \propto u_-$$

or $$\lambda_+ = ku_+ \qquad \lambda = ku_-$$

where k is a proportionality constant and u_+ and u_- represent the ionic velocities at infinite dilution. Since 1 equivalent of an ion under unit potential gradient carries a charge of 1 Faraday per sec., the proportionality constant $k = 96500$ coulombs.

Therefore,

$$u_+ = \frac{\lambda_+}{k} \quad \& \quad u_- = \frac{\lambda_-}{k} \qquad \text{.....(10.32)}$$

$$\boxed{\text{Ionic mobility} = \frac{\text{Ionic conductivity}}{96500}} \qquad \text{.....(10.33)}$$

Ionic conductivity is expressed in S cm^2, while ionic mobility is expressed in cm s^{-1}. Calculated ionic mobilities of few common ions at 25 ^oC are given in Table 10.7.

Table 10.7 Calculated ionic mobilities of few common ions at 25 ^oC

Cations	**u_+ (cm s^{-1})**	**Anions**	**u_- (cm s^{-1})**
H^+	36.2×10^{-4}	OH^-	20.5×10^{-4}
K^+	7.61×10^{-4}	SO_4^{2-}	8.27×10^{-4}
Ba^{++}	6.60×10^{-4}	Cl^-	7.91×10^{-4}
Na^+	5.19×10^{-4}	NO_3^-	7.40×10^{-4}

Exceptionally high mobilities of H^+ and OH^- ions is attributed great affinity of these ions for water molecules, which can initiate a chain of transfers from one to other as shown below;

Mechanism of H^+ ion transfer :

$$H-\overset{\overset{H}{|}}{\underset{+}{O}}-H \; + \; \overset{\overset{H}{|}}{O}-H \longrightarrow H-\overset{\overset{H}{|}}{O} \; + \; H-\overset{\overset{H}{|}}{\underset{+}{O}}-H$$

Mechanism of OH^- ion transfer :

$$\overset{\overset{H}{|}}{O}-H \; + \; \overset{\overset{H}{|}}{O^-} \longrightarrow \overset{\overset{H}{|}}{O^-} \; + \; H-\overset{\overset{H}{|}}{\underset{+}{O}}-H$$

This additional mobility/conductivity is referred to as **Grotthus conductivity**.

10.6.2 Ionic Conductivity and Transport Numbers

Eq. (10.21) gives the relation between transport number and ionic mobilities

$$t_+ = \frac{u_+}{u_+ + u_-} \quad \text{and} \quad t_- = \frac{u_-}{u_+ + u_-}$$

Substitute the values of u_+ and u_- given in eq. (10.32), we get;

$$t_+ = \frac{\frac{\lambda_+}{\kappa}}{\frac{\lambda_+}{\kappa} + \frac{\lambda_-}{\kappa}} = \frac{\lambda_+}{\lambda_+ + \lambda_-} \qquad .....(10.34)$$

Similarly,

$$t_- = \frac{\lambda_-}{\lambda_+ + \lambda_-} \qquad .....(10.35)$$

We also know $\Lambda_4 = \lambda_+ + \lambda_-$

Therefore,

$$t_+ = \frac{\lambda_+}{\Lambda_\infty} \quad \text{or} \quad \lambda_+ = t_+ \Lambda_4 \qquad .....(10.36)$$

and $$t_- = \frac{\lambda_-}{\Lambda_\infty} \quad \text{or} \quad \lambda_- = t_- \lambda_4 \qquad .....(10.37)$$

In general

Ionic conductivity of anion = Equiv. conductivity of electrolyte at infinite dilution × transport no .of the ion.

Thus a knowledge of transport number and equivalent conductance at infinite dilution allows one to calculate its ionic conductivity and vice versa.

10.6.3 Applications of Kohlrausch's law

The introduction of concept of ionic conductivity opened the vistas towards numerous applications. A few of them are discussed below.

I. Estimation of equivalent conductivity at infinite dilution of weak electrolytes

Method 1 : The extrapolation technique used for the determination of λ_4 of strong electrolytes is not useful for weak electrolytes. This is due to the steepness of the curve (Fig. 10.7) at the approach of zero concentration or infinite dilution. Λ_4 of weak electrolytes is calculated using Kohlrausch's law. Using acetic acid as an example, we have at infinite dilution

$$\Lambda_{\infty(CH_3COOH)} = \lambda_{(CH_3COO-)} + \lambda_{(H^+)}$$

Adding Λ_4 of the salt of a strong acid and strong base, eg., NaCl

$$\Lambda_{\infty(CH_3COOH)} = \Lambda_{\infty(CH_3COO^-)} + \lambda_{H^+} + \lambda_{Na^+} \; \lambda_{Cl^-}$$

on rearranging the ions on R.H.S, we may write

$$\Lambda_{\infty(CH_3COOH)} = \Lambda_{\infty(NaCl)} + \lambda_{(CH_3COONa)} + \lambda_{\infty(HCl)}$$

or $$\Lambda_{\infty(CH_3COOH)} = \Lambda_{\infty(CH_3COONa)} + \Lambda_{\infty(HCl)} - \Lambda_{\infty(NaCl)}$$

Because all the electrolytes on the right hand side are strong electrolytes, Λ_∞ values of these electrolytes can be obtained by the extrapolation of a Λ_v versus $\sqrt{C}$ plots and there by Λ_4 value of acetic acid.

Method 2 : Ionic conductance of H^+ and CH_3COO^- ions may be calculated by using transport numbers of these ions and Λ_4 of any strong electrolyte that contains these ions.

Using HCl we determine

$$\lambda_{H+} = t_+ \times \Lambda_{4(HCl)}$$

using CH_3COONa we determine

$$\lambda_{(CH_3COO^-)} = t_+ \times \Lambda_{\infty(CH_3COONa)}$$

Adding ionic conductances of H^+ and CH_3COO^- ions we get $\Lambda 4$ of CH_3COOH.

II. Calculation of degree of dissociation and dissociation constant of a weak electrolyte

The ratio of equivalent conductivity at any dilution (Λ_v) to that at infinite dilution ($\Lambda 4$) is called **conductance ratio.**

$$\boxed{\alpha = \frac{\Lambda_v}{\Lambda_\infty}} \qquad \text{.....(10.38)}$$

In dilute solutions of strong electrolyte, the conductance ratio is almost independent of the nature of the salt and is a constant value close to unity at any given concentration. For weak electrolytes this ratio is quite small and is a characteristic quantity called **degree of dissociation**. This in fact represents the fraction of molecules dissociated out of unit concentration of the electrolyte.

	CH_3COOH	$\rightleftharpoons$ $CH_3COO^- +$	H^+
Initial conc.	1	0	0
Equilibrium conc.	$1-\alpha$	α	α
When C moles are taken initially	$(1-\alpha)C$	αC	αC

Dissociation constant

$$K_D = \frac{[CH_3COO^-][H^+]}{[CH_3COOH]} = \frac{(\alpha C)(\alpha C)}{(1-\alpha)C}$$

K_D of uni-uni valent electrolyte

$$= \frac{\alpha^2 C}{(1-\alpha)} \quad \text{where} \quad \alpha = \frac{\Lambda_v}{\Lambda\infty}$$

where C is the concentration of the weak electrolyte and Λ_v is the corresponding equivalent conductivity.

III. Calculation of Ionic Product of Water

Water is a weak electrolyte, which dissociates partially to give H^+ and OH^- ions.

	H_2O	$\rightleftharpoons$ H^+ +	OH^-
For 1 mole at equilibrium	$1-\alpha$	α	α
For C moles	$C(1-\forall)$	$C\alpha$	$C\alpha$

Ionic product of water $K_w = [H^+]\,[OH]^- = (C\forall)\,.\,(C\forall) = C^2\,\forall^2$

Weight of 1000 mL of water = 1000 × 0.997 = 997 g

18 g (1 mole) of water dissolved in 1000 mL of water gives 1 M solution.

So 997 g of water present in 1000 mL of water has a concentration

qual to $= \frac{997}{18} = 55.4\text{ M} = 55.4\text{ N}$

Equivalent conductivity of water

$$\Lambda_v = \frac{1000\kappa}{C} = \frac{1000 \times 5.54 \times 10^{-8}}{55.4}$$

$$= 1 \times 10^{-6}\ \text{S cm}^2\ \text{equiv}^{-1}$$

Limiting equivalent conductivity $\Lambda_4 = 349.8 + 197.6 = 547.4$

Degree of dissociation

$$\alpha = \frac{\Lambda_v}{\Lambda_\infty} = \frac{1 \times 10^{-6}}{547.4} = 1.827 \times 10^{-9}$$

$$[OH^-] = [H^+] = \alpha C$$

$$= 1.827 \times 10^{-9} \times 55.4 = 1.01 \times 10^{-7}$$

Therefore, Ionic product of water

$$= [H^+]\,[OH^-] = (1.01 \times 10^{-7})^2 = 1.02 \times 10^{-14}$$

IV. Determination of Solubility Products

Another application of conductance measurements is the determination of solubility products of sparingly soluble salts. Salts like AgCl, $PbSO_4$, AgI, $BaSO_4$, PbI_2, etc., are *soluble to a negligible extent* and whatever that gets into solution dissociate completely. If ***S*** is the solubility, in gram equiv. per litre, of a given salt and **κ** is the specific conductance of the saturated solution, then Λ_v of thc solution is given by;

$$\Lambda_v = \frac{1000\kappa}{S}$$

These solutions are usually so dilute that their equivalent conductivity, Λ_v, is almost equal to Λ_4. Therefore,

$$\Lambda_4 = \frac{1000\kappa}{S}$$

$$\boxed{S = \frac{1000\kappa}{\Lambda_\infty}} \qquad \text{.....(10.39)}$$

Since the solutions are extremely dilute, the conductance contribution due to pure water is appreciable part of the total conductance and needs to be subtracted from the total value to get the contribution of dissolved salt.

Therefore,

$$\boxed{S = \frac{1000(\kappa - \kappa_w)}{\Lambda_\infty}} \qquad \text{.....(10.40)}$$

where κ_w is the specific conductance of pure water. In case of uni-univalent electrolyte like AgCl.

The solubility product $K_{sp} = (a_{Ag^+})\ (a_{Cl^-})$

The solution is dilute enough to regard the activity coefficient are unity.

Therefore, $a_{Ag+} = a_{Cl^-} = S$

Therefore $K_{sp} = S^2$

In case of AgCl

$$\Lambda_{4(AgCl)} = \lambda_{Ag+} + \lambda_{Cl^-}$$

$$= 61.90 + 76.35$$

$$= 138.25 \text{ S cm}^2$$

k of AgCl solution $= 3.41 \times 10^{-6}$ S cm^{-1}

k of pure water $= 1.60 \times 10^{-6}$ S Cm^{-1}

k of dissolved AgCl $= (3.41 - 1.60) \times 10^{-6}$ S cm^{-1}

$$= 1.81 \times 10^{-6} \text{ S cm}^{-1}$$

Substituting the above results in equation (10.39)

$$\text{Solubility of AgCl} = \frac{1000 \times 1.81 \times 10^{-6}}{138.25}$$

$$S = 1.31 \times 10^{-5} \text{ eq L}^{-1}$$

$$\text{Solubility product} = K_{sp} = S^2 = (1.31 \times 10^{-5})^2$$

$$= 1.72 \times 10^{-10} \text{ eq}^2 \text{ L}^{-2}$$

Problem 18.

The equivalent conductivities of sodium benzoate and HCl at 25°C are 82.4 and 426.16 S cm^2 equiv^{-1} respectively. If the transport numbers of benzoate and H^+ ions are 0.392 and 0.821 in the above mentioned compounds, find the equivalent conductivity of benzoic acid at 25°C.

Given : Λ_4 (Sodium benzoate) $= 82.4$ S cm^2 equiv^{-1}

Λ_4 (HCl) $= 426.16$ S cm^2 equiv^{-1}

t_+ (in HCl) $= 0.821$ & t_-(in Sodium benzoate) $= 0.392$

$$\lambda_+ = t_+ \Lambda 4 \quad \& \quad \lambda_- = t_- \Lambda_4$$

$$\lambda_{H+} = 0.821 \times 426.16 \qquad = 349.9 \text{ S cm}^2 \text{ equiv}^{-1}$$

$$\lambda_{benzoate} = 0.392 \times 82.4 = 32.3 \text{ S cm}^2 \text{ equiv}^{-1}$$

$$\lambda_{4\,(Benzoic\ acid)} = \lambda_{benzoate} + \lambda_{H+}$$

$$\lambda_{4\ Benzoic\ acid} = 32.3 + 349.9 \qquad = 382.2 \text{ S cm}^2 \text{ equiv}^{-1}$$

Problem 19.

The equivalent conductivity of NaCl at infinite dilution is 126.46 S cm^2 $equiv^{-1}$ and the ionic conductivity of chloride ion is 76.35 S cm^2 $equiv^{-1}$. Calculate transport numbers and ionic mobilities of Na^+ and Cl^- ions.

Given : $\lambda_4\text{(NaCl)} = 126.46 \text{ S cm}^2 \text{ equiv}^{-1}$

$\lambda_-\text{(NaCl)} = 76.35 \text{ S cm}^2 \text{ equiv}^{-1}$

$$t_- = \frac{\Lambda_v}{\Lambda\infty} = \frac{76.35}{126.46} \qquad = 0.604$$

$$t_+ = 1 - t_- = 1 - 0.604 \qquad = 0.396$$

So transport numbers of Na^+ and Cl^- are 0.396 & 0.604

$$\lambda_{Na+} = \Lambda_{4\,(NaCl)} - \lambda_{Cl-}$$

$$= 126.46 - 76.35 \qquad = 50.11$$

$$\text{Ionic mobility of Na}^+ (u_+) = \frac{\lambda_+}{F} = \frac{50.1}{96500} = 5.2 \times 10^{-4} \text{ cm S}^{-1}$$

$$\text{Ionic mobility of Cl}^- (u_-) = \frac{\lambda_-}{F} = \frac{76.35}{96500} = 7.9 \times 10^{-4} \text{ cm S}^{-1}$$

Problem 20.

The equivalent conductance values of 0.0005M (at 25^oC) NaCl, NH_4Cl, NaOH and NH_4OH are 124.50, 149.1, 245.6 and 9.38 mho cm^2 $equiv^{-1}$ respectively. Calculate the dissociation constant of NH_4OH.

$$\Lambda_{\infty(NH_4OH)} = \Lambda_{\infty(NH_4Cl)} + \Lambda_{\infty(NaOH)} - \Lambda_{\infty(NaCl)}$$

$$= 149.1 + 245.6 - 124.50$$

$$= 270.2 \text{ S cm}^2 \text{ equiv}^{-1}$$

Degree of dissociation

$$\alpha = \frac{\Lambda_v}{\Lambda_\infty} = \frac{9.38}{270.2} = 0.0347$$

Dissociation constant $K = \alpha^2 C$

$$= (0.0347)^2 \times (1/100)$$

$$= 1.2 \times 10^{-5}$$

Problem 21.

The ion conductances of H^+ and butyrate ions are 349.8 and 32.6 S cm^2 equiv^{-1} respectively. When same conductance cell is employed 0.1 m KCl and 0.1M butyric acid recorded a conductance of 1.4×10^{-2} and 5.85×10^{-4} S cm^2 respectively. If the conductivity of KCl soluiton is 0.01121 S cm^{-1}, calculate the dissociation constant of butyric acid.

A. *Calculation of cell constant.*

$$\text{Cell constant } X = \frac{\text{Conductivity}(\kappa)}{\text{Conductance}}$$

$$X = \frac{0.01121}{1.4\times10^{-2}} = 0.8$$

B. *Calculation of conductivity of butyic acid soluiton :*

Conductivity of butyric acid solution

$$\kappa = 5.85 \times 10^{-4} \times 0.8$$

$$= 4.68 \times 10^{-4} \text{ S cm}^{-1}$$

C. *Calculation of equivalent conductivity of butyric acid solution :*

$$\Lambda_V = \frac{1000\kappa}{C} = \frac{1000\times4.68\times10^{-4}}{0.1} = 4.685 \text{ cm}^2 \text{ equiv}^{-1}$$

D. *Calculation of limiting conductance of butyric acid solution :*

$$\Lambda_4 \text{ (Butyic acid)} = \lambda_+ + \lambda_-$$

$$= 349.8 + 32.6 = 382.4$$

E. *Calculation of degree of dissociaton of butyric acid solution :*

$$\alpha = \frac{\Lambda_v}{\Lambda\infty} = \frac{4.68}{382.4} = 0.012$$

F. *Calculation of dissociation constant :*

$$K_a = \alpha^2 C \quad = (0.012)^2 \times 0.1 \quad = 1.44 \times 10^{-5}$$

Problem 22.

Calculate the conductance of pure water at 25°C, when measured employing a cell with a cell constant 1.0. The ion conductivities of H^+ and OH^- are 349.8 and 197.6 S cm^2 equiv^{-1} respectively. Density of water at 25°C is 0.997 g cm^{-3}.

$$\Lambda_4 = \lambda_{H+} + \lambda_{OH^-} \quad = 349.8 + 197.6 = 547.4$$

Ionic product of water $K_w = 1 \times 10^{-14}$

$$H_2O \rightleftharpoons H^+ + OH^-$$

Initial	1	0	0
At equilibrium	$C(1-\alpha)$	$C\alpha$	$C\alpha$

$$[H^+] = [OH^-] \quad = C\alpha \quad = 1 \times 10^{-7}$$

Therefore,

$$\text{Degree of dissociation} = \alpha \quad = \frac{1\times10^{-7}}{C}$$

Concentration of water C

$$= \frac{997}{18} = 55.4 \text{ M}$$

Degree of dissociation

$$\alpha = \frac{1\times10^{-7}}{55.4} = 1.8 \times 10^{-9}$$

But $\qquad \alpha = \dfrac{\Lambda_v}{\Lambda_\infty}$

Therefore,

$$\Lambda_v = \alpha\, \Lambda_4 \quad = 547.4 \times 1.8 \times 10^{-9} \quad = 9.85 \times 10^{-7}$$

$$\Lambda_v = \frac{1000\kappa}{C}$$

$$\kappa = \frac{9.85 \times 10^{-7} \times 55.4}{1000} = 5.46 \times 10^{-8} \text{ S Cm}^{-1}$$

$$\text{Conductance} = \frac{\text{Conductivity}}{\text{Cell constant}}$$

$$\text{Conductance} = \frac{5.46 \times 10^{-8}}{1} = 5.46 \times 10^{-8} \text{ mho}$$

Problem 23.

The conductivity of a saturated solution at 298 K of a sparingly soluble salt MX is 2.5×10^{-6} S cm^{-1}, while that of pure water is 1.61×10^{-6} S cm^{-1}. If the equivalent conductivity of MX at infinite dilution is 156 S cm^2 equiv^{-1}, calculate the solubility product of the salt.

Conductivity of a saturated solution $= 2.5 \times 10^{-6}$ S cm^{-1}

Conductivity of pure water $= 1.61 \times 10^{-6}$ S cm^{-1}

Conductivity contribution of only salt (k) $= (2.5 - 1.61) \times 10^{-6}$

$= 0.89 \times 10^{-6}$ S cm^{-1}

Λ_4 of the salt = 156 S cm^2 equiv^{-1}

Solubility

$$S = \frac{1000\kappa}{\Lambda_\infty} = \frac{1000 \times 0.89 \times 10^{-6}}{156} = 5.7 \times 10^{-6} \text{ mol L}^{-1}$$

MX being a uni-uni valent salt

	MX	$\rightleftharpoons$	$M^+ + X^-$	
Initial	1 mole		0	0
at equilibrium	$1-x$		x	x

Solubility product $(K_{sp}) = [M^+][X^-] = x^2$

$= (5.7 \times 10^{-6})^2 = 32.5 \times 10^{-12}$ mol^2 L^{-2}

Solubility product of MX $= 32.5 \times 10^{-12}$ mol^{-2} L^{-2}.

Problem 24.

The conductivity of pure water and saturated solution of $Cu(OH)_2$ at 25°C are 1.622×10^{-6} and 1.705×10^{-6} mho cm^{-1} respectively. If the ion conductivities of Cu^{2+} and OH^- are 54 and 197.6 mho cm^2 equiv^{-1}, find the solubility product of Cu $(OH)_2$.

Given : $\Lambda_{\frac{1}{2}Cu^{2+}} = 54 \text{ mho cm}^2 \text{ equiv}^{-1}$

$\Lambda_{OH^-} = 197.6 \text{ mho cm}^2 \text{ equiv}^{-1}$

$\Lambda_4 \{\frac{1}{2} Cu(OH)_2\} = 54 + 197.6 = 251.6$

Conductivity due to ions dissolved in saturated solution (κ)

$$= 1.705 \times 10^{-6} - 1.622 \times 10^{-6}$$

$$= 0.083 \times 10^{-6} \text{ mho cm}^{-1}$$

$$\text{Solubility} = \frac{1000\,\kappa}{\Lambda_\infty} = \frac{1000 \times 0.083 \times 10^{-6}}{251.6} = 3.3 \times 10^{-7} \text{ mol L}^{-1}$$

	$Cu(OH)_2 \rightleftharpoons$	Cu^{2+} +	$2OH^-$
Initial	1 mole	0	0
At equilibrium	$1-x$	x	$2x$

where x is the solubility of the salt

$$\text{Solubility product } K_{sp} = [Cu^{2+}][OH^-]^2$$

$$= x(2x)^2 = 4x^3$$

$$= 4(3.3 \times 10^{-7})^3 = 1.44 \times 10^{-19} \text{ mol}^3 \text{ L}^{-1}$$

10.7 Conductometric Titrations

This is another interesting application which exploits the variation of conductance of a solution during a titration. This technique can be used advantageously with,

(a) colored solutions where selection of proper indicator is difficult;

(b) acids and bases and their mixtures;

(c) ions and their mixtures that are readily precipitated by the addition of titrant.

A grave disadvantage of conductivity based titrations is their non-applicability with solutions contaminated with high concentrations of electrolyte impurities. Presence of excess of electrolyte (ions) and hence high electrolytic conductivity masks the minor changes in conductivity, taking place as a result of addition of titration. Few examples of acid-base and precipitation titrations are discussed below.

10.7.1 Titration of a strong acid (HCl) with a strong base (NaOH)

Strong acids dissociate completely to give hydrogen ions. The ionic conductance data, Table 10.6 suggest that protons (H^+) are almost two times more conducting than hydroxy ions (OH^-), which themselves are more conducting than all the other ions. When NaOH is added to a solution of HCl we have the reaction.

$$(H^+ + Cl^-) + (Na^+ + OH^-) \longrightarrow (Na^+ + Cl^-) + H_2O$$

During the course of neutralization initially high conducting H^+ ions present in the acid are replaced by sodium ions having a much lower conductance. Therefore, addition of alkali is accompanied by a fall (curve AB) in the conductance till all the H^+ ions are totally neutralized. Further addition of alkali may be interpreted as only an addition of hydroxyl ions which are no longer used up in the acid-base reaction. Hence, after neutralization conductance of the reaction solution increases linearly (curve BC) with the volume of the alkali added. The titration graph will have the form shown in Fig. 10.12. Interpolation of both the linear segments AB and BC to meet at B gives the end point of the titration.

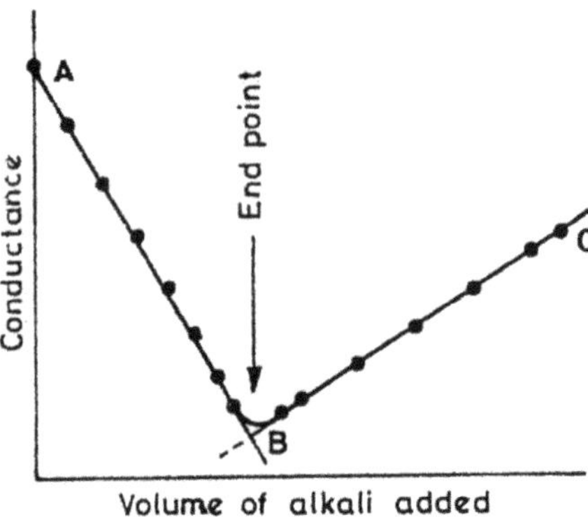

Fig. 10.12 Conductometric titration of a strong acid Vs strong base.

Precise measurements leave a small rounding near equivalent point B and is attributed to water dissociation.

10.7.2 Titration of Weak Acid (CH_3COOH) by Strong Base (NaOH)

A weak acid, like acetic acid, is characterized through its insignificant dissociation. Therefore, at the beginning of titration the acid solution has a low conductance. Addition of alkali to this moderately weak acid solution, initially result in a small fall of conductance owing to the replacement of free H^+ ions by acetate ions. If the acid is very weak, e.g., boric acid and phenol, the initial decrease of conductance upon neutralization is not observed due to insignificant dissociation. After the addition of a few drops of the alkali, the conductance increases in a regular fashion. This is due to suppression of ionization equilibrium of acetic acid by the acetate ions formed (buffering action) and increase in ionic population (sodium acetate is a strong electrolyte) in the solution.

$$(CH_3COOH) + (Na^+ + OH^-) \rightleftharpoons (CH_3COO^- + Na^+) + H_2O$$

Beyond neutralization point, addition of alkali result in a steep raise in conductance, the strong base being a better conductor than the salt.

Small rounding of the graph near equivalence point B is due to water dissociation.

10.7.3 Titration of moderately weak acid (CH_3COOH) by weak base (NH_4OH)

When a weak acid is titrated conductometrically with a weak base, the first part (line AB) of titration graph, is similar to that for strong base. Beyond equivalence point (B), the conductance remain unaltered and a straight line (BC) almost parallel to X-axis is obtained. This is due to insignificant dissociation of weak base.

The conductometric end point of moderately weak acid with a moderately weak base, is sharper than with a titration of a weak acid by a strong base. Conductometric titration of a very weak acid with a weak base can not be performed owing to extensive hydrolysis of their salt while this is possible with a strong base.

10.7.4 Conductometric titration of a mixture of a strong (HCl) and a weak acid (CH_3COOH) with a strong base (NaOH)

One of the valuable features of conductometric titrations is the titration of a mixture of a strong acid and a weak acid in one titration without employing any indicator. In the presence of a strong acid, the dissociation of a weak acid is completely suppressed due to common ion effect. Therefore, addition of a base to such a mixture will first result in the neutralization of the strong acid. The weak acid starts reacting only after neutralization of the strong acid. In the titration graph, Fig. 10.13, the line segment AB represents the neutralization of strong acid and the line segment BC represents the neutralization of weak acid.

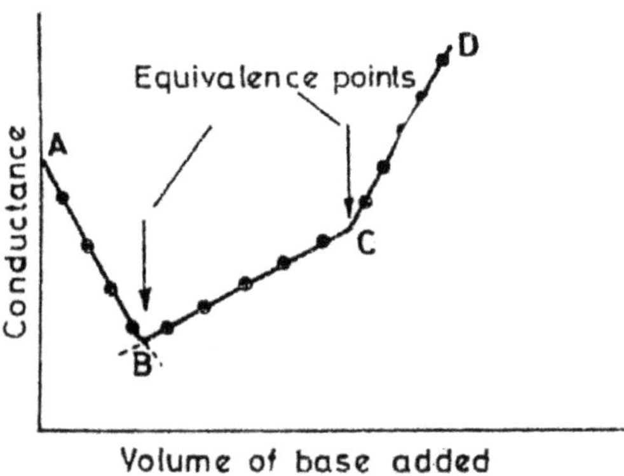

Fig. 10.13 Conductometric titration of a mixture of strong and weak acid *Vs* strong base.

10.7.5 Titration of salts of weak acids and bases

Conductometric titration of salt of a weak acid and a strong base, e.g., sodium acetate by a strong acid (HCl) or a salt of a weak base and a strong acid, e.g., ammonium chloride by strong base (NaOH) is another popularly accepted analytical tool. Corresponding reactions are given below.

$$(CH_3COO^- + Na^+) + (H^+ + Cl^-) \longrightarrow CH_3COOH + (Na^+ + Cl^-)$$

$$(NH_4^+ + Cl^-) + (Na^+ + OH^-) \longrightarrow (NH_4OH) + (Na^+ + Cl^-)$$

In the former titration prior to completion of neutralization, acetate ion is replaced by chloride, while in the latter sodium ion replaces the ammonium ion. Since all these ions differ very little in their ionic conductance, the conductance of the solution is almost unchanged up to neutralization. After crossing the neutralization point, however, there will be a marked increase in conductance owing to the addition of strong acid/base, Fig. 10.14.

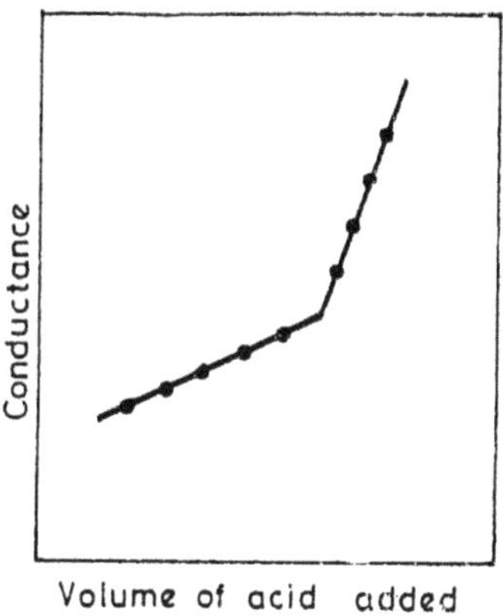

Fig. 10.14 Conductometric titration of CH_3COONa by HCl

It is possible to titrate a mixture of weak acid (CH_3COOH) and a salt of a weak base with strong acid (NH_4Cl) with a strong base, where, the first break corresponds to the neutralization of the acid and the second break to the completion of displacement of the salt.

10.7.6 Conductometric Titration involving Precipitation Reactions

Titration of silver nitrate solution with potassium chloride and the reverse titration are the typical examples of this class. In the titration of silver nitrate solution by KCl solution, represented by the reaction;

$$(Ag^+ + NO_3^-) + (K^+ + Cl^-) \longrightarrow AgCl\downarrow + (K^+ + NO_3^-)$$

before equivalence point the Ag^+ ions are replaced by slightly more mobile K^+ ions. Therefore, before equivalence point the conductance shows a small but gradual linear rise on the addition of KCl solution. After crossing the equivalence point, addition of potassium chloride solution simply raises the ionic strength (due to K^+ and Cl^- ions) of the medium and there by causes a sharp rise in the conductance, Fig. 10.15.

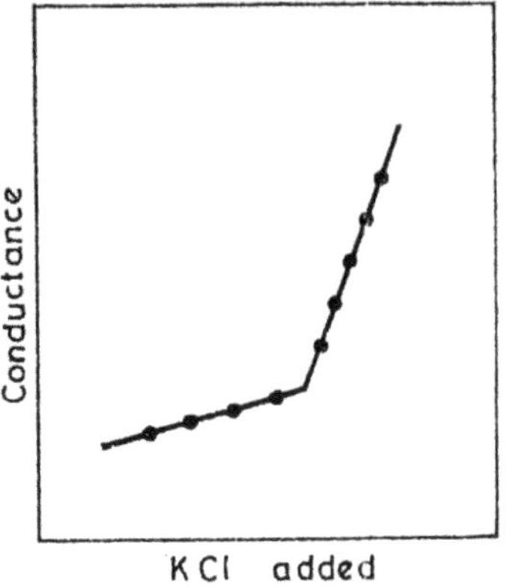

Fig. 10.15 Conductometric titration of $AgNO_3$ by KCl.

If the same titration is repeated with the sodium chloride solution, we notice a small initial fall in conductance followed by a sharp rise after equivalence point. This can be once again explained on the basis of ionic mobilities of Ag^+ and Na^+ ions, the former being more mobile than the later Fig. 10.16.

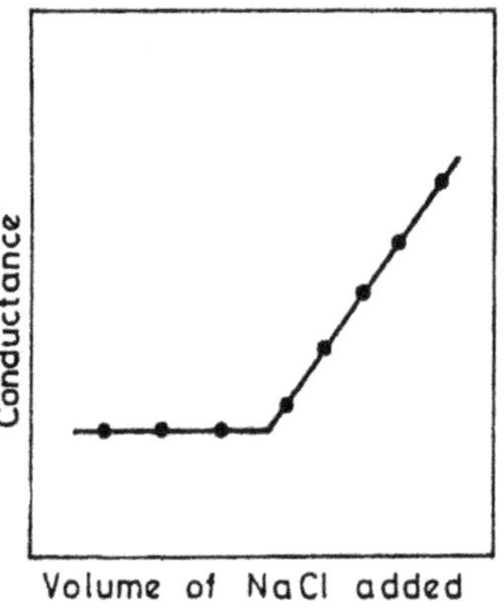

Fig. 10.16 Conductometric titration of $AgNO_3$ by NaCl.

Problem 25.

The equivalent conductivity of very dilute solution of $NaNO_3$ at 18^oC is 105.2 mhos cm^2. Being given that ionic conductivity of NO_3^- ions in the same solution is 61.7, Calculate the transport number of sodium ions.

Equivalent conductance of very dilute $NaNO_3$ = 105.2 mhos cm^2.

The mention of very dilute solution of $NaNO_3$ for a strong electrolyte indicates that this value correspond of Λ_4.

Given $\lambda_{NO_3^-} = 61.7$ mho cm^2 eq^{-1}

So $\lambda_{Na^+} = 105.2 - 61.7 = 43.5$ mho cm^2 eq^{-1}

$$t_{Na^+} = \frac{\lambda_+}{\Lambda_\infty} = \frac{43.5}{105.2} = 0.413$$

Problem 26.

At 18°C the conductivities at infinite dilution of NH_4Cl, NaOH and NaCl are 129.8, 217.4 and 108.9 mhos respectively. If the equivalent conductivity of N/100 solution of NH_4OH is 9.33 mhos, calculate the degree of dissociation of NH_4OH

$$\Lambda_{\infty(NH_4Cl)} = \lambda_{NH^{4+}} + \lambda_{Cl^-} = 129.8 \text{ S cm}^2 \qquad \text{I}$$

$$\Lambda_{4(NaOH)} = \lambda_{Na^+} + \lambda_{OH^-} = 217.4 \text{ S cm}^2 \qquad \text{II}$$

$$\Lambda_{4(NaCl)} = \lambda_{Na^+} + \lambda_{Cl^-} = 108.9 \text{ S cm}^2 \qquad \text{III}$$

I + II – III

$$\Lambda_{\infty(NH_4OH)} = (129.8 + 217.4) - 108.9 = 238.3$$

Degree of dissociation

$$\alpha = \frac{\Lambda_v}{\Lambda_\infty} = \frac{9.33}{238.3} = 0.039$$

Dissociation constant

$$K = \frac{\alpha^2 C}{1-\alpha} = 1.58 \times 10^{-5}$$

Problem 27.

At 25°C the equivalent conductivity of 0.1M benzoic acid is 8.9 ohm^{-1} cm^2 equiv^{-1}. The ionic conductivities at infinite dilution, of the hydrogen ion and benzoate ions are 349.8 and 32.3 ohm^{-1} cm^2 respectively. Calculate the degree of dissociation of 0.1 M benzoic acid.

$$\Lambda_v = 8.9 \text{ ohm}^{-1} \text{ cm}^2 \text{ eq}^{-1}$$

$$\Lambda_4 = \lambda_{H^+} + \lambda_{\text{benzonate}} = 349.8 + 32.3 = 382.1$$

$$\alpha = \frac{\Lambda_v}{\Lambda_\infty} = \frac{8.9}{382.1} = 0.023$$

Degree of dissociation of benzoic acid = 0.023

Problem 28.

The equivalent conductivity of NH_4Cl at infinite dilution is 149.7 ohm^{-1} cm^2 equiv^{-1} and the ionic conductances of OH^- and Cl^- are 198 and 79.3 ohm^{-1} cm^2 respectively. Calculate Λ_4 of NH_4OH.

$$\Lambda_{\infty(NH_4Cl)} = \lambda_{NH_4^+} + \lambda_{Cl^-} = 149.7 \text{ ohm}^{-1} \text{ cm}^2 \text{ eq}^{-1} \quad \text{I}$$

$$\lambda_{OH^-} = 198 \quad \text{II}$$

$$\lambda_{Cl^-} = 79.3 \quad \text{III}$$

$$\text{I} + \text{II} - \text{III} = 268.4$$

Λ_4 of NH_4OH is 268.4 ohm^{-1} cm^2 eq^{-1}.

Problem 29.

Calculate the cell constant of a cell where the area of the electrode is 2.5 sq cm and distance between them is 10 cm

$$\text{Cell constant} = X = \frac{l}{a} = \frac{10}{2.5} = 4 \text{ cm}^{-1}$$

Problem 30.

The equivalent conductivity at infinite dilution of HCl, CH_3COONa and NaCl are 426, 91 and 126 ohm^{-1} cm^2 equiv^{-1} respectively. Calculate the equivalent conductivity at infinite dilution of acetic acid.

$$\Lambda_{4(HCl)} = \lambda_{H+} + \lambda_{Cl^-} = 426 \quad \text{I}$$

$$\Lambda_{\infty(CH_3COONa)} = \lambda_{CHCOO^-} + \lambda_{Na^+} = 91 \quad \text{II}$$

$$\Lambda_{4(NaCl)} = \lambda_{Na^+} + \lambda_{Cl^-} = 126 \quad \text{III}$$

$$\Lambda_{\infty(CH_3COONa)} = (\text{I} + \text{II} - \text{III}) = 391$$

Problem 31.

An aqueous solution of copper sulphate was electrolyzed between copper electrodes. On the cathode 0.3 grams of copper was deposited. The solution in the anode compartment contained 1.43 grams of copper after electrolysis, and the same weight of water contained 1.214 grams of copper before electrolysis. Calculate the transport number of copper (eq. wt. of Cu is 31.77).

Amount of copper in anode compartment before electrolysis = 1.214g

Amount of copper in anode compartment after electrolysis = 1.43g

Increase in the concentration around anode = 1.43 – 1.214 = 0.216g

Amount of copper deposited on cathode = 0.3g

If there was no migration of copper from anodic compartment the actual raise would have been = 0.3 g

Fall in concentration of anodic compartment = 0.3 – 0.216 = 0.084g

$$t_{Cu^{2+}} = \frac{0.084}{0.3} = 0.28$$

$$t_{SO_4^{2-}} = 1 - 0.28 = 0.72$$

Problem 32.

Calculate equivalent conductivity from the following data :

(i) Cell constant = 1.15 cm^{-1}

(ii) its 0.1N solution has a resistance of 2.5×10^2 ohms

Given

$$R = 2.5 \times 10^2 \text{ ohm}, C = 1/R = 4.0 \times 10^{-3} \text{ mhos}$$

Cell constant $X = 1.15$

$$\kappa = C \times X = 4.0 \times 10^{-3} \times 1.15 = 4.6 \times 10^{-3} \text{ mho cm}^{-1}$$

$$\Lambda_v \text{ (Equivalent conductivity)} = \frac{1000\kappa}{C} = \frac{1000 \times 4.6 \times 10^{-3}}{0.1}$$

$$= 46 \text{ mho cm}^2 \text{ eq}^{-1}$$

Problem 33.

Specific conductance of a 0.02 M KCl solution at $25^{O}C$ is 0.002768 mhos m^{-1}. If the resistance of a cell containing this solution is 400 ohms, what is the cell constant.

Cell constant = Specific conductance x Resistance

$$= 0.002768 \times 400 \quad = 1.1$$

Problem 34.

When a certain conductance cell was filled with 0.02 M KCl, it had a resistance of 82.4 ohms at $25^{O}C$. When it is filled with 0.005 N K_2SO_4, it had a resistance of 326 ohms. Find the equivalent conductivity of the given K_2SO_4 solution. The specific conductivity of 0.02 M KCl is 0.002768 ohm^{-1} cm^{-1} at $25^{O}C$.

$$X = \kappa R$$

$$X = 0.002768 \times 82.4 = 0.2281 \text{ cm}^{-1}$$

$$\kappa_{K_2SO_4} = \frac{0.2281}{326} = 0.0006996 \text{ mho cm}^2 \text{ eq}^{-1}$$

$$\kappa_{K_2SO_4} = \frac{1000\kappa}{C} = \frac{1000 \times 0.0006996}{0.005}$$

$$= 139.9 \text{ mho cm}^2 \text{ eq}^{-1}$$

Problem 35.

The equivalent conductivity of 0.05 N solution of propionic acid at $25^{O}C$ is 6.32 $equiv^{-1}$ ohm^{-1} cm^2 and the value increases to 385.6 cm^2 $equiv^{-1}$ ohm^{-1} at infinite dilution. Calculate the degree of dissociation of the acid at that temperature.

Degree of dissociation of propionic acid

$$\alpha = \frac{\Lambda_v}{\Lambda_\infty} = \frac{6.32}{385.6} = 0.0164$$

Problem 36.

Calculate the ionic conductivity of H^+, given Λ_4 of HCl = 383.5, $t_{H^+} = 0.841$.

$$t_+ = \frac{\lambda_{H^+}}{\Lambda_{\infty(HCl)}} \quad \text{or} \quad \lambda_{H^+} = t_+ \times \Lambda_\infty$$

$$\lambda_{H^+} = 0.841 \times 383.5 \qquad = 322.52$$

Problem 37.

In a conductivity cell 0.01 N KCl solution gave a resistance of 225 ohms, while a 0.01 N solution of HCl gave a resistance of 77.1 ohms. Specific conductance of 0.01N KCl is 0.00141 ohm^{-1} cm^{-1}. Calculate the cell constant and equivalent conductance of HCl.

Cell constant = sp. conductance × resistance

$$= 0.00141 \times 225 \quad = 0.31725\ cm^{-1}$$

$$\kappa_{HCl} = \frac{0.31725}{77.1} = 0.0041\ mho\ cm^{-1}$$

$$\Lambda_{(HCl)} = \frac{1000\kappa}{C} = \frac{1000 \times 0.0041}{0.01}$$

$$= 410\ mho\ cm^2\ eq^{-1}$$

10.8 Introduction to EMF

Chemical changes are always accompanied by energy changes, where the difference appears generally in the form of heat (either evolution or absorption). Thus, when a zinc rod is placed in chlorine water, electron exchange takes place between Zn atoms and Cl_2 molecules at the point of inter-phase and Zn dissolves accompanied by evolution of heat.

$$Zn + Cl_2 \longrightarrow ZnCl_2 + \Delta H$$

This is an oxidation - reduction reaction for which partial ionic equations are

$$Zn - 2e^- \longrightarrow Zn^{2+}$$

$$Cl_2 + 2e^- \longrightarrow 2Cl^-$$

$$Zn^{2+} + 2Cl^- \longrightarrow ZnCl_2$$

Since electrical energy is another form of energy, we could devise a reaction system in such away that all electron movement takes place through an external electrical conductor and there by tap electrical energy. Conversely, a chemical reaction could be initiated by passing electricity through a reaction system. *The chemical and electrical changes are carried in a physical entity called cell. A cell may be defined as a single arrangement consisting of two electrodes and capable of producing electricity due of chemical reaction and vice versa.* A battery is a combination of two or more cells arranged either in series or parallel.

There are two types of cells in which the above electrochemical processes are carried out.

(i) **A galvanic or voltaic cell** uses the energy released from a spontaneous chemical reaction to generate electricity.

(ii) **An electrolytic cell** uses the electrical energy to cause a non-spontaneous chemical reaction to occur, and the process is commonly called **electrolysis**.

Modern society depends on the galvanic cells for running cars, radios, calculators, medical instruments, computers, on the field analytical devices, emergency flash lights and so on. The electrolytic cells are used in the chemical and metallurgical industries for the extraction, synthesis and purification of various chemicals. In this chapter our interest lies mainly with the galvanic cells.

10.9 Galvanic Cells

Under normal circumstances a redox reaction occurs when the oxidizing agent comes in contact with the reducing agent as in the reaction between Zn and Cl_2 described above. The electrons are transferred directly from the reducing agent to the oxidizing agent. In order to tap the energy of the reaction in the form of electrical energy, the reactants must be physically separated so as to prevent direct transfer of electrons. Electrons are then allowed to go from Zn to Cl_2 through an external circuit, made out of a wire or metallic conductor. In the current example both Zn rod and Cl_2 taken in two separate compartments which are separated either through a porous partition or a salt bridge. The porous partition or salt bridge prevents direct mixing of the reactants and at the same time are permeable to ions, Fig. 10.17.

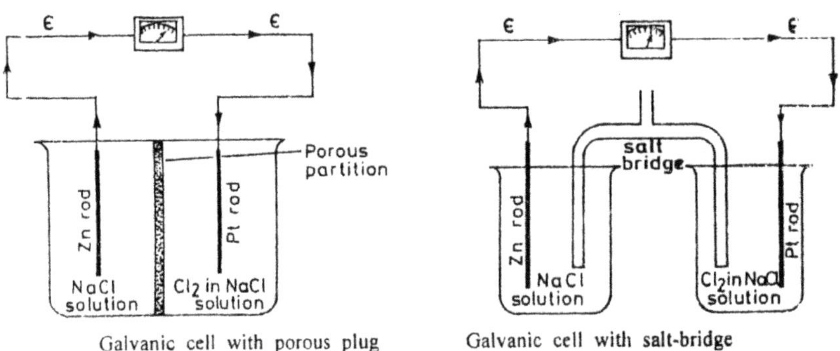

Fig. 10.17 A Galvanic cell.

Since the Cl_2 itself is a non-conductor unlike Zn rod, to conduct electricity in the Cl_2 compartment an inert metal wire like Pt is placed. The Zn rod and Pt wire are called electrodes. By *definition the electrode at which oxidation occurs is called the anode (Zn electrode); the electrode at which reduction occurs is known as cathode* (electrode is also called chlorine gas electrode).

For this system the *individual oxidation* and *reduction reactions at the electrodes are called* **half-cell reactions.**

Half-cell reaction of Zn electrode (anode) :

$$Zn(s) \longrightarrow Zn^{2+}(aq) + 2e^-$$

Half-cell reaction of Pt electrode (cathode) :

$$Cl_2(g) + 2e^- \longrightarrow 2Cl^- (aq)$$

Total cell reaction :

$$Zn(s) + Cl_2(g) \longrightarrow Zn^{2+}(aq) + 2Cl^- (aq)$$

In order to maintain electrical neutrality of both the compartments the Cl^- and Zn^{2+} move across the porous partition. For the movement of each Zn^{2+} ion from anodic compartment to cathodic compartment there is a concurrent migration of two Cl^- ions from cathodic compartment to anodic compartment. In the present example addition of NaCl is only to facilitate easy movement of ions before sufficient concentration of ions is built up with the course of time. A cell operates only so long as the chemical reaction, on which the cell is based, proceed. The electricity ceases as soon as one of the reactant is exhausted or the reaction reaches equilibrium.

Electric current flows from negative to positive electrode through external circuit, but long standing convention defines that current flows from positive to negative electrode (taking the medium into consideration). The passage of electrons from one electrode to the other suggests the presence of a potential difference between them. *The difference of potential, which causes the current to flow from an electrode at higher potential to the one of lower potential is called the electromotive force (emf) of the cell.*

10.9.1 Reversible & Irreversible Cells

As long as the products and reactants of a chemical reaction are within the cells, i.e., in a closed system, the reaction can be reversed by applying higher external emf and such cells are called **reversible cells.** Daniell cell is a common example of a reversible galvanic cell. It is shown diagrammatically in Fig. 10.18. A Zn strip is dipped in a zinc sulphate solution taken in a porous pot and connected to a copper sulphate solution having a Cu strip in it.

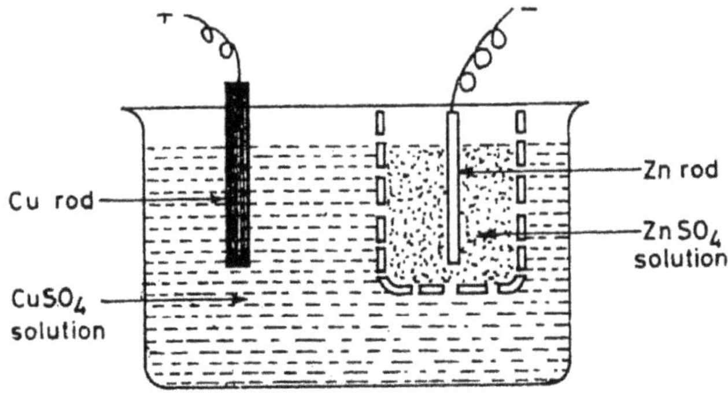

Fig. 10.18 A Daniel Cell.

The displacement of copper by the more active Zn, forms the basis of Daniell voltaic cell :

At anode	$Zn \longrightarrow Zn^{2+} + 2e^-$
At cathode	$Cu^{2+} + 2e^- \longrightarrow Cu$
Cell reaction	$Zn + Cu^{2+} \longrightarrow Zn^{2+} + Cu$

In case, one of the reactant or product escapes the cell, as in a cell constructed by dipping Ag and Zn rods in a solution of sulphuric acid;

Zn anode : $Zn(s) + H_2SO_4 \longrightarrow ZnSO_4 + H_2(g)\uparrow$

Ag cathode : $2Ag(s) + H_2SO_4 \longrightarrow Ag_2SO_4 + H_2(g)\uparrow$

the cell reaction cannot be reversed by applying higher external emf and hence are called **irreversible cells.**

10.9.2 Single Electrode Potential

The foregone conclusions *suggest the presence of two half cell reactions and the emf of a cell gives only the difference in the driving forces of the electrodes.* In the Daniell cell we can find more rational answer for the question of why copper does not displace zinc ion from the solution by understanding more about this driving force. When the cell reaction takes place, what we actually observe as **emf** or **cell potential** or **cell voltage** is a kind of tug-of-war between these driving forces. The species with greater tendency to reduce acquires an electron supplied by the less reducing species that gets oxidized. In the zinc-copper cell, therefore, copper ion must have a larger reducing nature or **reduction potential** than zinc. Thus;

$$E_{cell} = \textbf{Reduction potential of reduced species} - \textbf{Reduction potential of oxidized species}$$

or $$E_{cell} = \textbf{Reduction potential of reduced species} + \textbf{Oxidation potential of oxidized species}$$

or in short

$$E_{cell} = E_{cathode} - E_{anode} \qquad .....(10.41)$$

(Remember that when no specific mention is made the electrode potential is always the reduction potential)

or $$E_{cell} = E_{right} - E_{left} \qquad .....(10.41)$$

In the Daniell cell, therefore;

$$E^0_{cell} = E^0_{Cu^{2+}|Cu} - E^0_{Zn^{2+}|Zn} \qquad .....(10.41)$$

where $E^0_{Cu^{2+}|Cu}$ is the standard reduction potential for copper ion and $E^0_{Zn^{2+}|Zn}$ is the standard reduction potential for zinc ion.

Since $E^0_{Cu^{2+}|Cu}$ is larger than $E^0_{Zn^{2+}|Zn}$, E^0_{cell} should be positive for a spontaneous reaction. Here the prefix standard implies experimental conditions where the **half-cell potential** or **single electrode potential** are measured at 25°C, maintaining unit concentration of reversible ions. The **standard cell potential**, E^0_{cell}, represents the magnitude of difference between reduction potentials of two half cells of a redox reaction or sum of the standard oxidation potential and standard reduction potential of respective electrodes. *The emf is an intensive property like temperature and therefore 'independent' of the volume of the solution and size of the electrodes.*

Knowing the standard electrode potential of one of the electrode, it is possible to determine electrode potentials of all the other electrodes by suitably coupling and measuring the cell voltage. Unfortunately, even to start with, it is impossible to determine the absolute values of a single electrode potential. So arbitrarily the emf of a standard hydrogen electrode is chosen as zero and relative to this standard electrode, electrode potentials of various electrodes were determined. A standard hydrogen electrode consists of a platinum foil encased in a glass sleeve through which hydrogen gas is passed at a pressure of 1atm and is placed in one molar acid solutions, Fig. 10.3. Platinum foil is coated with a layer of platinum black. In addition to providing a conducting surface, the platinum electrode adsorbs molecular hydrogen and helps in maintaining the following equilibrium.

$$2H^+(aq) + 2e^- \rightleftharpoons H_2(g)$$

Table 10.8 Standard Reduction Potentials at 25^0C*

Oxidising power increases ↓

Electrode	Electrode Reaction	E^0(V)
$Li^+ \mid Li$	$Li^+(aq) + e^- \longrightarrow Li(s)$	– 3.045
$K^+ \mid K$	$K^+(aq) + e^- \longrightarrow K(s)$	– 2.925
$Ba^{2+} \mid Ba$	$Ba^{2+}(aq) + 2e^- \longrightarrow Ba(S)$	– 2.90
$Sr^{2+} \mid Sr$	$Sr^{2+}(aq) + 2e^- \longrightarrow Sr(s)$	– 2.89
$Ca^{2+} \mid Ca$	$Ca^{2+}(aq) + 2e^- \longrightarrow Ca(s)$	– 2.87
$Na^{2+} \mid Na$	$Na^{2+}(aq) + e^- \longrightarrow Na(s)$	– 2.714
$Mg^{2+} \mid Mg$	$Mg^{2+}(aq) + 2e^- \longrightarrow Mg(s)$	– 2.37
$Be^{2+} \mid Be$	$Be^{2+}(aq) + 2e \longrightarrow Be(s)$	– 1.85
$Al^{3+} \mid Al$	$Al^{3+}(aq) + 3e^- \longrightarrow Al(s)$	– 1.66
$Mn^{2+} \mid Mn$	$Mn^{2+}(aq) + 2e^- \longrightarrow Mn(s)$	– 1.18
$OH^- \mid H_2(g) \mid Pt$	$H_2O + e^- \longrightarrow H_2(s) + OH^-(aq)$	– 0.828
$Zn^{2+} \mid Zn$	$Zn^{2+}(aq) + 2e^- \longrightarrow Zn(s)$	– 0.763
$Cr^{3+} \mid Cr$	$Cr^{3+}(aq) + 3e^- \longrightarrow Cr(s)$	– 0.74
$Fe^{2+} \mid Fe$	$Fe^{2+}(aq) + 2e \longrightarrow Fe(s)$	– 0.440
$Co^{2+} \mid Co$	$Co^{3+}(aq) + 2e^- \longrightarrow Cd(s)$	– 0.408
$SO_4^{2-} \mid PbSO_4(s) \mid Pb$	$PbSO_4 + 2e^- \longrightarrow Pb(s) + SO_4^{2-}(aq)$	– 0.31
$Co^{2+} \mid Co$	$Co^{2+}(aq) + 2e^- \longrightarrow Co(s)$	– 0.28
$Ni^{2+} \mid Ni$	$Ni^{2+}(aq) + 2e^- \longrightarrow Ni(s)$	– 0.25
$Sn^{2+} \mid Sn$	$Sn^{2+}(aq) + 2e^- \longrightarrow Sn(s)$	– 0.14
$Pb^{2+} \mid Pb$	$Pb^{2+}(aq) + 2e^- \longrightarrow Pb(s)$	– 0.126
$H^+ \mid H_2(g) \mid Pt$	$2H^+(aq) + 2e^- \longrightarrow H_2(s)$	0.000
$Sn^{2+}, Sn^{4+} \mid Pt$	$Sn^{4+}(aq) + 2e^- \longrightarrow Sn^2(s)$	+ 0.13

Contd... Table 10.8

Electrode	Electrode Reaction	E^0(V)
Cu^{2+} | Cu	$Cu^{2+}(aq) + 2e^- \longrightarrow Cu^+(s)$	+ 0.15
Cl^- | AgCl(s) | Ag	$AgCl(s) + 2e^- \longrightarrow Ag(s) + Cl^-(aq)$	+ 0.2223
Cl^- | $Hg_2Cl_2(s)$ | Hg $	$Hg_2Cl_2 + 2e^- \longrightarrow 2Hg(s) + 2Cl^-(aq)$	+ 0.268
Cu^{2+} | Cu	$Cu^{2+}(aq) + 2e^- \longrightarrow Cu(s)$	+ 0.337
OH^- | $O_2(g)$ Pt	$O_2(g) + 2H_2O + 4e^- \longrightarrow 4OH^-(aq)$	+ 0.401
Cu^+ | Cu	$Cu^+(aq) + 2e^- \longrightarrow Cu(s)$	+ 0.521
I^-, I_2 | Pt	$I_2(s) + 2e^- \longrightarrow 2I^-(aq)$	+ 0.5355
MnO_4, OH^- | MnO_2 | Pt	$MnO_4(aq) + 2H_2O + 3e^- \longrightarrow$ $MnO_2(s) + 4OH^-(aq)$	+ 0.59
Fe^{2+}, Fe^{3+} | Pt	$Fe^{3+}(aq) + e^- \longrightarrow Fe^{2+}(aq)$	+ 0.771
Ag^+ | Ag	$Ag^+(aq) + e^- \longrightarrow Ag(s)$	+ 0.7991
Hg_2^{2+} | Hg | Pt	$Hg_2^{2+}(aq) + 2e^- \longrightarrow 2Hg(l)$	+ 0.85
Hg_2^{2+} | Hg^{2+} | Pt	$2Hg^{2+}(aq) + 2e^- \longrightarrow Hg_2^{2+}(aq)$	+ 0.920
Br^-, Br_2 | Pt	$Br_2(l) + 2e^- \longrightarrow 2Br^-(aq)$	+ 1.0652
H^+ | O_2 | Pt	$O_2(g) + 4H^+(aq) + 4e^- \longrightarrow 2H_2O$	+ 1.229
$Cr_2O_2^{7-}$, H^+, Cr^{3+} | Pt	$Cr_2O_2(aq) + 14H^+(aq) + 6e^- \longrightarrow$ $2Cr^{1+}(aq) + 7H_2O$	+ 1.33
Cl^- | Cl_2 | Pt	$Cl_2(g) + 2e^- \longrightarrow Cl^-(aq)$	+ 1.3595
Au^{3+} | Au	$Au^{3+}(aq) + 3e^- \longrightarrow Au(s)$	+ 1.50
MnO_4^-, Mn^{2+} | H^+ | Pt	$MnO_4^-(aq) + 8H^+(aq) + 5e^- \longrightarrow$ $Mn^{2+}(aq) + 4H_2O$	+ 1.51
Ce^{3+}, Ce^{4+} | Pt	$Ce^{4+}(aq) + e^- \longrightarrow Ce^{3+}(aq)$	+ 1.61
H^+ | H_2O_2 | Pt	$Ce^{3+}(aq) + 2H^-(aq) + 2e^- \longrightarrow 2H_2O$	+ 1.77
Co^{2+}, Co^{3+} | Pt	$H_2O_2(aq) + e^- \longrightarrow Co^{2+}(aq)$	+ 1.82
H^+ | O_3, O_2 | Pt	$O_3(g) + 2H^+(aq) + 2e^- \longrightarrow$ $O_2(g) + H_2O(l)$	+ 2.07
F^- | F_2 | Pt	$F_2(g) + 2e^- \longrightarrow 2F^-(aq)$	+ 2.87

Oxidising power increases ↓

* All the ions are at unit activity and gases at 1 atm pressure.

$ *EMF* of the calomel electrode containing saturated KCl is 0.2415 V and with 1 N and 0.1 N KCl solutions are 0.280 and 0.334 V respectively.

Some more details of this electrode are given in 10.5.1. When hydrogen electrode is paired with another electrode (half-cell) in a galvanic cell, it can undergo either oxidation or reduction depending on the magnitude of potential of coupled electrode relative to hydrogen electrode. The cell *emf* so determined, directly gives the single electrode potential of the coupled electrode. Some of the half cell reactions and corresponding standard reduction potentials are listed in Table 10.8.

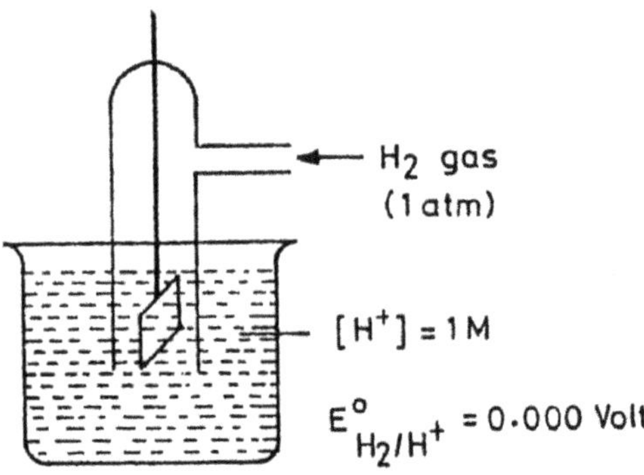

Fig. 10.19 Hydrogen gas electrode.

The standard reduction potential table serves many useful purposes.

1. From these tables, at a glance, *we can make out the good reducing agents and good oxidizing agents*. Thus fluorine tops the list of oxidizing agents as it has highest standard reduction potential. From a higher positive reduction potential we can understand that H^+ is a much powerful oxidizing than Zn^{2+}. Similarly lithium metal may be considered as best reducing agent.

2. *A knowledge of reduction potentials also helps in deciding whether a given redox reaction will proceed spontaneously in the forward direction or not.* Suppose if we consider the Zn-Cu cell reaction.

$$Zn + Cu^{2+} \rightleftharpoons Cu + Zn^{2+}$$

We have the two half cell reactions

Oxidation (anode) $\quad Zn \longrightarrow Zn^{2+} + 2e^- \quad E^0_{Zn^{2+}|Zn} = 0.76\ V$

Reduction (cathode) $\quad Cu^{2+} + 2e^- \longrightarrow Cu \quad E^0_{Cu^{2+}|Cu} = 0.34\ V$

$$E^0_{cell} = E^0_{Zn^{2+}|Zn} - E^0_{Cu^{2+}|Cu}$$

or $$E^0_{cell} = E^0_{oxidation} + E^0_{reduction}$$

$$E^0 = 0.76 + 0.34$$

$$= 1.10 \text{ V}$$

The positive E^0_{cell} value suggest that the redox reaction is spontaneous in the forward direction.

Let us now test the spontaneity of the reaction

$$Fe^{2+}(aq) + Ni(s) \longrightarrow Fe(s) + Ni^{2+}(aq)$$

The two half cell reactions are

$$Fe^{2+} + 2e \longrightarrow Fe \qquad E^0_{Fe^{2+}|Fe} = -0.44V$$

$$Ni \longrightarrow Ni^{2+} + 2e^- \qquad E^0_{Ni^{2+}|Ni} = -0.25 \text{ V}$$

$$E^0_{cell} = E^0_{Fe^{2+}|Fe} - E^0_{Ni^{2+}|Ni}$$

$$= -0.44 - (-0.25)$$

$$= -0.19V$$

Because E^0_{cell} computed is negative, we draw the conclusion that this reaction is not spontaneous under standard conditions.

3. *The standard electrode potential data is highly helpful in deciding the course of electrolysis reactions.* From the data given we see that the reduction potential for copper is 0.39V, for H_2O -0.83V and for aluminium -1.66V. During electrolysis of a Cu^{2+} solution, copper ions are more readily reduced (potential being more positive) than water and will plate on the electrode according to the half reaction.

$$Cu^{2+}(aq) + 2e^- \longrightarrow Cu(s)$$

In case of aluminium, due to more positive reduction potential (or less negative potential) of H_2O than aluminium, H_2O is preferentially reduced and no deposition of aluminium is seen.

$$2H_2O + 2e^- \longrightarrow H_2 + 2OH^-$$

$$2Al^{3+} + 6OH^- \longrightarrow 2Al(OH)_3$$

4. *The reduction potential data helps in designing cells through which required voltage can be drawn.* When zinc electrode is coupled with chlorine electrode, copper electrode and iron electrode we get cells of different voltages, Fig. 10.20.

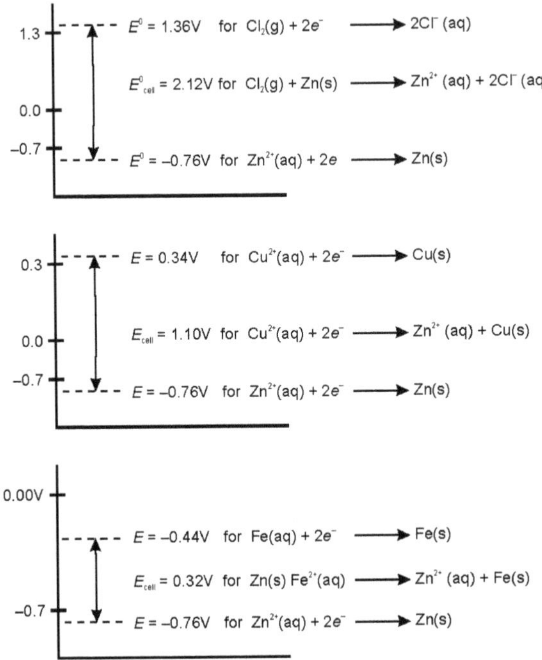

Fig. 10.20 A comparative graphical representation of E^0_{cell} values for various cells.

5. Finally from *a knowledge of these reduction potentials a non-spontaneous electrode/cell is made spontaneous by suitably altering the sign of the electrode.* Thus in the earlier example Fe-Ni cell may be made spontaneous by taking iron electrode as anode & nickel electrode as cathode. Then the reaction is

$$Fe(s) + Ni^{2+}(aq) \longrightarrow Fe^{2+}(aq) + Ni(s)$$

which has a cell potential of +0.19V.

10.9.3 Representation of a Cell

Sketching an electrochemical cell as in Fig. 10.17 and 10.18 is a tedious process. A symbolic representation is often used to describe a cell. This representation is called a **cell diagram**. Following rules are adopted in these representation. For example, a Daniell cell is represented as;

Anode Salt bridge Cathode

$Zn(s) \mid Zn^{2+}(0.1M) \parallel Cu^{2+}(0.1M) \mid Cu(s)$

Left Half cell Right Half cell

(i) The anode, where oxidation occurs is represented at the left and at cathode, where reduction takes place, is represented on the right of the cell diagram.

This order can be easily remembered by recollecting the Hindi phrase "**LAO SaR CaR**" or simply remembering the south asian country **LAOS**. Here the letters LAOS, stands for **L**eft **A**node **O**xidation **S**alt bridge.

In the Daniell cell $Zn \mid Zn^{2+}$ half cell is an oxidizing electrode and is written on the left.

(ii) A single vertical line represents the boundary between different phases that constitute an electrode.

(iii) A double vertical line indicates that the liquid junction potential has been decreased to minimum by suitable means like using a salt bridge.

(iv) The concentration of the dissolved substances are indicated in the parenthesis.

(v) In case of gases at equilibrium it is necessary to represent the pressure of the gas.

$Pt \mid H_2(1\ atm) \mid H^+\ (0.1\ M) \parallel Cu^{2+}\ (0.1\ M) \mid Cu(s)$

(vi) When two oxidation states of the same species are present in a homogenous medium, they are usually separated by a comma.

$Pt \mid H_2\ (1\ atm) \mid H^+\ (0.1\ M) \parallel Fe^{3+}\ (0.2\ M), Fe^{2+}\ (0.1\ M) \mid Pt$

10.10 Determination of Cell Potential

Approximate estimates of *emf* of a cell may be obtained by connecting an ordinary voltmeter across the electrodes. This method is inaccurate because the *current drawn by the voltmeter* disturbs the electrochemical equilibrium of the cell, thereby bringing some changes in the *emf.* At the most, a voltmeter is satisfactory only in measuring large *emf* values. For obtaining emf values without disturbing the electrochemical equilibrium **Poggendorff compensation principle** is widely used. In this technique the *emf* of a cell under examination is balanced by an equal and opposite emf from another source, Fig. 10.21.

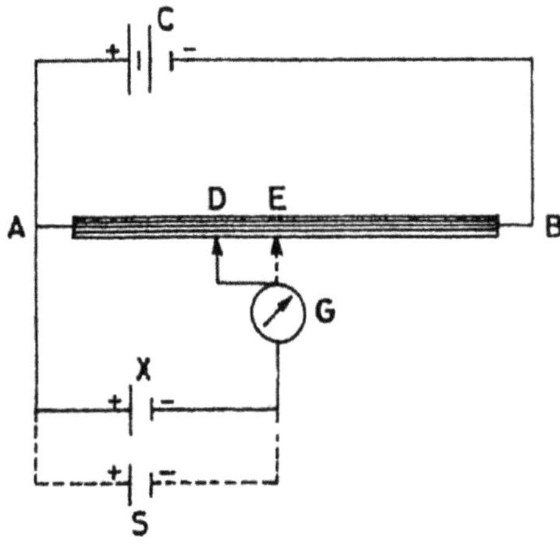

Fig. 10.21 Measurement of emf.

A storage cell C is connected across a uniform resistance wire AB. Now a cell X, whose *emf* is under study, is connected through a galvanometer G such that it opposes the emf of the storage cell. Now the position of the tap key D is adjusted until no current passes through the galvanometer. Under these circumstances

$$\frac{emf\ of\ X}{emf\ of\ C} = \frac{AD}{DB} \qquad .....(10.42)$$

Measurement of AD and BD enables us to calculate the *emf* of the unknown cell provided *emf* of the storage cell is known. But the emf of a storage cell is sensitive to minor changes in acid concentration, condition of the lead plates, temperature etc. To counter this problem the cell X is replaced by a standard cell S, whose emf is accurately known to a fraction of a millivolt and has negligible temperature coefficient. So if E is the balancing point when standard cell is used, then

$$\frac{emf\ of\ X}{emf\ of\ C} = \frac{AD}{AE} \qquad .....(10.43)$$

Therefore;

$$emf \text{ of } X = \frac{\text{AD}}{\text{AE}} \times emf\ of\ S \qquad .....(10.44)$$

Thus we get the *emf* of an unknown cell *X*.

10.11 Standard Cell

A standard cell is one whose potential is highly reproducible, accurate up to a fraction of a millivolt and has a very small temperature coefficient. **Weston standard cell** is best suited for this purpose. It is a H shaped tube. The negative electrode (C+.athode) consists of 12.5% cadmium amalgam, sprinkled with $3CdSO_4.8H_2O$ crystals over its top. The anodic limb (positive electrode) consists of mercury covered with solid mercurous sulphate. Now the H shaped tube is filled with a saturated cadmium sulphide solution, Fig. 10.22. The cell is represented as

12.5% Cd in Hg | $3CdSO_4.9H_2O(s)$ | Satd. soln. of $CdSO_4$ | $Hg_2SO_4(s)$ | Hg

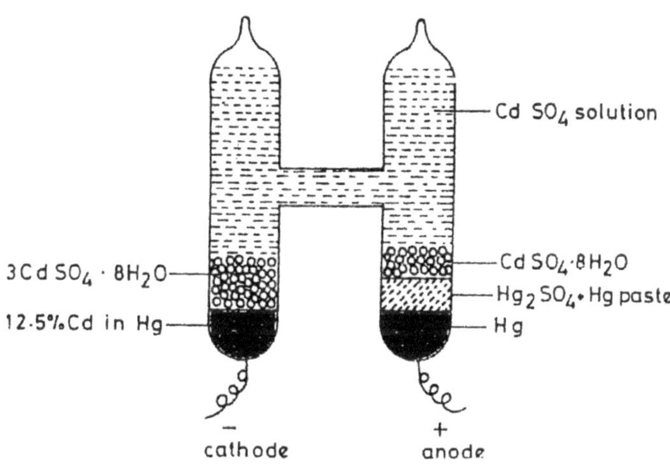

Fig. 10.22 The Weston standard cell.

The *emf* of this cell lies with in the range 1.0188 to 1.0198 V, depending on the exact activity of amalgam. The cell has a very small temperature coefficient as to be negligible.

10.12 Relationship between Cell Reaction and EMF

Nernst (1864-1941) studied the theoretical relationship between the electrochemical reaction and the corresponding cell emf. This relationship is generally known as **Nernst equation.**

Let us consider a galvanic cell whose cell reaction is written as

$$a\text{A} + b\text{B} \longrightarrow c\,\text{C} + d\,\text{D}$$

where lower case letter a,b,c,d represent the number of moles of A, B, C, D respectively at equilibrium. The Nernst equation for the cell is written as;

$$E_{cell} = \frac{RT}{nF}\ln K - \frac{RT}{nF}\ln\frac{[\text{C}]^c[\text{D}]^d}{[\text{A}]^a[\text{B}]^b}$$

where E_{cell} denotes the *emf* of the cell, R the gas constant, T the temperature, n number of faraday of current F passed, K the equilibrium constant and the quantities in the square brackets refer the respective concentration. The quantity $(RT/nF)\ln K$ is known as standard emf of the cell $E^{\text{o}}_{\text{cell}}$.

$$\boxed{E_{\text{cell}} = E^0_{\text{cell}} - \frac{RT}{nF}\ln\frac{[\text{C}]^c[\text{D}]^d}{[\text{A}]^a[\text{B}]^b}} \quad .....(10.45)$$

or

$$E_{\text{cell}} = E^0_{\text{cell}} - \frac{2.303\,RT}{nF}\log\frac{[\text{C}]^c[\text{D}]^d}{[\text{A}]^a[\text{B}]^b} \quad .....(10.46)$$

At room temperature (T = 298 K), taking R = 8.314J K^{-1}, F= 96457 C, the above equation simplifies to

$$\boxed{E_{\text{cell}} = E^{\text{o}}_{\text{cell}} - \frac{0.05916}{n}\log\frac{[\text{C}]^c[\text{D}]^d}{[\text{A}]^a[\text{B}]^b}} \quad .....(10.47)$$

From the equation (10.47) we notice that the standard cell *emf* is equal to the cell *emf* when the activities of both reactants and products is equal to unity.

10.12.1 Electrode Potential and Nernst Equation

Nernst equation is also applicable to half cell reactions. Few half cell reactions and corresponding Nernst's expressions are discussed below.

1. **Zn | Zn^{2+} electrode :** This a typical representative example of *Metal/Metal ion electrode.* This electrode can be constructed by bringing a zinc rod in contact with zinc ions. Oxidation of zinc atoms to give zinc ions is expressed by the following equation;

$$Zn \longrightarrow Zn^{2+} + 2e^-$$

In the reaction two electrons are exchanged between both the zinc species in the overall redox reaction making n=2. Therefore ,

$$E_{Zn|Zn^{2+}} = E^0_{Zn|Zn^{2+}} - \frac{0.05916}{2} \log \frac{a_{Zn^{2+}}}{a_{Zn}}$$

Since the activity of all the pure solids is unity (a = 1)

Therefore,

$$E_{Zn|Zn^{2+}} = E^0_{Zn|Zn^{2+}} - \frac{0.05916}{2} \log a_{Zn^{2+}}$$

This expression suggests that the oxidation electrode potential, *E*, decreases with increase in $\left[a_{Zn^{2+}}\right]$ and equals to standard electrode potential, E^o, at unit concentration of the electrolyte.

2. **Hydrogen electrode :** The electrode reaction of this electrode is written as;

$$\tfrac{1}{2} H_2 \longrightarrow H^+ + e^-$$

Applying Nernst equation

$$E_{H_2|H^+} = E^0_{H_2|H^+} - \frac{0.0592}{1} \log \frac{a_{H^+}}{p_{H_2}^{1/2}}$$

at 1atm pressure

$$E_{H_2|H^+} = E^0_{H_2|H^+} - 0.0592 \log a_{H^+}$$

$$E_{H_2|H^+} = E^0_{H_2|H^+} + 0.0592 \text{ pH} \qquad \text{.....(10.48)}$$

Advantages of Hydrogen Gas Electrode : It gives satisfactory results in the entire pH range of 0-14. The potential and pH are free from salt errors.

Limitations : The hydrogen gas used must be sufficiently pure. The electrode can not be used in the presence of gas impurities like air and oxygen and electrolyte impurities like oxidizing and reducing agents. It is very susceptible to poisoning by cyanides, arsenic, antimony compounds and alkaloids. It is cumbersome to handle in a normal laboratory.

3. **Redox electrode :** Platinum rod dipped in a mixture of two different oxidation states of a metal or two redox states of any compound is

called a **redox electrode**. Platinum wire immersed in a mixture of ferrous and ferric ion mixture is a typical example of this class.

Electrode reaction $\quad Fe^{2+} \longrightarrow Fe^{3+} + e^-$

$\qquad\qquad\qquad\qquad a_{Fe^{2+}} \qquad a_{Fe^{3+}}$

The electrode potential at any $[Fe^{2+}]/[Fe^{3+}]$ ratio is given by the expression,

$$E_{Pt|Fe^{2+},Fe^{3+}} = E^0_{Pt|Fe^{2+},Fe^{3+}} - \frac{0.0592}{1} \log \frac{a_{Fe^{3+}}}{a_{Fe^{2+}}}$$

4. **Quinhydrone electrode :** This is a redox electrode reversible to protons and often replaces the hydrogen electrode. Quinhydrone is a 1:1 molar mixture of quinone and hydroquinone. The electrode consists of a shiny platinum electrode dipped in a acid / base test solution, which is saturated with quinhydrone. The electrode reaction is given by

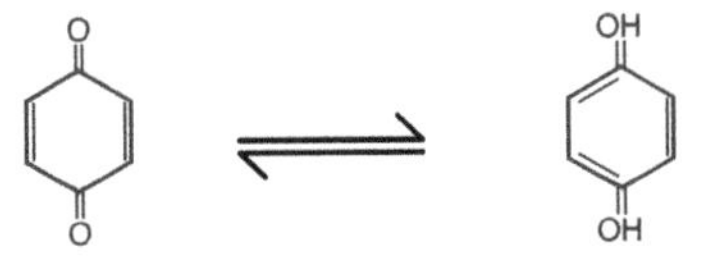

Quinone (Q) $\rightleftharpoons$ Hydroquinone (QH_2)

The electrode potential at 25°C is given by,

$$E_{Pt|Q,H^+,QH_2} = E^0_{Pt|Q,H^+,QH_2} - \frac{0.0592}{2} \log \frac{a_{QH_2}}{a_Q\, a^2_{H^+}}$$

Since quinone and hydroquinone are taken in equimolar amounts, i.e., $a_Q = a_{QH_2}$

We have

$$E = E^0 - \frac{0.0592}{2} \log \frac{1}{a^2_{H^+}}$$

$$E = E^0 + 2 \times \frac{0.0592}{2} \log a_{H^+}$$

$$E = E^0 + 0.0592 \log a_{H^+}$$

$$E_{Q,QH_2} = E^0_{Q,QH_2} - 0.0592\ \text{pH}$$

Quinhsydrone electrode can thus be used to measure pH of a solution. Due to instability of quinone in strong alkaline medium, this hydrogen ion indicator electrode is suitable only upto a pH of 8.

Advantages : The quinhydrone electrode is simple to set up and needs no removal of air. The reversibility equilibrium is achieved faster than hydrogen gas electrode thereby allowing a quicker measurement. pH values of solutions containing reducible substances like Cu^{2+}, Cd^{2+}, unsaturated acids, NO_3^- etc., and catalytic poisons can be measured using quinhydrone electrode, where hydrogen electrode can not be used.

Limitations : The electrode can not be used at pH values greater than 8. Even this electrode fails in the presence of strong oxidizing and reducing agents.

10.13 Cell Potential and Gibbs Free Energy Change

From thermodynamics, for a process carried out at constant temperature and pressure conditions we have,

$$\Delta G = W_{max}$$

where W_{max} is the maximum amount of work that can be done.

In a galvanic cell all the chemical energy is converted into electrical energy. Hence the maximum work becomes equal to the electrical work done by the cell.

$$W_{max} = W_{electrical}$$

The electrical work is given by the product of *emf* (volts) and the total change (nF coulombs)

Electrical work (Joules) = Total charge (coulombs) × *emf* (volts)

$$W_{electrical} = -nFE_{cell}$$

The negative sign indicates that the electrical work is done by the system.

Hence $\Delta G = -nFE_{cell}$ .....(10.49)

For reactions in which the reactants and products are in their standard states.

$$\Delta G^o = nFE^o_{cell} \quad .....(10.50)$$

But the standard free energy change for a reaction is related to the equilibrium constant as follows.

$$\Delta G^o = -RT \ln K$$

where K is the equilibrium constant comparing both the expression of standard free energy.

$$\boxed{nFE^{\circ}_{cell} = \frac{RT}{nF} \ln K} \qquad .....(13.51)$$

The Condition of spontaneity in terms of ΔG°, E°cell and K are summarized as in Table 10.2.

Table 10.2 Conditions of spontaneity.

ΔG^{o}	K	E^{o}_{cell}	Reaction under standard state conditions
Negative	> 1	Positive	Spontaneous
0	= 1	0	At Equilibrium
Positive	< 1	Negative	Non-spontaneous

10.14 Secondary Standard Electrodes (Reference Electrodes)

Although all the electrode potentials are measured relative to hydrogen electrode, whose potential is arbitrarily taken as zero, it is not always convenient to use this gas electrode in day to day potentiometric measurements. The main difficulties with this **primary reference electrode** is;

(a) Inconvenience in handling gases

(b) Maintenance of accurate pressures throughout the measurements

(c) Availability of sufficiently pure hydrogen gas.

(d) Necessity to platinize the platinum electrode with a solution of chloro platinic acid quite often.

For these reasons some **secondary reference electrodes** like silver chloride electrode and calomel electrode whose standard potentials are very accurately determined against a hydrogen electrode are widely used.

10.14.1 The Calomel Electrode

Calomel electrode is particularly very simple to construct, free from surface sensitivity and accurate to use even in a very normal laboratory.

A compact form of calomel electrode, as supplied by many commercial manufacturers is shown in Fig. 10.23.

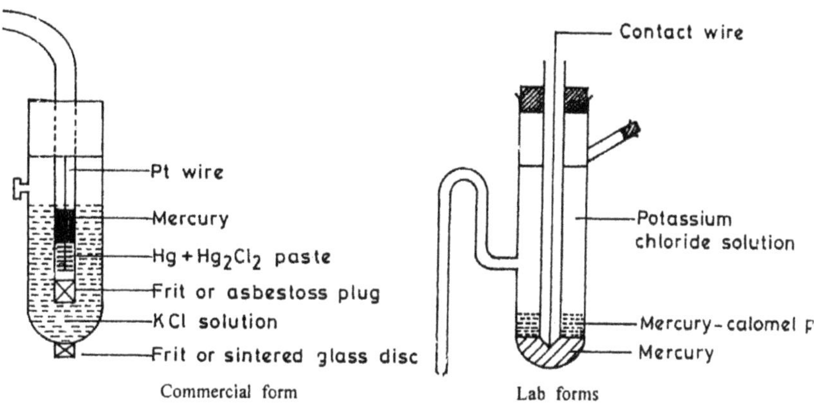

Fig. 10.23 The calomel electrode.

The calomel electrode consists of an inner glass tube and an outer jacket. In the inner glass tube a platinum wire is dipped into mercury which rests on a paste of mercurous chloride, Hg_2Cl_2 (commercially known as calomel) and mercury. This paste is in contact with KCl present in the outer jacket, through the glass frit plug fixed at the bottom of inner glass tube. The calomel electrode comes in contact with the experimental solution through a frit arranged to the outer jacket. The potential of this electrode depends on the concentration of KCl taken in the outer jacket. Some of the most popularly used concentrations of KCl and corresponding single electrode (reduction) potentials on the hydrogen scale at 25°C are given below;

0.1M KCl | Hg_2Cl_2 (s) | Hg, Pt	0.3338V
1.0M KCl | Hg_2Cl_2 (s) | Hg, Pt	0.2800V
Saturated KCl | Hg_2Cl_2 (s) | Hg, Pt	0.2415V

The corresponding electrode reaction is;

$$\tfrac{1}{2}\, Hg_2Cl_2 + 2e^- \longrightarrow 2Hg + 2Cl^-$$

and corresponding Nernst's expression is;

$$E_{Hg_2Cl_2|Cl^-} = E^0_{Hg_2Cl_2|Cl^-} - \frac{2.303}{F} \log a^2_{Cl^-}$$

10.15 Applications of Electrode Potentials

(i) **Determination of Ionic activities :** From the knowledge of standard cell emf and standard electrode potentials, it is possible to calculate the ionic activities of ions reversible to the electrode. Suppose the standard emf of the cell Hg(s), Hg_2Cl_2(s) | KCl (satd) || Cu^{2+} (a=x) | Cu is 0.036V.

$$E_{cell} = E_{right} - E_{left}$$

$$E_{cell} = E_{Cu^{2+}|Cu} - E_{calomel}$$

$$E_{Cu^{2+}|Cu} = E_{cell} - E_{calomel}$$

$$= 0.036 \quad + 0.242$$

$$E_{Cu^{2+}|Cu} = 0.278 \text{ V}$$

But
$$E_{Cu^{2+}|Cu} = E^0_{Cu^{2+}|Cu} - \frac{0.0592}{2} \log \frac{1}{a_{Cu^{2+}}}$$

$$0.278 = 0.337 + 0.0296 \log a_{Cu^{2+}}$$

or
$$\log a_{Cu^{2+}} = \frac{0.278 - 0.337}{0.0296}$$

$$a_{Cu^{2+}} = 0.01$$

Activity of Cu solution = 0.01

(ii) **Potentiometric titrations :** Potentiometric titrations is a very interesting application of electrode potentials. They involve a study of variation of *emf* with the volume of titrant added. Here we discuss the precipitation titration and acid base titration.

(a) **Precipitation titrations :** Consider the titration of Ag^+ ions with KCl. On addition of any chloride salt the silver ions fall out of the aqueous solution forming an insoluble precipitate of silver chloride. Hence we study the variation of silver electrode potential with the change in concentration of silver ions.

$$E_{Ag|Ag^+} = E^0_{Ag|Ag^+} - \frac{0.0592}{2} \log a_{Ag^+}$$

As the $E_{Ag|Ag^+}$ varies in a logarithmic scale of the concentration of silver ions, initially even for a hundred fold variation of silver ion concentration the *emf* variation is only 0.118V. With the approach of neutralization point, the E_{Ag^+} gets smaller and smaller, while the *emf*, a log value of Ag^+, increases rapidly.

Thus we notice a sigmoid curve with the steep portion indicating the neutralization point.

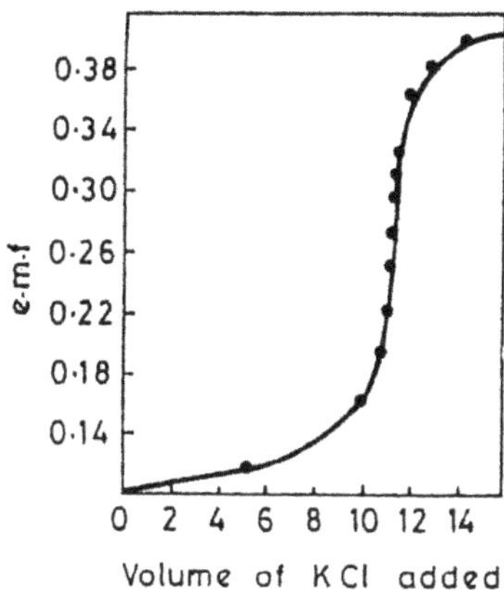

Fig. 10.24 Potentiometric titration of Ag^+ ions with KCl.

The Ag | Ag^+ electrode which is reversible to silver ion concentration is termed as the **indicator electrode**.

(b) **Acid-base titrations :** In case of acid - alkali titration a hydrogen electrode may be used as an indicator electrode. The suitable cell may be represented as

$$(Pt), Hg, Hg_2Cl_2 (s) \mid KCl (satd) \parallel H^+ \mid H_2 (1 atm), (Pt)$$

The emf of indicator electrode is given as

$$E_{H_2|H^+} = E^0_{H_2|H^+} + 0.0595 \log[H^+]$$

Since the standard emf of hydrogen electrode is zero, we have;

$$E_{H_2|H^+} = -0.0595 \log[H^+]$$

$$E_{H_2|H^+} = -0.0595 \text{ pH}$$

From the above expression we draw the conclusion that the emf decreases with decrease in the concentration of hydrogen ions or increase in the pH of the solution. On plotting the electrode potential or cell potential against the volume of titrant added, a sigmoid curve is obtained, Fig.10.25. The volume corresponding to zero emf gives the titre value of the acid solution.

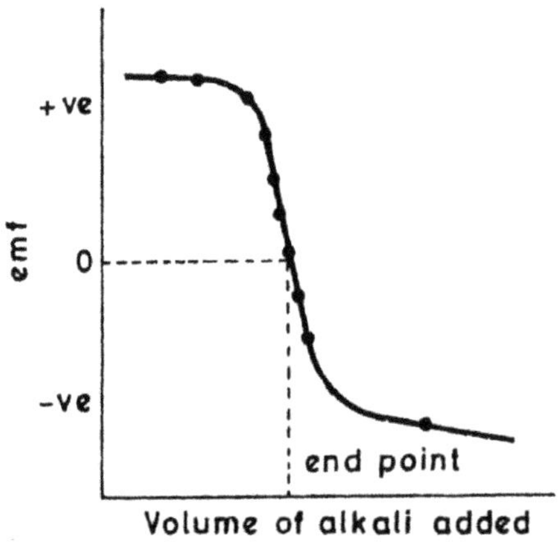

Fig. 10.25 Potentiometric acid-alkali titration curve.

In the laboratory titration generally the quinhydrone electrode is used in place of SHE.

(iii) Determination of pH : The pH of a solution could be determined by employing any electrode reversible to H^+ ions. The hydrogen electrode and quinhydrone electrode discussed above and glass electrode are popularly employed for the estimation of pH of a solution. The hydrogen indicator electrode could be coupled either to a standard hydrogen electrode (SHE) or to a standard calomel electrode (SCE). The cell diagram, when SHE is used as reference electrode, is given by

$$H_2 \text{ (1 atm)} \mid H^+ (a{=}1) \parallel H^+ (a{=}x) \mid H_2 \text{ (1 atm)}$$

The emf of the indicator electrode is given by

$$E_{H_2|H^+} = -\frac{RT}{F} \ln \frac{a_{H^+}}{p_{H_2}}$$

When $p_{H_2} = 1$ atm and T = 298 K

Since by definition pH = –log a_{H^+}, the above expression in terms of pH is

$$E_{H_2|H^+} = 0.0592 \text{ pH}$$

Because the electrode potential of the reference SHE is zero, for the present cell diagram

The cell diagram when SCE is used as reference electrode is given by

$$H_2 \mid H^+ (a=x) \parallel \text{Calomel(saturated)}$$

For this arrangement of room temperature

$$E_{H_2|H^+} = E_{\text{cell}} - E_{\text{SCE}} = 0.05916 \text{ pH}$$

(iv) **Determination of standard free energy change and spontaneity of a cell reaction :** The standard free energy change of a reversible cell given by the expression

$$\Delta G^o = -nFE^o_{cell}$$

The standard emf of the cell, E^o_{cell}, can be obtained from a knowledge of E_{cell} and activity of reduced and oxidized species in a cell reaction. E^o_{cell} could also be obtained by proper addition of standard reduction or oxidation potentials cxtracted from the data given in Table 10.1. Then the spontaneity of a cell reaction is established adopting ;the rules given in Table 10.2.

(v) **Determination of equilibrium constant of a reversible chemical reaction :** Suppose we are interested to find the equilibrium constant of the oxidation of Cu^+ by Ce^{4+}

$$Cu^+ + Ce^{4+} \rightleftharpoons Cu^{2+} + Ce^{3+}$$

The cell diagram representing this cell reaction

$$Pt \mid Cu^+, Cu^{2+} \parallel Ce^{4+}, Ce^{3+} \mid Pt$$

The standard oxidation potential of the left hand (oxidation) electrode, i.e., Pt | Cu^+, Cu^{2+}; is –0.153V (Table 10.2) where as the standard oxidation potential of the right hand (reduction), electrode, i.e., Pt | Ce^{4+}, Ce^{3+}; electrode is –1.61V (Table 10.3)

$$E^o_{cell} = (-0.153) - (-1.61)$$

$$E^o_{Cell} = -0.153 + 1.61 = 1.457\text{V}$$

Since $$\Delta G^o = -nFE^o_{\text{cell}} = -RT \ln K$$

$$E^o_{\text{cell}} = \frac{RT}{nF} \ln K$$

Therefore $(RT/nF) \ln K = 1.457$ V

At room temperature

$$\frac{0.05916}{1} \log K = 1.457$$

$$\log K = \frac{1.457}{0.05916} = 24.63$$

Equilibrium constant

$$K = \text{antilog } 24.63 = 4.25 \times 10^{24}$$

Problem 38.

Verify whether the following cell, as diagrammed, can function as a voltaic cell (Use reduction potentials given in Table 10.1, where ever necessary).

$$Ag \mid Ag^{+}(aq) \parallel Cu^{+}(aq) \mid Cu$$

As per the convention the cell reaction involves oxidation of silver at the left hand electrode (anode) and reduction of copper ions at the right hand electrode (cathode).

Oxidation	$Ag \longrightarrow Ag^{+}(aq) + e^{-}$		$E^{o}_{oxid} = -0.799V$
Reduction	$Cu^{+}(aq) + e^{-} \longrightarrow Cu$		$E^{o}_{redn} = 0.52$ V
Net	$Cu^{+} + Ag \longrightarrow Ag^{+} + Cu$		$E^{o}_{cell} = -0.279V$

Since the above cell diagram gives a negative emf, it is not a spontaneous reaction and the cell as diagrammed cannot function as a voltaic cell.

Problem 39.

Find the *emf* of the cell represented by the cell diagram $Zn \mid Zn^{2+} \mid H^{+}(1\ M) \mid H_2\,(1atm) \mid (Pt)$.

Given that the oxidation potential of $Zn \mid Zn^{2+}$ is 0.7618V

Emf of the cell

$$E_{cell} = \underset{(Oxid)}{E_{Zn|Zn^{2+}}} - \underset{(Oxid)}{E_{H_2|H^{+}}}$$

The hydrogen gas electrode is a standard hydrogen electrode, hence has a zero electrode potential.

Therefore $E_{cell} = 0.7618 - 0 = 0.7618V$

Problem 40.

The standard oxidation potential of Cd | Cd^{2+} electrode is 0.403V. Calculate the Cd^{2+} concentration if the electrode potential is 0.5V at 25°C.

$$Cd \longrightarrow Cd^{2+} + 2e$$

$$E = E^0 - \frac{RT}{nF} \ln \frac{[Cd^2]}{[Cd]}$$

$$0.5 = 0.403 - \frac{2.303 \times 8.314 \times 298}{2 \times 96500} \log [Cd^{2+}]$$

$$0.097 = -\frac{0.05916}{2} \log [Cd^{2+}]$$

$$= -0.02958 \log [Cd^{2+}]$$

$$\log [Cd^{2+}] = -\frac{0.097}{0.02958} = -3.279$$

$$[Cd^{2+}] = 5.26 \times x\ 10^{-4}$$

Therefore concentration of Cd^{2+} = 5.26 x 10^{-1} M

Problem 41.

Calculate the *emf* of the cell at 25°C

$$(Pt) \mid H_2 \ (1\ atm) \mid HCl \ (0.1\ M) \mid AgCl, Ag$$

Given the standard *emf* of the cell is 0.223V.

Cell reaction is

Oxidation	½ H_2	$\longrightarrow$	$H^+ + \bar{e}$
Reduction	$AgCl + e^-$	$\longrightarrow$	$Cl^- + Ag$
Cell reaction	$AgCl$ + ½ H_2	$\longrightarrow$	$H^+ + Cl^- + Ag$

Using the Nernst equation

$$E^o_{cell} = E^o_{cell} - \frac{RT}{F} \ln \frac{a_{HCl}}{a_{AgCl} \cdot p_{H_2}^{1/2}}$$

Since $a_{AgCl} = 1$, and $p_{H_2} = 1$ atm, we have;

$$E^{\circ}_{\text{cell}} = 0.223 - \frac{0.05916}{1} \ln \frac{0.1}{1}$$

$$= 0.223 + 0.0592$$

$$E^{\circ}_{\text{cell}} = 0.282\text{V}$$

Problem 42.

The Poggendorff method was used for measuring the *emf* of an electrochemical cell at 25°C. With the Weston standard cell (*emf* = 1.0182 V) the circuit balance point was obtained at 60cm while with the cell under examination it was 40cm. Calculate the emf of the cell.

$$\frac{\text{Balance length with std. cell}}{\text{Balance length with unknown cell}} = \frac{emf \text{ of Weston cell}}{emf \text{ of unknown cell}}$$

$$\frac{60}{40} = \frac{1.10182}{emf \text{ of unknown cell}}$$

emf of unknown cell

$$= \frac{1.10182 \times 40}{60} = 0.679 \text{ V}$$

Problem 43.

Calculate the emf of the following cell

$$\text{Fe} \mid \text{FeCl}_2\,(0.1\text{ M}) \mid \text{Cl}_2\,(1.5\text{ atm}) \mid \text{Pt}$$

Given the standard emf of the cell is 1.8005V

The cell reaction is $Fe + Cl_2 \longrightarrow Fe^{2+} + 2Cl^-$

or $\quad Fe + Cl_2 \longrightarrow FeCl_2$

(1.5atm) 0.1M

Using Nernst equation

$$E_{\text{cell}} = E^0 - \frac{RT}{nF} \ln \frac{a_{\text{FeCl}_2}}{p_{\text{Cl}_2}}$$

$$E = 1.8005 - \frac{0.05916}{2} \log \frac{0.1}{1.5}$$

$$= 1.8005 + 0.0348$$

$$E_{\text{cells}} = 1.8353 \text{ V}$$

Problem 44.

Arrange the following species in order of their increasing oxidizing capability, MnO_4^-, Li^+, Zn^{2+}, Ag^+, Co^{3+}, Ce^{4+}. Use the data given in Table 10.4.

According to data given in Table 10.4.

$$MnO_4^- (aq) + 8H^+ (aq) + 3e^- \longrightarrow Mn^{2+} + 4H_2O \quad E^o = 1.51V$$

$$Li^+(aq) + e^- \longrightarrow Li(s) \quad E^o = -3.05V$$

$$Zn^{2+}(aq) + 2e^- \longrightarrow Zn(s) \quad E^o = -0.76\ V$$

$$Ag^+(aq) + e^- \longrightarrow Ag(s) \quad E^o = 0.8V$$

$$Co^{3+}(aq) + e^- \longrightarrow Co^{2+}(aq) \quad E^o = 1.82V$$

$$Ce^{4+}(aq) + e^- \longrightarrow Ce^{3+}(aq) \quad E^o = 1.61V$$

Higher the standard reduction potential higher is the oxidizing capability. Therefore the order of oxidizing strengths are;

$$Li^+ < Zn^{2+} < Ag^+ < MnO_4^- < Ce^{4+} < Co^{3+}$$

Problem 45.

Predict what happens if bromine water is added to NaF, NaI and NaCl

Bromine water is a weak oxidizing agent

$$Br_2(l) + 2e^- \longrightarrow 2Br^-(aq) \quad E^o = 1.07V$$

$$Na^+(aq) + e^- \longrightarrow Na(s) \quad E^o = 2.71V$$

$$F_2(g, 1atm) + 2e^- \longrightarrow 2F^-(aq) \quad E^o = 2.87V$$

$$Cl_2(g, 1atm) + 2e^- \longrightarrow 2Cl^-(aq) \quad E^o = 1.36V$$

$$I_2(s) + 2e^- \longrightarrow 2I^-(aq) \quad E^o = 0.53V$$

Since the reduction potential $Br_2 \mid Br^-$ couple is less than the reduction potential of $F_2 \mid F^-$ and $Cl_2 \mid Cl^-$ couples, Br_2 cannot oxidize the F^- and Cl^- ions. Na^+ is already in the highest oxidation state and does not participate in any redox reaction with Br_2.

Hence Br_2 oxidizes I^- to give I_2 according to following reaction

$$Br_2 + 2I^- \rightleftharpoons 2Br^- + I_2$$

$$E_{Cell} = 1.07 - 0.53 = 0.54V$$

The positive E_{cell} value indicates that the process is spontaneous.

Problem 46.

The *emf* of the cell

$$Ag(s) \mid AgI(s) \mid I^-(aq, 1M) \parallel Cl^-(aq,1M) \mid Cl_2\ (g,\ p_{Cl_2} = 1.5\ atm)$$

is 1.5154 V at 25°C. The reduction potential of Ag | AgI | I^-(1 M) electrode is –0.1522V. Find the standard electrode potential of chlorine electrode.

Emf of the cell

$$E_{cell} = E_{Cl_2|Cl^-} - E_{Ag|AgI|I^-(1M)}$$

$$1.5154 = E_{Cl_2|Cl^-} - (-0.1522)$$

$$E_{Cl_2|Cl^-} = 1.5154 - 0.1522$$

$$= 1.358V$$

Chlorine electrode being cathode, the reduction reaction is

$$\frac{1}{2}Cl_2 + 2e^- \rightarrow Cl^-$$

$$E_{Cl_2|Cl^-} = E^0_{Cl_2|Cl^-} - \frac{0.05916}{1}\log\frac{[Cl^-]}{(p_{Cl_2})^{1/2}}$$

or

$$E^0_{Cl_2|Cl^-} = E_{Cl_2|Cl^-} + 0.05916\log\frac{[Cl^-]}{(p_{Cl_2})^{1/2}}$$

Therefore,

$$E^0_{Cl_2|Cl^-} = 1.358 + 0.05916\ \log\frac{1}{1.5} = 1.3476$$

Problem 46.

The reduction potentials of tin and copper electrodes are –0.136V and 0.340V respectively. Predict the reaction between these metals and 1N sulphuric acid.

Metals react with acids liberating hydrogen.

(a) Reaction between Sn and H_2SO_4

$$Sn + H_2SO_4 \longrightarrow SnSO_4 + H_2 \uparrow$$

or

$$Sn + 2H^+ \longrightarrow Sn^{2+} + H_2 \uparrow$$

Suitable cell diagram is

$$Sn \mid Sn^{2+} \parallel H^{+}(1N) \mid H_2\ (Pt,\ 1atm)$$

$$E^{o}_{cell} = E^{0}_{Sn|Sn^{2+}} + E^{0}_{H^{+}|H_2}$$

$$= 0.136 + 0$$

$$E^{o}_{cell} = 0.136V$$

The positive potential indicates that the reactions is thermodynamically feasible.

(b) Reaction between Cu and 1N H_2SO_4

$$Cu + H_2SO_4 \rightleftharpoons CuSO_4 + H_2$$

or $$Cu + 2H^{+} \rightleftharpoons Cu^{2+} + H_2$$

The cell diagram representing this cell reaction is

$$Cu \mid Cu^{2+} \parallel H^{+}(1N) \mid H_2\ (g,\ 1\ atm) \mid Pt$$

$$E^{o}_{cell} = E^{0}_{Cu|Cu^{2+}} + E^{0}_{H^{+}|H_2}$$

$$= -0.340 + 0$$

$$E^{o}_{cell} = -0.340V$$

The negative potential indicates that the reaction is not thermodynamically feasible.

Problem 47.

Use the standard reduction potentials given in Table 10.1 to find the equilibrium constant for the reaction.

$$2Ce^{4+}(aq) + 2Cl^{-}(aq) \rightleftharpoons Cl_2(g) + 2Ce^{3+}(aq)$$

The cell diagram corresponding to this cell reaction is

$$Pt \mid Cl_2(g,\ 1\ atm) \mid Cl^{-} \parallel Ce^{4+},\ Ce^{3+} \mid Pt$$

$$E^{o}_{cell} = \underset{\text{Oxid}}{E^{0}_{Cl_2|Cl^{-}}} + \underset{\text{Redn}}{E^{0}_{Pt\,|\,Ce^{4+},Ce^{3+}}}$$

Substitute the standard electrode emf values

$$E^{o}_{cell} = -1.3595 + 1.61$$

$$E^{o}_{cell} = 0.2505$$

$$\Delta G^{o} = -nFE^{o} \quad = -RT \ln K$$

$$\log K = \frac{nF}{2.303\, RT} E^{o}$$

$$\log K = \frac{1}{0.5916} E^{0} = \frac{1}{0.05916} \times 0.2505$$

$$= 8.4686$$

So $K = 2.94 \times 10^{8}$

Numerical Problems

1. Determine the gram atomic weight of the metal, whose monovalent salt solution deposits 0.15 g of the metal on passing a current of 134 coulombs.
2. A current of 0.05 amps is passed through a solution of $FeCl_3$ for 5 hours. Find the amount of iron deposited in the coulometer, (At wt of Fe 55.85).
3. Find the volume of hydrogen liberated by passing 460 coulombs of current through water. Hydrogen is collected and measured at 2 atm pressure and 300 K.
4. A solution of HCl is electrolyzed in a cell arranged in series with a copper coulometer. If 0.05g of copper is deposited in the coulometer at the end of electrolysis find the volume of hydrogen evolved at NTP (At wt. of copper = 63.54)
5. 0.1 M solution of KCl has a resistance of 97.36 ohms at 25°C. If KCl solution is replaced, in the same conductivity cell, by NaCl solution it recorded a resistance of 117.18 ohms. Calculate the conductivity and equivalent conductivity of NaCl solution. (Given that the conductivity of 0.1M KCl is 0.012856 mho cm^{-1})
6. At 25°C the equivalent conductivities at infinite dilution of HCl, CH_3COONa and NaCl are 426.16, 91.0 and 126.45 ohm^{-1} cm^{2} equiv^{-1} respectively. If the equivalent conductivity of a given acetic acid solution is 48.15 ohm^{-1} cm^{2} equiv^{-1} at 25°C, calculate the degree of dissociation of acetic acid.

7. Adoptng graphical method prove that the following data is in accordance with Kohlrausch equation.

[NaOH) mol L^{-1}	0.1	0.05	0.01	0.005	0.001	0.0005
Λ_v S cm^2 eq^{-1}	221	227	237	240	245	248

8. Calculate Kohlrausch constant and Λ_4 of $AgNO_3$. The equivalent conductivity values of 0.1, 0.02, 0.001 N solutions are 109.1, 121.4 and 130.5 S cm^2 $equiv^{-1}$ respectively.

9. Two electrolytic cells, one containing $CuSO_4$ with Cu electrodes and the other containing Ag_2SO_4 with Pt electrodes, were connected in series and current is passed. If 6g of O_2 is liberated in the Ag_2SO_4 cell calculate the amount of material deposited in both the cells.

10. In a Hittorf experiment involving $CuSO_4$ electrolysis between Cu electrodes 0.1836 g of Cu was deposited at the anode. The electrolyte in the anodic compartment before and after electrolysis contain 0.948 g and 0.987 g of Cu respectively. Calculate the transport number of participating ions.

11. 0.54434 molal H_2SO_4 in anhydrous methanol at 25^0C was electrolyzed keeping a Ag coulometer in the series. At the end of experiment the concentration of H_2SO_4 in anodic compartment was found to be 0.53846 molal, and coulometer was found to contain 1.7096 g of Ag metal. Calculate the transference number of participating ions.

12. The speed ration of Fe^{3+} to Cl^- in 0.4 m $FeCl_3$ is 0.8182. Calculate the transference numbers of Fe^{3+} and Cl^- at the given concentration.

13. 0.08 m solution of $CuSO_4$ was electrolyzed using 360 coulombs of current. At the end of electrolysis 50 g of anodic solution contains 0.626 g of $CuSO_4$. Calculate the Hittorf's numbers of Cu^{2+} and So_4^{2-}. The atomic weights of Cu, Ag, S and O are 63.5, 108, 32 and 16 respectively.

14. A solution of $AgNO_3$ is electrolyzed between Ag electrodes. 50 g of anodic solution before and after electrolysis contains 0.0901 g and 0.1563 g of $AgNO_3$ respectively. If 0.0792 g of silver is deposited in the Ag coulometer connected in series, calculate the transport numbers of Ag^+ and NO_3^- Atomic weights of Ag, N, O are 108, 14,16 respectively.

15. In problem 14, when Ag electrodes are replaced by Pt electrodes and electrolysis was performed keeping same initial concentration and current rating, 50 g of anode solution at the end of electrolysis has 0.0242 g of $AgNO_3$. Prove that the Hittorf numbers of participating ions are unaltered.

16. A 0.1 M KCl (specific conductance = 0.01321 mho cm^{-1}) recorded a conductance of 0.014 mhos. When the same conductivity cell is dipped in a 0.1 m chloroacetic acid solution it recorded a conductance of 5.84×10^{-3} mhos. If the ion conductance values of $ClCH_2COO^-$ and H^+ are 39.8 and 349.8 mho cm^2 $equiv^{-1}$, find its degree of dissociation, and dissociation constant.

17. The ion conductivities of Ca^{2+} and SO_4^{2-} are 59.50 and 80.02 respectively, the specific conductance of saturated solution of calcium sulphate at 25° is 1.159×10^{-5} S cm^{-1} and that of water is 1.60×10^{-6} S cm^{-1}. Calculate the solubility and solubility product of $CaSO_4$ at 298 K.

18. Predict whether the following reactions can occur spontaneously in aqueous solution at 25°C. The initial concentration of all aquated species are kept at 1.0 M.

(a) $K (s) + Ce^{4+} (aq) \longrightarrow Ce^{3+} (aq) + K^+ (aq)$

(b) $Ca^{2+} (aq) + Fe (s) \longrightarrow Fe^{2+} (aq) + Ca (s)$

(c) $Ni^{2+} (aq) + Cu (s) \longrightarrow Cu^{2+} (aq) + Ni (s)$

(d) $Cd^{2+} (aq) + Pb (s) \longrightarrow Cd (s) + Pb^{2+} (aq)$

(e) $MnO_4^-(aq) + I^-(aq) \longrightarrow Mn^{2+}(aq) + I_2(s)$

19. Predict whether Tl^+ can oxidize I^- to I_2 under standard conditions.

$$\epsilon^o = -0.36$$

20. Predict which metal among the following liberate hydrogen when dipped in 1N sulphuric acid.

(a) Nickel (b) Silver (c) Calcium,

(d) Zinc (e) Magnesium

21. For each of the following cell diagrams write the electrode reactions, cell reactions and corresponding Nernst equations.

$Hg \mid Hg_2Cl_2 \mid Cl^- \parallel Ag^+ \mid Ag$

22. The Ca | Ca^{2+} has a standard reduction potential of –2.76V.
    (a) Write the corresponding electrode reaction.
    (b) Write the Nernst expression for this system.
    (c) Does this half cell reaction have a greater or lesser tendency to occur under the standard conditions than $H_2 | 2H^+$ system.
    (d) Which is a stronger reducing agent under the standard conditions, Ca^{2+} or H^+ ?

23. If I_2, I^-, Fe^{3+} and Fe^{2+} are brought together, What spontaneous chemical reaction can occur and why.

24. Verify whether the following reactions occur spontaneously
    (a) $Cu + 2Ag^+ \longrightarrow Cu^{2+} + 2Ag$
    (b) $PbO_2(s) + 4H^+ aq + 2F^-(aq) \longrightarrow Pb^{2+}(aq) + F_2(g) + 2H_2O$
    (c) $2Cr^{3+}(aq) + 3Zn(s) \longrightarrow 2Cr(s) + 3Zn^{2+}(aq)$
    (d) $Fe^{2+}(aq) + Ni(s) \longrightarrow Ni^{2+}(aq) + Fe(s)$
    (e) $Cu (s) + 2H^+(aq) \longrightarrow Cu^{2+}(aq) + H_2(1atm)$

25. Calculate the standard cell potential for the following reactions.

    $Fe(s) + Br_2(aq) \longrightarrow Fe^{2+}(aq) + 2Br^-(aq)$

    $Sn^{2+}(aq) + Ni(s) \longrightarrow Sn(s) + Ni^{2+}(aq)$

    $Ce^{4+}(aq) + Fe^{2+}(aq) \longrightarrow Ce^{3+}(aq) + Fe^{3+}(aq)$

26. Draw the cell diagram that could use the following redox reaction to generate electricity;

    $Fe^{2+}(aq) + Mg(s) \longrightarrow Fe(s) + Mg^{2+}(aq)$

    $Ce^{4+}(aq) + Fe^{2+}(aq) \longrightarrow Ce^{3+}(aq) + Fe^{3+}(aq)$

    $Mg(s) + I_2 \longrightarrow Mg^{2+} + 2I^-$

27. Calculate the *emf* of the following cells at 25°C under given experimental conditions.

    $Zn(s) | Zn^{2+} (0.40M) \,||\, Cu^{2+} (0.02M) | Cu(s)$

    $Cd(s) | Cd^{2+}(0.0013M) \,||\, Ag^+(1.0M) | Ag(s)$

    $F_2(g, 1atm) | F^-(0.5M) \,||\, Li^+ (1M) | Li(s)$

    $Fe(s) | Fe^{2+} (0.2M) \,||\, H^+ (1.0M) | H_2 (1atm)$

28. Using the Nernst equation calculate the p^H of the following cell
    (Pt) $H_2 | H^+ || Cu^{2+} | Cu$
    whose emf was found to be 0.48V at 25°C

29. Calculate the equilibrium constants for the following cell reactions

$$Cl_2(aq) + 2Br^-(aq) \rightleftharpoons Br_2(aq) + 2Cl^-$$

$$Fe^{2+}(aq) + Ag^+(aq) \rightleftharpoons Ag(s) + Fe^{3+}(aq)$$

30. Calculate E^o_{Cell}, E_{Cell} and ΔG (in kilo joules) for the following reactions;

(a) $Ni(s) + Sn^{2+}\ (0.01M) \longrightarrow Sn(s) + Ni^{2+}(0.01M)$

(b) $2Au(s) + 3Ca^{2+}(aq,1M) \longrightarrow 2\ Au^{3+}(aq,1M) + 3Ca(s)$

[*Hint :* Here n=6 as six electrons are transferred during reaction]

(c) $2Zn(s) + 2Cr^{3+}(aq,\ 0.01M) \longrightarrow 3Zn^{2+}\ (aq,\ 0.01M) + 2Cr(s)$

31. Standard reduction potential of $Cd \mid Cd^{2+}$ and $Pb \mid Pb^{2+}$ electrodes are –0.408 volt and –0.126 volt respectively. State whether the following cell set up is feasible. Reason your answer $Pb \mid Pb^{2+} \parallel Cd^{2+} \mid Cd$

32. Write the cell reaction and calculate E at 25°C for the cell;

$$Cd \mid Cd^{2+}\ (aq,1M) \parallel Cu^{2+}\ (aq,1M) \mid Cu$$

$$Cd^{2+} \mid Cd(s) = -0.403V,\ Cu^{2+}(aq) \mid Cu = 0.337V$$

33. What is the cell reaction and what is the voltage at 25°C of the cell

$$Pb(s) \mid Pb^{2+}(a{=}1)\ \mid Ag^+(a{=}1) \mid Ag(s)$$

$$E^0_{Pb^{2+}+Pb} = -0.126V,\ \ E^0_{Ag^+|Ag} = 0.799V$$

34. Calculate the *emf* of the cell

$$Sn \mid Sn^{2+}\ (1M)\ \parallel Cu^{2+} \mid Cu$$

Given that $E^0_{Sn \rightarrow Sn^{2+}} = 0.136\,V,\ \ E^0_{Cu \rightarrow Cu^{2+}} = -\,0.345\,V$

35. The standard electrode potential of Zn is –0.76V. Determine the *emf* of the Zn electrode immersed in 0.1M $ZnSO_4$ solution at 25°C.

36. Calculate the *emf* of the cell $Cd \mid Cd^{2+} \parallel Cu^{2+} \mid Cu$ at 25°C. The standard reduction potential of cadmium and copper electrodes are -0.40 and 0.34 volts respectively.

37. The *emf* of the cell $Zn \mid Zn^{2+}\ (0.1M) \parallel Cu^{2+}(1M) \mid Cu$ is 1.1Volt at 298K. What is the *emf* of the cell

$$Cu \mid Cu^{2+}\ (1M) \parallel Zn^{2+}\ 1M \mid Zn$$

38. Calculate the *emf* of the following cell at 25°C

$Zn \mid Zn^{2+}$ (0.1M) || Cu^{2+} (0.01M) | Cu

Given that standard reduction potential for zinc and copper electrodes are –0.7623V and 0.337V respectively at 25°C

39. The standard *emf* of $Zn \mid Zn^{2+} \parallel Ni^{2+} \mid Ni$ is 0.51 volts. If the standard electrode potential of $Ni \mid Ni^{2+}$ is 0.25V, calculate standard potential of $Zn \mid Zn^{2+}$

Self-Test Questions

1. The resistance of a conductor is directly proportional to

   (A) Current passed (B) Area of cross section of conductor

   (C) Length of the conductor (D) Equivalent conductance

2. Sodium chloride is a poor conductor of electricity but transforms into a good conductor in molten state because

   (A) The covalent bonds are broken in molten state

   (B) The degree of dissociation increases

   (C) It form liquid crystals

   (D) The ions get mobility

3. The transport number of an ion represents

   (A) The number of ions of its kind (B) Current carried by an ion

   (C) The movement of the ion (D) Charge on the ion

4. If l is the distance between the electrodes of a cross sectional area a, then cell constant is given by the ratio

   (A) l/α (B) α/l (C) $\alpha \times l$ (D) $\alpha - l$

5. The resistance of a conductor is 5×10^{-2} ohms. Conductance is the equal to

   (A) 200 mhos (B) 20 mhos (C) 500 mhos (D) 50 mhos

6. The conductivity of 0.1 N KCl is 0.01120 mho cm^{-1}. If the cell constant is equal to 0.5 cm^{-1}, the conductance is equal to

   (A) 2.24×10^{-2} eq^{-1} (B) 5.6×10^{-3} mhos

   (C) 224 mhos (D) 560 mhos

7. The units of equivalent conductivity is

   (A) ohm^{-2} cm^{-2} eq^{-1} (B) ohm^{-1} cm^{-2} eq^{-1}

   (C) Ohm^{-2} cm^{-2} eq^{-1} (D) Ohm^{-1} cm^{2} eq^{-1}

8. The transport number of sodium ion in NaCl is 0.486. Ionic conductivity chloride ions is given by the expression

   (A) $0.486 \times \Lambda_4$ (B) $0.514 \times \Lambda_4$

   (C) $\alpha = \frac{0.486}{\Lambda_\infty}$ (D) $\frac{0.514}{\Lambda_\infty}$

9. In the electrolysis an electrolyte A^+B^- the anion moves three times faster than the cation. The number of ions discharged at each electrode are

   (A) Equal (B) More at cathode

   (C) More on anode (D) None of the them

10. The ionic conductivity of cation and anion of a salt are 30 and 70 mho cm^2 respectively. The transport number of cation is

    (A) 2.33 (B) 0.7 (C) 0.47 (D) 0.3

11. With increase in dilution conductivity of a solution,

    (A) Increases (B) Decreases

    (C) Does not change (D) Increase first and the decrease

12. Which of the following statements is correct

    (A) Transport number of cation is equal to the equivalent conductivity of cation

    (B) Transport number of cation is equal to transport number of anion.

    (C) Transport number of cation plus transport number of anion is equal to Faraday number

    (D) Transport number of cation plus transport number of anion is equal to unity

13. According to Kohlrausch law

    (A) $\alpha = \frac{\Lambda_v}{\Lambda_\infty}$ (B) $t_+ = \frac{\Lambda_v}{\Lambda_\infty}$

    (C) $\Lambda_+ + \Lambda_- = \Lambda_4$ (D) $\Lambda_v = KV$

14. Equivalent conductivity of an electrolytic solution

    (A) Increases with dilution

    (B) Decreases with dilution

    (C) Independent of dilution

    (D) None of the above

15. Cell constant of a conductometric cell

    (A) Increases with dilution

    (B) Decrease with dilution

    (C) Depends on the nature of electrolyte

    (D) Independent of the nature of electrolyte.

16. Equivalent conductivity of any electrolyte is algebraic sum of ________ conductivities.

17. Transport number is defined as the fraction of ______ carried by participating ions.

18. Degree of dissociation = $\frac{\Lambda_v}{.....}$

19. Specific conductance = Cell constant x ________________

20. In the electrolysis employing unattackable electrodes in a Hittorf apparatus the concentration of __________ compartment falls.

21. In the electrolysis employing attackable electrodes in a Hittorf apparatus the concentration of __________ compartment raises.

22. If the standard potential of Ca | Ca^{2+}, Ni | Ni^{2+} and Au | Au^{2+} couples are 2.87, 0.25 and -1.5V respectively, then

    (A) Ni can reduce Ca^{2+} and Au^{3+} (B) Ca and Ag can reduce Ni^{2+}

    (C) Ni^{2+} can oxidize Au and Cu (D) Au^{3+} can oxidize Ca and Ni

23. The standard oxidation potentials at 25°C of Ag, Cu and Zn electrodes are -0.8, -0.34 and 0.76V respectively. Which one of the following combinations of the electrodes gives a cell with a spontaneous reaction ?

    (A) Silver anode & Copper cathode

    (B) Copper anode & Zinc cathode

    (C) Silver anode & Zinc cathode

    (D) Zinc anode & Silver cathode

24. The standard reduction potentials of Mg, Zn, H_2, Cu and Ag electrodes are -2.4, -0.76, 0.00, 0.34, 0.80V respectively. If copper is made anode, which electrode should it be coupled to give a spontaneous cell reaction.

    (A) H_2 (B) Mg

    (C) Ag (D) Zn

25. In a galvanic cell

    (A) Chemical energy is converted into electrical energy

    (B) Electrical energy is converted into chemical energy

    (C) Chemical energy is converted into heat energy

    (D) Electrical energy is converted into heat energy

26. The cell reaction $Zn + Cu^{2+} \longrightarrow Zn^{2+} + Cu$ is correctly represented by

    (A) $Cu \mid Cu^{2+} \parallel Zn \mid Zn^{2+}$

    (B) $Cu \mid Cu^{2+} \parallel Zn^{2+} \mid Z^{4}$

    (C) $Zn \mid Zn^{2+} \parallel Cu^{2+} \mid Cu$

    (D) $Zn \mid Zn^{2+} \parallel Cu \mid Cu^{2+}$

27. If the oxidation potential of Zn metal is 0.76V, the E_{cell} for the redox reaction

    $Zn + 2H(aq,1M) \longrightarrow Zn^{2+} + H_2(1atm)$ is

    (A) 0.76 (B) –0.76

    (C) +0.38 (D) -0.38

Short and Essay Type Questions

1. Describe the measurement of "Transport numbers" by Hittorf's method.
2. State and explain Kohlrausch' law. Give its applications.
3. Write an account of acid base conductometric titrations.
4. State Kohlrausch law. Explain how it can be used to determine the Λ_4 of acetic acid.

5. What is transport number? Describe Hittorf method of determining transport number of silver ions.

6. What is the principle underlying conductometric titrations. Discuss titration curve obtained for titration of a mixture of HCl and CH_3COOH with NaOH. What are the advantage of conductometric titrations.

7. Define, the terms specific conductance and equivalent conductance. Explain clearly why on dilution the specific conductance decreases but equivalent conductance increases.

8. Show that the fall in concentration around an electrode in an electrolytic cell is proportional to the velocity of the ion moving away from it.

9. During conductometric titration of a strong acid against strong base, the conductivity decreases up to the neutralization point and then increases sharply. Explain why this happens

10. Explain the terms cell constant, equivalent conductivity. Give their units. Draw a graph showing the variation of equivalent conductance with concentration for both weak and strong electrolytes.

11. How does Kohlrausch's law of independent migration of ions help in determining the equivalent conductivity of a weak electrolyte like acetic acid at infinite dilution.

12. (a) Define the terms : (i) Resistivity (ii) Conductivity and (iii) Equivalent conductivity.

    (b) Describe a method for determining the equivalent conductivity of an electrolytic solution.

    (c) What is the effect of dilution on conductivity.

13. What is meant by cell constant of a conductivity cell? How is it determined ? What are its units.

14. Explain the following terms

    (a) Single electrode potential

    (b) Reference electrode

    (c) Standard hydrogen electrode.

15. What is meant by pH of a solution ? How is the pH of a solution determined.

16. How is the *emf* of a cell determined from a knowledge of single electrode potentials ? Give a suitable example.

17. What is an electrochemical cell ? Explain hosw you would setup an electrochemical cell with zinc and copper electrodes ? Write the reactions at each electrode and the overall cell reaction.

18. What are redox potentials ? Give their applications?

19. Write the construction of Weston standard cadmium cell

20. What are the reversible and irreversible cells

21. Write the cell reaction at 298K of the cell

    $Zn \mid ZnCl_2\ (a=0.5) \parallel AgCl \mid Ag$

22. (a) Explain what do you understand by single electrode potential and redox potential ?

    (b) What is hydrogen electrode ? How is constructed ?

Chapter 11

Corrosion and its Prevention

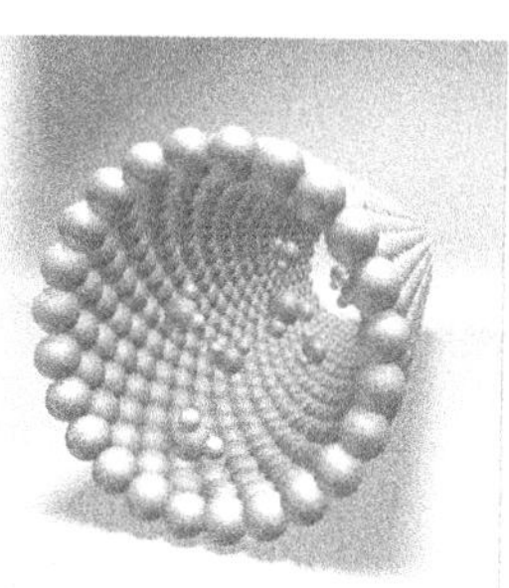

11.1 Introduction

One of the most widely accepted definitions of corrosion is that corrosion is the 'destruction of a metal (or alloy) by chemical or electrochemical reaction with its environment'. Thus corrosion involves chemical or electrochemical reaction and the use of the term is restricted to the destruction of metals or alloys. Destruction of wood, for instance, by rotting is not referred to as corrosion of wood. Physical or mechanical wearing away of a metal is not called corrosion but is called erosion. If corrosion and erosion take place together the destruction of the metal takes place very much faster.

Corrosion has also been defined as the deterioration of solids by liquid electrolyte. This definition covers the deterioration of refractory lining by molten glass, deterioration of concrete, etc. A still more comprehensive definition is that corrosion is an unintentional attack on a material by reaction with a surrounding medium. In this chapter corrosion will be considered mainly from the electro-chemical point of view.

11.1.1 Magnitude of the Problem

It is difficult to correctly assess the loss caused by corrosion. According to certain estimates, the direct loss due to corrosion in the U S A amounts to about Rs. 250 crores per annum. This does not include effects such as loss of lives, injuries to workmen, shut down of plants for repairs, etc. According to another estimate, the money spent in controlling corrosion is of the order of Rs. 350 crores in India, 2,500 crores in U K, Rs. 4000 crores in Australia and Rs. 10,000 crores in the U S A each year. In India due to its tropical climate the corrosion problem is more serious than in the colder countries.

The corrosion problem should be very seriously considered in all those cases where the structure or the equipment is meant to last a number of years, say 20 years or more. In some cases, like in some chemical equipment, the equipment is meant to last a few years only because of the changes in the process or in the demand for the product. In such cases it may not be economical to spend too much money on corrosion control or use expensive material for construction.

11.2 Theories of Corrosion

In atmospheric corrosion the metals tend to revert to the states in which they occur in nature, i.e., the ore, because that is the most stable state. Such compounds always have less free energy than the pure metal. Fundamentally this is the cause of corrosion. Several theories of corrosion have been proposed from time to time.

11.2.1 The Acid Theory

In the early nineteenth century it was demonstrated that for rusting of iron in the presence of both oxygen and water is necessary. It was also noticed that rusting takes place faster in acid solutions than in neutral or alkaline solutions. This was further confirmed by the detection of ferrous and ferric carbonates (due to carbonic acid in water) in iron rust which consists mostly of hydrated ferric oxide. This led to the acid theory of corrosion which considers acids to be the main cause. However, this is not true for all the metals. For instance, in the case of zinc the corrosion rates in alkaline and acidic solutions are comparable. Appreciable rate of corrosion in absence of carbon dioxide has also been reported in many cases.

11.2.2 Chemical Attack Theory

Direct attack of dry gases on metals does not take place to any appreciable extent at atmospheric temperatures. When it does take place, a thick film of the solid corrosion product formed on the surface protects the remaining metal from further attack. If a volatile product is formed the attack proceeds rapidly, as happens when dry chlorine attacks tin sheets, cans, etc., (which are made of low carbon steel sheet coated with tin) above, say, 120° causing escape of volatile stannic chloride. Ferric chloride being solid protects the iron surface.

The important industrial example of corrosion by direct chemical attack is the formation of scales during hot working of steel, as in the rolling of steel ingots into various sections. Attack of dry oxygen on iron does not take place at any appreciable rate below 800°, hence is not a major cause of destruction of iron in common usage.

It has been found that those metals whose oxides have lesser volume than the original metals corrode faster than those whose oxides are more voluminous. Alkali and alkaline earth metals belong to the former group. As the oxide has less volume, the deposit is porous and open space is left through which oxygen diffuses to the metal surface. Thus corrosion proceeds. The oxides of some other metal have larger volumes, therefore they cover the metal surface and prevent diffusion of oxygen. If the metal is heated then the protective oxide layer breaks, more especially if there is temperature fluctuation, and corrosion proceeds.

Addition of resistant metals to other metals or alloys makes them more corrosion-resistant for the same reason. Addition of chromium to iron causes, on exposure to oxygen, the formation of a film which is richer in chromium oxide. It is this continuous resistant film which is responsible for the heat-resisting properties of chromium steels.

11.2.3 The Electrochemical Theory

It is now accepted that most of the common cases of corrosion can be explained on the basis of the electrochemical theory. At ordinary temperatures the corrosion process involves oxidation. Very few cases of non-oxidative corrosion are met with under common conditions. Oxidation of a metal involves its loosing of electrons, thereby acquiring a positive charge.

If a piece of metal is dipped in water or a solution, it has a tendency to go into solution. Some of its atoms may loose electrons and go into solution as positively charged ions. These ions are held near the metal piece because of electrostatic attraction and the process of dissolution stops unless this balance is disturbed. Therefore, for the progress of corrosion, anodic and cathodic areas must exist in electrical contact with each other. Thus wherever corrosion is in progress, a number of tiny galvanic cells are active. In a sense metallic corrosion is opposite to electroplating.

A galvanic cell is formed when two dissimilar metals, electrically connected, dip in an electrolyte. The principle of their working can be explained by considering a Daniel cell, Fig. 11.1 which consists of zinc and copper plates dipping in solutions having zinc and copper ions, the solutions

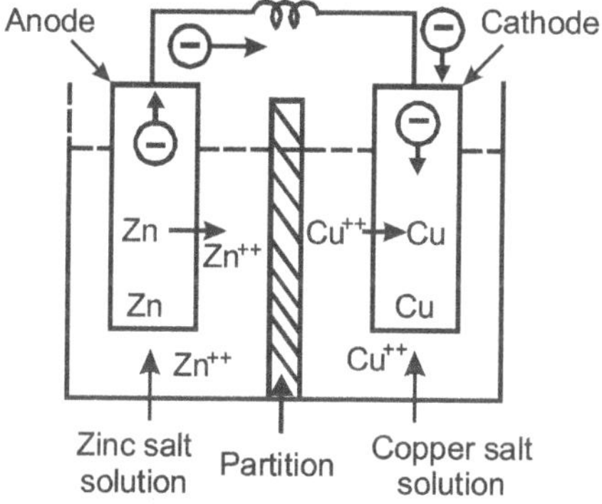

Fig. 11.1 Daniel cell.

being separated by a porous partition. The plates are connected by a metallic conductor. The emf developed is due to two separate reactions taking place at the electrodes. Zinc atoms have a tendency to go into solution as Zn^+ making two electrons available. This is an oxidation reaction:

$$Zn\ (solid) = Zn^{2+}\ (aq) + 2e$$

The electrons move along the conductor and are responsible for the reduction reaction at the copper plate.

$$Cu^{2+}\ (aq) + 2e = Cu\ (solid)$$

Thus zinc dissolves and copper is deposited at its cost. Exactly similar reactions occur in corrosion. The interest of the corrosion engineer is in preventing this kind of reaction.

In Daniel cell the zinc plate is called an *anode* because oxidation reaction, i.e., liberation of electrons, takes place at the zinc plate. Similarly, because of reduction reaction, i.e., consumption of electrons, takes place at the copper plate, the plate is called a *cathode*.

The concept of corrosion reaction can be developed on the basis of reaction occurring in a Daniel cell. If the two electrodes of the cell, instead of being kept in different solutions and connected together through an external circuit, are short circuited (connected internally) and kept in a solution containing cupric ions, there will be no change in the nature of chemical reactions. Since, in such a system the resistance is made very small, the reactions take place with maximum current and the rate of dissolution of Zn becomes much faster than in a simple Daniel cell. This is illustrated in the Fig. 11.2. The current in the system can be increased to a still higher value if an arrangement of alternate strips of zinc and copper is considered to be dipped in the Cu^{2+} solution. Such galvanic cells can be formed even if one considers a zinc sheet containing microscopic inclusions of copper kept in contact with a solution containing the same ions.

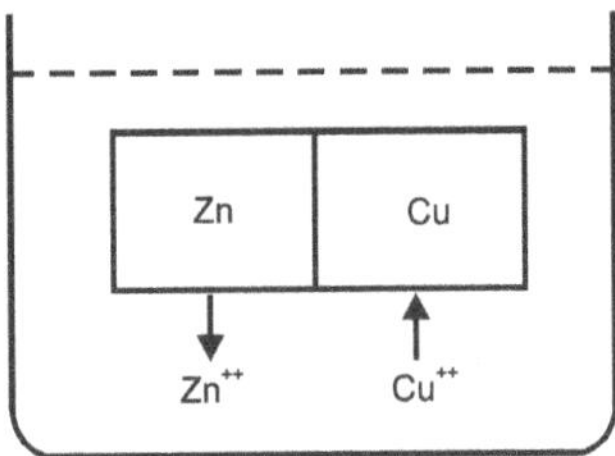

Fig. 11.2 Electrochemical reactions, when two bars of zinc and copper are welded together and immersed in cupric ion solution.

This is the most idealized case. In general, for any metal containing impurities and being in contact with a conducting solution or moisture, tiny galvanic cells are formed. The type of chemical reactions may be different in different cases. At anode there will always be the dissolution of a metal. The cathodic reaction may be the reduction of hydronium ion or oxygen.

The greater the potential difference between the two metals in actual contact, the faster will be the corrosion. Therefore, it is helpful to be able to determine the emf. The emf developed is the difference between the electrode potentials of cathode and anode. The electrode potential is a measure of the tendency of a metal to go into solution and depends upon several factors, like the metal, the concentration of the solution and the temperature. It is measured with reference to a hydrogen electrode whose potential has been taken as zero (A hydrogen electrode is prepared by bubbling hydrogen at 760 mm over platinum black plate immersed in a solution having 1 g of hydrogen ion per litre). The *normal electrode potential* of a metal is the potential between a hydrogen electrode and the metal immersed in a normal solution of its own ions (usually measured at 25°). When the elements are arranged in the order of their electrode potentials the well-known *electrochemical series* is obtained. A metal higher in the series, i.e., a more electro-positive metal, will displace the metal lower in the series, i.e., a less electro-positive metal. For example, iron will precipitate nickel and copper from their salt solutions.

The electrochemical series, though useful, does not give a basis for predicting, in a particular environment, whether a metal will or will not corrode. In actual circumstances some side reactions may take place which affect corrosion reactions. For example, copper will not corrode in acetic acid solution, but if dissolved oxygen is also present, corrosion will take place. There may be factors opposing corrosion reactions. Considering the conditions met with under many conditions of exposure, another series, known as the *galvanic series*, has been prepared. This series is of more direct help in corrosion processes than the electrochemical series. The galvanic series listing the more common metals and alloys is shown in the Table 11.1, from which the relative corrosion tendencies can be judged.

In the galvanic series chemically more active metals are given higher place. These metals easily react with oxygen, sulphur etc., hence, corrode readily. The metals at the lower end do not easily react, and therefore do not corrode so easily under ordinary atmospheric condition. When two metals are in electrical contact and an electrolyte is present, the one higher up in galvanic series corrodes and the other is protected. The greater the difference in their positions in the galvanic series the faster will be the corrosion. Metals close to each other in the galvanic series show less tendency to corrode when in contact with each other.

If to the Daniel cell just described a small electromotive force from all external source is applied in the reverse direction, the rates of copper deposition and zinc going into solution are reduced. As the applied emf is increased the reaction rates continue to decrease and when it is equal to the original emf of the cell neither current flow nor any reaction takes place, i.e., there is no corrosion. This principle is important as it provides a method for preventing corrosion. If a higher emf is applied the reverse reaction, i.e., deposition of zinc and corrosion of copper, can take place.

Table 11.1 Galvanic series (Anodic or corroded end)

Magnesium
Magnesium alloys
Zinc
Aluminum
Steel
Cast iron
Stainless steels (active)
Lead-tin solder
Lead
Tin
Muntz metal
Brasses
Copper
Bronzes
Copper nickel alloys
Stainless steel (passive)
Silver
Graphite
Platinum

There is another mechanism by which the reactions slow down. In the cell the concentration of zinc ions increases and that of cupric ions decreases as the current is withdrawn. This reduces the tendencies of zinc to ionize and copper ions to deposit. Thus gradually the cell is run out or *polarised* and the emf generated decreases.

It appears, at the first instance, from the above discussions that for a metal to corrode it must be in contact to some other metal which is cathodic to it and that a pure metal should not corrode, at all. But, the electrochemical theory also provides an explanation for the corrosion of a single metal through the formation of *concentration cell.* When two plates of the same metal dip in a salt solution of the same strength and are connected externally, no current flows, because both the plates have the same potential. If the concentrations of the solution around the plates are different there will be: the tendency for the metal around which concentration is lower, to go into solution, and metal will be deposited at the other plate. The surface exposed to a solution of lower concentration looses electrons hence becomes anode and dissolves, and the surface exposed to a solution of higher concentration becomes cathode as it receives electrons. This is clear from Fig. 11.3. Thus an emf will be produced. This shows that if a metal piece is in contact with a solution whose concentration is different at different ends of the piece, a concentration cell will be set up and corrosion will take place where the concentration of the electrolyte is lower. Thus, according to the electrochemical theory corrosion of a metal occurs either through the formation of a galvanic cell or concentration cell. The galvanic cells are formed due to inhomogeniety at the surface of metal whereas the heterogeneity of the environment at different parts of the structure causes the formation of concentration cell.

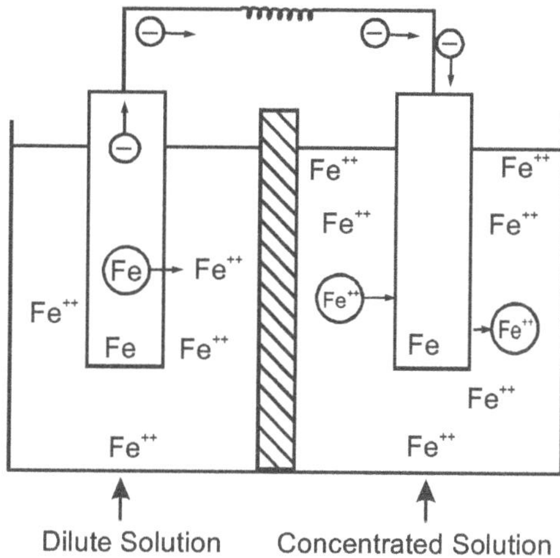

Fig. 11.3 Formation of a concentration cell.

From the foregoing, it is clear that for p rotection against corrosion all the factors that increase the resistance of the galvanic cell need to be promoted. The effect of such factors is called *polarization.* Polarization provides another approach to corrosion control. Polarization is due to several factors. The circuit resistance is one of these. As the current is withdrawn the ionic concentration around the anode increases and around the cathode it decreases. This causes concentration polarization. Yet another common cause of polarization is the formation of insoluble products on the cathode and anode surfaces. If the product forms a continuous, non-porous and adherent film the corrosion reactions may be completely stopped.

Among the metals that are used on a large scale the case of aluminium may be mentioned as a good example of protection against atmospheric corrosion. It readily reacts with oxygen of the air but a thin, non-porous film which adheres to the surface is formed. This film makes aluminium passive. Similar is the reason for the chromium and nickel plated surfaces being corrosion resistant. Passivity may be defined as a situation in which a thin film prevents a metal from corroding even in an environment in which it is unstable (i.e., it would have corroded if the protective film was not present).

11.3 Corrosion Reactions

Under most of the common conditions of corrosion, according to the electrochemical theory, a number of small galvanic cells are formed. It has just been explained that in a galvanic cell the anode dissolves and some material is liberated at the cathode. Often the formation of anodic and cathodic areas takes place in quite an unexpected manner, hence the equipment fails at unsuspected points. It is, therefore, necessary to understand the corrosion reactions somewhat clearly.

On a corroding surface, at the anode, the metal atoms give up their valence electrons to the environment and are oxidized to positive ions. The reaction taking place in iron in the presence of moisture is typical and can be represented thus :

$$Fe \rightleftharpoons Fe^{2+} + 2 \text{ electrons} \qquad .....(11.1)$$

If the ions and the electrons are not removed from the environment the forward reaction will stop and corrosion will not proceed. When corrosion is taking place the iron ions move away and come across the hydroxyl ions of the solution and form less soluble ferrous hydroxide:

$$Fe^{2+} + 2OH^{-} \rightleftharpoons Fe\,(OH)_2 \qquad .....(11.2)$$

In an oxidizing environment ferrous hydroxide is oxidized to ferric hydroxide, which is more insoluble, and appears as oxides of iron found in iron rust.

The electrons released at the anodes are conducted to the cathode and are responsible for cathode reactions which are more complicated than anode reactions. In the absence of oxygen, in acid media, the following reaction takes place:

$$2e + 2H^{+} = H_2\,(gas) \qquad .....(11.3)$$

In neutral or alkaline media and absence of oxygen, the reaction is:

$$2e + 2H_2O = H_2\,(gas) + 2OH^{-} \qquad .....(11.4)$$

As in the above reactions hydrogen gas evolves, this type of corrosion is known as *hydrogen type corrosion*.

In the presence of dissolved oxygen, in acid media, the reaction taking place is

$$4e + 4H^{+} + O_2 = 2H_2O \qquad .....(11.5)$$

In neutral or alkaline media the reaction is :

$$4e + 2H_2O + O_2 = 4OH^{-} \qquad .....(11.6)$$

It is noteworthy that electrode potentials of reactions (11.5) and (11.6) are different. In the former case it is +1.229 V and in the latter it is +0.401 V. The electrode potential of reaction (11.1) is -0.440V. This means that corrosion of iron in the presence of dissolved oxygen in acid medium will generate an emf of 1.669 V whereas in neutral or alkaline medium only 0.841 V.

Only in a few cases hydrogen peroxide is formed thus:

$$2e + 2H_2O + O_2 = H_2O_2 + 2OH \quad .....(11.7)$$

In many cases several of these reactions take place simultaneously but dominant one in a particular medium is the one mentioned above. Corrosion process involving oxygen is known as *oxygen type corrosion.*

Although theoretically it is possible to predict the rates of these reactions, in actual practice it has not been possible to predict the exact rate of corrosion reaction because there are several polarizing (i.e. opposing) reactions. For example, if an insoluble corrosion product is formed as a porous and non-adherent film on the metal surface, the polarizing effect is small. On the other hand if an insoluble product is deposited as an impervious, thin, continuous and adherent film, the corrosion reaction is completely stopped because polarization is complete. This condition is known as the *passivity* of metals. It is this process which is responsible for the corrosion resistance of some of the common metals and alloys like aluminium, tin, stainless steels, chromium, nickel, etc., on the surfaces of which oxide films are formed. If the environment is changed the passivity may disappear as happens when aluminium is used in sea water or stainless steel is in contact with hydrochloric acid solution.

The primary products of the electrochemical reactions are usually soluble substances. Sometimes insoluble products are later produced but generally they are porous and loose. Thus electrochemical reactions are more dangerous than direct chemical attack. The insoluble primary products of electrochemical reactions generally form impervious and adherent coatings and they prevent corrosion but such products are rarely formed.

To summarize, for electrochemical corrosion the three essentials are: 1. formation of anodic and cathodic areas; the anode corrodes by forming ions; at the cathode reduction takes place, 2. electrical contact between anode and cathode for the conduction of electrons, and 3. electrolyte, usually provided by the presence of moisture.

11.3.1 Formation of Anodic Areas

It is clear from the foregoing that surface or part of the equipment which becomes anodic gets dissolved or corroded. It is, therefore, necessary to know where the anodic areas are formed so as to safeguard against eventual failure. In many cases it is not easy to know the exact spots but the following illustrations will go a long way in locating the corrosion spots.

When more than one metal is used in the structure, with the help of the galvanic series it is easy to find the metal which is most Anodic. And it is this metal that will corrode, others will be saved. In an equipment using zinc and iron parts, zinc will corrode and not iron because zinc is anodic to iron. Iron will corrode in contact with copper because it is anodic to copper. The same principle applies when impurities are present. For example, impure zinc dissolves in acid faster than pure zinc because of the cathodic impurities. From the electrode potentials an idea of relative corrosion tendencies can be formed. For example, iron in contact with copper will corrode faster than in contact with lead. In drawing conclusions the importance of the polarizing effects, already discussed, should not be lost sight of.

A little more complicated is the case of a single metal being used, corrosion may take place due to the formation of concentration cells. A metal may be in contact with a solution whose concentration is not uniform, as can happen in the storage tank of a solution. A more common cause is dissolved oxygen whose concentration is not the same throughout the entire mass of the solution. This is known as *differential aeration*.

Differential aeration is quite a common cause of corrosion at many unsuspected spots. In the section on 'Corrosion Reactions' eqn. 11.5, 11.6 and 11.7 illustrate what happens at the cathode in the presence of oxygen. Of these the most common reaction is eq. 11.6 which is:

$$4e + 2H_2O + O_2 = 4OH^-$$

This shows the part played by oxygen. The cathode is the surface at which electrons are consumed. Oxygen helps consumption of electrons. Therefore, the surface which is more exposed to oxygen is cathodic to the surface less accessible to oxygen. Consequently, the surface on which oxygen concentration is lower gets corroded. Common examples of this type of corrosion are illustrated in Fig. 11.4. When there is a crack, in a metal there is lower concentration of oxygen in the crack hence it becomes anodic and corrosion proceeds in the crack increasing its size. The surface covered with dirt is also less accessible to oxygen and is anodic. In the same way there is less oxygen where the wires of a fence cross and also under a rivet head. When once corrosion starts on a metal surface at certain spots they become less open to oxygen with the result that corrosion proceeds faster at those very spots making deeper and deeper cavities. This is known as *pitting*. Pitting is easy to explain on the basis of differential aeration. The surface under the rust or scale is less accessible to oxygen, therefore, is anodic and corrodes. As corrosion proceeds, a cavity or pit is created and as it becomes deeper its bottom becomes still less accessible to oxygen which makes it still more anodic.

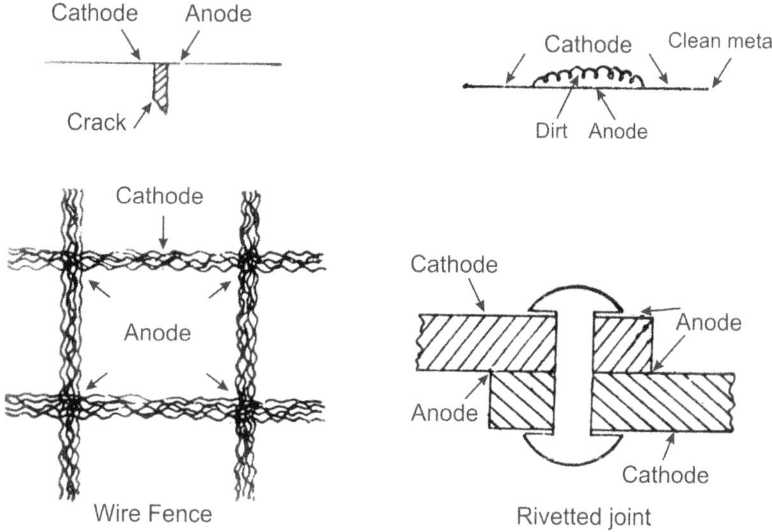

Fig. 11.4 Illustration of corrosion by differential aeration.

Corrosion by differential aeration has also been explained as being due to the initial formation of a thin metal-oxide film on the surface. A galvanic cell is now set up between the metal-oxide surface, which is anodic, and the metal surface. Therefore, the part which has higher oxygen concentration being cathodic is saved. The phenomenon has also been attributed to the formation of a protective hydroxide layer at the cathode. It seems thal corrosion due to differential aeration is quite a complicated process and several factors together are responsible for it.

A drop of water on an iron or steel plate (Fig. 11.5) is an interesting illustration for the study of the formation of an oxygen concentration cell. The concentration of oxygen at the center of the drop is less than at its surface because as oxygen is used up it is replaced more easily at the surface of the drop. Thus, there is a difference in oxygen concentration from the centre to the surface of the drop and consequently, the centre becomes anodic and the surface region becomes cathodic.

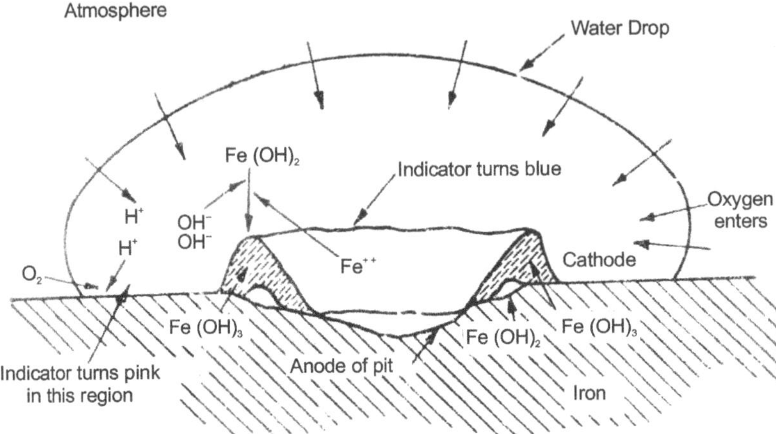

Fig. 11.5 A drop of water on a piece of iron sets up an oxygen-concentration cell. The reactions involved are indicated by a ferroxyl indicator.

(J.F. Young, Materials and Processes, Courtesy, John Wiley and sons)

As corrosion progresses ferrous hydroxide is formed which gradually gets oxidized to ferric hydroxide and is precipitated as rust, as has already been explained. The rust around the anode further prevents the entry of oxygen from the surface of the drop. The electrical potential of the cell is thereby increased accelerating the rate of corrosion. Again the surface under the rust is also excluded from oxygen and another oxygen concentration cell is formed. Some ferrous ions diffuse through the rust and deposit as green ferrous hydroxide under the red rust.

The reactions taking place in the drop can be observed by adding ferroxyl indicator solution prepared by mixing solutions of phenolphthalein (which indicates the presence of excess hydroxyl ions) and potassium ferricyanide (which indicates the presence of ferrous ions). For demonstration purposes it is helpful to add a trace of sodium chloride to the indicator so that corrosion may be accelerated and the experiment may be easily shown. On placing a drop of this solution on a polished iron surface a concentration cell is quickly formed because the oxygen is used up at the centre. The centre becomes anode and the solution in the centre turns blue because of ferrous ions. Cathode is formed at the edge of the drop and the solution there turns pink because of hydroxyl ions.

Water line corrosion is quite common in ships, iron tanks used for storing water, etc., in which a portion of the metal surface is under and the other is above water. The concentration of oxygen dissolved in water is greater at the water surface than deeper down. When a steel tank is regularly used for filling water up to a certain level the metal at the water surface is in contact with water having a higher concentration of oxygen than the metal under the surface. This causes the formation of a concentration cell and the anode is the lower portion and the cathode is at the water level. Because of the poor conductivity of water the ions just below the water level are more readily available for reaction, therefore, the metal corrodes just below the water level. Ships sunk for many years deep under water are known to be free from corrosion whereas water line corrosion of ocean-going ships is a practical problem.

Certain *physical factors* like surface conditions and internal or external stresses also affect the corrosion process. A rough surface is more likely to collect dirt than a smooth and polished surface. As has already been explained, the surface under dirt is anodic and will corrode. Many structures are known to have failed due to *stress corrosion*. The internal stresses may have been produced during manufacture or fabrication or heat treatment. The external stress is due to service conditions. From thermodynamic calculations it has been shown that the atoms under stress are at higher energy level. Consequently the points under stress, as well as the grain boundaries, are anodic and there is a large stress-free surface which is cathodic. The difference in potential, however, is very very small but the path of stress is very narrow, hence corrosion is concentrated in a narrow path which is anodic with a large cathodic surface, and helps to open up cracks or crevices which widen due to the stress. The cracks have been noticed in areas under tensile stress and not under compressive stress. Stress corrosion appcars quite suddenly and its reason is totally different from that of creep failure which is due to elongation under steady load. Caustic embrittlement of iron and steel, which is sometimes the cause of boiler failure, is similar to stress corrosion.

Another way the anodic areas are formed is by *scratches* on a protective coating which is cathodic to the metal. When a scratch is formed on the common 'tin sheet', which is in fact a mild steel sheet coated with tin, severe corrosion is noticed at the scratch. Tin is cathodic to iron and when it covers iron completely, iron does not come in contact with the corroding atmosphere. When a scratch is formed, the small iron surface exposed is anodic with respect to the large cathodic area around it. Therefore, the entire intensity of corrosion is concentrated on the area exposed by the scratch which often results in the formation of a hole in the sheet. In galvanised iron sheets, which are iron sheets coated with zinc, the case is the reverse. Zinc is anodic to iron and it is zinc that continues to be attacked even when a scratch is formed because the exposed iron surface is cathodic to it. Therefore care has to be exercised while handling metals protected by coatings.

When an organic coating like paint, is scratched, often the edges of the coating become loose and oxygen can enter to a small extent. If this happens, there will be corrosion inside the scratched edges due to the lower concentration of oxygen as compared to the concentration at the exposed surface.

In certain alloy systems corrosion takes place due to preferential attack of the environment, because one constituent gives rise to anodic and another to cathodic areas. This is known as preferential attack corrosion. In brass, zinc is anodic to copper and in corroding environments removal of zinc takes place. In cast iron there are quite large areas of graphite which are cathodic to iron. This helps in the removal of iron; another type of preferential corrosion is intergranular corrosion. It is due to impurities or inhomogeneties at the grain boundaries of the metals and alloys.

11.3.2 Factors affecting Corrosion Rate

It has already been explained that in spite of the rates of individual corrosion reactions being predictable, it has not yet been possible to predict the rate of corrosion in an actual situation because it is often a result of many complicated and not very well understood factors. However, there are several common environmental conditions whose effects in general are worth taking note of.

One of the factors that greatly affects the corrosion rate is the *relative proportion of anodic and cathodic areas*. If there is a large cathodic area with only a small anodic area, intense corrosion at the anodic area takes place, as happens when a protective coating, like tin, nickel or chromium on iron, breaks at a few points.

In the corrosion reaction described H^+or OH^- are also involved. Therefore the affect of pH is obvious. It can be easily seen from the chemical equation for a reaction the direction in which it will shift by change in concentration of H^+ or OH^-. As a general rule, acids are more corrosive than neutral or alkaline solutions. Exceptions to the rule are amphoteric metals, like aluminium, zinc and lead, which form anions as well as cations. In acid solutions these react quickly like other metals but in alkaline solutions they form complex ions and go into solution. Consequently these metals most resistant to corrosion in neutral solutions. In the case of other metals the corrosion rate is higher in acidic solutions than in alkaline solutions. The corrosion of iron or steel is quite slow in alkaline solutions and gradually increases as acidity increases and becomes rapid when the pH value is below, say, 4.5 to 3. This is the reason why the pH value of water for boiler feed or for cooling systems is kept in the alkaline range.

Oxidizing agents may increase or decrease corrosion rates. Systems handling water, e.g., boilers, heat exchangers, etc., are faced with serious corrosion problems due to dissolved oxygen. In high pressure boilers less than 0.005 mg/l of dissolved oxygen is permissible. The oxidizing agents oxidize materials and are themselves reduced at cathodes and, thereby support cathode reaction and promote corrosion. It is noteworthy that on some metals, like stainless steels, aluminium and magnesium, a thin film of oxide is deposited on the surface rendering the metal resistant to corrosion and makes it passive. Such metals and alloys are more resistant in oxidizing environments. Monel metals (Ni-Cu alloys) corrode rapidly in the presence of air whereas 18-8 stainless steel (Cr-Ni-Fe alloy) corrodes rapidly in the absence of air. It has been reported that in an experiment using 5% sulphuric acid at 30° the corrosion rate of Monel metal was 240 mdd (mg per sq. decimeter per day). When the acid solution was saturated with air, but on exposure to the same solution free from air corrosion was 40 mdd. Under similar conditions the corrosion rates of stainless steel were 2 and 300 mdd. This illustrates the sharp difference in the behaviours of Monel and stainless steel both of which are known for their resistance to corrosion. Hence, care needs to be taken while using materials

The effect of temperature on the corrosion rate is complicated because of the fact that it affects the various factors in different ways. The rate of chemical reaction increases, with rise in temperature but the solubilities of gases, like oxygen which affect corrosion, decrease. The temperature may affect the protective coatings in different ways. In general, the rate of corrosion due to oxygen or oxidizing agents is decreased with rise in temperature but the rate of hydrogen type corrosion is increased.

A mild steel equipment handling water gets corroded at a faster rate as the temperature increases, because of a higher rate of chemical reaction, reaching the maximum rate at 80°. Above this temperature the effect of lower solubility of oxygen in water becomes dominant and the rate of corrosion falls sharply. Similar results have been reported for the corrosion of Monel metal in dilute sulphuric acid in the presence of air. The effect of temperature on the corrosion of zinc in distilled water is interesting. Up to 50° the protective coating formed by the corrosion protects zinc. Above 50^{0} the protective coating gives way and rapid corrosion takes place up to about 65°. Above 65° the effect of decreasing solubility of oxygen becomes pronounced and the corrosion rate falls.

A number of *other factors* also affect corrosion rates. A liquid flowing with high velocity past a metal surface reduces the tendency to form a protective film and may even remove films already formed. It may even cause erosion. Consequently, at higher velocities the corrosion rates are higher. The dissolve salts may increase or decrease corrosion rates. The corrosion of stainless steel and of aluminium is increased in the presence of chloride ions as these react with the protective oxide films. In general, dissolved salts increase corrosion rates as they make the solution a better conductor, thereby facilitating diffusion of ions.

Mention may be made here of *inhibitors*, i.e., substances that hinder the reactions when added in small amounts to the environment, a number of organic as well as inorganic substances have been suggested. Some of these are polarizers and act by retarding cathodic reactions. Their exact action is varied e.g., formation of an insoluble salt, oxide on the metal surface, absorption on the metal surface, etc.

11.4 Corrosion Protection

Corrosion Protection means not allowing corrosion reactions to take place. In most practical cases the cost of a metal or alloy or of fabrication is such that the most resistant material cannot be used and some steps have to be taken to prevent corrosion, e.g., noble metals do not corrode but they cannot be used for common purposes. In some cases protection can be done directly, e.g., by painting, and in some case indirectly, e.g., by attaching blocks of magnesium to underground pipelines. The various measures taken for corrosion protection can be studied under: 1. design and material selection, 2. cathodic protection and 3. protective coatings.

11.4.1 Design and Material Selection

Design consideration opens up a vast field of study. From the corrosion point of view it means proper selection and proper use of the metals. Sometimes even a small amount of an impurity in the common metals and alloys may cause severe corrosion, as for example, 0.003% iron dissolves in magnesium forming a solid solution. If iron is present in even slightly greater amounts it appears as small particles which are highly cathodic to the surrounding magnesium surface and cause very severe corrosion around these particles. This type of corrosion is known as *pitting*. Similar is the case of lead as an impurity in zinc diecasting alloys (whose solubility is about 0.002%). In such cases it is necessary to use high purity metals.

Sometimes a difficulty like the above can be avoided by alloying. Iron may be an impurity in aluminium. The difference between their potentials is about 0.8 volts which causes severe corrosion of aluminium. If small amounts of chromium or manganese are added, alloys like Al-Cr-Fe or Al-Mn-Fe are formed whose potentials are very close to that of aluminium.

In case a metal is under external stress, *heat treatment* like annealing, which relieves the stress, will help corrosion resistance. It is rather difficult to generalize the effect of heat treatment on corrosion. A quenched steel has one phase, viz. martensite. If it is tempered at low temperatures, carbide and ferrite phases appear and rapid corrosion takes place. High temperature tempering causes the carbide phases to come together, thereby reducing the number of galvanic couples. Annealing, which reduces the corrosion rate in many cases, increases corrosion of cutlery grades of ferrite steel. The reason is that annealing causes the precipitation of chromium carbide and thus reduces the amount of carbon in the neighbouring areas which get corroded because they are anodic to chromium carbide.

A major factor in the corrosion failure of a component or structure is a faulty design. The design may satisfy all the basic mechanical requirements but it may incorporate features that tend to promote corrosion. This will lead to either early failure or unnecessary maintenance throughout its life.

A simple modification in design in many cases may protect them from corrosion to a large extent.

There are many equipments and structures which are meant to be exposed to corrosive environments, e.g., equipment handling acids, a bridge, and hull of a ship. In such cases often slight changes in their designs can be made so that there is no accumulation of corrosive material at a particular place.

Fig. 11.6 shows the design of drain pipe in a tank for chemicals. Fig. 11.6(a) illustrates a poor design because some solution will be left after draining and the life of equipment will be reduced because of corrosion. Fig. 11.6(b) shows a very slight modification in the design but the effects of the modification are far-reaching. In this case the solution can be drained completely. Cause of leaking drain pipes, taps, fittings, etc., can be traced to such defects in the design.

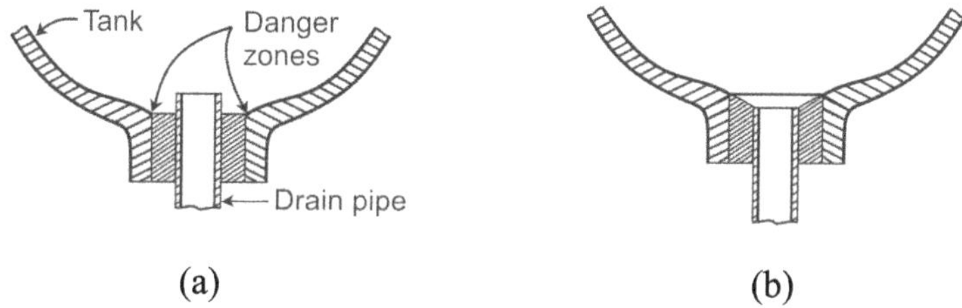

Fig. 11.6 Drain pipe in a tank for chemicals, (a) Poor design, (b) Correct design (L. P Mondolfo and O. Zmeskal, Engineering Metallurgy, *Courtesy, McGraw Hill Book Co., Inc.*)

Fig. 11.7 shows an electrical box. The design shown in Fig 11.7(a) is such that rain water collects at the top and also seeps between the bolt and the housing and remains there because of capillary action. The rain water at the top is not drained off and as water evaporates, concentration of the salts increases with the result that corrosion takes place. If the housing and the bolt are of different materials, as is quite often the case, the formation of galvanic couples is still easier. Fig. 11.7(b) shows how the defects can be corrected by slight modifications in the design.

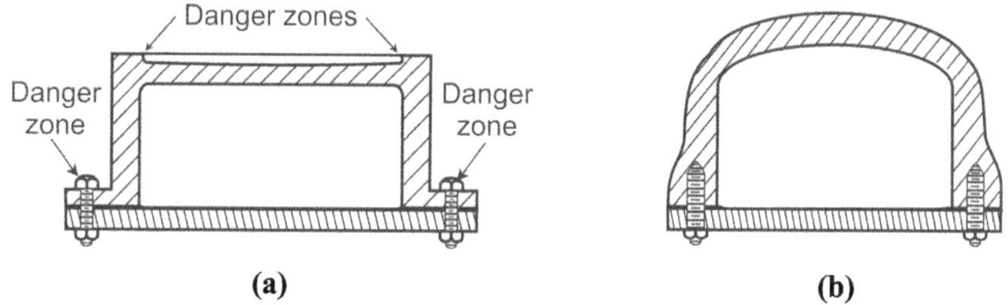

Fig. 11.7 Electrical box (a) Poor design, (b) Correct design.
(L. F. Mondolfo and O Zmeskal, *Engineering Metallurgy*), (*Courtesy, McGraw Hill Book Co., Inc*)

In the tropical countries roofing of corrugated iron sheets is quite common. Fig. 11.8 (b) shows a very bad way of joining the sheets as the rain water cannot be drained away. Fig. 11.8 (a) shows how the sheets should be assembled so that corrosion is not facilitated.

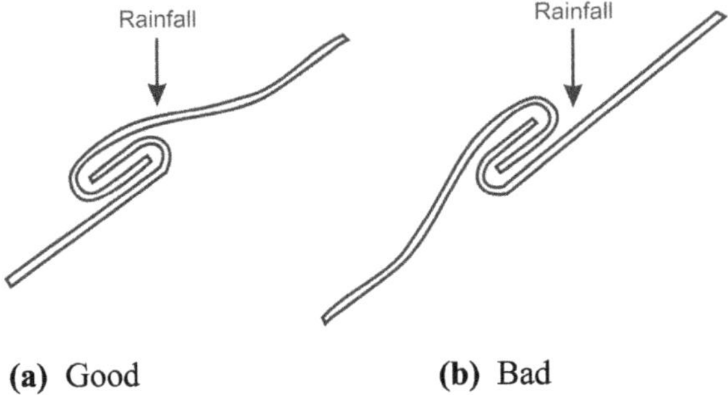

(a) Good **(b)** Bad

Fig. 11.8 Roofing sheet. (a) correct assembly (b) wrong assembly leading to rapid corrosion.

(L. F. Mondolfo and O Zmeskal, *Engineering Metallurgy*),
(*Courtesy, McGraw Hill Book Co., Inc*)

Materials like wood, paper, cloth, asbestos, magnesia, etc., which absorb and retain moisture, should not be used in contact with metals. Care should be exercised in the use of soft metal (like lead) washers because of the formation of galvanic couples. It is better to use rubber or plastics washers because they do not absorb moisture and moreover they insulate the metals. Sharp corners and projections should be avoided as they cause accumulation of materials.

The use of more than one metal in equipment often becomes unavoidable. This opens up the danger of formation of galvanic couples. Since the rate of corrosion depends upon, besides difference in potentials, the relative sizes of the anodic and cathodic areas it is desirable that the smaller parts be made of the noble metals. For example, a steel bolt in a copper equipment is undesirable because the effect of corrosion will be concentrated on the small steel bolt. Engineering News (1915) describes the failure of the yacht "Sea Call" whose hull plates were made of Monel metal. These were fastened with steel rivets to the steel frame. Within three months the Steel rivets were destroyed and the frame became so weak that the vessel had to be scrapped. If more than one metal has to be used in the construction, much of the corrosion problem can be avoided if (i) the metals are so chosen that they have as little difference in their potentials as possible, (ii) the smaller part is of nobler metal, and (iii) the metals are electrically insulated from each other.

A note of warning may be sounded. Corrosion is a complicated problem with many aspects whose effects are difficult to assess. Undue haste should not be exercised in changing designs that have stood the test of time. For example, aluminium sheets fastened by iron nails using lead washers are known to stand for years, Fig. 11.9. Electrochemically, it is undesirable because the difference between the electrode potentials of aluminium and lead is about 2 volts. The resistance to corrosion in this case is attributed to an insulating film of aluminium oxide which prevents the formation of the galvanic couple.

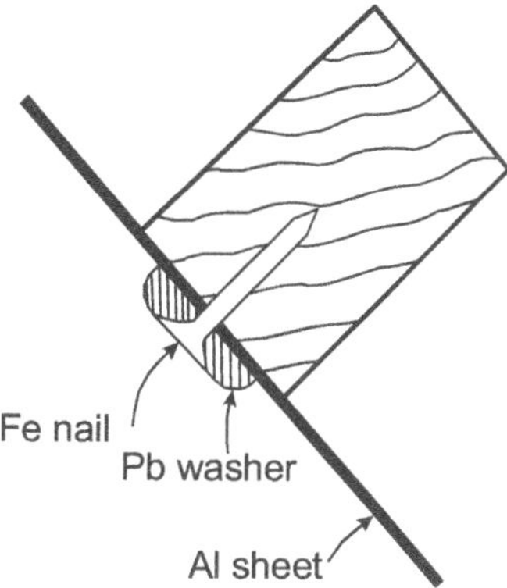

Fig. 11.9 Method of fastening roofing sheet.
(L. F. Mondolfo and O Zmeskal, *Engineering Metallurgy*),
(*Courtesy, McGraw Hill Book Co., Inc*)

Stray currents cause anodic corrosion, especially in areas where electric traction is used. Because of defective insulation stray currents pass through metallic conductors, like poles and pipes, supported on the ground. The point at which the current leaves the conductor becomes anodic and gets corroded. This can be checked by proper insulation.

11.4.2 Cathodic Protection

Interest in cathodic protection has greatly increased in recent years. It can be defined as a method of protecting metals and alloys from corrosion by making them completely cathodic and no site of the metallic object is allowed to act as an anode. Obviously, there cannot be a cathode without an anode. Therefore, to achieve this, an auxiliary anode is provided in the corroding medium which is connected to the structure to be protected through an external source of current. This is illustrated by the simple diagrams shown in Fig. 11.10.

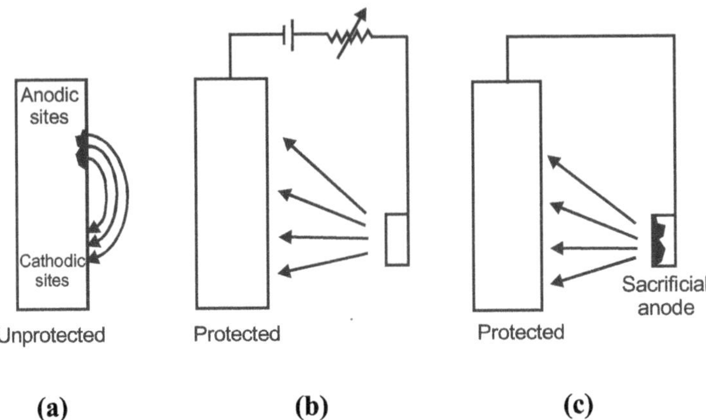

Fig. 11.10 Cathodic protection metal (a) piece being corroded (b) It can be protected by applying an impressed current or (c) by connecting a sacrificial anode.

The cathodic protection is applied either by using sacrificial anode or by applying direct current from a rectifier or a battery. In the former case a more active metal, like zinc, magnesium, aluminium or their alloys, than the metal to be protected is used as the anode which gets corroded in the environment and the metallic object which has now been made a cathode is protected. (See Fig. 11.11). The electrons for protection in this case, are supplied by the dissolution of sacrificial anode. This method is advantageous in the sense that it requires low installation cost and minimum maintenance cost after installation. In the impressed current method the auxiliary electrode is made of noble metal and is kept in contact with corroding environment. The anode, therefore, remains unaffected and electrons are supplied by rectifier or battery.

Although, the method of cathodic protection appears to be an ideal method for the prevention of corrosion, there are some demerits in this method. Firstly, this method is expensive as it requires very high current, especially in the case of impressed current method. Secondly, it is difficult to maintain uniform current over the entire metal surface and as a result localized corrosion may occur. Finally, as the metal under protection is made cathodic, hydrogen evolution reaction may occur increasing the pH of the environment which may cause the corrosion of base metal. The continuous production of H_2 at cathode may also lead to hydrogen embrittlement.

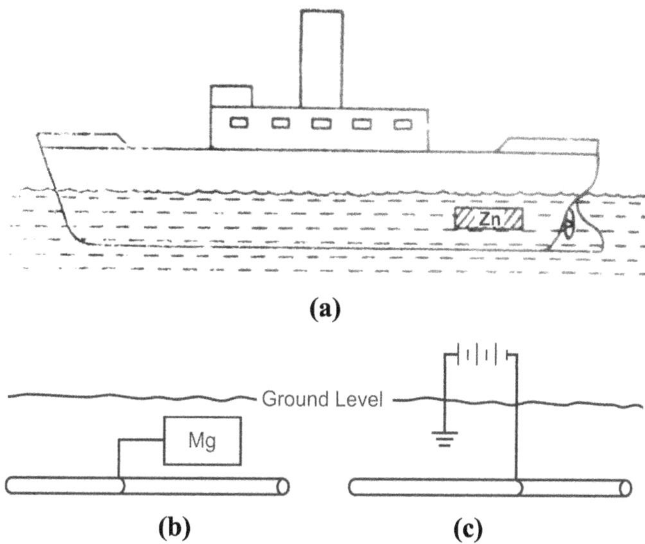

Fig. 11.11 Application of cathodic protection. (a) Protection of ship's hull below waterline by fixing zinc plates, (b) Protection of underground pipe-line by attaching magnesium plates. (c) Protection of underground pipe-line by applying a small impressed voltage. In all these cases the part to be protected has been made a cathode.

11.5 Protective Coatings

The need to apply protective coatings arises from the practical difficulties in using the most resistant metal and avoiding all the various factors that support corrosion. These difficulties are due to cost and availability of materials and fabrication problems. After taking all due care in the selection of the materials and using them in the best possible manner it often becomes necessary to cover the metal surface with a coating which is intended to protect it from corrosion. Such coatings are called protective coatings. These may be of metallic, inorganic or organic materials. Most of these coatings act by cutting off the contact between the metal and the corrosive surroundings. Before any coating can be applied, it is necessary to prepare properly the metal surface.

11.5.1 Surface Preparation

If the metal surface has not been properly prepared, the protective coating will not stick to the surface and not only will the purpose be defeated but still more severe corrosion may results at points exposed to the corrosive environment. The surface imperfection, besides physical defects, is generally due to the presence of two kinds of substances: oil or grease, and rust or scale. Oil or grease is removed by washing with soap or alkali solution in water. Several patented compositions containing alkalis, phosphates and silicates are marketed for removing oil. The use of solvents, mostly chlorinated hydrocarbons, has become quite wide spread in the U S A.

Rust or scale is generally removed by dipping the material in dilute (5 to 10%) sulphuric acid and by mechanical rubbing with a wire brush. Large equipment or structures like bridges are cleaned by the use of a wire brush or by sand blasting. Another method of cleaning is by applying flame from a kerosene burner. The old paint, oil or grease burns and the scale is removed because of sudden expansion of the metal.

A surface may also be cleaned by making it a cathode and passing high density current through it when the evolved hydrogen gas does the cleaning.

If the surface is not smooth, it is made smooth by mechanical polishing or by electropolishing. The clean surface catches dirt and gets corroded quickly. Hence it should not be left in that condition and should be quickly protected.

11.5.2 Metallic Coatings

The metals used for coatings may be placed under two categories: (1) metals which are anodic to the metal to be protected, i. e. metals standing above it in the galvanic series, or (2) cathodic metals, which stand below it in the galvanic series. Metals like zinc, magnesium, cadmium and aluminium, when applied on iron and steel fall under the first category. These being anodic get themselves corroded, saving the iron which is to be protected. As explained under 'Formation of Anodic Areas', in such cases even if the surface is scratched, the metal to be protected does not get corroded. These metals may be applied on the surface by hot dipping or spraying.

Metals like tin, nickel and chromium on iron and steel surfaces fall in the second category. These are cathodic to iron or more resistant to corrosion and offer protection only so long as the surface is completely covered. If the coating breaks or is applied in a manner so that pinholes are left, the points at which the iron surface is exposed gets very severely corroded because they are anodic to the large cathodic surface. Hence great care has to be exercised while applying these. Generally, the cathodic metal coating is applied by electroplating because by this method a more uniform coating is produced which completely covers the surface.

11.5.3 Galvanizing

Zinc coating on iron by hot dipping, is the most widely used metallic coating. The metal sheet is cleaned by alkali wash followed by pickling in dilute sulphuric acid and finally washed with water. It is then passed through a bath of molten zinc at about 450°. Smaller articles like bolts, nuts, etc., are kept in a perforated container and dipped in molten zinc and the excess of zinc is removed by centrifuging. A layer of an alloy of zinc on iron is formed to which the outer layer of zinc sticks The article should not be dipped in zinc at a very high temperature or for a long period as a thick layer of alloy is formed which is brittle.

Galvanizing is widely used for protecting iron exposed to the atmosphere, as is the case with roofs, wire fences, pipes and articles fabricated from galvanized sheets like buckets, tubes, etc. Galvanized ware is not used for keeping eatables because of the solubility of zinc. The popularity of galvanizing is due the low cost of zinc, easy application and the anodic protection offered by the zinc.

Tinning of mild steel plates is done mostly for the requirements of the foodstuff industry. Tin is a nobler metal than iron, therefore, it is more resistant to chemical attack. As has already been explained, if the coating formed does not cover the surface completely during use and leaves iron exposed, more rapid corrosion of iron will take place.

A cleaned iron sheet is passed through a bath of molten flux, like zinc chloride, then through molten tin and finally through a suitable vegetable oil (to protect the hot surface against oxidation). Lastly it is passed between rolls to adjust the thickness of the tin layer, which may be about 0.002 mm thick.

Nickel and chromium plating on steel are expensive but are commonly used for costlier equipments. Both the metals are cathodic to steel and care should be taken that the coating is continuous, uniform and firmly adherent to the surface, otherwise rust spots appear. Nickel coating deteriorates more quickly than chromium. Chromium gets covered with a strongly adherent oxidc film which is extremely stable. Chromium plating is more difficult and is more expensive. During plating, hydrogen evolves and makes the metal brittle. The layer should be thin otherwise it cracks. Steel is first plated with copper which is nickel plated and is then highly polished. Finally chromium plating is given. If due care is taken such a plating should remain bright for a long time.

Terne plate is the name given to the plate which has a coating of an alloy of tin and lead.

11.5.4 Inorganic Coatings

Inorganic coatings are produced either by converting the surface layer into a resistant compound or by applying a vitreous (i.e., glass like) enamel coating on the surface.

The most common examples of the former group are the anodized aluminium articles, often beautifully coloured, like plates, soap cases, etc., sold in the market. The article is made an anode in a chromic acid bath when a very thin and uniform coating of the metal oxide is formed on the surface. This resists attack by corrosive environments. Another method of forming this kind of coating is to dip the clean metal in a phosphate or a chromate solution when a compound is formed which resists the formation of corrosion products. In this method it becomes necessary to paint the surface after forming the inorganic coating.

The pans, basins, etc., commonly used in hospitals are common examples of articles made of iron sheet coated with vitreous enamels. The iron sheet is given the final shape, cleaned, and a suitable composition of silicates is applied and then fired. The silicates fuse to form a glass-like layer which protects the metal surface.

11.6 Organic Coatings

Organic coatings are more widely used than inorganic coatings. Generally these have two functions. Like inorganic coatings they protect the surface principally because they act as physical barriers between the metal to be protected and the corroding environment. Therefore, a continuous, adherent, impervious film must be formed which can withstand exposure for a reasonable period. It is worth repeating that if the protective coating is discontinuous or breaks, more severe corrosion will take place on the exposed surface. The second function of organic coatings is decorative. Various coloring matters are added which give a pleasing effect to the eye.

The formulations of organic coatings put on the market seem to be ever increasing. They may be outlined in the following groups.

11.6.1 Paints

Paint is a term which has been used to signify several types of painting materials but mostly it refers to oil paints. It is in this sense that the term will be used here. It essentially consists of a solid constituent, called pigment, which obscures the surface and provides a decorative colour effect, and a liquid which binds the pigment to the surface and protects it from decay. As it is desirable that the paint should dry soon, a third constituent, called a drier, is added which reduces the drying time. These three constituents remain in the paint film in one form or the other. A fourth constituent, a thinner, which is a volatile solvent, is often added. It helps to adjust the consistency of the paint. The thinner being volatile evaporates.

Unlike thinners, plasticiser remains permanently in paints and varnishes. They improve flexibility or elasticity of the film without impairing its other properties. They should be compatible with the other constituents, non-volatile and chemically inert and stable. Most of the plasticizers used are high molecular weight esters.

Often colourless inorganic substances like aluminium silicate, barium carbonate, and barium sulphate are added to the paints. These are called *extenders*. In oil paints the extenders have only an auxiliary function. They are added in limited quantities for adjusting the working qualities of the paint because they help to prevent the settling of heavier pigments. The extenders also increase the mechanical strength of the paint film and help the application of subsequent coats of paint. They reduce the cost of the paint.

Pigments usually comprise 60 to 80% of the weight of the paint film. Pigments increase the life of paint film because they prevent the penetration of the ultra violet rays which deteriorate the oil film. Generally, mixture of pigments is used so as to incorporate several essential properties. Many properties need to be looked for in a pigment. It should be chemically stable and resistant to the common corrosive agents. It should not be soluble in the vehicle. The particle size affects the shade of the colour. The pigment should have good hiding power, i.e. it should be opaque so that the surface underneath is not visible; otherwise all the dirty spots, surface defects etc., would be seen. Hiding power of a pigment, which depends primarily on its opacity, is its power of reflecting the light rays which fall on it instead of transmitting them. Opacity depends on the difference between the refractive power of the pigment and that of the medium which surrounds it; the greater this difference is, the greater is the opacity, if the two are practically equal the pigment becomes transparent. For example, magnesium carbonate and barytes have low refractive power and are considerably opaque in the dry state or when mixed with water. They become practically transparent in linseed oil. Zinc oxide has much greater refractive power, hence remains opaque in the oil giving high hiding power. Another important property that influences its use in paints is its spreading power which represents the surface area on which the pigment made into a paint can be spread. A paint made with a pigment of high spreading power can be applied on a larger surface than a paint made with the same quantity of a pigment of lower spreading power. As the spreading power depends upon the readiness with which the particles slide over one another, it is influenced by the size and more particularly by the shape of the particles. In practice it is determined by the oil absorption of the pigment which is the quantity of oil required to convert a given quantity of dry pigment into stiff paste. The property depends upon the size and shape of the pigment and with a few exceptions, gives an indication of its spreading power.

Among the *white pigments*, white lead (basic lead carbonate, $2PbCO_{11}$. $Pb(OH)_2$) has been considered to be important for the last few centuries. It is prepared by the corrosion of lead in acetic acid vapours in the presence of carbon dioxide. In the older processes the time taken was a few months. In the newer processes the time has been greatly reduced. White lead has a high hiding power which has been superseded only recently by titanium dioxide and antimony oxide pigments. Because of its high specific gravity the spreading power is low (1 kg of white lead made into a paint spreads over about 8 sq. m of the surface). White lead is a very good pigment but has two disadvantages. It is toxic and while stable in pure air it discolours in the presence of hydrogen sulphide due to the formation of black lead sulphide. Basic lead sulphate is another white pigment of lead.

Zinc oxide is made in large amounts either by heating high grade $ZnCO_3$ ores, like franklinite, or by heating commercial zinc in air. Zinc oxide clouds given off are condensed. It gives brilliant white paints. Its hiding power is comparable to that of white lead. It does not discolor in impure air but the paint dries rather slowly and tends to be brittle and crack. Therefore, combinations of white lead and zinc oxide are preferred. *Lithopone* is prepared by mixing solutions of zinc sulphate and barium sulphide when barium sulphate and zinc sulphide get precipitated.

Titanium dioxide is very resistant to chemical attack. It is highly opaque and is a very good white pigment. Because of the cost and high opacity it is used after mixing with other inert materials, called extenders, like calcium sulphate, barium sulphate, zinc oxide and magnesium silicate.

Red pigments are mostly iron oxide or lead oxide pigments. Iron oxide pigments are very suitable for giving the first coat on steel structures. Some deposits of clay contain various oxides of iron and have yellow, brownish yellow or greenish brown colours. They are called ochre, sienna or umber, respectively.

Red lead, Pb_3O_4 is an important red pigment. It is used on steel work more for protective than for decorative purpose. It is highly corrosion resistant.

Lead chromate is an important *yellow pigment.* Basic lead chromate is orange-coloured. Lead chromates are inexpensive but, like other lead pigments, tend to darken on exposure to air. *Zinc chrome* is a complex pigment of varying compositions. It is yellow in colour, does not discolor in air, non-toxic and has anticorrosion properties. The chromates, in general, are brilliant in colour, have high hiding and staining powers and are corrosion resistant.

Green pigments are prepared by mixing yellow and blue pigments. Oxides of chromium are also used as green pigments. Among the *blue pigments* mention may be made of ultramarine, which is a complex silicate of sodium and aluminium having a sulphur compound, and of prussian blue. *The black pigments* owe their colour to the presence of carbon in one form or the other.

11.6.2 Oils

The term 'oil' is used to signify materials which chemically differ widely. The mineral oils have been discussed ill detail under 'Lubricants'. For paint-making fixed oils are used which are derived from vegetable and animal sources. These are esters of higher fatty acids with glycerol and are called glycerides. The fatty acids may be saturated or may have one, two, three or even more double bonds. Besides the molecular weight, the number and position of double bonds present in the acid influences the properties and uses of the oils. On this basis the oils are classed under three groups. The non-drying oils have mostly saturated and monosaturated fatty acids and are used in lubricant manufacture. They do not easily absorb oxygen from air. The drying oils have fatty acids with two or three or more double bonds.

They readily absorb atmospheric oxygen at ordinary temperature and are covered with a solid film of oxidation product. They are used as vehicles in the paints. The composition of the semi-drying oil is in between drying and non-drying oils. Linseed oil is the most widely used oil in paint making. India is an important producer of linseed oil.

Drying of an oil like linseed oil has been a subject of extensive investigation. It is a rather complex process. In greatly simplified form it may be said that on exposure to air the oil absorbs oxygen at the double bonds, slowly for the first three days, then rapidly and again slowly and in about four weeks it dries completely. The initial delay in the absorption of oxygen is due to the presence of natural 'antioxygenic' materials. The drying process does not involve evaporation but is due to chemical changes. Oxygen is absorbed at the double bonds resulting in the formation of peroxides which decompose giving keto or hydroxyketo compounds. To some extent polymerization and condensation reactions follow initial oxidation. The results of oxidation, and also to some extent of polymerization and condensation is that large three-dimensional molecules are formed and the oil becomes progressively thicker and thicker until it passes from a gel stage (which makes it dry to touch) to hard dry film. This process makes the film insoluble in the organic solvents in which the original oil was soluble. This hard, continuous, impervious film is able to prevent contact between the surface to be protected and the environment; thus it protects the surface.

The time taken for the completion of these changes is too long. For practical use, the time needs to be reduced, otherwise the painted equipment will be out of use for a long period, and also it may gather atmospheric dirt. The time is shortened by adding some compounds, like linoleates of lead, manganese and cobalt which are added in amounts between 0.5 to 2%. These are called *driers*. These shorten the drying time by destroying certain natural compounds present in oils which resist the oxidation process and also act as catalysts for oxidation. Thus driers also become essential constituents of paints.

The driers are mostly prepared by heating the oxides of lead, cobalt acetate, manganese sulphate, borate or oxalate with linseed oil when linoleates are formed. Sometimes a salt may by directly used. Lead driers are efficient but are liable to darken on exposure to impure air, as in industrial towns, owing to the formation of lead sulphide. Manganese compounds are more powerful driers than lead but they tend to become pink, which is a disadvantage in white paints. Their action is not very regular. Cobalt gives the most powerful driers. It acts as a catalyst promoting both oxidation and polymerization. As it is used in much smaller quantity, the effect of discolouration is negligible but its action is erratic and is influenced by weather conditions. Its powerful action may cause rapid drying on the surface leaving the interior of the paint film soft. A combination of 10 per cent of cobalt and 90 per cent lead (calculated as metals) gives the best overall results. Naphthenate driers are more expensive but are better than linoleates and rosinates as they are formed more easily and form more stable solutions with the oil with the result that there is less discolouration and drying action is more dependable. They are the salts of naphthenic acid–this is the name given to a complex mixture of acids of high molecular weight contained in crude petroleum rich in aromatic compounds.

Instead of using the raw linseed oil, 'boiled' linseed oil is commonly used in paints and varnishes. In the older method of preparation, the raw linseed oil was heated to 260°, oxides of lead and manganese were added and heating was continued for a few hours. This gave a homogeneous solution of the drier in the oil. In the modern process the oil is heated to about 90^0 by steam coils and agitated by passing a current of air. Soluble driers–resinates or linoleates of lead and manganese, or of cobalt, are added in small amounts (e.g., up to about 0.1 per cent of lead). The temperature is increased to about 130° and agitation is continued for approximately two to three hours. The oxidation reaction being exothermic it becomes necessary to cool it by circulating water. The colour of the oil becomes darker if it is heated to a higher temperature. The oil is cooled, allowed to stand for some time and filtered. Paint made from boiled oils dries in 10 to 24 hours. Depending on the colour and viscosity, boiled oils of several grades such as extra pale, pale boiled and double boiled are marketed.

The term 'boiled' oil is a misnomer because the fixed oils do not boil–they decompose on heating strongly. Linseed oil as obtained from the seeds, called raw linseed oil, takes a number of days to dry. In the preparation of the boiled oil several changes take place in the oil. The oil becomes clearer due to settling of some suspended matter. The driers get dissolved in the oil. The natural antioxidants which retard drying get destroyed. In the preparation of the boiled oil little polymerization takes place and the oxidation of the oil is also small. Because of these changes boiled oil is used to a large extent for making paints, enamels and varnishes.

'Stand oil' is prepared by heating the oil to a high temperature without adding driers and is used in corrosion resistant paints, glossy paints, varnishes, etc. In boiled oils oxidation has taken place whereas in stand oils polymerization is the main reaction.

'Blown oil' is prepared by passing air at 120° after dissolving driers in the oil. Oxidation takes place to a considerable extent in the process. It is used in the preparation of certain paints (because it wets certain pigments better) and for some varnishes.

To lower the viscosity of a paint or a varnish and to make its application easier, certain volatile solvents are added. These are called *thinners*. They are mostly various petroleum fractions boiling from 100 to 250° and are known in trade as white spirits.

Manufacture of paints essentially consists of two stages. The pigment and the extender are mixed with drying oil to form a stiff paste. In the second stage the remaining oil and thinner are added to obtain the final paint of the desired consistency. To attain homogeneity of paint and uniformity of the film after application, it is necessary that the oil should wet the surface of the pigment and of the extender. Therefore, the pigment and the extender should be dry, otherwise oil will not wet them. The amount of pigment in a paint or enamel affects its gloss. In general, the greater the amount of pigment the lesser is the gloss. Therefore, pigment volume concentration (called P. V. C.), calculated as per cent by dividing the volume pigment by the volume of pigment plus volume of non volatile vehicle, is carefully adjusted. Several types of machinery have been employed for making the paste. Roller mills are quite common for the purpose. Oil pigment and extender are ground by passing them through hollow rollers made of hardened steel. Water is circulated through the hollow rollers to prevent over-heating during use. The pressure between the rollers is carefully adjusted. The paste is then agitated with the remaining oil, thinner and drier in a pan fitted with agitating gears.

Instead of carrying out the operation in two stages, the entire operation can be done in a ball mill. A ball mill consists of a cylinder lined with hardened steel, filled with hardened steel balls. The pigment and the extender with a portion of oil are first charged in the ball mill which is then rotated. The balls falling through the mass grind it into a past. The remaining material is then added and mixed thoroughly and the final paint is discharged. It is strained through a fine wire gauze to remove extraneous matter and lumps that have escaped grinding.

11.6.3 Water Paints

These have less protective power than the oil paints and are made with a medium miscible with water. They are mostly used for interior decoration, although the composition of the oil-bound paint can be so adjusted as to make the paint sufficiently durable for exterior use on concrete and similar finishes. As compared to oil paints they are cheaper but have less gloss and less protective value.

In this category oil-bound water paints, popularly known as emulsion paints, are most widely used. The emulsion of linseed oil and water is prepared by agitating them in the presence of an emulsifying agent, like soap or sulphonated castor oil, and a stabilizer which is usually a protein, like casein or glue. As the protein will purify on storage an antiseptic like phenol or mercuric chloride is added. The pigment is ground with the emulsion. Often an oil varnish is added so as to give gloss. The material is supplied in the form of a paste. Before use it is mixed with water to give a paint of the required consistency.

On applying, the water evaporates to give a porous film which is temporarily bound by the glue. Gradually oil undergoes oxidation and polymerization giving hard film.

11.6.4 Varnishes

Varnishes are transparent viscous liquids made by dissolving resins, or resins and drying oils in volatile solvents. The varnish dries by the evaporation of the solvent and in case of oil varnish by the oxidation and polymerization of the drying oils. It produces a hard, glossy and transparent film which protects the surface on which it is applied, but dot's not obscure it. Some colouring matter may also be present in the varnish but the colour effect is secondary. In these respects, varnish differs from a paint. Varnishes are mostly applied on wooden furniture giving a shining appearance and at the same time making visible the fine fibers of wood.

Varnishes are mainly of two types. The *spirit varnishes* or lacquers are commonly used on wooden furniture. They are solutions of resins, like shellac in volatile solvents, such as denatured (or methylated) spirit. On application, the solvent evaporates leaving a coating of the resins. The coating is hard but is soluble in the original solvent. It is noteworthy that the paint film does not dissolve in solvents because of the chemical changes that have taken place in the drying oil.

Oil varnishes, or oleoresinous varnishes contain drying oils in addition to resins and volatile solvents. The volatile solvent evaporates and then chemical changes take place in the oil on exposure to air. The coating produced is durable and dissolves only in the powerful solvents.

In varnish making both natural and synthetic resins are used. The natural resins are basically exudations from plants. They are distinguished from gums in being insoluble in water and soluble in organic solvents. The most common natural resin is shellac. It is extracted by certain insects from the sap of trees in the tropics. India is the largest producer of shellac. Fossil resins like copal, amber and damar are hard and more stable. Due to lack of uniformity and homogeneity of the natural resins some synthetic products, known as synthetic resins, are also used. They are mainly produced by polymerization and condensation reactions. The raw materials are heated at proper temperature and pressure and water and other reaction products are removed by distillation. Phenolic alkyd, maleic and urea resins are the common synthetic resins.

For making spirit varnishes or lacquers, the resins like shellac, and the solvent, like methylated spirit, are placed in a cylindrical drum. Often a plasticizer (to give elasticity to the film) and colouring matter are also added. The resin is dissolved either by agitating the mixture or by rotating the drum. The operation can be hastened by warming the mixture. The solution is removed from the drum and stored in tanks when dirt is deposited. It is then filtered or centrifuged and marketed.

Oil varnishes are more difficult to manufacture. Resins are high molecular weight substances and are not easily soluble in oil. The resin is taken in an aluminium vessel, known as a kettle, of about 25 kg capacity and heated over a fire pit or in a small furnace. The resin melts and the temperature is slowly increased to about 300°. In this process, which is known as gum running, some cracking or depolymerization of the resin takes place and about 25 per cent of the resin is lost in the form of pungent fumes. The required quantity (about 25 per cent of the weight of the resin) of boiled oil or linseed oil along with driers is separately heated to 200 to 220° and is slowly added to the heated resin with constant stirring until thorough combination has taken place. This operation is known as cooking. Heating the mixture to the correct temperature and for sufficient time is essential. If thorough combination of resin and varnish does not take place, on keeping the varnish the resin separates. Overcooking causes discolouration of the varnish.

The kettle is removed from the furnace and allowed to cool. White spirit, which is a petroleum fraction boiling between 150° and 200°, is the most common thinner. The thinner is added to varnish when the temperature of the varnish is below the flash point of the thinner. The material is constantly stirred during the addition.

The varnish is stored in tanks for some days for maturing. Foreign matter and also particles of resin, which have not dissolved, settle during this period. During maturing, the colour of the varnish also improves and it becomes more homogeneous. The varnish is either filtered or centrifuged and packed for marketing.

In the preparation of an oil varnish the resin may be replaced by a bitumen found in nature or obtained as a residue in petroleum distillation. Bitumens are mixtures of high molecular weight hydrocarbons, together with complex nitrogen and sulphur compounds. As the bitumens are soluble in oil it is not necessary to heat them. The boiled oil or raw oil with the driers is heated and the bitumen is slowly added and agitated until it is dissolved. Resins may also be added to give strength and durability to the film. Bitumens are cheaper than resins, are black in colour, and give black varnishes. Such compositions are popularly known as *Japans*.

For coating articles like bicycles, sewing machines, etc., on a mass scale *stoving* or *baking varnishes* are used. These are oil varnishes having synthetic resins and take advantage of the thermosetting properties of the synthetic resins. They are applied on the surface while hot, thus are more uniform, and thinner coatings can be applied due to the reduced viscosity of the varnish. After application the article is heated up to 160° or exposed to infrared rays. The coating so obtained is hard, flexible, glossy and resistant to abrasion.

11.6.5 Enamels

The term enamel has been used for such a variety of products that it no longer refers to a particular product. However, it has been mostly used to refer to what is now called a hard gloss paint. In a sense it is a combination of paint and a varnish incorporating the advantages of both and giving a harder, smoother and glossier finish than ordinary paint.

11.6.6 Lacquers

The term lacquer has been used to refer to several kinds of coatings. The term was originally derived from lac which was dissolved in spirit to give spirit varnish. Thus spirit varnishes are also called lacquers. In recent times the term lacquer is more commonly used for combinations of cellulose derivatives, like cellulose nitrate or acetate, in a volatile solvent. On applying a simple combination like this the solvent volatalizes, leaving a film which is brittle and does not adhere well to the surface. To remove these defects plasticizers, like camphor, blown castor or blown linseed oil, are added. These make the film smoother and flexible. Natural or synthetic resins are also incorporated to add hardness and gloss to the film. Colouring matter may also be added. The solvents used in lacquers are rather expensive. Common solvents for cellulose derivatives are ethyl acetate, amyl acetate, butyl acetate, acetone, etc. Some thinners or diluents, like benzene, toluene, xylene, etc., are added to reduce the cost and to dissolve the resins. These do not dissolve the cellulose derivatives. By such a combination a lacquer dries quickly giving a hard and glossy film. In this process of film formation, no chemical change takes place, therefore the film is again soluble in the solvents. This defect can be removed by incorporating alkyd resin which undergoes oxidation on exposure.

Since lacquers dry very quickly, they are applied by spraying. These are used in large quantities by the automobile and furniture industries.

11.7 Inhibitors

In electrochemical corrosion although the anode dissolves, the reactions take place at the same rate both at the anode and the cathode, but one of the two reactions may control the rate of corrosion of the system as a whole. If the rate of either of the reaction is diminished or stopped completely, the other reaction is also affected and corrosion is prevented. The National Association of Corrosion Engineers, U. S. A., has accepted the following definition of inhibitors. A substance which retards corrosion when added to an environment in small concentration. From this point of view the inhibitors may be classed as (i) Adsorptive and (ii) Bulk film formers and (iii) Neutralizer. As an inhibitor affects either anode or cathode reaction, another, array of classifying the inhibitors is as (i) anodic (ii) cathodic and (iii) mixed inhibitors.

Anodic Inhibitors react with the metal ions produced at the anode and deposit sparingly soluble compounds on the anode surface. If an insufficient amount of inhibitor is added, although the total amount of the metal dissolved is reduced, there are active corrosion spots on the anode at which severe corrosion takes place. This becomes particularly dangerous if the corrosion is under cathodic control and the equipment would fail earlier than it would have failed if no inhibitor had been added. The amount of anodic inhibitor added therefore must be sufficient if not in excess.

It has already been explained that several types of cathodic reactions take place depending upon environmental conditions like the pH, presence or absence of oxygen. Accordingly, *cathodic inhibitors* function in several ways. They may reduce the diffusion of hydrogen ions (by the absorption of organic compounds on the cathode surface), remove dissolved oxygen by the action of a reducing agent (like sodium sulphite) or eliminate the hydroxyl ion (by the precipitation of insoluble hydroxide). The cathodic inhibitors become effective only after an initial induction period when the cathode reaction has been slowed down to such an extent that it indirectly begins to decrease the rate of the anode reaction. If insufficient amounts of cathodic inhibitors are added, the corrosion rate will be reduced but not stopped. In this respect they are safer than anodic inhibitors which aggravate the damage when added insufficient amounts.

Among the *inorganic inhibitors* the action of oxidizing agents like chromates, nitrites and even oxygen itself may be mentioned. They act by oxidizing ferrous hydroxide and depositing it as ferric oxide, or by forming an oxide film. It is noteworthy that the oxidizing agents promote cathode reaction by depolarising hydrogen at the cathode. Nitrites and chromates in particular are very effective on iron and steel in neutral and alkaline mediums. In the presence of chloride a large amount of chromate must be added. Alkaline inhibitors like sodium hydroxide, sodium carbonate and sodium phosphate precipitate the metal ions and control the anode reaction. Sodium silicate is effective in pure water.

Inorganic inhibitors are generally effective in neutral and alkaline solutions. They are ineffective in the presence of acids, acidic brines, reducing agents and microbiological growths. For such conditions *organic inhibitors* like organic bases, polar compounds and colloids are useful. The organic base forms an ionisable salt with the corrroding acid. The positive inhibitor ion is attracted to the cathode and absorbed at the surface forming a protective layer. The layer is a few molecules thick. The most important inhibitors are the nitrogen compounds followed by the oxygen and sulphur compounds. Some organic inhibitors are absorbed at the surface by van der Waals' forces and increase the electrical resistance of the cathodic area. A mention of sodium benzoate which has wide use in preventing the corrosion in cooling systems containing glycerol or glycol, as in automobile radiators, may be made. Paper used for wrapping metals is also coated with it.

Colloidal inhibitors like waste sulphite liquor from paper manufacture, tanneries, etc., are effective in the presence of the growth of micro-organisms.

Besides preventing corrosion in general, organic inhibitors are found to be useful in the various metal-cleaning operations like pickling of metals, cleaning of boilers, condensers, heat exchangers, water pipes and chemical equipment. In such operations without the use of the inhibitor a considerable amount of the metal would be dissolved in the process of removing the rust or the deposit.

Some organic phase (amines, nitrites, etc.) have been used in a vapour phase. They are very convenient for use in closed spaces and during storage and packing.

Questions

1. Define corrosion.
2. Explain passivity.
3. Difference between galvanic and electro chemical series.
4. Explain pitting corrosion.
5. Explain galvanic corrosion.

6. What is the chemical formula of the black rust?
7. How Fe corrodes in neutral or alkaline medium.
8. 'Rusting of iron is more in saline water', explain.
9. Differentiate galvanic and concentration cell corrosion.
10. Corrosion of gases will depend on which factors.
11. Write important applications of protective coatings.
12. Differentiate oil varnish and sprit varnish.
13. What is the function of extender in the paint.
14. Mention the important application of lacquers.
15. What is the purpose of adding solvent to the paint.
16. Define anodizing.
17. Explain cathodic protection.
18. Give the function of adding anti skinning agent to the paint.
19. Distinguish varnish and enamel.
20. Write importance of vitreous coatings.
21. Why should nickel plated articles be free from pores and pit holes
22. What is corrosion? Discuss different methods of corrosion control.
23. What is cathodic protection? How is it done by using sacrificial anode method?
24. Explain why an iron pipe does not rust when connected to zinc metal but rusts rapidly when connected to copper metal.
25. Write a brief note on dry corrosion
26. What is metallic corrosion? Write a note on water line corrosion
27. Explain the factors affecting chemical corrosion
211. Explain the steps involved in pre-treatment of a metal surface prior to being given a protective coating
29. Write short note on
    (a) Cathodic protection against corrosion
    (b) Anodizing
    (c) Electroplating

30. How would you explain the corrosion of zinc rod partially immersed in a solution of sodium chloride.

31. List any three methods employs for corrosion control and explain their features.

32. What is a paint? What are the constituents and functions of a paint?

33. Describe the following metallic coatings
   (a) Tinning
   (b) hot dipping
   (c) paints

34. Explain the following methods of corrosion control
   (i) proper designing (ii) protective coatings

35. (a) What is the difference between pitting and stress corrosion?
   (b) Describe the important factors that influence the process of corrosion.

36. (a) Explain the process of differential aeration corrosion.
   (b) What is sacrificial anode and how does it work?
   (c) Write a note on metallic coatings that prevent corrosion.

37. (a) What is electrochemical corrosion?
   (b) Explain the process of galvanizing and tinning.
   (c) What are organic paints? Describe their uses.

311. (a) Explain the difference between wet and dry corrosion.
   (b) What are organic paints and what are their constituents. Explain their functions.

39. (a) Explain the electrochemical theory of corrosion.
   (b) Discuss the important factors influencing corrosion.

40. (a) Explain the process of
      (i) tinning
      (ii) electro plating.
   (b) What are organic paints? What are their important constituents and their respective functions?

41. (a) What is galvanic corrosion and how it can be prevented?

    (b) Explain the phenomenon of stress corrosion with suitable examples.

42. (a) Discuss the phenomenon of

    (i) Metal cladding

    (ii) Electroplating used to prevent corrosion.

    (b) Write a note on underground/soil corrosion.

43. Write a note on paints.

44. What is the role of surface preparation in corrosion prevention and how it is done?

45. (a) Explain the process of

    (i) tinning

    (ii) galvanizing.

    (b) Describe the method of anodic protection.

46. (a) Discuss the differences between varnishes and paints.

    (b) What are the constituents of paints and what are their functions ?

47. (a) Explain the process of pitting corrosion and differential aeration corrosion.

    (b) What are the important factors that influence the corrosion phenomenon?

411. (a) Describe the important applications of varnishes and lacquers.

    Write briefly, on

    (i) phosphate coatings

    (ii) chromate coatings.

50. Write short note on organic paints

    (a) Discuss the electro chemical theory of corrosion.

    Write notes on:

    (i) sacrificial anode

    (ii) corrosion inhibitors.

51. (a) What are varnishes? Mention their important uses.
    (b) Explain the process of tinning and electroplating.
52. Write a note on enamels.
53. How do sodium sulphite and hydrazine act as corrosion inhibitor.

Objective Questions

1. Which of the following acts as corrosion inhibitor by removing oxygen from aqueous solutions.

   (A) $ZnSO_4$ (B) Na_2CrO_4

   (C) N_2H_4 (D) $NaNO_2$

2. Galvanizing is the process for coating iron with

   (A) Ni (B) Zn

   (C) Cu (D) Sn

3. During galvanic corrosion the more active metal acts as

   (A) Anode (B) Cathode

   (C) Anode as well as cathode (D) None of the above

4. The principle in electroplating is

   (A) Electrolysis (B) Oxidation

   (C) Reduction (D) Anodization

5. The metal which is protected by a layer of its own oxide

   (A) Cu (B) Fe

   (C) Au (D) Al

6. During galvanic corrosion, the most noble metals act as

   (A) Anode (B) Cathode

   (C) Anode as well as cathode (D) Corroding metal

7. Corrosion of zinc metal containing an impurity of copper is called ______ corrosion.

   (A) water-line (B) galvanic

   (C) drip (D) crevice

8. The type of corrosion occurring in a wire fence is ______ corrosion
   (A) galvanic (B) differential aeration
   (C) inter granular (D) water line
9. A metal will be easily corroded if it is
   (A) less acidic (B) more acidic
   (C) less basic (D) more basic

Fill in the Blanks

1. Evolution of hydrogen type corrosion occurs in _____ environment.
2. The metal _____ (higher/lower) in electrochemical series undergoes corrosion.
3. _____ is an example for stress type of corrosion.
4. Presence of CO_2 in air _____ the rate of corrosion.
5. The smaller the grain size of metal _____ will be the corrosion.
6. Corrosion is a reverse of _____ process.
7. If the cathodic area is small and anodic area is large the rate of the corrosion will be _____.
8. _____can be used as cathodic inhibitor.
9. The purpose of adding chromium in steel is to prevent _____.
10. The main constituent of paint is _____.
11. Function of pigment in paint is _____.
12. Binder provides _____ to the paint.
13. An acid which is used in the phosphate coatings is _____.
14. Chromate coatings are mainly applied to _____ Metal.
15. The function of extender in paints is _____.
16. _____ is example for inorganic coatings.
17. _____ is used as drying agent in paints.
18. Function of thinner in paints _____.
19. In chromium electroplating electrolyte used is _____.

20. Opacity to the paint is provided by _____.

21. An acid which is used in electroplating of chromium is _____.

22. Coating of tin over iron is example for _____ coating.

23. Coating of zinc over iron is example for _____ coating.

24. In Anodic coatings the metal which is used for coating is _____ to the base metal.

25. _____ is an example for cathodic coating.

26. _____ is used as a main constituent in the lacquers.

27. _____ is an example for the drying agent.

28. _____ is an example for the synthetic coating agent.

29. _____ can be used as sacrificial anodic protection.

Chapter 12

Molecular Spectroscopy

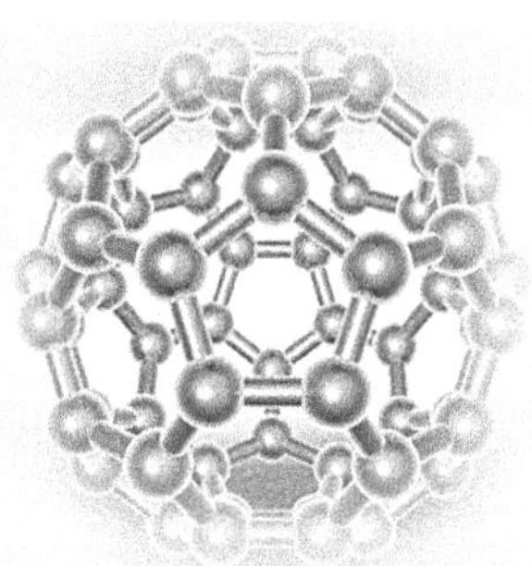

12.1 Introduction

Until around 1940, the structure determination or identification of a newly isolated organic natural product/newly synthesized organic compound was done entirely by chemical methods. The chemical approach involved determination of molecular weight and molecular formula of the organic compound, study of its functional group and degradation reactions. Based on this study a structure was proposed for the compound in question. The proposed structure was finally confirmed by the unequivocal synthesis of the compound. The chemical methods of structure determination, no doubt irreplaceable, were very tedious, time consuming and required large quantities of organic compound.

During twenties and thirties of 20th century, physicists all over the world were busy studying the interaction of light with matter. This led to the rapid growth of the technique of **spectroscopy** in the later years. Spectroscopy as an analytical technique and has several advantages over chemical analytical methods, like;

(i) The spectroscopic methods are relatively very simple.

(ii) Th e spectroscopic methods are very fast and require minimum time for structure determination.

(iii) The spectroscopic methods require very little of the pure compound for structure determination.

(iv) The spectroscopic techniques are generally non-destructive and the substance is recoverable after analysis.

12.2 Nature and Origin of Molecular Spectra

Molecular spectra is the study of the absorption or emission of electromagnetic radiation by molecules. Electromagnetic radiation is a broad term comprising of radiations starting from highly energetic cosmic rays to power waves having extremely small energy, Fig. 12.1.

ν (Hz)	$3 \times 10^6 - 3 \times 10^{10}$	$3 \times 10^{10} - 3 \times 10^{12}$	$3 \times 10^{12} - 3 \times 10^{14}$	$3 \times 10^{14} - 3 \times 10^{16}$	$3 \times 10^{16} - 3 \times 10^{18}$	10^{18}	$3 \times 10^{19} - 3 \times 10^{20}$
λ	10m – 1 cm	1cm – 100μm	100μm – 1μm	1μm – 10μm	10nm – 100pm	100pm	100pm – 1pm
Energy (J/mol)	0.001 – 10	$10^2 - 10^3$	10^4	10^5	$10^6 - 10^9$	10^9	$10^9 - 10^{11}$
Region	Radio	Microwave	Infra-red	Vis & UV	Vacuum UV	X-ray	γ-ray
Spectra	NMR	Rotational (esr)	Vibrational	Electronic	Loss of valence electrons	Core $\bar{e}$ transition	Nuclear transition

Fig. 12.1 Electromagnetic Spectrum.

Though all these radiations fall under a common class name electromagnetic radiation and propagate due to interaction between electric and magnetic components, they differ markedly in their method of production, handling, detection, wave length, and energy of their quanta. The radio waves, microwaves and T.V. waves are safely used in the transmission of information through electronic media and telecommunication media. Visible light is responsible for our vision. The electromagnetic rays in the form of ionizing radiations, viz., x - rays, γ - rays, cosmic rays etc., are helpful in medicine in obtaining x - ray photographs, in killing unwanted cell growths like cancerous tumors. Ionizing radiations are also posing the threats of killing or altering biosphere by arresting physical and mental growth and disease resisting capabilities by altering their genomes.

As mentioned already electromagnetic radiations have both *electrical* and *magnetic* components. All the material present in this universe are formed by the combination of atoms, which are mere description of electrons revolving in fixed orbits around the nucleus. This type of movement of the negatively charged electrons result in the electrical and magnetic fields. Hence *both the matter and electromagnetic radiation have electrical and magnetic components.* An interaction between these two forms of energy is expected only when they oscillate or resonate with the same frequency. A spectrum of a compound is a plot that reflects the amount of electromagnetic radiation absorbed or transmitted at each frequency of light. A **spectrum** is a finger print of a compound, which implies, no two compounds can have same electromagnetic spectrum. Therefore the spectroscopic techniques offer the most powerful tool in the elucidation of molecular structures. Interpretation of molecular spectra is largely based on the empirical correlation with the standard experimental data. An adsorption or emission data of a particular group or molecular arrangement taken for reference and suitable correlation of unknown compound, gives a reasonable assurance to the molecular structure.

12.3 Molecular Energy States & Classification of Spectroscopy

Quantum mechanical theories postulated by de Broglie, Schrodinger and Heisenberg and structural theories postulated by Pauli and Hund discuss about various electronic distributions and energies associated with them. In addition to these electronic energies a free molecule is made up of translational, rotational and vibrational energy components. Molecular spectra is the result of transitions in which one or more types of molecular energies are involved. Always a higher energy transition is accompanied by some lower energy changes also Fig. 12.2. Thus an electronic transition is always accompanied by suitable changes in vibrational and rotational energies. Similarly a change in vibrational energy levels show effects related to rotational changes also. The details of the individual energies are given in the following sections.

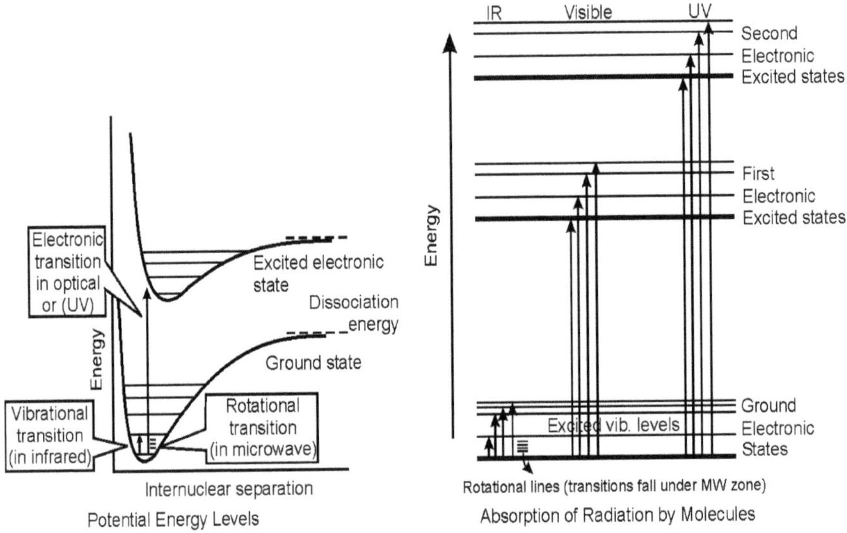

Fig. 12.2 Molecular energy levels (Rotational energy levels are shown only for few vibrational states only).

12.4 Translational Energy

Translational energy is possessed by virtue of movement of the molecules. All gaseous molecules have large kinetic or translational energy. It is given by the expression ½ mv^2 where m is the mass of the body and v the velocity. This energy decreases considerably in the solid state. Pressure of a gas is the consequence of this translational motion. *Quantum mechanics places no restriction on the amount of translational energy a molecule may possess.*

The continuously colliding gas molecules exchange this energy with one another during collissions. It is invariably raised with temperature. *Change of translational energy can not be observed spectroscopically.*

12.5 Rotational Spectra (Microwave Spectra)

A molecule acquires rotational energy by virtue of its rotation about its molecular axes. Rotational transition require very little energy. Pure rotational spectra falls in the microwave region of electromagnetic spectrum and for this reason rotational spectra is also called **microwave spectra**. Rotational energy of a molecule is rationalized by the quantum mechanics and is given by the formula;

$$E_{rot} = \frac{h^2}{8\pi^2 I} J(J+1) \text{ Joules} \quad (\text{where } J = 0, 1, 2, \ldots) \quad \ldots\ldots(12.1)$$

changing the above energy from Joules in to frequecy,

$$E_{rot} = BhJ(J+1) \text{ cm}^{-1} \quad \ldots\ldots(12.2)$$

where B (equal to $\frac{h}{8\pi^2 I}$ sec^{-1} or $\frac{h}{8\pi^2 IC}$ cm^{-1}) is a **rotational constant** which depends on the size and moment of inertia (I) of the molecule and J is the **rotational quantum number**. Here C represents the velocity of light and h is the plancks constant.

Thus considering a rotational transition from rotational quantum number J' to J'' the energy for the transition is

$$E_{J''} - E_{J'} = \Delta E \quad \ldots\ldots(12.3)$$

Therefore, $\Delta E = BhJ''(J''+1) - BhJ'(J'+1)$ .....(12.4)

But according to quantum mechanics the rotational transitions from one energy level to another follow the selection rule $\Delta J = \pm 1$.

Considering an excitation from J' we have

$$J'' - J' = 1 \quad or \quad J' = J'' - 1$$

Substituting for J' we have

$$\Delta E = BhJ''(J''+1) - Bh(J''-1)(J''-1+1)$$

$$\Delta E = BhJ''(J''+1) - BhJ''(J''-1)$$

$$\Delta E = BhJ'' \{J'' + 1 - J'' + 1\}$$

$$\Delta E = 2BhJ'' \text{ cm}^{-1} \quad (J'' = 1, 2, 3, ...) \qquad .....(12.5)$$

Since B is a function of moment of inertia and is equal to $\frac{h}{8\pi^2 IC}$ cm^{-1}

We have, $\Delta E_{rot} = 2 \times \frac{h}{8\pi^2 IC} \times J'' \quad = \frac{h}{4\pi^2 IC} J'' \text{ cm}^{-1}$

As this rotational transition occurs by the absorption of a photons of frequency ν, we have

$$\Delta E = h\nu \quad = 2\,B\,h\,\text{Jc}'' \text{ cm}^{-1} \qquad .....(12.6)$$

or Rotational frequency $\nu = 2BJ''$ .....(12.7)

or $$\boxed{\nu = 2BJ'' = \frac{h}{4\pi^2 IC} J''} \qquad .....(12.8)$$

(where J'' can assume values 1, 2, 3,...but not 0)

Hence the frequencies of rotational lines of any rotational spectrum are $2B$, $4B$, $6B$, ... i.e., integral multiples of $2B$.

This shows that the rotational lines are equally spaced as shown in Fig. 12.3.

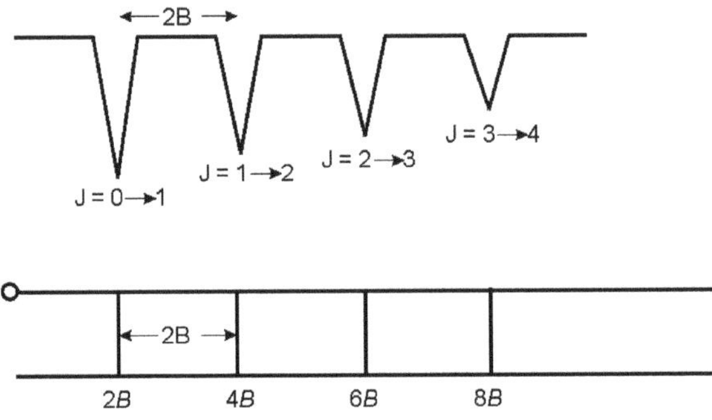

Fig. 12.3 Rotational lines (lines are equally spaced and separated by $2B$ cm^{-1}).

Rotational spectra is generally noticed in gaseous and liquid phases where a molecule has unrestricted rotational freedom. In the solid phase, where the molecular rotations are freezed, the rotational spectra is normally not observed.

Problem 1.

The rotational spectrum of AB has lines separated by 16.9 cm^{-1}. Calculate the moment of inertia.

Given frequency of rotational lines = $2B = 16.94\ cm^{-1}$

We know $2B = \frac{h}{4\pi^2 IC}\ cm^{-1}$

or $$I = \frac{h}{4\pi^2 C \times 2B}$$

Substituting the values of h, C, and $2B$

$$I = \frac{6.627 \times 10^{-27}}{4 \times (3.14)^2 \times 3 \times 10^{10} \times 16.94}$$

$$= 3.3 \times 10^{-40}\ g.cm^2$$

Problem 2.

Calculate the moment of inertia of C = O having a bond length of 1.128 $\overset{0}{A}$.

1 g atom (Avagadro number of atom) of carbon weighs 12 g

Weight of 1 carbon atom = $\frac{12}{6.023 \times 10^{23}}$

Similarly,

Weight of 1 oxygen atom = $\frac{16}{6.023 \times 10^{23}}$

Therefore;

$$\mu = \frac{m_1 m_2}{m_1 + m_2} = \frac{\left[\frac{12}{6.023 \times 10^{23}}\right]\left[\frac{16}{6.023 \times 10^{23}}\right]}{\frac{[12+16]}{\left[6.023 \times 10^{23}\right]}}$$

$$: = \frac{12 \times 16}{12 + 16} \times \frac{1}{6.023 \times 10^{23}}$$

$$= 1.14 \times 10^{-23}\ g/molecules$$

$$l = 1.128\ \overset{0}{A} = 1.128 \times 10^{-8}\ \text{cm}$$

But, $I = \mu l^2$

$$I = 1.14 \times 10^{-23} \times (1.128 \times 10^{-8})^2$$

$$= 1.45 \times 10^{-39}\ \text{g.cm}^2$$

Problem 3.

The first line ($J = 0 \rightarrow J = 1$) of a rotational spectrum of CO appears at 1.53×10^{11} cycles s^{-1}. From this data determine the bond length of CO bond.

Rotational frequency

$$\nu = \frac{1}{4\pi^2 I} J''\ \text{cycle sec}^{-1}$$

$$= \frac{1}{4\pi^2 \mu l^2} J''\ \text{cycle sec}^{-1}$$

For the first line $J = 0 \rightarrow J = 1, J'' = 1$

Therefore,

$$\nu = \frac{h}{4\pi^2 \mu l^2}$$

$$l^2 = \frac{h}{4\pi^2 \mu \nu}$$

Substituting the values of h, ν and μ (obtained from preceding problem), we have;

$$l^2 = \frac{6.62 \times 10^{-27}}{4 \times (3.14)^2 \times (1.53 \times 10^{11}) \times (1.14 \times 10^{-23})}$$

$$l^2 = 9.62 \times 10^{-17}$$

$$l = 0.981 \times 10^{-8}\ \text{cm}^{-1}$$

$$= 0.98\ A^{O}$$

So internuclear distance of CO is $0.98\ \overset{0}{A}$.

12.5.1 Applications of Rotational Spectra

(i) **Determination of molecular constants :** Microwave spectra are highly helpful in determining the molecular constants like bond length, bond angle, dipole moment, moment of inertia, etc. A detailed account of procedure involved in case of a diatomic molecules is discussed already in the preceding numerical problems.

(ii) **Detection of isotopes and isotopic abundance in a molecule :** The presence of any isotope of a given atom in a molecule under investigation and its abundance is clearly indicated in a microwave spectra. For example, if we consider NaCl, the chlorine atom has two isotopes, viz., Cl^{35} and Cl^{37}. This minute mass difference results in a frequency shift of rotational lines. Hence in the microwave spectrum we observe two peaks corresponding to both the isotopes, Fig. 12.4.

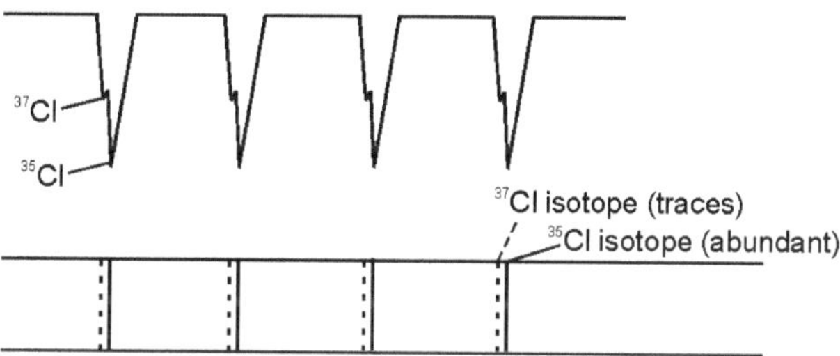

Fig. 12.4 Microwave spectra when two isotopes of an atom are present.

The intensity (area under the peak) of each peak of the rotational spectrum gives the isotopic abundance.

(iii) **Shapes of molecules :** Microwave spectra is observed by virtue of moment of inertia and can not pinpoint molecular groupings like –OH, $–CH_3$ etc. It gives a clear picture of over-all arrangement of the groups. With the help of microwave spectra we can distinguish between linear, symmetric top, spherical top and asymmetric top arrangements of the molecules.

(iv) **Detection of conformational isomers :** A unique application of rotational spectra is the detection of conformational isomers, such as the one formed in 1, 2-dichloroethane, Fig. 12.5.

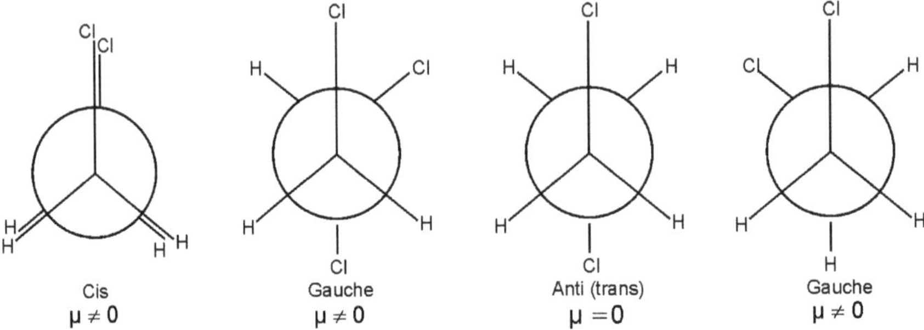

Fig. 12.5 Conformational isomers of 1, 2-dichloroethane.

(v) **Chemical examination of interstellar space :** One fascinating area where microwave analysis is being used is in the chemical examination of interstellar space. Some 30 or so chemical compounds like water, ammonia and formaldehyde were characterized from the analysis of microwave signals coming from interstellar space. This lead to the speculation regarding the origins of biological molecules of life on various stars and planets. The intensity of microwave signals also give an accurate estimate of the temperature of interstellar material.

12.6 Vibrational Spectra (IR Spectra)

The vibrational motion of a molecule takes place by the absorption of infrared radiation (300–1μm). Hence the **vibrational spectra** is often referred to as **IR spectra**. Vibrational spectra of a molecule indicates its flexibility i.e., the ease with which the **bond lengths** and the **bond angles** can be distorted from an equilibrium configuration of a molecule.

A diatomic molecule may be compared to two balls joined through a spring. In this analogy the electrostatic forces responsible for the bond formation are compared with the spring and the balls represent the binding atoms. Similar to the spring model, a diatomic molecule also undergoes expansion and compression.

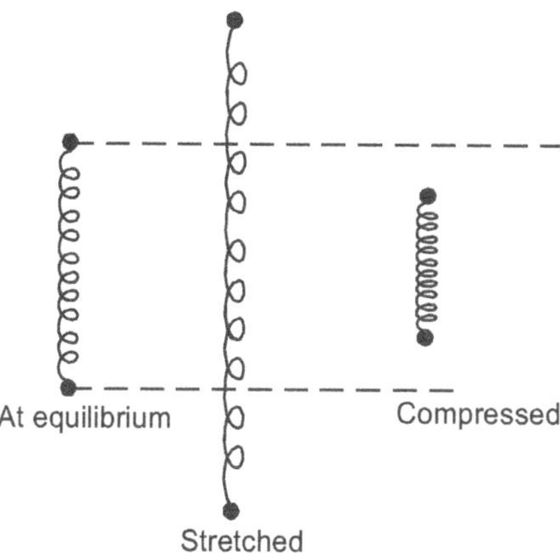

Fig. 12.6 Stretching vibrations of a diatomic molecule.

The oscillations of a molecule between expansion and compression modes of a bond are called **stretching vibrations**. Thus diatomic molecule like HCl undergoes characteristic stretching vibrations Fig. 12.6. Frequency of the vibrational transitions (oscillations) depend on the bond length (strength of the spring in the ball and spring model) and the mass of the atoms (mass of the balls). The vibrational frequency (ω) is given by the expression;

$$\omega = \frac{1k}{2\pi\sqrt{\mu}} \text{ Hz} \qquad \text{.....(12.9)}$$

$$\omega = \frac{1k}{2\pi C\sqrt{\mu}} \text{ cm}^{-1} \qquad \text{.....(12.10)}$$

or

$$k = \sqrt{4\pi^2 C^2 \omega^2 \mu} \text{ Nm}^{-1} \qquad \text{.....(12.11)}$$

where C is the velocity of light, μ the reduced mass $\frac{m_1 m_2}{m_1 + m_2}$ and k the force constant (in dyne cm^{-1} in CGS units and Newton.m^{-1} or Nm^{-1} in SI units; Conversion 1000 dyne cm^{-1} = $1Nm^{-1}$).

It is seen from the above expression that the vibrational frequency is directly proportional to the force constant and inversely proportional to the square root of reduced mass. In other words vibrational frequency increases with increase in the number of bonds but decreases with increase in the reduced mass. This is amply demonstrated in the representative groups given in Table 12.1 & 12.2.

Table 12.1 Spectroscopic data of hydrogen halides.

Molecule	ω (cm^{-1})	*k* (Nm^{-1})
H))) F	4138.5	966
H))) Cl	2990.6	516
H))) Br	2649.7	412
H))) I	2309.5	314

If we consider the hydrogen halides given in the Table 12.1, the hydrogen and halogen atoms are bound together through a single bond. Except HF the rest have force constants of similar magnitude. The exceptional behavior of HF is due to increased ionic character of the HF bond. The drop in the vibrational frequency (ω) is mainly due to increased reduced mass as we move from HF to HI.

Table 12.2 contains spectroscopic data of some single, double and triple bonds. Here the bond frequency changes depending on the mass of the constituent atoms, multiplicity of the bond and on the nature of the bond.

Table 12.2 Spectroscopic data of some single, double and triple bond.

Bond	Frequency range(cm^{-1})	Force constant ($Nm^{-1} \times 10^{-2}$)
→ C – O –	1000 - 1200	4.9
> C = O	1700 - 1750	12.80
→ C – C ←	1000 - 1200	4.5
> C = C <	1550 - 1650	10
– C º C –	2170 - 2270	16
>C – N <	1000 - 1200	4.8
→ C = N –	1650	12.10
– C ≡ N	2250	17.50

From the data we notice that there is a marked increase in the force constant and frequency values with the increase in the number of bonds. The force constant values given in Table 12.1 and 12.2 lead to the following generalizations :

(i) The forces constant is a measure of bond strength, the bond energy increases with increase in force constant value

(ii) The force constant increases with multiplicity of a chemical bond. The force constant values increase from a single bond to a double bond to a triple bond.

(iii) The force constant increases with increase in the ionic character of a bond given multiplicity.

Like rotational energy levels, vibrational energies are also quantized and are given by the expression,

$$E_v = (\nu + ½)\, h\omega \quad .....(12.12)$$

where ν is called the **vibrational quantum number**.

The selection rule for allowed vibrations is $\Delta\nu = \pm 1$

Applying the rule for a transition from ν +1 to ν

$$E = (\nu + 1 + ½)h\,\omega - (\nu + ½)\,h\,\omega \quad .....(12.13)$$

$$(\nu + 1) \rightarrow \nu$$

$$\Delta E_{\text{Vibration}} = h\,\omega \quad .....(12.14)$$

A Linear triatomic molecule (e.g., CO_2) can undergo two types of stretching vibrations and two types of bending vibrations, Fig. 12.7.

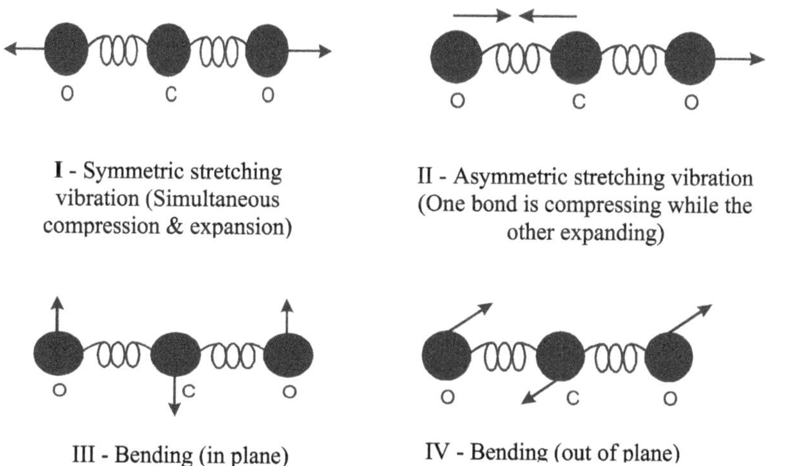

Fig. 12.7 Vibrational modes of a linear triatomic molecule.

A non-linear triatomic (e.g., H_2O, SO_2, NO_2, etc.) can under go three types of vibrations, Fig. 12.8.

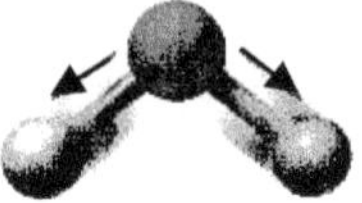

I - Symmetric stretching vibration (no change in bond angle)

II – Asymmetric stretching vibration (no change in bond angle)

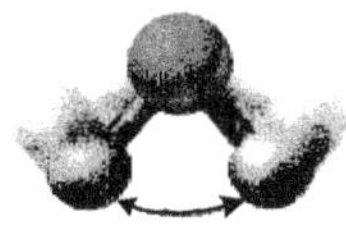

III – Scissoring or bending vibration (bond angle changes but no change in bond length)

Fig. 12.8 Vibrational modes of a non-linear triatomic molecule.

Experimental and theoretical spectral studies of different polyatomic linear and nonlinear molecules led to the following conclusions :

(i) A polyatomic molecule having *N* atoms has in total 3N degrees of freedom of which 3 degrees of freedom are needed to describe the translational motion in a three dimensional space. For example, a triatomic molecule has $3 \times 3 = 9$ degrees of freedom out of which 3 degrees of freedom correspond to translational motion and remaining 6 degrees of freedom describe the rotational and vibrational degrees of freedom.

(ii) All the linear polyatomic molecules have *3N–5* vibrational transitions and 2 degrees of rotational freedom. For example, a linear triatomic molecule has $3 \times 3 - 5 = 4$ vibrational degrees of freedom and 2 rotational degrees of freedom.

(iii) All the non-linear polyatomic molecules have *3N–6* vibrational degrees of freedom and 3 rotational degrees of freedom. For example, a non-linear triatomic molecule has $3 \times 3 - 6 = 3$ vibrational transitions and 3 rotational transitions.

(iv) Though *3N–5* degrees of freedom describe the vibrational motions in case of linear polyatomic molecules and 3N–6 in case of the non-linear olyatomic molecules, only those vibrations that produce a change in dipole moment of the molecule during the vibrational motion are detectable in a vibrational spectrum (also referred to as IR spectrum). For example, considering the stretching and bending vibration of CO_2, the symmetric stretching vibration produces no change in the dipole moment of the molecule and hence is inactive or not observed in IR spectrum Fig. 12.9.

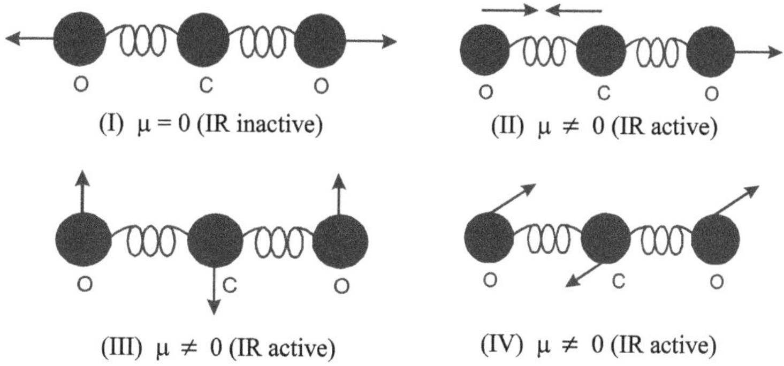

Fig. 12.9 Dipole moment change and IR activity.

In the asymmetric stretching vibration, although dipole moment is zero in the normal equilibrium mode, it assumes a change during vibration due to one of the oxygen atom approaching the carbon atom while the other is moving away. This makes an asymmetric vibration IR active.

Similarly, both in-plane and out of plane bending vibrations produce a change in dipole moment and the extent of deformation is one and the same. Therefore they produce only one peak at 667 cm^{-1} with double the probable intensity. In case of non-linear polyatomic molecules (e.g. H_2O), all the three vibrational transitions, viz., symmetric stretching vibration, asymmetric stretching vibration and bending vibration, produce a change in the dipole moment and are IR active.

For the same reason explained above H_2, Cl_2 and other homonuclear diatomic molecules are *IR* inactive while all heteronuclear diatomic molecules (HCl, HF, etc) *IR* active.

Problem 9.

Calculate the vibrational frequency of HF. Given the force constant of HF bond is 9.7×10^2 Nm^{-1}.

We know vibration frequency

$$\omega = \frac{1k}{2\pi C\sqrt{\mu}} \text{ cm}^{-1}$$

$$\mu = \frac{m_1 m_2}{m_1 + m_2} = \frac{\left(1/6.023\times10^{23}\right)\left(19/6.023\times10^{23}\right)}{\left[(1+19)/\left(6.023\times10^{23}\right)\right]}$$

$$\mu = \frac{19}{20} \times \frac{1}{6.023 \times 10^{23}} = 1.578 \times 10^{-24}$$

$$k = 9.7 \times 10^{2}\ \text{Nm}^{-1}$$

$$= 9.7 \times 10^{5}\ \text{dyne cm}^{-1}$$

Now substituting the values of h, k and : in the vibration frequency expression

$$\text{vibration frequency } \omega = \frac{1}{2 \times 3.14 \times 3 \times 10^{10}} \times \frac{9.7 \times 10^{5}}{\sqrt{1.578 \times 10^{-24}}}\ \text{cm}^{-1}$$

$$\omega = 4161.5\ \text{cm}^{-1}$$

Problem 10.

Calculate the force constant of ICl which vibrates at a frequency of 384.2 cm^{-1}.

$$\mu = \frac{\frac{m}{I} \times \frac{m}{Cl}}{\frac{m}{I} \times \frac{m}{Cl}} = \frac{127 \times 35}{127 + 35} \times \frac{1}{6.02 \times 10^{23}}$$

$$= 4.56 \times 10^{-23}\ \text{g molecule}^{-1}$$

Force constant

$$k = 4B^2 C^2 \omega^2 \mu\ \text{Nm}^{-1}$$

$$= 4 \times (3.14)^2 \times (3.0 \times 10^{10})^2 \times (384.2)^2 \times 4.56 \times 10^{-23}$$

$$= 2.39 \times 10^{5}\ \text{dyne cm}^{-1}$$

$$= 239\ \text{Nm}^{-1}$$

12.6.1 Infrared spectrophotometer

The essential parts of an infrared spectrophotometer are (i) a source of *IR* radiation (ii) a monochromator (iii) a sample slot (iv) a detector cum analyzer. The schematic diagram for the construction of an IR spectrophotometer with the above parts is shown in Fig. 12.10.

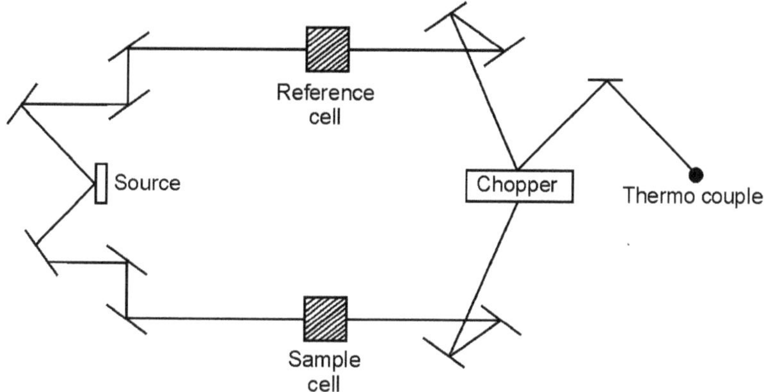

Fig. 12.10 Schematic diagram of an infrared spectrophotometer.

The source of infrared radiation is usually a Nernst lamp, which consists of a molded rod of ZrO, YtO and EbO-that is heated electrically to about 1500C. The monochromator consists of a series of gratings and prisms made out of alkali halide salts. It is used for obtaining a thin beam of light of single wavelength. A thin beam of intra-red light is initially split into two so that one passes through a sample solution and the other through the reference. The detector detects both the halves of the two transmitted rays. The analyzer picks up these signals and provides automatically provides a plot of wave number versus transmittance after subtracting the reference value from the sample value. The relationship between absorbance (A) and transmittance (T) is given by the equation,

$$A = \log (I/T)$$

In the sample slot, a suitable container is required to hold the compound in the path of the beam of infrared light. Regarding material for the container, glass or quartz cannot be used since these materials absorb radiation very strongly in the infrared region. However, metal halides such as NaCl or CaF_2 are transparent to air radiation. Hence, the cells for IR examination are usually made of NaCl or CaF_2.

Sample Techniques

The infrared spectrum of a compound may be recorded with the sample either in solid phase, liquid phase (neat liquid or in solution) or in the vapor phase. For recording the IR spectrum of a compound it is necessary that compound should be completely free of moisture because moisture absorbs very strongly around 3600 cm^{-1} of the infrared region.

The IR spectrum of a compound present in the solid phase may be obtained by taking the sample in the form of a pellet. The pellet is prepared by mixing around 1 mg of organic compound with around 100 mg of dry KBr and grinding together. The ground mixture is them pressed strongly in a hyraulic press to yield a pellet. This pellet is a kind of solid solution of the organic compound in KBr. This pellet is then placed in the path of the IR beam and examined for its spectrum. Since KBr does not absorb IR radiation, a compound is obtained.

For a solid compound, the IR spectrum may also be recorded with the sample in the form of a mull. The latter is prepared by grinding together, a mixture of the solid compound with a paraffin oil such as **nujol** to yield a mull. A drop or two of the mull is then placed between two sodium chloride plates and examined by IR radiation. Nujol shows strong absorption around 3000 cm^{-1} due to C-H stretching and so information in this region is lost when the compound is examined by IR as a nujol mull.

The liquid phase, spectrum of a compound can be obtained by examining the compound as neat liquid (between the two sodium chloride plates) or as a solution (about 1-2% in $CHCl_3$ or CCl_4) using glass cells having NaCl windows.

12.6.2 Interpretation of IR spectra of organic compounds

The most important fact for the interpretation of IR spectrum of an organic compound is that similar functional groups in similar compounds (i.e., in similar envIRonments) absorb at identical or closely related frequencies. Furthermore there are no well established relationships between concentration of the compound and intensity of absorptions. The intensities of absorptions are merely expressed in a qualitative manner as strong (s), medium (m), moderate strong, weak (w) and as very weak absorptions.

Some of the characteristic group frequencies are listed below

Characteristic group frequencies :

(I) –OH Stretching ≈ 3600 cm^{-1} (Free OH)

–OH Stretching ≈ 3300 – 3000 cm^{-1} (Hydrogen bonded –OH, broad, strong)

–NH Stretching ≈ 3300 cm^{-1} (medium intensity band)

$–NH_2$ Stretching ≈ 3300 and 3200 cm^{-1} (unequal doublet of medium intensity)

–CH Stretching ≈ 3000 cm^{-1} (medium to strong, usually a series of absorptions with shoulders etc.)

–SH Stretching ≈ 2700–2800 cm^{-1} (single band of moderate intensity)

(II) –C ≡ C–

–C ≡ N 2300–2000 cm^{-1} (single band usually as singlet)

–N = N–

$–N^+$ ≡ N

(III) Carbonyl groups

R–CO–R or R–CO–H ≈ 1715 cm^{-1} (srtong absorption band usually a singlet)

R–CO–OH or R–CO–OR ≈ 1800 cm^{-1} (strong absorption band, some times broad due to H-bonding)

$R–CO–NH_2$ or R–CO–NH–R ≈ 1650-1640 cm^{-1} (strong broad band)

IV –C = C–, –C = N– } ≈ 1600-1620 cm^{-1} (usually a moderate absorption, not diagnostic)

V. $–NO_2$, $–SO_2–$ } ≈ 1000-1300 cm^{-1} (usually a doublet due to asymmetric and symmetric stretchings of moderate intensity)

The infrared spectrum of a compound may be divided into three region namely,

(i) Functional group region (3600-1600 cm^{-1})

(ii) Finger-print region (≈ 1500-1000 cm^{-1}) and

(iii) Bending region (≈ 1000-600 cm^{-1}).

Most of the important functional groups in an organic compound usually show up in the functional group region and so its name. The finger-print region normally contains a large number of peaks serving like finger-print for compounds. Even if organic compounds contain identical functional groups and possess identical absorption bands in the functional group/bending region, there will be discernible differences in the spectra of compounds in the finger region. No two structurally different compounds, will show superimposable *IR* spectra.

12.6.3 Applications of IR spectra

(i) *Establishing identity of compounds :* If the IR spectrum of an unknown compound, matches completely with the IR spectrum of a known compound, then the identity of two compounds is completely established.

(ii) *Detection of functional groups in a compound :* This is done simply by examining the functional group region in the IR spectrum of the compound.

(iii) *Study of chemical reactions :* The progress of a chemical reaction can be studied by examining carefully, reaction mixtures and by looking for disappearence of peaks due to functional groups in starting material or appearance of new peaks due to products. For example; in the oxidation of an alcohol by selenium dioxide.

$$> CH - OH \xrightarrow{SeO_2} > C = O$$

one can monitor the change occurring in the reaction by recording the IR spectrum of the reaction mixture at different time intervals and look the disappearance of the peak due to –OH group or appearance of peak due to carbonyl functional group and thereby either understand the mechanistic aspects or know the extent of completion of the reaction.

(v) *Study of H-bonding :* H-bonding in an organic compound decreases the wave number of the absorption due to weakening of the functional group of the compound that is responsible for it. For example, in amines the stretching frequency of –NH– decreases due to H-bonding.

12.7 Vibration-Rotation Spectra

Vibration-rotation spectra is found in the same spectral range as vibration spectra (IR region). The general features of this spectra are listed below.

(i) Compounds like homonuclear diatomic molecules that do not give pure rotation spectra are also silent towards vibration-rotation spectra.

(ii) The general selection rules of vibration and rotation spectra $\Delta v = \pm 1$ and $\Delta J = \pm 1$ are applicable to vibration-rotation spectra also.

(iii) Every bond corresponds to a transition from one vibrational level to another

(iv) Fine structure of every vibration-rotation spectra is due to rotational changes that accompany the simultaneous vibrational changes.

(v) In the solution and solid states due to restricted rotations, the observed bands are smooth Fig. 12.11, while the same molecules in the gaseous state exhibit the fine structure Fig. 12.13.

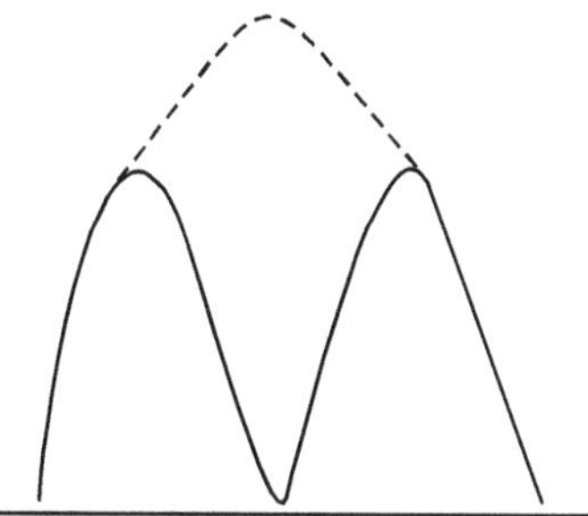

Fig. 12.11 Vibration-rotation spectra in condensed phase.

The total energy of absorption is equal to the sum of vibrational energy component and the rotational energy component.

$$E_{\text{vib-rot}} = E_{\text{vib}} + E_{\text{rot}} \qquad \text{.....(12.15)}$$

Substituting the E_{vib} (Eq. 12.12) and E_{rot} (Eqn. 12.2) values;

$$E_{\text{ vib-rot}} = h\omega\,(\text{v} + ½) + J(J+1)\,\frac{h^2}{8\pi^2 I}\text{ joule} \qquad \text{.....(12.16)}$$

For a transition $\text{v}' \rightarrow \text{v}''$ and $J' \rightarrow J''$

$$E''_{\text{vib-rot}} - E'_{\text{vib-rot}} = h\omega(\text{v}''-\text{v}') + \{J''(J''+1) - J'(J'+1)\}\,\frac{h^2}{8\pi^2 I} \qquad \text{.....(12.17)}$$

Applying the selection rules of allowed molecular energy transitions $\Delta\text{v} = \pm 1$ and $\Delta J = \pm 1$;

When $\Delta J = +1$ (i.e., $J'' = J' + 1$) & $\Delta\text{v} = +1$, Fig. 12.12 :

$$\Delta E_{\text{vib-rot}} = h\omega + \frac{h^2}{8\pi^2 I}\,[(J+1)(J+2) - J(J+1)] \qquad \text{.....(12.18)}$$

$$\Delta E_{\text{vib-rot}} = h\nu = h\omega + \frac{h^2}{8\pi^2 I}\,(J+2) \qquad \text{.....(12.19)}$$

or

$$h\nu = h\omega + Bh\,(J+2)\text{ Joule} \qquad \text{.....(12.20)}$$

or $\nu = \omega + B\,(J+2)\ \text{cm}^{-1}$ .....(12.21)

where ν is the frequency of light absorbed and $J = 0, 1, 2, ...$

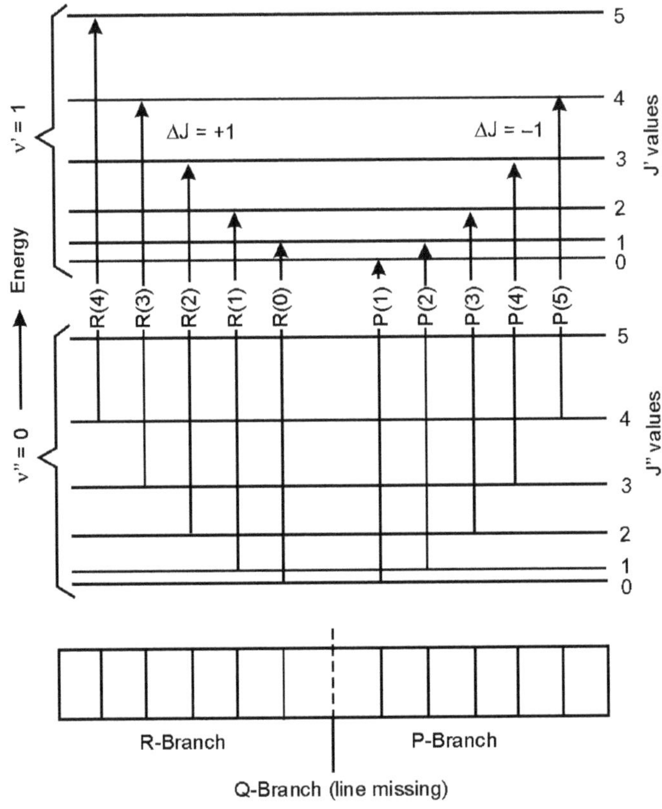

Fig. 12.12 $\Delta J = +1$ & $\Delta v = +1$ $\Delta J = -1$ & $\Delta v = +1$

When $\Delta J = -1$ (i.e., $J'' = J' - 1$) & $\Delta v = +1$, Fig. 12.12 :

$$\Delta E_{\text{vib-rot}} = h\omega + \frac{h^2}{8\pi^2 I}\,[(J-1)\,J - J\,(J+1)] \quad .....(12.22)$$

$$\Delta E_{\text{vib-rot}} = h\nu = h\omega + \frac{h^2}{8\pi^2 I}\,(-2J) \quad .....(12.23)$$

$$\Delta E_{\text{vib-rot}} = h\nu = h\omega - \frac{h^2}{4\pi^2 I}\,J \text{ Joule} \quad .....(12.23)$$

or $h\nu = h\omega + 2Bh\,(J+2)$ Joule .....(12.24)

or $\nu = \omega + 2B(J + 2)\ cm^{-1}$ .....(12.25)

where ν is the frequency of light absorbed and J = 1, 2, ...

A typical very fine vibration-rotation spectrum having transitions corresponding to $\Delta J = \pm 1$ is shown in Fig. 12.13.

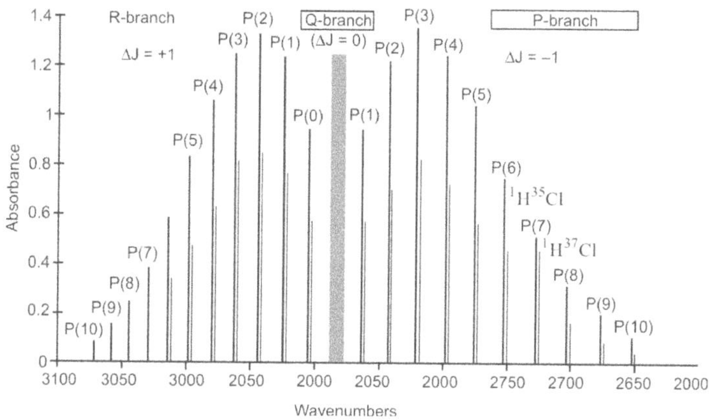

Fig. 12.13 A typical vibration rotation spectrum.

The vibration-rotation spectrum consist of equally spaced lines (spacing $\Delta\bar{\nu} = \Delta\bar{B}$) forming two branches. Lines corresponding to ΔJ = 1 are referred as P–branch and those correspond to ΔJ = +1 are called R-branch. As ΔJ = 0 will be missing from the spectrum and the position which this line would occupy ($\nu = \nu_0$) is known as the band origin or band center or Q-branch.

12.8 Electronic Spectra

Electronic spectra is the result of excitation or emission of electrons from one electronic level to another. Electronic excitation requires far higher energy compared to vibrational and rotational transitions. Electronic spectral bands, specially in case of gases, contain the fine structure due to simultaneous vibrational and rotational transitions, Fig. 12.2. Generally the fine structure of electronic spectra is too complicated and is not useful for obtaining information about the vibration and rotational transitions. Due to simplicity of production and detection techniques employed in obtaining electronic spectra it is popularly used in many analytical labs. Interpretation of fine lines of not too complicated spectra give a very valuable data concerning vibrational and rotational transitions. Electronic spectra are obtained in the visible and ultra-violet regions of electromagnetic spectrum and assumes the name UV-Visible spectra. The electronic energy levels get closer and closer together towards the shorter wavelength. This makes the bands in an electronic spectrum to converge and ultimately merge into a **continuous spectrum**. In an atomic spectrum this convergence limit gives the ionization potential and the continuous region in a molecular spectrum indicates an **ionization zone**. In case of a molecular spectrum the continuous spectral region is attributed to **dissociation of the molecule** into separate molecule fragments joined by the bond under consideration. Applying the **Born-Oppenheimer approximation**, which states that the electronic, vibrational and rotational energies of a molecules are completely independent of each other we have

$$E_{Total} = E_{Electronic} + E_{Vibrational} + E_{Rotational}$$

or $$\Delta E_{Total} = \Delta E_{Electronic} + \Delta E_{Vibrational} + \Delta E_{Rotational}$$

The approximate orders of magnitude of the changes are

$$\Delta E_{Electronic} \cdot \Delta E_{Vibrational} \times 10^{3} \cdot \Delta E_{Rotational} \times 10^{6} \qquad .....(12.26)$$

Rotational spectra are exhibited by molecules possessing a permanent dipole moment while the vibrational spectra require a change in dipole moment during the motion. On the contrary electronic spectra is accompanied by corresponding vibrational and rotational transitions for all molecules owing to the fact that every electronic excitation or emission is invariably accompanied by dipole moment change.

Thus the homonuclear diatomic molecules like H_2, Cl_2 etc., which show no rotation or vibration or vibration-rotation spectrum show electronic spectrum having rotational and vibrational fine structure. Rotation and vibration spectra accompanying the electronic spectra is highly useful in the calculation of molecular parameters like bond angle, bond length etc.

12.8.1 Construction of a UV–Visible Spectrophotometer

The important parts of a spectrophotometer are a source, a monochromator, a cell/cuvette and a detector cum analyzer Fig. 12.14.

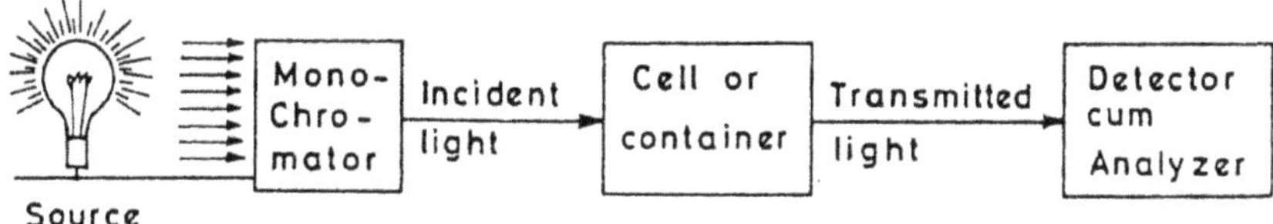

Fig. 12.14 Schematic diagram of the construction of a UV-visible spectrophotometer.

The source provides rays of ultraviolet light or visible light depending on region to be scanned. The monochromator allows filtering of the different wavelengths of radiation and allows only one radiation of a particular wavelength to pass through starting from 220 nm to higher wavelengths. The container, a quartz tube cell or a cuvette, hold the experimental solution in the path of the light. The detector cum analyzer detects and analyzes the transmitted light with reference to incident radiation. In the double beam spectrophotometers the primary source of light is split into two beams, one of which passes through a cell containing the reference solvent the other experimental solution. The spectrophotometer electronically subtracts any absorption due to the solvent, thus, nullifying effects due to absorption of light by the solvent. Most of the modern spectrometers provide straight away plots of wavelength versus absorbance Fig. 12.15.

The ordinary uv-visible spectrometer provides acceptable spectra in the region 200-800 nm. For recording of spectra in the region below 200 nm, the major limitation is the presence of air in the instrument. Oxygen absorbs strongly at about 185 nm. Therefore, the instrument has to be made free of air/oxygen. That is, the instrument has to be evacuated. The spectra obtained below 200 nm with evacuation of air/oxygen are called **vacuum ultra-violet spectra**.

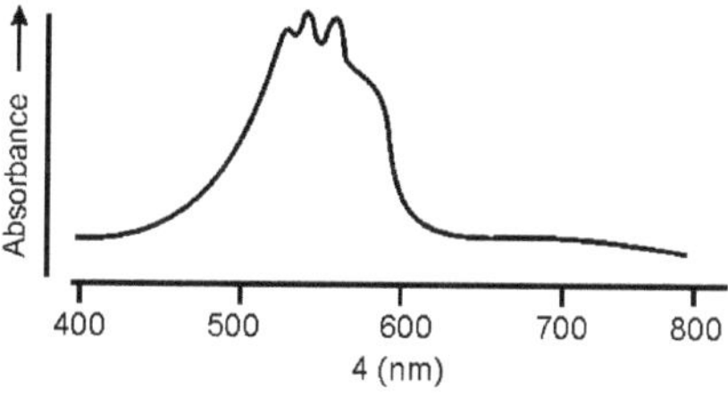

Fig. 12.15 A typical UV-visible spectra (Spectra of $KMnO_4$).

12.8.2 Shape of UV absorption spectra-Band spectra of organic compounds

The UV spectra arise as a consequence of the electronic transitions that take place in organic molecules. Since all these transition require fixed/quantized amounts of energy, an ultra-violet spectrum of a compound may be expected to consist of one or more lines or well-defined sharp-peaks. However, most or rather all the UV spectra consist of broad humps. The peaks are always broad because each electronic transition in a molecule is accompanied by vibrational and rotational transitions. The latter require smaller amounts of energy which is usually obtainable through the energy supplied to bring about electronic transitions. The rotational and vibrational spectra usually appear as sharp peaks.

12.8.3 Theory of Chromophores

In the early stages of study of spectra, compounds were discussed from the point of view of chromophores and auxochromes. It was recognized quite early that compounds appear colored because they contain functional groups which absorb radiation of certain wavelength when ordinary light falls on them. For example, a compound appears blue in ordinary day light (λ = 400-800 nm) because, it has a functional group/s that absorb yellow light (i.e. radiation corresponding to (570-590 nm). The transmitted light, which is now deficient in yellow light gives the effect of blue color. The transmitted light will have the remaining six colors i.e. white light minus yellow color, but the human eye can see only the dominant complimentary color i.e., blue. Such *functional groups conferring color on organic compounds were named as* **chromophores** [Greek : meaning color forming]. Most chromophores have multiple bonds such as C=O C=N, N=N, etc. Functional groups such as hydroxyl (–OH), amino (–NH_2) or halo confer color on substances but have the *ability to increase the coloring power of a chromophore. Such groups were named as* **auxochromes.** This term is obsolete.

In modern work a chrmophore is defined as any functional group which can absorb in the UV/visible region (200-800 nm) of the electromagnetic radiation. *An auxochrome is defined as any functional group which by itself does not absorb in the UV/visible region but when attached to a chromophore increases its absorption.*

Thus, for example, benzene has an absorption at 254 nm, $\varepsilon_{max} = 204$. However, when an auxochrome such as –OH is attached to the benzene ring (i.e., phenol) it shows λ_{max} of 270 nm and $\varepsilon_{max} = 1450$. The UV spectra of these two compounds are shown in Fig. 12.16.

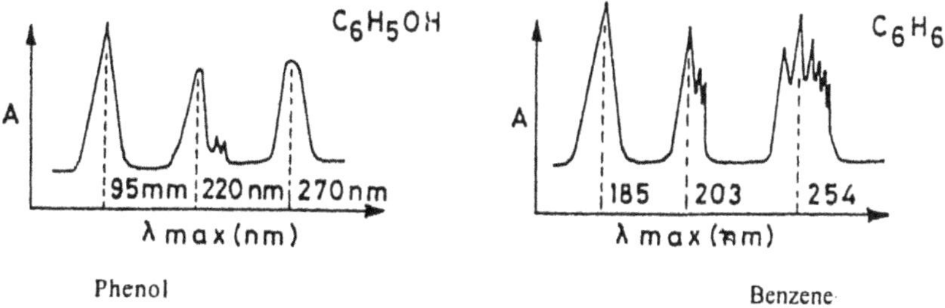

Fig. 12.16 Absorption spectra of Phenol and Benzene.

(i) A change in absorption peak a chromophore from a standard value to higher wave length is called the **bathochromic shift**; also called **red shift** since the movement is towards the red end of the spectrum.

(ii) A change in absorption peak a chromophore from a standard value to shorter wavelength is called the **hypsochromic shift**; also called **blue shift** since the movement is towards the blue end of the spectrum.

(iii) The increase in the intensity of a spectral band due to substituents or interaction with the molecular environment is called **hyperchromic effect**.

(iv) **Hypochromic effect** is the opposite of hyperchromic effect where the intensity of a band decreases due to substituent or environment.

Consider a chromophores such as N=N, –N=O, C=O etc. These chromophores contain, *n*(non-bonding), *s*(bonding) and p(bonding) electrons. The excited levels corresponding to these molecular orbitals in an organic compound would be p^* and s^*. Therefore, the probable transitions that can occur in these compounds/functional groups when irradiated with UV/visible light are

(a) $n \longrightarrow \pi^*$

(b) $n \longrightarrow \sigma^*$

(c) $\pi \longrightarrow \pi^*$

(d) $\pi \longrightarrow \sigma^*$

(e) $\sigma \longrightarrow \pi^*$

(f) $\sigma \longrightarrow \sigma^*$

Of these possible six electronic transitions, the most probable ones are $n \longrightarrow \pi^*$ (around 200 nm) $\pi \longrightarrow \pi^*$. (around 200 nm but can go to higher wavelength when an auxochrome is present).

When two or more chromophoric groups are present in a molecule separated by several single bonds such that there is no interaction between individual chromophores, the UV spectrum of the compound is the sum total of the UV spectral absorptions of individual functional groups. However, if the different chromophores are separated only by one double bond such that there is a large resonance interaction between functional groups, then there will be a large resonance interaction between functional groupS, then there will be a large increase in λ_{max} as well as in the value of extinction coefficients of corresponding peaks.

The explanation for these observations is that conjugation brings about delocalization of p electrons. Therefore, the ease with which these electrons can be excited to higher energy levels also increases thereby decreasing the wavelength of absorption maxima.

12.8.4 Interpretation of *UV* spectrum

The most important role of UV spectrum in structure determination is to identify the chromophore present in the organic compound. This is done with the help of a set of rules, derived with the help of the previous history of the compounds and with the help of a comparison of the spectrum of the compound with that of several authentic/model systems already available in the literature.

12.8.5 Applications of UV-vis spectroscopy

The following are some of the important applications of UV-vis absorption spectroscopy.

(i) *Determination of chromophoric groups :* This is done using the technique already mentioned above.

(ii) *Study of stereochemistry :* In geometrical isomerism, two types of isomers are encountered, namely, cis-isomers and trans-isomers. In general, the trans-isomers absorb at longer wavelength giving higher γ_{max} value than their cis-isomers. Hence, by comparing the UV spectra of any pair of cis and trans compounds, it is possible to distinguish between the two.

For example :

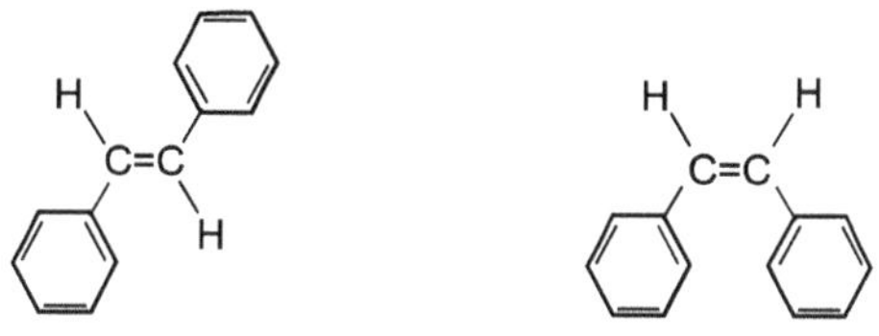

(trans-Stilbene) (cis-Stilbene)

(iii) *Study of tautomerism* : UV spectroscopy has been used extensively to study the presence and amounts of tautomers in a compound. For example, benzoylacetanilide was suspected to exist as an equilibrium mixture of two readily inter convetible keto and enol tautomers.

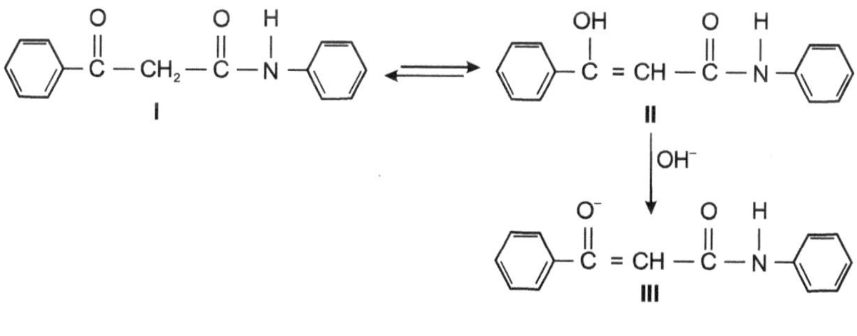

$C_6H_5-C(=O)-CH_2-C(=O)-NH-C_6H_5$ (I) $\rightleftharpoons$ $C_6H_5-C(OH)=CH-C(=O)-NH-C_6H_5$ (II)

II $\xrightarrow{OH^-}$ $C_6H_5-C(O^-)=CH-C(=O)-NH-C_6H_5$ (III)

The presence of two forms I and II was shown by the too maxima in the UV spectrum at 245 and 308 nm. The latter absorption peak shifted to 323 nm when a drop of alkali was added to the test solution and spectrum is recorded at pH 12.0. It is inferred that,

(i) The maxima at 245 nm is due to keto tautomer, I.

(ii) The maxima at 308 nm is due to enol tautomer, II.

(iii) The maxima at 323 nm was due to enolate anion, III.

From the relative heights of the maxima at 245 nm and 308 nm in the original UV spectrum, the relative ratio of tautomers I and II under different experimental conditions was calculated.

(iv) *Detection of ionic species :* The λ_{max} recorded for certain types of compounds, such as carboxylic acids, phenols etc., change to higher wave length side on addition of alkali.

$$Ar{-}OH + HO^- \longrightarrow Ar{-}O + H_2O$$

$$Ar{-}\underset{\underset{O}{\|}}{C}{-}OH + HO^- \longrightarrow Ar{-}\underset{\underset{O}{\|}}{C}{-}O^- + H_2O \text{ (Bathochromic shift)}$$

However, opposite is the effect, when the spectra of basic compound are recorded after addition of a drop of aqueous acid to the test solution in the quartz cell.

$$Ar{-}NH_2 + H^- \longrightarrow Ar{-}NH_2 \text{ (Hypsochromic shift)}$$

By calculating spectral areas of different peaks it is possible to determine the concentration of all the species on both sides of equilibrium. Thus, the value of equilibrium constant, rate constants, of acid-base dissociation etc., can be evaluated.

(v) *Assay of biologically and commercially useful products :* UV spectra can be used to determine the assay or purity of a biologically or commercially useful product. Since the height of a λ_{max} in the UV spectrum of a compound is directly proportional to the concentration of the species, a calibration curve plotting absorbance versus molar concentration can be constructed. The concentration of the unknown is then determined by simply recording absorbance of the compound. The routine assay of a large number of compounds belonging to the class of drugs, vitamins, antibiotics, pesticides, industrial intermediates can be done using this technique.

Numerical Problems

1. Estimate the moment of inertia and rotational frequency of NO whose internuclear distance is 1.151 Å.

   [*Hint* : $I = \mu l^2$ where μ is reduced mass and l the internuclear distance Rotational frequency $\nu = \frac{h}{4\pi^2 I}$]

2. Find the rotational spacing for HCl^{35}. Assume the inter nuclear distance is equal to 1.275 Å

   [*Hint* : Fist calculate the I_{HCl} and then calculate B using the expression $B = \frac{h}{8\pi^2 I}$ rotational spacing is equal to 2B].

3. Calculate the force constant of NO which exhibits a vibrational frequency of 1904 cm^{-1}. Given that the atomic weights of N ad O are 14 and 16 respectively.

   [*Hint* : Force constant $k = 4B^2 C^2 \omega^2 \mu$]

4. Find the vibrational frequency of HI with the help of its force constant 314 N m^{-1} taken from the literature. Given that the atomic masses of H and I are 1 and 127 respectively.

5. The rotation bands of CO in a microwave spectrum are spaced by 3.858 cm^{-1}. Estimate its moment of inertia and equilibrium internuclear distance.

   [*Hint* : The rotational bands in a microwave spectrum are spaced by $2B$, where $B = \frac{h}{8\pi^2 I}$]

Self Test Questions

1. A diatomic molecule is polar if the atoms involved are _________
2. CO_2 has zero dipolemoment because the molecule is _________
3. The number of translations degrees of freedom are

   (A) 3*N*-5, (B) 3*N*-6,

   (C) 3, (D) 2

4. The color property of a substance is due to

   (A) Vibrational transitions (B) Rotational transitions

   (C) Translational transitions (D) Electronic transitions

5. The number of vibrational modes of a non-linear molecule are

   (A) 3N-5, (B) 3N-6,

   (C) 3, (D) 2

6. Which of the following compounds is ir-inactive

   (A) H_2, (B) HCl,

   (C) HBr, (D) HI

7. Which of the following compounds show microwave spectra

   (A) O_2, (B) CO_2,

   (C) H_2O, (D) Cl_2

8. Total vibrational modes of CO_2

   (A) 2, (B) 3,

   (C) 1, (D) 4

9. Which one of the following statements is true

   (A) The force constant increases with the multiplicity of the bond.

   (B) The force constant decreases with the increases in the weight of molecule.

   (C) The force constant is a measure of polarity of a bond

   (D) None of the above

10. The bond length of Cl_2 molecule could be estimated from its

    (A) Translation energy changes (B) Rotational spectrum

    (C) Vibrational spectrum (D) Electronic spectrum

11. The line spectra is often produced by _________

12. The band spectra is a characteristic of _________

13. Pure rotational spectra is observed in _________ region.

14. CO_2 is IR insensitive because it has _________ dipole moment.

15. Electronic spectra is noticed in the _________ and _________ spectral regions.

Short and Essay Type Questions

1. Write any three applications of IR in the determination of structure of molecules.

2. What is a polar bond ? Explain "all molecules containing polar bonds need not be polar" with a suitable example ?

3. In which region does the transition due to vibrational energy change occur ? Calculate the number of modes of vibration expected for water molecules. Which type of vibrations do not show absorption bond.

4. (a) Give two possible structures for a molecule of the type AB_2. How can dipole moment values help in deciding the two structures.

   (b) In acetone and benzophenone which one absorbs at longer wave lengths? Explain.

5. What are the applications of UV spectroscopy.

6. (a) Discuss the principle underlying infrared spectroscopy.

   (b) How is the information from rotational spectral lines, used in the calculation of the bond length of a molecule.

7. Give regions at which IR, UV spectra fall. How does IR spectroscopy show whether a hydroxyl group in an organic compound is "Free or Associated".

8. Explain why CO_2 is linear while H_2O is non-linear molecule ?

9. Give applicatins of IR spectroscopy ?

10. Discuss important applications of UV & IR spectroscopy.

11. Which type of molecules show rotational spectra ? Which of following show rotational spectra

    (a) H_2 (b) HCl

    (c) H_2O (d) Br_2,

12. Using IR spectra how would you distinguish between

    (i) CH_3COCH_3 and CH_3 CN

    (ii) CH_3COCH_3 and CH_3 CH_2OH

    (iii) Cis and trans CH_3 $CH_2 = CH_2CH_3$

    (iv) o– and p– Chlorophenol

13. Using electronic spectra how would you distingish between.

    (i) Benzene and aniline

    (ii) Ethylene and butadiene

    (iii) Benzene and anthracene

    (iv) Neutral molecule and anion of phenolphthalein.

14. Indicate whether you would use IR and UV spectroscopy for distinguishing between,

    (a) $CH_2 = CH\text{-}C\text{-}CH_3$ and CH_3 CH_2 C CH_3

    5 5

    O O

    (b) CH_3 O CH_3 and CH_3CH_2OH

Chapter 13

Chemical Kinetics

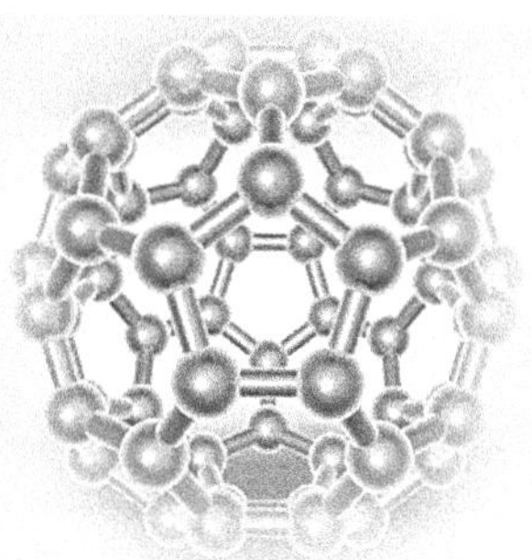

13.1 Introduction

Movement and change are the prime characteristic manifestations of life. A life cycle is in turn is a sequence of innumerable chemical reactions. Study of changes occurring in chemical reactions with time, is the life of a new branch of chemistry called chemical kinetics. It mainly deals with the changes occurring in the chemical reactions on account of changes in pressure, temperature, volume and concentration; about the routes traced in bringing these changes and stipulating conditions for optimization of desired products.

A change in a chemical reaction may occur million times in a wink of eye and give a feeling that it is instantaneous or may be observable in a reasonable time period, or may be so slow that it takes years to notice. Thus addition of silver nitrate to sodium chloride solution instantaneously gives a precipitate, while rusting is an extremely slow process where it is easy to realize the consequence but not time to time change.

As we already discussed, thermodynamics is mainly concerned with the system at equilibrium and is successful in predicting the spontaneity of a reaction. It lacks the *time sense* and therefore fails to predict *how long* a chemical reaction can occur. Thus combination of hydrogen and oxygen to give water is thermodynamically feasible, but kinetically not feasible at ordinary temperatures. Kinetics is concerned with *how fast* a chemical reaction occurs.

13.2 Rate of Reaction

With the elapsing of time, the concentration of reactants gradually falls while the concentration of products raises, Fig. 13.1.

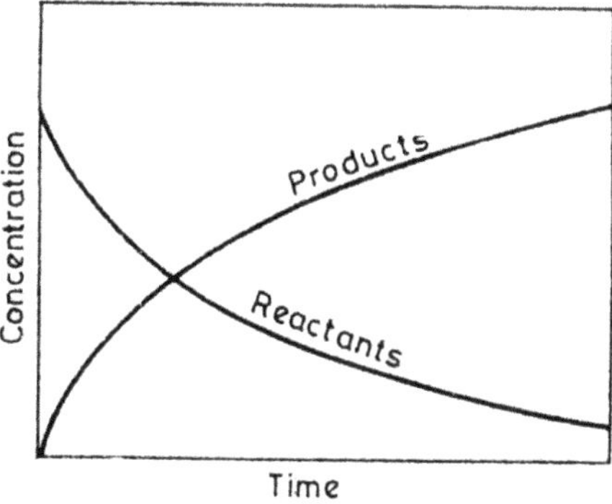

Fig. 13.1 Change in the concentration of reactants and products with time.

Rate of a reaction is defined as *the change in the concentration of reactant or product with time.*

$$\text{Rate of reaction} = \frac{-\Delta\, Reactant}{\Delta\, Time} = \frac{\Delta\, Product}{\Delta\, Time} \qquad .....(13.1)$$

The *negative sign* before Δ *reactant* implies that the amount of reactant is falling with time. The rate of a chemical reaction can be expressed in terms of rate of consumption of any reactant or rate of formation of any product. For example, oxidation of iodide by persulphate is represented by the equation,

$$K_2S_2O_8 + 2KI \longrightarrow 2K_2SO_4 + I_2 \qquad .....(13.2)$$

When persulphate and iodide solutions are mixed, originally colorless reaction mixture slowly turns brown due to formation of iodine. The reaction *rate*, or simply *rate* or *velocity* or speed may be expressed in terms disappearance of persulphate or iodide or in terms of appearance of sulphate or iodine with change in time.

For a reaction represented by the equation,

$$A + 2B \longrightarrow 2P + Q \qquad .....(13.3)$$

the rate of consumption of B is twice that of A and the rate at which Q is produced is one half the rate at which B is consumed. Therefore,

$$\text{Rate of reaction} = \frac{-dn_A}{dt} = \frac{-1}{2}\frac{dn_B}{dt} = \frac{dn_P}{dt} = \frac{dn_Q}{dt} \qquad .....(13.4)$$

Where n_A, n_B, n_P and n_Q are the number of moles of A, B, P and Q respectively.

For a general reaction,

$$aA + bB + ... \longrightarrow pA + qA + \;..... \qquad .....(13.5)$$

rate of reaction may be written as;

$$\frac{-1}{a}\frac{dn_A}{dt} = \frac{-1}{b}\frac{dn_B}{dt} = \frac{1}{p}\frac{dn_P}{dt} = \frac{1}{q}\frac{dn_Q}{dt} \qquad .....(13.6)$$

This rate expression is true for any open or closed homogenous and heterogeneous systems.

The homogeneous closed system, especially liquid phase reactions are carried at constant volume and are most convenient for kinetic measurements. For such systems,

$$-\frac{1}{V}\cdot\frac{dn_A}{dt} = -\frac{d[A]}{dt} \qquad .....(13.7)$$

where [A] is the molar concentration of any species A. Rewriting equation (13.6) in terms of concentration we have,

$$\text{Rate} = -\frac{1}{a}\frac{d[A]}{dt} = -\frac{1}{b}\frac{d[B]}{dt} = \frac{1}{p}\frac{d[P]}{dt} = \frac{1}{q}\frac{d[Q]}{dt} \qquad .....(13.8)$$

According to this formulation, the rate must have units of concentration divided by time. Conventionally rate is expressed in mol L^{-1} s^{-1} but S.I system requires the expression mol dm^{-3} s^{-1}.

Table 13.1 Decomposition of N_2O_5 (in CCl_4)

Time (s)	**[N_2O_5]** [M]	$rate = -\frac{d[N_2O_5]}{dt}$ **(Ms^{-1}) × 10^4**	$k = \frac{rate}{[N_2O_5]}$ **(s^{-1}) × 10^4**
0	1.4	8.0	5.7
200	1.24	7.0	5.6
400	1.10	6.0	5.5
600	0.98	5.5	5.6
800	0.87	5.0	5.7
1000	0.77	4.5	5.8
1200	0.68	4.0	5.9
1400	0.60	3.5	5.8
1600	0.53	3.0	5.7
1800	0.47	2.5	5.3
2000	0.42	2.4	5.7

Rate of reaction involves concentration term and therefore is not the same throughout the reaction. The rate is initially rapid and maximum due to high reactant concentration, and falls with the increase in time due to simultaneous fall in reactant concentration. Therefore rate is expressed at any given time or at any given concentration of a particular reactant. Rate of a reaction at any given time is readily obtained from the slope of tangent drawn at that particular instant in a concentration versus time graph. For example, in the decomposition of N_2O_5, we notice a gradual fall in the rate of reaction, Fig. 13.2 and Table 13.1.

$$2N_2O_5 \text{ (in } CCl_4) \longrightarrow 2N_2O_4 \text{ (in } CCl_4) + O_2 \text{ (g) with time.}$$

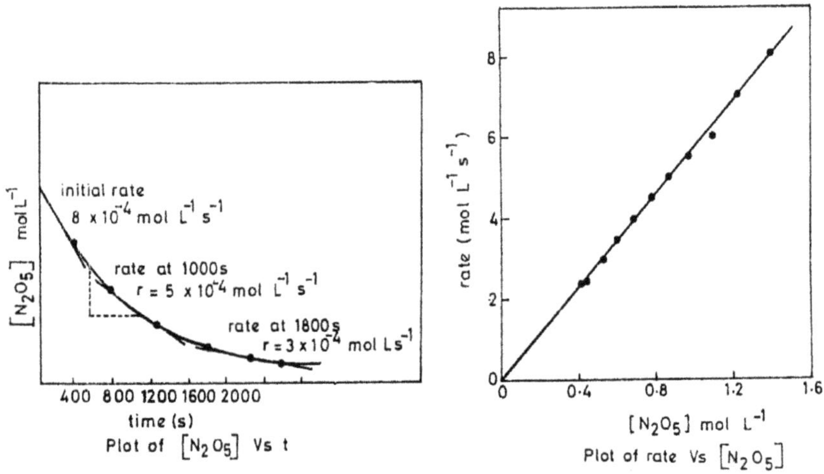

Fig. 13.2 Variation of rate of reaction.

13.3 The Rate Law

In the example cited above, the rate of decomposition of N_2O_5 (data given in Table 13.1) when plotted against concentration of N_2O_5 gives a straight line, Fig. 13.2. This suggests that the rate is directly proportional to the concentration of N_2O_5.

$$\text{Rate} \propto [N_2O_5]$$

$$= k\,[N_2O_5] \qquad .....(13.9)$$

The proportionality term k is called **the rate constant**. Rearranging the equation gives;

$$k = \frac{rate}{[N_2O_5]}$$

The k values obtained from such a rearranged equation are given in last column of Table 13.1. These values suggest an important feature that *the rate constant is independent of concentration of reactants* (minor variations in k found in Table 13.1 are only due to experimental errors).

Rate constant is thus defined as the rate of reaction when all the reactants are present at unit concentration. Hence rate constant is also called **specific reaction rate or velocity constant**. Rate constant, however, varies with change in temperature of the reaction medium.

Similar to expression (13.8), it possible to express the rate of reaction through a mathematical equation containing concentration terms of all reactants. Consider a general reaction,

$$aA + bB + ... \longrightarrow \text{products}$$

where a, b, ... are the coefficients in the balanced equation.

Since at any given temperature rate usually depends on concentration of each reactant, we can write

$$\text{rate} \propto [N_5O_5]$$

$$\text{or} \quad \text{rate} = k\,[A]^{\alpha}\,[B]^{\beta} \qquad .....(13.10)$$

Equation 13.10 which relates the rate with concentration of the reactants is called **differential rate law, rate law or rate equation**. The concentration exponents α, ß ..... are generally small integers, although in some cases they may be fractional or negative. *Sum of these concentration exponents in the rate equation is called* ***order of the reaction***. It is important to note that there is no relation between the concentration exponents α, ß ..... and corresponding coefficients *a*, *b*, ... What ever resemblance that may be found between these two values should be taken as only a matter of chance.

A rate equation serves three objectives ;

(i) Rate law enables the prediction of rate, given the values of specific reaction rate and the concentration of the reactants.

(ii) Rate law groups the reactions into various orders, viz., zero, first, second, etc.,

(iii) A rate equation helps in arriving at the reaction mechanism.

Some substance do not appear in the stoichiometric equation but affect the rate of reaction are called **catalysts** or **inhibitors**. The former increases the rate, while the later retards it. *Rate equation is purely an empirical relation*, i.e., it is purely obtained from experimental results. Table 13.2 gives few reactions and corresponding rate equations derived from the experimental data.

Table 13.2 Few reactions and corresponding Rate equations

Stoichiometric equation	Rate equation
(I) $CH_3COOC_2H_5 + H_3O \rightarrow$ $CH_3COOH + C_2H_5OH$	$r = \frac{d[CH_3COOH]}{dt} = k[CH_3COOC_2H_5][H^+]$
(II) $K_2S_2O_8 + 2KI \rightarrow$ $2K_2SO_4 + I_2$	$r = \frac{d[I_2]}{dt} = k[K_2S_2O_8][KI]$
(III) $CH_3-\overset{\overset{O}{\parallel}}{C}-CH_3 + I_2 \rightarrow$ $CH_3-\overset{\overset{O}{\parallel}}{C}-CH_2I + HI$	$r = \frac{d[I_2]}{dt} = k[CH_3-\overset{\overset{O}{\parallel}}{C}-CH_3][H^+]$
(IV) $H_2 + Br_2 \rightarrow 2HBr$	$r = \frac{1}{2}\frac{-d[HBr]}{dt} = \frac{k_1[H_2][B_r]}{1+k_2([HBr]/[B_{r_2}])}$

Analysis of rate laws and reactions given in Table 13.2 leads to the following conclusions.

1. *The rate law bears no simple relation to stoichiometric equation.* In reaction (II), although 1 molecule of $K_2S_2O_8$ reacts with 2 molecules of KI, we find only one molecule of each in the rate equation. Similarly reaction (IV) seems to be a simple 1:1 reaction but the rate law is complicated with $[Br_2]$ showing a fractional order and product molecule seen in the denominator.
2. *The rate law may not reflect the concentrations of all reactants.* In case of reactions (II) and (IV), the rate law contains the concentrations of every reactant but rate law (I) and (III) show independence of concentration of some of the reactants. For example, in equation III, iodine is a reactant but is not seen in the rate law.

3. *Generally a rate law is devoid of product concentration but may contain them in few reactions*, where product is not a silent spectator but sets some secondary reactions that interfere with the primary reaction.
4. *The rate law may contain the concentration of species which do not appear in stoichiometric equation.* Thus in equation I, the concentration of H^+ ions is not seen in the overall stoichiometric reaction but finds a place in the rate equation.
5. *The rate law is purely empirical and truly valid only for a given set physical parameters under which conclusions like rate constant are drawn.* The rate law may change by changing any one or more such physical parameters.

Problem 1.

Write all possible rate equations in terms of each of the reactant and product of the following reaction.

$$A + 2B \longrightarrow 3C + D + 2E$$

Adopting the convention indicated by equation (13.6) we can write the rate equations as follows.

$$\frac{-d[A]}{dt} = \frac{1}{2}\frac{d[B]}{dt} = \frac{1}{3}\frac{d[C]}{dt} = \frac{1}{2}\frac{d[D]}{dt} = \frac{1}{2}\frac{d[E]}{dt} = k\,[A]\,[B]^2$$

Problem 2.

Write the rate equation for the formation of D in each of the following elementary reactions.

(i) $2A + B \longrightarrow 2C + D + E$

(ii) $2A + B \longrightarrow C + 2D + E$

(iii) $A + B \longrightarrow C + 3D + 2E$

(iv) $2A \longrightarrow B + 2C + D$

Rate of formation of a product is always given a positive sign. Since the reactions are elementary reactions, the concentration coefficients represent order with respect to the reactant.

(i) $$\frac{d[D]}{dt} = k\,[A]^2\,[B]$$

(ii) $\frac{1}{2}.\frac{d[D]}{dt} = k[A]^2[B]$

(iii) $\frac{1}{3}.\frac{d[D]}{dt} = k[A][B]$

(iv) $\frac{d[D]}{dt} = k[A]^2$

13.4 Order of Reaction

The concentration exponents α, β, γ, ... *in the rate equation* (13.10) *decide the sensitivity of rate on the concentration of corresponding reactant and are called order of that particular reactant or individual order. Sum of the individual orders gives the overall order of the reaction.*

Overall order $n = \alpha + \beta + \gamma + ...$ .....(13.11)

Thus *order of a reaction may be defined as the sum of concentration exponents shown in the rate equation.*

The *order* with respect to a reactant or a product or any other foreign species that effect the reaction rate *could assume a positive, negative, integral or fractional value.* These individual and overall *orders are dimensionless.* The substances that appear in the stoichiometric equation but do not appear in the rate equation, i.e., when the rate is insensitive to a particular substance, are assigned a zero individual order.

Any foreign species (substance that does not appear in the rate equation) or product molecules that retard the reaction always determine experimentally and it is sinful to come to any conclusion on any theoretical considerations.

13.5 Molecularity

Some reactions like oxidation of oxalic acid by permanganate have complex stoichiometric equations.

$$5C_2H_2O_4 + 2KMnO_4 + 3H_2SO_4 \longrightarrow K_2SO_4 + 2MnSO_4 + 8H_2O + 10CO_2$$

In order to have the reaction in a single step 5 molecules of oxalic acid, 2 molecule of $KMnO_4$ and 3 molecules of H_2SO_4 are expected to combine in a single instant. Statistically such a combination is quite unlikely. These reactions were found to take place in a series of steps called **elementary reactions** and the reaction is said to follow a complex reaction mechanism.

Molecularity is defined as the number of reactant molecules participating in an elementary reaction. For a general elementary reaction;

$$a\,\mathrm{A} + b\,\mathrm{B} + c\,\mathrm{C} \rightarrow p\,\mathrm{P} + q\,\mathrm{Q} + r\,\mathrm{R} + \text{etc} \qquad .....(13.15)$$

the molecularity of the reaction $m = a + b + c$. Since collision of more than the three molecules/atoms at any given instant is statistically improbable, we have the boundary condition $m \leq 3$, i.e., $(a + b + c) \leq 3$. *Molecularity, m, is always given by the sum of stoichiometric coefficients of an elementary reaction and therefore is always a small positive integer less than or equal to 3 and cannot be equal to zero or fraction.*

The order and molecularity of an elementary step are equal. Elementary reactions with molecularity 1, 2, or 3 are termed as unimolecular, bimolecular or termolecular reactions respectively. *A uni-, bi-, or ter-molecular elementary reaction will invariably follow first, second or third order kinetics respectively but the reverse is not true.* Thus a first order reaction can involve more than one reactant molecule but a unimolecular reaction is always a first order reaction. Order can take positive, fractional and negative values but molecularity is always restricted to only three positive integer values of 1, 2, or 3.

When we talk about molecularity of a complex reaction, comprising two or more elementary steps, it is often defined as the number of molecules or ions that participate in the rate determining step. This definition is not quite precise, as a reaction may have two or more number of slow steps of comparable rates and it is difficult to pin point the rate determining step. So molecularity of a reaction may change on the basis of additional arguments which may subscribe to the view that some other elementary reaction is the rate determining step, but order, which is deduced directly from experimental results, is always precise for a given set of experimental conditions.

The main points of difference between the order and molecularity of a reaction are summarized in Table 13.3.

Table 13.3 Differences between order and molecularity

Order	Molecularity
1. Order is an empirical quantity	1. Molecularity is a theoretical quantity
2. It is the sum of concentration exponents in a rate equation.	2. It is the sum of Stoichiometric coefficients of an elementary reaction.
3. It can be a positive or negative integer or fractional number including zero.	3. It is always a positive integer less than or equal to 3 and can not be zero or fraction.
4. Order refers to overall reaction. Needs no knowledge of separate elementary steps.	4. Molecularity refers to elementary reaction.
5. Always certain for a given set on experimental conditions.	5. Molecularity can change on the basis of additional arguments which may support the existence of some other rate determining step.

Problem 3.

Find the molecularity and order of the following elementary reaction.

$$A + 2B \longrightarrow C + D + 2E$$

when (i) B is present in large excess and

(ii) when the concentration of A and B are comparable.

Number of molecules participating directly gives molecularity of the reaction. Therefore,

$$\text{Molecularity} = 1 + 2 = 3$$

(i) When B is present in excess rate of the reaction is not dependent on B and order with respect to B is zero.

$$\text{Order} = 1 + 0 = 1$$

(ii) When concentration of A and B are comparable

$$\text{Order} = 1 + 2 = 3$$

In this case order and molecularity are same.

13.6 First Order Reactions

First order reactions are one of the most widely encountered kinetic forms. All the nuclear decay reactions follow first order kinetics. In a first order reaction the rate is directly proportional to the concentration of one reactant. For the reaction,

	A $\longrightarrow$	Product/s	.....(13.16)
Initial	a	0	
at time t	a-x	x	
at infinite time	0	a	

Assuming that the reaction is following first order kinetics the rate law of this reaction may be written as,

$$\text{Rate} = \frac{-d[\text{A}]}{dt} = k[\text{A}] \qquad .....(13.17)$$

where [A] is the concentration of reactant A at time t and k is the first order rate constant.

Suppose that the initial concentration of A is a mol L^{-1} and (a-x) mol L^{-1} is the concentration of reactant at time t. By suitable substitution the rate of equation (13.17) may be written as,

$$\text{Rate} = \frac{dx}{dt} = \frac{-d(a-x)}{dt} = k\,(a-x) \qquad .....(13.18)$$

Rearranging the equation (13.18) we get,

$$\frac{dx}{(a-x)} = kdt \qquad .....(13.19)$$

Equation (13.19) can be integrated between the concentration limits of product i.e., 0 and x corresponding to the time limits 0 and t.

$$\int_0^t \frac{dx}{(a-x)} = \int_0^t kdt$$

The integrated rate expression is

$$-\ln(a-x)\Big]_0^x = kt\Big]_0^t$$

Applying the limits we have

$$-\ln(a-x) + \ln a = kt$$

or

$$\boxed{kt = \ln \frac{a}{(a-x)} \quad \text{or} \quad kt = 2.303 \log \frac{a}{a-x}} \qquad .....(13.20)$$

This is known as the **kinetic rate equation for a first order reaction.**

13.6.1 Characteristics of First Order Rate Equation

(i) **Units of first order rate constant :** In the first order rate equation (13.20), the concentration terms appear both in the numerator & denominator

$$k = \frac{2.303}{\text{time}} \log \frac{\text{mol L}^{-1}}{\text{mol L}^{-1}}$$

$$k = time^{-1}$$

In the kinetic runs, time is usually expressed in seconds. Therefore unit of k is s^{-1}.

(ii) **Half-life period :** The time required to convert any initial concentration, $[A_o]$, to half its value, $\frac{1}{2}[A_o]$, is called the **half life period**. The half-life period of a first order reaction may be obtained by substituting in equation (13.20) the initial and final concentrations of the reactant a and a/2 respectively.

$$t_{\frac{1}{2}} = \frac{2.303}{k} \log \frac{a}{(a/2)} = \frac{2.303}{k} \log \frac{2a}{a}$$

$$\boxed{t_{\frac{1}{2}} = \frac{0.693}{k}} \qquad .....(13.21)$$

Therefore, *half life of a first order reaction is a constant is independent of initial concentration of the reactant.*

(iii) **Kinetic plots :** Straight lines are obtained by plotting log (*a-x*) vs time (Fig. 13.3(a)) or log (*a/a-x*) vs *time* (Fig. 13.3(b)).

A plot of log (*a-x*) versus t gives a straight line with a slope of (–k/2.303) and intercept of log (*a*). A plot of log (*a/a-x*) versus *t* gives a straight line passing through origin with a positive slope of *k*/2.303 . In either case the first order rate constant is obtained by multiplying slope with 2.303.

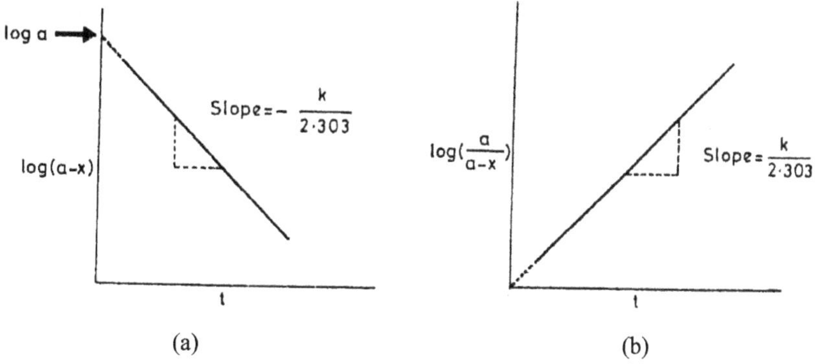

Fig. 13.3 Straight line relationship of first order rate equation.

13.6.2 Examples of First Order Reactions

(a) Decomposition of Nitrogen Pentoxide : Decomposition of N_2O_5 in gaseous phase and in solution follow first order kinetics.

$$2N_2O_5\,(g) \longrightarrow 2N_2O_4(g) + O_2(g)$$

$$2N_2O_5\,(\text{in } CCl_4) \longrightarrow 2N_2O_4\,(\text{in } CCl_4) + O_2\,(g)$$

(b) Molecular Rearrangements : Conversion of cis-2-butene to trans 2-butene and cyclopropane to propene follow first order kinetics.

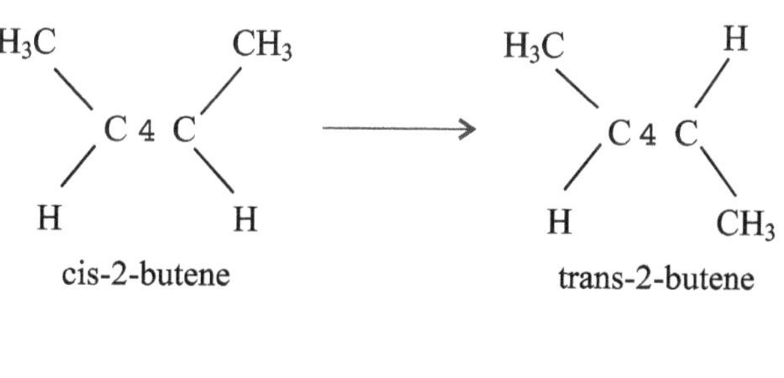

CH_2

H_2C))) CH_2 ⟶ CH_3) $CH = CH_2$

Cyclopropane Propene

(c) **Hydrolysis Reactions :** Hydrolysis of esters, diazonium salts and sucrose follow first order kinetics. These reactions are usually very slow in neutral medium and are catalyzed by mineral acids.

$$\underset{\text{Ethyl acetate}}{CH_3COOC_2H_5 + H_2O} \xrightarrow{H^+} \underset{\text{Ethyl alcohol}}{C_2H_5OH} + \underset{\text{Acetic acid}}{CH_3COOH}$$

$$\underset{\text{Benzene diazonium chloride}}{C_2H_5N_2Cl + H_2O} \xrightarrow{H^+} \underset{\text{Phenol}}{C_6H_5OH} + N_2 + HCl$$

$$\underset{\text{Sucrose}}{C_{12}H_{22}O_{11} + H_2O} \xrightarrow{H^+} \underset{\text{Glucose}}{C_6H_{12}O_6} + \underset{\text{Fructose}}{C_6H_{12}O_6}$$

In all the above hydrolysis reactions, we notice a combination of an organic molecule with a molecule of water to give hydrolysed products. Concentration of catalyst (H^+ ions) is unchange throughout the reaction. These reactions are expected to follow second order kinetics–first order with respect to each reactant, but water being in large excess the concentration is virtually constant and the reaction rate is controlled only by the concentration of organic molecule that is getting hydrolysed. These reactions, where a second order reaction behaving like a first order reaction is therefore called a **pseudo first order reaction**. Similarly a third order reaction (termolecular reaction) transforms into a pseudo first order reaction by taking two of the reactants in excess. In *general any higher order reaction, by virtue of placing one or more of the reactants in large excess, behaving as a lower order reaction is called a pseudo order reaction.*

(d) **Radioactive decay and Carbon dating :** An element is said to be radioactive if its nuclei disintegrates spontaneously, thereby transforming in to a different element. During the course of this process it emits nuclear particles and/or radiation. For example,

∀-Decay $\quad {}^{226}_{88}Ra \longrightarrow {}^{222}_{86}Rn + {}^{4}_{2}He$

ß$^{-}$–Decay $\quad {}^{32}_{15}P \longrightarrow {}^{32}_{16}S + \beta^{-1} + \text{neutrinos}$

ß$^{+}$–Decay $\quad {}^{22}_{11}Na \longrightarrow {}^{22}_{10}Ne + \beta^{+} + \text{neutrinos}$

Electron Capture $^{55}_{26}Fe \longrightarrow ^{55}_{25}Mn$ + neutrinos

(Electron capture is usually accompanied by β^+-decay)

The rate of decay of a radioactive element at any instant is proportional to the number of atoms (concentration) present at that instant. Therefore, all the radioactive decay reactions obey first order kinetics. This fact is advantageously used in finding ages of archeological remains of biological origin. This method is called **radiocarbon dating** technique. Earths atmosphere is constantly subjected to bombardment by cosmic rays. ^{14}C isotope is produced as a result of this bombardment. This unstable isotope along with stable ^{12}C isotope enters the biosphere through photosynthesis. From time to time ^{14}C lost through radioactive decay is replenished by cosmic ray bombardment, where by ^{14}C to ^{12}C ratio remains same in the atmosphere and in the living matter supported by it. In the dead plants and animals and biomatter formed by them like coal, paintings, etc., ^{14}C isotope is no longer replenished and decay constantly with a half life of 5730 years. Since the decay follow first order kinetics,

the rate constant $k = \dfrac{0.693}{5730} = 1.21 \times 10^{-4}\ \text{yr}^{-1}$

If N_o and N_t are the number of ^{14}C nuclei present at $t = 0$ and $t = t$ respectively, we have

$$kt = \ln \frac{N_0}{N_t} \qquad \text{.....(13.22)}$$

Since the rate of decay at any instant is directly proportional to the ^{14}C nuclei present at that instant, the first order rate equation (13.22) takes the form,

$$\boxed{t = \frac{1}{1.21\times10^{-4}} \ln \frac{\text{decay rate of fresh sample}}{\text{decay rate of old sample}}} \qquad \text{.....(13.22(a))}$$

Knowing decay rates of atmospheric ^{14}C and of old sample, t can be calculated which gives the age of old archeological sample.

Problem 4.

The first order reaction is found to be 15% completed in 20 minutes. Calculate the time taken to complete 75% of the reaction.

Let the initial concentration of the reactant be a

For 15% of completion of the reaction

$$x = \frac{15}{100} a$$

and $$a - x = a - \frac{15a}{100} = \frac{85a}{100}$$

Substituting the values of a, x and $(a–x)$ in the first order rate equation, we have,

$$k = \frac{2.303}{20} \log \frac{a}{(85a/100)}$$

$$k = \frac{2.303}{20} \log \frac{100}{85}$$

$$k = 0.0081 \text{ min}^{-1}$$

For 75% of completion

$$x = \frac{75a}{100} \text{ and } (a–x) = a - \frac{75a}{100} = \frac{25a}{100}$$

Since the value of k is independent of concentration terms

$$0.0081 = \frac{2.303}{t} \log \frac{a}{(25a/100)}$$

$$t = \frac{2.303}{0.0081} \log \frac{100}{25}$$

$$t = 170.6 \text{ min}$$

Problem 5.

A certain first order reaction is 75% complete in 1000 seconds. Calculate its half-life period and rate constant.

Let the initial concentration of the reactant be a

After 75% completion concentration of reactant

$$= a - (75a/100) \quad = 25a/100$$

$$k = \frac{2.303}{1000} \log \frac{a}{(25a/100)}$$

$$= \frac{2.303}{1000} \log \frac{100}{25} = 1.386 \times 10^{-3}\ s^{-1}$$

Half life period

$$t_{1/2} = \frac{0.693}{k} = \frac{0.693}{1.386 \times 10^{-3}} = 500\ s$$

Problem 6.

Show that for a first order reaction the time required for 99.9% completion is three times the time required for 90% completion.

For a first order reaction

$$k = \frac{2.303}{t} \log \frac{a}{a-x}$$

For 99.9% completion

$$a\text{-}x = 1 - \frac{99.9}{100} a = \frac{0.1}{100} a$$

therefore $t_{99.9\%} = \dfrac{2.303}{k} \log \dfrac{100a}{0.1a} = \dfrac{2.303}{k} \times 3$

For 90% completion

$$a\text{-}x = 1 - \frac{90}{100} a = \frac{a}{10}$$

$$t_{99\%} = \frac{2.303}{k} \log \frac{1}{(a/10)} = \frac{2.303}{k} \times 1$$

Therefore, $t_{99.9\%}$ of the reaction is 3 times $t_{90\%}$.

Problem 7.

For a reaction A $\longrightarrow$ B + C, the following data were obtained.

t (in secs)	0	900	1800
conc.of A (a-x)	50.6	19.7	7.62

Prove that the reaction is of first order.

At zero time (a-x) is nothing but the initial concentration a.

Therefore $a = 50.6$

$(a\text{-}x)$ at 900 s $= 19.7$

Substituting the data in first order rate equation

$$k = \frac{2.303}{900} \log \frac{50.6}{19.7} = 10.5 \times 10^{-3}\ \text{s}^{-1}$$

Similarly at 1800 s rate constant value is given by

$$k = \frac{2.303}{1800} \log \frac{50.6}{7.62} = 1.05 \times 10^{-3}\ \text{s}^{-1}$$

Since k value obtained by substituting the data in first order rate equation is same at both time intervals the reaction follows first order kinetics.

Problem 8.

The half life period of a first order reaction is 60 min. Calculate the time required for 87.5% completion of the reaction.

For the first order reaction

$$k = \frac{0.693}{t_{1/2}} = \frac{0.693}{60} = 1.155 \times 10^{-2}\ \text{min}^{-1}.$$

$$= 1.155 \times 10^{-2}\ \text{min}^{-1}$$

Let a be the initial concentration

For 87.5 % completion,

$$x = \frac{87.5}{100} a \text{ and } (a\text{–}x) = \frac{12.5}{100} a$$

Substituting the values in first order rate equation

$$t = \frac{2.303}{1.155 \times 10^{-2}} \log \left[\frac{\text{a}}{(12.5/100)\,a}\right] = 180 \text{ min}$$

Problem 9.

At 45 $^{\circ}$C a solution of N_2O_5 in CCl_4 gives 5.02 cc of oxygen in 1198 seconds and 7.33 cc of oxygen in 2315 seconds. The volume of oxygen evolved finally is 9.58 cc. From the above data, show that the decomposition of N_2O_5 is a first order reaction.

Decomposition of N_2O_5 is given by the equation

$$N_2O_5 \text{ (in } CCl_4) \longrightarrow 2N_2O_4 \text{ (in } CCl_4) + O_2 (g)$$

In this reaction amount of oxygen generated at the end of reaction corresponds to the initial concentration of N_2O_5. So we can take a = 9.56. Similarly the oxygen volumes 7.33 cc and 5.02 cc correspond to concentration of dissociated N_2O_5 at time intervals 2315 s and 1198 s respectively.

Therefore $(a-x)$ at 1198 s = 9.56 – 5.02 = 4.54 cc

$(a-x)$ at 2315 s = 9.56 – 7.33 = 2.23 cc

k for data at 1198 s

$$= \frac{2.303}{1198} \log \frac{9.56}{4.54} = 6.22 \times 10^{-4} \text{ s}^{-1}$$

k for data at 2315 s

$$= \frac{2.303}{2315} \log \frac{9.56}{2.23} = 6.28 \times 10^{-4} \text{ s}^{-1}$$

The constancy of k values at both the time intervals indicate that thc reaction follows of first order kinetics.

Problem 10.

Show that in a first order reaction the time taken for 99.9% completion of the reaction is 10 times greater than its half life period.

Let a be the initial concentration.

For 99.9% completion

$$(a-x) = a - \frac{99.9}{100} a = \frac{0.1a}{100}$$

Time for completion of 99.9% completion

$$t = \frac{2.303}{k} \log \frac{a}{(0.1a/100)} = \frac{2.303}{k} \times 3 = \frac{6.909}{k}$$

$$\text{Time 50\% completion} = \frac{0.693}{k}$$

$$\text{Ratio of times } \left(\frac{6.909}{k}\right) / \left(\frac{0.693}{k}\right) \approx 10$$

Problem 11.

In the decomposition of H_2O_2, 5 mL of reaction mixture needed 46 mL of $KMnO_4$ solution at the initial stage, while 5 mts later the same volume of reaction mixture required 37 mL of $KMnO_4$ solution. Assuming the reaction to be of 1st order, calculate the specific reaction rate.

46 mL of $KMnO_4$ corresponds to initial concentration of H_2O_2, say a.

37 mL of $KMnO_4$ corresponds to concentration of H_2O_2 at 5 mts, say $(a-x)$

Substituting these values in the first order rate equation

$$k = \frac{2.303}{5} \log \frac{46}{37} = 4.35 \times 10^{-2} \text{ min}^{-1}$$

Specific reaction rate = 4.35×10^{-2} min^{-1}

Problem 12.

A piece of wood found in a Egyptian pyramid has 70% as much ^{14}C as the recently grown piece of wood. If the half life of ^{14}C is 5600 years, how old is the pyramid ?

Radioactive decay follows first order kinetics.

Decay rate constant

$$k = \frac{0.693}{t} = \frac{0.693}{5600} = 1.2375 \times 10^{-4} \text{ year}^{-1}$$

When 70% ^{14}C is left; $(a-x) = 70$, a = 100

$$k = \frac{2.303}{t} \log \frac{a}{a-x} \qquad 1.2375 \times 10^{-4} = \frac{2.303}{t} \log \frac{100}{70}$$

Age of the pyramid = t = 2882.7 years

13.7 Second Order Reactions

In a second order reaction the rate of reaction depends on square of concentration of a single reactant or product of concentrations of two different reactant molecules each raised to the power of unity.

13.7.1 Second order reactions involving one kind of Molecules

A reaction of this kind may be written as,

	A	$\longrightarrow$ Product/s
Initial	a	0
at time t	$(a-x)$	x
at Infinite time	0	a

Rate law of this reaction is;

$$\text{rate} = \frac{dx}{dt} = k\,(a-x)^2$$

Rearranging the equation

$$\frac{dx}{(a-x)^2} = k\,dt \qquad .....(13.23)$$

Equation (13.23) may be integrated between product concentration limits 0 to x and corresponding time limits 0 to t. Integrated rate expression is,

$$\left.\frac{1}{a-x}\right]_0^x = kt\Big]_0^t$$

Applying the limits

$$kt = \frac{1}{a-x} - \frac{1}{a}$$

or

$$\boxed{k = \frac{x}{ta(a-x)}} \qquad .....(13.24)$$

13.7.2 Characteristics of second order rate equation involving one kind of molecules

(i) **Unit of k :** The units of second order rate constant k are obtained by substitution of concentration and time units in rate equation (13.24)

$$k = \frac{x}{ta(a-x)} = \frac{\text{mol}\,\text{L}^{-1}}{\text{s.mol}\,\text{L}^{-1}.\text{mol}\,\text{L}^{-1}} = \text{L. mol}^{-1}\,\text{s}^{-1}$$

(ii) **Half-life period :** At half-life period $x = a/2$, $(a-x) = a/2$ substituting these values in equation (13.24), we have;

$$t = \frac{a/2}{k\,a(a/2)} = \frac{1}{ka} \qquad .....(13.25)$$

Therefore the half life time (t) of a second order reaction involving one kind of molecules is inversely proportional to the initial concentration of the reactant.

(iii) **Kinetic plots :** Straight line relationships are obtained by plotting either $1/(a-x)$ versus t or $x/\{a(a-x)\}$ versus t.

Plot of $1/(a-x)$ vs t has an intercept of $1/a$ and the slope directly gives the rate constant k (Fig. 13.4(a))

Plot of $x/\{a(a-x)\}$ vs t also has a positive slope equal to k, but passes through origin (Fig. 13.4(b))

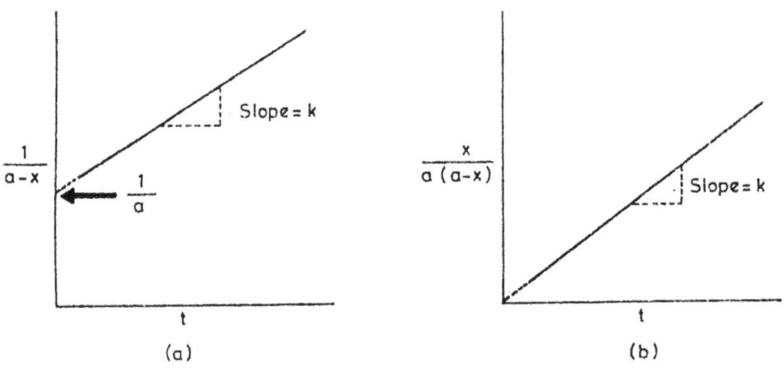

Fig. 13.4 Kinetic plots of a second order reaction.

13.7.3 Second order reactions with two different reactants of the same strength

When both the reactants have the same initial concentration i.e., [A] = [B] or $a = b$ the rate equation may be written as;

$$\frac{-d[\mathrm{A}]}{dt} = \frac{-d[\mathrm{B}]}{dt} = k\,[\mathrm{A}]\,[\mathrm{A}] = k\,[\mathrm{A}]^2$$

or

$$\frac{-d[\mathrm{A}]}{dt} = \frac{-d[\mathrm{B}]}{dt} = k\,[\mathrm{B}]\,[\mathrm{B}] = k\,[\mathrm{B}]^2$$

Either expressions are akin to above discussed second order reaction involving one kind of molecules. *The kinetics of such reactions are also identical and therefore the rate law is given by eq.* (13.24).

13.7.4 Second order reaction with two different reactants of different strengths

Let A and B be two different reactants of initial strengths a and b respectively.

	A	+ B	$\longrightarrow$ Product	where $a > b$
Initial	a	b	0	
at time t	$(a-x)$	$(b-x)$	x	
at infinite time	$(a-b)$	0	b	

Rate of this reaction is

$$\text{rate} = \frac{dx}{dt} = k\,(a-x)\,(b-x)$$

$$\frac{dx}{(a-x)(b-x)} = k\,dt$$

Splitting into partial fractions,

$$\frac{1}{(a-b)}\left[\frac{dx}{(b-x)} - \frac{dx}{(a-x)}\right] = k\,dt$$

Integrating we have,

$$\frac{1}{(a-b)}\left[-\ln(b-x) + \ln(a-x)\right] = kt$$

Applying concentration & time limits,

$$\frac{1}{(a-b)}\left\{-\ln(b-x) + \ln(a-x)\right\}\Big]_0^x = kt\Big]_0^t$$

$$\frac{1}{(a-b)}\{-\ln(b-x) + \ln b + \ln(a-x) - \ln a\} = kt$$

$$kt = \frac{1}{(a-b)} \ln \frac{b(a-x)}{a(b-x)}$$

$$\boxed{k = \frac{2.303}{t(a-b)} \ln \frac{b(a-x)}{a(b-x)}} \quad .....(13.26)$$

13.7.5 Characteristics of second order reaction with two different reactants of different strengths

(i) Unit of *k* : The units of *k* are obtained by substitution of concentration and time units in equation (13.26).

$$k = \frac{1}{\text{s.mol L}^{-1}} \ln \frac{\text{mol L}^{-1}.\text{mol L}^{-1}}{\text{mol L}^{-1}.\text{mol L}^{-1}} = \text{L. mol}^{-1}\ \text{s}^{-1}$$

(ii) Half-life period : On substitution of half life conditions the equation (13.26) does not reduce to any simple expression that merits discussions. So it is not considered here.

(iii) Kinetics plots : A plot of $\frac{2.303}{(a-b)} \log \frac{b(a-x)}{a(b-x)}$ versus *t* gives a straight line passing through origin. The slope of the line is equal to *k*.

(iv) When one of the reactant is present in large excess : Suppose the reactant A of initial concentration *a* is present in large excess. Under these conditions one can take $(a-x)$. a and $(a-b)$. a. The rate equation (12-26) may then be modified as

$$kt = \frac{1}{a} \ln \frac{ab}{a(b-x)}$$

Since concentration of *a* is constant throughout *ka* may be replaced by another constant *k*′.

$k't = \ln (b/b-x)$

This equation is identical to first order rate eq. (13.20). This explains why a second order reactions behaves like a first order reaction when one of the reactant is present in excess.

13.7.6 Examples of second order reactions

All the bimolecular elementary reactions reactions follow second order kinetics but multi-step second order reactions could have different molecularities. If one of the three reactants in a third order reaction is taken in excess, it also obeys second order kinetics and such reactions may be termed as **pseudo second order reactions**. The following are few examples of second order reactions.

(a) **Thermal dissociation of gaseous molecules :** Thermal dissociation of gases like ozone, nitrogen dioxide, chlorine monoxide, etc., and organic vapors of formaldehyde, acetaldehyde, etc., follow second order kinetics.

Dissociation of ozone at 100 ^{o}C :

$$2O_3 \longrightarrow 3O_2$$

Dissociation of nitrogen dioxide :

$$2NO_2 \longrightarrow 2NO + O_2$$

Dissociation of chlorine monoxide at 200 °C :

$$2Cl_2O \longrightarrow 2Cl_2 + O_2$$

Dissociation of acetaldehyde :

$$CH_3CHO \longrightarrow CH_4 + CO + \text{other minor products}$$

(b) **Combination reactions :** Combination of ethylene and hydrogen to give ethane and recombination of iodine atoms to give molecular iodine are typical examples of this category.

$$\underset{\text{ethene}}{C_2H_2 + H_2} \longrightarrow \underset{\text{ethane}}{C_2H_4}$$

$$I(g) + I(g) \longrightarrow I_2\,(g)$$

(c) **Saponification of esters :** Hydrolysis of an ester in the presence of an alkali is commonly called saponification. These reactions follow second order kinetics.

$$\underset{\text{ethyl acetate}}{CH_3COOC_2H_5} + \underset{\text{alkali}}{NaOH} \longrightarrow \underset{\text{sodium acetate}}{CH_3COONa} + \underset{\text{alcohol}}{C_2H_2OH}$$

(d) **Persulphate - iodide reaction :** This reaction is popularly used in the college laboratories to demonstrate a second order reaction.

This reaction is followed by estimating the iodine liberated by standard hypo-iodine titration.

$$K_2S_2O_8 + 2KI \longrightarrow 2K_2SO_4 + I_2$$

(e) **Organic displacement reactions** : Reaction between 2, 4-dinitrochlorobenzene (designated as N-Cl) and aniline in benzene and reaction between ethylene bromide and KI in methanol follow second order kinetics.

$$N\text{-}Cl + C_6H_5NH_2 \longrightarrow C_6H_5NHN + HCl$$

$$C_2H_4Br_2 + 2KI \longrightarrow C_2H_4 + 2KBr + KI_3$$

Problem 13.

50% of a second order reaction is complete in 40 minutes. What would be the time required for 75% of the reaction to complete.

Half life period of a second order reaction is given by the expression,

$$k = \frac{1}{t_{1/2}a}$$

Therefore; $k = \dfrac{1}{40a}$

For 75 % completion :

$$x = \frac{75a}{100}, \quad (a-x) = \frac{25a}{100}$$

Substituting these values in second order rate equation

$$t_{75\%} = \frac{1}{(1/40a)} \times \frac{(75a/100)}{a(25a/100)}$$

$$t_{75\%} = 40 \times \frac{75a}{25a} = 120 \text{ min}$$

Problem 14.

A second order reaction where $a = b$ is 20% completed in 500 seconds. How long will it takes for the reaction to go to 60% completion.

For the second order reaction where $a = b$, rate equation is

$$k = \frac{1}{t}\frac{x}{a(a-x)}$$

For 20% completion

$$x = \frac{20}{100}a, \quad (a-x) = \frac{80}{100}a \quad \text{and} \quad t = 500$$

substitute the values in the above rate expression

$$k = \frac{1}{500} \frac{(20/100a)}{a(80/100a)}$$

$$= \frac{1}{500} \times \frac{20a}{100} \times \frac{1}{a} \times \frac{100}{80a} = \frac{5\times10^{-4}}{a}$$

Similarly for 60% completion

$$x = \frac{60a}{100}, (a-x) = \frac{40a}{100} \; a \text{ and } t_{60\%} = ?$$

$$t_{60\%} = \frac{1}{(5\times10^{-4}/a)} \cdot \frac{(60a/100)}{a(40a/100)}$$

$$= \frac{a}{5\times10^{-4}} \times \frac{6}{4a} = 3000 \text{ s}$$

Problem 15.

In a second order reaction time taken for half change is 100 min. Calculate time taken for 90% change.

For second order reaction

$$k = \frac{1}{t_{1/2}a} = \frac{1}{100\text{a}}$$

for 90% change

$$x = \frac{90a}{100}$$

$$a - x = \text{a} - \frac{90a}{100} = \frac{10a}{100}$$

Substitute in second order rate equation

$$\text{k} = \frac{1}{t} \frac{x}{a(a-x)}$$

$$\frac{1}{100a} = \frac{1}{\text{t}} \frac{90a/100}{a(10a/100)}$$

or $$t = 100a \times \frac{90}{10a} = 900 \text{ min.}$$

Problem 16.

At 518 °C acetaldehyde decomposes to give methane and carbon monoxide. At constant volume the pressures recorded at different time intervals are as follows.

Time (min)	0	42	105	242	480
Pressure (mmHg)	363	397	437	497	557

Find the second order rate constant.

Given $CH_3CHO \longrightarrow CH_4 + CO$

1 vol 1 vol 1 vol

This is a second order reaction of the type $2A \longrightarrow$ Product(s). In this reaction 1 volume of acetaldehyde is dissociating to give 2 volumes of the product. So P_o is proportional to a and the pressure change, $(P_0 - P_t)$, is proportional to the concentration of the product (x).

Time (min)	0	42	105	242	480
$x \forall (P_o\text{-}P_t)$	0	34	74	134	194
$(a–x) \forall (P_o–x)$	363	329	289	229	169
$k = x/\{a(a–x)\}$		6.79×10^{-6}	6.72×10^{-6}	6.66×10^{-6}	6.59×10^{-6}

The average second order rate constant = 6.69×10^{-6} M^{-1} s^{-1}

13.8 Zero Order Reactions

All the zero order reactions proceed through more than one elementary step. Acid catalyzed reaction between acetone and iodine is a typical example of this class, where the rate of reaction is independent of iodine concentration. The mechanism involves the following steps.

$$\underset{\text{keto form}}{CH_3 - \overset{\overset{\displaystyle O}{||}}{C} - CH_3} \xrightarrow[\text{slow}]{H+} \underset{\text{enol form}}{CH_2 - \overset{\overset{\displaystyle OH}{|}}{C} = CH_2} \quad \text{Enolization}$$

$$CH_3 - \overset{\overset{\displaystyle OH}{*}}{C} = CH_2 + I_2 \xrightarrow[\text{fast}]{} CH_3 - \overset{\overset{\displaystyle O}{||}}{C} - CH_2I + HI \quad \text{Iodination}$$

When a reaction is described by a sequence of elementary steps, **the slowest step** controls the overall rate of reaction. This can be well understood by considering two analogies.

(i) The strength of a chain is decided by the strength of the weakest link. It is immaterial, how strong are the other links when one of the link is made out of an easily breakable material like thread or a plastic wire.

(ii) The rate of emptying of a packed theatre is decided only by the number and width of exit doors available to the audience but not on the time taken to reach the exit.

Since the slowest step controls the rate of overall reaction it is also called **the rate determining step**. In case of above example the enolization of ketone was found to be the slow step. Therefore, the rate of reaction is totally controlled by the presence of acetone and acid responsible for its enolization. Thus the reaction rate is independent of iodine concentration, or in other words, order with respect to iodine is zero.

Photochemical reactions are another important class that generally follow zero order kinetics. Photochemical reactions are initiated by the absorption of photons by ground state molecules. Since the excited molecules lose their energy and quickly return to the ground state, rate is controlled only by the concentration of excited molecules or in other words on the intensity of incident light rather than the ground state molecules that are available in plenty. Photochemical combination of H_2 and Cl_2 to give HCl is a typical example.

$$H_2\,(g) + Cl_2\,(g) \xrightarrow{\text{light}} 2HCl\,(g)$$

This reaction follows zero order kinetics with respect to reactants as long as the product (HCl) formed is removed simultaneously by dissolving in water. The product trapping is essential to avoid complications arising out of secondary reactions. Similarly photolysis of organic molecules also follow zero order kinetics with respect to reactant molecules.

Surface catalytic reactions are another group which follow zero order kinetics in reactants.

$$2N_2O \xrightarrow[(\Delta)]{\text{(Platinum)}} 2N_2 + O_2$$

$$2NH_3 \xrightarrow[\Delta]{\text{Tungsten}} N_2 + 3H_2$$

$$2HI \xrightarrow[\Delta]{\text{Gold}} H_2 + I_2$$

All these reactions essentially involve two steps,

(i) Adsorption of gas on the catalytic surface,

(ii) Decomposition of adsorbed gas and desorption of products.

Here, the rate is governed not by the amount of substance taken but by the small concentration of molecules adsorbed on to the surface of the heterogenous catalyst. That is, rate is directly proportional to the amount of the active surface area but not on the reactant concentration. Thus in the decomposition of ammonia on tungsten, the rate of reaction is proportional to the active surface area of tungsten and independent of ammonia concentration.

13.8.1 Derivation of integral form of zero order rate equation

Let the general reaction

$$A \longrightarrow \text{product}$$

be a zero order reaction in A.

$$rate = k\,[A]^0 = k$$

$$\frac{dx}{dt} = k$$

$$dx = k\,dt$$

On integrating between the time limits 0 to t we have

$$\int_0^x dx = k\int_0^t dt$$

$$x\Big]_0^x = kt\Big]_0^t$$

$$x = kt$$

$$\boxed{k = \frac{x}{t}} \qquad \text{.....(13.27)}$$

Eq. (13.27) is the zero order rate equation.

13.8.2 Characteristics of zero order reactions

(i) **Units of k :** Substituting the concentration and time units in equation (13.27), we get;

$$k = \frac{\text{mol L}^{-1}}{\text{s}} = \text{mol L}^{-1}\,\text{s}^{-1}$$

(ii) **Half life period :** At half life time $x = a/2$

$$k = \frac{a}{2t_{1/2}} \quad \text{or} \quad t_{1/2} = \frac{a}{2k}$$

Half life of a zero order reaction is directly proportional to the initial reactant concentration.

(iii) **Kinetics plots :** From equation (13.27) we can infer that straight lines passing through origin are obtained by plotting x versus t. The slope of the straight line directly gives the rate constant of the zero order reaction.

Problem 17.

A zero order reaction is 50% complete in 30 minutes. Find the percentage completion at the end of 45 min.

For a zero order reaction

$$k = \frac{a}{2t_{1/2}}$$

$$k = \frac{a}{2\times 30} = \frac{a}{60} \text{ concentration min}^{-1}$$

At the end of 45 min

$$x = kt = \frac{a}{60}\times 45 = \frac{3}{4}a$$

Percentage of completion

$$= \frac{x}{a}\times 100 = \frac{(3/4)}{a}\times 100 = 75\%$$

The differential and integral rate equations, units of k, proportionality of $t_{1/2}$ and suitable kinetic plots for reactions of different orders are listed in Tables 13.4 & 13.5.

Table 13.4 A summary of integral & differential rate equations.

Order n	Differential rate law	Integral rate law
0	$\frac{dx}{dt} = k$	$kt = x$
1	$\frac{dx}{dt} = k\,(a–x)$	$kt = ln\,\frac{a}{(a-x)}$
2	$\frac{dx}{dt} = k\,(a–x)^2$	$kt = \frac{a}{a(a-x)}$
2	$\frac{dx}{dt} = k\,(a–x)\,(b–x)^2$	$kt = \frac{1}{(a-b)} \ln \frac{b(a-x)}{a(b-x)}$
n	$\frac{dx}{dt} = k\,(a–x)^n$	$kt = \frac{1}{n-1}\left[\frac{1}{(a-x)^{n-1}} - \frac{1}{a^{n-1}}\right]$

13.9 Methods of Determination of Order of a Reaction

Various methods are available for the determination of reaction orders.

13.9.1 Method of Trial or Integration method

This method is most suited for simple reactions having integral positive orders. The method involves determination of changes in the concentration of reactants or products and substituting the data in standard rate law expressions (13.20), (13.24), (13.26) and (13.27). The equation giving constant value for the specific reaction rate, k, over a significant time interval gives the correct order of reaction.

Table 13.5 Characteristics of some rate equations.

Order	Units of k	$t_{½}$ is proportional to (value)	Straight line kinetic plots
0	mol L^{-1} s^{-1}	$a\ (t_{½} = a/2k)$	x vs t
1	s^{-1}	independent ($t_{½} = \frac{0.693}{k}$)	$ln\,\frac{a}{(a-x)}$ vs t
2	L mol^{-1} s^{-1}	$\frac{1}{a}\ (t_{½} = \frac{1}{ka})$	$\frac{1}{(a-x)}$ vs t

Contd... Table 13.5

Order	Units of k	$t_{1/2}$ is proportional to (value)	Straight line kinetic plots
where a = b			
2	$L\ mol^{-1}\ s^{-1}$	$\frac{1}{a}$	$ln\ \frac{b(a-x)}{a(b-x)}$ vs t
where a < b			
n	$L^{(n-1)}\ mol^{(1-n)}\ s^{-1}$	$\frac{1}{a^{(n-1)}}$	$\frac{1}{(a-x)^{n-1}}$ vs t

Alternatively graphical methods were developed based on the respective rate equations. For the first order reaction, $n = 1$, a plot of t vs $ln\ (a-x)$ gives a straight line while other nth order reactions having single reactant t against $\frac{1}{(a-x)^{n-1}}$. Different graphical schemes are already discussed under individual rate laws and listed in Table 13.5.

The trial method cannot be used for reaction with fractional order and negative order dependence.

13.9.2 Half-life method or Method of definite fraction

Half life period or period for completion of a definite fraction of reaction is

(a) directly proportional to the initial concentration of the reactant for a zero order reaction.

(b) independent of initial concentration of the reactant for a first order reaction and

(c) inversely proportional to the initial concentration of reactants for a second and higher order reactions.

For a reaction of nth order half life is given by the expression,

$$t_{1/2}\ \alpha\ \frac{1}{a^{n-1}} \quad \text{or} \quad t_{1/2} = \frac{k_n}{a^{n-1}}$$

where k_n is a proportionality constant.

Taking logarithms we have,

$$log(t_{1/2}) = log\ (k_n) + (1-n)\ log\ (a)$$

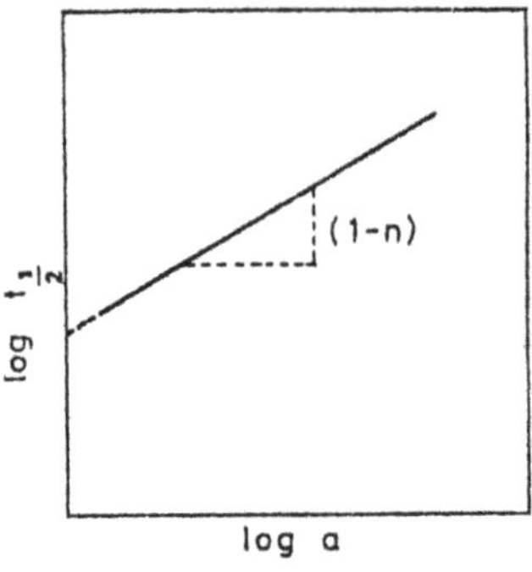

Fig. 13.5 Half-life method.

It is possible to directly find half life times $\left(t_{1/2}\right)$ for a series of initial concentrations *(a)*. When log $\left(t_{1/2}\right)$ is plotted against log *(a)* we get a straight line whose slope is equal to *(1–n)*, Fig. 13.5.

When we have only two or three sets of data points, we can calculate the order of the reaction employing a general formula.

$$n = 1 - \frac{\log t_{1/2} - \log t'_{1/2}}{\log a - \log a'} \qquad .....(13.28)$$

where $t_{1/2}$ and $t_{1/2}$ are half life period corresponding to initial concentration a and a' respectively. *This method is mainly useful for reactions with integer orders.*

Problem 18.

A substance with an initial concentration decomposes to half its concentration in 40 min. When the initial concentration is doubled the $t_{1/2}$ increases 80 min. Find the order of the reaction.

Let C be the initial concentration corresponding to $t_{1/2}$ of 40 min.

when initial concentration is 2C corresponding $t_{1/2}$ is 80 min.

we know

$$\frac{t_{1/2}}{t'_{1/2}} = \frac{(2C)^{n-1}}{(C)^{n-1}}$$

$$\frac{40}{80} = \frac{(2C)^{n-1}}{(C)}$$

$$\ln(1/2) = (n-1)\ln 2$$

$$\ln 1 - \ln 2 = n \ln 2 - \ln 2$$

$$n \ln 2 = \ln 1 - \ln 2 + \ln 2$$

$$n \ln 2 = 0$$

$$n = 0$$

Hence the decomposition reaction follows zero order kinetics.

Alternatively, direct proportionality between half life time and initial concentration also suggests a zero order dependence.

Problem 19.

For the thermal decomposition of an organic compound the following data was obtained.

Initial concentration (mol L^{-1})	0.12	0.18	0.24	0.40	0.5
Half life $t_{1/2}$ (min)	150	100	75	45	36

Find the order of the reactions.

This can be attempted in two ways

Numerical Method :

we know

$$\left[\frac{t_{1/2}}{t'_{1/2}}\right] = \left[\frac{C'}{C}\right]^{n-1}$$

$$\left[\frac{150}{100}\right] = \left[\frac{0.18}{0.12}\right]^{n-1}$$

$$n-1=\frac{\log(150)-\log(100)}{\log(0.18)-\log(0.12)}=\frac{2.18-2.00}{(-0.74)-(-0.92)}$$

By taking any other two point sets also we can prove $n = 2$

Graphical method : A plot of log $t_{\frac{1}{2}}$ vs log $t'_{\frac{1}{2}}$ gives a straight line with a slope equal to *(n–1)* from which *n* can be calculated.

13.9.3 Van't Hoff's Differential Method

For two different initial concentrations C_1 and C_2 the differential form of rate equation may be written as,

$$\text{rate} \quad r_1 = \frac{-dC_1}{dt} = k\,C_1^n$$

$$\text{rate} \quad r_2 = \frac{-dC_2}{dt} = k\,C_2^n$$

By taking logarithms

$$\log r_1 = \log k + n \log C_1 \qquad \text{I}$$

and

$$\log r_2 = \log K + n \log C_2 \qquad \text{II}$$

Now subtracting I from II, we get;

$$\log r_2 - \log r_1 = n\,[\log C_2 - \log C_1]$$

or

$$n=\frac{\log r_2-\log r_1}{\log C_2-\log C_1} \qquad \text{.....(13.29)}$$

or

$$\boxed{n=\frac{\log(-dC_2/dt)-\log(-dC_1/dt)}{\log C_2-\log C_1}} \qquad \text{.....(13.30)}$$

The estimation procedure involve plotting of concentration versus time plots and drawing tangents to the curve corresponding to two different concentrations C_1 and C_2 of reactant. The slope of these tangents give rates corresponding centralions, $-dC_1/dt$ and $-dC_2/dt$ and they can be readily used in the eq. (13.30) and order of that reactant may be calculated. Alternatively the slope of the linear curve obtained by plotting log *(–dC/dt)* against *log C* gives the order of the reactant.

This method is applicable even in case of reactions having more than one reactant. Suppose a reaction having two reactants A and B of same or different concentrations is taken and concentration C_A is varied from C_A to

$C_A{}'$ while keeping C_B constant. Corresponding differential rate equations are

Rate $r_1 = k\,(C_A)^{n}{}_1\ (C_B)^{n}{}_2$

Rate $r_2 = k\,(C_A{}')\ (C_B)^{n}{}_2$

Taking logarithms

$$\log r_1 = \log k + n_1 \log C_A + n_2 \log C_B \qquad \text{III}$$

$$\log r_2 = \log k + n_1 \log C_A' + n_2 \log C_B \qquad \text{IV}$$

Subtracting (IV-III) and rearranging, we get;

$$n_1 = \frac{\log r_2 - \log r_1}{\log C_A' - \log C_A} \qquad \text{.....(13.31)}$$

Thus order estimated corresponds to the reactant whose concentration is varied. To find the order in B, it needs only to keep C_A constant and vary C_B.

This method can be successfully employed even when the reaction does not fit in standard zero, first, second order rate expressions derived above but have fractional or negative order dependence on the reactants.

13.9.4 Ostwald's Isolation Method

W. Ostwald in 1902 found that if all reactants except one are taken in excess, the observed order will be determined by reactant isolated in this way. Suppose an elementary reaction involves ∀ molecules of A, β molecules of B and γ molecules of C, then the rate law is,

$$rate = k\, C_A^{\forall}\, C_B^{\beta}\, C_C^{\gamma}$$

and overall order of reaction = ∀ + ß + γ

If reactants B and C are taken in excess, according to Ostwald, the order of reaction will be equal to ∀. Similarly β and γ can be evaluated by studying the kinetics of reactions with A & C and A & B respectively in excess. Overall order of such a reaction then corresponds the sum of individual orders of the isolated reactants.

Problem 20.

The following data is given for the decomposition of phosphine. Find the order of the reaction.

Initial pressure	:	707	500	79	37.5
Half life	:	84	83.8	84	83.5

In this problem half-life is independent of initial pressure of phosphine. Therefore, decomposition of phosphine follows first order kinetics.

Alternatively : We know

$$n = 1 + \frac{\log t - \log t'}{\log C'' - \log C}$$

Considering the first two data points

$$n = 1 + \frac{\log 84 - \log 83.5}{\log 500 - \log 707} = 1$$

Similarly for the other data points also we get $n = 1$

Therefore, order in phosphine is unity.

Problem 21.

For a particular reaction in an equimolar mixture of reactants at 336 mm Hg initial pressure was half changed in 108 seconds and when the pressure was 288 mm Hg it changed in 147 seconds. Determine the order of the reaction.

$t_{1/2}$ corresponding to initial pressure of 336 mm Hg = 108 s

$t_{1/2}$ for an initial pressure of 288 mm Hg = 147 s

We know,

$$\left[\frac{t_{1/2}}{t'_{1/2}}\right] = \left[\frac{a'}{a}\right]^{n-1}$$

$$n = 1 + \frac{\log t_{1/2} - \log t'_{1/2}}{\log a' - \log a}$$

$$n = 1 + \frac{\log 108 - \log 147}{\log 288 - \log 336} = 3$$

Problem 22.

Find the order and velocity constant of inversion of sucrose in 0.9 M HCl at 25 ^{O}C from the following data.

Time(min)	0	18	27	36.8	46	56	68	∞
Angle of rotation.	24.09	17.735	15.00	12.40	10.02	7.80	5.455	–10.74

In this reaction concentration of unreacted reactant is directly proportional to the change of rotation still due to occur.

$$a = r_o - r_{\forall} = 24.09 - (-10.74) = 34.83$$

$$a - x = r_t - r_{\forall}$$

Time(min)	18	27	36.8	46	56	68	∞
$(a-x)$ values	28.475	25.74	23.14	20.76	18.54	16.155	0.000

Assuming that the reaction follows first order kinetics and substituting above data in first order rate equation, we have,

k	0.0112	0.0112	0.0112	0.0113	0.0113	0.0113

Constancy of k values suggest that the reaction follows first order kinetics.

13.10 Effect of Temperature on Reaction Rate

An easy way to understand the effect of temperature on reaction rates is to peep into our kitchen room. In one corner of the room we find the refrigerator packed with fresh vegetables, fruits and other food items. Spoiling and decay of food, a chemical reaction, is almost stopped by storing the food in the freezer. In the other corner, we find the rice being cooked. Rice cooks faster in a pressure cooker due to raise of boiling point of water. Therefore we can generalize that the rates of almost all chemical reactions increase with increase in temperature.

Another interesting example is that of reptiles. Snakes and turtles being cold blooded animals, cannot maintain a fixed body temperature. Their body temperature is affected severely by the temperature of the surroundings. Their body metabolism slows in the winter months making them sluggish and must seek shelter in holes. We therefore don't hear any croaking on the cold winter evenings, a common feature of summer days.

13.10.1 Temperature Coefficient

The first attempt to rationalize the effect of temperature on reaction rates was made in terms of **temperature coefficient**. The ratio of rate constants at two different temperatures separated by 10^0 is defined as temperature coefficient. For reactions that occur at room temperature these temperatures are conventionally fixed at 25 ^{O}C and 35 ^{O}C.

$$\textbf{Temperature coefficient } T_{V} = \frac{k_{35^{o}C}}{k_{25^{o}C}} . 2 \qquad .....(13.32)$$

In most cases this ratio was found to be around 2, i.e., the rate is doubled by increasing the temperature of a system from 25 ^{O}C to 35 ^{O}C. However, later studies of other systems revealed higher and lower temperature coefficient values. For example the decomposition of N_2O_5 at room temperature gave a temperature coefficient of 3.8. Further, in case of gaseous reactions like decomposition of acetaldehyde at 160 ^{O}C, the reaction rates were found to increase few hundred times for just an increase of 10 °C. These startling revelations led to the disappearance of phenomena of temperature coefficient into thin air.

13.10.2 Arrhenius Equation

With the rise in temperature the kinetic energy of the molecules increases. Therefore the chemical reactions were thought to be due to collisions between the reacting molecules. At a milli molar concentration (10^{-3} mol L^{-1}) level a molecule undergoes 3.5×10^{28} collisions per liter per second at 500 ^{O}C. If all the collisions were effective, all the reactions including biological processes would have been over instantaneously and the universe would not have existed. One of the intervening factors that restrict the rates of chemical reactions is the **activation energy** (A detailed account of nature of collisions and other intervening factors will follow the current discussion). In any chemical system the molecules are distributed with kinetic energies spread over a wide range. At ordinary temperature majority of the molecules have energies well below the activation energy. However, a small fraction have energies equal to or greater than activation energy. When these molecules are involved in a collision, called **fruitful collision** or **activated collision**, can lead to chemical reaction. At higher temperatures the distribution energy curve broadens accommodating more molecules with energies greater than activation energy. Therefore the reaction rate increases with increase in temperature, Fig. 13.6.

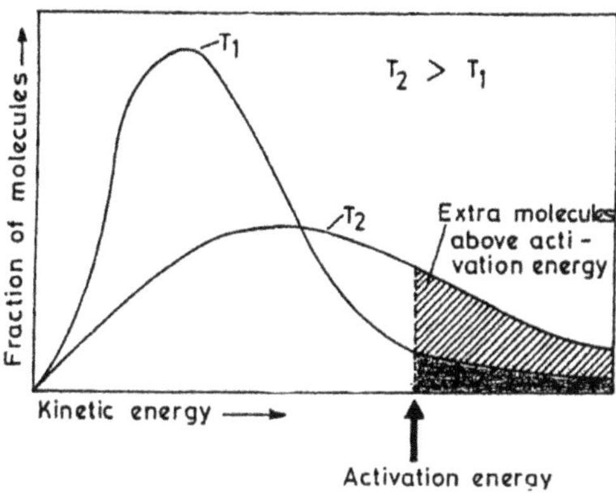

Fig. 13.6 The effect of temperature on reaction rates.

In 1869, a Swedish chemist, Svante Arrhenius, pointed out that the temperature dependence of the rate constant can be derived from the van't Hoff's eq. (13.33).

$$\frac{d\ln K}{dT} = \frac{\Delta H}{RT^2} \quad \text{.....(13.33)}$$

where K is the equilibrium constant, ΔH is the heat of reaction. Since $K = k_1/k_{-1}$, where k_1 and k_{-1} are the rate constants for forward and reverse reactions of an equilibrium respectively, Equation (13.33) may be modified as,

$$\frac{d\ln k_1 - d\ln k_{-1}}{dT} = \frac{\Delta H}{RT^2} \quad \text{.....(13.34)}$$

Again, heat of reaction ΔH may be expressed as a difference of two energies E_1 and E_{-1} related to forward and reverse reactions, i.e., $\Delta H = E_1 - E_{-1}$.

$$\frac{d\ln k_1 - d\ln k_{-1}}{dT} = \frac{E_1 - E_{-1}}{RT^2} \quad \text{.....(13.35)}$$

Splitting equation (13.35) into two equations we get,

and $$\frac{d\ln k_1}{dT} = \frac{E_1}{RT^2} + Z$$

$$\frac{d\ln k_1}{dT} = \frac{E_1}{RT^2} + Z$$

where Z is a constant. The generalized equation can be written as,

$$\frac{d\ln k}{dT} = \frac{E_a}{RT^2} + Z \quad \text{.....(13.36)}$$

Integrating eq. (13.36),

$$\ln k = \frac{E_1}{RT} + \ln A \qquad .....(13.37)$$

where ln A is an integration constant.

Removal of natural logarithms gives a quantitative expression relating rate constant (k) and activation energy (E_a), now known as **Arrhenius equation**.

$$k \; = Ae^{-E_a|R_T} \qquad .....(13.38)$$

where e is the base of natural logarithm, R is the gas constant and T is the absolute temperature. The quantity, A, called **frequency factor**, is related to the collision frequency and also to the importance of molecular orientations during collisions. Since number of collisions increase only by 2% for a raise of 10 ^{O}C, the frequency factor A can be treated as constant over a fairly wide temperature range. Further, because of the minus sign associated with the exponential factor E_a / RT, the rate constant increases with increase in temperature and decreases with increase in activation energy. The expression $e^{-E_a|RT}$ is called the **Boltzmann factor**, gives the fraction of molecules that have an energy E_a or more.

Equation (13.37) can be compared to the equation of a straight line ($y = mx + c$).

$$\ln k = \left(\frac{-E_a}{R}\right)\left(\frac{1}{T}\right) + \ln A$$

$$y \qquad m \qquad x \qquad c$$

Thus, plotting ln k *vs (1/T)*, we get a straight line with slope equal to $-E_a / R$ and intercept ln A, Fig. 13.7. Alternatively, when log k (logarithm with base 10) is plotted against *1/T*, the straight line obtained has a slope of $-E_a / 2.303\, R$.

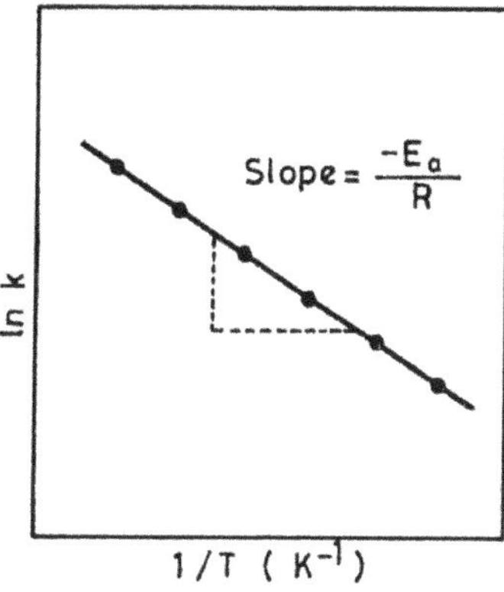

Fig. 13.7 Graphical determination of activation energy.

E_a can also be computed directly from a knowledge of rate constants k_1 and k_2 corresponding to temperatures T_1 and T_2 respectively. Required equation for this purpose can be derived from equation (13.37).

$$\ln k_1 = \ln A - \frac{E_a}{RT_1} \qquad \text{(V)}$$

$$\ln k_2 = \ln A - \frac{E_a}{RT_2} \qquad \text{(VI)}$$

Subtracting V from VI gives

$$\ln k_2 - \ln k_1 = \frac{-E_a}{R}\left(\frac{1}{T_2} - \frac{1}{T_1}\right)$$

$$\ln\left[\frac{k_2}{k_1}\right] = \frac{-E_a}{R}\left(\frac{1}{T_2} - \frac{1}{T_1}\right) \qquad \text{.....(13.39)}$$

Thus E_a can be determined from a knowledge of rate constants of a reaction at any two temperatures. Once E_a and k value at any temperature are known, eq. (13.39) can be used to predict rate constant at any other desired temperature.

The relation between heat of reaction and the activation energy can be illustrated with the aid of **energy barrier concept**. For this consider a reactant R forming product P. To form the product, the reactant must first acquire an excess energy E_1 and cross an energy barrier, Fig. 13.8. This excess energy E_1 is nothing but the activation energy E_a of the forward reaction.

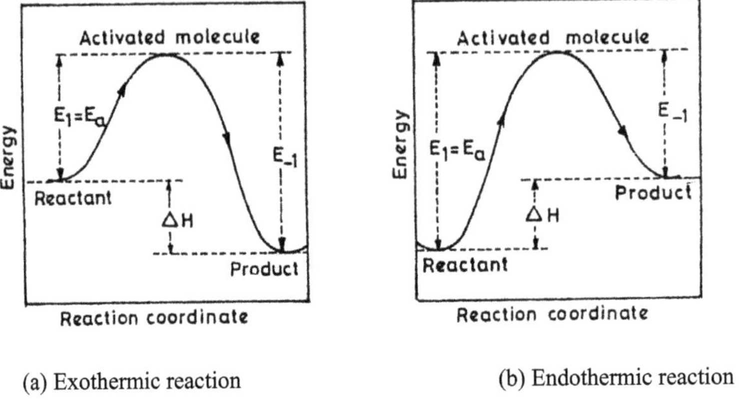

(a) Exothermic reaction (b) Endothermic reaction

Fig. 13.8 Activation energy barriers.

A reactant molecule possessing this energy E_a is generally called as **excited** or **activated molecule**. According to Arrhenius, the activated molecules form a product by rolling through a down hill process. While rolling down, if the product(s) is(are) at a lower energy level than reactants, the difference in energy ΔH is thrown out in the form of heat. The reaction is then termed as **exothermic reaction**. In case, the product(s) is(are) at higher energy level, the excited molecule retains some of the energy it got during the process of excitation and there by results in cooling of the reaction system. This type of reaction is called **an endothermic reaction**. The activation energy barrier also suggests that the activation energies of forward and reverse reactions are entirely different.

Problem 23.

The temperature coefficient of a reaction between 25 and 35°C is 1.8. Calculate the activation energy of this reaction.

$$T_\infty = \frac{k_{35}}{k_{25}} = 1.8 \quad \& \quad \log \frac{-E}{2.303R}\left[\frac{1}{T_1} - \frac{1}{T_2}\right]$$

So
$$\log(1.8) = \frac{-E}{2.303 \times 8.314}\left[\frac{1}{298} - \frac{1}{308}\right]$$

$$E = 44.85 \text{ kJ}$$

Problem 24.

A second order reaction has a rate constant of 6.0×10^{-6} at 17 ^{O}C and 9.0×10^{-3} at 50 ^{O}C. Calculate the energy of activation for this reaction.

Given $k_1 = 6.0 \times 10^{-6}$ $T_1 = 17 + 273 = 290\ K$

$k_2 = 9.0 \times 10^{-3}$ $T_1 = 50 + 273 = 323\ K$

Substituting this data in the Arrhenius equation

$$\ln\left[\frac{k_1}{k_2}\right] = \frac{E}{R}\left[\frac{1}{T_1} - \frac{1}{T_2}\right]$$

$$\ln\left[\frac{6.0 \times 10^{-6}}{9.0 \times 10^{-3}}\right] = \frac{E}{8.314}\left[\frac{1}{290} - \frac{1}{323}\right]$$

$$E = \frac{8.314\ \ln\dfrac{6.0 \times 10^{-6}}{9.0 \times 10^{-3}}}{\dfrac{1}{290} - \dfrac{1}{323}} = 172.6\ \text{kJ}$$

Problem 25.

The specific reaction rate of a certain reaction doubles, when the temperature of the reaction is increased from 20 ^{O}C to 40 ^{O}C. Calculate the activation energy of the reaction.

Let rate constant at 20 ^{O}C (20 + 273 = 293 K) be k_1

Then rate constant at 40 ^{O}C (40 + 273 = 311 K) = $k_2 = 2k_1$

$$\log\left[\frac{k_1}{k_2}\right] = \frac{-E}{2.303 \times 8.314}\left[\frac{1}{293} - \frac{1}{313}\right]$$

$$\log\left[\frac{1}{2}\right] = \frac{-E}{2.303 \times 8.314}\left[\frac{1}{293} - \frac{1}{313}\right]$$

$$E = \frac{(2.303 \times 8.314 \times \log 2)}{\left(\dfrac{1}{293} - \dfrac{1}{313}\right)}$$

$$E = 26.43\ \text{kJ}$$

Problem 26.

Calculate the activation energy of a reaction with a rate constant of 2.35×10^{-3} s^{-1} at 80 ^{O}C. Given the pre-exponential factor of the reaction is 10^{11}.

$$k = A\, e^{-E/RT}$$

$$2.35 \times 10^{-3} = 10^{11}\ e^{-E/(8.314 \times 353)}$$

$$\frac{2.35 \times 10^{-3}}{10^{11}} = e^{-E/(8.314 \times 353)}$$

$$\ln (2.35 \times 10^{-3}) = -\frac{E}{(8.314 \times 353)}$$

Activation energy $E = 92.1$ kJ

13.11 Theories of Reaction Rates

To account for the observed kinetics of reactions in terms of the molecular behavior, several theories have been proposed. Out of them two attempts, viz., collision theory and activated complex theory stand in the fore front. We are discussing here only collision theory.

13.11.1 Collision Theory

The basic requirement of the collision theory are :

(a) Each molecule must collide with other molecule.

(b) The colliding pair of molecules must have sufficient energy to climb over the energy barrier of the reaction.

The first requirement of collision theory emphasizes collision between two molecules. Calculations based on the kinetic molecular theory show that in gaseous phase, at ordinary pressure (1 atm) and temperature (300 K), there could be 1×10^{27} binary collisions in 1mL of volume per second. These collisions increase with increase in temperature. If all these collisions are successful in transforming into product, then all the reactions would be complete instantaneously. But this is against the experience. This necessitated the second requirement.

The second requirement imposes a condition that the molecules must possess sufficient energy to overcome the activation energy barrier. For this molecule must have sufficient kinetic energy and must collide at a proper angle. This can be well understood by considering an analogy of a moving car. A car running in good condition will not disintegrate or break-apart on its own while moving on the road. When it is moving slow enough, like in a busy city center and bumps against another car they simply bounce from each other. When the same car is moving at high velocity, like on a high-way, the car may break into pieces when it meets another car head-on and may escape scratched when it grazes another car. Similarly incase of molecular model also a molecule by itself doesn't undergo any disintegration unless hit by another molecule at a proper angle and with a proper velocity (energy).

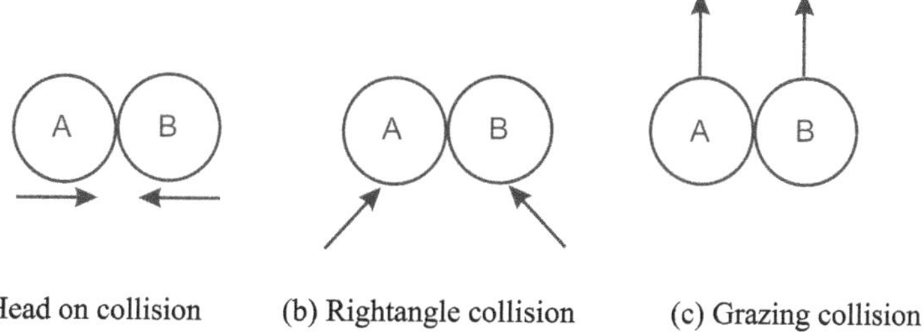

(a) Head on collision (b) Rightangle collision (c) Grazing collision

Fig. 13.9(a) Types of collision.

Two molecules are regarded as having collided if they approach a distance equal to the sum of their radii. Consider two molecules A and B participating in a chemical reaction and let the corresponding velocities be U_A and U_B.

Mean average velocity $\overline{U_r^2} = U_A^2 + U_B^2$

or $$U_r = \sqrt{U_A^2 + U_B^2} \quad .....(13.41)$$

From Maxwell's distribution law, square average velocity of any molecule is equal to $8R_T/\pi M$ where M is the molecular weight and R, the gas constant.

$$\overline{U_A^2} = \frac{8RT}{\pi M_A} \text{ and } \overline{U_B^2} = \frac{8RT}{\pi M_B}$$

Therefore equation (13.41) may be written as

$$\overline{U_r} = \sqrt{\frac{8RT}{\pi M_A} + \frac{8RT}{\pi M_B}} = \sqrt{\frac{8RT}{\pi}\left(\frac{1}{M_A} + \frac{1}{M_B}\right)}$$

$$\overline{U_r} = \sqrt{\frac{8RT}{\pi}\left[\frac{M_A + M_B}{M_A M_B}\right]} = \sqrt{\frac{8RT}{\pi\mu}} \qquad \text{.....(13.42)}$$

where : = $\frac{M_A + M_B}{M_A M_B}$ and is called **reduced** mass of the reactant pair.

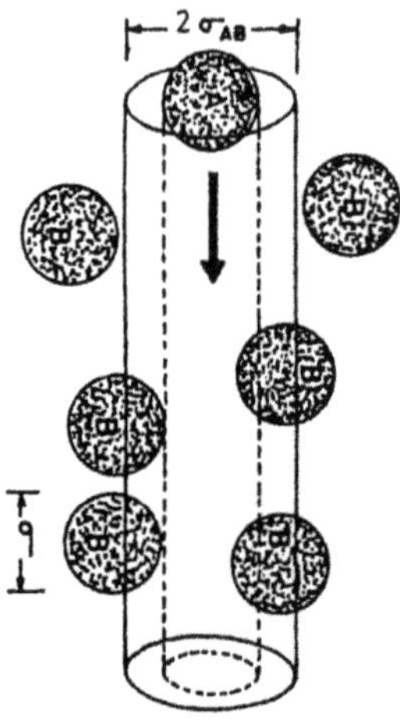

Fig. 13.9(b) Movement of A through a cylinder containing B.

For the purpose of derivation of an expression for calculating number of collisions made by a molecule per unit time, let us consider only the movement of A keeping B stationary Fig. 13.9(b). In unit time molecule A, with a molecular diameter d_A and mean velocity $\overline{U}_r$, would encounter collisions with all those B molecules present in a cylinder of radius $\sigma_{AB} = \frac{d_A + d_B}{2}$ and length $\overline{U}_r$.

Total volume swept by molecule A per unit time = $B\,(\Phi_{AB})^2\,\overline{U}_r$

If N_B is the number of molecules of B present per unit volume, then,

Total number of collisions per unit time $Z_B = N_B\,B\,(\Phi_{AB})^2\,\overline{U}_r$

In case N_A molecules of A are also present per unit volume in the cylinder discussed above,

Total number of collisions between A & B per unit time $Z_{AB} = N_A Z_B$

$$Z_{AB} = N_A\,N_B\,B\,(\Phi_{AB})^2\,\overline{U}_r$$

Substituting the value of U_r from eq. (13.42),

$$Z_{AB} = N_A\, N_B\, B\, (\Phi_{AB})^2 \left[\frac{8RT}{\pi\mu}\right]^{1/2}$$

$$Z_{AB} = N_A\, N_B\, (\Phi_{AB})^2 \left[\frac{8\pi RT}{\mu}\right]^{1/2} \qquad .....(13.43)$$

From the first requirement of collision theory, Rate of reaction is also proportional to number of collisions made by participating molecules per unit time.

$$Rate\ of\ reaction\ \forall\ Z \qquad .....(13.44)$$

But the second requirement of collision theory opinions that all the collisions are not fruitful and they must possess a minimum activation energy. Fraction of activated collisions is given by the expression $e^{-E/RT}$

Including this correction term in the rate equation (13.44),

$$Rate\ of\ reaction = Z\, e^{-E/RT} \qquad .....(13.45)$$

From the fundamental concepts of reaction kinetics, rate of reaction is directly proportional to the concentration of reactants

$$Rate\ of\ reaction = k\, N_A\, N_B \qquad .....(13.46)$$

Equating both the rate expressions (13.45) & (13.46)

$$k\, N_A\, N_B = Ze^{-E/RT}$$

or $$k = \frac{Z}{N_A N_B}\, e^{-E/RT} \qquad .....(13.47)$$

$$k = Z_0\, e^{-E/RT} \qquad .....(13.48)$$

where $$Z_0 = \frac{Z_{AB}}{N_A N_B} = \frac{1}{N_A N_B}(\sigma_{AB})^2 \left[\frac{8RT}{\mu}\right]^{1/2} \qquad .....(13.49)$$

For Like Molecules : $N = N_A = N_B,\ M = M_A = M_B$ and $d = \Phi_{AB} = d_A$

Reduced mass $= : = \left[\frac{M+M}{M^2}\right] = \frac{2}{M}$

Substituting these values in eq. (13.49);

$$Z_0 = \frac{1}{N^2}\, d^2\, \frac{(8\pi RT)^{1/2}}{(2/M)^{1/2}}$$

or $$Z_0 = \frac{2}{N^2} d^2 \left[\frac{\pi RT}{M}\right]^{1/2} \quad .....(13.50)$$

13.11.2 Success of Collision Theory

(i) Collision theory was successful in predicting the rate constants of reactions involving simple molecules and there by the activation energy of the reactions. A good agreement was observed between the experimental and theoretical activation energy values, Table 13.5.

Table 13.5 Experimental & Theoretical activation energy values

Reaction	*Activation energy in kcal mol⁻¹*	
	Calculated from collision theory	*Experimental*
$H_2 + I_2 \longrightarrow 2HI$	40.0	39.5
$2HI \longrightarrow H_2 + I_2$	44.0	44.0
$2N_2O \longrightarrow 2N_2 + O_2$	56.0	58.0

(ii) The expression of rate constant obtained from collision theory is also widely used for estimating molecular diameters. This procedure uses the experimental activation energy.

13.11.3 Failure of Collision Theory

Exhaustive collection of kinetic data of various reactions pointed out many limitations for the collision theory. The reactions considered in Table 12-6, where activation energy values derived from collision theory are in good agreement with the experimental values, are simple molecules behaving like hard and inelastic spheres. But in reality many molecules are compressible and products are formed only by the collision of molecules in particular orientation. Thus bimolecular polymerization of ethylene was found to be 10^3 times slower than the predicted value, while 1, 3-butadiene polymerization was slower by 10^4 times. In general larger the size/length of an organic molecule slower was the rate of reaction.

On the other extreme chain reactions, e.g., photochemical combination H_2 and Cl_2 to give HCl, were reported to be 10^7 times faster than the rate predicted by collision theory.

In order to accommodate these deviations, a new term P called **probability factor**, was introduced in to the equation (13.46) and the modified equation is written as;

$$k = PZ_O\, e^{-E/RT} \qquad \text{.....(13.51)}$$

In the above equation P is simply a ratio of theoretical rate constant & observed rate constant and has no special meaning.

The value of P ranged from 10^{+9}, for reactions needing particular orientations before an activated collision, to 10^{-9} for the chain reactions. Though this modification was quite attractive in the beginning, its charisma fades soon for want of theoretical justification. Its introduction served no purpose except to deceiving one's own self. Reasons for the failure of collision theory are summarized below.

(i) Collision theory considers the reactant molecules as hard, inelastic spheres but in reality majority of the molecules are compressible and elastic.

(ii) Collision theory does not take steric factors into account. But many molecular reactions need a particular orientation before collision. For example, in the bromination of p-nitrophenol, a

$$HO-C_6H_4-NO_2 + Br^- \longrightarrow Br-C_6H_4-NO_2 + OH^-$$

collision is fruitful only when Br^- strikes at the aromatic carbon where OH^- group is attached. All the other collisions, though energetically rich, will not lead to any product formation.

(iii) Collision theory failed to account for higher reaction rates as found in case of chain reactions than anticipated.

(iv) There is no way to explain the phenomena of catalysis, where introduction of small amount of catalyst, increases the reaction rate many fold without increasing the number of collisions per second per unit volume.

(v) Similar to general catalysis discussed above, collision theory is also incapable of explaining surface catalysis of gas phase reactions. In these reactions the rate is proportional to the active surface area of the catalyst and independent of the concentration terms of the reactants and the collisions between them.

(vi) It is possible for the excited molecule to distribute its activation energy among all the bonds than localize it in a particular bond that may lead to a spontaneous reaction. Such a situation is not discussed while deriving an expression of k in collision theory.

(vii) With the increase in temperature by 10 ^{O}C, the number of collisions increase by only 2% but the rate increases many fold. There is no explanation in collision theory for such an irrationality.

13.11.4 Activated Complex Theory or Theory of Absolute Reaction Rates

In 1935 Eyring chose an entirely new and radical approach to explain dependence of reaction rates on concentration, temperature and other parameters and explain various other kinetic features which the collision theory failed to accommodate. The approach involves postulation of an equilibrium between reactants and a transient called activated complex. The activated complex is regarded as a normal molecule in all respects except for its transient existence and capability of converting itself into product(s) at a definite rate in a particular vibrational mode.

Thus in the biomolecular decomposition of hydrogen iodide, for example, the activated complex may be formulated as follows;

$$\begin{matrix} H & & H \\ | & + & | \\ I & & I \end{matrix} \rightleftharpoons \begin{matrix} H & \cdots & H \\ \vdots & & \vdots \\ I & \cdots & I \end{matrix} \longrightarrow \begin{matrix} H & - & H \\ & + & \\ I & - & I \end{matrix}$$

Reactants Activated complex Products

Generalizing the above reaction we can represent a reaction scheme as follows;

$$A + B \overset{K^{\neq}}{\rightleftharpoons} (AB)^{\#}$$

$$(AB)^{\#} \overset{\nu}{\longrightarrow} \text{Product}$$

In case of unimolecular reactions, the activated complex may be visualised as a suitably rearranged reactant capable of giving product(s). The activation energy, hither to thought only as energy required for fruitful collisions, is now defined as;

> "*the difference in the energy content of reactants and activated complex*"

or

> "*the energy which the reactants must acquire in order to form activated complex of the reaction*".

That is to say, the reactant molecule must possess enough energy to climb over an activation energy barrier, Fig. 13.10, in order to form activated complex. Once the activated complex is formed, it is left with two options, viz.,

(i) to revert back to reactants,

(ii) to form the products.

Thus, the rate of any chemical reaction depends on,

1. The concentration of activated complex.
2. The rate (frequency) at which the activated complex disintegrates/rearranges to give product(s).

   Therefore, Rate of reaction $= [AB^{\#}].\ \nu$

But $K^{\#} = \dfrac{[AB^{\#}]}{[A][B]}$ or $[AB^{\#}] = K^{\#}\,[A]\,[B]$

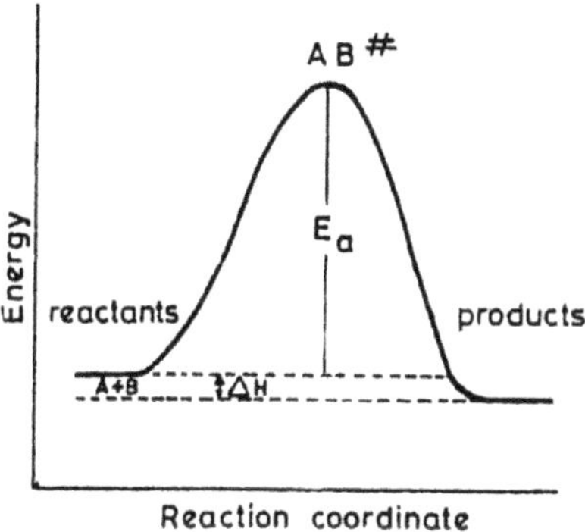

Fig. 13.10 Activation energy barrier.

Substituting for $[AB^{\#}]$ in the above rate equation,

$$\text{Rate of reaction} = K^{\#}\,[A]\,[B] \times \nu$$

Frequency, ν, with which an activated complex converts itself into products was found to be equal to RT/Nh.

$$\text{Rate of reactions} = K^{\#}\,[A]\,[B] \times \frac{RT}{Nh}$$

or

$$\text{Rate constant } k = K^{\#}\,\frac{RT}{Nh} \qquad \text{.....(13.52)}$$

From thermodynamics we know,

$$\Delta G = -RT\,ln\,k \quad or \quad k = e^{-(\Delta G/RT)}$$

and

$$\Delta G = \Delta H - T\Delta S$$

∴

$$-RT\,ln\,K^{\#} = \Delta G^{\#} = \Delta H^{\#} - T\Delta S^{\#}$$

$$ln\,K^{\#} = \frac{-\left(\Delta H^{\#} - T\Delta S^{\#}\right)}{RT}$$

$$K^{\#} = e^{\frac{-\left(\Delta H^{\#} - T\Delta S^{\#}\right)}{RT}} = e^{\frac{-\Delta H^{\#}}{RT}} \times e^{\frac{\Delta S^{\#}}{R}}$$

Substituting the value of $K^{\#}$ in equation (13.52)

$$K^{\#} = \frac{RT}{Nh}\,e^{-\Delta H^{\#}/RT}\,e^{\Delta S^{\#}/R} \qquad \text{.....(13.53)}$$

Here rate expression is associated with an enthalpy term and an entropy term. Enthalpy change, $\Delta H^{\#}$, between reactants and activated complex is nothing but Arrhenius activation energy, E_a.

$$k = \frac{RT}{Nh}\,e^{-E_a/RT}\,e^{\Delta S^{\#}/R} \qquad \text{.....(13.54)}$$

Similarly, $\Delta S^{\#}$ in the rate equation takes care of the **steric factors** that determine the rate of reaction. For a reaction which requires a particular orientation, we can expect a more orderly arranged activated complex, i.e., ΔS is +ve. This accounts for a lower K value than expected by collision theory. The enhancement of reaction rate in the presence of a catalyst was explained as due to lowering of activation energy. Thus the theory of absolute reaction rates corrects the inadequacies of collision theory, simultaneously providing a thermodynamic base for its arguments.

13.11.5 Applications of Chemical Kinetics

Chemical kinetics is a powerful tool in arriving at the mechanism of a reaction, for this one needs to study the reaction trends at various concentrations, temperatures, pressures and a variety of solvents and solutes. The reaction mechanism arrived should never be concluded as ultimate and final as collection of new data or determination of old data with a new precision could some times lead to an entirely new thinking and thereby a new mechanism. However, in case of simple reactions the mechanism could be almost absolute. *The strength of reaction kinetics in arriving at the mechanism is well projected in the nucleophilic substitution reactions.* In these reactions the substitution species seeks out a positive toward which it can push electrons. Our present discussion is mainly centered around the reactions between alkyl halides and the hydroxide ions. In aqueous ethanol medium the over all reaction is given as

$$RX + OH^- \longrightarrow ROH + X^-$$

Where OH^- is a nucleophilic reagant and X is halide replaced by the hydroxyl group.

There are two schemes by which this substitution can occur.

Scheme – I : By a one step biomolecular process

$$RX + OH^- \longrightarrow ROH + X^-$$

Scheme – II : By a two step process involving unimolecular ionization of alkyl halide followed by attack by nuclephile.

$$\text{(a) } RX \xrightarrow{\text{slow}} R^+ + X^-$$

$$\text{(b) } R^+ + OH^- \xrightarrow{\text{fast}} ROH$$

In the first scheme the rate of reaction is proportional to the concentration of both RX and OH^-. In the second process the ionization of alkyl halides is a slow step and there by rate determining. The ion R^+ produced in step (a) is known as carbomium ion. The nucleophilic reagant appears only to the concentration of alkyl halide. Hughes and Ingold referred the mechanistic scheme-I and mechanistic scheme-II as S_N^2 (substitution, nucleophilic, bimolecular) and S_N^1 (substitution, nucleophilic, unimolecular). Kinetically S_N^1 reactions will have an order of one and S_N^2 reaction will have an order of two, unless one of the reactant is present in excess (pseudo order conditions) when the reaction will have an order of one. The criteria for S_N^1 or S_N^2 mechanisms is discussed below.

(i) **Ionization Order :** The order of different alkyl halides is;

Tertiary > Secondary > Primary

This is because of the stabilization of the tertiary carbomium ion by the three electron releasing (+I) alkyl groups over the primary carbomium ion. This ionization order favors a S_N^1 mechanism in case of hydrolysis of t-alkyl halides and S_N^2 mechanism for primary alkyl halides.

(ii) **Effect of medium :** The dielectric constant of the medium has a significant effect in deciding the nature of nucleophilic substitution. The ionizing solvents solvate and stabilize the carbomium ions formed and will therefore strongly favor reactions by S_N^1 mechanism.

(iii) **Nucleophilicity of Nucleophilic reagant :** In S_N^1 reactions the nucleophilic reagant appears in the fast step and therefore the rate will be independent of whether the reaction involves OH^- or H_2O or $NO3^-$, i.e., nucleophilicity of the nucleophilic reagant. The S_N^2 reactions on the contrary are influenced greatly by the nuclephilicity of the nucleophile. For example, in the substitution reaction.

$$R-I + \left\{\begin{matrix} F^- \\ Cl^- \\ Br^- \end{matrix}\right. \longrightarrow R\left\{\begin{matrix} F \\ Cl \\ Br \end{matrix}\right.$$

the rate of reaction follows the order

$$F^- > Cl^- > Br^-$$

Which is in accordance with the nucleophilicity of the nucleophile. Similarly the rate of hydrolysis of a t-alkyl halide following S_N^2 mechanism higher with OH^- than H_2O, since hydroxide ion is a much more powerful base than water.

Numberical Problems

1. The rate constant of a first order reaction is $1 \times 10^{-3}\ s^{-1}$. Calculate the half life of the reaction.

2. For a reaction k at $7^{O}C$ was 7×10^{-7} and at $57^{O}C$ was 9×10^{-4}. Calculate energy of activation.

3. When one gram of ethyl acetate was hydrolysed with one gram molecule of sodium hydroxide, the concentration of NaOH noted at different times is as follows.

t (in secs)	0	600	1200
$(a-x)$	8.04	3.50	2.22

Show that the reaction is of second order.

4. Find the value of specific reaction rate for the hydrolysis of methyl acetate from the following data :

Time (min) :	0	20	∀
Vol. of NaOH (ml) :	24.5	25.85	47.15

5. The specific rate constant of a first order reaction is 4.0×10^{-5} min^{-1} at 25°C, and 8.0×10^{-4} min^{-1} at 45°C. Calculate the energy of activation of this reaction.

6. The following data is obtained for the thermal decomposition of N_2O_5 in CCl_4 at 25°C.

Volume of O_2 generated (in mL)	258	493	694	874	1030	1186
Time (min)	200	400	600	800	1000	1200

Determine the order of the reaction.

7. N-chloroacetanilide changes to p-chloroacetanilide and the reaction can be monitored by withdrawing a known amount of reaction mixture at different time intervals, adding KI solution and titrating the liberated iodine with a standard hypo solution. The titre value, therefore, corresponds to unreacted N-chloroacetanilide at any given instant. One such typical data is given below.

Time (min)	0	60	120	180	240	360	480	∀
Titre (cc)	49.3	35.6	25.8	18.5	13.8	7.3	4.8	0.0

Find the order of the reaction adopting Trial and error method and Graphical method.

8. The following data is obtained for the decomposition of N_2O_5 at different temperatures.

K (s^{-1})	7.87×10^{3}	3.46×10^{5}	4.98×10^{6}	4.87×10^{7}
T (k)	273	298	318	338

Draw arrhenius plot and estimate the activation energy of the reaction.

9. Peroxodisulphate-iodide reaction gave the following rate constants at several temperatures.

T (^{O}C)	3	13	24	33
K (L mol^{-1} s^{-1})	1.4×10^{-3}	2.9×10^{-3}	6.2×10^{-3}	1.20×10^{-3}

(a) Construct Arrhenius plot

(b) Calculate the value of k at 50 ^{o}C

(c) Calculate the activation energy.

10. A certain first order reaction A $\longrightarrow$ products gave k value of 4×10^{-3} min^{-1} at $25^{o}C$. Calculate the rate constant at $50^{o}C$, if the activation energy is 25 kcal mol^{-1}.

11. For a reaction 2A $\longrightarrow$ Products, the heat of reaction is 52 kJ mol^{-1}. The activation energy of the forward reaction is 123 kJ mol^{-1}. Find the activation energy of the reverse reaction.

12. Find the order w.r.t hydrogen peroxide decomposition

$2H_2O_2 \longrightarrow 2H_2O + O_2$ from the following data;

$[H_2O_2]$ (M)	0.10	0.20	0.30	0.50	0.70
Rate of H_2O_2 disappearance (mol L^{-1} min^{-1})	$0.6{\times}10^{-4}$	$1.2{\times}10^{-4}$	$1.8{\times}10^{-4}$	$3.0{\times}10^{-4}$	$4.2{\times}10^{-4}$

13. The rate constant of the reaction A + B $\longrightarrow$ products is 1.5×10^{10} dm^3 mol^{-1} s^{-1}. Calculate the half life of the reaction, if [A] = [B] = 10^{-2} mol dm^{-3}.

[*HINT* : From the units of *k* it is clear that the reaction follows second order kinetics. For a second order reaction with equal reactant concentrations, $t_{½} = 1/(ka)$]

14. The rate constant for the formation of HI at $283^{o}C$ and $430^{o}C$ are $1.2{\times}10^{-4}$ dm^3 $mol^{-1}s^{-1}$ and $1.7{\times}10^{-1}$ dm^3 mol^{-1} s^{-1} respectively. What is the value of E_a and A for this reaction. If the bond dissociation energies of H_2, I_2 and HI are 103.4, 36.2 and 71.4 kcal respectively, find the energy of activation for the decomposition of HI.

15. The activation energy of the reaction between ICl and H_2 is 168 kJ mol^{-1}. If the velocity constant at $230^{o}C$ is 0.163, find the frequency factor and *k* at $280^{o}C$.

16. A piece of wood buried under a glacier was found to have 30% as much ^{14}C as present in a fresh piece of wood. Estimate the age of wood. Given $t_{½}$ of ^{14}C is 5600 years.

17. Temperature coefficient for the persulphate-iodide reaction between 25 and $35^{o}C$ is 3.5. Find the specific reaction rate at $50^{o}C$. Given frequency factor $A = 1 \times 10^{11}$.

18. A certain reaction represented by the equation

$A + B \longrightarrow 2C + D + E$

was studied at 25 ^{O}C at different concentrations of A and B and the following data was obtained.

[A$_o$] (mol L^{-1})	**[B$_o$] (mol L^{-1})**	***Initial rate of formation* of D (mol L^{-1} s^{-1})**
0.01	0.01	1.5×10^{-3}
0.02	0.01	3.0×10^{-3}
0.03	0.01	4.5×10^{-3}
0.03	0.02	9.0×10^{-3}
0.03	0.03	1.35×10^{-4}

(a) What are the orders with respect to A and B

(b) What is the rate law for the reaction

(c) If [A$_o$] = 0.02 and [B$_o$] = 0.03, What will be the rate of formation of D.

19. Half life period of disintegration of ^{32}P is 14.28 days. How long will it take for the disintegration of 90% of a given sample.

(*HINT :* Nuclear disintegration processes always follow first order kinetics)

20. Radium (^{226}Ra) disintegrates with a half life of 1600 years. Find the disintegration constant and time needed for 75% disintegration.

Self Test Questions

1. For the reaction $N_2 + 3H_2 \longrightarrow 2NH_3$ at a particular instant the rate of disappearance of hydrogen is 0.6 mol L^{-1} s^{-1}. The rate of formation of ammonia in mol L^{-1} s^{-1} is

(A) 0.6 (B) 0.4 (C) 0.2 (D) 0.1

2. Unit of rate constant of second order reaction 2A $\longrightarrow$ Products is,

(A) L mol^{-1} s^{-1} (B) mol L^{-1} s^{-1} (C) L mol^{-1} s (D) mol L^{-1} s

3. The rate constant of a first order reactions is 6.93 s^{-1}. The half-life of the reaction is

(A) 0.001 s (B) 0.01 s (C) 0.693 s (D) 0.1 s

4. Arrhenius equation is written as

(A) $\frac{d\ln k}{dT} = e^{-E/RT}$ (B) $\frac{d\ln k}{dT} = \frac{E}{RT} + Z$

(C) $\frac{d\ln k}{dT} = \frac{E}{RT^2}$ (D) $\frac{d\ln k}{dT} = \frac{E}{RT}$

5. The unit of zero order rate constant is

   (A) s^{-1} (B) L mol^{-1} s^{-1} (C) mol L^{-1} s^{-1} (D) Dimension less

6. Rate constant of a reaction changes with changes in

   (A) Concentration of reactants (B) Temperature

   (C) Time (D) Concentration of products

7. 20 g of a radio-active substance decays to 10 g in 10 minutes. The total time required for 20 g of this sample to become 5 g is

   (A) 40 min (B) 20 min (C) 10 min (D) 5 min

8. The half life period of a second order reaction $2A \longrightarrow$ products is

   (A) directly proportional to the initial concentration.

   (B) directly proportional to the square root of initial concentration.

   (C) independent of initial concentration

   (D) inversely proportional to the initial concentration.

Short and Essay Type Questions

1. (a) Describe any two methods for the determination of order of a chemical reaction.

   (b) Deduce the units for the zero, first and second order rate constants.

2. (a) What are pseudo unimolecular and zero-order reactions.

   (b) Explain briefly "The collision theory of bimolecular reactions".

3. (a) Derive an expression for the rate constant of a first order reaction.

   (b) What are the units of the first order rate constant? Will a change in concentration units affect the rate constant ? Explain.

   (c) Explain how one can determine half-life period for a particular reaction.

4. (a) Explain the terms rate constant and half-life period.

   (b) Give the relation between half-life period and rate constant for first and second order reactions.

5. Derive the rate equation for a second order reaction when the concentration of reactants are equal.

6. Give the characteristics of a 2nd order reaction. Explain one method for determining the order of a reaction.

7. What is the effect of temperature on a reaction rate? Give Arrhenius equation. Plot the graph showing the variation of k with temperature. What can you calculate from the slope ?

8. Write equation for 2nd order reaction, when initial concentrations of the two reactants are equal. What are the units of k ? Explain how the specific reaction rate for the reaction between persulphate and iodide can be obtained experimentally.

9. What is meant by a reaction being zero, first or second order? Derive an expression for the rate constant of a first order reaction.

10. Show that the time of half change for a first order reaction is independent of initial concentration.

11. Distinguish order and molecularity of a reaction.

12. What is activation energy.

13. Define the terms (i) Molecularity (ii) Order (iii) Half life period of a reaction. Describe any one method for the determination of the order of a reaction

14. Write the differential rate equation for the following reactions in terms of the disappearance of the reactants and appearance of the products.

    (a) $N_2(g) + 3H_2(g) \longrightarrow 2NH_3(g)$

    (b) $H_2O_2 + 3KI\ (aq) + 2HCl \longrightarrow KI_3(aq) + 2H_2O(aq) + 2Kcl\ (aq)$

    (c) $2N_2O_5 \longrightarrow 4NO_2 + O_2 + 2KCl(aq)$

    (d) $CH_3COOC_2H_5 + NaOH \longrightarrow CH_3COONa + C_2H_5OH$

15. Explain the dependence of rate constant on the following physical quantities

    (a) Concentration (b) Nature of reactants

    (c) Temperature (d) Volume

16. The reaction $NO_2(g) + CO(g) \longrightarrow CO_2(g) + NO_2(g)$ follows the following mechanistic scheme

$$NO_2 + NO_2 \xrightarrow{slow} NO_3 + NO$$

$$NO_3 + CO \xrightarrow{fast} NO_2 + CO_2$$

Explain why the reaction is zero order in CO.

17. "Order of a reaction is generally decided by the slowest step while molecularity is decided by only the individual elementary steps" explain the statement.

18. Chemical reaction between A and B in the presence of acid traces the following is sequence of steps.

Step 1 $A + H^+ \longrightarrow P_1$

Step 2 $P_1 + B \longrightarrow P_2 + H^+$

(a) Write the overall reaction

(b) If step 1 is the rate determining step what could be the order with respect to B.

(c) In case step 2 is the slow step, which concentration terms determine the rate of the reaction.

19. Write the differential rate equations of the following elementary reactions.

(i) $A + 2B \longrightarrow$ Products

(ii) $A + B \longrightarrow 2C + D$

(iii) $3A \longrightarrow 4P + Q + 3Rs$

(iv) $2A + B \longrightarrow$ Products

Chapter 14

High Polymers

(Plastics, Rubbers, Adhesives and Laminates)

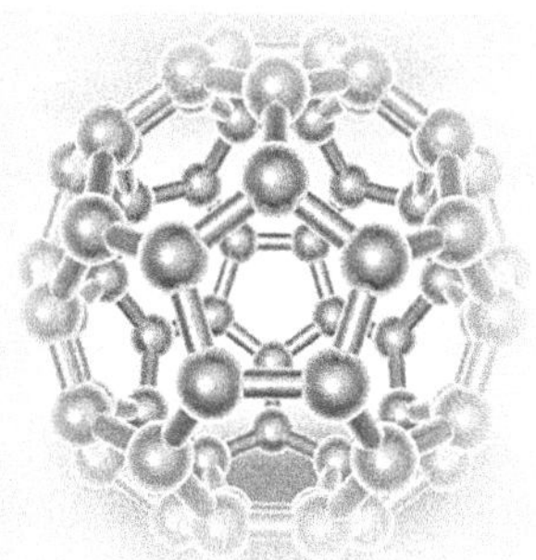

14.1 Polymers

14.1.1 Introduction

Although the large number of macromolecules, containing few hundred to few thousand atoms) all linked to one another firmly, have evolved along with the universe, their science remained obscure and emerged suddenly into lime light, to name this technology era as *polymer or plastic age*. Their actual importance may be realized from the statistics that half the American chemists and chemical engineers are employed in industries concerned with polymeric materials. A common man can readily realize their importance by attempting to recognize the polymeric materials around him and within him. The naturally occurring macromolecules, viz., cellulose, wool and rubber are used to drape ourselves, to live, to walk and to roll on them. Majority of the food items contains polysacharides, proteins and fibers and the very existence of life and its replication is the result of biopolymers called DNA and RNA. Similarly most of the synthetic materials existing around us in the form of plastics, fibers, gums, laminates, paints, gadgets, etc., are all different forms of macromolecular materials. Some of the synthetic polymers have gone to the extent of replacing ailing hearts and other very sensitive organs.

Polymers, often called as *macromolecules*, are very large molecules formed by linking together small, simple chemical units called *monomers*. The resulting structure comprises of repetition of relatively small molecular fragments usually referred to as *repeating units*. For example, polyethylene is a polymer made up of inter linking of a number of ethylene monomer molecules. In this process the double bonds of ethylene molecules open up and bind with the adjacent C_2H_4 unit. Here $\left(C_2H_4\right)$ is the repeating unit.

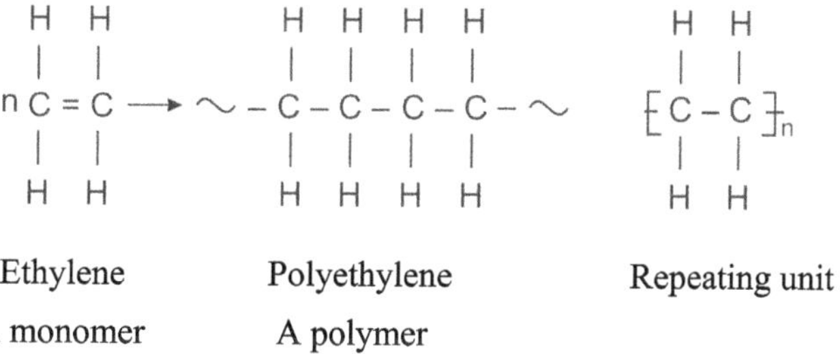

Ethylene A monomer

Polyethylene A polymer

Repeating unit

The physical and chemical properties of the polymer are much different from the monomer. In fact polyethylene behaves much like saturated hydrocarbons with their inactivity towards chemical reactions characteristic of double bonds like facile bromination.

Although polymeric materials were discovered by early chemists, polymer science may be thought to have entered this new era largely because of untiring efforts by Staudinger through his macromolecular hypothesis presented in 1920, which earned him the 1953 Noble Prize. He proposed and substantiated the molecular formula of polystyrene, rubber and other polymers. This soon paved way for commercial exploitation of polymers like PVC, Nylon, Neoprene, Styrene -Butadiene synthetic rubber, etc. Ziegler and Natta (1955) developed new polymerization catalysts with stereo-regulating powers and obtained the second Nobel Prize of polymer chemistry in 1963. Flory elucidated the mechanisms of polymerization processes and added one more Noble Prize to the field in 1974. As a result of these discoveries the polymer industry currently produces around 30 billion pounds of polymer materials annually.

14.1.2 Classification

The polymeric materials can be broadly classified on the basis of a physical verification of origin, elements involved in the backbone chains and physical properties.

(a) **Natural and Synthetic Polymers :** Depending on their origin, polymer can be grouped into Natural and Synthetic polymers. The natural polymers are either naturally occurring materials or their chemical modifications. In spite of discovery of synthetic routes, for example - natural rubber from isoprene, these materials still fall under natural polymers. Cell constituents like DNA, proteins etc., also fall under this category. Some of the commercially used natural polymers are cotton, silk, leather, wool, rubber, cellulose, acetate and cellophane.

The synthetic polymers are relatively of lower molecular weight when compared to their natural counter parts but find use in variety of applications. Classifications and uses of polymers of this class is provided in the next section.

(b) **Organic and Inorganic Polymers :** Organic polymers, as expected, have carbon back bone structures while inorganic polymers usually have back bone structures made out of non-metals other than carbon. Elements like Si, Ge, N, P, As, O, S, Se etc., with limited catenations (inter linking) usually make the back bone of inorganic polymers. In either case the side valencies of the back bone elements are satisfied by elements like hydrogen, halogens, oxygen, nitrogen, etc. Almost all the polymers of biological origin and majority of the synthetic origin belong to the group of organic polymers. Therefore, when we mention simply polymers, we generally mean organic polymers. As we already stated, organic polymers have a variety of applications and does not fit in any simple frame work. However, inorganic polymers have some specific and unique applications, where organic polymers are not of much use. The typical and most useful polymers of this class are glass and silicones (-Si-O-Si-back bone). Depending on the size and cross linking silicones may be fluids, greases, resins or rubber. Silicone rubbers retain elasticity over a wide range of temperatures and silicone lubricants are mixed in mineral oils.

(c) **Thermosetting and Thermoplastic Polymers :** This classification depends on the differences in physical properties on heat treatment. Polymers that soften on heating and regain original physical properties on cooling are known as Thermoplastics. On the contrary, Thermosetting plastics are deformed irreversibly on heat treatment. Linear chain polymers fall under the first category of thermoplastics. The process of heating, reshaping and retaining the same on cooling can be repeated several times on this class of polymers. Polyethylene, PVC, Nylon, etc., are typical examples of thermoplastic polymers. In case of cross linked polymers, heat treatment brings changes in the cross linkages and converts them into infusible mass.

(d) **Fibers, Elastomers, Plastics and Liquid Resins :** This nomenclature is based on the ultimate form which in turn depends on dominant physical property. Fibers, elastomers, plastics and liquid-resins differ in the extent of the crystallinity and physical state and are referred as crystalline, semi-crystalline, amorphous and liquid respectively.

(a) ***Fibers :*** *Being highly crystalline, possess high tensile strength. Because of this property these materials are drawn into fibers used in weaving and making ropes. eg: Nylon, Terelene.*

(b) ***Elastomers :*** *These polymers possess only partial crystallinity and therefore can be stretched to one and half times its original length, but regain original shape on releasing. Because of their rubbery characters they are also used as synthetic rubbers. They exhibit good strength and elongation. eg., Silicone rubber, Neoprene, Natural rubbers.*

(c) ***Plastics :*** *Polymers that can be shaped into hard and tough utility articles by the applications of heat and pressure are called plastics. Eg : PVC, Teflon, Polymethylmethacrylate, Polystyrene, etc.*

(d) ***Liquid Resins*** **:** Polymers existing in liquid state are used as adhesives, lubricants, sealants and resins. Eg. Some Silicones, Polysulphide Sealants, Epoxy resins.

14.1.3 Types of Polymerization

W.H. Carothers in 1929 categorized the polymerization reactins into *Addition polymerization and Condensation polymerization.*

14.1.4 Condensation Polymerization

Condensation polymers are those in which the repeat unit the polymer lacks certain atoms present in corresponding monomer(s). Two like or unlike monomers join each other by the elimination of small molecules, such as water, to give the polymer chain.

When the monomers of the same kind join together, the polymer obtained is called monopolymer. Nylon-6 is formed by self condensation of ε-aminocaproic acid

$$2n\,H_2N-(CH_2)_5-COOH \rightarrow -[HN-(CH_2)_5-CO-HN-(CH_2)_5-CO]_n-OH + n\,H_2O$$

ε-aminocaproic acid Nylon-6

Polymerization of two or more different monomers gives a copolymer. Nylon-6,6 is a best example of condensation polymers. It is a copolymer, formed by the combination of a dicarboxylic acid and a diamine.

Polyamide (Nylon 6, 6)

$$n\,H_2N(CH_2)_6-NH_2 + n\,HOOC-(CH_2)_4-COOH \xrightarrow{} [-NH-(CH_2)_6-NH-CO(CH_2)_4-CO-]_n + 2n\,H_2O$$

Hexamethylene diamine + Adipic acid → Nylon

A general mechanistic scheme for condensation polymerization involves a simultaneous combination of any two monomers to give dimers, which in turn combine to give tetramers and so on and finally grow into a polymer in a step-wise manner.

$$H-\overset{H}{\underset{}{N}}-(CH_2)_6-\overset{H}{N}-H + HO-\overset{O}{\overset{\|}{C}}-(CH_2)_4-\overset{O}{\overset{\|}{C}}-OH$$

Hexamethylene diamine + Adipic acid ↓

$$H-\overset{H}{N}-(CH_2)_6-\underbrace{\overset{H}{N}-\overset{O}{\overset{\|}{C}}}_{\text{Amide bond}}-(CH_2)_4-\overset{O}{\overset{\|}{C}}-OH$$

↓

$$...\overset{H}{N}-(CH_2)_6-\underbrace{\overset{H}{N}-\overset{O}{\overset{\|}{C}}}_{\text{Amide bond}}-(CH_2)_4-\underbrace{\overset{O}{\overset{\|}{C}}-\overset{H}{N}}_{\text{Amide bond}}-(CH_2)_6-\underbrace{\overset{H}{N}-\overset{O}{\overset{\|}{C}}}_{\text{Amide bond}}-...\text{ etc}$$

Poly (Hexamethylene adipamide) or NYLON – 6,6

Molecular weight of condensation in polymers is determined by the number of reactive functional groups present on the monomer. Monofunctional monomers give only low molecular weight condensation products. Bi and poly functional monomers give high molecular weight (longer chain) condensation polymers. Others most useful condensation polymers are Polyesters (trade name Dacron) and Polyurethane (trade name PUF).

Polyester (Dacron or Terylene) is another typical example of a condenser polymer

$$n\,HO-CH_2-CH_2-OH + n\,H_3CO-OC-C_6H_4-CO-OCH_3$$

Ethyleneglycol + Dimethylterephthalate ↓

$$\left[-\overset{O}{\overset{\|}{C}}-C_6H_4-\overset{O}{\overset{\|}{C}}-C-O-CH_2-CH_2-O-\right]_n + n\,CH_3OH$$

Polyester (Polyethyleneterephthalate)

Polymers having more than one atom in their back bone structure are called *heterochain polymers or heteropolymers.* Because Condensation polymers have functional groups as a part of polymer back bone, they invariably have more than a single element in the backbone structure and therefore are always heterochain polymers.

Table 14.1 Condensation Polymers with commercial uses.

Polymer name	Monomer(s)	Polymer	Use
Polyamides (nylon)	$HOC(=O)(CH_2)_nC(=O)OH$ $H_2N(CH_2)_nNH_2$	$-C(=O)(CH_2)_nC(=O)-NH(CH_2)_nNH-$	Fibres, molded objects
Polyesters (Dacron, Mylar, Fortrel)	$HOC(=O)-C_6H_4-C(=O)OH$ $HO(CH_2)_nOH$	$-C(=O)-C_6H_4-C(=O)-O(CH_2)_nO-$	Linear polyesters, fibers, recording tape
Polyesters (Glyptal resin)	Phthalic anhydride $HOCH_2CHCH_2OH$ (OH on central C)	C_6H_4 with $-C(=O)-$ and $-COCH_2CHCH_2O-$ (C=O, OH; O–)	Cross-linked polyester, paints
Polyesters (Casting resin)	$HOC(=O)CH=CHC(=O)OH$ $HO(CH_2)_nOH$	$-C(=O)CH=CHC(=O)-O(CH_2)_nO-$	Cross-linked with styrene and benzoyl peroxide, fiberglass boat resin, casting resin
Phenol-formaldehyde (Bakelite)	C_6H_5OH (OH), $CH_2=O$	$-CH_2-$, OH, CH_2, OH, CH_2-; CH_2, CH_2	Mixed with filler, molded electrical cases, adhesives, laminates, varnishes
Cellulose acetate (cellulose is a polymer of glucose)	CH_2OH, O, O–, –O, OH, OH CH_3COOH	CH_2OAc, O, O–, –O, OAc, OAc	Photographic film

Contd... Table 14.1

Polymer name	Monomer(s)	Polymer	Use
Silicones	$Cl-Si(CH_3)_2-Cl$ H_2O	$-O-Si(CH_3)_2-O-$	Water-repellent coatings, temperature-resistant fluids and rubber
Polyurethanes	$CH_3C_6H_3(N=C=O)_2$; $HO(CH_2)_nOH$	$CH_3C_6H_3(NHC(=O)-O(CH_2)_nO-)_2$	Foams, rigid and flexible, fibers

14.1.4 Addition Polymerization

Addition polymers are those in which the repeat unit and monomer(s) have same composition. Additional polymers are formed by simply adding monomer units together w*ithout undergoing any loss of atoms or groups.*

$$n\,CH_2 = CH_2 \longrightarrow \left[CH_2 - CH_2 \right]_n$$

Ethene polythene

$$n\,CH_2 = CH-Cl \longrightarrow \left[CH_2 - \underset{\displaystyle Cl}{\underset{|}{CH}} \right]_n$$

Vinylchloride polyvinylchloride

Addition polymerization is usually initiated by a radical or a cation or an anion. Most of the vinyl monomers undergo addition polymerization. For example ethylene polymerizes into polyethylene at high temperatures (100-200 ^{0}C), high pressures (100-300 atm) and in the presence of benzoyl peroxide. The reaction is split into three parts.

Initiation : Homolytic cleavage of initiator leads to formation of free radicals. Thus Benzoyl peroxide decomposes to give benzoyloxy free radicals C_6H_5-CO-O$^{\cdot}$ (denoted by $R^{\bullet}$).

$$2\,R\text{-}O\text{-}R \xrightarrow{heat/h\nu} 4R^{\bullet} + O_2$$

Propagation : Benzoyloxy radical readily combines with ethylene, causing opening of double bond and yielding another highly reactive radical. This combines further with another ethylene molecule and this process goes on till a very long chain of CH_2 groups in built.

$$\dot{R} + CH_2 = \dot{C}H_2 \rightarrow R - CH_2 - \dot{C}H_2$$

$$R - CH_2 - \dot{C}H_2 + CH_2 = CH_2 \rightarrow R - CH_2 - CH_2 - CH_2 - \dot{C}H_2$$

$$\rightarrow R\,(-CH_2 - CH_2)_n - CH_2 - \dot{C}H_2 \rightarrow$$

***Termination* :** Eventually, the reaction is terminated by the recombination of final free radicals.

$$R\left[CH_2 - CH_2\right]_{n-1} CH_2 - \dot{C}H_2 + R\left[CH_2 - CH_2\right]_{m-1} CH_2 - \dot{C}H_2$$

$$\longrightarrow R\left[CH_2 - CH_2\right]_{n+m} R$$

Here *n* gives the total monomer molecules combined to give the polymer.

The above mechanism suggests a gradual but fast disappearance of monomer to give a long polymer chain. Since the process is akin to interlinking of rings one by one to give a chain, addition polymerization is also called *Chain polymerization*.

Table 14.2 Addition Poymers with commercial uses

Polymer name	Monomer(s)	Polymer	Use
Polyethylene	$CH_2 = CH_2$	$-CH_2-CH_2-$	Most common polymer. Used in bags, wire insulation, and squeeze bottles
Polypropylene	$CH_2 = CH$ with CH_3 substituent (1 - propene)	$-CH_2-CH-$ with CH_3 substituent	Fibers, indoor-outdoor carpts, bottles
Polystyrene	$CH_2=CH$ with phenyl ring substituent (styrene)	$-CH_2-CH-$ with phenyl ring substituent	Styrofoam, molded objects such as table ware (forks, knivesand spoons) trays, video-cassette cases.
Poly (vinyl chloride (PVC)	$CH_2=CH$ with Cl substituent (Vinyl chloride)	$-CH_2-CH-$ with Cl substituent	Clear food wrap, bottles, floor covering,synthetic leather, water and drain pipe.
Polytetrafluoro-ethylene (Teflon)	$CF_2 = CF_2$ (tetraflouroethene)	$-CF_2-CF_2-$	Nonstick surfaces, plumbing tape, chemical resistant containers and films

Contd... Table 14.2

Polymer name	Monomer(s)	Polymer	Use
Poly (methyl methacrylate) (Lucite, Plexiglas)	$CH_2{=}C(CO_2CH_3)(CH_3)$ (Methyl methacrylate)	$-CH_2-C(CO_2CH_3)(CH_3)-$	Glass replacement, paints, and house-hold products
Polyacrylonitrile (Acrilan, Orlon, Creslan)	$CH_2{=}CH(CN)$ (acrylonitrile)	$-CH_2-CH(CN)-$	Fibers used in knit shirts, sweaters, blankets, and carpets
Poly (vinyl acetate) (PVA)	$CH_2{=}CH(OOCCH_3)$ (vinyl acetate)	$-CH_2-CH(OOCCH_3)-$	Adhesive (Elmer's glue), paints, textile coatings, and chewing gum
Natural rubber	$CH_2{=}C(CH_3)-CH{=}CH_2$ (2-methyl-1, 3-butadiene)	$-CH_2-C(CH_3){=}CH-CH_2-$	Rubber bands, gloves, tires, conveyor belts, and house hold materials
Polychlorprene (neoprene rubber)	$CH_2{=}C(Cl)-CH{=}CH_2$ (2-methyl-1, 3-butadiene)	$-CH_2-C(Cl){=}CH-CH_2-$	Oil and gasoline resistant rubber
Styrene butadine	$CH_2{=}CH(C_6H_5)$; $CH_2{=}CH-CH{=}CH_2$ 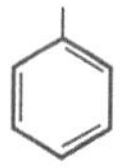	$-CH_2-CH(C_6H_5)-CH_2-CH-CH-CH_2-$ 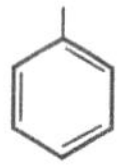	Non-bounce rubber rubber (SBR) used in tires

The distinguishing features of addition and condensation polymers is given in Table 14.3.

Table 14.3 Distinguishing features of Addition and Condensation polymers.

Condensation Polymers	Addition polymers
1. Repeat units and monomers are different.	1. Repeat units and monomers are same.
2. Polymerization is accompanied by loss of molecules or atoms.	2. Polymerization involves no loss of molecules or atoms.

Contd... Table 14.3

Condensation Polymers	Addition polymers
3. Almost all the monomer disappears in the first 10% of the reaction.	3. Monomer concentration decreases steadily.
4. Molecular weight of the polymer rises throughout reaction.	4. All the polymer is formed at once and molecular weight of the polymer changes little throughout
5. Low molecular weight polymers are formed. Longer reaction times are essential for obtaining high molecular weight polymers.	5. High molecular weight polymers are formed all at once.
6. Mainly heterochain polymers often have functional group atoms as a part of chain.	6. Mainly homochain polymers often have carbon chains.
7. Monomer needs to be bi- or poly- functional.	7. Monomer could be mono-functional.

14.1.6 Mechanism of polymerization

There are three main types of mechanisms encountered in polymerization reactions :

(a) ***Free radical Mechanism.*** This type of mechanism, as discussed under chain polymerization, operates in the presence of peroxide catalyst, light or heat energy.

(b) ***Anionic Mechanism***. This type of mechanism is brought into play by the use of coordination catalysts i.e. Zeigler-Natta catalysts. Here the catalyst transfers a negative charge to the monomer giving a *carbanion*, which undergo initiation, propagation and termination to yield the polymer.

(c) ***Cationic Mechanism.*** In this type of mechanism, the polymerization is initiated by protonic acids, metal halides and carbonium salts e.g. tertbutyl perchlorate, which dissociates to give carbonium ion.

14.1.7 Polymer Branching

There are three main types of polymer branching :

(a) ***Linear Chain*** exhibited in the case of polymerizations of bifunctional groups (Fig. 14.1 a).

(b) ***Branched Chain*** present in the case where branches grow from the parent chain (Fig. 14.1 b).

(c) ***Cross linked Chain*** shown in the case of polymerizations from bi and tri-functional groups (Fig. 14.1 c).

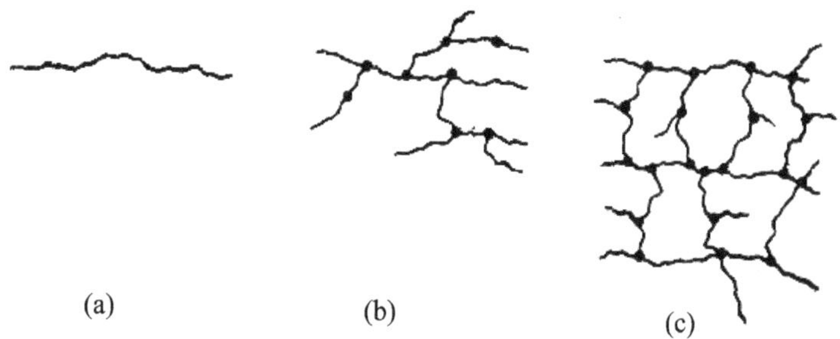

Fig. 14.1 Formation of plastics (a) Linear polymer, (b) branched polymer, (c) cross linked polymer.

14.2 Plastics

A substance that can be easily formed or moulded into a desired shape is plastic. A plastic material can be permanently formed into a desired shape at some stage in its manufacture or processing by means of mechanical force, or under combined effect of mechanical force and heat. In most cases the temperatures required are not very high and steam heating is adequate. The common plastics are organic high polymers 'which possess the property of plasticity at some stage during their fabrication, and which are substantially rigid after fabrication'.

The origin of modern plastics may be traced back to the discovery of nitrocellulose in 1846. Articles made of nitrocellulose were displayed in 1862. The history of industrial plastics begins with the preparation in 1869 of celluloid from cotton linters, wood pulp, camphor and nitric acid. The second landmark in industrial plastics was the preparation of casein plastics in 1890 from formaldehyde and sour milk. Real incentive to the industry came in 1909 from the synthesis of Bakelite: named after its discoverer Dr. L. H. Bakeland. It is prepared by the reaction between phenol and formaldehyde. During the last half century, a large number of plastics have been developed and are manufactured on a very large scale.

In the manufacture of plastics raw materials like coal, petroleum, cellulose (prepared from wood and cotton), salt, sulphur, limestone, air, water, etc, are used. These are subjected to several processes of chemical decomposition and recombination giving intermediates which are then made into plastics.

The Indian plastics industry is of recent origin. The production of phenol formaldehyde started in 1946, polystyrene and polyethylene in 1957, polyvinyl chloride in 1961 and cellulose acetate in 1963. The development of the plastics industry depends to a considerable extent on the petro-chemical industry. As the latter is firmly established in the country the production of plastics will increase.

14.2.1 Plastics as Engineering Materials

The chief engineering advantages of plastics over other engineering materials result from their light weight low fabrication cost, low thermal and electrical conductivities, high resistance to corrosion and solvents and transparency. In addition they have a decorative effect. A further advantage is that the plastics can be used in conjunction with metals.

The combined use of metals and plastics has enabled engineers to overcome many design problems. Many plastics (sp. gr. 1 to 2) have greater strength than the metals per unit weight. The tensile strength of plastics (average strength about 700 kg/ cm^2) is about one fourth of aluminium and gravity makes up for low strength. The stress-strain-relationship of plastics is similar to that of the metals. Just like the metals the strength decreases with increase in temperature and also with increase in time under a constant load (i.e., plastics show creep). Humidity also affects the strength of the plastics.

Some of them have greater wear resistance and are used for making parts that rub against other parts, e.g., bearings, gears, etc. Remarkably low coefficient of friction of nylon against metals enables it to be used for making bearings and gears. Another advantage is that plastics reduce noise and vibration in machines. Plastics parts can be easily and cheaply made with accurate dimensions. Complicated shapes can be produced by moulding. Plastics, particularly thermosetting plastics, have a tendency to shrink.

Most plastics cannot be used at elevated temperatures, e.g., thermoplastics generally cannot be used above 100° and thermosetting above 150°. Plastics being bad conductors of heat are useful for making handles used for hot objects, and for the same reason in cold climates the plastics feel warm to the touch. Most plastics are inflammable, but some are fire-proof.

Most plastics are electrical insulators and find large scale use in the electrical industry.

In general, plastics are resistant to chemicals. Thermosetting plastics are more insoluble in organic solvents than the thermoplastics.

Most plastics are clear and transparent. They can be given beautiful and brilliant colours. Optical clarity combined with strength makes them suitable for making wind screens for automobiles, aircrafts, etc.

Many failures in the usage of plastics materials in engineering have occurred because of inadequate knowledge of their properties under working conditions. Many of the plastics have poor weather resistance and their properties deteriorate and dimensional changes take place. Generally they are soft, have low strength and low modulus of elasticity. It is difficult to repair parts made of plastics.

14.2.2 Types of Plastics

Plastics have been classified in several ways. From the engineering point of view it is most satisfactory to classify them as : 1. thermoplastics and 2. thermosetting plastics. Some of the more common plastics are classed as under:

Types of Plastics

Thermoplastic Resins	Thermosetting Resins
Acrylics	Alkyds
Cellulose derivatives (e.g., celluloid cellulose acetate)	Melamines
Polyamides (e.g., Nylon)	Phonolics (e.g., Bakelite)
Polyethylenes	Polyesters (e.g., terylene)
Polystyrenes	Silicones
Polyvinyls	Ureas

The behaviour of a few plastics is in between that of the two extreme types but most plastics fall under one or other of these types. The load-carrying capacity of thermoplastics is appreciably affected by the working conditions, because the material becomes soft and plastic with an increase in temperature. They have higher compressive strengths than thermosetting plastics materials, although the actual strength of the latter is usually greater than that of the former.

14.3 Thermoplastic Resins

As already explained, the thermoplastics soften or melt when heated. On cooling they become rigid. On reheating they again soften or melt and can be given the desired shape by moulding. Their behaviour can be compared to that of the common sealing wax which softens on heating and takes the desired shape. On reheating it softens again and a new shape can be given. The thermoplastic resins consist mostly of linear polymers. Heating or moulding opeations do not change the chemical structure or the molecular weight of the material. The thermoplastics are smaller molecular weight compounds than the thermosetting plastics. In them the inter-molecular forces are weaker than in thermosetting plastics. This explains why these swell or dissolve in suitable solvents. This property enables them to be mixed with plasticizers which keep the material soft and pliable.

On the basis of the raw materials used, the thermoplastics may be subdivided into the *cellulosic* and the *non-cellulosic* types.

14.3.1 Cellulose Resins

The important cellulose derivatives are its nitrate, acetate, acetate-butyrate, propionate and methyl cellulose. Strong alkalis decompose most of the cellulosic plastics. Weak acids and weak alkalis affect them slightly. Cellulosic plastics are very good electrical insulators. They have considerable strength and toughness. Most of them have poor fire resistance. Rayon is regenerated cellulose in the form of filament. It has low gravity and great tensile strength.

Cellulose molecule, $(C_6H_{10}O_5)_n$ has a large chain of $C_6H_{10}O_5$, groups joined together, each of which has three hydroxyl groups which may be esterified or etherified Cellulose derivatives in general and some other thermoplastics are not workable. Substances like camphor, glycolesters, phthalates, etc. known as *plasticizers*, are added to these plastics. The plasticizers improve the flow properties of the plastics. They enable the plastics to be worked at lower temperature, make them resistant to water and temperature change, and improve toughness, flexibility and non-inflammability. Plasticizers usually decrease the strength.

Cellulose acetate is made by first preparing cellulose triacetate by acetylating cellulose, e.g., purified cotton linter is treated with acetic anhydride, glacial acetic acid and sulphuric acid. When the acetate content is about 62.5% it is hydrolysed, precipitated, washed and dried. The product is kneaded in a mixer with a plasticizer. Cellulose acetate plastics are stable, transparent and slow burning. They are not very resistant to moisture, but are tough, have high impact strength, and are available in attractive colours. On account of its mechanical strength, toughness, wear resistance, ease of fabrication and transparency, cellulose acetate has found wide application in the automotive industry. Cellulose acetate plastics can be more easily moulded into the desired shape than cellulose nitrate by extrusion and especially injection moulding. Cellulose acetatebutyrate has similar properties but has better resistance to moisture and better dimensional stability than the acetates.

Cellulose nitrate is quite hard and brittle. It can be used as a plastic material only after incorporating a plasticizer like camphor. Rods, sheets and tubes of cellulose nitrate are available in various colours. Cellulose nitrate plastics are highly inflammable, unstable at higher temperatures, sensitive to sunlight and are affected by acids, alkalis and organic solvents like alcohols, ethers and ketones. In spite of these drawbacks, they are extensively used because of their toughness, strength, dimensional stability and water resistance. They are used mostly as sheets or blocks and are shaped by sawing, turning, drilling or pressing the pre-cut solid blocks. They cannot be processed by injection moulding as they do not withstand the temperature required for the process. Unlike most other plastics materials, granules of cellulose nitrate are not used for moulding because they do not weld together.

Ethyl cellulose is the most important of the cellulose ethers. It is chemically stable and is resistant to alkalis and to dilute acids. It has a low softening temperature (120°). Ethyl cellulose plastics are very tough, strong, resistant to moisture and are good insulators. They have good dimensional stability and do not distort on heating. Thin sheets of this material are widely used for packaging. Because of its low softening point, it is easily fabricated by compressing, extrusion and injection moulding. Ethyl cellulose is largely used in coatings on adhesives.

14.3.2 Non-Cellulose Resins

The second type of thermoplastics is derived from *non-cellulosic materials*. Plastics of this type are in general very tough, chemically inert, non-inflammable or slow-burning and have good electrical and water resistant properties.

Poly crystals : A large quantity of non-cellulosic plastics are derived from ethylene. Acrylic esters are prepared from ethylene by converting it first to ethylene chlorohydrin, then ethylene cyanohydrin which on esterification gives the acrylates :

$$\underset{\text{Ethylene}}{CH_2 = CH_2} \xrightarrow{HOCl} \underset{\text{E. Chlorohydrin}}{CH_2OH.CH_2Cl} \xrightarrow{NaCN} \underset{\text{E. cyanohydrin}}{CH_2OH.CH_2CN}$$

$$CH_2OH.CH_2CN\ CH_3OH \xrightarrow{H_2SO_4} \underset{\text{Methylacrylate}}{CH_2 = CHCOOCH_3} + NH_3$$

The esters obtained in such a manner are polymerized to give the plastics called polyacrylates. The **acrylates** stand as a class because they are water-clear transparent. Age, weather and sunlight have little effect on their transparency. Because of these properties, numerous applications follow, e.g., aircraft and cockpit enclosures, windshields, 'Plexiglas', tubes, sheets, rods, etc.

The acrylates also have toughness, strength, flexibility, low specific gravity and permanence of dimensions. Acrylates are also avaibable as staple fibres (e.g., orlon).

14.3.3 Teflon

Teflon belongs to the *fluorocarbons*. These have very good chemical, electrical and thermal resistance and can be used from –70 to 250° temperature range. Because of these properties, the fluorocarbons are used for handling hot and corrosive liquids and as insulators in equipment using high voltage and high frequency electricity.

The condensation of diamines and dibasic acids gives linear products. Properties can be varied greatly (from hard and tough products to soft and rubbery products) by varying the nature of the acid and the amine. The use of nylon, which is a *polyamide*, in the textiles is very well-known. It is tough, and heat and abrasion resistant. It is also moulded to produce gears and bearings.

14.3.4 Polyethylenes

These are prepared by polymerizing ethylene under high pressures and moderately high temperature. The polymer is then subjected to substitution and other reactions to give the plastics. These are resistant to chemicals, have good electrical resistance, low specific gravity, low water absorption and flexibility. They are much cheaper than the other plastics.

Their resistance to chemicals falls drastically at higher temperatures. They are adversely affected by sunlight. Above 60° oils, greases, aromatic hydrocarbons, etc., affect polyethylenes.

14.3.5 Polystyrenes

These are prepared from styrene (i.e., phenylethylene). They have high electrical resistance and also resistance to acids and alkalis. These properties combined with low water absorption make polystyrenes very suitable materials for making electrical and electronic appliances. They also have permanence of dimensions and are hard.

14.3.6 PVC

Vinyl plastics are numerous. They are basically derived from acetylene. The vinyl group, CH_2=CH - is very reactive and readily gives acetal, acetate, alcohol and chloride. These materials, on polymerization give the corresponding plastics. In general, these plastics have high electrical resistance and resistance to chemicals and to corrosion, combined with flexibility. These properties make them very suitable for coating wires and cables. They are strong, machinable, do not absorb moisture and the shapes of the finished products do not change. Vinyl plastics have a wide range of properties, e.g., from soft and flexible to hard rigid.

The largest use of vinyl resins is in the manufacture of safety glass. They are also used in large quantities in the manufacture of chemical resistant fabrics and for electrical insulation.

14.3.7 Epoxy Resins

They are widely used as adhesives (like Araldite). These are ether polymers containing phenylene groups as well as aloholic groups. These are either solids of low melting point or viscous liquids. On mixing with various organic acids, amines, etc., cross-links are formed and the material sets. These are thermosetting resins and set slowly at the room temperature and rapidly on warming.

14.3.8 Nylons

Among the Polyamides the most widely known are the **nylons** of which nylon 6,6 is the most common. It is prepared by the condensation reaction of hexamethylene diamine with adipic acid. It has remarkably low coefficient of friction against many metals which makes it a good material for making bearings and gears. It is self extinguishing when it comes in contact with a flame. It has low water absorption and good strength which does not deteriorate even when used for a long period of time.

Nylon can be shaped by casting, extrusion, blowing and moulding giving products of excellent durability and toughness.

14.4 Thermosetting Resins

As already explained, the thermosetting resins are produced when the molecular weight and chemical structure of the resin changes on heating. When the material is moulded, additional cross links are formed between the long chains and the molecular weight increases. In the formaldehyde resins the chemical reaction is condensation, in the glyptal resins it is esterification. In all such cases cross linking is very important. In the partially set state these are usually soluble in solvents but when cross linkings are formed, the plastics acquire some of their characteristic properties, e. g., they dissolve or swell less in solvents and soften less on heating, and hard, tough, and highly insoluble products are obtained. The thermosetting resins have direct covalent links between the long chain molecules, giving a threedimensional molecule. In the fibres like nylon, which are thermoplastics, there are strong intermolecular forces of attraction which bind the molecules, and the material consists of many smaller (as compared to thermosetting resins) molecules held together by these forces. This accounts for the differences in the behaviour of the two groups of plastics.

14.4.1 Alkyd or Glyptal Resins

The earliest plastics of this group were formed by esterification of glycerol with phthalic anhydride, hence the name 'glyptal' resins was given to the group. Now a number of other resins have been prepared by the combination of other polyhydric alcohols and polybasic acids and the name 'alkyd' (from alcohol-acid) is more appropriate. The alkyds are resistant to heat, moisture and weathering conditions, hence are used in protective coatings. They find applications for electrical insulations. The alkyds are transparent, can be easily coloured, are tough and flexible and can be joined to metals. Therefore, these are used for aircraft, automobiles, and for making sheets, rods, tubes. etc., which find a number of engineering uses.

14.4.2 Melamine Resins

Melamine is a polymer of cyanamide ($NH_2C = N$) obtained from calcium cyanamide which is manufactured by heating calcium carbide in nitrogen. Melamine, on reacting with formaldehyde, forms a resin. The intermediate product of reaction is soluble in water. The moulded resins have high resistance to heat and dielectric etc. Their high dielectric strength and low water absorption makes them suitable for making moulded electrical parts. They are colourless, hard, brittle, rigid and very resistant to abrasion.

14.4.3 Phenolic Resins

Resins of this group are more widely used than those of any other group. The most important ones are formed by the condensation of phenol and formaldehyde. The properties of the product depend upon several factors, like the concetrations of reactants, the catalyst, the time and temperature of the reaction. Phenol-formaldeyde resin, known as Bakelites, have outstanding electrical and mechanical properties and the resistant to heat, water and many chemicals. Strong caustic alkalis and concentrated Sulphuric and nitric acids decompose them. They are hard and infusible and retain their shapes. Usually various fillers and fabrics are combined with the resins for different purposes. To increase mechanical and dielectric strength and lightness, wool floor is commonly used as a filler. The resulting product is used where light material is desired, as for aircraft, automobile and telephone industries. Resistance to chemicals and heat is increased by using asbestos as a filler. Moulding sheets, which have many applications, are made by impregnating paper with the resin. Thus by a suitable filler the range of usefulness of phenol-formaldehyde plastics is greatly increased.

Phenol also forms a condensation product with furfural. The product has properties similar to phenol formaldehyde plastics and can be used under difficult and complicated moulding conditions which require a long time of flow. Thus articles of thick sections can be made out of it.

14.4.4 Polyesters

These are prepared by the condensation of a dialcohol and a diacid. Many of them have long chain molecules which are characteristics of thermoplastics. A common polyester is made from ethylene glycol and phthalic acid. The molten material is extruded through an orifice and stretched to give a strong fibre known as Terylene or Dacron. The same material in the form of film is known as Mylar. It has remarkable properties, e.g., it has excellent electrical and mechanical properties. The polyesters on combination with fillers give plastics which combine lightness with hardness, resistance to abrasion and heat and good electrical and mechanical properties. These have a strength to weight ratio which is superior to other plastics.

14.4.5 Silicones

These have silicon-oxygen chain to which are linked various organic groups. They are available in a wide range of consistency. The thermosetting silicone compositions are resistant to heat and water and have high electrical resistance. Therefore besides other applications, they are used to make insulations for high frequency equipment. (Some more information can be found under rubbers).

14.4.6 Urea-Formaldehyde Resins

Urea and formaldehyde on condensation give a variety of products which are used as plastics, adhesives and laminates. The properties of the plastics are similar to those of phenolic plastics. They have good mechanical and excellent electrical properties. Paper pulp and wood are the common fillers for these plastics. Besides other uses, they are used as laminating agents. They are impregnated in sheets of absorbant paper, dried and heated to give a sheet suitable for making ,wall panels, table tops, lighting reflectors, etc. (Some more information can be found under rubbers).

14.5 Compounding of Plastics

The materials described above are called resins or binders and are the chief ingredients of plastics objects. They have plasticity or binding property but need other ingredients to be mixed with them for fabrication of useful shapes.

Almost invariably plasticizers have to be added to the thermoplastics because the thermoplastics are inflammable and lack flexibility and toughness. Plasticizers reduce these defects and also improve the flow-property of the plastics. The plasticizer should dissolve the resin at the working temperature and should not evaporate. Plasticizers are not used with thermosetting resins because the two are incompatible and a plasticizer gradually sweats out.

Thermosetting resins generally lack elasticity and they crack under stress, To remove this defect *fillers* or *extenders* like wood flour, cotton or cloth fibres, asbestos, mica and clay are added, These are usually cheaper than the resins and the amount added may be more than that of the resin. Fillers impart some very useful properties e.g., wood flour reduces the specific gravity and absorbs shocks, cotton and other fibers impart strength and toughness, asbestos and mica are especially useful for plastics for electrical appliances.

To give decorative effect, *dyes* and *Pigments* are added. To prevent the plastics from sticking to the moulds lubricants like oils, waxes and soaps are added.

14.5.1 Fabrication of Plastics

To give suitable shapes to plastics, several methods of fabrication are in use. The method chosen depends upon the type of the plastics, i.e., thermoplastics or thermosetting plastics, and the shape of the finished product.

Compression Moulding

This is the most widely used method, both for thermosetting and thermoplastic resins. The quantity of the resin, filler and other ingredients placed in the mould should be such that the volume of the compact mass is slightly more than the volume of the cavity of the mould. The mould is usually preheated. It is closed after placing the ingredients and heat and pressure are applied, then the ingredients soften and fill the cavity of the mould. For thermoplastic resins, the mould is cooled before releasing the pressure, otherwise the object will distort. Thus because of alternate heating and cooling, compression moulding is not the most suitable method for thermoplastics. The thermosetting resin has to be kept in the mould for a period of time sufficient for the completion of the chemical reactions and the formation of the permanent shape. Cooling of the mould is not necessary. The temperature ranges from 100 to 200° and the pressure from 100 to 500 kg/cm^2.

Cold Moulding

It is mostly used for phenol-formaldehyde resins for electrical fixtures. The products are of inferior strength and surface finish. The moulding composition is shaped by pressure alone. In the case of a thermosetting resin, the article is cured, i.e. heated in an oven. The chief advantage of the method is that the output is large because the time taken in the mould is short, say, 10 seconds.

Injection Moulding

It is the most widely used method for moulding thermoplastic resins. In compression moulding of thermoplastics the difficulty is that the mould has to be cooled before removing the object from it. In injection moulding use of hot plastics in a cold mould removes this difficulty. The material is passed through a heating chamber where it softens. See Fig. 14.2. It is then forced by a ram into a cold mould and left there to become hard. The process is usually made automatic and the time required is 10 seconds to 1 minute. *Jet moulding* is a modification of injection moulding so as to make it suitable for thermosetting plastics. As the material is forced through the nozzle, intense heat is applied by induction heating equipment. The material hardens in the mould. The advantage of this process is a high rate of production.

Injection and jet moulding methods are similar to compression moulding methods but the latter method is slower. However, more complex shapes can be formed and a variety of raw materials can be used in compression moulding.

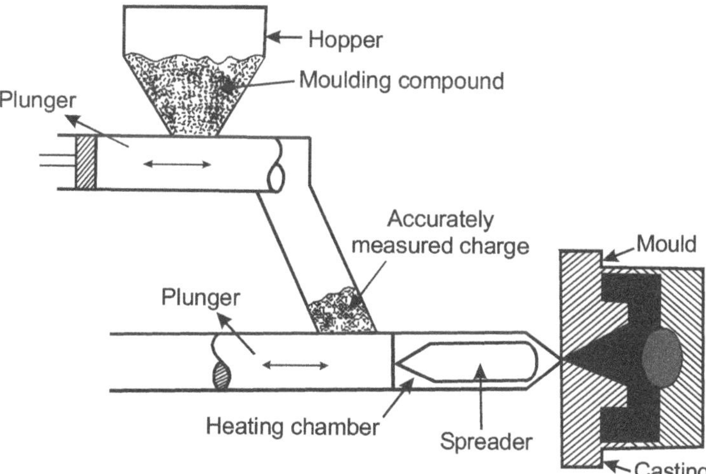

Fig. 14.2 Injection moulding.

Transfer Moulding

This method is almost exclusively used for thermosetting plastics and has been developed to make compression moulding applicable to more complicated shapes and permits the use of delicate inserts. The material is heated in a chamber to melt it and then is forced into a hot mould. The mould is opened after the plastics has set.

Extrusion Moulding

It is similar to injection moulding and is applicable to thermoplastics only. The material is forced by a screw drive into a heated chamber where it softens, and is forced out through a die. The finished product coming out from the die is cooled by blowing air or spraying water. Extrusion moulding is used for making continuous shapes like rods, tubes, sheets, etc.

Casting

Casting is a very simple method but is only used to a limited extent. The mould is usually made of lead. The molten resins are poured in the mould and cured at about 70° for several days. The mould is removed, remelted and a new mould is again made. This method is used for making special shapes. The products are water-white in colour, free from internal stresses and can take high polish. The cost of production is high.

Blowing

This method is similar to that used in the glass industry. The thermoplastic material is softened and then blown by air or steam into a closed mould.

14.6 Elastomers–Rubbers

Polymers having molecules arranged in the form of zig-zag or spring-shaped chains are commonly called elastomers. Most of the commercial rubbers are thermoplastic materials of this type. In recent times the distinction between rubbers and plastics has become very obscure, because a large number of copolymers now being used have structures between rubber and simple thermoplastic resins.

14.6.1 Natural Rubber

Natural rubber is a high molecular weight hydrocarbon polymer represented by the formula $(C_5\,H_8)_x$. It is obtained from a milky emulsion, called latex, by tapping the bark of the tree *Hevea brasiliensis*. This tree was originally found along the Amazon river in South America and its seedlings were planted in Ceylon.

Rubber plantations flourish best in the equatorial climate. In India about 95% of the plantations are in Kerala. The remaining are in Tamil Nadu, Karnataka. Goa, Assam and Maharashtra States. The total production nearly 2 lac tonnes.

When the trees are 6 years old spiral grooves are cut in the bark, the sap flows slowly and is collected in cups. It has about 35% rubber. The latex is diluted with water and filtered to remove coagulated rubber, barks etc., and acetic or formic acid is stirred in it when the rubber coagulates. Rubber is separated from the liquid by squeezing between rolls and is then washed with water.

The sheets so obtained are dried either in the air when 'crepe' rubber is obtained, or in hot smoke when 'smoked' sheet rubber is obtained.

The product so obtained is a high polymer of isoprene.

$$\left[CH_2 - \overset{\overset{\large CH_3}{|}}{C} = CH - CH_2 \right]_n$$

In a normal position the molecule is like a coiled helix. It uncoils in tension. When tension is removed, the molecule recoils and returns to the original coiled state. This accounts for the elasticity of rubber. If rubber is stretched too much, one molecule chain slips over another and a permanent deformation takes place. In this process of uncoiling and recoiling some energy is lost. This gives rubber the property of absorbing shock and reducing vibrations in machines. Raw rubber has some serious defects, e.g., it becomes soft and sticky with heat and stiff and brittle in cold. It is not very tough and deteriorates through slow atmospheric oxidation giving a bad small. These defects are removed by vulcanization.

14.6.2 Vulcanization

When raw rubber is passed between hot rolls both physical and chemical changes take place and the mass becomes plastic. Before vulcanization proper substances are added in order to control properties like strength, toughness and hardness and to make it resistant to abrasion, chemicals, solvents, oxygen, light and oils. Some fillers like barytes ore, cotton, etc. are also added. Zinc oxide and zinc soaps are added to decrease the vulcanization time and are called accelerators. The vulcanized rubber deteriorates with age, hence some age-resisters, mostly organic nitrogen compounds, are also added. Antioxidants are used to give resistance to oxygen. Carbon black increases abrasion resistance and is the important reinforcing agent. Raw rubber, when in a plastic state during hot rolling, is mixed with these ingredients and with sulphur, which is the most important vulcanizing agent. Soft rubber is obtained when 0.5 to 5% sulphur is used. Addition of over 30% sulphur produces hard rubber or ebonite. For uniformity of composition the material is passed many times between the rolls. This operation, called milling, requires considerable power and the temperature rises. Rubber is then cut into pieces of suitable size, placed in moulds, and formed into the desired shape and vulcanized in a single operation by the simultaneous application of heat and pressure. The rubber flows and fills the cavities and sets by vulcanization giving the desired shape. The moulds are opened and the finished products are obtained.

Vulcanization involves complex and not too well understood, chemical reactions. The helix-like molecules of unvulcanized rubber can be stretched but they slide over each other and raw rubber deforms readily, i.e., does not regain the original shape. In vulcanization, cross links are formed between the long unsaturated hydrocarbon chains of the original molecules. The cross links provide anchorage and prevent inter-molecular movements with the result that when the vulcanized rubber is subjected to stress, elastic strain is produced and it regains the original shape on removing the stress. The cross links may be directly from carbon to carbon or through a sulphur atom or through an oxygen atom or through all the three. Thus the vulcanized rubber is of higher molecular weight than the raw rubber and is less unsaturated.

On vulcanization, the properties of rubber change considerably. For instance, the unvulcanized rubber deforms, very readily becomes plastic and soft on working between hot rolls, and becomes viscous in hydrocarbons like benzene; but vulcanized rubber is elastic, crumbles between hot rolls and retains its form in many solvents. Besides bringing about these changes in properties, vulcanization enables the manufacturer to manipulate the material to produce the desired shapes.

14.6.3 Artificial Rubber

Strictly speaking, synthetic rubber is not produced in the sense that the synthetic product differs in chemical composition and properties from those of the natural rubber. 'Synthetic' rubbers are only rubber-like products. Therefore the term 'artificial' rubber is more appropriate than 'synthetic' rubber. The development of the automobile industry brought a revolution not only in the petroleum industry but also in the rubber industry. Rubber from the plantations could not meet the demand and attempts were made to synthesise natural rubber. As this did not succeed, rubber-like materials were synthesised. Further, due to several specific and specialized applications, need was felt to retain elastomeric properties over a wider range of condition than natural rubber has. As natural rubber is produced in tropical regions, the industrialized countries felt the urgency for having their own resources of rubber. With the development of the petrochemical industry, it is planned to increase the artificial rubber production. In India in 1980 its production was 28,000 tonnes.

At present the following five principal types of rubber are in large scale use :

Buna S. (SBR, GRS) Rubber : It is a copolymer of butadiene., $CH_2 = CH - CH = CH_2$ (75% by weight) and styrene, $C_6H_5 - CH = CH_2$ (24% by weight). In the early days of its synthesis sodium was used as the catalyst, hence the name, bu (for butadiene), na (symbol, Na, for sodium) and S (for styrene). It is also called GR-S (Government Rubber Styrene) or SBR. Buna S rubber can be combined with natural rubber, thus combining the useful properties of both, and is used for the manufacture of tyres belting, shoe roles, coated fabrics, floor tiles, tank lining, carpet backing, etc. This rubber has lower strength and flexibility than the natural rubber.

An emulsion of butadiene and styrene is made in water with the help of soap-like emulsifying agents. A catalyst is added and the mass is heated under pressure when a latex-like liquid is obtained. It is then steam distilled to remove the uncombined hydrocarbons. Rubber is coagulated and processed in a manner similar to the natural rubber using sulphur as the vulcanizing agent. Buna S rubber is synthesised more than the other types of rubber.

At Bareilly, U. P., Synthetics and Chemicals Ltd. have an artificial rubber plant of 20,000 tonnes a year capacity. The plant annually uses about 70 million litres of ethyl alcohol, which is supplied by the various distilleries, and 9 million litres of benzene obtained from the coke oven plants of the steel industry. Alcohol is dehydrogenated in the presence of a catalyst, to give acetaldehyde. Acetaldehyde mixed with more alcohol is passed over another catalyst, butadiene is obtained :

$$CH_3CH_2OH \longrightarrow CH_3CHO + H_2$$

$$CH_3CHO + CH_3CH_2OH \longrightarrow CH_2 = CH - CH = CH_2 + 2H_2O$$

Dehydration of ethyl alcohol gives ethylene which, with benzene, in the presence of catalysts, gives styrene:

$$C_6H_6 + CH_2 = CH_2 \longrightarrow C_6H_5CH_2CH_3$$

$$C_6H_6CH_2CH_3 \longrightarrow C_6H_5CH = CH_2 + 2H_2O$$

Copolymerization of butadiene and styrene gives Buna S rubber which has a structural unit of

$$-[CH_2 - CH = CH - CH_2 - \underset{\displaystyle C_6H_5}{\underset{|}{CH_2 - CH}}]_n -$$

Buna N (GR-A and GR-N). It is a copolymer of butadiene 75% and acrylonitrile, (CH_2 = CHCN), 25% and its structure is similar to that to Buna S rubber, styrene being replaced by more polar acrylonitrile. The stable nitrogen component makes Buna N rubber highly resistant to solvents, oils and chemicals. It also has good resistance to abrasion, aging and to low temperatures and these properties increase with increase in acrylonitrile percentage. It is than natural rubber. The rubber is vulcanized with sulphur and is used for hose, conveyor belts, tank linings and for high altitude aircraft. It is also used in matching adhesives, oil resistant foams and automobile parts. Further, in the form of latex it is used for impregnating paper, textiles and leather. In India a plant with installed capacity of 2,000 tonnes of nitrile rubber was started in 1976 which produces about 1000 tonnes.

Butyl Rubber (GR-I). 98% of this rubber consists of isobutylene which polymerizes to give a saturated polymer:

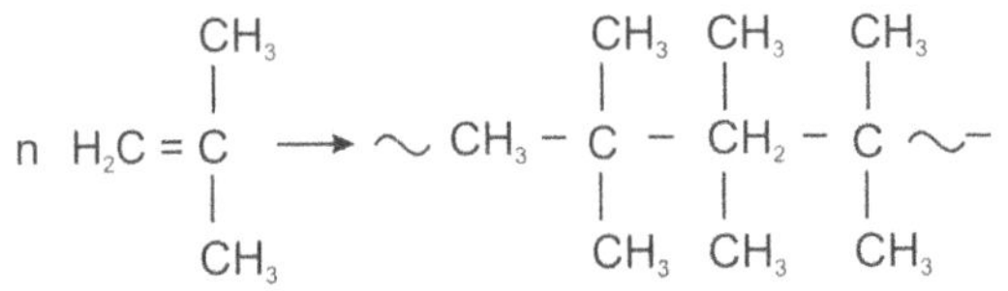

$$n\ H_2C = \underset{\displaystyle CH_3}{\overset{\displaystyle CH_3}{\overset{|}{\underset{|}{C}}}} \longrightarrow \sim CH_3 - \underset{\displaystyle CH_3}{\overset{\displaystyle CH_3}{\overset{|}{\underset{|}{C}}}} - \underset{\displaystyle CH_3}{\overset{\displaystyle CH_3}{\overset{|}{\underset{|}{CH_2}}}} - \underset{\displaystyle CH_3}{\overset{\displaystyle CH_3}{\overset{|}{\underset{|}{C}}}} \sim$$

Polybutene (Butyl rubber)

Because it is saturated no cross links can be formed, hence it cannot be vulcanized. To overcome this difficulty, about 2% of butadiene is copolymerized with isobutylene. Butadiene provides centers for cross links and thus the rubber can be vulcanized. A plant for 20,000 tonnes per annum capacity of polybutadiene is working in India since 1978.

Butyl rubber is cheaper than other artificial rubbers. Another advantage of this rubber is that it is impermeable to gases, hence is satisfactory for making cycle and automobile tubes. It is also resistant to aging, ozone, acids or alkalis. Its other uses are for hose, tank linings and electrical insulation. A limiting factor of butyl rubber is that because of very low unsaturation it cannot be hardened much.

Neoprene (GR-M). It is a polymer of chloroprene :

$$CH_2 = CH - \underset{\displaystyle Cl}{\underset{|}{C}} = CH_2$$

Starting materials for it are acetylene and hydrogen chloride. It is valcanized by zinc oxide or magnesium oxide. It has very useful properties and resembles natural rubber but is more expensive than rubbers. It has very good resistance to aging and to oils and is used for hose, wire insulation and belts.

Thiokol (GR--P). Several types of rubbers of this group are made by copolymerization of organic dichlorides, like ethylene dichloride, with inorganic polysulphides, like, Na_2S_4.

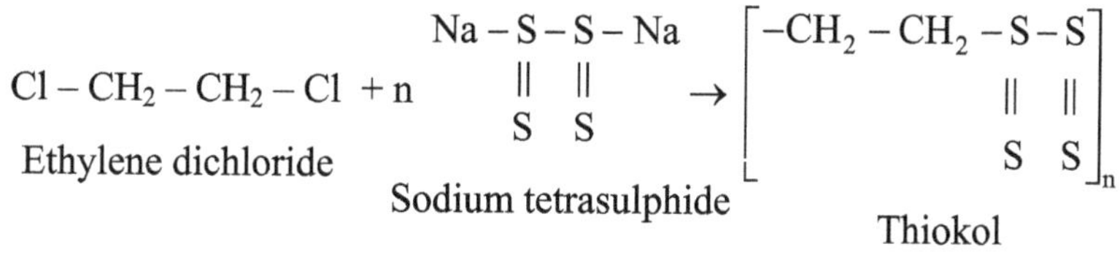

$$\underset{\text{Ethylene dichloride}}{Cl-CH_2-CH_2-Cl} + n \underset{\text{Sodium tetrasulphide}}{\begin{matrix} Na-S-S-Na \\ \quad \| \quad \| \\ \quad S \quad S \end{matrix}} \rightarrow \underset{\text{Thiokol}}{\left[\begin{matrix} -CH_2-CH_2-S-S \\ \qquad\qquad\quad \| \quad \| \\ \qquad\qquad\quad S \quad S \end{matrix}\right]_n}$$

These rubbers neither resemble natural rubber in structure nor in properties they have very good resistance to solvents and can withstand cold but not heat. They are cheap and can easily be shaped, but have an unpleasant smell. Because of their being practically unaffected by petrol, fuel oils and lubricating oils they are mostly used for making hose pipe and tank linings for handling these materials.

Vulcanizing of Thiokol using sulphur is not done since it does not yield the hardened rubber. Vulcanization is done using active metal oxides like ZnO.

Silicone rubber *:* Silicone rubbers are polymers composing alternate silicon and oxygen atoms in which the silicon atoms are joined to organic groups.

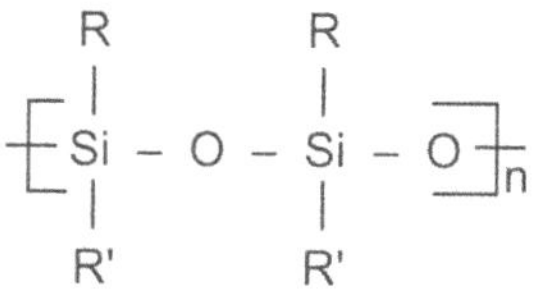

$$\left[\begin{matrix} R \\ | \\ Si \\ | \\ R' \end{matrix} - O - \begin{matrix} R \\ | \\ Si \\ | \\ R' \end{matrix} - O \right]_n$$

Silicone polymers were developed in 1930s during searches for heat resistant electrical insulating materials. Dicholrosilanes readily react with water to give silanols, which are unstable and condence to form silicones.

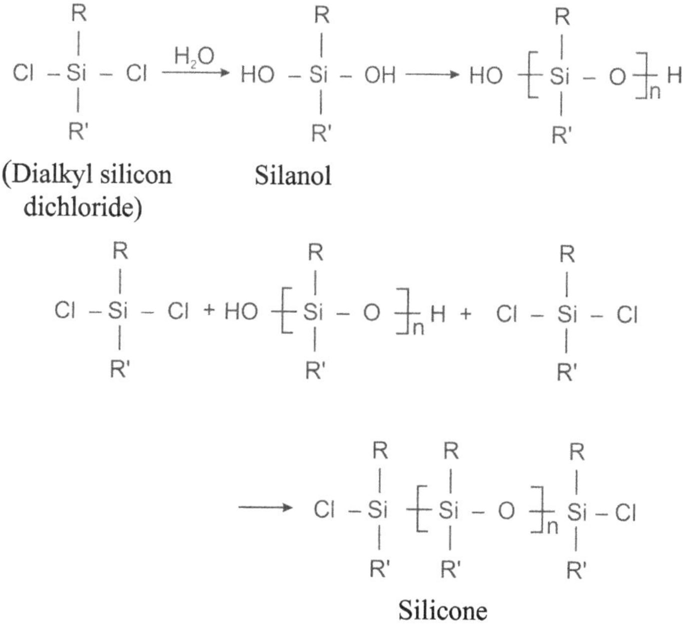

$$\underset{\text{(Dialkyl silicon dichloride)}}{Cl - \begin{matrix} R \\ | \\ Si \\ | \\ R' \end{matrix} - Cl} \xrightarrow{H_2O} \underset{\text{Silanol}}{HO - \begin{matrix} R \\ | \\ Si \\ | \\ R' \end{matrix} - OH} \longrightarrow HO \left[\begin{matrix} R \\ | \\ Si \\ | \\ R' \end{matrix} - O \right]_n H$$

$$Cl - \begin{matrix} R \\ | \\ Si \\ | \\ R' \end{matrix} - Cl + HO \left[\begin{matrix} R \\ | \\ Si \\ | \\ R' \end{matrix} - O \right]_n H + Cl - \begin{matrix} R \\ | \\ Si \\ | \\ R' \end{matrix} - Cl$$

$$\longrightarrow \underset{\text{Silicone}}{Cl - \begin{matrix} R \\ | \\ Si \\ | \\ R' \end{matrix} \left[\begin{matrix} R \\ | \\ Si \\ | \\ R' \end{matrix} - O \right]_n \begin{matrix} R \\ | \\ Si \\ | \\ R' \end{matrix} - Cl}$$

Silicones have an outstanding thermal stability and can be heated in air to about 200 °C. Silicone rubber is also water repellent. So silicone rubbers are mainly used in making lubricants, electric blankets, protective coatings for fabrics, types for fighter aircrafts that could with stand high frictional heat and for making thermal impulsion for washing machines.

Polyurethane Rubber : These are obtained by condensation polymerization of short chain polyalcohols. The short chain hydroxyl terminated polymers like polyesters and polyethers are treated with diisocyanate to give the polyurethane rubbers

$$HO \sim\sim R \sim\sim OH \;+\; OCN - R' - NCO \quad \left[O - R - O - \underset{\underset{O}{\parallel}}{C} - NH - R' \right] NCO$$

Hydroxyl terminated Short polymer   Diisocyanate   Polyurethane

The properties of polyurethane rubber are decided by the short chain polymers and the final molecular length. Required strength for the rubber is achieved by vulcanisation using dissocyanate.

Urethane rubbers are highly resistant to oxidizing environments and to many organic solvents. However, they are attacked by acids and alkalies particularly in hot conditions. They form light forms and are often used as mattresses and insulators. They are also used in making surface coatings.

Rubber as an Engineering Material. Rubber has several properties which make it useful as an engineering material. Rubber has large deformability. Most structural materials, if deformed more than 10%, exceed their elastic limit but some rubbers can be stretched 10 times without rupture and regain almost the original shape after the removal of the stress. This constitutes both an advantage as well as a disadvantage depending upon the service requirement. Rubber has *low modulus,* i.e., low stresses produce deformation or change in shape. This property enables rubber to absorb shocks and to be used for cushioning and sealing. Because rubber regains its original shape after large deformation it can store energy more than most other materials. When rubber is subjected to cyclic (i.e., alternating) loading, as in automobile tyres which are cyclically stressed when the automobile is moving, not all the energy is recovered. This results in some *hysteresis loss* of energy which appears as heat energy and the rubber gets heated. This shortens the life of rubber. This loss of energy enables rubber to reduce vibrations. Rubber continues to deform under steady load with passage of time i.e., it has high creep. This behaviour limits the use of rubber as structural material.

Rubber is a bad conductor of electricity. Soft rubber is used as an insulator and hard rubber or ebonite is used as a high grade dielectric. Rubber is impermeable to water and to a large degree to gases.

Natural rubber is superior to synthetic rubber in mechanical, properties but is inferior in resistance to solvents and oils.

14.7 Adhesives

14.7.1 Introduction

Adhesion is attraction between unlike surfaces. Cohesion is attraction between like surfaces, usually due to primary or secondary forces of attraction. Materials, like the common gum or glue, that can hold surfaces together are called adhesives. Adhesives join the surfaces in three distinct ways. They may keep the surfaccs together by valence or intermolecular forces of attraction. This is known as *specific adhesion*. The adhesive may fill the voids of porous or rough surfaces and hold the surfaces by interlocking action, ie. by *mechanical adhesion*. The third mechanism, which is less common, is by *fusion* or the surfaces being partially dissolved in the adhesive or its solvent.

The function of an adhesive is similar to that of a nail or rivet used for keeping the surfaces together. A nail or rivet may not be satisfactory in many cases. The advent of adhesives made of synthetic resins, which are unaffected by moisture, has made possible wide scale use of laminates which have distinct engineering advantages

Types of Adhesives. The adhesives have been classified in several ways. They can be conveniently studied as protein adhesives, starch adhesives and synthetic adhesives.

Protein Adhesives. A large quantity of the protein glues are made from hides obtained from the leather industry or animal tissues, hides and blood from the slaughter houses or from the bones. The animal tissues and hides are treated with lime to remove hair and are washed. The bones are heated with water when the grease which floats is removed. The bones are crushed and treated with hydrochloric acid when the mineral matter dissolves leaving behind the organic matter. The stock from the hides or bones is then heated with water from 60 to 100° when the glue is extracted from the stock. The liquor is filtered hot, concentrated under reduced pressure and solid glue is obtained. The glue so obtained is commonly used for joining wood. It makes strong joints but is not resistant to water.

Another source of glue is casein obtained by curdling skimmed milk by the addition of dilute acid. Casein is separated, mixed with lime and preservatives, and sold. While using, the glue is mixed with water and the paste is applied to the surfaces to be joined. As water evaporates, the glue solidifies joining the surfaces by specific as well as mechanical adhesion. A cheaper casein glue is obtained by mixing calcium and sodium hydroxides with the soyabean meal left after the oil has been extracted.

Starch adhesives. These are mostly made by heating starch suspension in water with dilute alkali until the starch granules burst. Starch adhesives are also prepared by treating starch with dilute acids, enzymes and peroxides. These adhesives are cheaper and have lower strength than the casein glues.

Synthetic Adhesives. These are mostly resins used in the plastics industry and are classed as thermosetting or thermoplastic adhesives. The adhesives of the former group are more widely used. Of this group, i.e., the thermosetting, the most important is phenol and formaldehyde but it is not cured. It is sold as a solution, a solid, or as a film prepared by impregnating paper with the resin which is applied between the surfaces and cured by pressing and heating (110 to 170°) the surfaces. This results in the formation of cross links and the joint has very good strength and is unaffected by water. Some of the thermosetting resins, like urea formaldehyde resin, are partly formed and while applying are mixed with water and a catalyst which helps in the formation of cross links. They can set without heating but better strength is obtained by slightly heating them after joining.

Some of the cellulose derivatives, like cellulose acetate, are dissolved in solvents and used as adhesives. These can easily be used but have poor water resistance and are inflammable. Raw rubber, dissolved in a suitable solvent, gives a waterproof adhesive but it deteriorates with aging. This defect can be removed by curing either by chemicals or by the application of heat and pressure.

The animal glues give strong joints but swell in moisture and shrink in dry atmosphere. Casein and soyabean glues are more water resistant. Starch adhesives are much cheaper but are weaker and have low resistance to water. The thermoplastic adhesives have high strength and water resistance and their joints are unaffected by weather.

14.8 Laminates

14.8.1 Definition

According to the American Society for Testing Materials, a product obtained by joining two or more layers of a material (or materials) is defined as a *laminate.* When some layers are arranged at right angles to others with respect to the grains (or the strongest! direction in tension) it becomes cross *laminated.* It is called *parallel laminated* when the grains of the layers are approximately parallel to each other. The material which is bonded to another material by an adhesive is called *adherend.*

14.8.2 Common Laminates–Plywood

Thin sheets (0.5 to 10 mm thick) of wood are called *veneers.* Plywood is obtained by bonding together an odd number of veneers so that the grains of alternate sheets are at right angles to each other. This causes a redistribution of the various properties of wood along the length and width of the panel. The properties of the panel become almost the same in all directions and shrinkage of wood is not only minimised but is equalised in both directions. Splitting of wood is eliminated. Since veneers can be dried in a very short time, dry plywood can be produced from green wood in a day.

The alternate layers should be of the same wood and of equal thickness. In lumber-core plywood the central ply is a thicker panel. The plywood on the exterior should be waterproof and is bonded with a phenolic resin by hot pressing. The plywood for interior use is bonded with an extended resin glue or a protein glue, the former is more waterproof than the latter. It may be pressed hot or cold.

To make plywood more resistant to abrasion and weathering action the surface is coated with a plastics. Plastic-surfaced plywood has a glossy finish and a longer life than the ordinary plywood. The adhesive used is phenolic resin and several sheets of paper impregnated with resin are placed on the surface. It is then hot pressed when the veneers of wood and the plastic-facing cure at the same tune. Sometimes a metal sheet is bonded at the surface thus increasing its strength and abrasion resistance properties.

The essential difference between plywood and *laminated beam* is that in the latter the grains of all the sheets are parallel to each other instead of grains of alternate sheets being at right angles. The laminate so prepared has greater strength than the original wood because selected veneers are joined, the defective portions being rejected. Large cross-sections and longer pieces can be made by this method from wood of a smaller size. Such laminates are used when the stresses are mostly in the direction parallel to the direction of the grain.

Plywood is particularly useful whenever a material is required to cover large areas with strong and rigid but light sheeting. Such important applications are for cabinet making, house-building (for false ceilings, doors, cupboards, etc.) box-making, railway coachbuilding and aircraft-building.

14.8.3 Laminated Plastics

Laminated plastics are prepared by impregnating sheets of paper, cloth, wood, asbestos fibre, nylon, glass or metal foil with a resin solution. Mostly thermosetting resins are used. The roll of paper, cloth, etc., is passed through a solution of the resin (e.g., phenol formaldehyde resin dissolved in alcohol) and dried at a temperature at which the resin is not cured. The sheets are cut in suitable sizes, piled over one another and subjected to heat and pressure when the resin becomes hard and joins the material.

In the laminated plastics the fibre or the filler is continuous whereas in the moulded product the filler is not continuous but is dispersed in a continuous phase of the resin. Therefore, the laminated plastics are stronger but such complicated shapes cannot be made as are possible by moulding. The laminated plastics find maximum use in the electrical industry and for making panels for table tops, walls, etc.

14.8.4 Laminated Glass

Glass sheets or plates may be joined with a plastic sheet between them. Polyvinyl butyral resin is used if the laminate is to remain transparent. By laminating, the strength of glass is increased. For making safety glass a dry sheet of plastics is fixed between two glass plates by moderate heat. It is then subjected to pressure and heated in an oven. Such a glass laminate does not shatter to pieces on receiving an impact. High grade safety glass finds applications in making windshields for cockpits of the aircraft. Bulletproof glass is made in a similar manner. It has several sheets of glass joined together and is much thicker, say, I to 10 cm thick.

Questions

1. Distinguish Thermoset and Thermoplast polymers.
2. Important uses of Nylons.
3. What are the ratio of phenol and formaldehyde in novolac and resol.
4. Differentiate plasticized and unplasticized PVC.
5. Mention the conditions of HDPE manufacturing.
6. Explain the term plasticity.
7. Define co-polymerization
8. What are the advantages of polymers over metals?

14. Give the chemical structure of butyl rubber.

10. What are the important uses of Teflon?
11. What are the advantages of vulcanized rubber?
12. Define functionality of monomer and give the requirement for a molecule to act as monomer
13. What is vulcanization and its chemistry?
14. Mention any two synthetic rubbers and indicate how they are prepared. How are their properties compared to natural rubber?
15. What are the main differences between addition and condensation polymerization? Give an example in each case?
16. What is meant by thermosetting and thermoplastic polymers? Give at least two differences with examples.
17. Write the preparation and two properties each of

    (a) polythene (b) polyurethane

18. What is Buna-S. Mention its important uses.

114. Write short notes on the following
    (i) Phenol-formaldehyde resins
    (ii) Silicones
    (iii) Thiokol
    (iv) Teflon

20. What are the drawbacks of natural rubber?

21. Taking suitable example give the mechanism of free radical addition polymerization

22. Write informative notes on –
    (i) Fabrication of plastics (ii) Compounding of rubber
    (iii) Injection moulding (iv) Vulcanization
    (v) Adhesives (vi) Laminates.

23. What are elastomers? Give the preparation and application of
    (i) Thiokol
    (ii) SBR
    (iii) Neoprene

24. Write a note on compounding and fabrication of plastics.

25. Discuss the preparation, properties and uses of the following:-
    (a) Nylon (6,6)
    (b) Buna–S
    (c) PVC

26. (a) Explain the differences between thermoplastics and thermoset plastics.
    (b) Write a note on fabrication and moulding of plastics.

27. (a) Describe "Addition" and "Condensation" polymerizations with suitable examples.
    (b) What is vulcanization and why is it done?
    (c) Describe the preparation, properties and uses of Teflon

28. (a) Give an account of the compounding of natural rubber.

 (b) Outline the preparation, properties and uses of

 (i) Bakelite (ii) Polymethyl methacrylate.

214. (a) Differentiate between thermoplastics and thermoset plastics.

 (b) Write a note on compounding and fabrication of plastics.

30. (a) What is the constitution of natural rubber?

 (b) Describe the preparation, properties and uses of buna-N rubber.

 (c) Write briefly on the compounding of rubber and its applications as an engineering material.

31. Discuss the preparation, properties and applications of the following polymers:-

 (a) Nylon

 (b) Bakelite

 (c) Polythene

32. (a) What is the constitution of natural rubber?

 (b) What is Buna-S rubber? How is it manufactured? What are its applications?

 (c) Describe the preparation, properties and applications of PVC.

33. (a) Explain the differences between thermoplastics and thermo set plastics.

 (b) Discuss the preparation, properties and uses of different grades of polythenes such as LDPE, HDPE, and LLDPE.

34. Write a note on bakelite

35. (a) What is the constitution of natural rubber?

 (b) How is crude rubber obtained from latex?

 (c) Describe the uses of rubber as an engineering material.

36. (a) What is polymerization? Explain the different types of polymerizations with appropriate examples.

 (b) What is vulcanization and why it is done?

 (c) Write a note on preparation, properties and uses of Teflon.

37. Outline the preparation, properties and applications of the following polymers :

    (a) Nylon (6,6)

    (b) Bakelite

    (c) PVC

38. Discuss the preparation, properties and applications of

    (i) Silicone rubber

    (ii) Buna-S rubber

    (iii) Thiokol rubber

314. (a) Explain the difference between "addition polymerization" and "condensation polymerization" with suitable examples.

    (b) Write a note on fabrication and molding of plastics.

40. Write a short note on polythene

41. (a) What are high polymers? Describe their classification with suitable examples.

    (b) Describe the manufacture, properties and applications of

    (i) PVC

    (ii) polymethyl methacrylate.

42. (a) What is natural rubber?

    (b) How is raw rubber obtained from latex?

    (c) Explain the process of vulcanization and its uses.

Objective Questions

1. Which one among the following is a co-polymer

    (A) Nylon 6,6 (B) Teflon

    (C) PVC (D) Polybutadiene

2. A molecule is eliminated in

    (A) Addition polymerization

    (B) Condensation polymerization

    (C) Free radical polymerization

    (D) Ionic polymerization

3. Bakelite is prepared by the condensation of

(A) Phenol-acetaldehyde (B) Cresol-formaldehyde

(C) Phenol-formaldehyde (D) Urea-formaldehyde

4. Polyesters belong to the _____ type of polymer

(A) Addition, thermoplastic

(B) Addition, thermosetting

(C) Condensation, thermoplastic

(D) Condensation, thermosetting

5. Natural rubber is a _____ polymer of _____ monomer

(A) Addition, chloroprene (B) Condensation, chloroprene

(C) Addition, isoprene (D) Condensation, isoprene

6. _____ is the compound used in refrigeration instead of sulphurdioxide and ammonia

(A) Teflon (B) sulphur hexafluoride

(C) Freon-12 (D) Boron trifluoride

7. Chloroprene is the repeating unit in

(A) PVC (B) SBR

(C) Neoprene (D) Teflon

8. Polymer obtained by the condensation polymerization is

(A) PVC (B) Teflon

(C) Nitrite rubber (D) Bakelite

14. Which of the following is a copolymer

(A) SBR (B) Buna-S

(C) both a & b (D) none

10. Cellulose acetate is a

(A) thermoplastic (B) thermosetting plastic

(C) both (D) none

11. Which of the following is the film forming substance in oil varnish

(A) Linseed oil (B) Methyl alcohol

(C) Turpentine (D) Phenol-formaldehyde resin

12. Coagulation of natural rubber is carried out by adding

(A) CH_3COOH (B) HCl

(C) alum (D) detergent

Fill in the Blanks

1. _____ is example for theromoset resin.
2. _____ plastic will become rigid on cooling.
3. _____ are permanent setting resins.
4. Thermoplastic resins are formed by _____ polymerization.
5. Thiokol rubber is made by reaction between _____and _____.
6. Butyl rubber is a copolymer of isobutene and a small amount of _____.
7. _____ will be formed by co-polymerization of butadiene and Acrylonitrile.
8. _____ polymer is used for making of motor tyres.

14. Nylon is obtained by polymerization of _____.

10. Natural rubber is polymer of _____.
11. _____ can be used as vulcanizing agent.
12. Styrene rubber is a polymer of styrene and _____.
13. _____can be used for making gaskets.
14. _____ rubber is used for making tyre tubes.
15. Silicone resins contain alternating _____ and _____.
16. Vulcanization of rubber improves_____.
17. For making polyurethane rubber, ethylene di-isocyanate and _____ are used.
18. _____ is used for making electric switches.

114. Nylon-6 is used for making _____.

20. In SBR-Rubber the percentage of styrene is _____ and percentage of 1,3 Butadiene is _____.
21. Phenol and Formaldehyde reacts in presence of Acidic catalyst and forms _____.
22. _____ Polymer is having excellent insulating properties.
23. _____ can be used as free radical initiator.

Chapter 15

Ceramics

(Refractories, Glasses, Pottery & Abrasives)

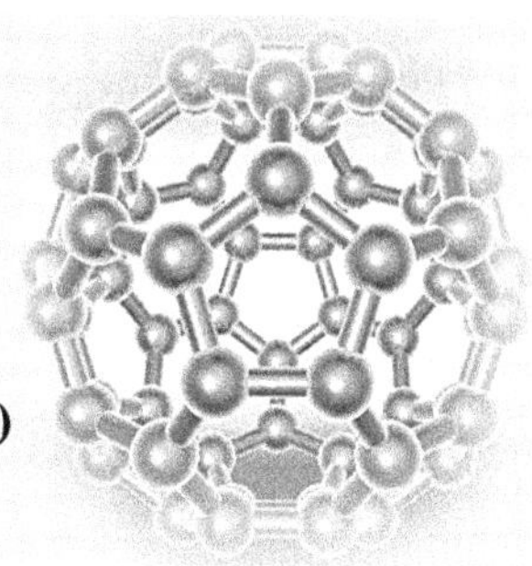

15.1 Introduction

The word ceramics comes from the Greek word 'keramos' meaning "burnt stuff". At present it is used to mean refractories, glass, enamels, potteries and porcelain, cements, bricks, abrasives, etc.

Earlier, the term 'silicates' was used for these products because silicates were the predominant constituents. As the newer materials include many non-silicates, the term 'ceramics' has now replaced 'silicates'. Broadly speaking, these include inorganic materials and products, except the metals, whose manufacture or preparation requires high temperature. In one sense ceramics may be considered to be high temperature technology. Till recently high temperature materials mostly referred to those materials which could withstand a temperature of, say, 1500 to 2000^0 but advancing technology requires materials to withstand a temperature of the order of 5000^0 (Thermonuclear reactions employ entirely different technique for handling very high temperature problems).

In this chapter all these materials, except cements and bricks, are covered. Manufacture of the common bricks has not been considered in this book. On account of its importance, a separate chapter has been devoted to Portland cement in which chapter are included other related materials like limes, pozzolan cement, plaster of Paris, etc.

15.2 Refractories

15.2.1 Definition and Functions of Refractories

Any substance that is difficult to fuse is a refractory, but when this term is used for engineering materials, several other properties are implied. Refractories retain their chemical and physical identity and also their mechanical function at elevated temperatures. Many engineering, more particularly metallurgical and ceramic, operations require the use of high temperatures for which materials are needed which can withstand the high temperatures. Common building materials like the ordinary brick, if exposed to such high temperatures, would fuse and the structure would fail. Besides resistance to high temperatures, the material is required to resist other destructive factors like heavy load, fluxes, corrosive fluids like slags, thermal shock due to rapid changes in temperature, abrasion (rubbing) due to movement of materials or dust-carrying gases. In addition, a refractory should be satisfactory for use as a lining for a furnace or equipment handling hot metals. Accordingly, a number of properties need to be considered in a refractory material. In 1981 about 9 lakh tonnes of refractories were made in India.

Properties

The following properties need to be considered while selecting a refractory material.

1. *Refractoriness :* This is a measure of the fusibility of the refractory. It indicates the temperature at which the refractory softens and is generally indicated in terms of pyrometric cone equivalent (P. C. E.) value. As the primary function of a refractory is to withstand high temperatures, it is of prime importance that the softening (or melting) temperature should be well above the operating temperature. It is noteworthy that in a furnace the inner end of the refractory wall is at a much higher temperature than the outer. Therefore, unless the brick melts away completely, it can often be is used to withstand a temperature higher than its softening temperature. The outer end will be at a lower temperature and still in a solid state, thus giving strength.
2. *Strength* : It must be strong enough to resist physical wearing away and to take load. *Refractoriness under load* is an important consideration because usually a refractory fails at a lower temperature when subjected to load. This is supposed to be due to the fact that some crystals which have become fluidic act as a lubricant and deformation becomes easier when subjected to load. In many cases, especially in taller furnaces, the refractory has to support a heavy load, hence strength under the combined effect of temperature and load is an important criterion.
3. *Specific gravity :* The specific gravity is an important routine test for refractories as it gives an idea of its true porosity. Bulk density is represented by weight per unit volume of the refractory, the volume includes pore also. It is also a criterion while buying the raw material for making bricks or for lining, etc. Raw materials of higher specific gravity will give fewer bricks per unit weight. This means that the cost of a brick of higher specific gravity is more than that of lower specific gravity (taking the price of the raw material to be the same per unit weight). While using a refractory material in powdered form for making furnace lining or for repairs, lesser weight of a material with lower specific gravity will be required. Materials of lower specific gravity are also preferable because bricks in the lower part of a full structure will not be subjected to a heavy load.

4. *Thermal expansion :* In furnace design, allowance has to be made for thermal expansion. This becomes very important factor if the refractory is subjected to rapid changes in temperature as a furnace door is, which on opening is suddenly exposed to air at room temperature. Within a single brick in a wall there may be a temperature drop of a few hundred degrees from the hot face to the cold face. This causes differences in thermal expansion in different sections of the brick. There may be changes in the internal structures of the material with the result that there are changes in volume, as will be explained under properties of silica refractories. All these result in the development of high internal stresses.
5. *Thermal conductivity :* In most cases the thermal conductivity should be low so that the heat of the furnace may not be lost. Exceptions to this rule are not many, e.g. coke oven walls, muffle furnaces and retorts are heated from outside, and should be made of refractories of high thermal conductivity. To prevent heat loss sometimes a refractory is backed by an insulating material like asbestos. It is to be noted that under these conditions the refractory will be maintained at a higher temperature and consequently its life will be shorter.
6. *Porosity :* If the refractory has open pores the gases and slags will enter the material more easily and to a greater depth and may react and reduce the life of the refractory.
7. *Electrical conductivity :* For electrical furnaces the electrical conductivity of the refractory should be low. Except graphite, all other refractories are bad conductors of electricity. The electrical resistivities of the refractories drop rapidly with rise in temperature.
8. *Chemical composition :* It gives the true identity of the particular type of the refractory. The chemical composition of the refractory should be such that the surroundings do not chemically attack the refractory and corrode it.

15.2.2 Classification

The most commonly used classification of refractories is on the basis of the chemical nature of the principal oxides (or compounds) present in them. Accordingly, the refractories are classified.

1. *Acid Refractories :* Acid refractories are those which consist of acidic materials and therefore are not attacked by acidic materials present in the furnace but are easily attacked by basic materials. These are of two types : (a) Siliceous materials have silica as the principal constituent as in sand, ganister, quartz, etc.; (b) Alumino-silicates consist of Al_2O_3 and SiO_2. Fire clays are the most important refractories of this group.

2. *Neutral Refractories :* Neutral refractories are not neutral in the strictly chemical sense. These consist of feebly acidic or feebly basic materials with the result that they resist the action of acidic as well as basic materials. Examples of this group are chromite, graphite, silicon carbide, etc.

3. *Basic Refractories :* Basic refractories are those which consist of basic materials, therefore they are not attacked by other basic materials but are easily attacked by acidic materials. Magnesite, chrome-magnesite, dolomite, etc. are the most widely used refractories of this class.

Some of the common refractories are described below :

15.2.3 Silica Refractories

The raw materials for silica refractories are the various forms of natural deposits of SiO_2 which contain only small amounts of other oxides. Such deposits are quartz, quartzite, ganister, sand, sandstone, etc. The presence of an excess of basic oxides is particularly undesirable as they lower the fusion temperature. One great difficulty with silica refractories is that silica occurs in several allotropic forms which are stable at different temperatures and have different specific gravities. These changes will be discussed presently. Therefore, a raw material is chosen which can be easily converted into the more stable form. For this reason common sands and pure quartzite are not very satisfactory, the raw material should have some binding material preferably fine-grained silica particles themselves. About 4 lakh tonnes of quartz and quartzite are produced annually in India.

For the manufacture of silica bricks the rock is crushed and ground with 2% lime and water. At higher temperatures lime forms fusible calcium silicate which acts as a binding material. The mixture is then made into bricks by hand moulding or by machine pressing. The bricks are dried either in the sun or in heated rooms. The dried bricks need to be handled with care, otherwise they can crumble. They are then transferred to kilns which are heated by burning coal or gas. In the manufacture of refractory bricks on heating, mechanically and chemically held water is first removed. On heating further very often mineral transformation and volume changes take place. In the case of silica bricks another very important aim of heating is to convert quartz into cristobalite or tridymite which are the more stable allotropes of silica. If this change does not take place, while firing, the volume will change during use and the structure will become weak. The temperature is slowly raised, in about 21 hours, to say 1500^0 and maintained at this temperature for nearly 12 hours so as to allow quartzite to be converted into cristobalite which slowly changes to tridymite during cooling. Maintaining the temperature at 1500^0 for a longer period is desirable but is not economical. Cooling has to be done still more carefully and takes 1 to 2 weeks. This results in the formation of a mixture of cristobalite, and excess of tridymite.

Properties

Silica bricks are slightly yellowish in colour with brown specks distributed throughout the mass. The porosity is about 25% and the specific gravity is 2.3 to 2.4.

An important property is that quartzite is converted into tridymite at 870^0 which changes to cristobalite at 1470^0, each of which has α and β forms and different specific gravities. The changes may be summarized as follows :

Allotrope	Quartzite	$\underset{}{\overset{870^0}{\rightleftharpoons}}$	tridymite	$\overset{1470^0}{\rightleftharpoons}$	cristobalite
Sp. Gr.	2.65		2.26		2.32

Hence the transformation of quartz to tridymite causes a volume increase of 17.2 per cent; of quartz to cristobalite, 14.3 per cent. The linear expansion of silica bricks during firing, due to these transformations, is about 3.5 per cent. Therefore, if during firing of the bricks, quartzite is not converted into tridymite and cristobalite the bricks will expand on being heated in the furnace and the structure will fail. The reverse reaction, i.e. the formation of quartzite, takes place so very slowly that for all practical purposes tridymite and cristobalite may be taken as stable forms. That is why it is important that these forms should be produced while making the bricks.

Silica bricks have high refractoriness under load and can stand a load of about 3 kg/cm^2 at about 1600^0. No change in shape is noticeable up to 1350^0 under load. Bricks almost free from alumina do not deform under load even at 1680°. Below 600^0 silica bricks cannot withstand thermal shocks, (i.e. rapid changes in temperature). These properties make silica bricks very satisfactory for use in steel melting furnaces and for coke ovens. The thermal conductivity is better than that of fire clay bricks but not as good as that of the basic refractories.

Silica being an acid oxide, the bricks cannot be used in contact with basic fluxes and slags because of the chemical reactions between them. Silica bricks are relatively light and have high mechanical strength and rigidity.

Uses

The largest consumer of silica bricks is the iron and steel industry. They are widely used wherever the material in contact with them is not basic and there is a steady high temperature, i.e. the bricks are not exposed to temperature fluctuations. In addition, they are used in coke ovens, glass furnaces, etc. Silica bricks are used more than any other type of refractory except firebricks.

15.2.4 Alumina-Silica Refractories

In the alumino-slilicate refractories, as the name suggests, the two principal oxides present are alumina and silica. Fe_2O_3, TiO_2, CaO, MgO, Na_2O and K.O are the main impurities present and amount to 2 to 5 percent. Alumina percentage in the semi-silica type varies from 15 to 20% and in corundum type from 99 to 99.7%. In the silica refractories alumina is avoided because it may form an eutectic, which has a very low fusion temperature. A combination of 94.5% SiO_2 and 5.5% Al_2O_3 melts at 1545^0. The most widely used refractories of this group are the fire clay refractories.

Fire clays are hydrated aluminium silicates and their chemical composition may be represented by $Al_2O_3.2SiO_2.2H_2O$. At higher temperatures water is driven out. The common fire clays contain 18 to 47.5 per cent Al_2O_3 and 80 to 50 per cent SiO_2. Clays with higher Al_2O_3 contents are usually called high-alumina refractories. Fire clays are classified in several kinds of clays e.g. flint (hard clay), plastic clays, kaolin, etc. The term 'firebricks' refers to fire clay bricks.

In India most of the fire clay is obtained from deposits under the coal seams of Bihar, Bengal, Madhya Pradesh and Andhra Pradesh. Good deposits are also found near Jabalpur.

The fire clays can hold water mechanically in addition to the chemically held water. During firing, due to the expulsion of all this water and the fluxing effect of the impurities, the product shrinks considerably and becomes out of shape and size. To overcome this difficulty the fire clay is mixed with materials like silica that expand and/or large amounts of materials which do not shrink. This latter material is known as grog. Grog is fire clay which has already been heated to, say, 1300^0 so that no further contraction in volume takes place. Broken and used fireclay bricks are also used as grog. A small amount of plastic clay is added to facilitate moulding. A mixture is made by grinding raw fire clay, grog and plastic clay with water. There are several ways of giving the desired shape to the mixture. The bricks are mostly sun-dried. In colder and more humid climates they are dried in driers.

The bricks are then fired in a kiln. At the lower temperature water is expelled. At 980^0 an exothermic reaction takes place due to the formation of mullite ($3Al_2O_3.2SiO_2$) and the excess silica crystallizes as tridymite. Usually the temperature is raised to 1400^0 and maintained for a few hours to obtain uniform properties. For higher temperature use, the firing temperature may be 1500^0. The furnace is then closed and allowed to cool slowly. The total time taken is about a week.

Properties

Fire clay bricks of several grades are manufactured and their properties and prices differ to some extent. The bricks are light yellow in colour and depending on the content of iron oxide may be reddish in colour. While selecting the bricks for a particular use it is to be noted that fine-textured bricks are more resistant to corrosion by slags and gases because these cannot penetrate the bricks. The open-textured bricks are more resistant to thermal shock, i.e. changes in temperatures, and are to be preferred where such conditions exist.

Firebricks are slightly acidic in character. The extent to which they are acidic depends upon the SiO content. They cannot therefore resist highly basic slags. They are less refractory than silica bricks though bricks fired up to a high temperature can be used up to 1700^0 but ordinarily they can be used up to 1400^0. The refractoriness under load is much poorer than that of silica bricks and most grades begin to fuse at 1350^0 under a load of 2 kg/sq.cm. The crushing strength at atmospheric temperature is high. It is about 200 kg/sq. cm.

A very useful property of firebricks is resistance to thermal shocks which makes them very suitable for making a checker work of regenerative furnaces and those parts of furnaces, like the charging doors, which are subjected to temperature fluctuations.

Firebricks are cheaper than silica bricks.

In blast furnaces, for which firebricks are very extensively used, the disintegration of firebricks takes place by the deposition of carbon from carbon monoxide and is believed by some authorities to be retarded by the presence of iron oxide.

Uses

Fire clay bricks are more widely used than any other refractory bricks because of the cost factor and as they are satisfactory for most purposes. Under extreme conditions the guiding principle is that they are satisfactory if the temperature is not very high and the slag is not highly basic. As they can resist thermal shocks, furnaces like crucible furnaces, or parts which are frequently cooled are lined with firebricks. They are used in large numbers in lining blast furnaces, regenerators, boiler furnaces and for the general lining given to varied equipments like chimneys, flues, etc.

High-alumina Refractories

Refractories of this group are not used so extensively, because they are costly. Some of them are very good refractories. These contain 45 to 100 percent alumina. As a general rule the refractoriness increases with increase in the Al_2O_3 content of alumina-silicate refractories.

Kyanite and sillimanite are the more important refractories of this group and both can be approximately represented by the formula $Al_2O_3.SiO_2$. Both change on heating to mullite ($3Al_2O_3.2SiO_2$) and silica. In the case of kyanite this change takes place at 1310^0 and is accompanied by a great increase in volume whereas in sillimanite the change to mullite takes place at about 1600^0 without any appreciable increase in volume. Therefore, it is necessary to strongly heat kyanite and convert it to mullite and then make bricks and finally fire in the usual manner. Sillimanite can be made directly into bricks and fired as usual. In both cases, plastic clay is used as binding material.

Kyanite deposits are found in the Chhota Nagpur division (Bihar) and sillimanite in the Khasi Hills (Assam), Rewa and Bhandala (M.P.) and as beach sand in Kerala. India produces annually about 65,000 tonnes of kyanite and 11,000 tonnes of sillimanite, most of which are exported.

Due to the formation of mullite, both kyanite, and sillimanite are almost ideal refractories. Mullite contributes high refractoriness (can be used up to 1800^0), high refractoriness under load, chemical resistance and low coefficient of expansion, therefore resistance to thermal shocks. They are more or less neutral refractories and can be used in most environments. Both the materials, and sillimanite in particular, have high mechanical strength.

They are costly and hence are not used in very large quantities.

Uses

Kyanite and sillimanite refractories are widely used in glass-making furnaces.

15.2.5 Bauxite Bricks

Bauxite is the ore from which aluminium is extracted. Pure Al_2O_3 is a highly refractory material (m.p. 2010^0)) but ordinary bauxite contains iron and silica as impurities and gives bricks fusing at about 1000^0, but properly prepared bricks can be used up to 1750^0.

Besides containing iron oxide and silica, which lower m.p., bauxite contains water of crystallization. It is heated to 600^0 to remove water and can then be used directly for making inferior grade bricks. Usually the calcination temperature is 1400 to 1500^0 which prevents further shrinkage in the bricks. The calcined material is crushed and mixed with plastic refractory clay and made into bricks and fired at 1500^0.

Refractoriness under load is also good, being about 3 kg/sq.cm at 1625^0. Resistance to thermal shock is not as good as that of fireclay bricks but is better than that of silica bricks.

Bauxite bricks are classed under basic refractories but they behave to some extent as neutral refractories and can resist the action of even slightly acidic slags.

Uses

They are used for making the bottoms and hearth of basic steel making furnaces and as lining for cement and other kilns and other furnaces working at high temperatures.

15.2.6 Chromite Bricks

Chromite refractories are important neutral refractories. The raw material is chromite ore which is usually represented by Cr_2O_3FeO. It contains $MgO.Al_3O_3$ which acts as binding agent for the bricks. Cr_2O_3 content of the ore varies from 30 to 60%. Some of the Indian ores contain as high as 52% Cr_2O_3. The presence of large amounts of silica and iron oxide reduces the fusion temperature, hence these are undesirable. Chromite deposits in India are found in Karnataka, Salem district (Tamil Nadu). Singbhum (Bihar), Ratnagiri (Maharashtra) and Ladakh (Kashmir). The annual production is about 4 lakh tonnes.

The ore is ground with binding clay and water and made into bricks and heated slowly (to avoid swelling of the bricks and their becoming out of shape) at 1500 to 1700^0.

Chromite bricks can be used up to 1800^0 and their refractoriness under load is 3.5 kg/sq.cm at 1430^0. At the atmospheric temperature the crushing strength is 350 to 550 kg/sq.cm. They are important neutral refractory bricks. Chromite bricks are resistant to basic and moderately acid slags, but are affected by the slags in a reducing atmosphere. They are not resistant to thermal shocks. They have moderate thermal conductivity.

Uses

Chromite bricks find extensive use in steel-making and copper refining furnaces. They are particularly useful in separating acid and basic refractories and are good for any use so long as they are not subjected to thermal shocks, but they are expensive.

15.2.7 Chrome-Magnesite Bricks

Chromite bricks are very good refractory bricks but their serious drawback is lack of resistance to thermal shocks. Efforts to over come this difficulty have led to the addition of 10 to 30% of dead burnt magnesite (MgO) to chromite. (In chrome-magnesite the amount of chrome ore exceeds the amount of magnesite; those in which magnesite predominates are termed 'magnesite-chrome.' The mixture is bonded by organic materials (like molasses) and made into bricks and fired at 1350^0 to 1650^0). At a higher temperature the organic materials burn off and ceramic bond is formed.

In recent times chrome-magnesite bricks using a chemical compound as binding agent are being manufactured. The bricks are formed at high pressures with the chemical bonding agents. When used in the furnace under the influence of temperature the chemical bond is lost and ceramic bond is developed by sintering of the particles.

Chrome-magnesite bricks appear to have the good qualities of chromite as well as of magnesite bricks. They can be used up to 1700^0 and their refractoriness under load is 2 kg/sq.cm. at 1600^0. They are more resistant to thermal shocks than chromite bricks. Their chief defect is that they swell and burst in the presence of slags containing iron oxide. They behave as neutral bricks and are resistant to both acid and basic slags.

The defects of chrome-magnesite bricks have been further reduced by three improvements. The improved bricks are made by using coarse chromite mixed with fine calcined magnesite and using a chemical binding agent. The fine magnesite prevents slag penetration and thus reduces the bursting of the bricks. The chemical binder eliminates and saves the cost of firing. The third improvement is the use of alternate layers of these bricks with layers of the bricks enclosed in boxes made of thin steel plates. In actual use, due to the high temperature, these boxes fuse together and provide a strong structure. As the bricks are enclosed in boxes, these are known as metalkase or metalclad or steelclad bricks.

Uses

Chrome-magnesite bricks find extensive use in acid as well as basic open hearth furnaces in making the hearth, roofs, side walls and ports.

15.2.8 Graphite Refractories

Graphite is a crystalline form of carbon and its natural deposits mixed with calcarious and siliceous materials occur in Bihar, Kerala and Karnataka States. The natural deposit is known as plumbago. Coke has been used as a cheap substitute for graphite but is not as satisfactory as the latter.

For obtaining superior products, graphite after grinding is treated with hydrochloric acid to remove the mineral impurities. It is mixed with a binding material like plastic clay, sodium silicate or tar and made into crucibles. Carbon bricks, which are also extensively used, are made from coke of very low ash content and employing tar as the binding material. During firing, contact of the bricks with air is minimized by filling the space in between the bricks with a mixture of sand and powdered coke.

Fireclay refractories coated with tar or carbon deposited on them, under controlled conditions, are also used. The carbon makes them more resistant and increases the life.

Graphite is infusible and sublimes at 3600^0. It is not attacked by any slag or chemical but in the presence of oxygen, at a higher temperature it oxidizes. It is resistant to thermal shocks and has high thermal conductivity.

Uses

Carbon bricks find the most extensive use in lining the hearth and the bosh of the blast furnace. They are also used in other furnaces and equipment where the temperature is high and the atmosphere is nonoxidizing.

15.2.9 Zirconia Refractories

Zirconia refractories are usually classed as neutral refractories but they are not quite resistant to acid slags and hence are on the border line between neutral and basic refractories.

The raw material is ZrO_2, which occurs as zirconite, zirconia and zircon mineral. In India it is found as beach sand in the Kerala State. The mineral undergoes volume changes on heating and cooling and is stabilized by adding MgO or CaO and firing at 1700^0. The most satisfactory binding material is colloidal zirconia prepared by drying zirconium nitrate solution and taking the residue into solution and again drying. Alumina can also be used as a cheaper binding material. The mixture is made into bricks, dried slowly and fired at 1750^0 or at an even higher temperature for still better bricks.

Zirconia bricks can be used up to 2000^0 and especially prepared bricks can be worked up to 2600^0. Refractoriness under load is 3.5 kg/sq.cm at 1600^0. They resist the basic slags but are not satisfactory for use with acidic slags. They are resistant to thermal shocks.

Uses

Zirconia bricks are very expensive and are used only when the temperature is very high as in high frequency electric furnaces.

15.2.10 Magnesite Refractories

Magnesite refractories are the most important and widely used basic refractories. Magnesite is $MgCO_3$ and occurs as a natural deposit but before it can be used as a refractory, it has to be heated strongly and the resulting product, MgO, is used as a refractory material. Magnesite deposits in India are found in the districts of Singbhum (Bihar), Kurnool (Andhra), Salem and Trichinapally (Tamil Nadu), Mysore and Hassan (Karnataka), Almora (U.P.) and several places in Rajasthan and Kashmir. Magnesite is being manufactured at Almora and Salem. Its annual production is about 33 lakh tonnes.

Preparation

Magnesite, $MgCO_3$, is heated when most of CO_2 evolves at about 800^0 giving caustic magnesia. Caustic magnesia is used as a cheaper refractory material. It is then heated to 1700^0 till the decomposition is complete. Then no further shrinkage will take place in actual use. It does not absorb water or carbon dioxide from the atmosphere. Thus from magnesite about 48% of dead burnt magnesite is obtained. Mineral or crystalline form of magnesium oxide (m.p. 2800^0) is called periclase. In recent times magnesia is also produced by heating magnesium hydroxide. Its deposits, mineral brucite, are rare. It is also prepared by precipitating magnesium hydroxide from sea water or brine from petroleum wells by adding slaked lime or slaked dolomite. Periclase is crushed to the proper size and mixed with caustic magnesia or sulphite lye (obtained as a waste product from paper factories) or iron oxide as binding material and ground with water and made into bricks. The bricks are slowly heated to 1500^0, kept at that temperature for 8 hours and slowly cooled. Chemically bonded bricks which do not require firing are also made.

Properties

Magnesite bricks can be used up to 2000^0 without load and up to 1500^0 under load (3.5 kg/sq.cm.). They are basic in nature and therefore should not be used in the presence of acidic slags. But they are not as basic as dolomite and the reaction with other acidic bricks is not so very rapid; therefore, they can be used in conjunction with slightly acidic bricks.

Magnesite bricks cannot resist thermal shocks, loose strength at high temperature, and are not resistant to abrasion. To overcome these defects the bricks may be packed in metal boxes and used as described under chromite-magnesite metalkase bricks.

Uses

Magnesite refractories are used widely in the iron and steel industry in the form of bricks and as 'pea' magnesite (about 1cm in diameter). Pea magnesite is used for making the bottoms of basic furnaces. The bricks are used in making basic furnaces, hot metal mixers, soaking pits etc.

15.2.11 Dolomite Refractories

Dolomite may be represented by $CaCO_3$. $MgCO_3$. On calcination it gives CaO.MgO. The difficulty in using it is that lime absorbs water and carbon dioxide at atmospheric temperature, and thus causes the disintegration of the bricks. This can be prevented by either coating the bricks with tar so that during storage contact with water and carbon dioxide is avoided, or stabilized by mixing serpentine ($MgO.SiO_2$) with dolomite and calcining, resulting in the formation of di or tri-calcium silicate which does not absorb water and carbon dioxide. After calcination it is mixed with water, a silicate binder and made into bricks. The bricks are dried and fired at 1500^0 for a day.

Dolomite refractories are more basic and less satisfactory than magnesite. Furnaces using unstabilized bricks should not be cooled during use because lime would slake and the bricks disintegrate. Dolomite bricks can withstand a load of 3.5 kg/sq.cm at 1650^0. They are not very resistant to thermal shocks.

Use

Stabilized dolomite bricks are replacing magnesite bricks because they are much cheaper. Besides being used as bricks, dolomite refractory is widely used in repairing basic furnaces.

15.2.12 High-Refractory Materials

There are many specialized service conditions for which, though the quantity of the material needed is small, resistance to very high temperatures is needed. Demand for such materials is increasing for use in gas turbines, nuclear reactors, rockets and missiles besides small furnaces for melting certain materials. Materials for this group come from pure metallic oxides, carbides, nitrides, silicides and sulphides. (In Fig. 15.1 are given the melting points of high melting materials. The melting points of copper and iron have been included for giving a comparative idea of the melting points). It is a matter of common knowledge that impurities lower the melting points. Therefore, great care should be taken to use pure materials and not to use any low melting binding agent. The shape is to be given by special techniques, like the combined effect of heat and pressure.

Oxides

The highest melting oxide is thoria (m.p. 3070^0) but its use is very limited due to its cost and difficulty in fabricating it. Widely used oxides are magnesia (2800^0), zirconia (2580^0), beryllia (2570^0) and alumina (2010^0). The pure oxide is finely ground with water and a temporary organic binder such as starch, wax, a high boiling alcohol, etc., is added. On firing, the temporary binder is destroyed and the fine particles sinter together. Solid surface reactions take place and gradually new tiny crystals are produced which hold the shape. As no material is used, the fusion temperature of the final shape is practically the same as that of the pure material.

Each of the oxides mentioned has some special property. For example, alumina is the hardest of all and is useful where wear resistance is required. Beryllia has high resistance to thermal shocks and is not reduced by carbon or carbon monoxide up to 2000^0. In spite of the high melting point of magnesia it cannot be used above 1700^0 in a reducing atmosphere and 2200^0 in an oxidizing atmosphere. As zirconia undergoes polymorphic changes, its resistance to thermal shock is low, above 1250^0. However, it can be stabilized by adding calcium or magnesium oxide.

Carbides

The carbides have higher melting points than the oxides but they are not resistant to oxidation at high temperatures. Carbides of silicon (decomposes above 2200^0)), boron (m.p. 2450^0), zirconium (m.p. 3750^0) and titanium (3160^0) are the more widely used carbides.

Silicon carbide, also known as carborundum, is the most commonly used carbide. It is made by heating, in an electric furnace, sand with coke and small amounts of sawdust and salt. The gases evolving from the sawdust make the mass porous. Salt forms volatile chlorides of iron and other impurities thus increasing the purity of the product. Several types of bonds have been used for giving the final shape. Silicon carbide decomposes before melting. Above 1000^0 it begins to oxidize in air. It has high thermal conductivity and low thermal expansion.

Boron carbide, B_4C, needs special reference due to its ability to absorb thermal neutrons (i.e. slow moving neutrons) which makes it a useful material for nuclear power plants. It is very hard and has high wear resistance. Above 1000^0 it readily gets oxidized in air and reacts with molten iron. This restricts its use in many furnaces. Other carbides also get oxidized in air much below their melting points and are very satisfactory for use in an atmosphere of nitrogen or in reducing atmosphere.

Borides

The industrially important compounds of this group are the borides of titanium (TiB_2 m.p. 2890^0), zirconium (ZrB_2 3040^0), chromium (CrB_2 2760^0), molybdenum (MoB_2, 2100^0) and tungsten (WB, 2860^0). The borides are very hard, more resistant to oxidation than the carbides and have good resistance to chemicals.

Nitrides

As a group the nitrides have somewhat higher melting points than the borides but have poorer resistance to oxidation (Boron nitride is soft like graphite and sublimes at 2730^0).

The *silicides* and *sulphides* have relatively much lower melting points. Silicides resist oxidation on account of the formation of a glass-like coating of silica on the surface.

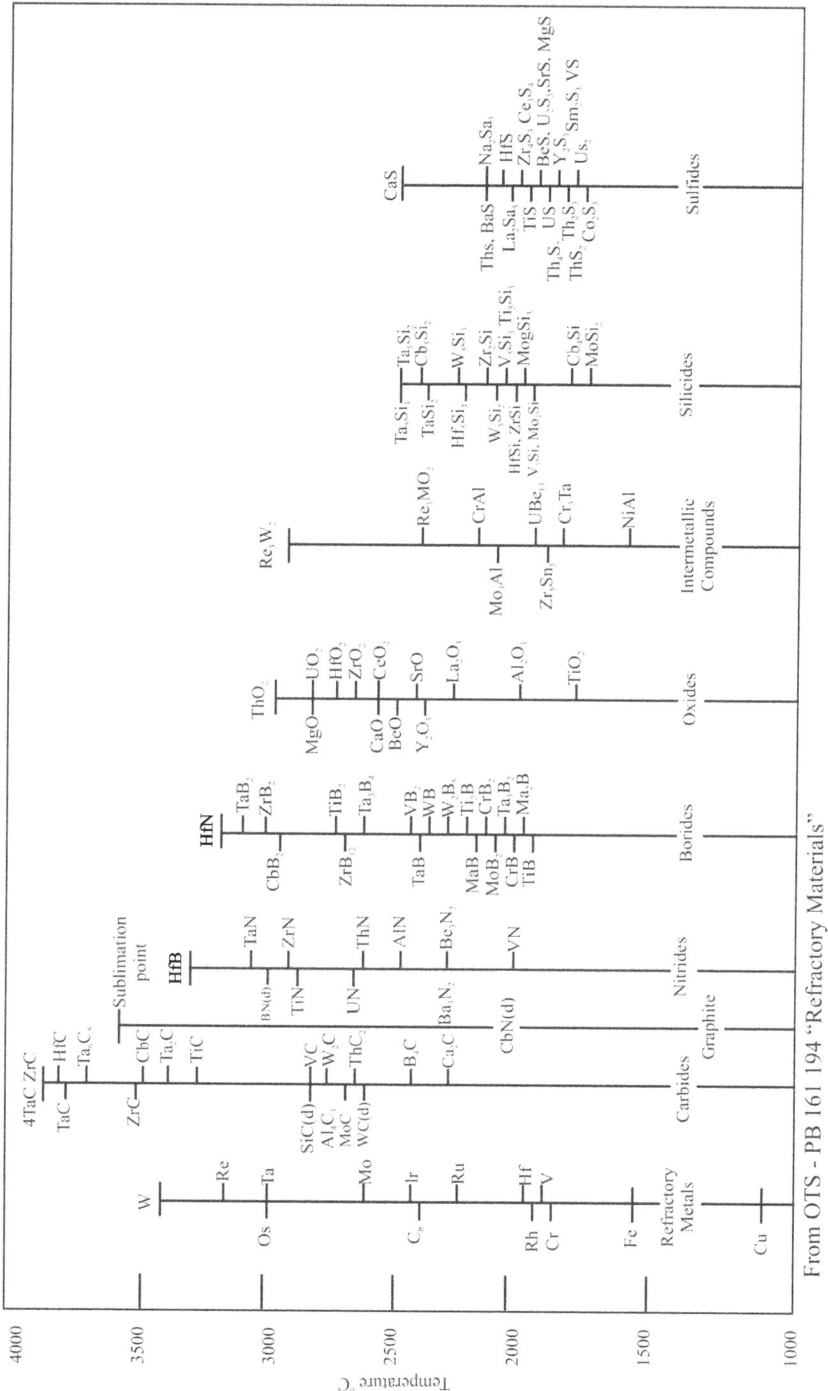

Fig. 15.1 Melting points of high refractory materials (Melting point of Cu and Fe are given for comparison).

15.2.13 Super-Refractories

In some cases like very high operating temperatures, severe corrosion due to turbulent flow of molten material, the conventional refractories prove unsatisfactory. For such applications super-refractories are used. The term is used to indicate the superior properties of the refractories.

Very pure raw materials like silicon carbide, mullite, alumina, zirconia, etc. are used. These are fused in an electric furnace when the porosity becomes negligible. The fused mass is broken into small grains, called fused grains. The fused grains of different sizes are taken so that the packing characteristics of the product are improved. These are then made into the required shape by the conventional methods of manufacture.

15.2.14 Failure of Refractories

The most common cause of failure of a refractory is chemical reaction with the environment in which it is working. It should certainly not be in a chemically dissimilar environment, e.g. an acid refractory should not be used in a furnace using basic fluxes and slags and vice versa. In spite of using proper fluxes there may be impurities which may react with the lining. In the chemical reaction the porosity of the refractory plays an important part. The more porous it is, the greater will be the depth to which the slag will penetrate and destroy the refractory. Porosity also affects the reactions with the gases. The deposition of carbon from carbon monoxide in fire clay refractories in a blast furnace is an important cause of its failure (Deposition of carbon sets up high internal stresses resulting in the disruption of the brick). The higher the temperature the faster does the chemical reaction take place. Sometimes rise in temperature beyond the safe limit quickly brings about the destruction of the refractory. Lastly, the agitation or movement of the materials is important. If there is agitation on the surface, new surface is quickly exposed to chemical attack and is destroyed.

The second important cause is "spalling" which has been defined as 'the loss of fragments or spalls from the face of a refractory brick or structure through cracking and rupture, with exposure of inner portions of the original refractory mass'. The cause of spalling may be thermal, mechanical, or structural. Thermal spalling may be due to unequal expansion or contraction caused by the difference in temperature at different parts or by rapid changes in temperature. Mechanical spalling is mostly due to carelessness in loading the furnace or in the removal of materials from the furnace thereby damaging the refractory. Structural spalling takes place because of the change in composition of the refractory due to reactions with slags, etc., with the result that its coefficient of expansion changes. Thus different parts expand and contract to a different extent. Spalling of a refractory can be reduced by proper design, construction and operation of the furnace.

Abrasion may be listed as the third factor. Abrasion means mechanical rubbing away of the material. In the furnace there is movement of materials and often hard and solid materials slide along the refractory. There are heavy molten slags and metals flowing. Dust-bearing gases often move at high speeds. All these rub off the refractories.

15.2.15 Selection of Refractories

The choice of a suitable refractory depends on a large number of factors. These factors may be classified into the following groups :

1. *Factors related to refractories :* The properties of the refractories available in the market need to be studied. The properties and their significances have already been discussed in this chapter.
2. *Factors related to the furnace design :* The choice of the refractory will depend upon the type of the furnace, conditions of heating (e.g., a coke oven is kept continuously at a high temperature for years but a cupola is daily cooled) and loading, degree of insulation, provision for thermal expansion made in the furnace, whether there are moving parts or not (as are in rotary kilns and in tunnel kiln cars).
3. *Factors related to operation :* The most important factor in this group is the chemical nature of the materials (like the ores, fluxes, fuels, etc.) in contact with the refractory, temperature in different parts of the furnace and temperature fluctuations, fluidity of the metals and slags, velocity of gases in contact with the furnace lining and abrasion due to moving charge, gases and the molten metals.
4. *Cost factors* : While considering the initial cost of different refractories their life during use should not be overlooked.

It may, therefore, be seen that selection of a suitable refractory is quite complicated. Much data is available from the working of the refractories in similar furnaces elsewhere. These will help in actual selection. Out of the large number of factors only a few are so essential that by comparision other factors are of minor importance. For example, in some cases exposure to high temperature is the chief problem. In other cases high load has to be supported, in only a few cases the refractory has to withstand both high temperature and high load. In some cases high insulating property is needed, in others, high thermal conductivity. Greatest attention, in the selection, should be given to the more essential properties for the particular use.

The cost of handling the refractory, for example, transportation, brick-laying cost, are the same whether superior or inferior bricks are used. If an inferior refractory is used and it has a shorter life, it will have to be replaced more often. This will involve additional handling costs, in addition to loss in production. Therefore the true cost of refractories should be considered on the basis of their effect upon the overall cost of production per unit output e.g., cost per tonne of steel produced.

15.2.16 Characteristic Properties of Refractory Bricks

Type	Characteristics
High-Alumina	High refractoriness, which increases with increasing alumina content. High mechanical strength at high temperatures. Excellent to fair resistance to spalling. Greater resistance than fire clay bricks to corrosion by most basic slags and fluxes.
Fire clay	The chemical and physical properties of fire clays vary between wide limits; hence fireclay bricks with widely varying combinations of properties are available. This fact accounts for their suitability for service under widely different operating conditions. Most fireclay bricks have relatively good spalling resistance and thermal insulation value. Fair resistance to acid slags and fluxes; lower resistance to basic slags and fluxes.
Silica	High refractoriness and resistance to abrasion. High mechanical strength at high temperatures. Thermal conductivity at high temperatures is about 25 per cent greater than that of fireclay bricks. High resistance to corrosion by acid slags. Fair resistance to attack by lime, magnesia and iron oxide. Readily attacked by basic slags and fluorine. Not subjected to thermal spalling at temperature above 650^0; poor resistance to thermal spalling at low temperatures.
Magnesite	Extremely high refractoriness and high thermal conductivity. Great resistance to corrosion by basic slags; poor resistance to slags containing high percentages of silica. The chemically bonded and metal encased bricks have marked resistance to spalling.
Chrome	High resistance to corrosion by basic and moderately acid slags and fluxes. In general basic slags not adhere to chrome bricks. Under certain unusual conditions iron oxide is absorbed and causes a damaging expansion. Thermal conductivity is lower than that of magnesite bricks but higher than that of fireclay bricks.
Magnesite Chrome	Mechanical strength and stability of volume at high temperature. Excellent resistance to spalling. High resistance to corrosion by basic slags.

Contd...

Type	Characteristics
Chrome Magnesite	Mechanical strength and stability of volume at high high temperatures. Excellent to good resistance to spalling. High resistance to corrosion by basic slags.
Forsterite $2MgO.SiO_2$	High refractoriness. Excellent strength at high temperatures. Marked resistance to corrosion by alkali compounds. Fair resistance to most basic slags; attacked by acid slags.

(Based on *Modern Refractory Practice*, *Courtesy*, Harbison Walker Refractories Co., Pittsburg).

Note : Basic refractories in general are less resistant to thermal shock than fireclay and high alumina refractories. However, resistance to spalling of some basic bricks or special composition is equal to that of super-duty fireclay bricks.

Basic bricks of any class may be metal-encased and internally plated. Such bricks have much better resistance to spalling than bricks without the casings. Most metal-encased bricks are magnesite chrome compositions.

15.3 Glass

15.3.1 Definition

Glass may be defined as "inorganic product of fusion which has been cooled to a rigid condition without crystallization". It is an under-cooled liquid of very high viscosity. The glassy state is a condition which is continuous with and analogous to the liquid state of that substance.

15.3.2 Importance of Glass

Generally glass is regarded as a weak material hence not very useful to the engineers. This view is not correct. Bullet proof and shatter proof glasses are well known. Glass has a unique place among the engineering materials because it has a combination of an unusual variety of properties. For example, it is hard and brittle when cold but on heating it softens and becomes so plastic that it can be given any desired shape by blowing, rolling or pressing. No other material lends itself to shaping in so many different ways. Glass is an excellent electrical and thermal insulator. It is resistant to many chemicals (which attack the metals) and organic solvents (to which organic materials are not resistant). Above all, its optical properties greatly increase its usefulness. Its refractive index can be controlled rendering it fit for making lenses.

15.3.3 Glassy State

In chapter I, one important characteristic of the solid state was described as the crystalline state in which the atoms are arranged in a geometrical or orderly manner. In the glassy state, (which for all practical purposes is a solid state) the atoms form a random three-dimensional network. The glasses are non-crystalline or amorphous.

X-ray diffraction techniques have thrown light on the structure of glass. Glass produces only a few diffused bands, showing that little order is present, whereas the X-ray spectrum of a crystal has many sharp lines. The crystals fracture along definite cleavage planes related to the structure of the crystal. On the other hand, when glasses break, the fracture cannot be predicted. Mostly it follows curves, except in flat glasses in which fracture lines are practically straight for considerable length. Thus the glassy state is a *vitreous* state. Appearance of tiny crystals in glass is called *devitrification*.

The most important behaviour of glasses, from a practical as well as theoretical point of view, is exhibited when they are heated or cooled. When they are cooled from the liquid state, viscosity steadily increases and the rigid, elastic condition of ordinary glass is reached without any sudden change. On heating from the solid condition, the glasses eventually soften and gradually become more and more fluidic. They do not have melting points (Fig. 15.2). Some arbitrary methods have been designed to fix some degree of mobility or ductility and to designate the corresponding temperature as *softening temperature* e.g. for the common soda-lime glasses it is about 700^0. In this respect also the glasses differ from crystalline materials which, at a certain fixed temperature, the melting point, suddenly absorb heat without rise of temperature.

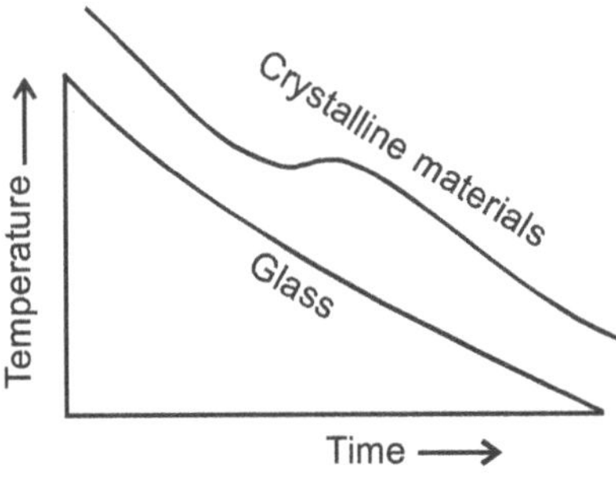

Fig 15.2

The glassy state, not being quite a stable state, tends to attain the crystalline state. If glasses are kept for a long period at a temperature slightly above the softening temperature crystals begin to appear or they *devitrify*. At the ambient temperatures also this process takes place but the rate is imperceptibly slow. The effect is seen only in glassware made centuries ago.

15.3.4 The Glass-Making Oxides

The glass-making oxides are classified into the following three classes.

1. *Network formers :* These can produce glass by themselves. These are oxides of such elements whose atomic diameters are such that they can be surrounded by four oxygen atoms each, in the form of a tetrahedron. They form chains and networks that may exist in a completely random manner. The rules stated by Dr. Zachriasen lay down four conditions : (i) each oxygen atom must not be linked to more than two cations, (ii) the number of oxygen atoms around any one cation must be small, not more than four, (iii) the oxygen polyhedra must share the corner positions, not the edges, to form a three-dimensional network, and (iv) at least three corners of each tetrahedron must be shared.

The most important network forming oxide is SiO_2. Other important ones are B_2O_3, GeO_2, P_2O_2 and As_2O_5. Other less important materials are Sb_2O_3, As_2O_3, Bi_2O_3 and BeF_2. All these on melting can give glass without the addition of any other oxide. Some others of this category are of purely theoretical interest. In general these are acidic oxides.

Silica may be considered to be the most important raw material for glass making, not only because it can produce glass by itself but also because of its availability at low cost. Deposits of glass-sand are available in different parts of India, in addition to other crystalline forms being available (see Refractories). Its melting point is high (1723^0) and molten silica is so viscous that the gas bubbles formed in melting are not easily moved. It is difficult to form many articles from molten silica because it becomes rigid so quickly that it is not easy to give the shape. Its high m.p. makes it suitable for making glass for use at high temperature. Its coefficient of thermal expansion is low, has good electrical properties and high transmission for ultraviolet light. Silica glass is expensive.

The most harmful and common impurity in glass sand is iron oxide. For glass of good colour Fe_2O_3 should be less than 0.04% and for optical glass less than 0.015%. Lime and alumina, once considered as impurities, are now not considered so because they are incorporated in glass compositions. It needs to be seen that they are uniformly distributed.

Boric oxide is another network former and about 20% of it is present in borosilicate glass, with a small percent of alkali, the remaining being silica. The glasses have low coefficient of thermal expansion, are resistant to corrosive action of acids, can be used at higher temperatures and are resistant to thermal shocks. These properties make the borosilicate glasses suitable for many industrial and laboratory applications. It is used in small amounts in other glasses mainly for giving lower coefficient of expansion.

2. *Network modifiers :* To this class belong a large number of oxides of elements whose atoms have larger diameters and, therefore, can have more than four oxygen atoms as their nearest neighbours. In other words their coordination numbers are higher. These melted alone cannot form glass as they form regular crystals but on melting with network formers can enter into the holes of the network. They loosen the bonds of the network formers with the result that the softening point as well as chemical resistance is lowered but the coefficient of thermal expansion is increased.

   Important network modifiers are the oxides of the alkali metals. Alkaline earths, lead, zinc and some other oxides are of lesser importance. In general these are basic oxides.

   Soda ash is the most widely used alkali metal oxide although for special purposes potash is also used. In tonnage its amount is next to silica. In soda of good quality, sodium chloride is less than 0.5% and iron oxide is negligible. For glass-making granular soda ash is preferred as it is not blown off easily by the furnace gases. The most important effect of soda is the fluxing action i.e. lowering the melting temperature. When it is used in larger amounts, it increases thermal expansion, and lowers tensile strength, elasticity and resistance to weathering.

   Lime and limestone come next to soda in glass-making on the basis of weight. Limestone of good quality is widely available in India. There are points in favour of both of them. Limestone is cheaper and is easier to store. Carbon dioxide evolved from it during glass-making helps to stir the material. Whereas use of lime saves heat of the furnace – which is a distinct advantage – and the transportation cost is less but the storage is more expensive. Good limestone should contain $< 0.1\%$ Fe_2O_3. Dolomite, which contains carbonates of calcium and magnesium, is also used.

   When added in correct amounts, lime contributes to glass stability or permanency, hardness, facilitates melting and refining.

   Intermediate glass formers : These also cannot form glass by themselves but on fusion with the network formers may act like network formers taking up their coordination numbers. Under suitable conditions they may even form glass with other non-glass forming oxides. For example, Al^{8+} may partially substitute for Si^{4+} in the network. The intermediates are amphoteric oxides.

Oxides of lead are prepared by heating lead in a current of air. Red lead, Pb_3O_4 is preferred to litharge PbO, because it helps to ensure oxidizing conditions and is free from metallic lead which is highly undesirable in glass. In most cases PbO content is less than 15% but can be as high as 80%. It increases brilliance and improves working quality. Lead glasses are used in electric lamps, neon signs, tubes, optical glasses and for absorbing X-rays.

Barium is introduced in the form of $BaCO_3$ which gives BaO in the glass melt. It is prepared by reducing $BaSO_4$ to BaS by heating with coke, which is then dissolved in hydrochloric acid and then $BaCO_3$ is precipitated by adding Na_2CO_3. It is used only in small amounts but is an important constituent of heat resistant and optical glasses.

Alumina is more widely used than any other intermediate oxides. Its deposits in the form of bauxite are abundant in India. It is purified by Bayer process (described in the Section on aluminium). The alumina so obtained is expensive but is free from iron oxide. The use of feldspar, $K_2O.Al_2O_3$, $6SiO_2$, and other clay minerals has increased in recent times not only because of lower costs but also because alkali, which is expensive, is supplied at the lower cost when a material like clay is used.

Up to 3% alumina can be advantageously used in the ordinary soda-glasses as it contributes to resistance to weathering decreases tendency to devitrification and lowers the thermal expansion. Alumina along with boric oxide is an important constituent of all types of low expansion glasses.

15.3.5 The Functions of the Glass-Making Oxides

In brief, it may he stated that silica is the best glass-former, but its melting point (1723^0) and also its viscosity on melting are very high. Alkalis act as flux and facilitate the melting of silica and also reduce the viscosity but resistance to chemicals, weathering, and even water are greatly reduced. For example, addition of 25% Na_2O to silica reduces its melting point to 850^0 but makes it water soluble. The alkaline earths, like lime, make glass resistant to water, weathering and chemicals. Thus silica, soda and lime are the most widely used raw materials in glass-making. The common glass is made of these raw materials and is known as soda-lime-silica glass. Other oxides are added to improve the properties or to incorporate certain desirable properties, like devitrification (i.e., prevention of crystallization or maintenance of the glassy state), reduction in coefficient of thermal expansion, modification of optical properties, high temperature properties, fluidity, durability, resistance to weathering and chemicals. Colourants may be added for making coloured glasses. Special glasses, of course, need special consideration. In the

Fig. 15.3 effects of common oxides used in glass making are represented in an over simplified manner. The effects are, however, not as distinct and sharply defined as the simple diagram may suggest but a fairly good idea of the role played by the common raw materials and also some idea about the guiding principles in choosing the raw materials can be had from it.

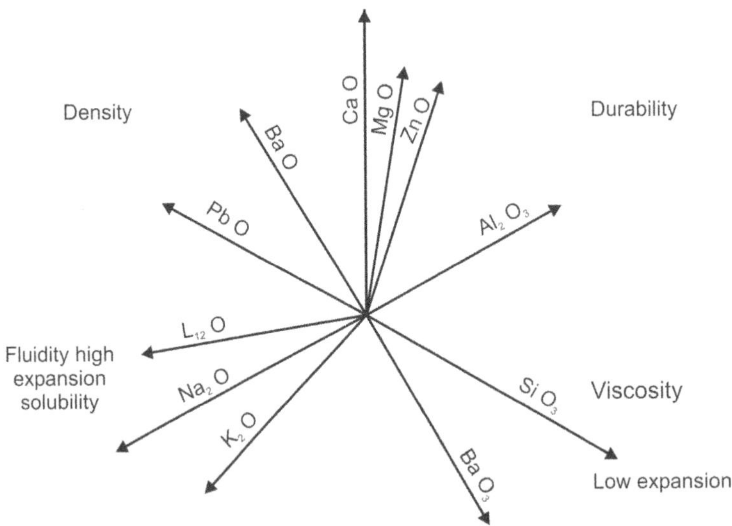

Fig. 15.3 Relative functions of glass-making oxides.
(Courtesy, *Modern Glass Practice* by S.R. Scholes)

15.3.6 Manufacture of Glass

From the above discussion it is clear that the *raw materials* for commercial glasses are of three types viz : (i) silica, (ii) alkaline fluxes (mainly soda and/or potash), and (iii) stabilizing oxides (one or more of the oxides of Ca, Mg, Ba, Al, Pb, Zn). Besides requirements for special purposes the raw materials are chosen to give the best conditions for rapid fashioning of the ware in the modern automatic machines and also taking into account : (a) easy fusibility, (b) freedom from crystallization, and (c) permanence of the product.

The raw materials are analysed, especially with a view to judge the purity and to take care of the objectionable impurities. They are then taken in the correct proportions. It may be noted that even in the case of simple common glass whose raw materials are sand, soda and lime, small amounts of several other oxides have to be added. For example, ferrous oxide, which is a common impurity, gives a greenish tinge. An oxidizing agent like sodium nitrate needs to be added to oxidize it to ferric oxide which gives a yellowish tinge and is less noticeable to the eye. At higher temperature the ferric iron dissociates to ferrous iron which gives a greenish tinge to glass. Oxides of arsenic and antimony upset the ferrous-ferric equilibrium and keep the proportion of ferrous low. Alternatively, manganese oxide is introduced which gives a pink colour complimentary to the green colour of ferrous iron. Molten glass will have a large number of bubbles due to the air entrapped between the grains of the raw material and also to the CO_2 liberated. Reduction in the viscosity helps to eliminate them but refining agents like arsenic oxide need to be added to help the bubbles to rise to the surface. Small amounts of feldspar may be introduced in the batch with a view to give certain desirable properties, which have already been discussed.

Besides purity of the raw materials, the particle size is of importance. Large grains may not react fully and remain behind as “stones" ("Stones" is the name given to any non-glassy material embedded in glass). Too finely divided material causes dusting. Materials should be fine and of uniform grain size. Waste or broken glass is always available in the factory. It is called "cullet" and forms, say, one-third to three-fourths of the total charge. Cullet has to be carefully sorted out before adding.

The *furnaces* used for glass-making are of two types. In the first type, separate pots, may be from 1 to 20 pots, are kept in the furnace. The pots may be open in which case the glass is heated from the surface also. The covered pots protect the glass from the furnace atmosphere. The pots of capacity up to two tonnes are in use, but for special glasses pots of smaller capacities are use.

The second type of furnace is the tank-furnace which has an open hearth over which flames pass from one end to the other. In the continuous tank furnace the raw materials are fed at a rate equal to that at which molten glass is withdrawn. This type of furnace is used where mechanical or continuous working on large scale is required.

The common fuels used are coal, producer gas and oil. The furnace is worked on a regenerative or recuperative system of heat economy, which has been explained in detail in the Section on open-hearth furnace for steel-making.

The refractory materials used in the construction of the pots and the furnaces are carefully chosen.

The batch, along with the cullet, fining agents and colourants is introduced in the furnace either by shovelling by hand or mechanically. During the melting, a series of reactions takes place as the temperature rises. Around 800^0 the alkali silicates are formed and limestone begins to decompose. Lime and other basic oxides gradually begin to enter the solution forming double silicates with the alkali silicates. As the temperature increases and viscosity decreases the excess of silica begins to dissolve. The gases, mostly CO_2, liberated due to the reactions with the carbonates violently agitate the molten mass. The reactions continue for some time after the maximum temperature is reached. In fact, molten glass is heated to 1400 to 1500^0 which are higher than is needed for the completion of the reactions.

This is done so as to reduce the viscosity of the melt and thereby facilitate the escape of the entrapped gas bubbles. Once the fusion process starts, it should proceed rapidly. Therefore the design and working of the furnace should be such that there is good transfer of heat to the charge.

The molten glass is cooled to several hundred degrees below the maximum temperature so that it has satisfactory viscosity for working. It is then given the final shape by various methods of glass working. The principal methods are: pressing in mould, blowing for making bottles and jars etc, drawing into sheets, tubes, rods etc., and spinning for making fibers.

Glass is a bad conductor of heat, Therefore on cooling the finished product great care has to be exercised. Rapid cooling will cause the surface to become hard and rigid whereas the interior still remains soft. Severe thermal stresses will result and the ware will be weak and break easily. It is therefore kept for annealing i.e., for slow cooling at a controlled rate, either in an oven or placed on a belt moving slowly in an annealing furnace. As the glassware moves, the temperature gradually drops to ambient temperature,

Glass is easily affected by the weathering action of the atmosphere and the chemicals. Its chemical stability is increased by : (i) the increase of lime and decrease of alkali, (ii) the addition of aluminium, (iii) replacement of small amounts of alkali with B_2O_3 and BaO or ZnO, (iv) the substitution of PbO for part of the lime, and (v) the use of a combination of soda and potash.

15.4 Pottery and Porcelain

Pottery and porcelain are taken together as their principal raw material is clay and also that their method of manufacture involves similar processes. *Pottery*, also known as earthenware, refers to all ceramic bodies that are not vitrified. Consequently they are slightly porous and coarser. The vitrified clay product is known as *stoneware* or *porcelain*. The distinction between these last two is very vague and in this section the term porcelain will be used. All the products may either be glazed or unglazed.

The igneous rocks contain *feldspar*. The common feldspar is the potash feldspar which may be approximately represented as K_2O, Al_2O_3. 6 SiO_2. In place of K_2O soda, Na_2O, or lime, CaO, may be present to give soda or lime feldspar. In the course of ages its slow chemical decomposition by the action of carbon dioxide and water, and mechanical disintegration gave *kaolinite* Al_2O_3. 2SiO. $2H_2O$. It occurs mixed with particles of feldspar, sand and other impurities and is washed to give white clay, *koolin or china clay,* which is the principal raw material for the pottery and porcelain industry.

Clay has two important properties. Firstly, when ground and mixed with water, it is plastic i.e. it can be moulded into the desired shape which it retains. Secondly, on firing it forms a brittle product which is highly resistant to corrosive agents that attack most metals and organic materials. The clay pots, when dried in the sun, become soft and collapse in contact with water but on heating to about 600^0 are unaffected by water. China clay is a refractory material and does not deform at high temperature, therefore it forms the framework of the porcelain or chinaware body. On heating to very high temperature the clay vitrifies or fuses to give a glassy, nonporous body. The temperature can be reduced by mixing with more readily fusible materials which seal the pores of the unfused material.

The earthenwares are made of cheaper raw materials and are fired to lower temperature (1000^0 to 1550^0). These are not vitrified.

Of all types of clays it is only koolin which yields a white translucent, vitreous ceramic when fired. For making porcelain 40 to 50% of it is mixed with fusible materials like feldspar and silicas of various types. To give an example, a composition containing equal parts of clay and feldspar vitrifies (i.e. fuses to form a glassy substance) at 1200^0; becomes translucent at 1250^0 and melts at 1400^0. About ten to fifteen percent of ball clays are mixed with the raw materials. These clays have a high degree of plasticity and make the body more workable and help in fusing or vitrification. Flint or pure silica needs to be added to lower the shrinkage, cracking and warping of the body during firing. Feldspar acts as a flux and is the first to melt and dissolve the other materials and fuse them into a vitreous mass. On cooling the ware, feldspar hardens and acts as a binding agent that increases hardness and translucency of the fired ware.

The raw materials in proper proportions are mixed with the correct amount of water and ground in a ball mill. The muddy mass is passed through a screen to remove particles of larger size and then between electromagnets so as to remove iron impurities which cause discolouration. It is passed through a filter press to remove extra water. This gives a soft, plastic mass of clay from which air bubbles are removed which could give defects in the finished product.

After giving the desired shape the wares are very carefully packed in the kiln for firing. Firing transforms the soft, dull-coloured clay body into a brilliant white translucent porcelain or chinaware. The temperature of firing required for maturing the body, glaze and decorations varies widely. For ordinary pottery it may be as low as 900^0 and for good porcelain 1500^0. The ware moves on kiln cars or movable trays through a kiln. It passes through zones of different temperatures, gradually increasing to the maximum and then cooling as the ware approaches the exit.

Industrial uses

Commonly one thinks of the potteries and porcelain mainly in terms of its use as domestic crockery. The annual production figures in India, in tonnes, are given after each item: crockery 15,700; sanitary wares 20,000; stoneware jars 4,200; stoneware pipes 47,200; glazed tiles 40,000; high tension insulators 18,000 and low tension insulators 1,300. Power generation and transmission requires special ceramic products and there is a rapid increase in the demand for increase for these uses.

15.4.1 Glazes

The surface of most of the ceramic bodies is uneven and difficult to clean, has pores and lacks luster and gloss. These defects can be removed by applying a *glaze* which is a continuous adherent coating or layer of a glassy material.

The glaze forming ingredients are made into a suspension in water. It is then applied on the surface of the piece and dried. On firing the piece, the ingredients react and melt to form a thin layer of glass. The firing of the glaze may be done at the same time as the body or the body is first fired, the glaze applied and fired for the second time.

The two widely used types of glazes are the *raw glazes* and the *fritted glazes*. The raw glazes contain ingredients which are insoluble in water. After grinding them in proper proportion they are made into a suspension in water and then applied on the body and fired. The compositions for fritted glaze contain some water soluble ingredients also. They are melted to give an insoluble glass and quenched in water. The material is ground with water to give a suspension ready for application.

The glazes are applied by brushing, dipping, spraying or pouring on the surface. A portion of the water is either absorbed in, the pores of the body or evaporated on a vitreous body. The composition of the frit should be such that it reacts with the body to form an intermediate layer between the body and the glaze itself.

15.4.2 Enamels

The term 'enamel' is used to designate a wide variety of materials, e.g. the enamel paints, the ceramic enamels. In this chapter by 'enamel' is meant a vitreous glaze or combinations of vitreous glazes which are used on a metallic surface. They are durable, protect the metal surface excellently and are washable. They may be white or of various colours.

Depending on the kind and proportion of the other oxides mixed with silica, the enamel may be soft or hard. They are termed soft when the temperature required to fuse them is low. In general the atmospheric agencies easily decompose the soft enamels. The hard enamels which fuse at higher temperature, are more resistant.

As the enamels are mostly applied on dark surfaces it is necessary, more particularly for white or light coloured enamels, that the enamel should give opaque coating. Otherwise thick coating will have to be applied which will increase the cost and at the same time will reduce the mechanical strength.

For obtaining opacity small particles having refractive index different from that of the glass are distributed in the enamel glass. The greater the difference in their refractive index the greater is the scattering power, consequently the greater is the opacity. One method of doing this is to add finely divided particles of zirconium and titanium compounds to the batch and grind well. During the short period of heating while applying the enamel, these do not dissolve in the glass.

A large user of the enamels is the *sheet steel* industry. Steel of low carbon content is preferred. It is given the desired shape and cleaned. The ground coat is prepared by first preparing a frit. It is a glassy material and is prepared by melting in a furnace and is then quenched in water. It is mixed in a ball mill with about 50% of its weight of water and a suspending medium, so as to give a suspension. The ground coat is applied on the surfaces by spraying and heating for 2-3 minutes at 830 to 885^0. It is cooled and a cover coat is sprayed and heated for about three minutes at a slightly lower temperature (800 to 850^0). The cover coat is also prepared in a similar manner by preparing the frit and suspending it in water with suspending agents. The compositions of the ground coat and the cover coat differ slightly from each other. Application of the ground coat can be avoided by treating the surface with a soluble nickel salt and applying the cover coat directly and firing. Although this cuts the cost, the mechanical properties are poorer.

For *cast iron* enameling, grey cast iron, which contains more free graphite due to higher content of silicon in the metal is preferred. A ground coat is given as for sheet metal. For smaller articles a cover coat is also given by spraying the suspension. For larger pieces it is more convenient to heat the piece and sprinkle powdered material on it which fuses as it falls on the hot metal and sticks in place. It requires great skill to apply it uniformly.

15.5 Abrasives

15.5.1 Definition

An abrasive is a material that can be used for cutting, grinding, surfacing or finishing another material. Most of them are hard and mechanically resistant but some are relatively soft. From the point of view of preparation of the material, the abrasives have been classified as *natural and artificial abrasives*. From the users point of view they are classified according to their *hardness*. The mechanism of hardness has been indicated in Chapter I. On the Moh scale, the hardness of diamond has been taken to be 10 and of talc as 1. The common uses of the abrasives are in the preparation of sand paper, sand cloth and grinding wheels which are widely used in the workshops. Annual production of abrasives in India is about 2,500 tonnes.

15.5.2 Hard Abrasives

Diamond

It is the hardest material and the purest form of carbon found in nature. When pure and free from defects it is highly valued as a gem. The diamonds are crystalline substances. Those which are used as gems have layer formations while in the industrial diamonds the crystal growth has taken place in all directions. So these do not crack during use. Diamond dust obtained from the cutting of gem diamonds is used as an abrasive for hard steel, for making diamond wheels for grinding and for cutting other gems. The black diamonds are dark grey and even black in colour and are used to greater extent for industrial purposes than the gem diamonds. *Bort* and *carbonados* are black diamonds which are considered as intermediate forms between diamonds and graphite. They differ in quality and hardness.

The industrial uses of diamonds range from making tips for drills used for cutting rock to cutting glass pencils. They are also used as teeth for saws and on abrasive wheels.

Boron Carbide

Among the hard abrasives, boron carbide, B_4C stands next to diamond in hardness. It is an artificial abrasive and is prepared by strongly heating boron oxide and coke in an electric furnace. It is very hard, hardness being 9.5 to 9.7, and crystalline material, and can resist the action of chemicals and heat. It is used for cutting hardened steel and tungsten carbide and for making nozzles for sand blasting.

Silicon Carbide

It is also called *carborundum*. Its hardness, 9 to 10, is slightly less than that of diamond. It is an artificial abrasive and is prepared by fusing a mixture of coke and sand in an electric furnace. After cooling, the mass is treated with strong sulphuric acid, washed and dried. It is used as an abrasive powder for making grinding wheels and polishing cloth and paper. An advantage of carborundum is that it is resistant to chemicals and can withstand high temperatures.

Aluminium Oxide

Aluminium oxide found in nature as an extremely hard (hardness is 9) and crystalline material is known as *corundum*. The clear crystals, whether coloured or colourless, are valued gems like the ruby, sapphire, etc. Aluminium oxide abrasive is also prepared artificially. The calcined bauxite (i.e., aluminium ore) is fused in an electric furnace. Molten alumina is allowed to solidify under controlled conditions. The impurities present in the ore help to bind the particles of alumina and thus contribute to its hardness. Corundum is used in grinding wheels, especially for grinding iron castings, and for abrasive paper and cloth. Mixed with bonding material it is used to make emery wheels for grinding purposes.

Emery

It is a naturally occurring impure variety of corundum Al_2O_3 containing high percentage of magnetite, Fe_3O_4. It is dark grey in colour. Because of the presence of magnetite its hardness 7 to 9, is less than that of corundum. Emery is mostly used for making emery paper and cloth of various grades. The grades depend upon the fineness of the particles. Graded emery powder is glued to one side of the paper or cloth to give emery paper or cloth.

Garnet

The term 'garnet' refers to a group of naturally occurring trisilicates mainly used for making abrasive cloth and paper. The garnets are not quite as hard as emery. Their hardness on Moh scale is 6.5 to 7.

Quartz and Flint

These are almost pure forms of silica, SiO_2 and have a hardness of about 7. Quartz is crystalline but flint is not crystalline. They are crushed to suitable size and glued on thick paper to give abrasive paper. Quartz and flint pebbles are used for grinding pigments in ball mills.

Metallic Abrasives

Among the metallic abrasives mention may be made of steel balls whose hardness ranges from 5 to 8.5. Steel balls are used in the ball mills for grinding cement clinkers, pigments, etc. Steel shots of smaller size are used for grinding and polishing stone. Still smaller shots are used in place of sand for cleaning the surfaces of castings, forgings. etc., by blasting (i.e. by directing the shots along with compressed air on the surface to be cleaned).

Sandstones

Sandstone, oilstone and pulpstone are more or less similar in composition. The sandstone consists of grains of sand joined together by natural binding materials, like clay, into a compact rock. Several varieties of sandstones are available. The efficacy of the sandstone as an abrasive is due to the fact that the sand particles, whose sharp edges cut the material, are not held very firmly by the binding material with the result that before they lose sharpness they are removed and new particles are exposed which are sharp. It is, therefore, necessary that the stone has the same hardness allover, otherwise the stone will grind more at certain spots than at others.

Oilstone has finer grains of silica than the sandstone. It is bluish white in colour and is quite commonly used for sharpening razors. The name *pulpstone* is given to large blocks of sandstone which are made into circular blocks for grinding wood for making pulp in the paper mills. The pulpstone should be of medium hardness and uniform texture and should have tearing action more than grinding action on the wood. The stone is aged for a year or two before use, or heated to about 90° for some time, for improving the binding properties of the natural binders.

Grains of *sand*, free from clay, are widely used for making sand paper and for sand blasting.

Pumice

Pumice is produced as a result of volcanic eruptions. Because of the escaping gases it has a porous structure and is of light weight. Its hardness on Moh scale is 6. It is used for fine polishing and for making metal polishes.

15.5.3 Soft Abrasives

Diatomite

It is also known as kieselguhr and diatomaceous earth. The remains of aquatic organisms, called diatoms, deposited and produced the diatomite. In diatomite the particles of silica are loosely held due to which its hardness on Moh scale is low, being I to 1.5. It is used as an abrasive in metal polishes. Besides being used as an abrasive, it finds use as a filler in soaps, plastics and paints.

Metal Oxides

The most widely used metal oxide abrasives are the various forms of oxides of iron. *Rouge* is a widely used abrasive. It is hydrated iron oxide and is reddish in colour. The material is quite hard but its grains are round in shape, hence it is used for fine polishing and is classed under soft abrasives. *Black rouge* and *green rouge* are Fe_2O_4, and Cr_2O_3, respectively, and are also used for fine polishing.

15.5.4 Selection of Abrasives

Abrasives function primarily by cutting the other material. Therefore hardness is certainly an essential property to be considered in selecting an abrasive, but hardness only is not enough for satisfactory service. Abrasion also depends upon the shape of the particles. The particles should have sharp and durable cutting edges which depend upon the mode of fracture through the crystals of the abrasive, during use. For instance, diamonds with layer formations are more valued as gems whereas for industrial application diamonds with crystal growth radially from the central point are preferred because these do not crack so easily. The usefulness of carborundum, for example, besides its hardness, depends on the fact that small chips break off and thus it maintains a sharp cutting edge.

The bond strength between the particles of the abrasives is also important. However hard an abrasive may be, its sharp edges are bound to be rounded off in course of use. It is therefore necessary that the particles should be dislodged and new particles with sharper edges be exposed. This point should be remembered while bonding the abrasive wheels, papers and cloths.

The temperatures of the surfaces rise during cutting, grinding or polishing. Sometimes the rise in temperatures is considerable and the properties of the abrasive and the material being worked or are adversely affected. The rise in temperature can be controlled to some extent by regulating the speed of operation. The problem is also tackled by the use of a suspension or slurry of the abrasive in a suitable medium

15.5.5 Grinding Wheels

For making grinding wheels, abrasives are selected depending on the use of the wheel. The abrasives are then bonded into the desired shape. The common types of bonds are : ceramic, silicate, shellac, resin and rubber bonds. The *ceramic bond* is produced by mixing the abrasive with clay and water to make a paste. It is then moulded into the wheel, and heated in a kiln when clay forms glass-like material which holds the abrasive particles together. The ceramic bond is most commonly used for making wheels because the product is hard and can withstand the effects of chemicals and temperature. The abrasive may be made into a paste by mixing with sodium silicate solution and then moulded into shape and dried. On drying silicic acid gel is formed which hardens to give the *silicate bond*. The wheel is then baked at 250^0. The silicate bond is weaker than the ceramic bond, hence the wheel is softer than that produced by ceramic bond and can be used only up to a lower temperature.

The wheels can also be made by mixing the abrasive with molten shellac and giving the desired shape. The *shellac bond* is weaker still and can be only used up to a much lower temperature. Shellac bonded wheels are used for producing fine and sharp edges, as those on the razor blades. Some thermosetting resins are also used for producing the *resin bond*. The resin is mixed with the abrasive, filler plasticizer, etc., and given the shape by the application of heat and pressure. The bond so produced is quite strong and in some cases compares well with the ceramic bond. *Rubber bonded* wheels are prepared by mixing raw rubber with the abrasive, filler and the vulcanizing agent and subjecting it to heat and pressure in a mould when the rubber vulcanizes. The wheels are used for finer and high speed work.

15.5.6 Abrasive Papers and Cloths

Abrasive papers and cloths are prepared by first coating a stiff paper or cloth with an adhesive like glue. The abrasive of suitable particle size is then uniformly spread over it. The sheet is dried and a very thin adhesive is spread over the surface and dried again. The sheet is then cut into suitable sizes. About 2 lakh reams of coated abrasive are produced annually in India.

Chapter 16

Conductors and Insulators

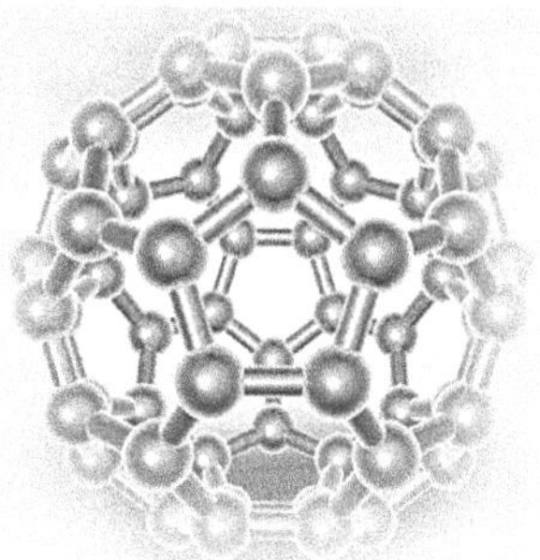

16.1 Electrical Conductors & Insulators (Dielectrics)

A metal rod held in hand and rubbed with fur will not seem to develop a charge. The same rod when held with a glass or rubber handle and rubbed without any human or metal contact gets charged. The explanation to this phenomena is that the metals and human body are *conductors* of electricity and glass and hard rubber are *insulators* or *dielectrics*. In a conductor, electric current can flow freely. Metals such as copper typify conductors (Fig. 16.1). Metals are classified as conductors because their outer electrons are not tightly bound. Experimental proof of *Hall effect* shows that only electrons are free to move and the positive charge is immobile in metals. For conductors like electrolyte, we find movement of both positive and negative charges.

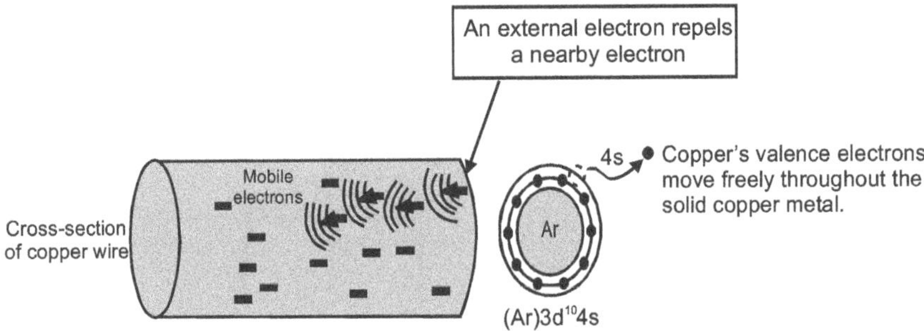

Fig. 16.1 Conduction in Copper.

Insulators offer very large resistance to the flow of electric current since their outermost electrons are so tightly bound that there is essentially zero electron flow through them at ordinary voltages. Although there are no perfect insulators, the insulating ability of fused quartz is about 10^{25} times as great as copper.

Material	Resistivity (ohm m)
Glass	10^{12}
Mica	9×10^{13}
Quartz (fused)	5×10^{16}

Insulators (have high resistivity)

Material	Resistivity (ohm m)
Copper	1.7×10^{-8}

Conductors (have low resistivity)

The materials having intermediate properties between conductors and insulators in their ability to conduct electricity are called *semi-conductors*.

16.1.1 Charecteristics of Insulators (Dielectrics)

1. *Resistivity* It is the resistance offered by a unit cube of the material. Insulators have resistivities ranging between 10^6–10^{17} ohm.m at room temperatures.
2. *Dielectric Polarisation* In the presence of applied electric field the dielectrics undergo separation of charge called dielectric polarization. Here the dielectric is internally charged and non-equipotential but prevents any flow of charge through them. If Q is the charge collected in the presence of an applied voltage V, then the capacitance of the material is $C = Q/V$.
3. *Dielectric Constant* The capacity of a material to store electric charge in the presence of an applied electric field is called dielectric constant. It is the ratio of the electrical capacity of the condenser filled with the material to that of the electrical capacity when evaculated. A good electrical insulator should have a low dielectrical constant.
4. *Dielectrical Strength* The ability to resist puncture or rupture of insulating material towards applied field is called dielectric strength and is expressed in volts/mm. Dielectric strength decreases with increase in thickness of material, time of application of charge, humidity and temperature. In case of liquid insulators it also depends on the viscosity of the material and increases with increase in viscosity. Ageing and weathering of the material is the consequence of decrease in dielectric strength.
5. *Dielectric losses* These are caused by internal dipole friction called absorption of electric energy and by leakage of current through the medium. A good insulator should have minimum dielectric losses.
6. *Thermal Expansion & Contraction* An ideal insulating material should have minimum thermal expansion and contraction, since they lead to gaps in the insulator.

7. *Thermal & Chemical Ageing* – Breakdown of insulator with temperature and oxygen content and contact of other chemicals is called ageing and makes it less serviceable for long time periods.

8. *Porosity* – Porosity increases moisture holding capacity of the insulator, thus leading to unwanted results.

9. *Mechanical Requirements* – Lower surface tension of the insulator permits greater wetting of the electrical component and thus giving better cooling. The surface tension of dielectrics is generally modified by the addition of wetting agents. In case of solid insulators surface roughness as well as hardness are unwanted properties of dielectrics.

16.1.2 Classification of Insulating Materials

The insulators are classified depending on their physical state into Gaseous, Liquid and Solid insulators.

(i) *Gaseous Insulators* All the gases and air are good insulators of heat and electricity. Air is a good insulators when the voltages are not very high. For these applications generally inert gases and nitrogen are used. Power cables are filled with nitrogen at pressures ranging from 10-200 psi depending on the voltage. Presence of oxygen and halogens damage the electrical equipment through their chemical reactions and are generally avoided. Hydrogen is a light gas and has good coolant capacity. It is therefore used used in large turbo engines and synchronous condensers. Organic gases are preferred owing to their chemical reactivity and combustibility. Gases like hexafluoro sulphur , SF_6, are non-explosive, non-inflammable and non-toxic and have better cooling capacity than air and nitrogen. Its dielectric strength is 2.35 times greater than air. It has corrosive properties and is not suitable under corrosive working conditions and temperatures.

(ii) *Liquid Insulators* These insulators have better heat transfer properties than gas dielectrics but suffer from chemical breakdown. So they are often used in conjunction with solid stabilizers. We have different dielectrics of this class covering different temperature ranges (Table 16.1).

Table 16.1

Insulating liquid	Operating temperature range
Mineral oil	–50 to 110 °C
Askarels	–50 to 110 °C
Silicone liquids	–90 to 220 °C
Fluorinated liquids	–50 to 200 °C
Synthetic hydrocarbons	–50 to 110 °C
Organic esters	–50 to 110 °C
Vegetable oils	–20 to 100 °C

(a) ***Mineral oils*** Their insulating properties are influenced by the presence of hydrocarbons, oxygen, sulphur, etc. They have low dielectric constants (2.5). Used in the transformers.

(b) ***Askarel*** These are noninflammable, synthetic insulating liquids. They have excellent arc resistance and have high dielectric strength and low dielectric constant (4-6). They have excellent thermal, electrical and chemical stability.

(c) ***Silicone fluids*** They are non-corrosive to metals up to 200^{0}C. They are water like with oily consistency. On break down they yield considerable amount of gases and substances like silicon carbide, silicon dioxide. Because of this disadvantage the silicones are not suitable as switch gear oils. They are used in radio, aircraft and pulse transformers.

(d) ***Fluorinated liquids*** They are non-inflammable and chemically inert, and their vapors are non-explosive. They are highly volatile and decompose in the presence of moisture. They find use in small electric and radio devices.

(e) ***Vegetable oils*** These are cheap and oldest insulating liquids. The drying ois such as linseed oil, tung oil are used in making insulating varnishes used in the transformers. The non-drying oils such as castor oil, coconut oil, etc., are popular plasticizers in insulating resin components.

(iii) Solid Insulators Most of the non-metals and compounds like mica, polymers, silk, paper, asbestos, etc, fall under this category. These are capable of retaining their insulating properties up to a fixed temperature called limiting temperature.

(a) ***Cellulosic Materials*** Cellulosic materials like cotton, paper, etc., in their raw form have thermal stability with limiting temperatures of 90°C. When they are impregnated with moderate heat resistant materials this value increases to 105 °C.

(b) ***Fibrous Materials*** Natural fibers like wool, jute, etc., and synthetic fibres like rayons, nylons, fiberglass, etc., show variable limiting temperatures depending on the impregnating and bonding materials as shown in the Table 16.2.

Table 16.2

Insulating Materials	**Limiting Temperature**
Cellulosic materials like cotton, silk, paper that are not impregnated.	90 °C
Silk, paper, cotton, nylons, etc., impregnated with moderate heat resistant materials.	105 °C
Fibrous materials like Asbestos, fibre glass, etc., impregnated with varnishes of less heat resistance materials.	130 °C
Mica, marble, ceramics, glass, teflon	>180 °C
Enamelled wire insulation on polymer resin bases	120 °C
Polymer resins bonded with varnishes made of high heat resistant materials	155 °C
Fibre glass, mica and asbestos impregnated with silicones and silicone rubbers.	180 °C

(c) ***Synthetic Polymers*** Polymers have low dielectric constants ranging from 2 to 5. Normal polymers like polyethylene and polystyrene are excellent insulators for low temperature operations but undergo softening at higher temperatures. They are used as thin films and find their use in chokes, capacitors, electrical bushings and telephones.Teflon is a good synthetic polymer that could withstand temperatures above 180^0C and has a dielectric constant of 2. Epoxy resins possess excellent electrical and mechanical properties and are widely used in making insulators, bushings, etc., for high voltage operations. Laminates and insulating varnishes are also prepared using epoxy resins.

(d) ***Rigid Insulators*** Glass, ceramics and refractories fall under this category. Glass has excellent electrical properties. Pyrex glass even has good chemical stability. Their major application is found in lighting systems. Glasses have similar coefficient of expansion as metals and are therefore used as insulating materials and metal claddings. Normal porcelains are used for low voltage applications while at high voltages glazed ceramics are used. Zircon and alumina porcelains are used in spark plug cores. High dielectric constant porcelains (15-5000) are prepared using titanates and are used for high capacity condensers with small spaces. Mylax is a mica bonded with lead glass that can be used for operations up to 600 ^{0}C and has good insulating properties with mechanical stability.

(e) ***Impregnating Coatings and Bonding Materials*** These are synthetic high molecular weight organic and organo-metallic compounds. Bitumen, resins, drying oils etc., are known as insulating varnishes. They not only enhance the insulating properties but also help in stopping ageing of cheaply available materials like paper and cotton and reducing hygroscopicity of materials. Bonding materials are generally solutions of phenolic resins in alcohol and are used in low capacity electrical units. Other adhesives too could be used as bonding materials for low voltage and low temperature applications.

16.2 Thermal Conductors

Heat transfer by conduction involves transfer of energy within a material without any motion of the material as a whole. The rate of heat transfer depends upon the temperature gradient and the thermal conductivity of the material. Thermal conductivity is a reasonably straight forward concept when you are discussing heat loss through the walls of your house, and you can find tables which characterize the building materials and allow you to make reasonable calculations.

More fundamental questions arise when you examine the reasons for wide variations in thermal conductivity. *Gases* transfer heat by *direct collisions between molecules*, and as would be expected, their thermal conductivity is low compared to most solids since they are *dilute media. Non-metallic solids* transfer heat by *lattice vibrations* so that there is no net motion of the media as the energy propagates through. Such heat transfer is often described in terms of "*phonons*", quanta of lattice vibrations. Metals are much better thermal conductors than non-metals because the same mobile electrons which participate in electrical conduction also take part in the transfer of heat.

Conceptually, the thermal conductivity can be thought of as the container for the medium-dependent properties which relate the rate of heat loss per unit area to the rate of change of temperature.

$$\frac{\Delta Q}{\Delta t\, A} = -k\frac{\Delta T}{\Delta x}$$

where, $\Delta Q/\Delta t\, A$ is power per unit area transported;

$K =$ k is thermal conductivity;

$\Delta T/\Delta x$ is temperature gradient.

For *an ideal gas* the heat transfer rate is proportional to the average molecular velocity, the mean free path, and the molar heat capacity of the gas.

$$K = \frac{n(\mathrm{v})\,\lambda\, C_V}{3N_A}$$

where, $K =$ Thermal conductivity;

$n =$ Particles per unit volume;

$\lambda =$ Mean free path;

$C_V =$ Molar heat capacity

$N_A =$ Avogadro's number

$v =$ Mean free path

For *non-metallic solids*, the heat transfer is viewed as being transferred *via lattice vibrations, as atoms vibrating more energetically at one part of a solid transfer that energy to less energetic neighboring atoms*. This can be enhanced by cooperative motion in the form of propagating lattice waves, which in the quantum limit are quantized as phonons. Practically, there is so much variability for non-metallic solids that we normally just characterize the substance with a measured thermal conductivity when doing ordinary calculations.

For *metals*, the thermal conductivity is quite high, and those metals which are the best electrical conductors are also the best thermal conductors. At a given temperature, the thermal and electrical conductivities of metals are proportional, but raising the temperature increases the thermal conductivity while decreasing the electrical conductivity. This behavior is quantified in the Wiedemann-Franz Law:

$$\frac{k}{\sigma} = LT$$

or $$L = \frac{k}{\sigma T} \quad \text{Wiedemann-Franz law}$$

where, k = thermal conductivity

σ = electrical conductivity

L = Lorenz number

where the constant of proportionality L is called the *Lorenz number*. Qualitatively, this relationship is based upon the fact that *in metals the heat and electrical transport both involve the free electrons present in the valence band.* The thermal conductivity increases with the average particle velocity since that increases the forward transport of energy. However, the electrical conductivity decreases with particle velocity increases because the collisions divert the electrons from forward transport of charge. This means that the ratio of thermal to electrical conductivity depends upon the average velocity squared, which is proportional to the kinetic temperature.

Thermal conductivity values of some materials Table 16.3.

Table 16.3

Material	Thermal conductivity (cal/sec.cm^3 ^{0}C/cm)	Thermal conductivity (W/m K)*
Silver	1.01	406.0
Copper	0.99	385.0
Brass	...	109.0
Aluminum	0.50	205.0
Iron	0.163	...
Steel	...	50.2
Lead	0.083	34.7
Mercury	...	8.3
Ice	0.005	1.6
Glass,ordinary	0.0025	0.8
Concrete	0.002	0.8
Water at 20 °C	0.0014	...
Asbestos	0.0004	...
Hydrogen at 0 °C	0.0004	0.14
Helium at 0 °C	0.0003	0.14
Oxygen	...	0.023
Snow (dry)	0.00026	...
Fiberglass	0.00015	0.04
Brick,insulating	...	0.15
Brick, red	...	0.6
Cork board	0.00011	0.04
Wool felt	0.0001	0.04
Rock wool	...	0.04
Styrofoam	...	0.01
Wood	0.0001	0.12-0.04
Air at 0 °C	0.000057	0.024

*From Young, Hugh D., University Physics, 7th Ed.

16.3 Superconductors

Onnes, a Dutch physicist, discovered in 1911 that mercury loses all resistance to electrical flow when cooled to approximately 4 K. A current, once started, will flow continuously and this phenomenon is known as superconductivity. At ordinary temperatures, metals have some resistance to the flow of electrons, due to the vibration of the atoms that scatter the electrons. As the temperature is lowered, the atoms vibrate less and the resistance declines smoothly, until a critical temperature, T_c, is reached. The value of this temperature depends on the material that is being cooled. At this point, the resistance drops abruptly to zero (Fig. 16.4).

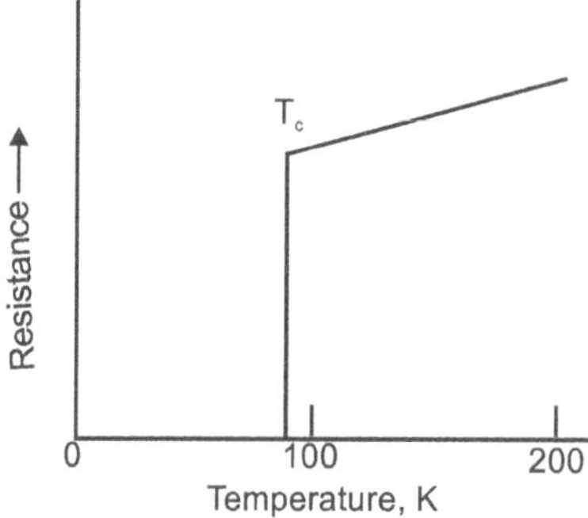

Fig 16.4 Electrical resistance of a superconductor.

If an electrical current is started in a superconducting ring, it will continue forever.

Superconductors are also perfectly diamagnetic and repel a magnetic field. This property was discovered in 1933 and is known as the Meissner effect. A magnetic field induces a current in a conductor; conversely, a current induces a magnetic field. When a magnet approaches a superconductor, it induces a current in the superconductor. Since there is no resistance to the current, it continues to flow, thus inducing its own magnetic field which then repels the magnet's field. If the magnet is sufficiently small and strong, the repulsion will be enough to counterbalance the pull of gravity and the magnet will levitate above the surface of the superconductor.

Until recently, the highest known critical temperature was 23 K, observed in the intermetallic compound Nb_3Ge. Superconductors nonetheless found a number of applications, the most common being superconducting magnets for nuclear magnetic resonance (NMR) machines. NMR's are used extensively in research and for medical imaging. These instruments require liquid helium as the refrigerant, which is expensive. In a major breakthrough in 1986 (Nobel prize in 1987), Bednorz and Müller at IBM Zurich discovered superconductivity in copper-containing oxides at temperatures above 30 K, instigating a massive worldwide research effort. This culminated in the discovery of a metallic oxide of yttrium, barium, and copper that was superconducting at about 92 K. Liquid nitrogen (BP 77 K), which is cheaper and easily handled in, for example, a Dewar flask, could therefore be used as a coolant rather than helium. Related oxides now superconduct up to 133 K.

The compound, which has the formula $YBa_2Cu_3O_{7-\delta}$, is readily prepared by heating an intimate mixture of yttrium oxide, barium peroxide, and cupric oxide at approximately 930°C for 10-12 hours. At this stage, the crystalline structure is formed by the interdiffusion of ions, but has an oxygen deficit ($\delta \sim 0.5$). Cooling the material to 500°C, and annealing at this temperature for 10-12 hours allows it to react further with oxygen from the air, reducing δ to less than 0.1. The overall reaction (with only the metal ions balanced) is given by:

$$0.5\ Y_2O_3 + 2\ BaO_2 + 3\ CuO \rightarrow YBa_2Cu_3O_{7-\delta}$$

This compound is often called the 1-2-3 material, from the molar ratios of Y:Ba:Cu.

Chapter 17

Cements and Limes

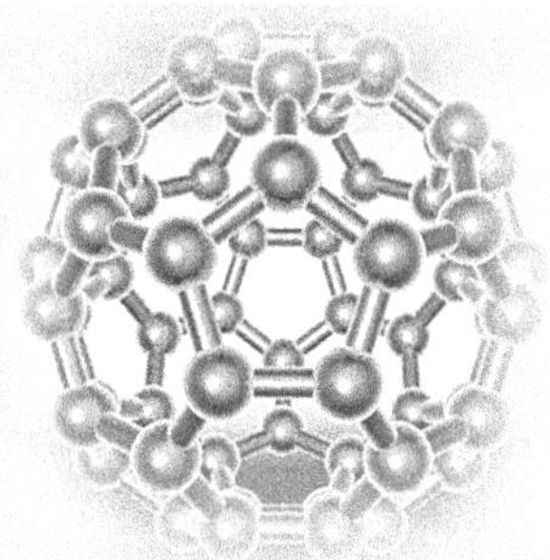

17.1 Introduction to Cements

The term 'cement' and 'concrete' are often loosely used. One frequently speaks of a cement road. It is really speaking about a concrete road, not a cement road. Portland cement is the most important ingredient of concrete. Concrete is used as a material of construction more than any other material, e.g., weight of concrete used in construction is 5 to 10 times the weight of steel used. About a tonne of concrete is used per human being per year. 33 million tonne of Portland cement was manufactured in India in 1985 as compared to 2.7 million tonnes in 1951. To meet the increased demand of the cement, attempts are being made to increase the installed capacity to 60 m.t.

Various limes and mortars have been used for construction since ancient days. The earlier materials did not set under water. It is said that the Romans were the first to prepare cement, a hydraulic lime, by mixing lime paste and volcanic ash, that would set under water. In some pyramids lime plaster used by the Egyptians 4,500 years ago is still in good condition. In 1824 Joseph Aspdin, an English bricklayer of Leeds, prepared a cement in his kitchen by strongly heating a mixture of limestone and clay. The product on powdering and mixing with water would set under water. Aspdin patented it and gave it the name Portland cement because its colour resembled the colour of stones from Portland, England.

Portland cement is a product obtained by calcining a suitable mixture of lime bearing (calcareous) and clay bearing (argillaceous) materials and adding small amounts of other approved materials (like gypsum) and finally powdering the mixture. This powdered mixture hardens into a solid mass on reacting with water and gives a water resistant product. Thus the basic raw materials for the cement manufacture are lime, silica and alumina. The use of Portland cement has increased so greatly because of its excellent properties, cheapness and abundance of the raw materials used in its manufacture.

The uses of concrete are too well known to be narrated here. Large structures for which no other material is available have been possible on account of the unusual and varied properties of concrete. To give just one example, the tallest chimney constructed in India, at Singrauli, is 220 m high. Concrete is such a versatile construction material that it is used in quantities larger than any other manmade construction material. An estimate puts its use in India in 1986 as 70 million cubic meters.

The biggest cement factory in India, with an annual capacity of 1.7 million tonnes of cement is at Chunar, Dist. Mirzapur, U. P. Clinkers are produced at Dhala, which is surrounded by limestone deposits and is very close to the Singrauli coalfields. The clinkers are transported to Chunar and ground with blast furnace slag (8 lakh tonnes per year) brought from Bokaro steel plant and gypsum from Rajasthan. There are already two other cement factories, in the public sector, producing 8 lakh tonnes of cement annually.

17.1.1 Manufacture of Portland Cement

Raw materials for the manufacture of Portland cement fall into three groups :

1. *Calcareous*, i.e., those which contain lime. Lime-stone is mostly used for this purpose. The limestone or chalk should not contain more than 5% of magnesium carbonate as it gives free MgO in the final product. Magnesium oxide very slowly reacts with water and expands causing cracking of the concrete. Presence of free silica in lime-stone is undesirable as it does not react easily but silica combined with alumina is helpful. Presence of sulphur is undesirable.
2. *Argillaceous*, i.e., those which contain silica, alumina and iron oxide. Mostly siliceous clay is used for this purpose. Bauxite is added if the proportion of alumina is not sufficient. Blast furnace slag is also used.
3. *Gypsum* is added during final grinding. It controls the ratio of setting and hardening.

The solid state reactions taking place between the raw materials during the manufacture of cement are promoted by iron oxide which chiefly acts as a flux. The reactions between the limestone and the clay would take place at a much higher temperature without the fluxing action of iron oxide. Therefore, presence of iron oxide in the raw materials is desirable.

For every tonne of iron 750 to 800 kg of slag is produced. Accumulation of this slag produces a serious disposal problem, hence manufacture of slag cement has been started. The molten slag from the blast furnace is poured in water when it gets granulated, i.e. broken into small porous pieces. Generally, basic slag containing about 50% CaO, 30% SiO_2 and remaining alumina plus iron oxide with less than 3% of MgO is used. Acid slags are used to a lesser extent. The Associated Cement Company is making cement at Jamshedpur using blast furnace slag. A cement factory with a capacity of 500,000 tonnes per annum is working under Bhilai plant of Hindustan Steel Ltd. using the blast furnace slag. As the slag from the Rourkela plant contains manganese, it was considered unsuitable for cement manufacture, but a scientific team of Geological Survey of India has reported its feasiblty. For making slag cement both the granulated slag and limestone are ground and calcined in the cement kiln or the slag and Portland cement clinkers are ground separately and mixed. For instance, at Durgapur 0.7 million tones of slag cement is manufactured by mixing nearly equal quantities of granulated blast furnace slag and Portland cement clinkers and grinding to fine powder in the usual way.

The proportion of calcareous and argillaceous material to be used depends upon their composition and the properties desired in the final product. It may be said that on an average for each tonne of cement produced about 0.2 tonnes of gypsum, 0.35 tonnes of coal and 3 tonnes of water and 80 KWH of electricity are needed besides limes and clay minerals. There are two processes of manufacture.

The *dry process* is used to a considerable extent in the U S A. but not in India or in Europe. In India wet process is most widely used. A few plants using dry process have been installed and the recent trend is in favour of the dry process mainly due to the economy in the fuel costs. In this process the raw materials are crushed, dried, ground and mixed in the proper proportion. The mixture is then finely powdered and fed into a rotary kiln 50 to 80 metres long and 3 m in diameter. The kiln is inclined and is rotated at the rate of about one revolution per minute. Powdered coal is blown from the lower end and combustion gases escape from the upper end. In the first half portion of the kiln limestone decomposes. Alumina and silica combine at 1400 to 1500^0 with lime giving the fused product. The clinkers are removed from the lower end, cooled, ground with about 4 to 5% gypsum and marketed. In the dry process the cost of fuel is less than in the wet process.

The *wet process* for manufacturing Portland cement is used to a very much larger extent than the dry process. In this process greater control is obtained over the composition and the product is more homogenous. The cost of grinding the raw materials is less than in the dry process but the kiln has to be longer and more water has to be removed, therefore fuel consumption is greater. The choice between dry and wet process depends mainly on the cost of fuel and the cost of grinding. If the raw materials available are hard, the wet process will cost less to grind the materials than the dry process. In many cases the overall advantages of the wet process are greater than those of the dry process.

The raw materials are carefully proportioned to give the required amount of lime, silica, alumina and iron oxide in the final product. Iron oxide and alkali oxides reduce the reaction temperature. Limestone after excavation from quarries is crushed and the size reduced to below 2 cm. It is fed along with washed clay in a grinding mill, water is added, and the materials are ground while wet. This gives a mud-like mixture, called slurry, containing 30 to 50% water. The slurry is collected in a tank and if necessary its composition is adjusted

As the temperature of the kiln is below the fusion temperature of the products, it is necessary to grind and mix the materials well so that chemical reactions may take place more easily. The slurry is fed directly into the cement kiln and calcined up to a temperature of about 1400 to 1500^0 when the materials react chemically and hard lumps called clinkers are obtained.

Rotary kilns (Fig. 17.1) are used for burning cement. A rotary kiln is cylindrical in shape, 2 to 4 metres in diameter and 70 to 150 m long. It is made of steel plates lined with firebricks. It is inclined (inclination 1 in 25, and mounted on heavy friction rollers and rotated at a speed of about one revolution per minute. At the upper end of the kiln is a flue through which the combustion gases escape after the removal of dust. At the lower end there is an opening through which the clinkers fall. There is a firing pipe at the lower end for blowing powdered coal or oil along with air. The hot gases leave through the flue at the upper end at a temperature of about 700° and the clinkers fall from the lower end at about 1200°.

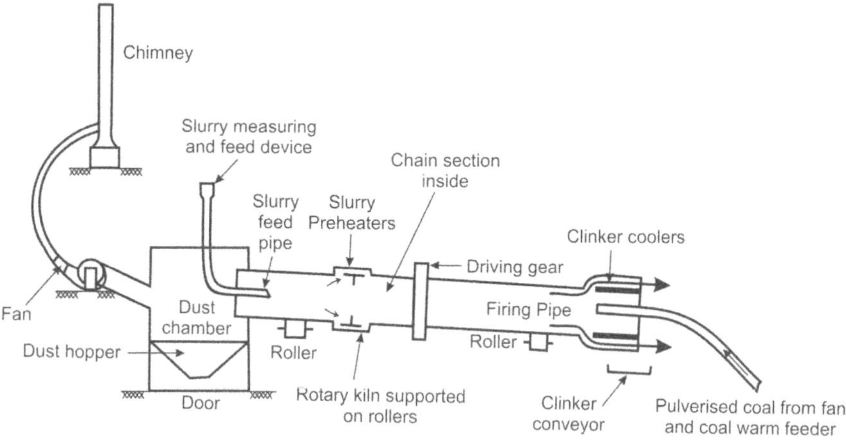

Fig. 17.1 A cement kiln.

The slurry is stored in a tank situated at a height so that it can be fed in the kiln by gravity. It passes through a measuring and feed device and a pipe (which is sloping towards the kiln because the slurry is viscous) leads it into the upper end of the cement kiln. At this end there are slurry pre-heaters. These are chambers riveted on flanges on the kiln body. These pre-heaters have inlet and outlet passages into the shell of the kiln with grates at the inlet and outlet ends. The slurry enters from the upper end and hot gases enter from the lower end. They come in contact inside the preheaters with the result that heat of the hot waste gases is transferred to the slurry. Water is thereby evaporated and the material gets heated. As the materials travel down the kiln they meet gases at higher and higher temperature, moving in the opposite direction. Decomposition of the carbonates takes place at about 900°. As the mass moves downwards, formation of the silicates and aluminates takes place above 1100°. To ensure complete reactions temperature is kept to about 1500° at the lower or firing end. The fused mass solidifies into small lumps, greenish black in colour. These lumps are called cement clinkers. At the lower end of the kiln 16 clinker coolers are attached to the kiln shell, all round the periphery. From the outlet of the kiln the clinkers enter the coolers. There are chains hanging inside the coolers. The clinkers drop down from the lower end of the coolers through which some air enters. Air helps to cool the clinkers and then enters the main kiln and there takes part in combustion reactions. Only partial cooling is accomplished in the clinker coolers. The clinkers drop on a shaker conveyer on which they cool further and are taken to ball mills for pulverising. The materials take a few hours to move from the feed to the discharge end of the kiln.

High efficiency dust filters need to be installed before discharging the gases into the atmosphere so as to control air pollution

The powdered clinkers on mixing with water will set very quickly and the cement mortar will unworkable. Therefore it is necessary to add some retarders to slow down the setting process, allowing time for placing cement mortar in the desired place. Gypsum, $CaSO_4.2H_2O$, is commonly used as the retarder. Usually about 4 to 5% of gypsum is mixed with the clinkers and the product is ground finely and stored.

Powdered coal is mostly used for heating but oil is also used. The coal received from the mines is crushed and taken by belt conveyer to ball mills and pulverised. This powdered coal is stored in coal bins from which it is fed by a worm conveyer into a pipe. Air is also forced into this pipe by a fan. This pipe, called firing pipe, feeds the air draft mixed with powdered coal in the kiln from the lower end. The flames and hot gases move upwards and supply the heat necessary for the reactions.

The reactions taking place during firing are complicated but may be represented thus :

$CaCO_3 + SiO_2 + Al_2O_3 + Fe_2O_3 + [H_2O]$
(limestone) (clay)

→ CO_2, H_2O

$3CaO.SiO$,
tricalcium silicate
$2CaO.SiO_2$
dicalcium silicate
$3CaO.Al_2O_3$
tricalcium aluminate
$4CaO.Al_2O_3.Fe_2O_3$
tetracalcium alumino ferrite

Cement for common purposes contains about 63% CaO; 22%, SiO_2; 6% Al_2O_3; 3% Fe_2O_3; 3% MgO, the balance being other constituents. As will be discussed presently, the composition may be widely different for certain uses. According to Meade, the ratio of the silica to the alumina content in the cement should be between 2.5 and 4 to 1 and iron should not be more than alumina. Magnesia and alkalis should each be less than 3%.

17.1.2 Vertical Shaft Kiln Technology

Cement plants using rotary kilns require large capital investment. *Mini Cement Plants* are being set up in the country for which vertical shaft kilns have been proposed. Such kilns are already functioning satisfactorily in some of the large cement factories as at Dalmiapuram in Tamil Nadu. These utilize low volatile coal, coke or char made of lignites or even the agricultural wastes. The capacity of the mini cement plants may range from 40,000 to 140,000 tonnes per year. A tiny cement plant of capacity 20 tonnes per day, installed on a plot of about 2000 m^2 near Jodhpur at an estimated cost of Rs. 20 lakhs, has gone in production in April 1982. It also employs the V S K technology.

The vertical shaft kiln Technology for manufacture of cement was in vogue prior to 1920 but with the advent of rotary kiln technology it went into oblivion. In 1950 with the invention of pan pelletizer or pan nodulizer or sometimes called dish granulator, where uniform pellets are made and fed into the kiln, vertical shaft kiln technology was once again revitalised for the manufacture of quality cement comparable to the rotary kiln cement. It is necessary to proportion the raw materials correctly to have uniform quality of raw meal in the nodules to achieve good burning conditions in the kiln and thereby improve the quality of clinkers produced. There are two processes in vogue in V S K technology : one is called the "black meal" process in which the raw materials, i.e., limestone, clay and any correcting material like bauxite or laterite, are ground along with coal in the desired proportion in a ball mill. The meal obtained from the ball mill contains coal particles which are black compared to raw meal powder, hence the name "black meal". It is then made into nodules on the pan nodulizer with about 15% water.

In another process called "fuel slurry" process the raw meal is ground separately either in a tube mill or vertical roller mill and coal is ground wet, with the addition of equal amount of water in another tube mill. The raw meal and the coal slurry are made into pellets on a nodulizer. The pellets or nodules prepared by either of the processes are then fed into the kiln through a rotating feeder spread over the periphery of the shaft which is lined with refractory, bricks and the nodules while passing through the shaft get well sintered and then discharged through a slowly rotating grate followed by cold air blown through the discharge unit to cool the clinkers. In doing so the cold air gets heated, and this air is utilised for combustion purposes in the kiln. Clinkers are then stored and fed to cement mills in the usual way and ground along with 4-5% of gypsum. Some of the plants adopting this technology claim the following advantages :

1. The plant is compact, the kiln being only about 10 metres tall, the requirements of buildings, foundations, structures, etc., are greatly reduced.
2. There is no dust discharged into the atmosphere.
3. Low melting eutectics are not formed. Therefore there is no clogging of clinkers to the firebrick lining in the hot zone. This saves manual labour and loss of production and reduces consumption of firebricks.
4. As the shape of the clinkers can be regulated there is saving in the cost of grinding and the wear of the grinding machinery is less.

17.1.3 The Setting and Hardening of Portland Cement

Portland cement on mixing with about one third its own weight of water is converted into a plastic mass, called cement paste, which slowly loses its plasticity and becomes stiff and ultimately a rock-like mass is obtained. The stages in the change are indicated by the terms *initial set* and *final set*. Initial set is said to have taken place when a standard needle does not penetrate the paste, made in standard manner, beyond a certain depth. When the needle does not penetrate at tall, the cement is said to have reached the final set. The time for the initial set should not be less than 45 minutes and for the final set not more than 10 hours. It is necessary that in the beginning the cement should set slowly to allow time for mixing and placing it in position and to ensure that it adheres well with adjacent batches of concrete. At the same time it is necessary that the concrete should quickly become hard after the initial set so that it may soon be loaded. Setting of cement is completed in a matter of hours, while hardening is not completed for months. As a matter of fact, it has been found that hardness and strength continue to increase for years. It may be added that there is no fundamental difference between setting and hardening.

The hydration reactions begin immediately on mixing water and continue till the anhydrous compounds that constituted the cement disappear. Hydrolysis reactions also take place. The reaction products, energy changes and speeds of reactions during setting and hardening of Portland cement have been studied for over half a century. The main constituents of the common cement are given in Table 17.1.

Table 17.1

Name	Abbreviation	Formula	Percentage
Tricalciun Silicate	C_3S	$3CaO.SiO_2$	45
Dicalcium silicate	C_2S	$2CaO.SiO_2$	25
Tricalcium aluminate	C_3A	$3CaO.Al_2O_3$	10
Tetracalcium Aluminoferrite	C_4AF	$4CaO.Al_2O_3..Fe_2O$	10

The remainder consists of free MgO and small amounts of K_2O and Na_2O in various combinations with CaO, Al_2O_3, SiO_2 and SO_3.

The amount of the water taken up, the amount of heat given up and the rate of setting of all the four principal constituents of cement are different. Therefore, by adjusting the relative amounts of the constituents of the cement its properties may be altered considerably.

On mixing with water the aluminates, like C_3A, quickly hydrate giving off a considerable amount of heat and give the initial set. To check the rate of initial set gypsum, $CaSO_4$, $2H_2O$ is added which forms crystals of calcium sulpho-aluminate which retard the initial set and allow time for placing the concrete in position. The exact mechanism of the action of gypsum is not well understood. Calcium sulpho-aluminate has a low specific gravity and occupies a large volume. Therefore, if excess of gypsum is added there is a danger of high stress being produced due to expansion after hardening of the concrete. This may lead to cracking and failure of the structure. The aluminates are mainly responsible for the strength during the first day. These reactions are complete in four weeks. Thus if C_3A content is increased the cement would set quicker and would attain greater strength during the first four weeks but the ultimate strength would be less.

Tricalcium silicate, C_3S reacts more slowly with water but this reaction is the most important reaction and may be represented thus :

$$2(3CaO.SiO_2) + 6H_2O + 3CaO.2SiO_2.3H_2O + 3Ca(OH)_2$$

Tricalcium silicate water Tobermorite gel Calcium hydroxide

(Calcium silicate hydrate crystals resemble a natural mineral tobermorite, hence the name tobermorite gel). This reaction is slower and is almost complete during the first week and is responsible for the early hardness and strength of cement. The large amount of calcium hydroxide produced in this reaction adversely affects corrosion resistance of cement. It is noteworthy that no calcium hydroxide is liberated in the hydration of C_3A.

The reaction of water with dicalcium silicate is similar to that with tricalcium silicate but is very slow and its effect on the strength and hardness continues to be felt for a year. Aluminoferrite also hydrates :

$$2\,(2CaO.SiO_2) + 4H_2O \rightarrow 3CaO.2SiO_2.3H_2O + Ca\,(OH)_2$$

The engineering properties of concrete are mostly due to tobermorite gel (calcium silicate hydrate gel), which is the main cementing agent. This gel has an enormous surface area. It has been calculated that if one gram of a cement clinker is ground to the fineness of cement its surface area will be 3,000 cm^2. One gram of tobermorite gel has a surface area of million cm^2. Therefore, the surface forces of even a small weight of this gel are enormous and cause adhesion of the gel between its own particles as well as with other crystals, grains of sand and pieces of aggregate present in the concrete. Calcium hydroxide has greater surface energy than tobermorite has, but its surface area is much smaller; therefore its binding effect is not felt to such an extent.

In addition to surface forces, the strength and dimensional stability of the concrete depends upon factors like porosity. The pores do not contribute strength and hardness and are, therefore, undesirable. Further, if water enters the pores and solidifies, as happens in colder climates, it exerts great pressure because ice occupies a greater volume than water, and cracks may form. Airentrained concrete, i.e., concrete having a very large number of tiny air bubbles, is used to relieve this internal stress. Changes in atmospheric humidity also affect the amount of water in the concrete and cause volume changes. Therefore, if pure cement is used for construction, besides the cost factor, the volume-changes will be great and the mass will crack and will be weak. Inert materials like sand (called fine aggregate) and gravel or coarse stone (called coarse aggregate) are mixed with cement and water so that the volume-changes, and also heat-changes may be minimised. Such a composition, which solidifies into a rock-like mass, is known as *concrete.* A mixture of cement, water and sand is called *mortar.* An inert material, like sand, gravel and stone pieces, is called *aggregate.* The voids, or the space, between the larger particles is filled by smaller particles. The remaining voids are filled by cement paste. The entire aggregate is completely embedded in cement paste which binds it firmly.

Cement concrete is used for structural purposes under compressive loading. It has very poor tensile strength.

17.1.4 Types of Cement

In this brief review of the changes taking place in setting and hardening, the heat-changes which take place have been omitted, but these affect the use of cement concrete to a great extent. The hydration reactions are exothermic. Tricalcium aluminate releases the maximum heat. In a massive structure, like a dam, heat release is undesirable as the temperature of the interior may rise, say, 50° above the outside temperature and set up high thermal stresses. For such purposes low-heat cement is needed. It has a smaller amount of tricalcium aluminate but large amount (more than 50%) of dicalcium silicate which has a low heat of hydration. For making concrete blocks it is desirable that the concrete sets quickly and gains strength and 'high early strength' cement is used which has about 60% of tricalcium silicate. Thus for specialized uses five general types of cements are made.

Without going into details of the types, the behaviour of the constituents can be summarised thus : Tricalcium aluminate gives early set and high heat of hydration (207 cal/g). It has poor resistance to sulphates, therefore cement for use in huge structures like dams or for use in sea waters should have a minimum of tricalcium aluminate. High alumina cement is also made which is resistant to sea water. It attains high strength in a short time and has high heat evolution. This does not contain tricalcium aluminate but other aluminates are present.

Tricalcium silicate gives early strength and makes the clinkers easy to grind. Its heat of hydration (120 cal/g) is high but is less than that of tricalcium aluminate. Dicalcium silicate has minimum heat of hydration (62 cal/g) and is therefore, an essential constituent of low heat cement. Its early strength is low and full strength is attained in a year or more. It is hard, therefore it makes clinkers difficult to grind.

'Rapid setting' cement contains a much smaller amount of gypsum. It reacts quickly with water and is used for emergency work like sealing water-tanks, repairing busy roads, etc.

17.1.5 Decay of Cement Concrete

Cement concrete, though mechanically very strong, is highly susceptible to chemical attack. Concrete contains some free lime. In acidic water lime dissolves making the concrete weak. Most of the natural waters have a pH value slightly above 7 (i.e., are slightly alkaline) and do not have any marked effect on concrete but as the pH value decreases (i.e., as the acidity increases) deterioration of concrete increases. Lime is more soluble in soft water than in hard water. Therefore, concrete deteriorates more quickly in contact with soft water because the lime dissolves.

Various sulphates and chlorides present in water also remove lime. Sodium carbonate removes silica. All these weaken the concrete. If it is soaked in mineral oil for some time, its resistance to abrasion decreases. Some cases of failure have been reported, the cause of which has been traced to the use of old sugar sacks for carrying sand or for covering the freshly laid concrete. It has been reported that even as low as 0.1% sugar when added to cement delays its setting time and greatly reduces its strength during the first four weeks.

The most serious type of damage to concrete takes place when a new compound which occupies more volume is formed. The common example of this type is in the presence of sulphates. Life of concrete is greatly reduced in the presence of sulphates because sulpho-aluminates are formed. This causes expansion and results in the destruction of the concrete structure. Such a danger is avoided by eliminating tricalcium aluminate from the cement composition. In cement containing tetra-calcium alumino ferrite, instead of tricalcium aluminate is formed. It is resistant to attack by sulphates.

A very inexpensive method of protecting cement concrete against corrosion is to give a coating of a bituminous material which will prevent contact between water and concrete. Concrete can also be protected by coating the surface with silicon fluoride in a soluble form with oxides of magnesium, zinc or aluminium. This precipitates insoluble calcium fluorides in the capillaries and prevents dissolution of lime.

17.2 Limes

17.2.1 Raw Materials

The commercial calcium oxide, CaO, is known as quicklime or common lime. It is obtained by the calcination of naturally occurring materials like limestone, chalk, marble, oyster shells, etc. which are rich in calcium carbonate, $CaCO_3$. Often the raw material contains magnesium carbonate, $MgCO_3$. It is not regarded as an impurity but its presence changes the properties of lime. If magnesium oxide is more than 25% the product is known as dolomitic lime and is put to different uses. Limestone is usually associated with silica, alumina and iron oxide. These constitute the chief impurities. If the limestone has sulphur compounds then its use in metallurgical industries, notably the iron and steel industry, becomes restricted.

Limestone deposits are found scattered in various parts of India. Satna and Katni in M. P. have been centres of its manufacture. Deposits are also found more or less around iron and steel centres of the country but due to the large consumption of the material in the iron and steel industry new deposits are being explored. Some such deposits are being worked in Mirzapur district, U. P., for the iron and steel industry. About 29 million tonnes of limestone is produced annually in India.

17.2.2 Manufacture

Calcium carbonate decomposes at 890° to give calcium oxide and carbon dioxide. The reaction is endothermic. In the manufacture the temperature of the hottest part of the kiln is kept from 900° to 1100°. Alumina, silica and iron oxide are the common impurities from limestone and from coal, in case coal and limestone mixture is fired in the kiln. These combine to form clinkers. The higher the temperature the greater is the danger of lime combining with these impurities to form fusible slag. Magnesium oxide, which is quite commonly formed, slakes slowly and is undesirable in large percentages in quicklime. Magnesium oxide heated to high temperature does not react with water and makes slaking difficult.

For small scale manufacture *intermittent kilns* are used. The kiln is a shell-like structure made of ordinary bricks and lined with fireclay bricks. At the bottom an opening is provided for the entry of air for combustion and for the removal of lime. Fire is lighted by burning wood, and 6 parts of limestone mixed with 1 part of coal are charged from the top. Coal burns and supplies heat for the decomposition of limestone. After calcination the kiln is allowed to cool and lime is removed. The product gets mixed with the various impurities present in limestone and coal. Some of the impurities combine to form hard lumps or clinkers. This kind of working is uneconomical on a large scale because of its small capacity, waste of heat and inferior quality of lime. Such kilns are still used in India, especially in smaller quarries, because the overhead expenses are eliminated and lime is sold in the nearby local market.

For large scale manufacture *continuous kilns* are used. In India, in the common kiln of this type limestone mixed with fuel is charged in the kiln. The kiln is a vertical structure lined with fireclay bricks. To facilitate continuous removal of lime, it is usually built on a slope. Lime is removed from an opening at the ground level. Fire is lighted in the kiln and alternate layers of limestone and coal (sometimes wood) are charged from the top. As calcination progresses, lime is continuously withdrawn and the charge is fed from the top. Air for combustion enters from the same opening through which lime is removed. Thus lime gets cooled and air gets heated. The lime gets mixed with the ash of the fuel but the furnace has a large capacity and thermal efficiency is higher than that of other furnaces.

Continuous separate feed kilns are used when lime of high purity is required. The kiln is similar to the kiln described above except that two to four fireplaces are provided by the side of the klin near the bottom, Fig.17.2. Coal or wood is burnt in the fireplace and the hot gases pass up the shaft of the kiln. Wood has the advantage that it burns with long flames and does not raise the temperature of limestone too much. When coal is burnt, some steam is also introduced. Steam keeps the temperature lower and forms water gas which burns and heats the kiln to a greater height. Limestone is charged from the top and lime is removed from the bottom as shown in Fig. 17.2. Thermal efficiency of this type of kiln is lower than the kiln in which limestone and fuel mixture is charged but the product is of superior quality.

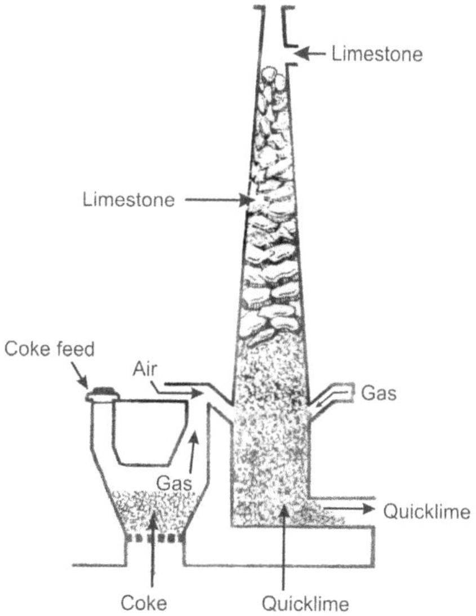

Fig. 17.2 A continuous kiln for making lime.

The rotary kiln for burning lime is similar to that used for cement manufacture but is much smaller and is worked at a lower temperature. Limestone has to be crushed to much smaller pieces and is fed from the upper end. Powdered coal, oil or producer gas is used as the fuel and is forced up, as is done in cement manufacture Heat from the hot gases escaping the rotary kiln is recovered by waste heat boilers or heat exchangers. The product obtained is of superior quality because of better control on the operation. Rotary kilns are quite widely adopted in the highly industrialized countries and for making lime for chemical industries.

Slaking

For making mortar or plaster or for white-washing . quicklime is mixed with water. Three parts of quicklime react with one part of water to form calcium hydroxide, $Ca(OH)_2$, or slaked lime. As the commercial quicklime contains impurities the amount of water needed is less than the theoretical amount. The hydration reaction, called slaking, is accompanied by considerable heat and the volume) of slaked lime becomes two to three times the volume of the quicklime. In slaking, lumps of quicklime disintegrate and crumble to fine powder. It is essential that the lime used for mortar be completely slaked before use, otherwise it will slake in the joints of the plaster and make the construction defective because of expansion and local rise in temperature. The time taken for hydration of lime depends upon several factors. The presence of impurity reduces the rate of hydration. Lime containing a larger amount of magnesia, i.e., dolomitic lime, hydrates slowly. Porous or finely divided lime hydrates quickly because water gets inside more rapidly. Hydration takes place slowly whether the temperature of burning has been high or low. High temperature during burning causes lime to react with the impurities and at a low temperature decomposition of limestone is incomplete, hence it is important to control the temperature in the lime kiln.

The temperature should not be allowed to rise very much during slaking, otherwise the mortar is of lower strength. For slaking, high-calcium quicklime should be added to water and the water should cover the lime completely. Care should be taken to avoid boiling of water. High magnesia or slow-slaking lime should be moistened "with water and more water should be gradually added so that the mass remains hot, thereby the hydration reaction is facilitated. Enough time should be allowed for the completion of slaking. From the foregoing discussion it is clear that slaking is not as simple as it is often thought to be. In some countries quicklime is carefully slaked and marketed in bags.

Quicklime on exposure to air slowly absorbs moisture, swells and disintegrates to fine powder giving slaked lime. This product is known a~ air slaked lime and is of no use for making mortar. The reason is that lime absorbs not only water but also carbon dioxide from the air, hence air-slaked lime contains mostly calcium carbonate and lime mortar made from it will not harden.

Lime Mortar

Lime mortar for joining masonry materials and for plastering is made by mixing 4 to 5 parts by volumes of sand with each part of hydrated lime. Sand should not be considered to be merely a cheapening material. It reduces shrinkage and prevents the formation of cracks. It helps in the hardening by allowing air to enter which provides the carbon dioxide needed for hardening. The mechanism of setting of lime mortar is different from that of Portland cement. Lime mortar gradually looses moisture and absorbs carbon dioxide from air and forms crystals of calcium carbonate which interlock sand particles and the mortar gradually sets. As carbon dioxide is needed for setting of lime mortar, it does not set under water. Thick layers of lime mortar or masonry joints take months or even years for completing the reaction because carbon dioxide reaches the interior slowly. Dolomitic lime is slower to set but gives harder mortar than high calcium lime.

This process of loosing water is accompanied by shrinkage. If less sand is 4sed, there will be excessive shrinkage and the mortar or plaster will become weak and may crack. If too much sand is used, the mortar will be difficult to work and there will be less of binding material in it, i. e. it will be weak. Therefore 'sand-carrying capacity' of a lime is a guide to the quantity of sand that should be mixed with the lime samples. Presence of magnesia reduces the 'sand carrying capacity', but enable's the mortar to be worked more easily with a trowel. High-calcium limes have lower plasticity and are likely to stick to the trowel.

Lime mortar is cheaper, weaker and slower to set than Portland cement.

17.2.3 Types of lime

On the basis of chemical composition, limes are classified into several types. From the foregoing discussion on slaking, it is clear that lime should be slaked properly, otherwise maximum strength will not be obtained, and that slaking depends upon the chemical composition of lime. It is therefore, necessary to consider the classification of lime so that these can be put to best use.

High-Calcium Lime or Fat Lime

This is the purest commercial grade of lime and contains more than 90% of calcium oxide CaO. It is also called rich lime or caustic lime. Average Ca content of limes of this grade is 95%. Alumina, silica and iron oxide content is less than 2%. The remainder consists of magnesia, MgO, water and carbon dioxide. High-calcium lime should be slaked carefully because it reacts vigorously with water and the temperature is likely to rise excessively making the lime inert. Mortar made from high calcium lime is sticky non-plastic and difficult to work under the trowel. Lime of this type has high sand-carrying capacity, i.e. can be mixed with a larger amount of sand than limes of other types but it produces less hard mortar. High-calcium lime is more expensive than other types of lime and the chemical, metallurgical and glass industries have a good demand for it.

Calcium Lime or Lean Lime

Limes of this grade have at least 75% CaO content, the remainder being mostly alumina, silica and iron oxide. Lean lime, also called *poor lime*, is cheap, is of inferior quality, and requires less water for slaking. It has low sand carrying capacity and mortar made from it is weak. It is widely used for white-washing.

Magnesian Lime

Magnesian lime contains not less than 20% of Mg') and the percentage of the impurities is small. It slakes slowly with the evolution of less heat, and the expansion in volume is also less than that of the high-calcium lime. For slaking magnesian lime, water should be added to the lime in small amounts at a time so that the paste does not cool. Enough time should be allowed for the completion of hydration. It has lower sand-carrying capacity but the mortar is plastic, is easy to work under a trowel, and sets more slowly but to a harder mass than mortar made from high-calcium lime. It finds valuable use as a flux in the metallurgical industry.

Dolomitic Lime

Dolomitic or high-magnesian lime contains at least 25% of MgO; often it has 30 to 40% of MgO. It is prepared by burning dolomite ore which is almost an equimolecular mixture of $CaCO_3$ and $MgCO_3$. The percentage of the impurities is small. Dolomitic lime slakes very slowly with comparatively least expansion and the least evolution of heat. Great care has to be exercised while slaking it, otherwise there is the danger of part of it remaining unslaked and slaking gradually in the mortar with the result that the buildings crack. It has low sand-carrying capacity but the mortar is very plastic, works easily and smoothly under the trowel and sets to a hard mass. Dolomitic lime is too expensive to be used as a common mortar. It fetches a good price from the metallurgical industry for use as a flux for preparing special slags and for repairing the basic refractories.

Hydraulic Lime

Hydraulic limes are like inferior grade natural cements. Their chemical composition and behaviour are intermediate between the common lime and natural cement. They are prepared by burning limestone containing at least 100;0 of silica, alumina and iron-oxide. The product after calcination contains calcium silicates, aluminates, etc., along with free lime. Manufacture of hydraulic lime is similar to that of quicklime except that calcination is done at a higher temperature. The calcined product obtained from the kiln is allowed to cool, spread in thin layers and just sufficient amount of water is sprinkled. Excess of water should not be used while slaking because the lime will harden. Free lime present in the product slakes and expands with the result that the lumps break up into fine powder.

Hydraulic limes are of several grades. Hydraulic property, i.e., capacity to set under water increases as the content of silica and alumina increases. *Feebly hydraulic lime* contain, less than 15 percent of silica and alumina. It is not truly hydraulic in the sense that it does not set easily under water. It slakes, slowly, taking 5 to 60 minutes to slake with small expansion in volume. It gives a good mol tar. Semi-hydraulic or moder.1tely hydraulic lime contains 15 to 25 per cent of silica and alumina. It sets under water but develops considerably less strength than hydraulic lime. It slakes very slowly with small expansion. It makes very good mortar. The mortar is stronger than the mortar made from *feebly hydraulic lime* and is used for good brick work and masonry. The true hydraulic lime is also called very hydraulic lime and contains 25 to 35 percent of silica and alumina. Chemically it is similar to Portland cement and sets and develops considerable strength under water. It is finely ground.

Lumps of hydraulic lime which do not break on the addition of water are separated and sold as *Groppier cement*. It consists mostly of calcium silicate and is of lighter colour than the Portland cement

Pozzolan Cement

Pozzolan is a material which on mixing with lime at ordinary temperature sets under water. It is prepared from volcanic lava called pozzolan, which contains alumino silicates. Two to four parts of pozzolan is mixed with one part of hydrated lime which combines with the siliceous material and sets under water. It is not of commercial significance.

Artificial pozzolans include granulated blast furnace slag, burnt clay and crushed bricks and tiles. Burnt clay or common bricks are crushed to give *surkhi*. Surkhi mixed with lime and water is quite commonly used in India as a building material.

The pozzolans are effective because of the presence of reactive silica and alumina which slowly react with lime at ordinary temperatures forming calcium silicate and aluminate hydrates. They consist only partly of such active materials and the remaining constituents are inactive. The reaction being slow, it is necessary to grind well lime and *surkhi*, so that they are in a finely divided state. Concrete made by using pozzolanic cement, for example from the surkhi mortar is more resistant to corrosive enviroments than Portland cement. The structures have been found to be quite resistant to sea water, which corrodes Portland cement. As the heat of hydration is small and the reaction proceeds slowly, the pozzolanic cement is especially suited for making massive structures like dams. Mixture of pozzolans and Portland cement have been used with beneficial effects.

Pozzolan cement is made from clinkers produced either in the rotary kiln or vertical shaft kiln. The clinkers are mixed with 10 to 25% of pozzolanic material and ground \with 4 to 5% of gypsum. Thus the proportion of clinker: pozzolanic material: gypsum is about 80 : 15 : 5. The setting times, strengths and other characteristics of pozzolan cements are similar to ordinary Portland cement, excepting that in the case of pozz6lan cement the 3.days strengths are not specified. However 7-days strengths for pozzolan cements are the same as that of ordinary Portland cement.

17.3 Plaster of Paris

Manufacture

Gypsum, $2CaSO_4.2H_2O$, is the raw material used for the manufacture of plaster of Paris. In India, gypsum deposits are found in Rajasthan, annual production being 13,00,000 tonnes. It is associated with about 5% of impurities. Gypsum rock is crushed and ground to powder. The kettles for heating it are steel vessels, about 3 m high and 2 m in diameter with bottoms made of cast iron. Each kettle is fixed in a masonry enclosure. The kettle gets heated from the sides and the bottom and the hot gases escape through the flues which pass through the kettles. Kettles are provided with stirring paddles, thus heat is uniformly distributed. Powdered gypsum is introduced in the kettle and is slowly and carefully heated. Mechanically held water escapes at about 1500 and the mass is stirred to avoid local heating. The temperature is raised to about 2000 when the plaster of Paris is formed.

$$2CaSO_4.2H_2O \longrightarrow (CaSO_4)_2.2H_2O + 3H_2O$$

The mass is removed through openings near the bottom. For making plaster board and gypsum products, the temperature is raised when anhydrous calcium sulphate is obtained which does not easily take water. The plaster is also prepared by heating gypsum in a rotary calciner in which the temperature is raised up to 300° for a short time. Plaster of Paris meant to be used for making hard wall plaster is mixed with a small amount of glue so that is sets slowly allowing enough time for applying it.

Setting and Hardening

Plaster of Paris on mixing with water takes back water of crystallization and forms $CaSO_4.2H_2O$ crystals. If the plaster is pure, it sets in 2 minutes and is used by dentists. For general purposes small amounts of retarders, like glue, are added so that .the plaster remains in a plastic state for some time. Setting of the plaster is accompanied by increase in volume of about 0.2% due to which it makes sharp impression.

Uses

Plaster of Paris is used for making moulds and models in the pottery industry. In the U S A. it is manufactured on a large scale and is used as a wall plaster for the interior because it can be easily mixed with sand and water and it sets and dries more quickly than other plasters. Hard plaster made by heating gypsum to 500° is used as flooring.

Questions

1. Mention the ways to reduce thermal spelling of a refractory in furnace
2. Enumerate the important properties a material should possess in order to be a good refractory material
3. What are fire clay bricks ? Describe the properties and uses of these bricks
4. What are ceramics ? How are they classified ? Give an example in each case
5. Why dolomite bricks are rarely used as direct refractories ?
6. How do earthenware differ from stoneware ?
7. What is a refractory material ? Explain the differences between acidic and basic refractories ?

8. What is meant by glazing? Explain why the ceramic white wares are essentially glazed.

9. What are insulators ? How are they classified? Give examples

10. Describe the manufacture and uses of silica bricks

11. Explain the following properties as applied to refractory materials.

    (i) Refractoriness

    (ii) Refractoriness under load

    (iii) Thermal spalling

17. Explain how the dimensional stability and porosity effect the properties of refractories ?

13. What is meant by parametric cone equivalent of a refractory ?

14. Examine the truth in the statement "silica refractories can be used in basic media".

15. What is meant by white wares ? Give their two uses. Explain vitrification.

16. Describe the chemical conversion that take place in the manufacture of ceramic products

17. Write note on polymorphism in ceramic materials.

18. What type material are used as transformer liquids ?

19. Define porosity of a refractory.

20. What is the main advantage of liquid insulating materials over solid insulating materials ?

21. How the thermal spalling of a refractory can be reduced ?

22. Define dielectric constant.

23. On what basis solid insulating materials are classified ?

24. How the strength of refractory can be measured ?

25. What are advantages of SF_6 as dielectric gas ?

26. What is the difference between insulator and dielectric materials ?

27. (a) What are thermal insulators ? What are their characteristics and engineering applications ?

28. Describe the chief applications of thermal and electrical insulators.

29. (a) What are insulators ? Mention the different types of insulators.

    (b) Describe the properties and applications of

    (i) thermal insulators (ii) electrical insulators.

30. What are the causes for failure of a refractory material ?

31. What are refractories ? Discuss their important characteristics.

32. Write a note on mono-and di-propellants used in rockets ?

Objective Questions

1. Graphite bricks are widely used as material for construction of _____.

   (A) Atomic reactors (B) Open hearth furnace

   (C) Coke oven (D) Blast furnaces

2. Porasity in an important property of refractory bricks and it increases

   (A) Chemical stability (B) resistance to corrosion

   (C) Thermal conductivity (D) Resistance to thermal spalling

3. High refractoriness under load is shown by

   (A) Dolamite (B) Magnesia

   (C) Alumina (D) Silica

4. Sewage (drainage) pipes are made of

   (A) earthen ware (B) stone ware

   (C) white ware (D) GI pipes

5. Dolamite is an example of _____ refractory

   (A) Acidic (B) Basic

   (C) Neutral (D) Amphoteric

6. Clays are formed by weathering of

   (A) Igneous rocks (B) Feldspathic rocks

   (C) Dolomitic rocks (D) Both a & b

7. Ceramic obtained from china-clay, feldspar and flink is –

   (A) Stone-ware (B) White ware

   (C) Brick (D) Tile

Fill in the Blanks

1. Permeability of a refractory depends upon_____.
2. Basic refractories are _____.
3. More pores in refractory _____is the thermal spalling.
4. _____ are gaseous insulators.
5. Refractoriness is the ability of material to withstand _____ without appreciable deformation.
6. In porous bricks supplied air act as _____.

Chapter 18

Composite Materials

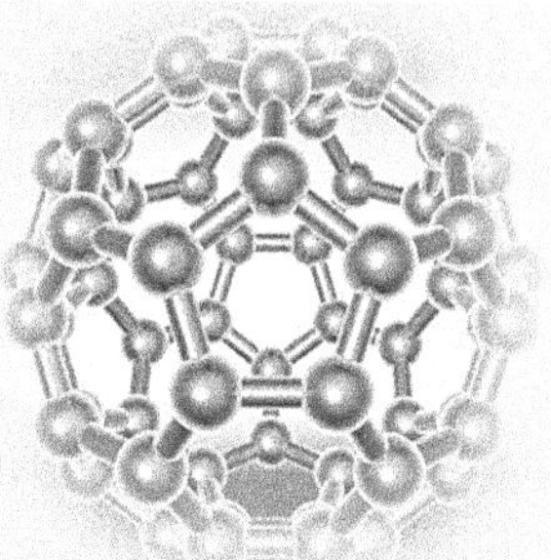

18.1 Introduction

Although the term 'composite materials' has attracted the engineers only in recent times, these materials have been used for many years. For decades he has depended, for instance on reinforced concrete for making strong structures like dams and large buildings. There are many smaller components in use for many years like the tyres-with cord embedded in rubber–made of composite materials. During the *last half* a century or so the materials technology has made advances beyond expectations in the fields of special steels and other metal alloys, polymeric materials and ceramics. The pace of progress seems to have slowed down during the *last decade or so* and the thrust has been in the use of combinations or mixtures of materials. Each class of engineering materials–metals, high polymers and ceramics–has its distinct and outstanding properties as well as limitations. However, very stringent requirements of supersonic aircraft, gas turbines, high temperature reactors, not to speak of missiles and spacecraft, have forced the technologists to think of such combinations of materials that weak points of one are covered by the other. In spite of the fact that no definition will be completely satisfactory, one may define a composite material as a materials system which consists of a mixture or combination of two or more microconstituents mutually insoluble and differing in form and/or material composition. The components form distinct phases. The combination not only has its own distinctive properties but its strength or some other desirable property is superior to either of the components or different from either of them.

The composite materials are generally made by placing the dissimilar materials together in such a manner that they work as a single mechanical unit. The properties of the new material so produced are different in kind and scale from those of any of the constituents. Thus it has become possible to incorporate and/or alter properties or, more than that, introduce a combination of properties like high strength and stiffness, especially at elevated temperatures, low specific gravity, desirable coefficient of thermal expansion, corrosion resistance and ability to withstand extreme temperature conditions. Such combinations of properties have not been within the easy reach of the engineers, either because the materials were not available or were too expensive. Metals, for instance, loose their strength at elevated temperatures. High-polymeric materials in general, can withstand still lower temperatures. Ceramics outstrip the metals and polymers in their favourable melting points, ability to withstand high temperatures, strength and thermal expansion properties, but due to their brittleness they are often unsatisfactory structural materials. This led to the exploration of combinations of metals and polymers with glass and other ceramics, resulting in composites having requisite properties. Thus the composite materials seem to be the hope for the future.

All this may seem 'high-sounding' and give the impression that the composites are unfamiliar and very special materials, not within the daily use. Wood, packing paper, impregnated with wax or bitumen, used to protect biscuits and books from dampness, the rain coat (not the simple plastics ones) made of cloth impregnated with water-proof material, the insulating tape for the electrical wiring, and concrete are common examples of composites in contact with which one so frequently comes.

There are definite advantages of composites over the widely used alloys and metals. These become clear from Fig. 18.1 in which properties of composites are compared with those of aluminum and steel.

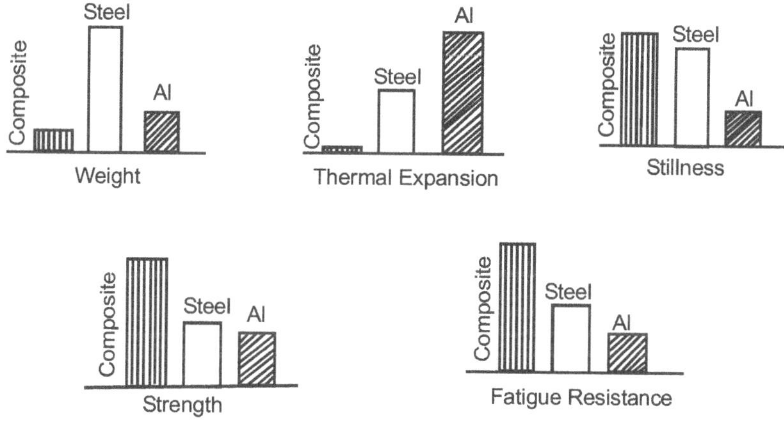

Fig. 18.1 Comparison of properties of composites with steel and aluminium

Before concluding these introductory remarks, it may be helpful to define a few of the important terms used in the study of the composites. ASTM has defined *fiber* as any material in an elongated form such that the ratio of its minimum length to its maximum average transverse dimension is 10:1, its maximum cross-sectional area is 4×10^{-4} cm^2 (corresponding to a circular cross-section of 0.025 cm diameter) and its transverse dimension is not greater than 0.025 cm. A *whisker* is any material that fits the definition of a fiber and is a single crystal. *Filament* is a continuous fiber and a wire is a metallic filament. When a number of filaments are so bundled that they can be handled as a single unit it is called *yarn*. The bonding material which contains and adheres to the reinforcing material like the fibers or the particles is known as *matrix*. Thermoplastic and thermosetting resins are most widely used for this purpose. Metals, glass or ceramic materials are also in general use as matrix.

18.1.1 Broad Principles

This survey will be restricted to the consideration of broad principles governing the behaviour of the special composites leaving aside for the time being materials like concrete. The production of fiber-reinforced composites began by the incorporation of strong fibers in a metal or polymer matrix. The objective was to obtain at room and elevated temperatures, superior mechanical properties like higher yield strength, fracture strength, creep strength and fatigue life. This could be achieved because the strong fiber inhibits gross plastic flow in the matrix and also takes up a proportion of the load. In Chapter I, in the Section on deformation of rigid bodies, the phenomena of slip and crack propagation were explained. The fibers come in the way of crack propagation and stop or inhibit it, thereby increase the above-mentioned mechanical properties. On the application of the load there is localized plastic flow in the matrix which transfers load from the matrix to the fiber embedded in it. In addition, the matrix not only binds the fibers but also protects their surfaces from damage due to reaction with the environment. The geometry, dimension and volume fraction which also play an important role will be considered at the appropriate place.

When hard reinforcement is present in a soft matrix the strength and modulus of matrix are increased. In the case of soft phase present in hard matrix, the shock resistance of composite is increased. The improvement in the properties is due to the movement of dislocations being obstructed by the reinforcement.

At this stage it is necessary to clarify the criterion for selection and comparison of the materials. The two properties that have been given paramount importance in the past are *specific strength* and *specific stiffness*. These refer to the ratio of the property to density i.e., ultimate tensile strength and Young's modulus divided by the specific gravity give specific strength and specific stiffness respectively. Specific strength is an important criterion for the constructional material, as in the aircraft, and specific stiffness is of special significance for the light weight structures. Alloys with greatly improved specific strength and stiffness have been produced but their production usually depends on sophisticated technology which is reflected in increased cost. It is interesting to note that while specific strength of the common metallic materials ranges from 1,700 kgf/cm^2 to 3,200 kgf/cm^2, their specific stiffness values are constant at around 25 × 10^4 kgf/cm^2. For example, the specific strength of steel plano wire, aircraft steel, duralumin and magnesium alloy in kgf/cm^2 are 3,200; 1,700; 1760 and 1,700 and their specific stiffness in kgf/cm^2 are 27 × 10^4, 27 × 10^4, 26 × 10^4 and 25 × 10^4 respectively. The strength and modulus of tungsten are relatively very high but its high specific gravity (19.3) brings down the specific values in the above range. Interestingly, the specific strength of oak wood is 1,000 to 1,400 kgf/cm^2 and specific stiffness is 14 to 21 × 10^3 kgf/cm^2.

Compared to these the special glass and silica fibers have remarkably high specific strengths, 11.2 to 20 × 10^4 kgf/cm^2 and 25 × 10^3 kgf/cm^2 respectively and slightly higher specific stiffness 29 to 35 × 10^4 kgf/cm^2 and 29 × 10^4 kgf/cm^2 rcspcctively. Thus the advantages of making composites are clear when one sees that the specific strength of glass fiber is almost ten times or even more, than that of the bulk metals, although the specific stiffness increases by about one fourth only.

It needs only to be added that different criteria need to be adopted for different conditions of loading and working.

One fundamental difference between ceramics on one hand and metals and polymers on the other hand may be mentioned. The former have saturated bonds and highly oriented forces. Once a crack is formed, it moves quickly through it. (See Fig. 18.2). Therefore the ceramics are used in powder form so that cracks do not find any continuous path through the material. The bonds in metals and polymers are non-directional and are unsaturated with the result that the atoms or molecules can readily form new bonds. A crack does not propagate readily through such a material. Therefore these are good for use as matrix.

In a composite material, ceramics may be used in the form of particles or fibres, bonded and separated by metal matrix. A crack initiated in a brittle ceramic phase does not have a continuous path and it is deflected by matrix as shown in Fig. 18.3.

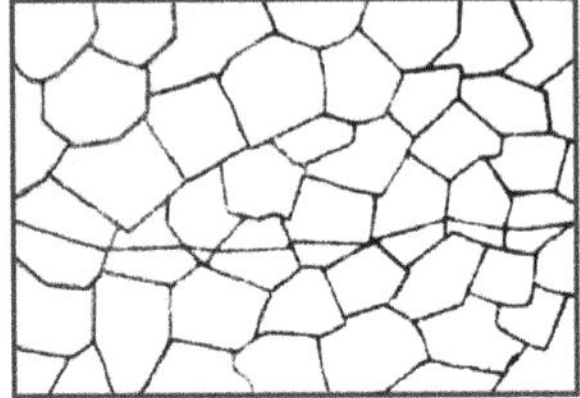

Fig. 18.2 Crack path is continuous in brittle matrix

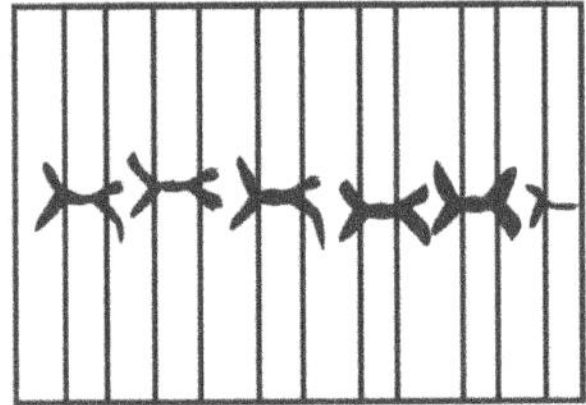

Fig. 18.3 Crack produced in a fiber is obstructed by the matrix

18.1.2 Advantageous Properties of the Composites

There are a number of materials from which the matrix and the reinforcing materials can be chosen. This gives a larger number of permutations and combinations for producing the composite materials. For practical purposes, of course, the choice is restricted. The important advantages of the composites over the common bulk materials can be summarised as follows :

1. They have higher specific strength.
2. The strength can be maintained up to higher temperatures, and loss of strength even near the melting temperature is not much.
3. They have higher specific stiffness.
4. Lower specific gravity.
5. They are tough, having good impact and thermal shock resistance.
6. Thermal expansion and conductivity, as well as electrical conductivity, are low and can be controlled to some extent.
7. Creep and fatigue strength are often better than for metals.
8. In many composites corrosion and oxidation resistance are much better than for metals.
9. Large parts can often be fabricated more easily and cheaply than from the metals.

18.1.3 Classification

On the basis of the structure of the reinforcing material, the composites are classified as 1. Fiber-reinforced composites, 2. Particulate composites and 3. Dispersion hardened composites.

1. ***Fiber-reinforced***

As the name suggests the reinforcement is provided by fibers which are embedded in a suitable matrix. For the reinforcement to be effective two conditions must be fulfilled. Firstly, the Young's modulus of the fiber must be higher than the matrix. This means that the fiber is stiffer and undergoes less change in dimensions than the matrix. If this does not happen the fiber will not carry the load as strain builds up. Secondly, there must be such bonding between the matrix and the reinforcement (i.e. fiber) that the stresses are transferred from the matrix to the reinforcement. This transfer is usually by shear.

The most familiar example of fiber-reinforced material is that of glass-reinforced plastics. It has already been shown that specific strength of glass fiber is about ten times that of the bulk metal (not metal fiber or whisker) and its specific stiffness is also higher. This happens because it contains fewer defects, which as explained in the first Chapter reduce the mechanical strength. The same holds true for other fibers and more especially the whiskers. These approach the strength of the perfect crystals. It is obvious that for obtaining from two materials a composite having the highest strength and modulus it is essential that there should be the maximum number of fibers contained per unit volume, and that each fiber takes its full share of the load.

The fiber composites have anisotropic properties i.e. the properties in different directions are different. The maximum strength is in the direction in which the fibers are aligned. If it is loaded in other directions much lower strength is found. This is advantageous in certain applications such as turbine blades for a jet aircraft engine in which loading is primarily in one direction. A pressure vessel, for example, is subjected to load in all directions and isotropic properties become essential. For such purposes fibers are oriented randomly within the matrix. Such is the case with ordinary fiberglass. It may be added that cost of laying fibers aligned in one particular direction is higher.

In addition to binding the fibers and help in transferring the load, the matrix helps to reduce the danger of sudden or catastrophic failure as normally happens with brittle materials used as fibers. Polymeric matrices are more widely used than metal matrices. The former have low densities and therefore, give higher specific strength. Polymeric matrices suffer from the disadvantage that usually there is only relatively weak physical bonding and interlocking rather than strong chemical bonding at the interface with the fiber. On the other hand they are more easily produced and fabricated and that too at a lower cost than the metal matrix. The metal-matrix composites have the great advantage of being able to withstand higher temperatures, besides generally having stronger chemical bonding at the fiber-metal interface.

From the foregoing considerations the reasons for the popularity of the composites with polymeric matrix having glass, carbon and boron filaments is clear. Of these, the glass fibers used to reinforce resins are the oldest materials in the field, in use since the 1930s, and continue to hold an important position. They have the advantages of high strength, low density and low cost in addition to being readily available according to the specifications. Glass fibers are easily affected by water which reduces the strength. They have to be protected by a coating, like a resin. The resins burn, char or flow at about 200^0 but certain types can withstand 300^0 or even up to 400^0. Demands of the aircraft industry–ability to withstand higher temperatures and having higher stiffness–has partially replaced its dominant position.

Requirements of higher temperatures turned the attention to ceramic whiskers which are essentially small single crystals practically free from defects, thereby having very high strength, stiffness and ability to withstand higher temperatures. This introduced the use of Al_2O_3, BeO, SiO_2, SiC and B_4C. In the form of whiskers these are about five times as stiff as the best glass fibers and also have superior tensile strengths.

Tungsten wires have been used in matrix of cobalt-nickel alloys. Tungsten retains its strength up to 1500^0 and cobalt-nickel alloys are resistant to oxidation. Such composites have been used up to 1000^0 and above. These have higher specific gravity than other composites.

Among the light weight elements fibers of boron and carbon have been used. Due to the difficulty of producing pure boron whiskers, it is deposited from its vapours on tungsten wire. It is thicker than glass fibers and has about the same strength but higher stiffness. Carbon, besides being very light, is very refractory and seems to be highly promising. Its whiskers are very expensive and are as yet materials for laboratory experimentations, which have demonstrated very good mechanical properties. On the other hand, simple techniques have been developed for producing carbon fibers by controlled and careful heating of organic fibers like the cotton or rayon fibers. These suffer from the disadvantage of low strength and low stiffness but are very much cheaper and more readily available than other fibers. It may be remembered that in many cases high strength and stiffness are really not needed.

Among the metal matrices aluminium is most commonly used. Boron filaments are popularly used in aluminium. Another approach is to produce the fibrous material in the metal by unidirectional solidification of eutectic alloys as is the case with TaC-NiCr, Al-Al_3Ni, Al-$CuAl_2$, Ta-TaC, etc. systems. Although such materials do not have very high strength or stiffness, because the fibers are not thin and also their volume in the total composite is small, they are several times stronger than reinforced materials. These are cheaper and easier to produce within the material (hence called in-situ composites) and their over-all advantages are considerable. However, in-situ composites with better properties can now be produced by controlled solidification of alloys other than eutectic compositions resulting in higher volume fractions of reinforcing phase (e.g. Al_3Ni; $CuAl_2$). The strength of such composites can further be improved by heat treatment which causes precipitation hardening resulting in strengthening of the matrix (Fig. 18.4).

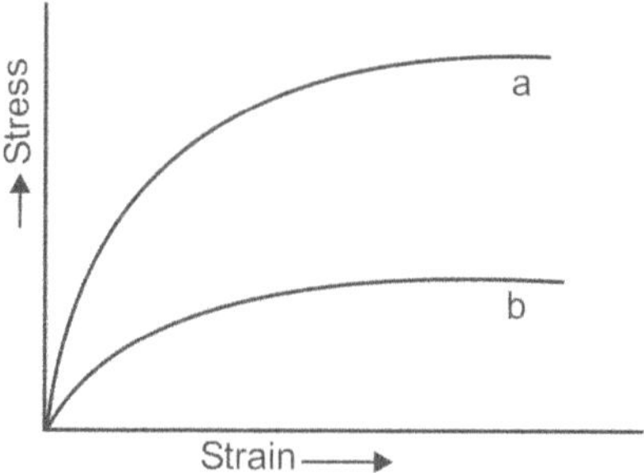

Fig. 18.4 Load taken for a particular strain is higher for directionally solidified alloy *(a)* than for an ordinary alloy *(b)*.

Finally, the volume factor needs to be considered. The improved mechanical properties are due to the properties contributed by the fibers. It is obvious, therefore, that the higher the volume percentage of the fibers in a particular matrix the better will be its mechanical properties. The volume depends on the fibers as well as on the matrix used. Laying the fibers in one direction gives better unidirectional strength but it costs more. Three-dimensional or random orientation is produced by adding chopped fibers (glass fibers being most widely used) in the matrix. In this way complex shapes can be cheaply produced but the composite has lower strength and stiffness.

In addition to the factors pertaining to strength, stiffness and bond between the fiber and the matrix, there are a few other essential requirements. The coefficient of thermal expansion of the fiber should closely match that of the matrix. The fiber and matrix should be chemically compatible with each other so that there is no undesirable chemical reaction between them. The fiber should be stable at room and elevated temperature and should retain a high percentage of its strength at the elevated temperature. Besides the fiber being laid in the direction of the load, it is essential that its ends should be tapered or rounded so as to reduce stress concentrations in the matrix at these points.

2. ***Particulate Composites :*** These have particles of a wide range of size, from microscopic to macroscopic, dispersed in a matrix. The dispersoid size exceeds about 1μ and the volume of the dispersoid is greater than 25%. Concrete, which is used in larger tonnage than any other material of construction, comes under the category of macroscopic particulate composites. On the other side, magnetic tape materials are microscopic particulate composites. Metal-ceramic combinations, called *cermets*, are refractory metal carbides dispersed in alloys for use as cutting tools for hardened steels, are new advances in materials falling under the broad group of particulate composites.

   One of the important and earliest cermets is *WC.Co*. In all the composites the volume fraction of the particulate and the matrix phases are important but in the cermets its effect is very marked. As the volume fraction of the dispersoid increases, the mechanical properties improve reaching an optimum value and then begin to fall. At higher volume fractions the brittle ceramic particles tend to come in contact with each other, thus a continuity of brittle phase is formed. This promotes premature failure, as once a crack is formed in a tungsten carbide particle, it propagates directly through other particles without being checked by the ductile metal matrix. (Fig. 18.5)

To give an example, in the case of tungsten carbide-cobalt composites the maximum strength is reached when volume of tungsten carbide is around 60%. In those cases in which the cermet is made primarily for giving a hard material and strength is secondary, 90% or even more of the hard brittle dispersoid has been used. In the cermets which are used for making cutting tools, dies for metal working operations and for similar purposes besides good abrasion, resistance and stiffness, toughness is essential. Such materials must have lower volume fraction of the ceramic so that they are less brittle. (Fig. 18.6).

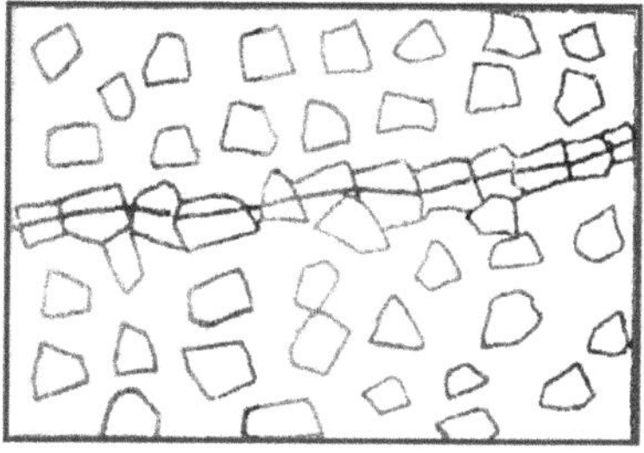

Fig. 18.5 With high volume fraction of the particles the crack path is likely to be continuous

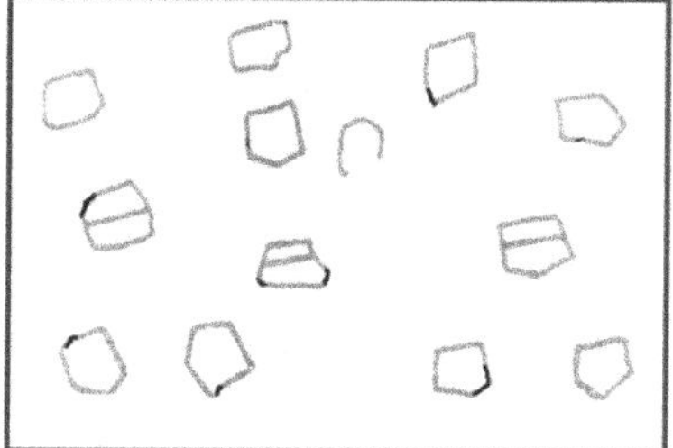

Fig. 18.6 With low volume fraction of the particles the crack path is checked by the ductile matrix.

In a macroscopic particulate composite like concrete, the difference in the size of the components is very large. Gravel of several centimeters in diameter is used in the concrete for dams. The space in between is filled with sand and the pores between the sand particles by cement-water paste. The binding action has been considered in detail under setting and hardening of cement. Concrete is brittle and can undergo only a limited amount of deformation prior to fracture. With a view to increasing its load bearing capability steel bars are incorporated in the concrete. The bars are interlocked with each other and are properly laid so that mechanical grip is ensured between the concrete and the steel bars. This arrangement overcomes the defect of the very weak bond between concrete and steel. It is an example of macroscopic fiber composite which gives it capability of supporting tensile loads also.

3. Dispersion-hardened Composites

These have still finer particles (0.01 to 0.1μ size) dispersed in the matrix. The volume percentage of the particles is also smaller, say, up to 15%. The most widely used materials of this class are the precipitation-hardened alloys. An important example is that of the age-hardening of aluminium alloys (discussed in that Section). The alloy of the proper composition (say. 95.5% Al and 4.5% Cu) is heated to the correct temperature so as to form solid solution and then quenched when a hard constituent, like $CuAl_2$ is precipitated in fine particle form. Nitriding (see Heat Treatment) is another example. The dispersoid obstructs the free movement of the dislocations, thereby giving strength and hardness.

Structure of precipitation-hardened alloys has been imitated by dispersing very fine ceramic particles. Important examples of this class are thoria, ThO_2, dispersed nickel and nickel-chromium alloys. In such artificial dispersions the particles are neither as fine nor as well-dispersed as in precipitation-hardening. Consequently the strengthening effect is not so great, but as the dispersoid does not dissolve in the alloy the artificial dispersion-hardened alloys retain their properties up to higher temperatures than the precipitation hardened alloys. In the latter case the precipitated hard constituent dissolves in the alloy at the elevated temperature and becomes soft.

18.1.4 Mechanism of Strengthening

Fiber Reinforced Composites

It has already been explained that the fibers and whiskers are stronger than the bulk material due to their freedom from defects. The matrix has lower mechanical strength. The two together act as a single mechanical unit and equalise their different strains. The fiber has higher modulus and carries most of the load and the matrix, besides several minor functions, transfers the stress to the fibers by shear. The matrix is weaker and tends to elongate. In the vicinity of the fiber, elongation of the matrix is restricted because the fiber elongates to a very much smaller extent. At some distance away from the fiber the matrix is free to elongate a little more. This non-uniform elongation produces a shear stress at the interface resulting in tensile stress in the fiber which is capable of withstanding high tensile stress. For efficient and successful working it is necessary, therefore, that the shear strength at the fiber-matrix interface should be higher than the shear stress developed there. In other words, there should be a strong bond between the fiber and the matrix and the matrix should be ductile so that the stress is redistributed after the fracture of the fiber.

As the fibers usually available are brittle, the failure may be caused by their fracture. Matrix plays an important role in preventing brittle fracture. It should be tough and ductile so as to be able to prevent propagation of crack across the composite.

There are two types of the reinforcing phase of the fiber. It may be continuous throughout the matrix, in which case the stress is constant over the entire fiber and the load is directly applied on it and little load is transferred to it by the matrix. In such a case the main function of the matrix is to bind the fibers together. If σ_C is the strength of the composite, σ_f is strength of the fiber, σ_m is stress in the matrix at the breaking strain of the fiber, vf and v_m are the volume fractions of the reinforcing phase and the matrix respectively, the strength of the composite is given by

$$\sigma_C = \sigma f\, vf + \sigma_m v_m$$

The reinforcement may be present in discontinuous form. When load is applied to the composite the stress developed at the ends of the fiber is zero and increases to a maximum value at the central portion of the fiber. In such a system the stress distribution in the fiber is given by

$$\sigma_f = \frac{2r_f\,(l/2 - Z)}{r}$$

where, Z = distance from the centre of the fiber and r_f shear flow stress of the matrix, l is the length of the fiber and r is its radius. The strength of the composite with discontinuous fiber is given by

$$\sigma_C = \sigma'f\, vf + \sigma'_m v_m$$

where $\sigma'f$ is average stress in the reinforcing phase and $\sigma'm$ is stress in the matrix at the breaking strain of the fiber.

The length of the fiber which allows the ultimate strength of the fiber to be reached at its center is known as its critical length (l_c). For this length $\sigma'f = \frac{\sigma f}{2}$. When the length of the reinforcing fiber l is greater, than the critical length l_c, the average stress in the fiber is given by

$$\sigma'f == \sigma f\left(1 - \frac{l_c}{2f}\right)$$

and the strength of the composite is

$$\sigma_C = \sigma f\left(1 - \frac{l_c}{2l}\right) v_f + \sigma'_m (1 - v_m)$$

From the foregoing it is clear that the strength of the discontinuous reinforcement composite increases as the length of the reinforcing phase, i.e. the fiber, increases. It approaches the strength of the continuous reinforcement as l/l_c, approaches infinity.

The strength of the fiber-reinforced composite depends on (i) continuity of the reinforcing phase, (ii) its strength and (iii) its volume fraction. The strength of the reinforcing phase increases as its diameter decreases and remains uniform provided the section is perfect and remains uniform. Presence of imperfections at the surface of the fiber helps initiation of crack and makes it weak. For maximum efficiency the distribution of the fibers in the matrix should be uniform and oriented in the direction of the applied stress. There should be good bonding between the fiber and the matrix and there should be no solubility of the fiber in the matrix for it causes surface imperfection.

Mechanism of strengthening of dispersion-hardened composites

It has already been explained that particulate and dispersion-hardened materials differ firstly in the size of the dispersoid and secondly the volume fraction, i.e., the percentage by volume of the dispersoid. In the dispersion-hardened composites the size as well as the volume of the particle is much smaller and mechanism of hardening is very different from that of the fiber reinforced composites and will be considered first. In the fiber reinforcement the fiber is the main load-bearing constituent whereas in the dispersion-hardened composites the matrix is the chief load-bearing constituent assisted by the dispersoid in such a manner that its deformation is highly restricted with the result that its strength is greatly increased.

Besides the shape, the dispersoid can be described by its average diameter, volume fraction and the mean free path between the particles or the inter-particle spacing. These three factors are related. Particle size remaining the same, an increase in volume fraction decreases the mean free path. Volume fraction remaining the same, decrease in particle size decrease the mean free path, because the number of particles increases. The mean free path, mfp, is given by the formula :

$$\text{mfp} = \frac{2d}{3v_p} \times (1 - v_p)$$

where, v_p = volume fraction and d = average diameter of the particles.

In Chapter I while briefly explaining the mechanism of strain hardening, it was explained that strengthening takes place by any factor which obstructs the movement of the dislocations. The formation of loops around the impurities was also referred to. When a stress is applied the deformation takes place by the movement of dislocations through the matrix. The dislocation comes across the dispersed particle which acts as a barrier and is unable to move forward. Thus no deformation takes place unless increased stress is applied. This causes the dislocation to bend like a bow between two fixed dispersoids. Increased stress breaks the dislocation, leaving a circular loop around the dispersoid and the remaining portion moving forward. When new dislocations approach the dispersoid, the loop exerts a stress in the opposite direction and restricts its movement to a still greater degree. The net result is that over increasingly higher shear stresses need to be applied for deformation by slip to take place. In other words, the matrix very rapidly becomes strain hardened.

If the yield strength of the material is equal to higher fraction of its theoretical shear strength, the required interparticle spacing for strengthening will be lower than in the case when yield strength is lower fraction of theoretical shear strength, the other factor remaining the same.

18.1.5 Mechanism of Hardening of Particle Reinforcement

It has already been explained that in the fiber-reinforced composites the load is mostly taken by the fiber and in the dispersion-hardened material by the matrix. In the particle-reinforced composite the load is shared by both the matrix and the particle. The particle restrains the deformation of the matrix. The magnitude of the restraint is a function of two factors, (i) the ratio of the mean distance between the particles and the diameter of the particles, (ii) the ratio of the elastic properties of the matrix and the particles.

On the application of a load the matrix begins to deform. The particles restrain the deformation at their rigid interfaces. As the load increases, a state of hydrostatic stress is developed. When its value is 3 to 3.5 times the yield strength of the matrix material a fracture takes place. If the particle is not deformed at a stress level of 3 to 3.5 times the matrix yield strength, fracture initiated by particle cracking occurs through the matrix.

18.1.6 Fabrication of the Composites

The different composite materials differ from each other mainly in the manner, besides the kind, of dispersion of the hard phase. The various techniques for the fabrication of the composite materials differ from one another in the method of distribution of the second phase. Although the principles for the successful production of the composites have already been discussed they may be enumerated as follows :

1. The second, i.e., the hard phase must be uniformly distributed in the matrix.
2. The dispersoids must be well-bonded with the matrix.

3. Although there must be intimate contact between the dispersoid and the matrix, there must be no direct contact between the dispersoids.
4. There must be no chemical interaction between the dispersoid and the matrix under the working conditions.
5. The orientation of the fibers in the fiber-reinforced materials should be in the direction of the applied stress.

Keeping the above points in mind the manufacturing techniques can be divided into two groups; embedding the separately prepared strengthening phase in the matrix, and, forming the second phase within the matrix.

The reinforcing phase can be incorporated in the matrix by the following methods :

In the *liquid metal infiltration* technique the reinforcing phase is arranged in a mould and the molten metal is poured in it and allowed to solidify.

Hot pressing can be done in several ways. One of the methods is to arrange the reinforcing fiber between the matrix foil (Fig. 18.7(a)) or to arrange the reinforcing fiber between the matrix foil or metal wire (Fig. 18.7(b)). A simple and common method is to coat the reinforcing phase with the matrix and arrange it in the desired manner. The bond is formed by applying pressure at the right temperature. This can also be achieved by arranging the fibers or filaments in position and then filling the matrix powder in the space between them followed by the application of temperature and pressure.

In the *electro deposition* process the filament is wound in such a manner that it has the desired orientation and the metal is electrodeposited simultaneously as the winding progresses. On entering the mandrel the two phases are bound together to give the composite. This technique can be used to produce both monolayer and multiplayer composites.

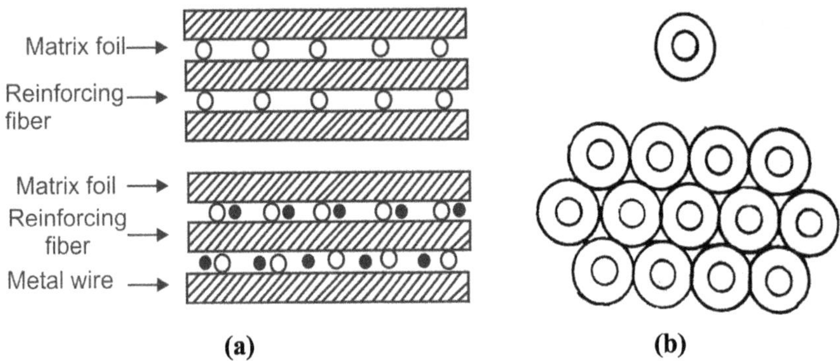

Fig. 18.7 (a) Reinforcing fiber between matrix foil (b) Fiber inside the matrix (top). Matrix coated fiber array (bottom).

Another technique is known as *plasma spraying*. A layer of the matrix is applied on the mandrel either in the form or a foil or by plasma spraying, the reinforcing filament or wires are now wrapped over it. A subsequent layer of the matrix is applied by spraying. Droplets of the molten matrix cover the reinforcement and quickly solidify. Several layers can be applied in this manner. When the matrix is a metal the operation has to be carried on in an inert atmosphere so as to prevent the oxidation of the droplets of the metal.

Chemical vapour deposition is not of very wide application. The metal to be used as matrix is carried along with another material in the vapour phase which decomposes on meeting the hot surface on which is wound the reinforcing phase. Thus the matrix gets deposited on the second phase.

Cold pressing sintering technique is popular for making discontinuous fiber and also particulate composites. Randomly oriented discontinuous fibers or the particles are mixed with the matrix powder, which may be metal or resin, and rolled into sheets or extruded through a die to give rods. If necessary, the operation can be carried on at higher temperatures so as to cause bonding by sintering.

In the directional solidification of eutectic alloys the strengthening phase is produced in the matrix. It requires high purity material. The volume fraction of the reinforcement is fixed by the composition of the eutectic. The molten eutectic alloy is solidified uni-directionally with the result that an inter-metallic phase, which is the strengthening phase, is formed in the direction of solidification. This avoids the need to handle two different kinds of materials. There is the additional advantage of good bonding at the interfaces of the strengthening phase and the matrix.

The above-noted methods are especially applicable for fiber reinforced composites but particulate composites can be prepared by liquid metal infiltration, hot pressing, cold pressing and sintering. Sintering can be used for producing small parts of simple shapes only.

Liquid metallurgy technique has been developed for making particulate composites. The technique is very versatile, using a variety of materials and is capable of producing a variety of shapes and of different sizes. Briefly speaking, the particles are introduced in the molten metal and held in suspension, otherwise they will segregate according to specific gravity, and then solidifying the mass in the desired shape. Various techniques adopted in this broad process differ from each other mainly in the method of keeping the particles uniformly dispersed in the molten metal. One such method uses a stream of a gas to inject the particles in the molten metal. It is then cast in the mould. In another method the molten metal is stirred in such a manner that a vortex is created in it and the particles are introduced. The vortex prevents the segregation of the particles. An interesting way consists of preparing a pellet or briquette of the particles along with the alloy. The concentration of the particles is high. It is then mixed with the molten metal. The pellet disintegrates and the particles get suspended in the melt. A somewhat similar technique is of vigorously agitating the particles with partially solidified metal and then allowing the mass to solidify. It is remelted and cast in moulds. Centrifugal casting is useful in making bearings of large diameter. When graphite or mica particles are dispersed and centrifuged in the molten metal, the particles segregate at the inner periphery of the cylindrical casting. This is helpful as the particles are needed at the inner periphery for lubrication.

Ultrasonic vibrations have also been used to mix and suspend the particles. This has the advantage that the particles get cleaned by the ultrasonic vibrations and become free from adsorbed gases and impurities. This results in stronger bonds between the metal and the particles.

Chapter 19

Solid State Chemistry

19.1 Introduction

The inter molecular forces enhance as we pass from gaseous state to condensed state. Liquids and solids, the constituents of the condensed state, possess a definite volume, definite density but differ in their rigidity or assuming a definite shape. While liquids take the shape of the vessel in which they are poured the solids does not change their shape. This rigidity of shape in solids cannot be attributed to any stronger intermolecular forces or compactness of the atoms or molecules in them as the liquid and solid states have comparable densities. Generally liquids have slightly lower density than solids and we are aware of the exceptions like ice and water where the solid ice has lower density than liquid water at a given temperature. Thus the rigidness of solids is ascribed to decrease in randomness of the components rather than any increase in the compact packing of the atoms or molecules.

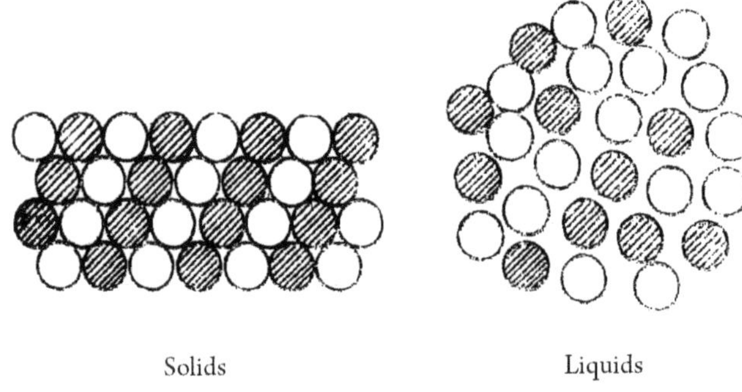

Fig. 19.1 Schematic view of structure in solids and liquids.

From Fig. 19.1 we see that in both solid and liquid states of a substance the densities are almost same, as density is only a measure of number of particles present per unit volume, but the arrangement of particles in the liquid is less orderly when compared to a precise orderly arrangement in the solid state, known as **long range order**. In a small region liquids may show some regularity, termed **short-range order**, but miss the long range order essential for maintaining a definite shape. When we powder a rigid solid, the long range order is broken and the powder assumes the shape of the container, a property of a liquid.

19.2 Classification of Solids

Solid substances are classified into two classes, depending on the randomness of particle arrangement, as crystalline and amorphous solids

19.2.1 Crystalline Solids

A crystalline solid is one in which the constituents have a long range order and are arranged to give a geometrical form characteristic of the substance. The common salt, mica sheets, ice flakes, sugar particles etc are all examples of the crystalline behavior of solids. On breaking a rock salt lump, it always breaks into small sharp-edged pieces and a mica sheet is comfortably pealed into thin sheets as decided by the inherent crystalline characters. Crystalline forms are obtained when the substances are cooled or created by precipitation reaction and the process of preparing the crystalline forms is called **crystallization**. *The branch of science which deals with the geometry, properties and structures of crystals and crystalline substances is called **crystallography**.*

19.2.2 Amorphous Solids

An amorphous solid is one in which the constituents are arranged to give structures having only partial regularity and do not have a definite geometrical arrangement. Glass, pitch, rubber, plastic, gelatinous precipitate of aluminium hydroxide etc., are some typical examples of amorphous solids. They have all the attributes of solids like definite shape, hardness, rigidity but only miss the geometric shapes or crystalline behavior. In true sense they are not solids and occupy an intermediate position between liquids and solids and are also called super cooled liquids of high viscosity. On long standing or heating or under certain conditions these substances achieve certain degree of crystallinity. Thus old glass pieces or heat treated glass show crystalline nature and turns brittle.

A distinguishing property of crystalline and amorphous solids is that the former gives definite, characteristic diffraction pattern for both crystals and their powders where as the later do not.

19.3 Characteristic Geometry of Crystalline Solids

19.3.1 Characteristic Geometry

As discussed already the main characteristic of crystalline solids is their regular arrangement which differentiates from amorphous solids which have partial regularity. In crystalline solids there is a long range order and in amorphous solids the constituents possess a short range order as in liquids. For example the alkali halides like NaCl have cubic structures, ice has a hexagonal, while $CaSO_4$ is rhombohedral (trigonal) in geometry.

19.3.2 Crystallization and Fusion

Translational velocities or kinetic energy of the molecules decrease with the decrease in temperature and on approaching freezing point, the attraction forces overcome the transnational energy and form a crystalline solid. When crystallization starts, heat is evolved, and this evolution arrests any temperature drop. The solid and liquid are at equilibrium at this temperature known as melting point or fusion temperature of the solid or freezing point of the liquid. Any attempt to lower the temperature only results in the freezing of some more liquid. *The amount of heat evolved for the crystallization of one mole of substance is called* ***heat of crystallization.***

The reverse of crystallization is *fusion or melting* of the solid. On heating a crystalline solid the vibrational forces increase, breaking completely the attractive forces and converting into liquid. As the liquid and solid are at equilibrium at fusion temperature any extra heat supplied is used for further fusion. *The amount of heat required for the fusion of one mole of a solid substance is called* ***heat of fusion***.

19.3.3 Melting Points

Owing to a definite arrangement, the crystal constituents are bound together by same type of cohesive forces throughout the crystal. Hence supply of heat is shared uniformly by all the bonds, followed by an increase of internuclear distance. At a particular temperature, called melting point, the translation forces are sufficient to over come the internuclear attractions and the solid converts to liquid. Thus *all the crystalline solids have a sharp melting points.* On the contrary an amorphous solid has only short range order resulting in different internuclear forces. Hence heat supplied will be able to snap some weak bonds, while still some other bonds are intact. Consequently *amorphous substances have no sharp melting points and extend over a temperature range.* Supply of heat initially softens followed by gradual conversion into a fluid.

19.3.4 Sublimations

In case of some solid substances like camphor and solid carbondioxide, *the vapor pressure of the solid crystalline substance reaches the* atmospheric pressure before reaching the melting point. In such cases the solid directly passes into vapour phase without touching the liquid form. This phenomenon where the solid and vapour of a substance are at equilibrium is known as *sublimation* and the equilibrium temperature as sublimation temperature. This category of substances can be melted by increasing the external pressure sufficient to raise the vapor pressure above the solid-liquid equilibrium vapor pressure.

19.3.5 Isotropy and Anisotropy

All solids that have *same properties, independent of direction are called* ***Isotropic***. In cubic crystals and amorphous solids, properties like thermal conductivity, elasticity, mechanical compressibility, refractive index etc, do not depend on the direction of measurement or applications. For example light travels with the same velocity through glass, irrespective of direction of observations.

The crystalline solids other than cubic crystals are said to be *anisotropic as the properties depend on the direction of observations*. When light is passed through an anisotropic crystal it is split into components which travel with different velocities and follow different paths. This phenomenon is called *double refraction*.

The dependence of properties on direction can be understood by considering a two dimensional, crystal lattice. On focussing light on a lattice, with two kinds of atoms, represented in Fig. 19.2, experiences at least two different environments.

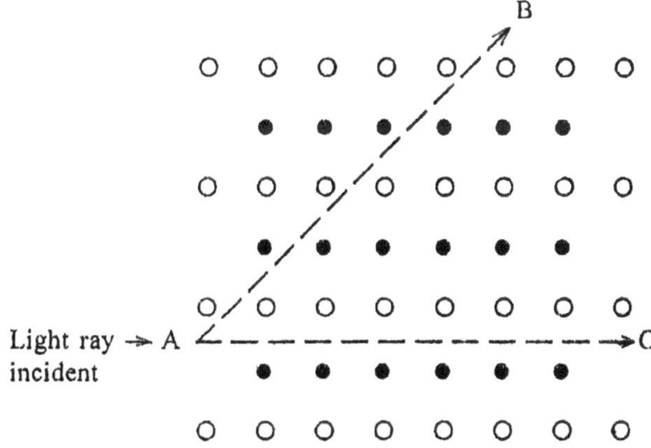

Fig. 19.2 Light passing through an anisotropic lattice.

The component AB passes through planes having alternately placed atoms while the component AC passes between two planes having each, having one kind of atoms. In case, the above lattice consists of one kind of atoms, it is called a cubic lattice. In the cubic lattice components AB and AC experience similar environment, suggesting the absence of directional properties in cubic crystals, Fig. 19.3.

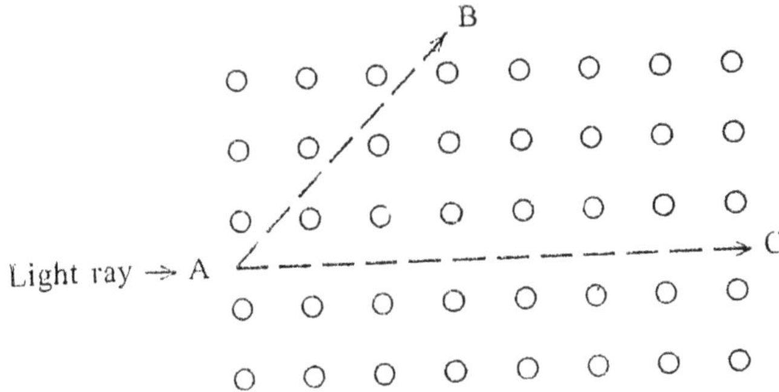

Fig. 19.3 Light passing through a cubic crystal lattice.

In case of an **amorphous solid,** the atoms are placed at random and any ray passing in any direction experiences always an average environment, explaining their **isotropicity.** For similar reason liquids are also isotropic.

Other characteristics of crystals are covered as three **fundamental laws of crystallography.**

19.4 Laws of Crystallography

The size and shapes of crystals depend on the experimental conditions employed for crystallization. For example quick cooling of a hot, saturated solutions of brine gives small crystals but a slow cooling gives crystals of few centimeter length. Traces of impurities, often greatly modify the relative rates of growth of various faces and then lead to different crystalline forms. Irrespective of differences in external geometric shapes in a three dimensional space, the crystal structures are based on three fundamental laws, (A) the law of constancy of interfacial angles (B) the law of rational indicies and (C) the law of symmetry.

A. The Law of Constancy of Interfacial Angles

Although the crystal appearance of a substance is variable the angles between the various faces are characteristic and distinctive features of a substance. Thus a crystal may appear, in a two dimensional space, as a perfect hexagonal (structure (a)) or an irregular hexagonal (structures (b) and (c)), Fig. 19.4.

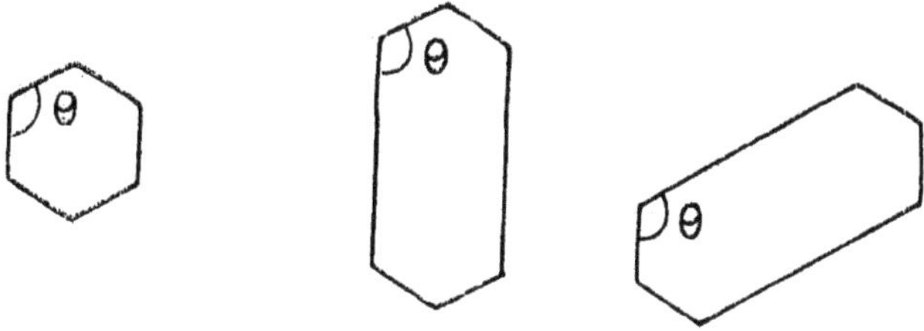

Fig. 19.4 Three crystal structures of the same substance.

Irrespective of the shapes or appearance, the angle of intersection of any two corresponding faces remain the same. With large crystals, a contact goniometer is used to measure these *crystal or interfacial angles*, but for tiny crystals a reflection goniometer is used. From the interfacial angles it is possible to derive a set of crystallographic axes that describe the crystal.

B. Law of rational indices (Haüy's law)

In 1784 Haüy proposed the law of rationality of indices or intercepts. It states *that the ratio of intercepts made by the crystal on the three coordinate axes remain the same without altering with the size of the crystal.*

Consider a crystal plane intersecting the x, y, z axes at A, B, C respectively such that OA = a, OB = b and OC = c, Fig. 19.5. The ratio of the intercepts is

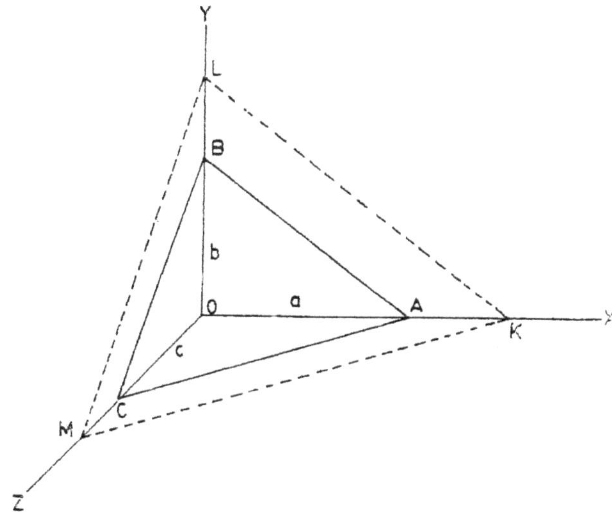

Fig. 19.5 Rationality of indices.

a:*b*:*c*. Considering one and half times larger crystal the same plane intersects the *x*, *y*, *z* axes at K,L,M respectively. Such that OK = $\frac{3}{2}$ *a*, OL = $\frac{3}{2}$ *b* and OM = $\frac{3}{2}$ *c* therefore OK:OL:OM = *a*:*b*:*c*. Hence the ratio of the intercepts, OK:OL:OM for the larger crystal is same as the intercept of a smaller crystal on the three coordinate axes.

*This law of rational indices may also be stated that it is possible to choose along the three coordinate axes unit distances (**a**, **b**, **c**), not necessarily the same length, such that the ratio of the three intercepts of any plane in the crystal is given by **ma:nb:pc**, where m, n and p are either integral whole numbers including infinity or fractions of whole numbers.* When a plane is parallel to a particular axis, the coefficient corresponding to that axis takes a value of infinity.

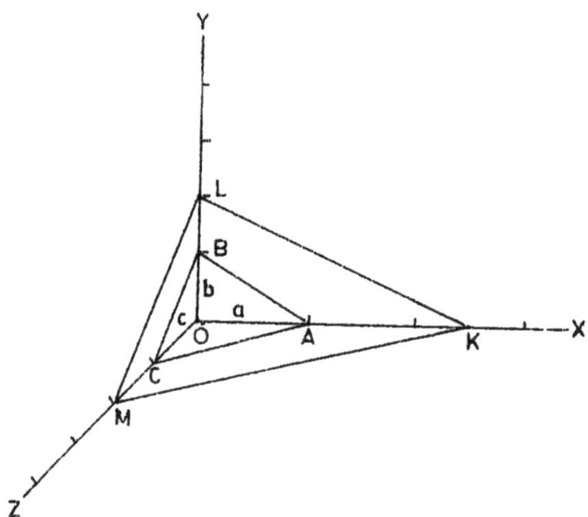

Fig. 19.6 Miller indices.

The Coefficients *m*,*n* and *p* are called **Weiss indices**. In 1938 W.H. Miller, modified the Weiss indices and introduced more convenient **Miller indices** to identify different planes of a crystal. In Fig. 19.6, let ABC be is a ***unit plane*** with distance *a*,*b* and *c* along *x*,*y* and *z* axes respectively. In order to describe any other lattice plane KLM in terms of the unit plane, first we measure the distances OK, OL and OM as multiple of *a*,*b* and *c*. In present example OK = $\frac{5}{2}$*a*, OL = 2*b* and OM = 2*c*. The Weiss indices of the plane

KLM are 5/2, 2, 2. *The Miller indices or* ***h k l values*** *of a plane are obtained by taking reciprocals of the Weiss indices and multiplying through smallest number that will express all the reciprocals as integers.* Hence Miller indices for K L M plane are obtained as follows.

(a) By taking reciprocals of Weiss indices : 2/5, 1/2,1/2

(b) Multiplying with 10, the least common multiple (LCM) of the denominators to get integers: 4, 5, 5. The KLM plane is therefore designated as (4, 5, 5) plane. Similarly for the unit plane ABC, the Miller indices are (111). In case of the plane ABC is cutting x and y axes but parallel to z axis, it is identified by Miller indices as (110) plane. If the plane ABC just intersects x axis, but parallel to other axes it is designated as (100) plane.

In addition to designating the planes, Miller indices could also be utilized to calculate the distance, d_{hkl}, between the parallel planes of a cubic crystal using the general formula

$$d_{hkl} = \frac{a}{\sqrt{h^2 + k^2 + l^2}} \qquad .....(19.1)$$

where a is the length of the side of a cube and h, k, l are the Miller indices of the plane.

C. The Law of Symmetry

This law states that *all crystals of the same substance possess the same elements of symmetry*. There are three possible types of symmetry, (i) plane of symmetry (ii) axis of symmetry and (iii) center of symmetry.

(i) **Plane of symmetry :** A crystal is said to have plane of symmetry, when an imaginary plane can divide it into two parts that are identical mirror images.

A cubic crystal can have two types of plane of symmetry.

(a) **Rectangular planes of symmetry :** A cube has six sides. Three rectangular imaginary planes, parallel to x, y and z axes, which can cut the cube into two identical halves, Fig. 19.7(a). Hence we say that a cube has three rectangular planes of symmetry.

(b) **Diagonal planes of symmetry :** The six imaginary planes passing through opposite edges and splitting the cube into two identical halves are called diagonal planes of symmetry, Fig. 19.7(b).

Thus we have total nine (3 + 6 = 9) planes of symmetry.

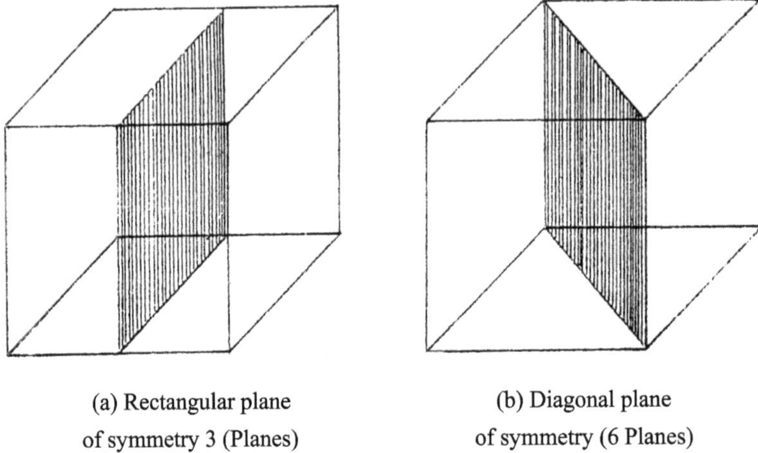

(a) Rectangular plane of symmetry 3 (Planes)

(b) Diagonal plane of symmetry (6 Planes)

Fig. 19.7 Planes of symmetry.

(ii) **Axis of symmetry :** A crystal is said to have an axis of symmetry, when an imaginary line passing through the crystal about which it may be rotated presents exactly the same appearance more than once in the course of one complete (360°) revolution.

(a) **Four fold axis of symmetry :** When a cube is rotated about an imaginary line passing through middle of opposite sides, the faces present themselves four times, Fig. 19.8(a). Hence the axis is said to have **four fold symmetry**.

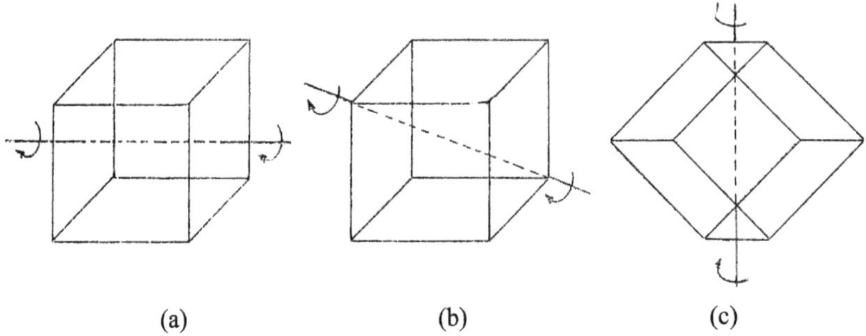

(a) (b) (c)

Fig. 19.8 Axis of symmetry.

(b) **Three fold axis of symmetry :** If a crystal is rotated about an axis that passes through diagonally opposite corners of a cube the sides repeat thrice in the course of rotation by 360°. Such axis is termed as axis of **three fold (triad) symmetry** (Fig. 19.8(b)).

(c) **Two fold axis of symmetry :** In case a crystal is rotated about an axis passing through centers of the two diagonally opposite edges of a cube, the cube presents the same appearance twice in the course of rotation of 360°. This axis is said to be one of **two fold (diad) symmetry** (Fig. 19.8(c)).

A cubical crystal has only the above three types of axes of symmetry. A regular cube has three axes of four-fold symmetry, passing through three pairs of opposite sides, four axes of three-fold symmetry passing through opposite corners and six axes of two-fold symmetry emerging from opposite edges making a total of thirteen (3 + 4 + 6) axes of symmetry.

(d) **Multi-fold axis of symmetry :** Depending on the geometry, a crystal is allowed to have even more than four-fold axis of symmetry.

Thus a hexagon can have six-fold axis of symmetry, Fig.19.9(a), and a pentagon can have five-fold axis of symmetry, Fig.19.9(b), etc.

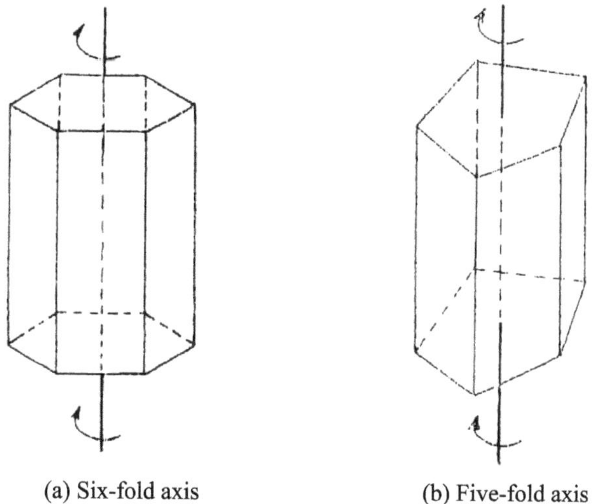

(a) Six-fold axis (b) Five-fold axis

Fig. 19.9 Multi-fold axis of symmetry.

(iii) **Center of symmetry :** A crystal may have a central point, known as center of symmetry, such that any line drawn through it will intersect the surface of a crystal at equal distances in both directions. The center of symmetry is a mass center and all the planes and axes passing through the crystal invariably pass through this point.

A cubic crystal has one center of symmetry as shown in Fig. 19.10.

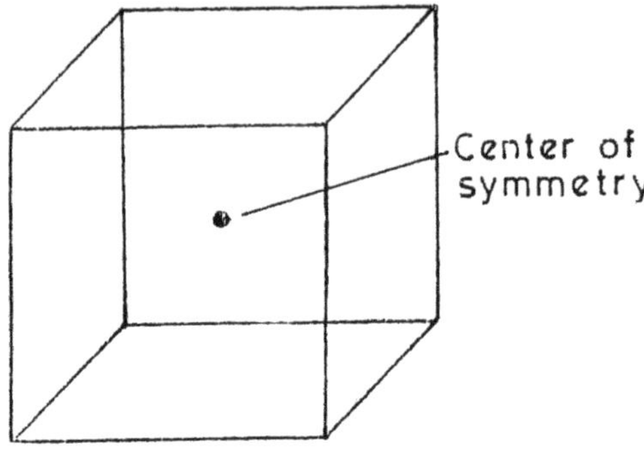

Fig. 19.10 Centre of symmetry.

All the three types plane, axis and center of symmetry, are collectively called **elements of symmetry**. A perfect cube has a total of twenty three (9 planes + 13 axes + 1 center) elements of symmetry.

19.5 Crystal Systems

Any repetitive pattern has a symmetry element associated with it, whether it is stacking of bricks in a wall or stacking of rice bags in a godown or stacking of fruits on a push cart, Fig. 7.11.

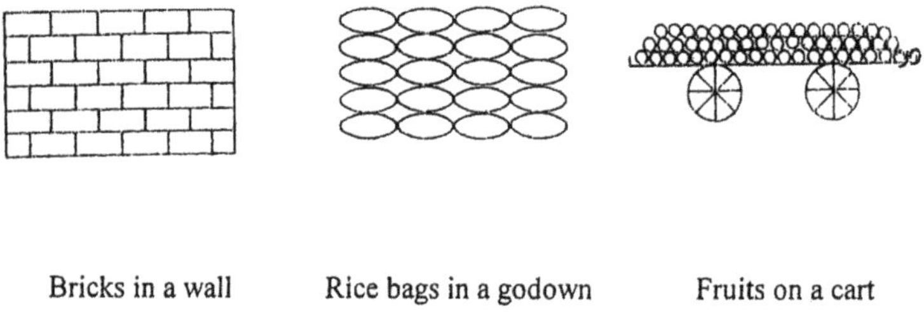

Bricks in a wall Rice bags in a godown Fruits on a cart

Fig. 19.11 Symmetry in some repetitive patterns.

Similarly the atoms or ions or molecules are packed in an orderly fashion to give crystals and are called **asymmetric units**. The pattern formed by points representing the asymmetric units is called **space lattice** or **crystal lattice**. Each point in the lattice has exactly same environment as any other point representing the same atom or ion. The space lattice, ineffect, is the scafolding for the crystal structure.

19.5.1 Unit Cell

In a crystal the number of particles is enormous and describing the positions of all these particles or their lattice points is impossible. Fortunately *it is sufficient if we describe the basic repeating unit of the lattice to completely understand the space lattice*. This basic repeating unit is called **unit cell**. It is like brick in the wall.

Fig. 19.12 illustrates a two dimensional lattice. In this a square lattice forms the basic unit and entire space lattice could be generated by repeated movement of this unit lattice in a two dimensional space. *An infinite number of unit cells can be described in the same lattice, but we normally choose the one with sides that have shortest lengths and are most nearly perpendicular to each other.* So in this example only a square lattice is considered.

Fig. 19.12 A square lattice (2D – lattice representation).

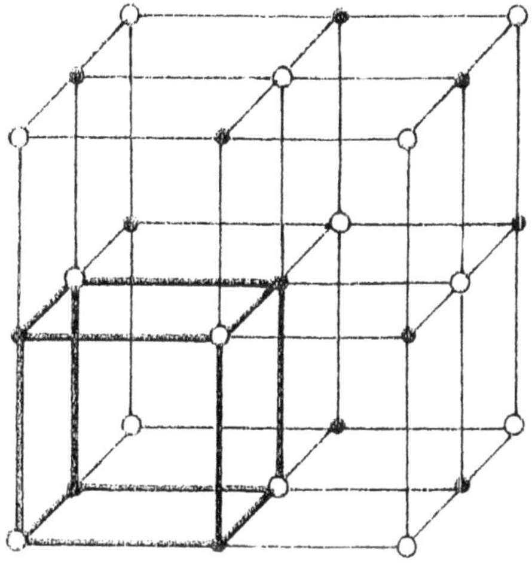

Fig. 19.13 A simple cubic lattice (3D – lattice representation).

Extension of above cited two dimensional lattice model, to a three dimensional space we get a simple cubic lattice as the unit cell, represented by dark lines in Fig. 19.13.

If the unit cells has one lattice point or asymmetric unit at its origin (i.e., only at the corners), it is called a **primitive cell**. If the lattice points are at the centers or on the face of the cell they are called non-primitive cells.

Irrespective of shapes of the crystals, the unit cells of all three dimensional lattices are similar in that they all have eight corners Fig. 19.14. Unit cells differ by the lengths of their edges (a, b and c) and angles between them (α, β and γ), Fig. 19.14.

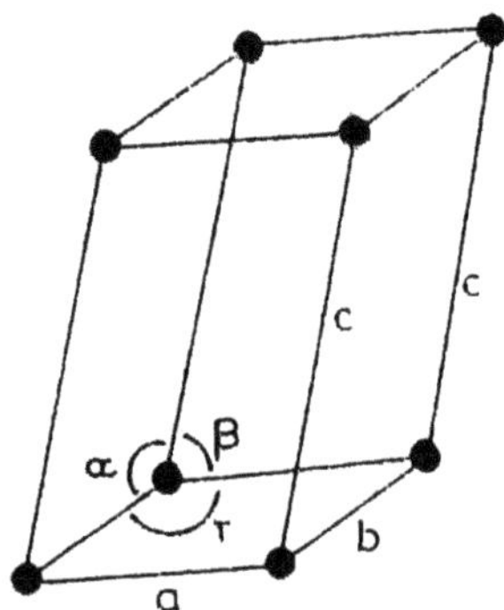

Fig. 19.14 A typical unit cell.

19.5.2 Bravais Lattice

In 1848 August Bravais showed that only 14 different kinds of lattices are possible, known as **Bravais lattices**, these were further divided into seven basic **crystal systems**. For example a cubic crystal system has one or more of three kinds of lattices known as (i) The simple cube (Fig. 19.15 (a)), (ii) The face centered cube (Fig. 19.15 (b)) and (iii) The body centered cube (Fig.19.15(c)). The other non-cubic Bravais lattices are also represented in Fig. 19.15.

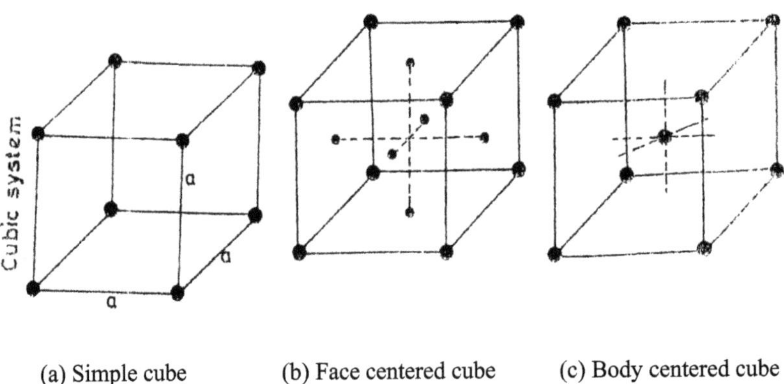

(a) Simple cube (b) Face centered cube (c) Body centered cube

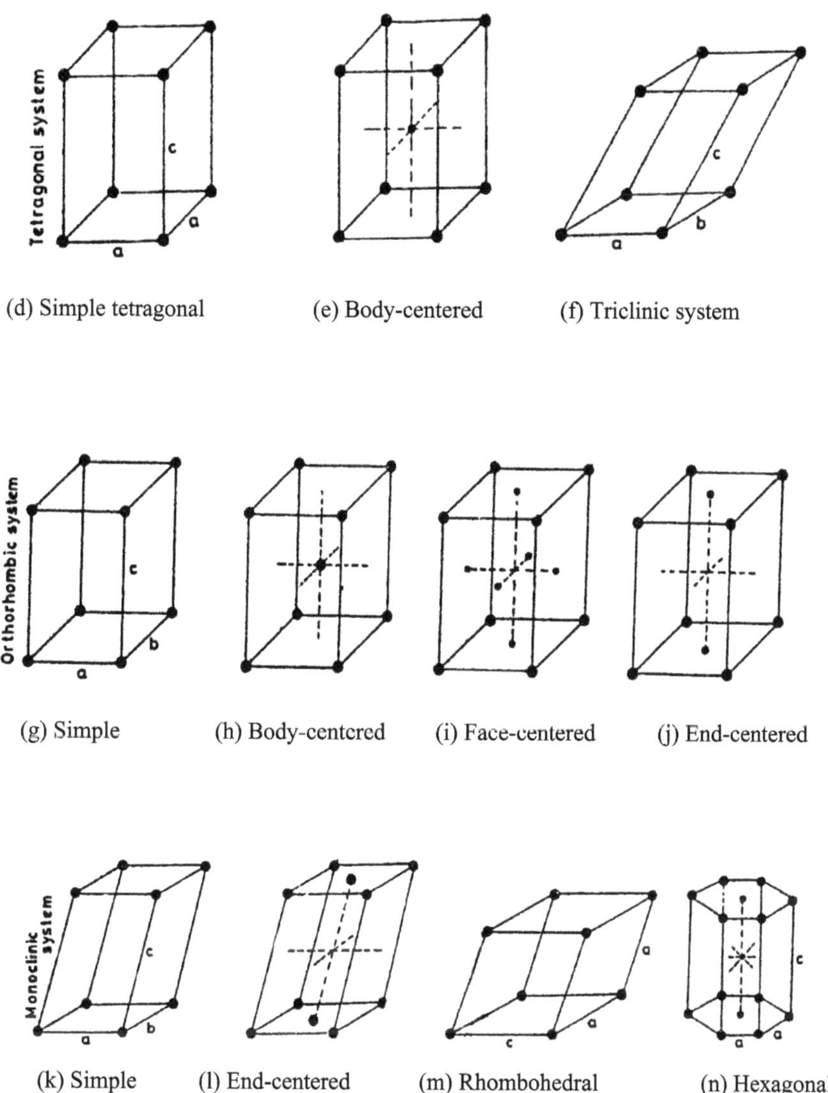

(d) Simple tetragonal (e) Body-centered (f) Triclinic system

(g) Simple (h) Body-centered (i) Face-centered (j) End-centered

(k) Simple (l) End-centered (m) Rhombohedral (n) Hexagonal

Fig. 19.15 The Bravais lattices.

A simple cube lattice has points only at the eight corners of a unit cell. A face centered lattice has in addition to the points at eight corners of the unit cell, it has a point at the center of each of the six faces. A body centered lattice has an additional point in the mass center of a simple cubic lattice.

The properties of the seven crystal systems are summarized in Table 19.1

Table 19.1 Properties of the crystal systems.

System	Edge Lengths	Angles	Examples
Cubic	$a = b = c$	$\alpha = \beta = \gamma = 90^o$	NaCl, CaF_2, ZnS, Diamond
Tetragonal	$a = b \neq c$	$\alpha = \beta = \gamma = 90^o$	SnO_2, TiO_2, $NiSO_4$
Orthorhombic	$a \neq b \neq c$	$\alpha = \beta = \gamma = 90^o$	KNO_3,K_2SO_4, S (rhombic)
Monoclinic	$a \neq b \neq c$	$\alpha = \beta = 90^o \neq \gamma$	$NaHCO_3$, $CaSO_4.2H_2O$ S (monoclinic)
Rhombohedral	$a = b = c$	$\alpha = \beta = \gamma \neq 90^o$	$CaSO_4$, $CaCO_3$, $NaNO_3$
Hexagonal	$a = b \neq c$	$\alpha = \beta = 90^o$, $\gamma = 120^o$	SiO_2, HgS, Graphite, Ice
Triclinic	$a \neq b \neq c$	$\alpha \neq \beta \neq \gamma \neq 90^o$	$K_2Cr_2O_7$, $CuSO_4.5H_2O$

19.6 Difraction of X-Rays

Sound waves bend round the corners. Sounds made in the adjoining rooms reach us after bending round the edges of the walls. Fresnels observed this type of bending for the radiation and named it as diffraction of light. But the wave length of light being very small in comparison to wave length of sound waves, the diffraction effects are not very prominent and require special apparatus to study these effects. To observe the diffraction effect of light one has to pass fine beam of light over a razor edge or a thin wire or a diffraction grating constructed by stretching a number of thin silver wires (200 per cm) across a frame work or ruling thousands of lines per inch on a glass plate. The primary condition for obtaining a diffraction pattern using a grating is that the separation of lines as the grating should be less than the wave length of the light employed. Thus manually prepared gratings were found suitable to visible radiation having wavelength in the range 4000-8000 $\overset{0}{A}$, but, failed to produce diffraction patterns when X-rays having a wave length of $\approx 10\overset{0}{A}$ are used, instead of visible radiation having wave length in the range 4000-8000 $\overset{0}{A}$. W.von Laue (1912) first explored successfully the possibility of using a crystal as grating, where the inter atomic distance is of the order of few $\overset{0}{A}$ units.

19.6.1 Laue diffraction pattern

Laue passed a beam of inhomogeneous X-rays, that is a beam consisting of X-rays of many wave lengths, through a crystal and the resulting X-rays were allowed to fall on a photographic plate. A pattern indicated by Fig. 19.16 was observed, which is called Laue photograph or Laue pattern. Though each type of crystals exhibits, a characteristic pattern interpretation needs advanced mathematical and geometrical knowledge.

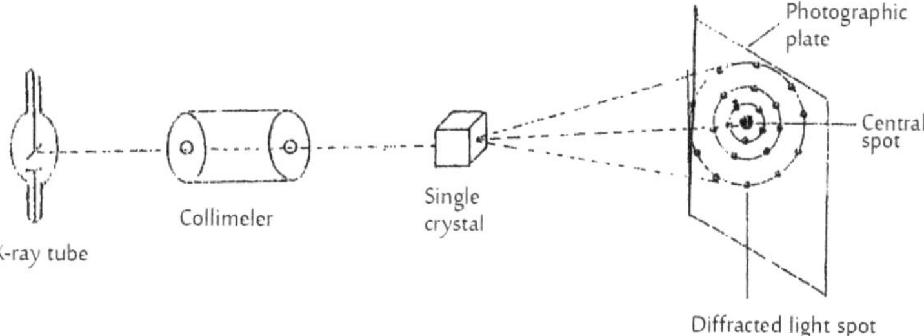

Fig. 19.16 The Laue Method.

The fact that the light coming out of the single crystal forms Laue pattern, consisting of spots lying on equally spaced circles indicates that the atoms in a crystals are spaced in a definite, repetitive manner.

19.6.2 Bragg's diffraction pattern

A clue from Laue pattern was taken by W.L. Bragg and W.H.Bragg, who obtained a reflection diffraction pattern, which is rather simple in the interpretation. They used homogenous X-rays (X-rays of single wavelength) and crystal as the diffraction grating. Consider that a crystal consists of different planes *AA*, *BB* etc., parallel to each other (Fig. 19.17).

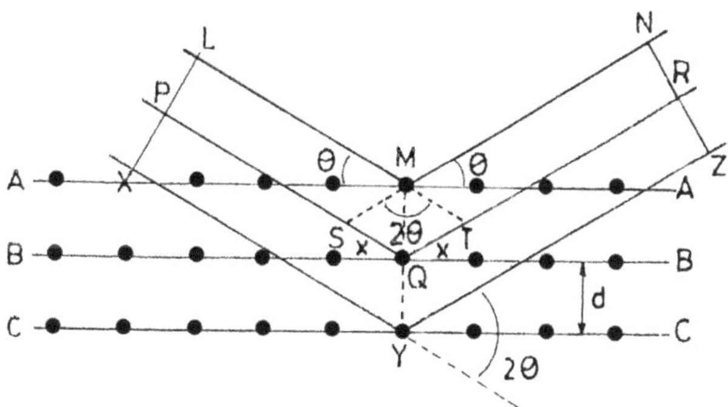

Fig. 19.17 Derivation of Bragg's Law.

The collimated X-ray beam strikes the crystal surface at a glancing angle θ. Part of the beam *LM* is reflected by the atoms of the first plane, *AA*, to yield reflected ray *MN*. Similarly another part *PQ* that could pass through the inter atomic spacing of first plane in reflected by the second plane *BB* to yield *QR*. If the rays *LMN* and *PQR* have to reinforce one another, they must have the same phase. This condition is met when the difference in lengths of the path traced by these two rays is equal to an integral number of wave lengths. The difference in the path length can be obtained by drawing perpendicular *MS* and *MT* to *PQ* and *QR* respectively.

If $\quad SQ = QT = x$

Difference in path length = SQ + QT = $2x$

From the above stated condition for reinforcement, the path difference;

$$2x = n\,\lambda, \qquad \text{.....(19.2)}$$

where n is an integer.

As $\angle LMS = \angle AMQ = 90^{O}$, we get the glancing angle

$$\angle LMA = \angle SMQ = \theta.$$

Similarly $\angle TMQ$ is also equal to θ.

In the right angle triangle $\angle SMQ$, the length of the side

$$SQ = MQ \text{ Sin } \theta$$

If the inter planar distance MQ is taken as "*d*" then;

$$x = d\ Sin\ \theta \qquad \text{.....(19.3)}$$

Substituting the value of x in eqn. 19.2 we have,

$$\mathbf{n\lambda = 2d\ Sin\ \theta} \qquad \text{.....(19.4)}$$

where $\quad n = 1, 2, 3....$

This result is known as **Bragg equation**.

For a given set of lattice planes the inter planar distance, *d*, has a fixed value and for homogenous X-rays, the wave length λ also has a definite value. Hence the possibility of obtaining reflection maxima or reinforcement depends on the glancing angle θ. This values gradually increases as defined by the integer values of *n* equal to 1, 2, 3.... in the Bragg equation.

The maximum value of Sin θ being unity for the maximum reflection angle or reflection of nth order;

The inter planar distance $d = n\,\lambda/2$ .....(19.5)

19.6.3 Bragg X-ray Diffractometer

In the Bragg X-ray diffractometer (or spectrometer), a collimated beam of X-rays are allowed to fall upon a crystal mounted on a rotating table and the angle of rotation may be measured on the scale instrument, Fig. 19.8. After reflection on the crystal surface the X-ray beam is allowed to pass into a detector, an ionizing chamber or a Geiger counter, the response of the detector being directly proportional to the intensity of the reflected X-ray beam. Reinforcement of the beam is indicated by the strongest reflection corresponding to the different integral values. By measuring the values of θ, the inter planar spacing d can be computed from Bragg equation which thus defines the plane of a crystal. This procedure may be repeated to identify all important planes of a crystal. This helps in the determination of crystal structure of the compound.

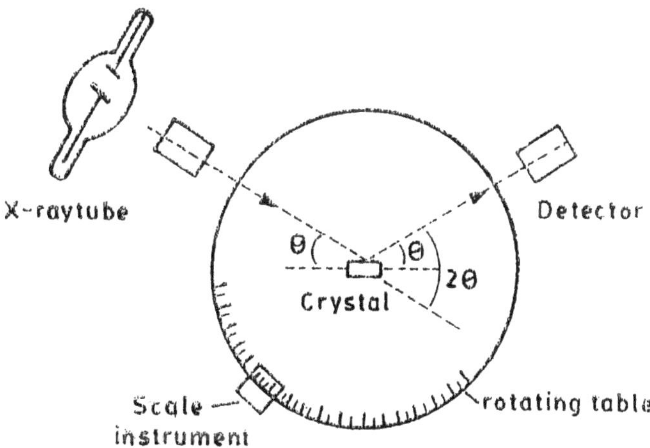

Fig. 19.18 Bragg X-ray diffractometer.

A typical diffraction pattern of tungsten is shown in Fig. 19.19. The number on the peaks are Miller indices of the planes which produce the peak.

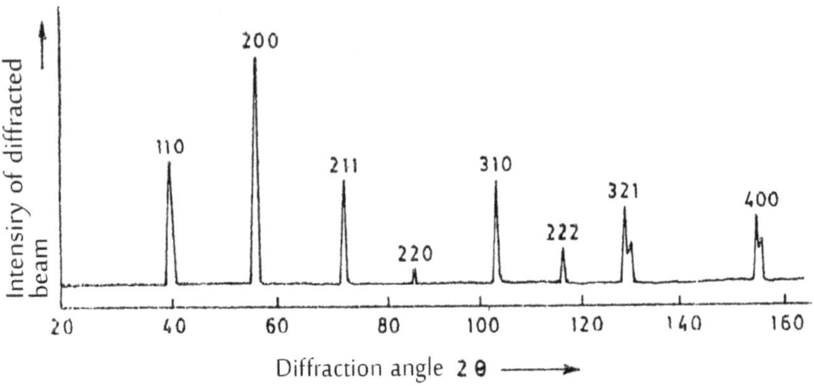

Fig. 19.19 Diffraction pattern of tungsten.

The fact that some planes are densely placed than others, produces a variation in intensity of the diffracted beam. The planes of high atomic density, scatter X-rays better and produce the more intense peak; example 200 peak in Fig. 19.19.

19.6.4 Interpretation of X-ray diffraction results

From the knowledge of X-ray diffraction studies we get a very useful quantity *d*, the distance between the lattice planes. As already discussed under Miller indices, a comparison of experimentally obtained *d* values and the values reached by theoretical consideration of various lattice planes help in arriving at the crystal structural of a substance.

For example;

A simple cube has three lattice planes described by the Miller indices (100), (110) and (111), Fig. 19.20.

Similarly in a face centered cube the lattice planes are described mainly by Miller indices (200) (220) and (111) and a body centered cubic lattice planes by (110), (200) etc. These are represented in Fig. 19.21, 19.22 and 19.23. The conditions for the appearance of X-ray diffraction pattern may be generalized as follow;

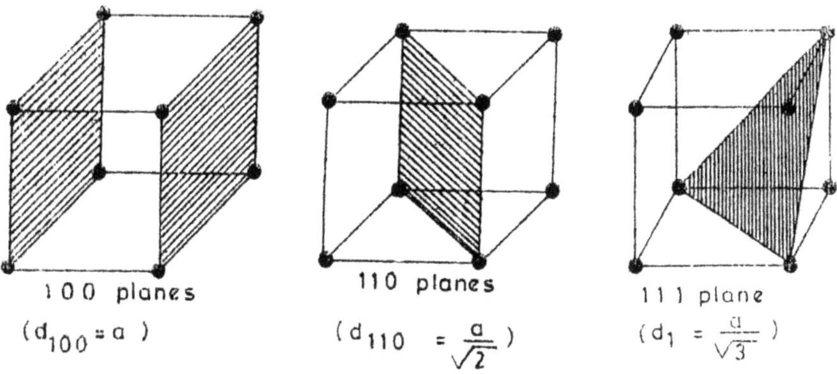

Fig. 19.20 Lattice planes of a simple cube.

(i) Simple cubic lattice : *h, k, l* may have any value.

(ii) Face-centered cubic lattice : *h. k. l* are odd or even (including zero)

(iii) Body centered cubic lattice: $h + k + l$ must be even.

Substituting possible *h k l* values in the equation 19.1 gives the ratio of distances of separation for lattice planes of any crystal.

For simple cube it is;

$$d_{100} : d_{10} : d_{111} = \frac{a}{\sqrt{1}} : \frac{a}{\sqrt{2}} : \frac{a}{\sqrt{3}}$$

$$= 1 : 0.707 : 0.577$$

The ratio for face centered cubic lattices is;

$$d_{200} : d_{220} : d_{111} = \frac{a}{2} : \frac{a}{2\sqrt{2}} : \frac{a}{\sqrt{3}}$$

$$= 0.5 : 0.354 : 0.577$$

For body centered cube;

$$d_{200} : d_{110} : d_{222} = \frac{a}{2} : \frac{a}{\sqrt{2}} : \frac{a}{2\sqrt{3}}$$

$$= 0.5 : 0.707 : 0.289$$

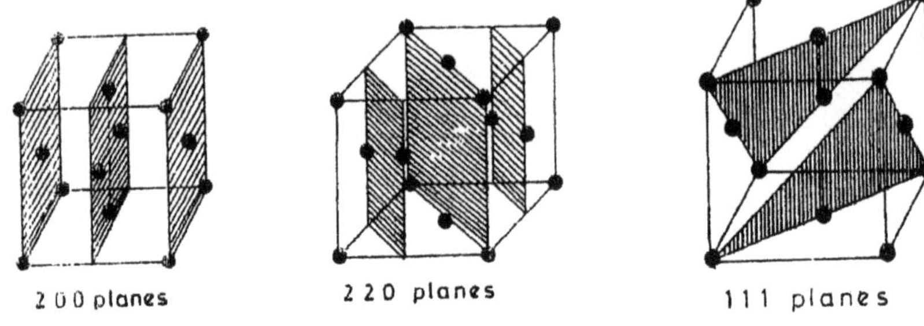

Fig. 19.21 Lattice planes of F.C.C.

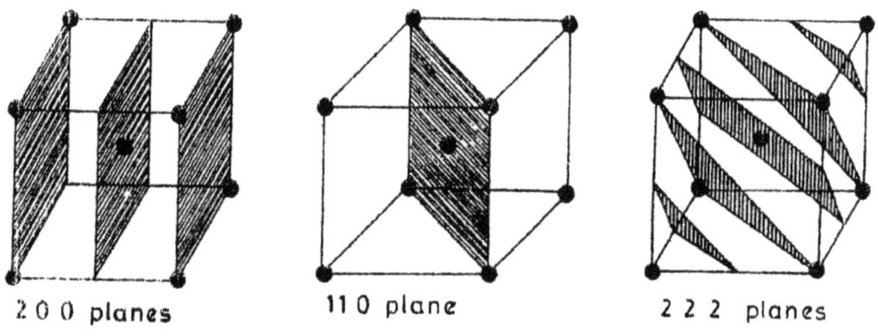

Fig. 19.22 Lattice planes of B.C.C.

Thus a comparison of experimental *d* values, and *d* values obtained from geometrical considerations lead to assignment of crystal structure.

19.7 Crystal types and properties

The geometry and physical properties of crystalline solids are determined by the types of particles present at the lattice points and the nature of attractive forces between them. Crystals, based on the nature of particles that form the space lattice, are classified into four categories, (A) Ionic crystal, (B) Covalent crystals, (C) Molecular crystals and (D) Metallic crystals.

19.7.1 Ionic Crystals

In an ionic crystal, the lattice points are occupied by positive and negative ions and the binding force between them is electro-static in nature.

(i) *Chemical units at lattice points :* As the definition states positive and negative ions are the chemical units at lattice points.

(ii) *Nature of forces operative between lattice points :* The electrostatic attractions are non-directional and each ion maintains an environment of oppositely charged ions. Because of non-directional property of electrostatic forces, the resultant lattice structure is mainly decided by the size of ions and the charge on them and the ions arrange themselves to maximize the forces of attraction and minimize the repulsions.

(iii) *Melting points :* Because electrostatic forces are strong, ionic crystals have large lattice energies. Hence ionic crystals melt at relatively high temperatures.

(iv) *Hardness :* Owing to strong electrostatic forces the lattice points are compactly arranged imparting hardness to the ionic crystals.

(v) *Brittleness :* Ionic crystals are very brittle and tend to shatter when struck with a hammer. Before striking, the stability of an ionic crystal is due to facing of ions of opposite charge to each other. When it is struck with a hammer the lattice planes slip from their position bringing like charged ions face each other, Fig. 19.23. The repulsive forces between like charged ions lead to the shattering of the crystal.

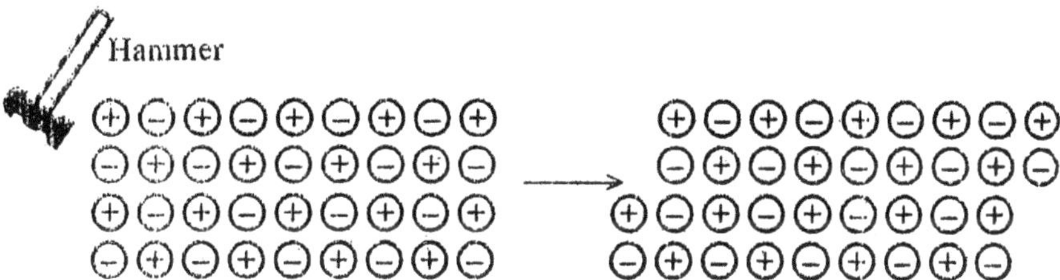

Stable due to attraction between oppositely charged ions facing each other.

Repulsions between like charged ions lead to shattering of the crystal.

Fig. 19.23 Ionic Crystals.

(vi) *Electrical conductivity :* In crystalline state the ionic crystals are poor conductors of electricity because the valence electrons are immobile and held rigidly. In the molten state or when dissolved in a solvent, the ions are free to move about and ionic substances become good conductors of electricity.

19.7.2 Covalent Crystals

In a covalent crystal the lattice points are occupied by atoms that are bound to the neighbouring atoms by covalent bonds.

(i) *Chemical units at the lattice points :* Covalently bound atoms form the chemical units at the lattice points.

(ii) *Nature of forces operative between lattice points :* Covalent bonds, which are highly directional, are operative between lattice points. Diamond is an example of this category where elemental carbon atoms are covalently bonded to four nearest, neighbouring carbon atoms. Carborundum (Silicon carbide) and quartz (silicon dioxide) are other typical examples of covalent crystals. The directional properties of these crystals result in the development of specific faces of a crystal.

(iii) *Melting points :* Because of strong covalent bonds covalent crystals have relatively high melting points.

(iv) *Hardness :* Because of inter locking network of the covalent bonds covalent crystals are very hard and used as abrasive materials and cutting tools.

(v) *Electrical conductivity :* The electrons in the crystals are tied up in covalent bonds and hence except graphite they are bad conductors of electricity.

(vi) *Solubility :* Covalent crystals are not easily soluble because it involves breaking of high energy covalent bonds.

19.7.3 Molecular Crystals

In molecular crystals, the lattice points are occupied either by molecules or atoms and are bonded by weak forces of attraction like random forces or dipole-dipole attractions or hydrogen bonds.

(i) *Chemical units at lattice points :* Atoms and molecules occupy the lattice points.

(ii) *Nature of forces operative between lattice points :* The atoms or molecules are bonded by forces much weaker than covalent bonds. Non-polar substances like O_2, naphthalene, CO_2 etc., are bonded by random forces. Polar molecules like SO_2 in their solid crystalline state are together due to dipole-dipole interactions. Polar molecules like H_2O (in ice), NH_3, HF etc., which have a possibility of formation of hydrogen bond are held together in the solid state by hydrogen bonds.

(iii) *Melting points* : The weak intermolecular forces present in the molecular crystals are responsible for low melting points of these solids.

(iv) *Hardness* : Molecular crystals are soft due to less compact structures resulted by weak molecular forces.

(v) *Electrical conductivity* : Molecular crystals are also poor conductors of electricity as the electrons bound to molecules are not able to move freely. In the molten state polar molecules conduct electricity but the non-polar solids remain poor conductors.

19.7.4 Metallic Crystals

In metallic crystals the positive metal ions are situated at lattice points, with valence electrons that belong to the whole crystal instead of any single element, around them. The negatively charged electrons hold the positively charged ions together. To maintain the electrical neutrality the number of positive and negative charges are exactly balanced. They assume one of the three following arrangements.

1. Hexagonal Close Packed - HCP
2. Cubic Close Packed (Face Centered Cubic) – CCP or FCC
3. Body Centred Cubic – BCC

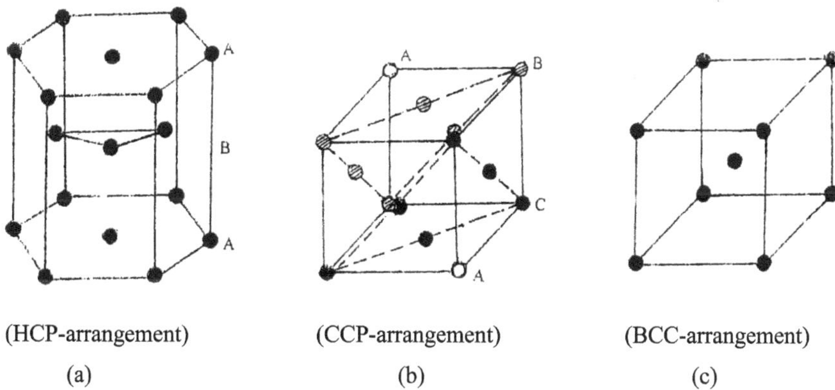

(HCP-arrangement) (a) (CCP-arrangement) (b) (BCC-arrangement) (c)

Fig. 19.24 Packing arrangements in metallic crystals.

In hexagonal close packed arrangement each atom has 12 closest neighbours (coordination number =12) and arranged in two different layers of atoms, each type appearing alternatively. This repeating pattern is called AB AB AB ..... pattern. The atoms occupy 74% of the available space.

In CCP or FCC arrangement the crystal is generated by three different layers A, B, C repeated and the pattern is represented by (ABC.ABC .....). This crystal too has a coordination number of 12. This class of crystals occupy 74% of the available space.

The BCC crystals are relatively less compact and has 8 neighbouring atoms. They fill 68% of the available space. The characters are summarized in Table 19.2.

Table 19.2 Comparison of HCP, CCP and BCC structures

Property	**hcp**	**ccp**	**bcc**
Coordination No.	12	12	8
Volume occupied	74%	74%	68%
Type of packing	(AB AB AB...)	(ABC.ABC...)	---
Malleability and ductitity	Less malleable hard & ductile	Malleable & ductile	Malleable & ductile
Metals	Be, Mg, Ca, Cr V, Mo & Zn	Cu, Ag, Au, Pt.	Alkali metals & Fe

Bonding in metal is discussed later.

(i) *Chemical units at lattice points :* The positive metal ions occupy the lattice points and give metallic crystals.

(ii) *Nature of forces operative between the lattice points :* The positive ions in the crystal lattice are held together by electrostatic attractions with a sea of electrons moving around them. The electrons can move freely in between the lattice planes and they act as a glue in holding any two positive ion lattice points, Fig. 19.25.

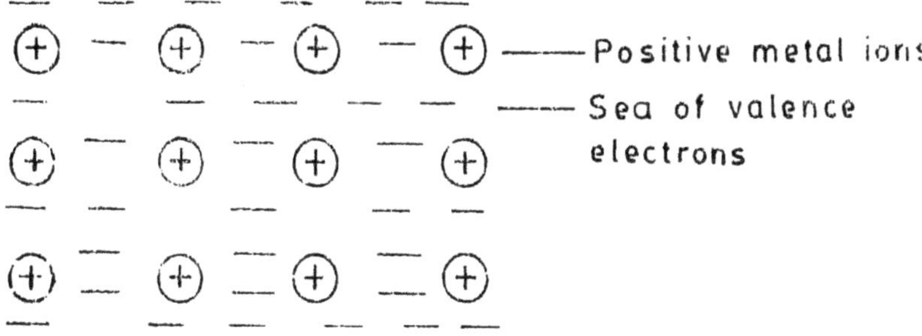

Fig. 19.25 Metallic Crystals.

The formation of sea of mobile electrons can be understood by considering the fact that the valence electrons possess almost identical lattice energy. When two metallic atom come into contact, due to identical energy states all the valence electrons move around them without any labelling of electrons to a particular atom. The strength of the metallic bonds (electrostatic attractions) depend on the number of free electrons available per atom, the atomic radii and nature of valence orbital (s, p, d or f).

(iii) *Melting points :* As the electrostatic attractions depend on nature of valence orbital, atomic radii and mobile electrons available per atom, and the bond strength vary from very weak to very strong. Hence metallic crystals show low and high melting points. For example sodium metallic crystal utilizing 2s valence electrons melt at a low temperature while copper or iron utilizing 3d valence electrons have high melting point.

(iv) *Hardness :* Due to dependence of electrostatic attractions on valence electrons the hardness of the metal crystal also varies from soft to very hard. Sodium metal is soft and can be cut with a knife while iron requires special cutting tools.

(v) *Electrical conductance :* Because of free mobile electrons all the metallic crystals are good conductors of electricity.

(vi) *Metallic Lustre :* The free electrons on the external surface of a metallic crystal are excited when light falls on them. The excitation energy is reemitted in the form of light giving an appearance of reflection of light which we call the metallic lustre.

(vii) *Malleability and ductility* : The metallic bond is non-directional and the lattice planes, containing the positive metal ions are held together by a mobile electron film. These two properties allow the flattening of metal by hammering into thin sheets, known as malleability,and drawing into thin wires, known as ductility.

(viii) *Tensile strength* : Even thin wires of metal bear heavy loads and resist stretching and break after crossing a particular load. This behavior is termed as tensile strength. The high tensile strength is due to strong electrostatic attractions operating between metallic planes.

All four types of crystal properties are summarized in Table 19.2.

Table 19.2 Summary of properties of crystalline solids

Property	Ionic	Covalent	Molecular	Metallic
1. Asymmetric units at lattice	Positive and negative ions	Atoms	Atoms or molecules	Positive ions
2. Nature of forces operating at lattice points	Electrostatic attractions between the ions	Covalent bonds	London forces or dipole-dipole attractions or hydrogen bonds	Electrostatic attractions between positive ions and valence electron sea
3. Melting point	High melting	High melting	Low melting	Low to high melting
4. Mechanical properties	Hard and brittle	Very hard	Soft	Soft to hard, malleable, ductile and high lustre
5. Electrical conductance	Poor conductance	Poor conductance	Poor conductance	Good conductance
6. Examples	NaCl (salt) $CaCO_3$ (lime stone), $MgSO_4$ (epsom)	C(diamond) SiC, SiO_2	CO_2 (dry ice) H_2O (ice), NH_3 HF, SO_2	Na, Cu, Fe, Pt Ni, etc.

19.8 Defects in Crystals

A crystal lattice is an idealization never achieved cent percent by a structure. Normally impurities and loose electrons generated by the absorption of heat or light cause these imperfections. Defects arise in all types of crystal lattices but the present discussion is restricted to ionic crystals of the type AB (e.g NaCl, FeO, etc).

There are three ways in which ionic crystals are defective or imperfect. (A) Defects in stoichiometric crystals, (B) defects in non-stoichiometric crystals and (C) Mechanical defects. Defects in stoichiometric and non-stoichiometric crystals (A & B) arise either due to missing of an atom or ion from the lattice site (Vacancy defect) or due to a foreign atom or ion entering the lattice site (Impurity defect). As these defects are restricted to few lattice points they are also called *Point defects*. The mechanical defects arise due to displacement of a complete lattice plane and hence are called *Line defects*.

A. Stoichiometric Crystals

The stoichiometric crystals are one in which the positive and negative ions are in the ratio represented by the chemical formulae of the compound. The stoichiometric compounds exhibit both (i) Frenkel and (ii) Schottky defects.

(i) **Frenkel defect :** *A Frenkel defect is a misplaced ion, generally a cation held interstitially in a region that would otherwise be perfect lattice. For every interstitial ion of this kind there is also a vacant ion site.* This defect arises due to vacant lattice point developed by the movement of one of the ion to another position, generally in between the lattice planes.

In Fig. 19.26, the cation A^+ moved between the two lattice planes, vacating its original position. In stoichiometric crystals, Frenkel defect occurs :

(a) When the cation is smaller than the anion, as in Ag Br, thus allowing the cation to slip into an interstitial position.

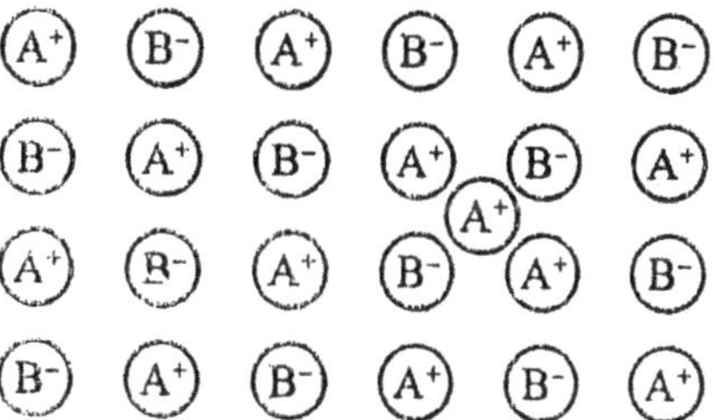

Fig. 19.26 Frenkel defect in a stoichiometric crystal.

(b) when the cation and anion have low oxidation numbers and larger sizes resulting in weak electrostatic attractive forces (e.g., Ag Br)

(ii) ***Schottky defect :*** A Schottky defect is a vacant ion site. Both anions and cations are missing from what might have been a perfect structure. The ions from the vacated sites are considered to have moved to the surface of the crystal. In crystals like NaCl, a cation and an anion completely miss from their lattice points causing Schottky defect.

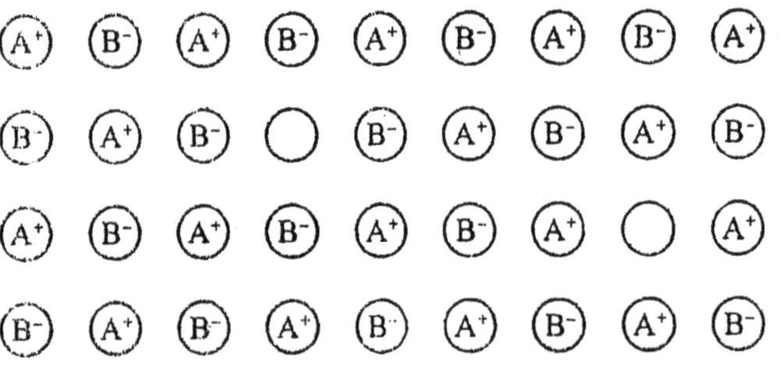

Fig. 19.27 Schottky defect in a stoichiometric crystal.

In the Fig. 19.27, a cation moved away along with an anion from the lattice points, maintaining electrical neutrality.

This defect is common among the stoichiometric crystals having :

(a) Ions of equal size and

(b) Ions with high oxidation numbers.

Schottky defect are generally one hundred times more probable than Frenkel defects. The energy requirement of Schottky defect site is about 25 kcal mole^{-1} of defect sites. Defects of both kinds may be present at one in fifty sites near the melting point, where the ions are more free to move, but at ordinary temperatures it is only one in one million possible sites.

B. Defects in non-stoichiometric crystals

Non-stoichiometric crystals are those in which the ratio of number of positive and negative charges do not coincide with the composition represented by chemical formulae. The defects in non-stiochiometric crystals is classifed into (i) metal excess defects and (ii) metal deficient defects.

(i) *Metal excess defect :* In ZnO, if the ratio of Zn^{2+} and O^{2-} is 0.95:1, the balance of positive charge is maintained by the higher valent metal ion or by the presence of an extra electron and an extra metal ion trapped in the interstitial planes. In either case there is an excess of metal ions in the crystal and hence called metal excess defect. The two ways of defects cited above are illustrated in Fig. 19.28.

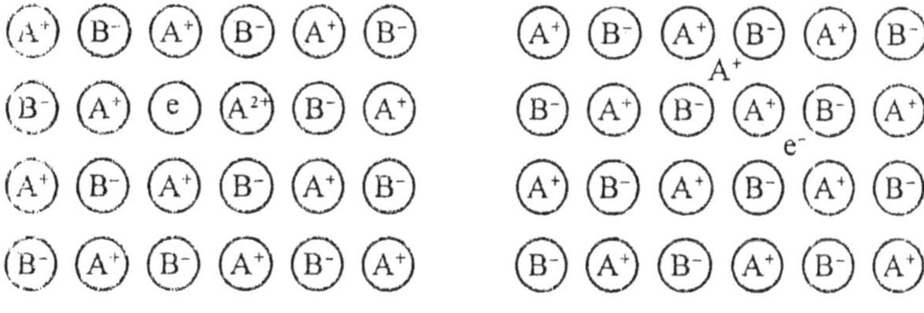

(a) Defect due to missing anion (b) Defect due to interstitial cation

Fig. 19.28 Metal excess defects.

(a) *Defect due to missing anion :* The illustration, Fig. 19.28(a), explains the defect due to missing anion. In this the loss of an anion from a lattice point creates an anion void and the charge balance is achieved by filling the void by an electron donated by a neighbouring cation which raises itself to a higher valence state. The missing anion leaves the crystal surface in the form of vapor or by combination with surrounding metal vapor. This defect arises in crystals susceptible for Schottky defects. When excess sodium metal is heated with chlorine gas and slowly cooled, yellow sodium chloride crystals are formed because of development of defects due to missing anion.

(b) *Defect due to interstitial cation :* In this a cation is trapped in interstitial sites along with the charge balancing electron as represented by Fig. 19.28(b). This defect is common with the crystals susceptible to Frenkel defects, eg., ZnO.

(ii) *Metal deficiency defect :* Metal deficiency defects are just the converse of metal excess defects. Here cations are either completely removed or shifted to interstitial sites amounting to an excess anions or deficient of metal ions in the crystal. We have two types of metal deficiency defects :

(a) Defects due to missing cation

(b) Defect due to interstitial anion.

(a) *Defect due to missing cation :* In the lattice plane illustrated by Fig. 19.29(a), a cation is missing from the lattice point. To compensate this loss and maintain the charge neutrality of the crystal, one of the cation reaches a higher oxidation state. Thus on the whole there is an excess of anion and the lattice remains deficient of cations. Considering a ferrous oxide, FeO, crystal, we notice some ferric, Fe^{3+}, ions in the defective crystals in addition to ferrous, Fe^{2+}, ions. The other compounds that show this defect are FeS, TiO, NiO etc.

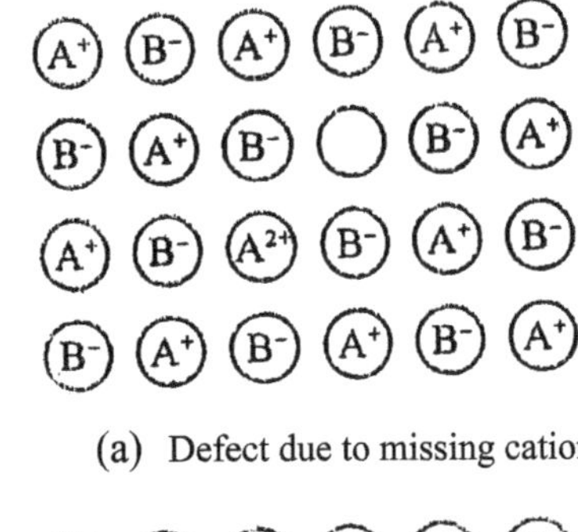

(a) Defect due to missing cation

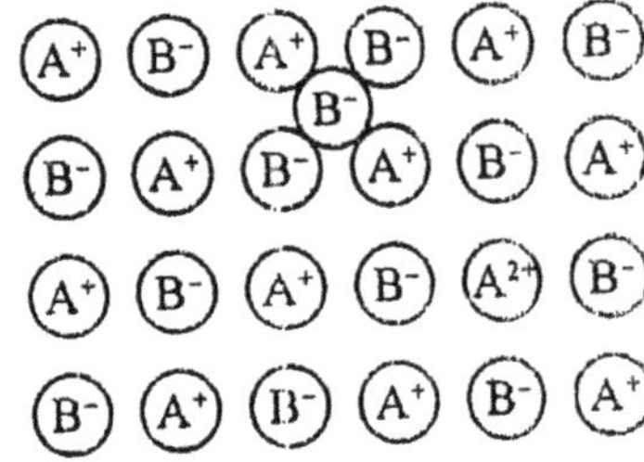

(b) Defect due to interstitial

Fig. 19.29 Metal deficiency defects.

(b) *Defect due to interstitial anion :* In this type of defect an anion is trapped in between the lattice planes of a crystal. The extra negative charge associated with the anion occupying interstitial site is compensated by an extra charge present in one of the cation, Fig. 19.29(b). In the crystals possessing this defect the anions are relatively small when compared to the size of cation, thus allowing the anion to find a place in the interstitial site.

C. Mechanical defects

When a stress or strain is applied on a localized plane, there is a deformity of crystal lattice. They have a large impact on the mechanical properties of the material and hence termed as mechanical defects. Suppose that a path is followed round the crystal from atom to atom or from ion to ion. If this path returns itself it is called a **Burgers circuit**. If the Burgers circuit returns exactly to the starting lattice point, the circuit is said to enclose no dislocations or defects. Instead if the Burgers circuit did not end on the starting atom or ion it is said to enclose one or more dislocations. There are two types of dislocations : (i) Edge dislocation (ii) Screw dislocation.

(i) *Edge dislocation :* When a strain because of compression or expansion is felt along a localized plane, part of the lattice gains extra half plane of atoms or ions. This dislocation is called edge dislocations. The deformity is produced in a plane perpendicular to the direction of applied strain, Fig. 19.30.

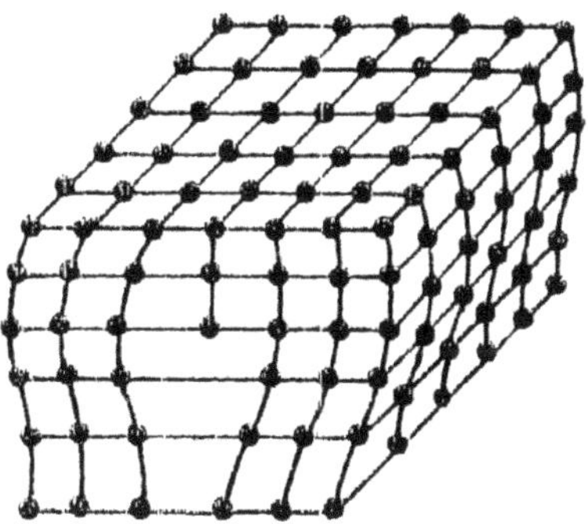

Fig. 19.30 Edge dislocation.

(ii) *Screw dislocation :* In screw dislocation, few lattice planes are dislocated by one or more lattice units relative to the other lattice planes by the applied stress, Fig. 19.31.

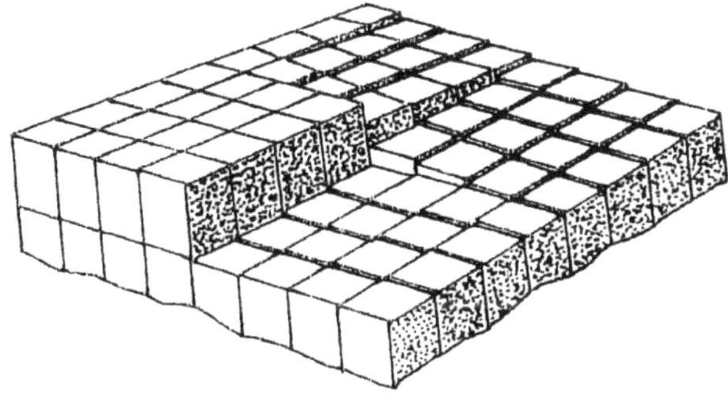

Fig. 19.31 Screw dislocation.

19.8.1 Better side of crystal defects

The free electrons available in imperfect lattice planes give special properties to the crystals like color and conduction properties.

(a) *Color Properties :* Depending on the extent of imperfection or in other words free electrons, crystals assume different colors. This is due to absorption of light by the free electrons and their excitation to higher energy levels. Thus we can prepare sodium chloride crystals that are yellow and blue. Similarly zinc oxide crystals turn yellow when hot and white when cold.

(b) *Semi-conductivity : Pure covalent and ionic crystals are non-conductors of electricity. Defects in crystals, like metal excess defect, put free electrons in their lattice planes and give them partially the capacity to conduct electricity known as semiconductivity and then the crystals are called semiconductors.* Generally quartz (SiO_2), TiO, etc., function as semiconductors and are used in the solar cells for the conversion of light energy into electrical energy.

Solved Problems

1. Designate (200) and (110) Miller planes of a *body centered cube.*

Solution :

For (200) Miller plane the Weiss indices are 1/2, $1/\infty$, $1/\infty$. Hence the plane is parallel to Y and Z axes and cuts the unit cell into two equal halves. Therefore the planes are indicated as in the following Fig. 19.32.1

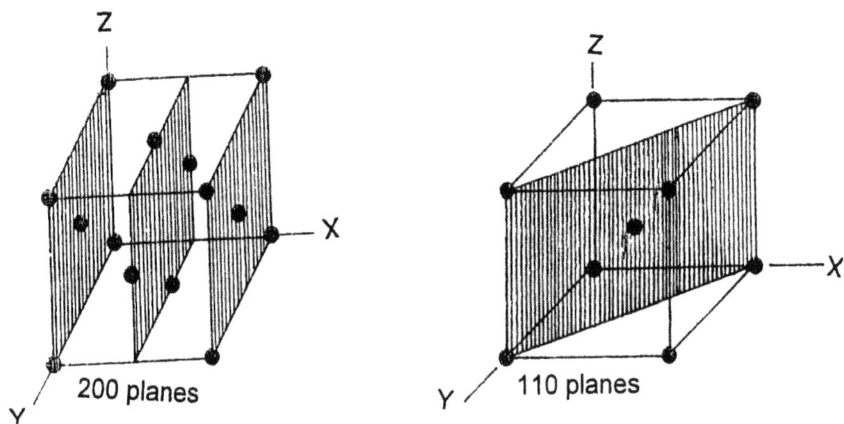

Fig. 19.32.1 Illustrate (200) and (220) planes of a body centered cube.

For (110) Miller planes the Weiss indices are 1, 1, $1/\infty$. Hence the plane is parallel to z-axis and intersects at the cell length.

2. Illustrate (200) and (220) planes of a face centered cube.

Solution :

For (200) planes, Weiss indices are 1/2, $1/\infty$, $1/\infty$. Hence the plane is parallel to y and z axes and divides the unit cell on x-axis into two halves.

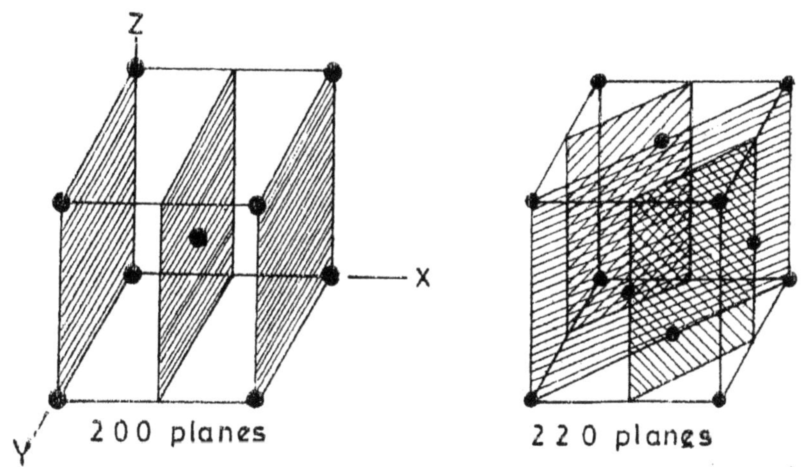

Similarly for (220) plane, Weiss indices ½, ½, $1/\infty$. Hence the plane is parallel to Z-axis but cuts the units cell side parallel to *X* and Y axes in to two halves (220) planes

3. Calculate the Miller indices of crystal planes which cut the crystal sides at 2a, b, 3c.

Solution :

	X	Y	Z
The planes cut the axes at	*2a*	*b*	*3c*
Coefficients	2	1	3
Therefore Weiss indices are	2	1	3
Inverse of Weiss indices are	1/2	1	1/3
Multiplying with 6 the LCM	3	6	2

Therefore Miller indices of the plane are (3 6 2)

4. Calculate the distance of separation of (222) planes of a body centered cube of the edge length 5 $\overset{0}{A}$.

The Distances of separation are given by the

formula $d_{(hkl)} = \dfrac{a}{\sqrt{h^2 + k^2 + l^2}}$

Therefore $d_{222} = \dfrac{5}{\sqrt{h^2 + k^2 + l^2}} = \dfrac{5}{\sqrt{h^2 + k^2 + l^2}} = 1.44 \overset{0}{A}$.

Self Test Questions

1. For crystalline solids, the physical properties like thermal conductivity, refractive index, mechanical strength are different in different directions. This phenomenon is called

   (A) anisotropy (B) Allotropy (C) Isotropy (D) polymorphism

2. In sodium chloride crystal, each chloride ion is surrounded by

   (A) 6 Na^+ (B) 4 Na^+ (C) 2 Na^+ (D) 8 Na^+

3. Phenomenon of a substance existing in more than one crystalline form, is called.

   (A) polymorphism (B) isotropy
   (C) anisotropy (D) isomorphism

4. Which type of bonding is present in diamond?

   (A) ionic (B) metallic (C) covalent (D) coordinate

5. A body centered cubic structure has

   (A) 8-8 coordination (B) 4-4 coordination
   (C) 6-6 coordination (D) 2-2 coordination

6. Which of the following substance is a molecular solid.

   (A) iodine (B) Li (C) diamond (D) NaCl.

7. A crystal assumes congruent position for every rotation of 120 degrees when rotated about an axis. The axis is

   (A) diad axis (B) triad axis (C) tetrad axis (D) hexed axis

8. Which of the following statements about Frenkel defect is correct ?

   (A) When this defect is present in the crystal, the stoichiometry is altered.

(B) This defect results in the change of electrical neutrality of the crystal.

(C) This defect arises because of shifting of cation into space between two layers.

9. The crystal systems are classified into

(A) 14 types (B) 7 types (C) 32 types (D) 3 types.

10. Transition temperature is

(A) temperature at which a substance melts

(B) temperature at which one allotrope changes into another

(C) temperature at which all the defects in the crystals disappear.

11. Number of Bravais lattices are ____________

12. Type of bonding present graphite is ____________

13. For a given substance angle between the corresponding faces or planes forming the external surface of the crystal remains constants, no matter how the faces develop. This statement is called ____________

14. The number of planes of symmetry of a cube are ____________

15. The number of axes symmetry for a cube ____________

16. A solid which has definite geometrical shape, sharp edges bounded by planes is called ____________

17. Allotropy is a special case ____________

18. Silica and quartz are examples of ____________

19. Arrangements in space of isolated points in a regular pattern, showing the positions of atoms, molecules, ions in the structures of the crystal is called ____________

20. A crystal is completely ordered at temperature ____________

Short and Essay Type Questions

1. Explain how Miller indices are useful in the identification of different crystal planes in a crystal.

2. Give a neat sketch of a cubic crystal showing the (100), (110) and (111) plane.

3. What are anisotropic and isotropic substances?
4. State the law of constancy of interfacial angles of a crystal.
5. Write a short note an Frenkel defects in crystals.
6. Define the following terms; Lattice, lattice plane and unit cell
7. What are the common defects in crystals? How do they influence the properties of materials.
8. Define unit cell and lattice planes. What are the three categories of crystalline cubic solids
9. Define the terms (a) unit cell (b) crystal lattice
10. Write notes on (a) Schottky defects (b) Frenkel defects.
11. (a) Define the term unit crystal lattice

    (b) Mention the types of defects that occur in crystals
12. State laws of crystallography.
13. What is crystalline solid and amorphous solid ?
14. Explain the point defects in crystals.
15. Explain the terms plane of symmetry, centre of symmetry and axis of symmetry with suitable examples.
16. Derive Bragg's equation.
17. Write two characteristics of crystalline solids.

Chapter 20

Chemical Aspects of Biotechnology

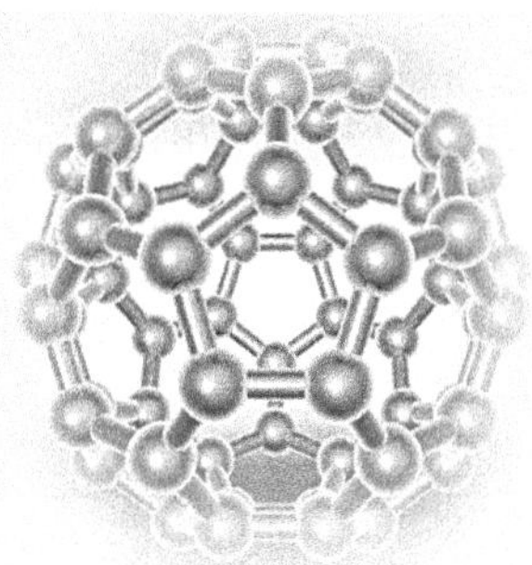

20.1 Introduction and Definition

Biotechnology is an interdeciplinary field of technology based on biology, chemistry and genetics, especially when used in agriculture, food science, and medicine. Biotechnologists tinker with living organisms. They use biochemistry and genetics to create new products for the agriculture and industry. These products include vaccines, medicines, growth hormones for plants, genetically modified seeds and animals and food additives.

Biotechnology can be defined in many ways

- *Biotechnology is the manipulation of living organisms to do practical things and to provide useful products.* Going by this definition we can put biotechnology to prehistoric times of 12000 BC, where man learnt the art of domesticating animals which are also living organism. The dogs may be mutants of wolves produced in the domestication process. During this time he also learnt the art of picking potential seeds, a mutant selection process.
- *Biotechnology is the use of cells and biomolecules to solve problems or make useful products.* Even this cellular definition is also prehistoric since man started using living cells like yeast, a microorganism, to ferment fruit or grains to produce alcohol. In a later period he also learnt the art of using microorganisms to make cheese, yogurt, kimchee, pickles, some sausages, etc., without having any knowledge about microscope or cell.

Jenner, 1795, is credited with invention of vaccination in Western medicine and the modern biotechnology. Jenner found that vaccination with cowpox virus makes people immune to smallpox, which is caused by smallpox virus (though of course the germ theory of disease and the concept of viruses did not come until much later). The word *vaccination* is derived from the Latin word for cow *(vacca*). In 1885, the French chemist Louis Pasteur showed that fermentation is caused by microbes, and concluded that these are different for different fermentation processes. Louis Pasteur is also known for invention of rabies vaccine. Ian Fleming, another chemist, discovered penicillin, the first antibiotic that led to discovery of many other antibiotics (an antibiotic is a substances made by one microorganism to inhibit the growth of, or kill, other microorganisms).

Basically before 1980s

- We could only move huge groups of genes by plant or animal cross breeding
- This meant genetic progress was slow, and not very specific

- Progress in plant breeding came in large part from identifying useful mutations
- Mutagenesis was accomplished through chemicals or radiation to "make" new genes in an organism.

Since the 1980's we have been able to:

- Find the genes for some specific traits
- Place genes in organisms where they did not originate
- Get those genes to work in their new location

The Modern biotechnology is totally based on our understanding of 'Genes' which determine the code of living systems and its technological applications like genetic engineering, cloning, etc. It is often associated with

- use of genetically altered microorganisms such as *E. coli* or yeast for the production of substances like insulin or antibiotics.
- refers to transgenic animals or transgenic plants, such as Bt cotton, Bt corn and animal cloning.
- Genetically altered mammalian cells, such as Chinese Hamster Ovarian (CHO) cells, that are widely used to manufacture pharmaceuticals.
- with breakthroughs in new medical therapies and diagnostic devices.

So the definition in the gene context is

"Biotechnology is the use of gene science to create new products from plants and animals"

However Biotechnology retains its classical flavour of

- use of organisms for the manufacture of organic products (examples include beer, milk products, and skin).
- Use of naturally occuring bacteria in the mining industry for bioleaching.
- Use in the recycling, treating bio-waste
- clean up of sites contaminated by industrial activities (bioremediation)
- use in new medical therapies and diagnostic devices
- on the negative side use in producing biological weapons.

There are also applications of biotechnology ***that do not use living organisms***. Examples are DNA microarrays used in genetics and radioactive tracers used in medicine.

The '*UN Convention on Biological Diversity*' has gone through various aspects biotechnological frontiers and formulated a definition that is one of the broadest:

"Biotechnology means any technological application that uses biological systems, living organisms, or derivatives thereof, to make or modify products or processes for specific use."

20.2 Benefits through Biotechnology

Biotechnology provides various benefits to mankind and some are listed below.

20.2.1 In Agriculture

Agricultural biotechnology is intended to address some of the challenges faced by farmers. These challenges include disease resistance (to viruses, fungi, and bacteria), insect resistance, environmental stress resistance (to heat, drought, salinity, flooding, soil pH), quality improvement (nutritional improvement, increased yield), and reduced post-harvest losses. Scientists can modify both

- agronomic traits (traits that determine how well a plant is suited for its environment, such as drought tolerance)
- quality traits (such as the oil content of a soybean) of crops.

Improved agronomic traits can help increase crop yields in numerous ways :

1. increasing the actual amount of food produced per plant;
2. reducing crop loss due to pests, disease, or weeds;
3. compensating for inhospitable environments that limit or prohibit planting;
4. extending growing and harvest seasons;
5. reducing production risks

In crops, two main kinds of genes have been commercialized

- Insect resistance genes, called "Bt"
  - Bt specific for corn borers
  - Bt specific for rootworms
- Herbicide resistant genes
  - Resistance to the herbicide "Roundup"
  - Resistance to the herbicide "Liberty"

One estimate indicates that the use of *Bt* cotton in China has reduced the amount of insecticide applied by 20 percent or as much as 78,000 tons of insecticides. In addition to creating new varieties of insect- and herbicide-resistant crops, biotechnology can develop new varieties of virus-resistant crops. Virus-resistant varieties of squash and papaya, for example, are among the GM foods approved in the United States.

20.2.2 In improving food quality

Biotechnology also has the capability to develop new crop varieties with improved qualities beyond those developed by conventional methods. Quality traits include delayed ripening of fruits enabling them to be available during off season, enhancement of oil content in oil seeds and improved nutrition in crops like carrots and potatoes. Dr. Ingo Potrykus at the Swiss Federal Institute of Technology and Dr. Peter Beyer at the University of Freiburg developed a strain of rice that produces beta-carotene (vitamin-A) in the grain itself and named it 'Golden Rice'. Work on commercializing this golden rice is ongoing under a collaboration between IRRI and the Indian Council of Agricultural Research for planting in 2006.

The other products include,

- Higher protein peanuts
- Longer shelf life for bananas and pineapples
- Sweeter bell peppers
- Tastier tomatoes
- Lower fat vegetable oils

20.2.3 In Medical Advances

Biotechnology has paved way for commercial production of penicillin from the mould pinicillium, the first antibiotic, which saved many lives. This process soon led way to manufacture a wide range of antibiotics like streptomycin used for the treatment of tuberculosis. To overcome the problem of development of resistance in disease causing bacteria, research is in progress to synthesize 'hybrid antibiotics' through genetic engineered microorganism that have required disease fighting capability but immunity can not be developed since such organism is not present in nature. Previously, porcine (from pig) or bovine (from cows) insulin was extracted and used, but these are not quite identical to human insulin. In some patients, they were recognized as foreign by the immune system. In 1982 the genetic engineers have introduces gene coding of human insulin recombinant human protein into a bacteria and are producing more reliable, inexpensive and easily acceptable human insulin. Other applications are :

- Rice - Enriched with beta-carotene (Golden Rice)and iron
- Bananas - Containing edible hepatitis vaccine
- Erythropoietin – Used by kidney patients to replenish the red blood cells that they lose during dialysis.

20.2.4 In Environmental Benefits

As already stated pest free crops produced by genetic modifications also reduce the usage of various pesticides. For example, 7% of the world's maize crop is destroyed by corn bore and Bt corn bore reduces the vulnerability of the crop to the pests thus saving 7% crop and the pesticides used to control the disease. Similarly biotechnology also helps in reducing the usage of herbicides for controlling the viruses, fungi and round worms. So the benefits can be summarized as :

- Reduced pesticide use
- Lower energy requirements
- Cleaner water
- Less soil erosion
- Waste management–Lipases as drain-cleaning agents

20.2.5 In animal Biotechnology

Although man is comfortably playing with the plant kingdom, when it comes to animals he is cautious about ethical and safety stand points. However methods involved without touching the cell mutations are accepted and used widely.

- **Blood clotting** is the same as **coagulation**. Most of the proteins involved are called **Factor Something**, where "Something" is a Roman numeral. Clotting factors are used as therapy for **hemophilia** (especially Factor VIII for hemophilia A and Factor IX for hemophilia B).
- Similarly in some cases dissolving clots is a problem. Tissue plasminogen activator (tPA) is a recombinant human protein that is given to heart attack victims (within a certain window of time) to dissolve the clot that caused the heart attack.
- **Hematopoietic stem cells** (HSCs) are in bone marrow and give rise to all blood cells. The purpose of bone marrow transplants is to transfer hematopoietic stem cells from the donor to the recipient, in hopes that the HSCs will become established in the recipient.

- Cloning of animals is currently a controvercial field. Cloning is aimed at producing and multiplying the preferred animal stock, producing human organs that are acceptable to immune system. People are afraid that it may lead to unethical practices raring humans for their organs, creation of monstorous animals that are difficult to be handled by nature.

20.2. 6 In Industry

- Food processing
    - Amylases in bread-making
    - Lipases in flavour development
    - Proteases in cheese making
    - Pectinases in clarifying fruit juices
- Textiles
    - Cellulases in treating denim to generate 'stone-washed' texture/appearance
- Grain processing
    - Conversion of corn starch to high fructose syrups
- Feed enzymes
    - Enzymes to assist in the digestibility of animal feeds (cellulase, xylanase, phytase)
- Diagnostic enzymes
    - Reporter enzymes (alkaline phosphatase, glucose oxidase, b-glucosidase) and diagnostic enzymes (DNA polymerase)

20.2.7 In Digestion of Food

The enzymes, Pepsin and Ptyalin originated from *saliva* work together to break down protein and starch molecules in to smaller fragments which are then converted into energy and required products by other enzymes like insulin produced by body. The irregularity in the production of insulin by pancreas is called diabetics.

Enzymes in the human alimentary canal and what they digest :

Enzyme	Substrate
Amylase	Starch
Maltase	Maltose
Sucrase	Sucrose
Lipase	Fats
Pepsin	Proteins

20.3 Fermentation

Definition : The term fermentation is derived from the latin word fermentare-*meaning boil.*

- ***Fermentation*** *typically refers to the conversion of sugar to alcohol using yeast.* The process is often used to produce or preserve food, typically wine or beer, but also includes the making of yoghurt (a form of milk curd). The other common examples of fermentation are putrification of meat and other food stuffs, curing of tobacco, making bread, idly, pickels, etc.
- *The science of fermentation is known as* ***zymology and the process zymosis***. It basically refers to anaerobic metabolic breakdown of a nutrient molecule, such as glucose, without net oxidation. Fermentation does not release all the available energy in a molecule; it merely allows glycolysis (a process that yields two ATP molecules per glucose) to continue by replenishing reduced coenzymes. Lactate, acetic acid, ethanol, or other reduced metabolites are also the products of fermentation.
- **Fermentation** is also used much more broadly to refer to the bulk growth of *microorganisms* on a *growth medium.* Fermentation does not make any distinction between aerobic and anaerobic metabolism. When fermentation stops prior to complete conversion of sugar to alcohol, a **stuck fermentation** is said to have occurred.

20.3.1 History

Since fruits and other food stuffs ferment naturally, fermentation precedes human history. However, humans began to take control of the fermentation process at some point. There is strong evidence that people were fermenting beverages in Babylon circa 5000 BC, ancient Egypt circa 3000 BC, pre-Hispanic Mexico circa 2000 BC, and Sudan circa 1500 BC. There is also evidence of leavened bread in ancient Egypt circa 1500 BC and of milk fermentation in Babylon circa 3000 BC. In *Vedic times* , circa 10000 BC, we read the conclusion of *yaga* with *somaras*, an alcoholic beaverage.

Scientific history starts probably with Louis Pasteur(1822) who considered fermentation to be a live process caused by micro-organism. Liebig considered it as a chemical process involving nitrogeneous substances. In 1896 Butchner proved that it is actually caused by **zymase**, a substance obtained by squeezing ground yeast. This was later recognized as an enzyme and is itself lifeless. Similarly other enzymes that cause fermentation are diastase and pepsin.

20.3.2 Conditions favourable for Fermentation

The conditions of enzymatic reactions are most suitble for fermentation processes. They are listed below :

1. **Temperature :** The enzymes are deactivated at high temperatures. The optimum temperature is 30-50^{0}C. Below this temperature the process occurs slowly.
2. **Presence of other substances :** Non-enzymatic substances like Co, Zn and some proteins act as coenzymes and substantially activate enzymes.
3. **Concentrations :** High concentrations of the solution render the enzyme inactive. So solutions used should be sufficiently diluted.
4. **Aeration :** Fermentation process is generally facilitated by the presence of air although we find some anearobic fermentation processes too.
5. **Absence of preservatives :** Excess salt, sugar, and oxidants are used as preservatives and they inactivate the enzymes. Substances like boric acid and mercury salts too deacivate the enzymes and should be avoided.

20.3.3 Types of Fermentation

Fermentations are of two types.

1. *Aerobic fermentation* occurs in the presence of air/oxygen. The free oxygen acts as a hydrogen acceptor. Eg., Fermentation of citric acid and acetic acid.
2. *Anearobic fermentation* process occurs in the absence of air/oxygen. Here substances like aldehyde act as abstractors of hydrogen. Eg., fermentation of alcohol, lactic acid, acetone and butanol.

20.3.4 The aerobic breakdown of glucose

The oxidation of glucose by removal of hydrogen atoms involves four phases :

1. *Glycolysis* – the breakdown of glucose to two molecules of *pyruvate* in the cytoplasm with no oxygen needed–yields 2 ATP.
2. *Transition reaction* – pyruvate is oxidized to a 2-carbon acetyl group carried by CoA, and CO_2 is removed–occurs twice per glucose molecule
3. *Citric acid cycle* – a cyclical series of oxidation reactions that give off CO_2 and produce one ATP per cycle–occurs twice per glucose molecule.

4. *Electron transport system*–a series of carriers that accept electrons and pass them from one carrier to the next until they are passed to the final electron acceptor, O_2, which is reduced to water with release of energy. Energy released is used to synthesize 32 to 34 ATP

Glycolysis occurs in the cytoplasm and is the breakdown of glucose to ***two*** pyruvate molecules. Glycolysis does not require oxygen. When oxygen is available, pyruvate enters the *mitochondria*, where it undergoes further breakdown giving energy. Pyruvate is converted to *acetyl CoA* and CO_2 is released. Each acetyl group received from the transition reaction is further oxidized to $2CO_2$ molecules. Thus a total of six carbon dioxide molecules are on the oxidation of glucose.

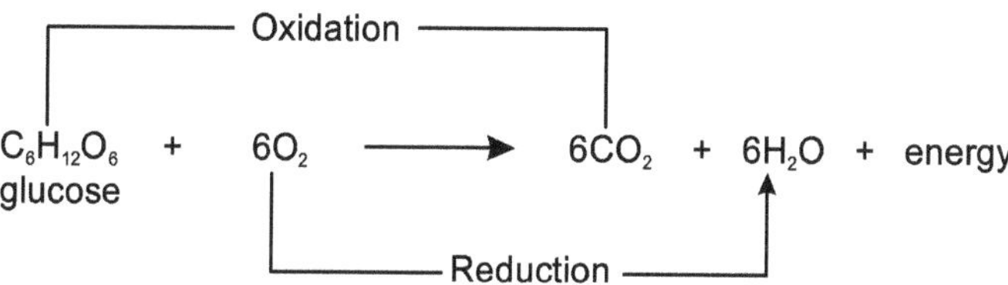

Electrons are donated to the Electron Transport System by NADH and $FADH_2$. As a pair of electrons is passed from carrier to carrier, energy is released and is used to form ATP.

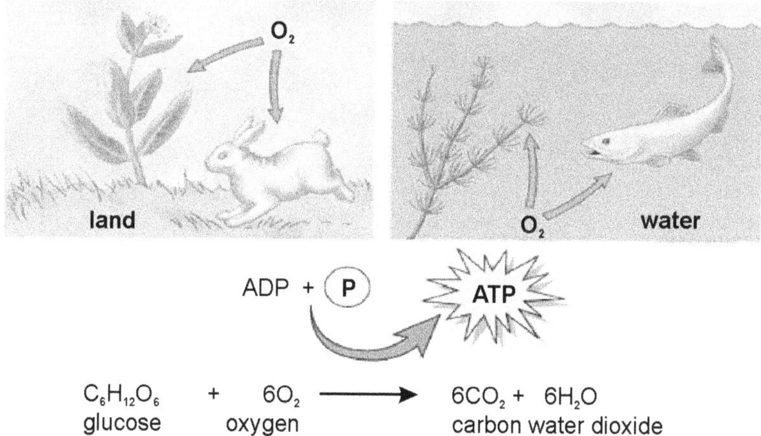

20.3.5 Anaerobic oxidation

If oxygen is not available, *fermentation* occurs and pyruvate undergoes *reduction*. In human systems, pyruvate is reduced to *lactic acid* during fermentation. When certain species of yeast (most importantly, *Saccharomyces cerevisiae*) metabolize sugar in the absence of oxygen, produce ethanol and carbon dioxide. The overall chemical reaction conducted by the yeast may be represented by the chemical equation.

$$C_6H_{12}O_6 \xrightarrow{\text{Zymase}} 2\ CH_3CH_2OH + 2\ CO_2$$

20.3.6 Advantages and Disadvantages of Fermentation in cells

Fermentation can provide a rapid burst of ATP in muscle cells, even when oxygen is in limited supply. Lactate, however, is toxic to cells. Initially, blood carries away lactate as it forms; eventually lactate builds up, lowering cell pH, and causing muscles to fatigue.

So now *Oxygen debt* occurs, and the liver must reconvert lactate to pyruvate.

20.4 Fermentation Process

20.4.1 Alcohol Beverages

These are of three types; Liquors (includes rum, gin, whiskey, brandy and vodka), Wines and Beers.

1. ***Liquors :*** Obtained by fermenting starch and sugars present in cane, beet, carrot, figs, berries, etc.
   - (i) Rum is prepared from beet sugar in France, carrot sugar in Portugal, strawberry in Brazil and cane sugar in Cuba. The flavour is mainly due to presence of ethyl acetate and ethyl butyrate.
   - (ii) Gin is obtained from plum spirit. After distillation it is flavoured by distilling with berries.
   - (iii) Brandy is obtained by distilling wine, and then suitably flavouring by distilling with fruit juices.
   - (iv) Vodka is prepared in Russia and obtained by distilling fermented rye starch.
2. *Wines :* Prepared by fermenting fruit juices, mainly grapes. The quality depends on sugar content in the fruits, color of grapes and other additives present. Grapes are macerated without crushing seeds by passing them through wooden rollers. The must obtained is fermented in still fermentors for 3-4 months. During this period yeast, tartaric acid and some salts separate out as argol. The clarified wine is now left for ripening for 2-4 years. During ripening tannins present in the skin of fruits are precipitated and the acids are converted to esters increasing quality of wine.

3. *Beers :* Malt is germinated barley and beer is product of fermented malt. For making killed malt is taken and mashed. A portion of the starched obtained is now converted in to dextrin by treating with an enzyme diastase. The proteins are converted into peptone by the proteolytic enzymes present in the malt.

20.4.2 Manufacture of Alcohol

Ethanol is made by converting the carbohydrate portion of biomass into sugar, which is then converted into ethanol in a fermentation process similar to brewing beer. Ethanol is the most widely used bio-fuel today with current capacity of 1.8 billion gallons per year based on starch crops such as corn. Ethanol produced from cellulosic biomass is currently the subject of extensive research, development and demonstration efforts. In countries like Cuba and India ethanol is often obtained by fermentation of molasses, a byproduct of sugar industry.

I. Ethyl Alcohol from molasses : Molasses ia a dark coloured viscous liquid left after the crystallization of sugar from cane juice. It contains 50-55% total sugars (sucrose + glucose + fructose) of which 35-40% is the sucrose. The different steps involved are:

   (i) *Dilution :* Molasses, because of its high sugar content does not support direct yeast fermentation. So molasses is first diluted to have 10-15% sugars.

   (ii) *pH adjustment :* The pH of the diluted solution is brought down to 4-4.5 by addition of sulphuric acid. This pH is not only favourable to the growth of yeast but also prevent growth of unwanted bacteria.

   (iii) *Nutrients :* Ammonium phosphate serves as nutrient to yeast and is added in small quantities.

   (iv) *Fermentation :* The above solution is pumped in to a steel fermentar and charged with yeast (about 5% of total volume) and temperature is kept at 30^0C. The fermentation sets in and a two step reaction takes place to give ethanol.

$$\underset{\text{Sucrose}}{C_{12}H_{22}O_{11}} + H_2O \xrightarrow{\textit{invertase}} \underset{\text{Glucose}}{C_6H_{12}O_6} + \underset{\text{Fructose}}{C_6H_{12}O_6}$$

$$\underset{\text{Glucose \& Fructose}}{C_6H_{12}O_6} \xrightarrow{\textit{zymase}} \underset{\text{Ethyl alcohol}}{2CH_3CH_2OH} + 2CO_2$$

The reactions are exothermic and hence cooling of fermentar is necessary to maintain the temperature. Fermentation is carried out for 48 hrs and is stopped by passing through the liquid. Yeast is filtered and the filtrate having 6-10% ethanol, called wash or wort, is collected.

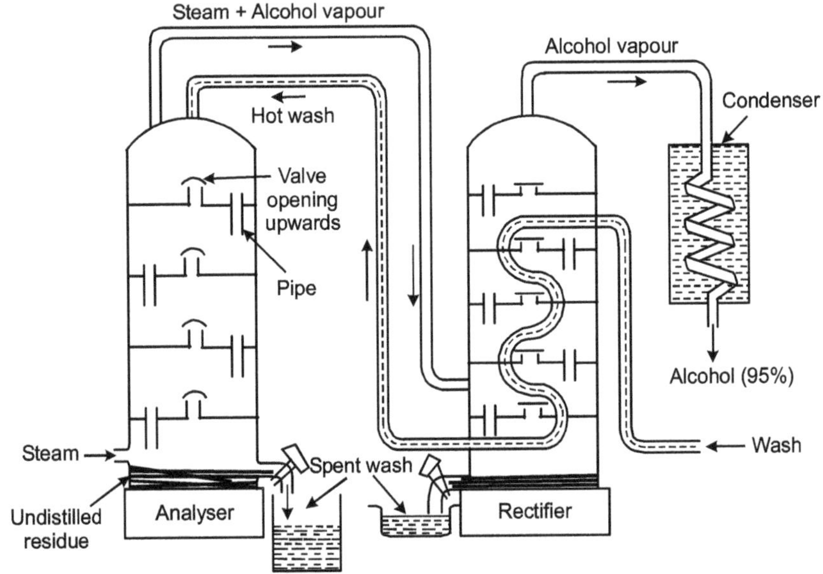

Fig. 20.1 Distillation Still.

(v) Distillation: In addition to alcohol the wash also has 3-5% glycerol, fusel oil (a mixture of propanol and higher alcohols), acetaldehyde, etc. The fractional distillation is carried out in a two towered arrangement called coffey still. The first fractionating tower is called *analyzer* and the second *rectifier* and these are provided with perforated plates having valves opening upwards. It works on the counter current principle where the steam and alcohol travel in opposite directions through the still. Steam passed from the bottom of the analyzer takes away the vapours of alcohol from the down coming wash. In the rectifier condensation takes place. High boiling water (b.pt.=100 ^{0}C) is condensed first followed by alcohol (b.pt.=78.3 ^{0}C) and the other volatile matter. The distillate is collected in three fractions. The *first fraction* contains low boiling liquids like acetaldehyde. The *second fraction* contains 95% ethanol called *rectified spirit* or *industrial alcohol.* The last fraction contains fusel oil (a mixture of n-propyl, n-butyl, amyl alcohols). The undistilled residue left in the still is used as cattle feed.

20.3.2 Ethyl Alcohol from Starches

Starch is generally produced from potato, corn, wheat, barley, etc. For fermentation following steps are adopted.

(i) *Sacharification :* This is a two step process.

(a) Malting : The grains are first cleaned and dried. They are now put in large steel water tanks (10000-13000 gallons capacity) for 3-5 days at 10-18 ^{0}C. Water is now drained out and the grains are spread on floor and allowed to germinate. The germination step is skipped in case of starch rich roots like potatoes. The germinated seeds are withered for 24 hrs by drying in a kiln at 70-95 ^{0}C. This stops germination without destroying enzyme amylase. The dead seeds are stored in ageing bins that enhance diastase content.

(b) *Mashing :* The crushed grains are taken to mashing tank where they are heated to 55 ^{0}C for about 15 minutes. This helps in solubulisation and gelatinization of mass in to a mixture of dextrose (glucose) and maltose.

$$\underset{\text{Starch}}{(C_6H_{10}O_5)_n} + nH_2O \xrightarrow{\textit{diastase}} \underset{\text{Maltose}}{nC_{12}H_{22}O_{11}}$$

(i) *Fermentation :* The maltysed starch solution is now treated with yeast for 2-3 days at 30 ^{0}C. The enzyme maltase is converts maltose into glucose which then follows the traditional path of conversion into ethanol.

$$\underset{\text{Maltose}}{C_{12}H_{22}O_{11}} + H_2O \xrightarrow{\textit{maltase}} \underset{\text{Glucose}}{C_6H_{12}O_6}$$

$$\underset{\text{Glucose}}{C_6H_{12}O_6} \xrightarrow{\textit{zymase}} \underset{\text{Ethyl alcohol}}{2CH_3CH_2OH} + 2CO_2$$

(iii) *Distillation & Rectification:* This is done in a similar way as described for molasses.

(iv) *Absolute alcohol :* The rectified spirit has 95.5% and the rest water. This is an azeotrope and can not be concentrated further by distillation. To get absolute alcohol (100% alcohol) the rectified spirit is treated with quick lime CaO and then distilled.

$$CaO + H_2O \longrightarrow Ca(OH)_2$$

This sample is now treated with anhydrous copper sulphate (white) which take out any reminiscent traces of water forming its blue colored hydrate.

$$CuSO_4 + 5H_2O \longrightarrow CuSO_4.5H_2O$$

20.4.3 Production of Acetic Acid

Wines become sour when exposed some time to the air. This souring is due to the oxidation of alcohol to acetic acid by certain aerobic fungi and bacteria like *mycoderma aceti* and *bacterium aceti*. Vinegar used in the ketchups and in the sotening of meat is a dilute solution (5.5-10%) of acetic acid. Liquors with 10-15% ethanol are fermented by these bacteria at 30-35 ^{0}C and the process is called acetous fermentation.

$$\underset{\text{Ethanol}}{CH_3CH_2OH + O_2} \xrightarrow[\textit{Bacterium aceti}]{\textit{Mycoderma aceti} \text{ or}} \underset{\text{Acetic acid}}{CH_3COOH + H_2O}$$

The oxidation is a two step process

(i) Oxidation of ethanol to acetaldehyde

$$\underset{\text{Ethanol}}{CH_3CH_2OH + O} \xrightarrow[\textit{Bacterium aceti}]{\textit{Mycoderma aceti} \text{ or}} \underset{\text{Acetaldehyde}}{CH_3CHO + H_2O}$$

(ii) Oxidation of acetaldehyde to acetic acid

$$\underset{\text{Acetaldehyde}}{CH_3CHO} + O \xrightarrow[\textit{Bacterium aceti}]{\textit{Mycoderma aceti} \text{ or}} \underset{\text{Acetic acid}}{CH_3COOH + H_2O}$$

The evidence comes from presence of traces of intermediate and aerobic oxidation of acetaldehyde to give acetic acid. Best results are obtained when the alcohol initially is contaminated with acetic acid, since sour media facilitates growth of bacteria. $MgSO_4$, $(NH_4)_2SO_4$ are used as nutrients for bacteria.

Anaerobic oxidation of both ethanol and acetaldehyde is also possible in the presence of methylene blue, which promotes the participation of molecular oxygen.

Preparation of Glacial acetic acid from Vinegar

Pure acetic acid called glacial acetic acid is treating vinegar with quick lime. Calcium acetate crystallized is distilled with concentrated sulphuric acid to get pure acetic acid.

$$2CH_3COOH + CaO \longrightarrow (CH_3COO)_2Ca + H_2O$$

$$(CH_3COO)_2Ca + H_2SO_4 \longrightarrow 2CH_3COOH + CaSO_4$$

20.4.4 Production of Lactic acid

This is a fermentation product of sugar with the enzyme *Lactic bacilli*. It is this type of bacteria that convert lactose into lactic acid in yogurt, giving it its sour taste. The fermentation is carried out at 37$\pm$2 ^{0}C in the presence of chalk powder. Chalk helps in keeping the free lactic acid concentration below 1%, since the enzyme is denatured above this concentration. Free acid is regenerated by treating the calcium lactate formed with dilute sulphuric acid. The calcium sulphate is filtered and the pure lactic acid is obtained by distillation under reduced pressures.

20.4.5 Production of Acetone

When starch or molasses are fermented using micro organism *Clostridium acetobutylicum* we get a 6:3:1 mixture of butanol, acetone and ethanol with evolution of carbondioxide and hydrogen. Pure acetone is obtained by fractional distillation.

20.4.6 Production of Vitamins

Mainly riboflavin (vitamin B_2), nicotinamide (Vitamin B_3), cyanocobalamin (vitamin B_{12}) and ascorbic acid (vitamin C) are prepared commercially through fermentation technique.

(i) *VITAMIN B_2 :* Riboflavin is prepared from butanol-acetone fermentation residue using *Clostridium acetobutylicum*

(ii) *VITAMIN B_3 :* Nicotinamide is an essential vitamin, and is widely used in the health-food and animal food-and-feed industries. Biological production, using *Rhodococcus* biocatalyst operates at about 5kT p.a.

$$\text{3-cyanopyridine} \xrightarrow[\text{cell biocatalyst}]{\textit{Rhodococcus}\text{ whole}} \text{nicotinamide}$$

(i) *VITAMIN B_{12} :* This is prepared using a culture medium having 20 parts of yeast, 20 parts of glucose, 0.2 parts of glycine, 0.08 parts of cobalt chloride dehydrate and 1000 parts of water. The fermentation is carried out at a pH of 7-7.5 at 30^{0}C for 4-5 days. The conversion is basically due to *propionibacterium-freudenreichil.*

(ii) *VITAMIN C :* The commercial production of ascorbic acid is developed by Prof. Yin of china in 1990. This is a partial fermentation process. First glucose is fermented using Bacillus thurigiensis to give 2-KGA (2-keto-gluconic acid). 2KGA is now chemically processed to give vitamin-C.

20.4.7 Production of Acrylamide

This is a typical example *involving release of ammonia by bacteria*

$$\underset{\text{Acrylonitrile}}{CH_2{=}CH{-}CN} \xrightarrow[\text{}]{\text{Nhase-containing } \textit{Rhodococcus} \text{ cells in stirred tank bioreactor}} \underset{\text{Acrylamide}}{CH2{=}CH{-}CONH_2} + NH_3$$

Acrylamide is widely used in industry as a precursor for the formation of acrylic polymers, in the construction, paint, and household products industries, and for laboratory use. The biological production of acrylamide has advantages over the chemical synthesis because of the absence of side-reactions, and the simpler recovery of the reaction product

20.4.8 Production of Aspartame

This is a typical example where two amino acids are fused into another compound. Aspartame (L-phenylalanyl-L-aspartyl-methyl ester) is a *low-calorie artificial sweetener* used widely in soft drinks and confectionary products (e.g., Diet Coke). It can be synthesised chemically, or biocatalytically by peptide synthesis using a thermostable protease enzyme – Thermolysin, *Bacillus thermoproteolyticus.*

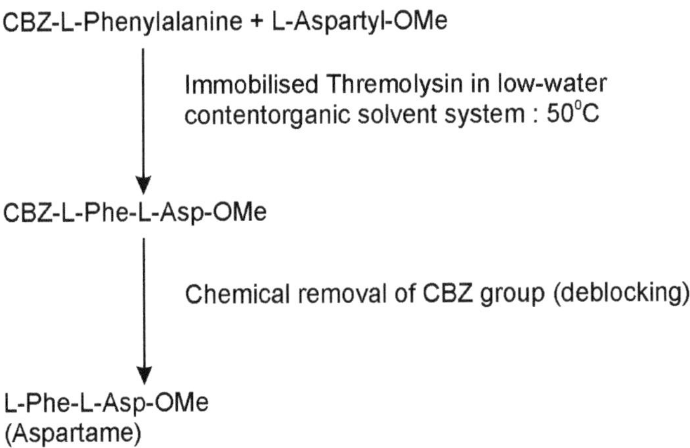

20.5 Industrial Applications of Enzymes

As already discussed enzymes are widely used in many industries. Some more applications are givn in the following Table.

Application	Enzymes used	Uses
Bio-surfactants	Primarily *proteases*, produced in an extracellular form from bacteria	Used for presoak conditions and direct liquid applications helping with removal of protein stains from clothes.
	Amylase enzymes	Detergents for machine dish washing to remove resistant starch residues.
Baking industry	Fungal *alpha-amylase* enzymes: normally inactivates about 50 degrees Celsius, destroyed during baking process	Catalyze breakdown of starch in the flour to sugar. Yeast action on sugar produces carbon dioxide. Used in production of white bread, buns, and rolls
	Protease enzymes	Biscuit manufacturers use them to lower the protein level of flour.
Baby foods	Trypsin	To predigest baby foods
Brewing industry	Enzymes from barley are released during the mashing stage of beer production.	They degrade starch and proteins to produce simple sugar, amino acids and peptides that are used by yeast to enhance fermentation.
	Industrially produced barley enzymes.	Widely used in the brewing process to substitute for the natural enzymes found in barley.
	Amylase,glucanases, proteases	Split polysaccharides and proteins in the malt
	Betaglucosidase	Improve the filtration characteristics.
	Amyloglucosidase	Low-calorie beer
	Proteases	Remove cloudiness during storage of beers.
Fruit juices	*Cellulases, pectinases*	Clarify fruit juices
Dairy industry	Rennin, derived from the stomachs of young ruminant animals (calves, lambs)	Manufacture of cheese, used to split protein

Contd... Table

Dairy industry	Microbially produced enzyme	Now finding increasing use in the dairy industry
	Lipases	Is implemented during the production of Roquefort cheese to enhance the ripening of the blue-mould cheese.
	Lactases	Break down lactose to glucose and galactose
Starch industry	*Amylases, amyloglucosideases* and *glucoamylases*	Converts starch into glucose and various syrups
	Glucose *isomerase*	Converts glucose into fructose (high fructose syrups derived from starchy materials have enhanced sweetening properties and lower calorific values)
Rubber industry	*Catalase*	To generate oxygen from peroxide to convert latex to foam rubber
Paper industry	*Amylases*	Degrade starch to lower viscosity product needed for sizing and coating paper
Photographic industry	*Protease* (ficin)	Dissolve gelatin off the scrap film allowing recovery of silver present

20.6 Other New Trends in Biotechnology Applications :

20.6.1 BIOFUELS

Biofuels include Biodiesel, Ethanol, Methanol, Methane produced by biological means and biocrude. Enzymatic production of ethanol is already discussed. It is currently used in the automobile industry after mixing with petroleum. Methane is obtained from fermentation of biowastes like Cow dung (Gobar) and is directly used by farmers for cooking and lighting. Biodiesel is produced through a process in which organically derived oils (triglycerides, fatty acids) are combined with alcohol (ethanol or methanol) in the presence of a catalyst to form ethyl or methyl ester. This process is called trans-esterification. The biomass- derived ethyl or methyl esters can be blended with conventional diesel fuel or used as a neat fuel (100% biodiesel). Biodiesel can be made from soybean or Canola (rapeseed)oils, animal fats, waste vegetable oils, or micro-algae oils.

20.6.2 Biosensors

A **biosensor** *is a device for the detection of an analyte that combines a biological component with a physicochemical detector component.*

It consists of 3 parts:

- The *sensitive biological element* (biological material (eg. tissue, microorganisms, organelles, cell receptors, enzymes, antibodies, nucleic acids etc), a biologically derived material or biomimic. The sensitive elements can be created by biological engineering.
- The *transducer* in between (associates both components)
- The *detector element* (works in a physicochemical way; optical, piezoelectric, electrochemical, thermometric, or magnetic.)

The most widespread example of a commercial biosensor is the blood glucose biosensor, *which uses an enzyme to break blood glucose down.* In so doing it transfers an electron to an electrode and this is converted into a measure of blood glucose concentration.

Recently, arrays of many different detector molecules have been applied in so called electronic nose devices, where the pattern of response from the detectors is used to fingerprint a substance.

Many of today's biosensor applications are similar, in that they use organisms which respond to toxic substances at a much lower level than us to warn us of their presence. Such devices can be used both in environmental monitoring and in water treatment facilities.

20.6.3 Biomembranes

*A **biomembrane** is a membrane which acts as a barrier within or around a cell.* It is, almost invariably, a lipid bilayer (except for Archaea which have isoprene membranes), being composed of a double layer of lipid-class molecules, specifically phospholipids, with occasional proteins inter-twined, some of which function as channels.

Such membranes typically define enclosed spaces or compartments, in which cells may maintain a chemical or biochemical environment that differs from the outside. For example, the membrane around peroxisomes shield the rest of the cell from peroxides, and the plasma membrane separates a cell from its surrounding medium. Most organelles are defined by such membranes, and are called membrane-bound organelles.

Probably the most important feature of a *biomembrane is that it is a selectively permeable structure*. This means that the size, charge and other chemical properties of the atoms and molecules attempting to cross it will determine whether they succeed to do so. Selective permeability is essential for effective separation of a cell or organelle from its surroundings.

If a particle is too large or otherwise unable to cross the membrane by itself, but is still needed by a cell, it could either go through one of the protein channels, or be taken in by means of endocytosis.

20.6.4 Biofertilizer

Biofertilizers include microorganisms and their metabolites that are capable of enhancing soil fertility, crop growth, and/or yield. These include both indigenous microbes and microbial inoculants, that is, microorganisms that replace fertilizers or increase a crops fertilizer use efficiency. Soil microorganisms such as bacteria, ectomycorhiza, arbuscular mycorrhizal fungi, and soil algae, especially the N_2-fixing cyanobacteria have potential as biofertilizers. Nitrogen-fixing inoculants based on *Rhizobium* species were among the first biofertilizers introduced into agro-ecosystems. Normally farmers grow legume plants that help in N_2-fixation and plough them into soil that drastically cuts the use of chemical fertilizers. Similarly conversion of bio-wastes into humus is done by earth-worm and the vermi-compost is now a big cottage industry. The farming done using biofertilizers is now called Organic farming. Biofertilizers provide an alternative to agricultural chemicals as more sustainable and ecologically sound practices to increase crop productivity. Biofertilizer sales forecasts in the United States for the years 2001 and 2006 represent $690 million and $1.6 billion, respectively. Examples of some biofertilizers currently in use worldwide are shown in following Table.

Organisms, mode of action, crops, and producers of biofertilizers currently in use for agriculture

Type	Mode of action	Crop	Used in
Rhizobium spp.	N_2 fixation	Legumes	Russia; several countries
Cyanobacteria	N_2 fixation	Rice	Japan; several countries

Type	Mode of action	Crop	Used in
Azospirillum spp.	N_2 fixation	Cereals	Several countries
Mycorrhizae	Nutrient acquisition	Conifers	Several countries
Penicillium bilaii	P solubilization	Cereals, legumes	Western Canada
Directed compost	Soil fertility	All plants	Several countries
Earthworm	Humus formation	Vegetables, flowers	Cottage industry

Question

1. Define Biotechnology and explain its role in modern times.
2. Explain past and present of Biotechnology.
3. Give an account of benefits and role of Biotechnology.
4. Explain the role of Biotechnology in

   (i) Agriculture (ii) Controlling pollution

   (iii) Animal functioning (iv) Industry

   (v) Improving food quality (vi) Digestion of food.
5. "Future food needs are met through Biotechnology" support the statement.
6. Define and Classify fermentation. Give conditions favourable for fermentation.
7. Write an account of types of fermentation taking suitable examples.
8. What are alcohol beverages. Classify and give their details.
9. Define Fermentation. Write the chemistry involved in the manufacture of alcohol.
10. Give method of preparation of acetic acid. Differentiate Vinegar and glacial acetic acid.

11. Explain the Fermentation chemistry in the preparation of following chemicals.

    (i) Acetate (ii) Vitamins

    (iii) Acrylamide (iv) Aspartame

12. Define "Enzyme". Give an account of its role in the digestion process.

13. Give an account of industrial application of Enzymes.

14. Write about

    (i) Biofuels (ii) Biofertilizers

    (iii) Biosensors (iv) Biomembranes.

Chapter 21

Phase Rule

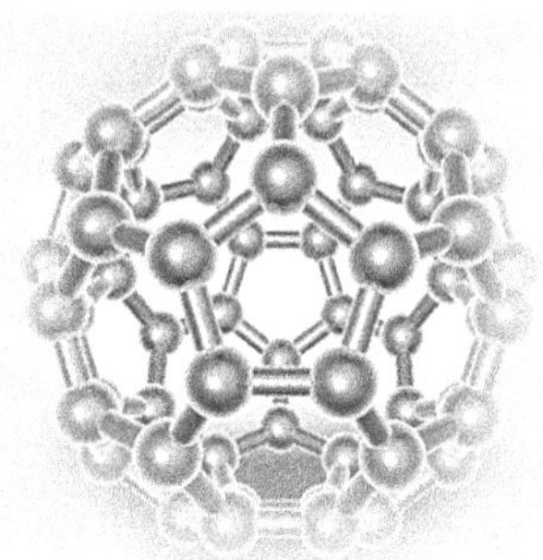

21.1 Homogeneity and Inhomogeneity of Systems

A system which exhibits identical physical and chemical properties through-out is called a ***homogenous system.*** Thus we say that copper sulphate solution is homogenous as it exhibits identical physical properties like uniform color, physical indistinguishability of copper sulphate even under ultra microscope, etc. Though milk possesses uniform white color, we classify it under heterogeneous solution, as milk is a colloidal solution and the presence of fat particles can be felt under ultramicroscope. *The particles of solute in a true solution have a diameter of few angstroms. The x-rays too have wavelengths of few angstroms, and therefore give characteristics diffraction patterns of these particles.* To solve this homogeneity and inhomogeneity of the true solutions, we have to restrict our *physical indistinguishability to visible light*, wherein the colloidal particles are large enough for detection. Keeping this boundary condition in view, we can comfortably identify all the constituents of a heterogeneous systems, which are generally termed as **phases**.

21.2 Phase Rule

Fusion, vaporization, sublimation, dissolution of a salt, liquefaction of a gas etc., all involve changes in the phases of the constituents. These phase changes are completely reversible and are said to make a **heterogeneous equilibrium**. A study of variations in the physical parameters like pressure, temperature, volume, concentration etc., of constituents on the heterogenous equilibria were discussed quantitatively by Henry's law, Clapayron-Classius equation, Raoult's law, Nernst distribution law, Le Chatelier principle, law of mass action etc. *Phase rule is a generalization of the above mentioned laws and gives an unified approach as regards to fixing the physical variables to decide the existence of the phases present in a heterogenous equilibria.* It is not meant for either to invalidate or supersede the existing laws, which mainly describe the quantitative relation among the variables.

The **phase rule** was deduced theoretically by J. Willard Gibbs (1876), but its application to physical chemistry was made by H.W.B. Roozeboom (1884). It is described by the mathematical expression;

$$F = C - P + 2 \qquad .....(21.1)$$

where F is called **number of degrees of freedom** of variables

C is the number of **components** of the system

P is the number of **phases** of the system.

The phase rule is defined as if an equilibrium is dependent only on temperature, pressure and concentration variations and is not influenced by gravity, surface forces, electrical or magnetic forces, then the **number degrees of freedom**, **(*F*)**, exceeds the difference of number of **components, (*C*)**, and the number of **phases**, **(*P*)** by 2.

21.2.1 Phase

A phase is defined as any homogenous, physically distinct with well defined boundaries and mechanically separable part of a system. Boiling, freezing, sublimation and conversion of one allotropic form to other, all involve changes of phase without change of chemical composition. Ice below $0^{O}C$ and normal pressure conditions exists in mere solid phase. As we raise the temperature, it melts to give liquid water and water vapor, thus forming a system of two or three phases that are homogenous, physically distinct and mechanically separable. As we cool they once again form solid ice. Hence we refer that these three phases of water are at a heterogenous equilibrium, Fig. 21.1. Addition of few ml of alcohol to this three phase equilibria, will not generate another phase, although, the liquid phase is now referred to as a solution of alcohol in water and the vapour phase is a homogenous mixture of alcohol vapour and water vapour. But carbon tetrachloride being immiscible with water, its addition to above three phase equilibria generate a fourth liquid phase, Fig. 21.2.

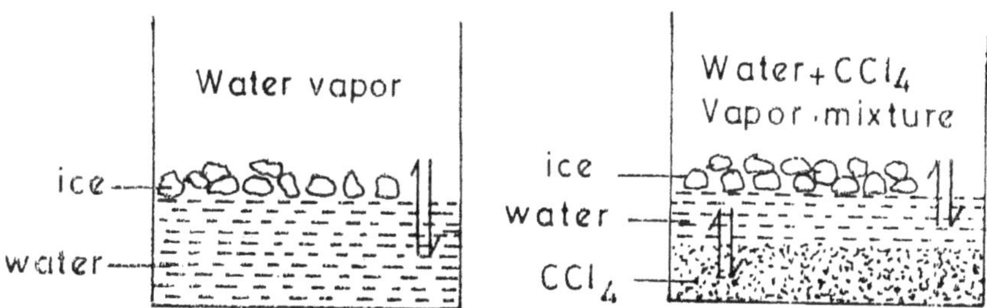

Fig. 21.1 Three phase equilibria. **Fig. 21.2** Four phase equilibria.

Similarly addition of every solid constituent generates a separate phase as they are once again physically distinct, defined by separate boundary conditions. Thus addition of naphthalene balls to the three phase water equilibria generates a fourth phase of solid naphthalene which then involves in an exchange of naphthalene with vapor and liquid water. On the other hand addition of sand to the three phase equilibrium under consideration is not considered as a creation of fourth phase as it lacks any exchange of chemical species. For the same reason, the beaker in which the constituents are present is also not a phase of the system.

Similar to a liquid solution, a solid solution (e.g., alloy) is also considered as a single phase, no matter how many individual substances are present in it.

A gas or mixture of gases always constitute a single phase as they are physically indistinguishable and always form a true solution.

21.2.2 Components

The number of components in a system is the minimum number of independently variable chemical species necessary to describe the composition of each phase. Thus at 100 $^{\circ}$C temperature when we take a mixture of H_2(g), O_2(g) and H_2O(g), it forms a three component system. At arc temperatures, if this mixture is allowed to equilibrate to give water vapor,

$$2H_2\ (g) + O_2\ (g) \longleftrightarrow 2H_2O\ (g) \quad K_{eq} = \frac{[H_2O]^2}{[H_2]^2[O_2]}$$

it then becomes a two component system as the description of any two components of the system, and a knowledge equilibrium constant allows one to calculate the third component. Instead, if we start with a known amount of water vapor and allow it to dissociate to give O_2 and H_2, it becomes a one component system as the knowledge of dissociation constant of water and concentration of any one of the component fully describes all the three constituents of the vapor phase.

Similarly in the decomposition of $CaCO_3$(s) to give CaO(s) and CO_2(g), we notice three phases viz., solid $CaCO_3$, solid CaO and gaseous CO_2.

$$CaCO_3(s) \longleftrightarrow CaO(s) + CO_2\ (g) \uparrow$$

At the first sight, it might appear, that this is a three component, $CaCO_3$, CaO and CO_2, system. In reality it is a single component system as one mole of decomposed $CaCO_3$ gives one mole each of CaO and CO_2 and the description of any one of the component is sufficient to describe the concentration of all the three phases present in the system. In the formation of $CaCO_3$(s) by the combination of CaO(s) and CO_2(g); we have an open

$$CO_2 + CaO(s) \longleftrightarrow CaCO_3\ (s)$$

option of taking different amounts of CaO and CO_2. Hence we have to know at least, the concentration of two of the components, to fully describe the equilibrium under investigation, making it a two component system.

Another interesting example is the sodium sulphate-water system. $NaSO_4$ and H_2O combine in different proportions to give, starting from anhydrous Na_2SO_4 to aqueous solution of Na_2SO_4. Some of the equilibria are.,

$$Na_2SO_4\,(s) + 0\ H_2O\,(l) \longleftrightarrow Na_2SO_4(s)$$

$$Na_2SO_4\,(s) + 7\ H_2O\,(l) \longleftrightarrow Na_2\ SO_4.7H_2O\,(s)$$

$$Na_2SO_4\,(s) + 10\ H_2O\,(l) \longleftrightarrow Na_2SO_4.\ 10H_2O\,(s)$$

$$Na_2SO_4\,(s) + x\ H_2O\,(l) \longleftrightarrow Na_2SO_4(aq)\,(l)$$

$$0\ Na_2SO_4\,(s) + \underset{(ice)}{H_2O\,(s)} \longleftrightarrow H_2O\,(s) \text{ or } (l) \text{ or}(g)$$

$$0\ Na_2SO_4\,(s) + H_2O\,(s) \longleftrightarrow H_2O\,(s) \text{ or } (l) \text{ or}(g)$$

$$0\ Na_2SO_4\,(s) + H_2O\,(g) \longleftrightarrow H_2O\,(s) \text{ or } (l) \text{ or}(g)$$

Although composition of certain phases (last three equilibria) may be stated in terms of only one of the constituents, certain others require the knowledge of the amounts of any two constituents to specify unambiguously the composition of each phase of the system. Thc smallest number by which the compositions of all the phases can be described being two, sodium sulphate water system is called a two component system.

Liquid water contains presumably enormous number of associated species; H_2O, $(H_2O)_2$, $(H_2O)_3$...$(H_2O)_n$

$$H_2O + H_2O \longleftrightarrow (H_2O)_2$$

$$H_2O + (H_2O)_2 \longleftrightarrow (H_2O)_3$$

$$\vdots \qquad\qquad \vdots$$

$$H_2O + (H_2O)_{n-1} \longleftrightarrow (H_2O)_n$$

The equilibria between various polymers of water shift very rapidly, within 10^{-11} seconds at the most. Ordinary measurements require much longer time. Of the **n** species, representing $(n-1)$ equilibria, only one species is chemically independent. Hence the system is effectively one component system.

In contrast to this behavior the dissociation of H_2O to give H_2 and O_2 in the presence of a catalyst is a three component system, while it reduces to one component system in the absence of catalyst as this dissociation equilibrium is not established for practical purpose (extremely slow).

From the above discussion it is clear that an accurate assignment of the number of components in a system pre-supposes some experimental knowledge of various equilibria involved in the system.

21.2.3 Degrees of Freedom

The number of degrees of freedom or variance (F) is the minimum number of ***intensive variables*** *(mass independent properties such as T, P, and concentration) that must be specified in order to fix the values of all remaining intensive variables.* The mass dependent properties like weight, volume, etc., are called **extensive properties** and they do not determine the equilibrium state of the phase. Physical properties like temperature, pressure, density, refractive index, viscosity, concentration (moles per unit volume), etc., are mass independent variables, and they only define the phases of a system.

Considering a system of pure water, to describe the phase and know whether water is present as solid, liquid or vapor it is just sufficient to know the pressure and temperature conditions as the other intensive variables (refractive index, viscosity etc) are automatically defined. Thus we have a freedom choice of only two variables, viz., the pressure and the temperature, of the various permitted intensive variables, and the water present in a single phase thus has only two degrees of freedom. Any change in either of these two variables could change the phase of the water. Thus water, existing in the vapor phase at 100 ^{O}C and 1 atm pressure, converts into liquid water either by increasing the pressure or decreasing the temperature.

Systems with $F = O$ are known as **invariant** or as having **no degrees of freedom**; with $F = 1$ they are called **univariant** (one degree of freedom), with $F = 2$ they are called **divariant** (two degrees of freedom); and so on. These are demonstrated in the following examples :

(i) **A pure gas :** A pure gas is completely described by just knowing the temperature and pressure. Hence it is divariant system or has two degrees of freedom. This conclusion could be arrived from the phase rule expression (21.1) as a pure gas is a one component, (C=1), one phase, (P=1) system.

Therefore;

Degrees of freedom, $F = 1 - 1 + 2 = 2$ (Divariant)

(ii) **A gaseous system consisting of two gases :** To describe the system fully, we need to define the composition of the mixture, temperature and pressure of the system. Hence it becomes a trivariant system. The same conclusion could be arrived as the gas mixture is a two component and one phase system.

Degrees of freedom, $F = 2 - 1 + 2 = 3$ (Trivariant)

(iii) **A gaseous system consisting of three gases :** Here the system could be fully described, if the composition of all three constituents, temperature and pressure are known. Now substituting the number of components as 3 and phases 1, we have from the phase rule expression.

Degrees of freedom, $F = 3 - 1 + 2 = 4$

(iv) **An undissolved solute in contact with its solution :** When CO_2 is filled in a jar containing water, the water gets saturated with CO_2 and the dissolved CO_2 is in equilibrium with free CO_2 gas. From Raoult's law, we know that the partial vapor pressure of a component at a given temperature is directly proportional to the mole fraction of that component. Just the description of temperature defines the solubility and the vapor pressure of the gas. Hence it is a univariant system. The same conclusion can be drawn from phase rule as the number of components is one and phases are two.

Degrees of freedom, $F = 1 - 2 + 2 = 1$

(v) **Ice-Water-Vapor system :** All the three phases of water, viz., ice, water and vapor, are at equilibrium at a temperature of 0.0098 ^{o}C, pressure 4.58 mm Hg. Any change in temperature or pressure disturbs the three phases equilibrium and we then get either one or two phases. Hence such a system has no degrees of freedom or invariant. As the number of components = 1, phases=3; it could be seen from the phase rule equation that;

Degrees of freedom $F = 1 - 3 + 2 = 0$

21.3 Phase Diagram

A diagram which illustrates the conditions of equilibrium between various phases of a substance is called a ***phase diagram or equilibrium diagram***. A polyphasic equilibrium which can be studied with the help of any two intensive properties, say temperature and pressure or temperature and composition or pressure and composition are represented by a two dimensional phase diagram. A trivariant system requires a three dimensional phase diagram. The importance of the phase diagrams could be understood by considering the following one component system.

21.4 One Component System

Pure water forms an example of one component system. Similarly the physical equilibria like melting, sublimation, evaporation etc., of pure substances are all example of one component heterogenous systems.

Solid ⟷ Solid (Allotrophic transition)

Solid ⟷ Liquid (Fusion)

Liquid ⟷ Vapor (Evaporation)

Solid ⟷ Vapor (Sublimation)

A one component system, should have at least one phase. Thus a one component system possess a maximum variance of two as seen by the phase rule expression.

Degrees of freedom $F = 1 - 1 + 2 = 2$

Hence a one component system is defined by at least two physical parameters viz., temperature and pressure. As we increase the number of phases of the one component system we can achieve a variance of 1 and zero.

(a) When two phase of the system are at equilibrium; degrees of freedom, $F = 1 - 2 + 2 = 1$

Hence it becomes a univariant system, allowing a variation of either temperature or pressure. This equilibrium vanishes by the variation of both the physical variables.

(b) When any three phases of the system are at equilibrium; degrees of freedom $= 1 - 3 + 2 = 0$

Hence it becomes an invariant system, allowing no variation of any of the intensive properties.

These results infer that it is not possible to have any four or any higher polyphasic equilibrium for a one component system.

21.4.1 Water System

Water system forms a typical example of one component system. Argument in support of water being a one component system has already been provided in section (21.2.2). *Water system comprises of three phases, viz., ice, liquid water and water vapor,* which show the following equilibria.

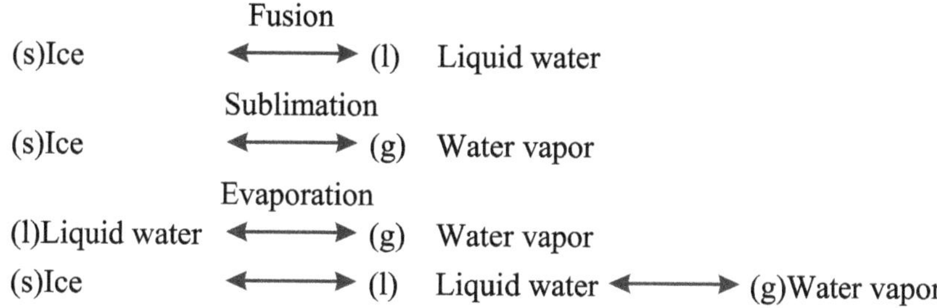

(s)Ice $\xleftrightarrow{\text{Fusion}}$ (l) Liquid water

(s)Ice $\xleftrightarrow{\text{Sublimation}}$ (g) Water vapor

(l)Liquid water $\xleftrightarrow{\text{Evaporation}}$ (g) Water vapor

(s)Ice $\longleftrightarrow$ (l) Liquid water $\longleftrightarrow$ (g)Water vapor

Similar to other one component systems, water can have 2, 1 and 0 degrees of freedom. Because the maximum variance is 2 we can build the phase diagram of water system by studying the effects of pressure and temperature on various phase equilibria, Fig. 21.3.

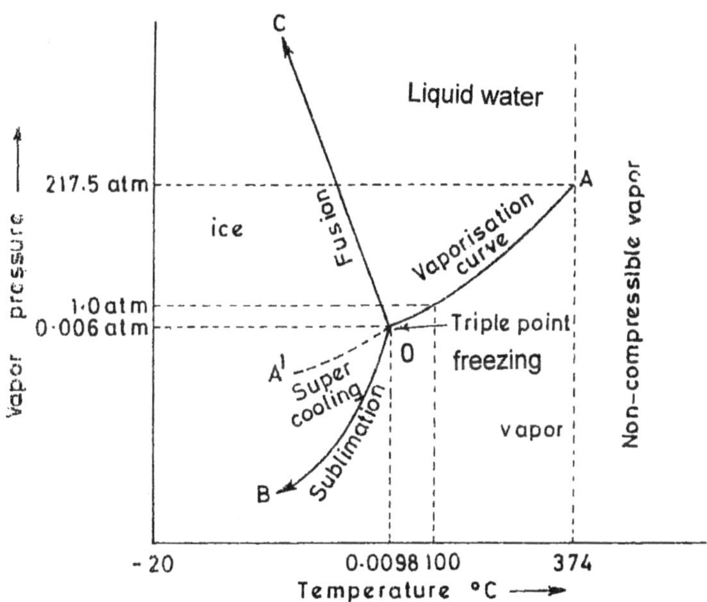

Fig. 21.3 Phase diagram of water system.

(i) Vaporization curve, OA : Liquid water at any given temperature has a precise vapor pressure. *With the increase in temperature we notice an increase in the vapor pressure*, but the rate of increase at higher temperatures is higher than at lower temperatures making *the vaporization curve, OA, non linear and slant upwards away from the* *temperature axis. This increasing trend continues till the system reaches a critical temperature* of 374°C *and a critical vapor pressure* of 217.5 atm. *Any attempt to pass beyond **A** destroys the liquid vapor equilibrium leaving the water totally in incompressible vapor phase.* It is not possible to liquify water vapor beyond 374°C critical temperature by the application of any amount of pressure.

Point O on the lower part of vaporization curve represents the freezing point of water, 0.0098^{O}C, under its own vapor pressure,4.58 mm of Hg or 0.006 atm. The lower value of normal freezing point of water (0°C) *measured under* laboratory conditions is due to :

(a) depression of freezing point by the dissolved air,

(b) the external pressure being 1atm, far higher than the critical vapor pressure of water vapor at the freezing point.

(ii) Sublimation curve,OB : *Below the freezing point the liquid phase totally converts into solid ice* and then we observe only an equilibrium between solid and vapor phases and *the curve representing this equilibrium is called **sublimation curve OB***. With the decrease in temperature initially *the vapor pressure drops steeply followed by slow decrease. This sublimation curve can be extended below B, but the immeasurably small values of vapor pressure bring more uncertainty than any use to the system and hence is left with an arrow head at B.*

(iii) Fusion Curve, OC : *The fusion equilibrium between liquid water and ice is represented by the curve OC.* The equilibrium may be represented

$$\underset{\text{(Occupies more volume)}}{\text{Ice}} \xrightleftharpoons{\text{heat}} \underset{\text{(Occupies less volume)}}{\text{Liquid water}}$$

as ice occupies more volume than liquid water, according to Le Chatelier's principle, any raise in pressure shifts the equilibrium in favor of liquid water, accompanied by a simultaneous drop in the fusion temperature. This is seen by the sloping of the curve OC towards the vapor pressure axis. *This curve is extendable, also beyond C and this in fact is indicated by the arrow head at C.*

(iv) Super cooling curve, OA' : *It is possible to cool the liquid water at 0, by gradually lowering the temperature without making any mechanical movements like stirring, scratching etc.*, and avoiding seeding by the dust particles or a tiny crystal of ice. Any mechanical shock, or the presence of a tiny crystal of ice result in shooting up of temperature to 0°C. *This* *cooling below the freezing point indicated by the broken line OA'* indicates a metastable equilibrium between ice and super cooled water and is called a super cooling curve. The vapor pressure of metastable phase curve OA' *at any temperature is greater than the vapor pressure of stable phase (curve OB). In the super cooled liquid, the solid and super cooled liquid are imagined to be separated by a thin vapor membrane and any slightest disturbance removes this insulatory vapor membrane leading to freezing of water and shooting of temperature to* $0^{0}C$.

(v) **Super heating :** Similar to super cooling curve, attempts to obtain super heating of a solid beyond the freezing point O were found futile. Hence no super heating curve is shown in the phase diagram.

After obtaining the phase diagram we can make an attempt to interpret the results obtained, in terms of number of degrees of freedom or variance of water system. The phase diagram contains four curves, a point of intersection of these four curves and areas separated by these curves.

Curves : Out of the four curves OA,OB,OC and OA' present in the phase diagram of water system, the former three curves (OA,OB & OC) representing phase equilibria of any two of the stable phases. Applying the phase rule equation for this *one component, two phase equilibria represented by these curves;*

The number of degrees of freedom $F = 1 - 2 + 2 = 1$

Thus the curves represent a univariant system, ie., along the curve on *the phase diagram fixing any one of the intensive properties* (either pressure or temperature), automatically fixes the other variables. *The broken curve OA' represents a metastable equilibrium of vapor and super cooled liquid and is also an univariant system.*

Areas : Areas in a one component phase diagram indicate the existence of single phase. Applying the phase rule equation for this one component, one phase equilibria represented by the areas AOB, BOC and COA;

The number of degrees of freedom, $F = 1 - 1 + 2 = 2$

Thus the areas represent a bivariant system, suggesting that in these areas for locating any point both temperature and pressure have to be fixed.

In these areas;

(a) area covered by AOB represents only vapor phase.

(b) area covered by BOC represents only solid ice phase and

(c) area covered by COA represents only liquid water phase.

Point of inter section of curves or Triple point : All the four curves intersect at O and this point of intersection is called the **triple point**. *At the triple point all the three phases coexist and are at equilibrium.* Applying the phase rule equation for the *one component, three phase equilibrium represented by point O*;

The number of degrees of freedom, $F = 1 - 3 + 2 = 0$

Thus the triple point represents an invariant system, ie., any variation in even one of the variable (either pressure or temperature) breaks the three phase equilibrium and transforms into a two phase equilibria represented by the curves. *The temperature and pressure of the triple point are fixed* at $0.098^{O}C$ and 4.58 mm Hg respectively.

21.5 Two Component System

In a two component system the composition of all the individual phases available in various phase equilibria are expressed with the help of not less than two components. *Systems involving fusion of any two metals, formation of freezing mixtures and salt hydrates are some of the typical examples of two component systems.* In a two component system, minimum number of possible phases could be taken as unity and this corresponds to three degrees of freedom ($F = 2 - 1 + 2 = 3$). Hence to illustrate fully a two component system we require to define at least three physical variables, viz., temperature, pressure and composition. As explained earlier the three possible variables could only be illustrated on a three dimensional phases diagram. To simplify the situation and bring it down to a two dimensional illustration, normally one of the variables in kept constant throughout and only two of the intensive properties are varied and corresponding curves are plotted. We now illustrate the phase diagrams of silver-lead system and KI-water system and explain the finer details of these phase equilibria.

21.5.1 Silver-Lead system

The four possible phases of the silver-lead system are (a) solid silver (b) solid lead (c) solution of silver and lead in molten state and (d) vapor. Silver and lead being very high temperature boiling metals, the vapor is practically absent, therefore pressure changes have a negligible impact on the phase diagram. Negligible pressure effects on the phase equilibria led to a practice of plotting the temperature against percentage composition of one of the metals. As discussed already, this fact helped in illustrating the system by a two dimensional phase diagram, Fig. 21.4.

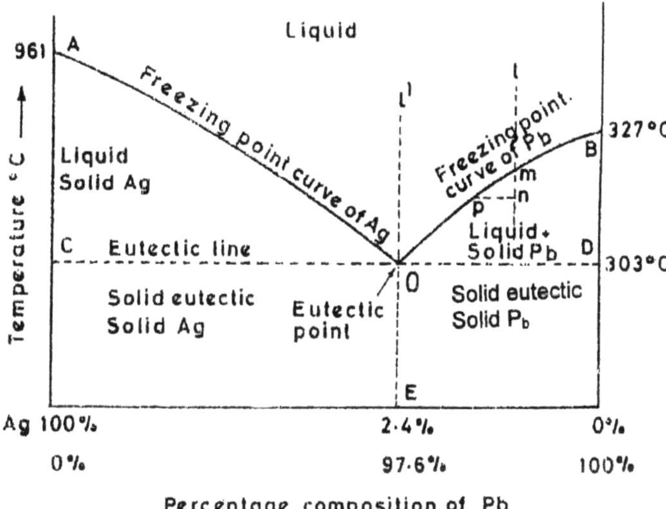

Fig. 21.4 Phase diagram of silver-Lead system.

In plotting the phase diagram, because the measurements are made *at a constant pressure of 1 atmosphere,* the variance of the system reduces by a factor of one and *corresponding phase rule equation, commonly called* ***reduced or condensed phase rule*** *equation may be written as;*

$$F' = C - P + 1$$

The freezing point curve OA : The point A represents the melting point of 100% pure silver (M.P = 961°C) where the solid silver, silver melt (liquid) and vapor coexist. The melting point of silver falls gradually by the addition of lead (depression of freezing point) till the silver melt gets saturated with lead as represented by the curve AO. At **O** the Ag solution is fully saturated with Pb and any further addition of lead is not accepted and it simply separates out as solid Pb. This saturation limit represented by O is called the ***eutectic point*** (Greek: eutectos = easy melting).

All along the curve AO, we notice three phases, solid silver, liquid and vapor in equilibrium. Applying the reduced phase rule equation for this two component, two phase equilibrium. (It is important to note that the vapor phase is already accounted for, while deducing the reduced phase rule equation);

The number of degrees of freedom $F' = 2 - 2 + 1 = 1$

Hence the curve AO represents a univariant system, where fixing any one of the variables automatically fixes the other physical variable and even a slight variation of one of the variables leads to the total disappearance of one of the three phases.

Freezing point curve of lead OB : In the freezing point curve OB, the point B represents the melting point (M.P = 327^{0}C) of 100% pure lead where the solid lead melt and vapor coexist. By the gradual addition of silver, the melting point of lead falls till the lead melt gets saturated by silver, as represented by the curve OB. Any further addition of lead beyond O, simply separate out as solid lead.

Similar to the freezing point curve of silver, OB also represents a phase equilibrium between solid lead, liquid and *vapor*. Applying the reduced phase rule equation for this two component, two phase equilibrium;

The number of degrees of freedom, $F' = 2 - 2 + 1 = 1$

Hence the curve BO also represents an univariant system where defining one variable amounts to fixing of the other variables automatically.

Eutectic Point O : At this point all the four phases, viz., solid Ag, solid Pb, liquid solution and *vapor* coexist. Applying the reduced phase rule equation;

The degrees of freedom, $F = 2 - 3 + 1 = 0$

Hence point O represents an invariant system, where any minor variation in one of the variables shatters the three phase equilibrium resulting in a two phase equilibrium. The temperature and composition of the eutectic mixture corresponding to eutectic point O at 1 atmospheric pressure are 303^{0}C and 2.4% Ag + 97.6% Pb respectively.

The salient features of eutectic point are;

(i) Eutectic point refers to lowest freezing point of a liquid mixture of any two metals at a given external pressure (conventionally fixed at 1atm).

(ii) Freezing any unsaturated liquid solution described by areas AOC and BOD leaves an eutectic mixture and pure solid metal corresponding to that area.

(iii) Similar to a molten metal, eutectic mixture freezes at a constant temperature. The solid eutectic so formed has a fixed composition and is called an alloy.

(iv) In spite of having a definite composition and fixed freezing point, eutectic mixture is never considered as a compound as the crystals of components are clearly observable under microscope.

The Areas : Drawing a horizontal line CD passing through the eutectic point, we notice three areas above the eutectic line, CD, and two areas below the eutectic line.

The Area AOB : The area AOB represents a liquid solution of Ag and Pb in equilibrium with its vapor. As the pressure is arbitrarily fixed at 1atm;

The number of degrees of freedom in the area AOB are $F' = 2 - 1 + 1 = 2$.

Hence AOB is a bivariant region. On cooling the liquid at any point l', vertically above the eutectic point, eutectic solid is separated only after touching the eutectic point O. The liquid at any other point, say, ***l*** when cooled, remain unchanged until it reaches a temperature corresponding to m on the freezing curve. At *m*, solid Pb will commence to deposit keeping the liquid composition at *m*. On further cooling it enters a univariant region BOD. At any point in this region, say *n*, the composition of the liquid is given by the point *p*, corresponding to the intersection of the tie line and BO curve. Further cooling to eutectic temperature, the point *q* on the eutectic line corresponds to solid eutectic and the solid Pb.

The Area AOC : The area AOC has solid silver and liquid melt in equilibrium. The composition of the liquid melt directly corresponds to the tie line intersection point on the AO curve. This area is a univariant region ($F' = 2 - 2 + 1 = 1$).

The Area BOD : The area BOD has solid lead and liquid melt in equilibrium. The composition of the liquid melt directly corresponds to the tie line inter section point on the BO curve. Thus is also a univariant region.

The Area COE : The area COE has solid eutectic and solid Ag in equilibrium. Applying reduced phase rule equation we find that the region is univariant ($F' = 2 - 2 + 1 = 1$)

The area DOE : The area DOE has solid eutectic and solid pb in equilibrium. This is also a univariant system just like area COE.

21.5.2 Pattinson's process for the desilverization of lead

Silver is often recovered from the work up of lead and copper ores. Galena is a lead ore containing 0.1% silver in it. The enrichment of silver in this ore to 2.4% was achieved from a knowledge of Ag-Pb phase diagram. This process is called **Pattinson's process or desilverization process**. It was already discussed under the area AOB, that a molten liquid ***l***, which is rich in lead,

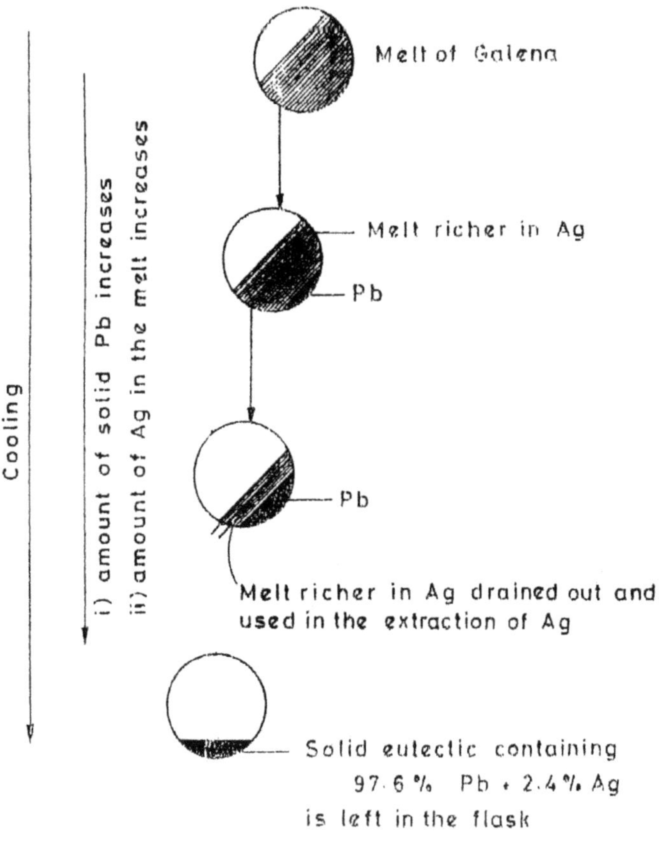

Fig. 21.5 Pattinson's process.

when cooled traces a path ***l m n q***, with a separation of lead metal. The concentration of the liquid at points ***m, n,*** and ***q***, correspond to the intersection point of BO curve with the tie lines passing through these points. Hence, as the liquid cools to **q**, the composition of the melt approaches the composition of the eutectic mixture (2.4% Ag and 97.6% Pb). This solid eutectic rich in Ag is profitably then used for the subsequent recovery of silver employing electrochemical techniques. The relative increase in the amount of lead in the solids and increase of silver in the melt on cooling argentiferrous lead is shown in Fig. 21.5.

21.5.3 Other applications of eutectic mixtures

(i) In the soldering of two metal pieces, often an alloy (solid eutectic) that has much lower melting point than the individual metal pieces to be joined is used and the process is called soldering.

(ii) The safety valves fitted in the pressure cookers to prevent bursting due to any malfunctioning are made up of the alloys that fuse as soon as a particular limiting temperature is reached.

(iii) Safety fuses made out of low melting alloys like Pb-Sn alloys, commonly called wood metal, (Bi 50%, Pb 25%, Sn 12½% and Cd 12½%) are used in protecting buildings against fire hazards. Wood metal melts at 65°C. It is used for plugging water sprinklers arranged in the building roof. In case of any fire in the building, the safety plug melts off, putting automatically the water sprayer in action to put off the fire.

Few eutectic systems with eutectic temperature and compositions are given in Table 21.1.

Table 21.1 Some Eutectic Systems.

Components (M.P.in brackets)		**Eutectic**	**Eutectic composition**
Component A	**Component B**	**temperature**	**composition(%ofA)**
Ag (961°C)	Pb (327°C)	303°C	2.6%
Bi (273°C)	Cd (323°C)	140°C	60%
Zn (419°C)	Cd (323°C)	270°C	67%
KCl (773°C)	LiCl (601°C)	361°C	42%
Sb (630°C)	Pb (326°C)	246°C	–

21.5.4 KI-Water System

This is another example of a two component system, and is a representative of freezing mixtures. Since *KI* does not form hydrates, the four possible phases of *KI*-water system are ice, solid *KI*, solution of *KI* in water and vapor. As in Ag-Pb system here also the pressure is maintained at 1 atm and the phase diagram is illustrated between temperature and composition of *KI*. Most of the treatment of this system is similar to Ag-Pb system.The phase diagram is given in Fig. 21.6.

Freezing point curve AO : The point A is the freezing point of water under normal conditions. Here the phase equilibrium exists between solid ice, liquid water and vapor. This freezing point is lowered by the addition of KI as represented by the freezing point curve AO. Point O represents lowest freezing point attainable by the addition of KI to ice-water equilibrium (freezing) and is called the eutectic temperature. The eutectic point also represents the maximum solubility of KI in water at the eutectic temperature. Any further addition to the eutectic mixture only leads to simple separation of solid KI.

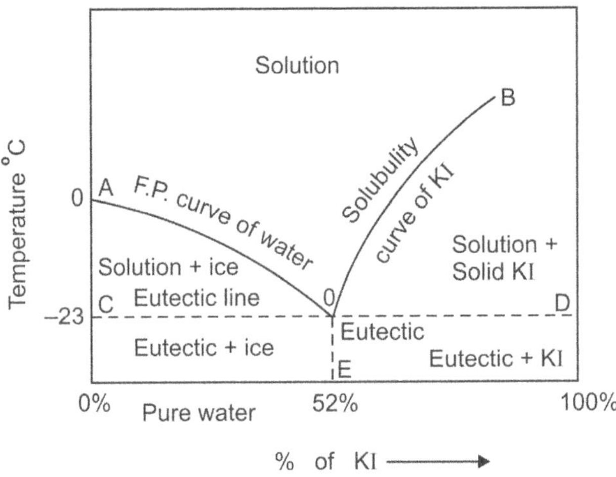

Fig. 21.6 Phase diagram of KI-water system.

Along the curve we observe a three phase equilibrium between ice, KI solution and *vapor*. Applying the reduced phase rule equation for this two component, two phase equilibrium curve AO;

The number of degrees of freedom F' = 2 – 2 + 1 = 1.

Thus curve AO represents a univariant system, where fixing any one of the variable automatically defines the other two variables and even a slight variation of one of the variable leads to the total disappearance of one of the two phases.

Solubility curve of KI, OB : The curve OB represents the effect of increase in solubility with increase in temperature. Along OB we find a three phase equilibrium between solid KI, saturated KI solution *and vapor.* This solubility curve can not be traced to 100% KI because the saturation of KI puts a limit at B on the boiling point. Thus the point B, owing to its constant temperature, pressure and composition conditions becomes an invariant system.

Applying the reduced phase rule equation for this two component, two phase equilibrium curve BO;

The number of degrees of freedom $F' = 2 - 2 + 1 = 1$

Thus curve BO represents a univariant system, where fixing any one of the variables automatically defines the other two variable and even a slight variation of one of the variables leads to the total disappearance of one of the two phases.

Eutectic point or Cryohydric point : The meeting point **O**, of the curves AO and OB is as usual called an eutectic point and in these freezing mixture equilibria in specific it is termed as **cryohydric point**. At this point all the four phases, viz., solid KI, ice, solution of KI and *vapor* coexist. Applying the reduced phase equation;

The number of degrees of freedom $F' = 2 - 3 + 1 = 0$

Hence cryohydric point, O, represents an invariant system, where any minor variation in one of the variables shatters the four phase equilibrium, resulting in the creation of a three phase equilibrium. The cryohydric temperature and composition are $-23^{O}C$ and 52% KI respectively at 1 atm pressure.

The Areas : The horizontal line passing through the eutectic point **O** is called the **eutectic line**. In the phase diagram we notice three areas above the eutectic line and two below the eutectic line.

The Area AOB : This represents the aqueous solution of KI in equilibrium with vapor. From the reduced phase rule equation we find that it is a bivariant equilibrium ($F' = 2 - 1 + 1 = 2$).

The Area AOC : The area AOC represents a phase equilibrium between ice, aqueous solution of KI and vapor. This area is an univariant region ($F' = 2 - 2 + 1 = 1$).

The Area BOD: The area BOD represents a phase equilibrium between solid KI, aqueous solution of KI and vapor. This is also an univariant region.

The Area COE : The area COE represents a phase equilibrium between eutectic, ice and vapor. This is also an univariant region.

The Area DOE : The area DOE represents the equilibrium between the eutectic, solid KI and vapor. This is an univariant region.

21.5.5 Applications of Freezing Mixtures

From the phase diagram of KI - water system it is clear that when we add a salt to ice, the freezing point curve AO is traced and we notice a depression of freezing point. Initially this fall in the freezing point is directly proportional to the amount of salt added and addition of excess of salt drops the temperature to touch the cryohydric temperature (eutectic temperature) and no further drop occurs. In the laboratory the freezing mixtures are used conveniently to bring down the reaction medium temperature to sub zero levels. Some of the commonly employed cryohydric mixtures and their temperature are given in Table 21.2.

Table 21.2 Some cryohydric temperatures

Cryohydric mixture	**Composition at cryohydric point**	**Cryohydric temperature**
KI + Ice	52%	$-23^{o}C$
NaCl + Ice	24%	$-22^{o}C$
$CaCl_2$ + Ice	15.2%	$-55.9^{o}C$
NH_4Cl + Ice	20%	$-17^{o}C$
KCl + Ice	–	$-11.4^{o}C$
KNO_3 + Ice	–	$-2.6^{o}C$
Alcohol + Solid CO_2	–	$-72^{o}C$
Chloroform + Solid CO_2	–	$-77^{o}C$

Self -Test Questions

1. Number of phases present in a mixture of nitrogen and oxygen are

   (A) 1 (B) 2

   (C) 3 (D) 4

2. Number of components present in a solution of sodium chloride is

   (A) 1 (B) 4

   (C) 3 (D) 2

3. The number of phases present in powdered tin and lead is

   (A) 1 (B) 4

   (C) 3 (D) 2

4. At melting point of ice, the system has two phase. The system is one component. The degrees of freedom of the system are

   (A) 1 (B) 2

   (C) 3 (D) 4

5. Water system is non variant at

   (A) melting point (B) boiling point

   (C) triple point (D) critical point

6. Mathematical statement of reduced phase rule is

   (A) $F' = C - P + 2$ (B) $F' = C - P + 1$

   (C) $F' = C - P + 3$ (D) $F' = C - P$

7. Maximum degrees of freedom for one component system are

   (A) 1 (B) 3

   (C) 2 (D) 4

8. Number of components in the system of Ag and Pb at eutectic temperature is

   (A) 1 (B) 2

   (C) 3 (D) 4

9. For Ag-Pb system, when it has three phases, the number of degrees of freedom is

   (A) 1 (B) 2

   (C) 0 (D) 3

10. The liquid mixture of any two components of definite composition, lowest freezing point, is called_________.

11. A homogenous and physically distinct part of the system which is separated from other parts of the system is called_____________.

12. The minimum number of constituents required in order to describe the composition of phases present in a system is called_______.

13. Eutectic temperature of Pb-Ag system is ________.

14. Phase diagram of Ag-Pb system is drawn between ________ and composition.

Short & Essay Type Questions

1. Draw the phase diagram for water and mention its parts.
2. Construct the phase diagram of silver - lead system and explain the importance of eutectic point
3. State phase rule and explain the terms involved. Apply the phase rule to silver-lead system.
4. What is a phase diagram ? Explain.
5. What is an eutectic ? Give one example.
6. Explain with suitable examples the following terms used in the phase rule study of heterogenous equilibrium

   (i) Triple point (ii) Eutectic point

   (iii) Univariant system
7. Discuss the application of phase rule to potassium iodide -water system.

8. (a) Write the phase rule equation. Discuss the application of the phase rule to a one component system.

   (b) Define the terms 'Eutectic point' and triple point.

9. Define the terms in the phase rule. Explain these terms in the following system

   (i) A gas in equilibrium with its solution in a liquid

   (ii) Ammonia gas

   (iii) $I_2(s) \longleftrightarrow I_2(g)$

10. (a) Define the terms (i) Component (ii) Phase of a system

    (b) Predict the degrees of freedom for the following

    (i) A one component system having two phases in equilibrium

    (ii) A one component system having a single phase.

    (iii) A one component system having three phases in equilibrium.

11. Explain the term component. How many components are present in the following system ?

    Water $\longleftrightarrow$ Water vapor

    KCl + water $\longleftrightarrow$ KCl hydrate.

12. Discuss a method to obtain the phase diagram of Ag-Pb system and give details of the number of degrees of freedom of the curves and areas in it.

13. (a) Draw the phase diagram of KI-H_2O systems.

    (b) Answer the following questions pertaining to the diagrams.

    (i) How many components are there in the system

    (ii) How many phases are in equilibrium along any one line

    (iii) What is the number of degrees of freedom along any line

    (iv) Name the phases which are in equilibrium at the eutectic point.

    (v) What is the number of degrees of freedom at the eutectic point.

14. Find the degrees of freedom and number of components in the following systems

    (a) $CaCO_3(s) \longleftrightarrow CaO(s) + CO_2(g)$

    (b) A two phase system made by mixing arbitrary amounts of C, O_2 and CO_2 at a temperature at which the reversible equilibrium establishes

    $C_{(s)} + O_2\ (g) \longleftrightarrow CO_2(g)$

15. State phase rule and apply to water system.

16. Plot the phase diagram of a one component system and identify all the lines, areas and phases.

17. Explain phase diagram of salt-water systems.

Chapter 22

Extraction of Metals

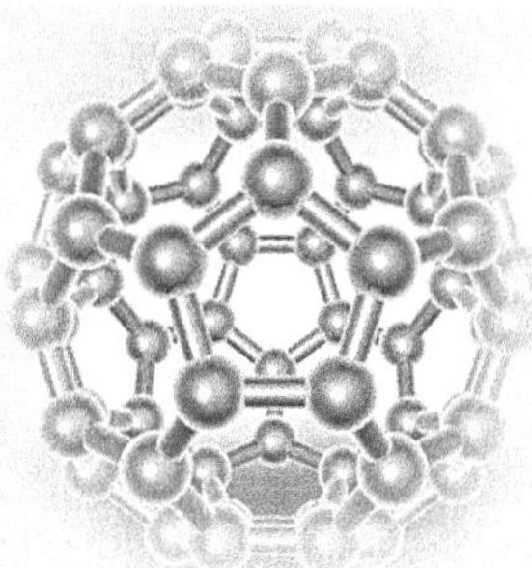

22.1 Occurrence of Metals

It is interesting to note that three-fourths of the elements are metals. The Table 22.1 shows the percentages of the various elements present in the earth's crust:

Table 22.1 Average analyses of igneous rocks.

Element	%	*Element*	%
Oxygen	46.59	Lead	0.002
Silicon	27.72	Thorium	0.002
Aluminium	8.13	Cobalt	0.001
Iron	5.01	Beryllium	0.001
Calcium	3.63	Molybdenum	0.000,n*
Sodium	2.85	Arsenic	0.000,n
Potassium	2.60	Tin	0.000,n
Magnesium	2.09	Antimony	0.000,0n
Manganese	0.90	Cadmium	0.000,0n
Titanium	0.63	Mercury	0.000,0n
Nickel	0.020	Silver	0.000,00n
Vanadium	0.017	Platinum	0.000,000,n
Copper	0.010	Gold	0.000,000,n
Uranium	0.008	Iridium	0.000,000,0n
Tungsten	0.005	Palladium	0.000,000,0n
Zinc	0.004	Radium	0.000,000,000,n

* The 'n' indicates the first significant Figure.

(From F. W. Clarke and H. S. Washington, Composition of the Earth's Crust, U.S. Geological Survey Professional Paper 127, 1924).

The table reveals that about 75% of the earth's crust consists of oxygen and silicon alone and only six metals occur in amounts of more than 2%. Of these six metals, only three are of industrial importance. The amount of an important metal like, copper is only 0.01%. Luckily, the metals are not uniformly distributed in the crust, otherwise it would have been next to impossible to recover them. Due to several factors, like soil movements, precipitation and crystallization, the metals are concentrated in the form of ores. It is from these concentrates that the metals are recovered.

Sea water is an unsuspected repository of mineral wealth and a metal like magnesium is already being extracted from it. It is estimated that a cubic kilometer of sea water contains one million tonnes of magnesium, 15,000 tonnes of strontium, 5 tonnes each of gold, copper, manganese, zinc and lead.

Only a few metals like gold and platinum are found in free state, and a few others, e.g., copper, silver and iron occasionally occur in free state. The vast majority of metals occur in nature as chemical compounds such as oxides, sulphides, carbonates, sulphates, silicates, etc. These compounds of metals, which occur in nature, are known as *minerals*. The minerals are almost always mixed with relatively valueless materials, like sand, clay, etc., which are called *gangue*. An *ore* is a mineral from which a metal can be profitably extracted. Valuable portions obtained after removing gangue from ore are called *values*.

Whether a deposit is mineral or an ore depends upon the cost of extraction and the market price of the metal. As prices have increased and methods of extraction improved, some of the minerals, which were discarded earlier, are now being used for extraction of metals and are now called ores. Generally, an iron ore must contain at least 45% iron, a copper ore at least 1% copper and a radium ore at least 5×10^{-9} % radium before the metal can be economically extracted. Because of the increased demands and improved methods, much poorer ores are also being worked.

22.2 Mining

The first step in the production of a metal is to bring the ore to the surface. This is done by mining which is of two general types, open-pit mining and underground mining. *Open pit mining* as the name suggests, is employed when the ore is at, or very close to, the surface, say, within 15 meters. In most cases the ore is covered with a layer of soil, called overburden. When the overburden is removed, the mining becomes a simple process of excavation or digging. When the ore is covered with water, the process is called *dredging*. The advantages of open-pit mining are employment of large and varied types of machinery, lower mining cost, large-scale production and better and safer working conditions.

Most of the deposits are deep underground and have to be brought to the surface by *underground mining*. First 'holes' or shafts or tunnels are drilled to reach the ore. This limits the use of the machinery and involves the problems of risk to life, ventilation, drainage, etc. Underground mining is more expensive but widely adopted because most of the ores are found underground. A reference here may be made to petroleum, which is brought to the surface by pumping the liquid to the surface. Certain materials are brought to the surface by forcing water, hot or cold, through the wells, which brings the material along with it to the surface.

22.3 Ore Dressing

Since rich ores are few and many of them have already been worked for some time, it has become increasingly necessary to eliminate the gangue from the values. The process of physically separating the particles of the mineral from the gangue, giving concentrates or values, which contain most of the mineral, is known as *ore dressing*. Ore dressing has two main advantages. Firstly, transport cost of the gangue is saved. Secondly, it causes considerable saving in cost while extracting the metal in the furnace. By using a concentrated ore the capacity of the furnace increases, heating cost becomes less because gangue has not to be heated, and less flux needs to be used for the removal of the gangue.

In case the metallurgical works are situated near the mine, the ore dressing is done at the factory. If the mine is at a distance, the ore dressing plant is located at the mine and the concentrate is transported to the main works for extraction of metal.

Ore dressing begins while mining the ore. The barren parts are not mined. Ore is obtained from the mine in pieces of different size and by *hand picking,* the rich portions is separated from clayey or sandy portions. Hand picking is adopted where labour is cheap.

To facilitate concentration it is necessary to reduce the size of the ore as mined. The final size depends upon the treatments that are to be given later. The primary crushers, like jaw crushers, and gyratory crushers, reduce the size to 5 cm to 20 cm.

The size is further reduced to 0.5 cm to 3 cm by crushing rolls and in hammer mills. In many cases further reduction in size is not necessary. If necessary, the ore is finely ground in grinding equipment such as a ball mill. The material is screened to remove the over-size particles.

Further concentration is done on the basis of difference in the properties of values and gangue. One of the simplest of the concentration methods is the *gravity separation*. The particles are suspended in water when the heavier particles settle more readily and are separated. The settling rate depends on the size and density of the particles and also on the shape.

A modification of the gravity method is the *sink and float* method. In this the powdered ore is suspended in a liquid whose specific gravity is intermediate between the densities of the gangue and values. The lower density material floats and the higher density material sinks. This method is used very widely in coal cleaning in various coal washeries in India. The specific gravity of coal is about 1.4 and that of most minerals is about 2.7. In this method the difficulty is of finding a liquid of the proper specific gravity. Solution of calcium chloride in water is often used. Suspensions of sand in water giving liquids of specific gravities up to 3.2 are used. *Jigging and tabling* methods are based on shaking or giving jerks to the ore in contact with water in different kinds of machinery.

Froth flotation method has revolutionized the ore dressing operations and very poor ores, which could not be worked earlier, are now concentrated by it. It is based on the difference in wettability of different materials. Certain reagents attach themselves to the values and do not allow them to become wet with water. The gangue is wetted and sinks, whereas the values float with air bubbles and produce froth. The ore is finely ground with water giving a thick pulp containing about 70% solids. It is stirred with the proper reagent, called a collector or *oiler* which attaches to the minerals. Xanthates, which are dithiocarbonates like $C_2H_5ON_aCS_2$ are the common collectors. The sulphide mineral attaches to the molecule and becomes water-repellent. To produce a dry and fairly stable froth, chemicals called frothers are added. The entire material is taken into an equipment and air is blown. Air bubbles adhere to oil-coated mineral particles and float them in the form of froth, which is collected. The gangue sinks.

Some ores contain more than one mineral and separation of one mineral from the other in addition to separation from the gangue is necessary. To achieve this, a depressing agent is added which suppresses the flotation of one of the minerals. An important example is that of lead-zinc ores. If the ore is floated without a depressor, both lead and zinc sulphides collect in the froth. On adding a small amount of cyanide or zinc sulphate and blowing air, only lead sulphide floats and flotation of zinc sulphide is depressed. After removing lead sulphide, copper sulphate, which acts as activator, is added and air is blown when zinc sulphide floats. This method is known as *differential flotation*. It has been reported that when a sample of ore containing 10% PbS and 15% ZnS was treated by this method, the first concentrate containing 93% PbS and second 90% ZnS was obtained. Lime is used as a depressor in the flotation of copper ores. It depresses the flotation of iron sulphide.

Magnetic separation is used when one of the constituents is magnetic. In some cases a constituent can be made magnetic by roasting. The crushed ore is passed over a belt conveyor which is subjected to a magnetic field, thus the constituent is separated.

22.4 Metal Extraction

Ore dressing is very helpful in winning the metals but it deals with the separation of impurities that are physically mixed with the mineral. In metal extraction the metal is liberated from chemical combination. Extractive metallurgy is classified under three broad divisions: 1. pyro-metallurgy, 2. hydro-metallurgy, and 3. electro-metallurgy.

22.4.1 Pyro-metallurgy

In pyro-metallurgical operations the ore is heated by burning fuel.

The objective of *drying* is to remove absorbed water. If an ore is to be transported, drying reduces freight charges. The *calcination* operation is similar to drying except that the ore is heated to a higher temperature when some decomposition takes place. Some common examples of calcination are:

$$CaCO_3 \rightarrow CaO + CO_2$$

$$FeCO_3 \rightarrow FeO + CO_2$$

$$2Al(OH)_3 \rightarrow Al_2O_3 + 3H_2O$$

Roasting is heating the ore in the presence of air and involves oxidation. It is mostly applied for sulphide ores which are oxidized to oxide or sulphates. Due to higher temperature employed, some impurities, like sulphides of arsenic and antimony, get oxidized and volatilized. Typical roasting reactions are:

$$4FeS_2 + 11O_2 \rightarrow 2Fe_2O_3 + 8SO_2$$

$$ZnS + 2O_2 \rightarrow ZnSO_4$$

$$2ZnS + 3O_2 \rightarrow 2ZnO + 2SO_2$$

Sintering is a process of heating finely divided solid material to a temperature approaching its melting point. The particles soften and join together to form lumps. The process need not necessarily involve a chemical change. Sintering is being widely practiced in India in agglomerating finer particles of iron ore. The dust and fines cannot be charged in a blast furnace, hence they are sintered to produce lumps which are charged in the furnace.

Smelting is a process of heating the ore, often with flux and fuel, to a high temperature at which the mass is in a molten state. In reduction smelting, as of iron ore, impure metal and molten slag are produced. In matte smelting, molten matte, i.e. mixture of purified sulphides, and molten slag are produced. When the material is heated by burning carbonaceous fuel, mostly coke (or coal), it simultaneously reduces the oxide ore thus:

$$MeO + C \rightarrow Me\ (Metal) + CO$$

At the temperature prevailing in most furnaces the reaction is:

$$MeO + CO \rightarrow Me + CO_2$$

(Above about 1000°, CO_2 gets reduced to CO by C)

An important accomplishment of *smelting* is the removal of gangue, which has so far escaped removal. Gangue has a high melting point. Therefore, certain material, known as *flux*, is added which combines with the gangue and produces an easily fusible mass known as *slag*. Acidic gangues, like silica and alumina, are removed by adding basic flux, like lime and magnesia, and vice versa. In some ores the gangue contains both acidic and basic materials in such proportions that on heating they react and form slag. They are called self-fluxing ores.

$$\underset{\substack{\text{(gangue or flux)} \\ \text{solid}}}{SiO_2} + \underset{\substack{\text{(flux or gangue)} \\ \text{solid}}}{CaO} \xrightarrow{\text{heat}} \underset{\substack{\text{(slag)} \\ \text{molten}}}{CaSiO_3}$$

Proper slag formation is an important metallurgical operation. Removal of impurities, which are not otherwise easily removed, is done by slags. The slag must have low melting point, so that it melts easily, and has low specific gravity so that it readily separates from the molten metal. It should also have low viscosity so that it may be easily removed from furnace. The slags are mostly silicates. The common basic oxides are CaO, MgO, FeO, Na_2O, etc. The fusibility of a slag depends upon the basic radical. Alumina slag is less fusible than lime slag and soda slag is the most fusible. The physical properties of slags also depend upon the ratio of oxygen in base and in acid components.

An important principle to note is that all substances in a reduced state tend to go together and those in an oxidized state also goes together. Therefore the impurities, which are reduced collect in the metal and those that remain in oxidized state collect in the slag.

A metal from sulphide ore cannot be directly obtained by reduction smelting. The sulphide has to be first oxidized to oxide and then reduced. The most important sulphide smelting is that of copper ore. It occurs as sulphide along with a considerable amount of iron sulphide. The smelting is done in a neutral atmosphere and molten slag and molten sulphide mixture known as *matte* are drawn off. The matte is treated by 'converting' to obtain the metal.

In *converting*, the molten metal or molten matte is taken in a converter and air is blown in. Heat from an external source is not supplied. When a metal is taken, the object of converting is oxidation of the impurities and the process is a refining process. When a matte is taken, the metal is produced, as in the extraction of copper.

Distillation is applicable to metals having low boiling points. Sometimes a low-boiling metal is obtained by reducing the ore above the boiling point of the metal. The ore is mixed with the reducing agent and heated and the vapours are collected. For example, ZnO is mixed with carbon and heated in retorts. ZnO gets reduced to zinc, which distils over. Distillation is adopted for mercuric sulphide also but in this case no reducing agent is needed. Mercuric sulphide is heated in air when mercury distils over. Distillation is also used for refining the metals. An interesting example is that of refining nickel by distillation. Finely divided nickel is heated to about 60^0 in a current of pure carbon monoxide when nickel carbonyl $Ni(CO)_4$, distils over and the impurities are left behind. At a higher temperature nickel carbonyl decomposes giving pure nickel and carbon monoxide. In some cases distillation under reduced pressure is adopted.

When the melting point of a metal is low and that of the impurities is high, the metal can be purified by *liquation*. It is heated on the sloping bed of a furnace above the melting point or the metal but below the melting point of the impurities. In the case of tin most of the impurities, like copper, iron, arsenic and antimony are removed in this way.

Another example of liquation is the desilverising of lead (Parkes's process). In molten lead, zinc is added which is completely soluble in lead above 800°. As the mixture is cooled, the solubility of zinc decreases and its crystals containing silver and gold separate. Zinc is distilled off and the noble metals are left behind.

22.4.2 Hydro-Metallurgy

The principal application of hydro-metallurgy is in the case of low-grade ores which cannot be economically concentrated by ore dressing. Leaching is an essential operation in hydrometallurgy. The powdered ore is treated with a chemical, which dissolves the mineral leaving behind the gangue. In some cases the ore is leached after giving a treatment which makes it soluble in the chemical.

An important example is of leaching oxidized copper ore by dilute sulphuric acid. Gold is leached by a dilute (0.05%) sodium cyanide solution in the presence of air. Recovery of alumina from bauxite by the Bayer process is adopted on a very large scale. Bauxite is ground and heated with caustic soda solution under pressure when aluminium goes in solution as sodium aluminate. The filtered solution is hydrolysed to give precipitate of aluminium hydroxide.

The metal is obtained from the solution mostly by precipitation by adding a less noble metal, e.g., copper is precipitated by adding iron filings to the solution, or by electrolysis. Ion exchange and solvent extraction methods are also being employed.

22.4.3 Electro-Metallurgy

Quantitatively electro-metallurgy is adapted to a lesser extent than pyro-metallurgy chiefly because of the high cost of electricity. The chief advantage of electro-metallurgical methods is that the purity of the metal produced is much higher than by other methods. They, also, have lower maintenance expenses, higher efficiency and less difficult working conditions.

In electro-metallurgy, electricity has two distinct functions. It is used for electrolysis and also for heating by resistance or by arcing. In some cases both the functions take place simultaneously. These methods are used for extraction as well as for refining the metals.

Electrolysis of an electrolyte solution is widely used for copper. In the extraction of copper the insoluble anode, made of a lead alloy is used and the copper sheet is the cathode. On passing direct current, copper is deposited from copper solution at the cathode. Sulphuric acid is regenerated and used again for leaching copper ore. The refining process is similar except that impure copper blocks form the anodes and the cathode is of pure copper sheets, with dilute sulphuric acid as electrolyte. Anode dissolves and pure copper is deposited at the cathode.

Aluminium, magnesium and sodium are extracted by the electrolysis of fused salts. Electrolysis in fused state is done in the case of highly active metals. They require high temperature for liberation of the metals from the compounds. The high activity of the metal also presents another problem. It gets easily oxidized at the high temperature. In some cases, e.g., sodium, the metal reacts with water and electrolysis of sodium in water is not possible. A large scale example of electrolysis in fused state is that of aluminium extraction. Alumina is dissolved in a fused mixture of fluorides with carbon blocks as anodes and molten aluminium as cathode.

Use of electric furnaces is on the increase. High-grade steel is manufactured in electric furnaces because of freedom from contamination and the superior control the process offers.

Chapter 23

Iron and Steel

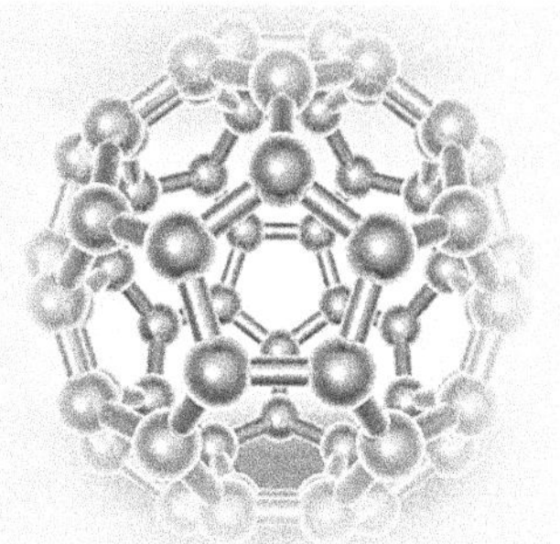

23.1 Introduction

Among the metals, iron comes next only to aluminium in abundance on the earth's surface. The following three chemical properties of iron are the most significant ones for practical purposes (i) Iron oxidizes easily in moist air, thus it gets rusted; (ii) it can be reduced from its oxides at temperatures above 260°; thus it can easily and cheaply be extracted from its ores; (iii) it has a great affinity for carbon; this is responsible, directly as well as indirectly, for most of its engineering uses. Likewise strength, magnetism and adaptability are the three important physical properties of iron. It is adaptable in the sense that by proper manufacturing and treatment, involving addition of not more than 5% of foreign elements, it can be made very strong or very weak, very hard or very ductile or very brittle. Its corrodability, conductivity, coefficient of expansion, etc., can also be altered greatly. Added to all this is its low cost. Thus it is easy to realize why iron and steel play such an important role in the present civilization.

In India, the chief producers of iron and steel have been Tata Iron and Steel Co. Ltd., Jamshedpur, India Iron and Steel Co. Ltd, Burnpur, and Bhadravati Iron and Steel works Ltd., Bhadravati. In 1954, the Government of India set up a corporation in the public sector known as Hindustan Steel Ltd. which set up three plants, situated at Rourkela (Orissa), Bhilai (M. P.) and Durgapur (W. Bengal). Each of these has its own coke ovens which supply the large amounts of coke needed for steel manufacture. Also, the coke ovens have associated with them by-product plants which are essential for the growth of chemical industries. In the public sector the fourth steel plant has been set up at Bokaro. It has an initial capacity of 1.7 million tonnes of ingot iron to be expanded to 4 million tonnes. The plant uses well-prepared and blended burden comprised of a highly fluxed sinter percentage and no limestone is used as flux in the bast furnace. Oxygen converters are used for steel making. The plants at Bhilai and Durgapur employ the basic open hearth process of steel making. The Rourkela plant is unique in the fact that it was the first to employ a process, viz. L. D. process, for manufacturing three quarters of its steel. The remainder is produced by the basic open hearth process.

The first phase of Salem Steel Plant (Tamil Nadu) has been commissioned. The fifth steel plant in the public sector has been commissioned at Visakhapatnam. Its special features are–blast furnace of 3200 m^3 capicity, 7 m tall coke ovens and the adoption of continuous casting machine instead of conventional ingot casting

It would be apparent from the figures appended below, the present growth of iron and steel industry in India :

Year	*Pig Iron* (In Million Tonnes)	*Crude Steel* (In Million Tonnes)
1958	2.00	2.1
1975-76	10.00	8.0
1976-77	10.06	9.62
1979-80	9.57	9.60
1980-81	9.81	9.07

The existing steel plants are under various stages of expansion and technological improvements.

Steel industry is facing a serious challenge. The quality of performance is clear from a report about the working of a plant of Steel Authority of India Ltd. (SAIL) which has 10,000 m^3 of blast furnace and 11,000 tonnes of steel making (converter) rated capacity for 1985-36 of 3 million tonnes of steel, the actual production being much less. Pohang Iron and Steel Works in Korea, with 11,000 m^3 of blast furnace capacity and only 900 tonnes of converter capacity has a rated capacity of 9.4 million tones and actually produced 9.1. m. t. of steel.

One of the reasons given is poor quality of coke (22-23% ash). The ash content in most other countries is around 10%. Since each one per cent of ash reduces the output of the blast furnace by 3 to 4%, a 12% reduction in ash content in the coke, say, by better coal washing or blending with imported coking coal, is imperative for improving the performance of steel plants.

23.2 Pig Iron

23.2.1 Manufacture of Pig Iron

Pig iron is the metallic product obtained by the reduction of iron ore smelted in a blast furnace. Most of this material, while still in a molten state, is converted into steel. The remainder is cast into small shapes or pigs. A few decades back the molten metal was run into cavities in a sand floor and allowed to cool. The resulting shape looked like a pig, hence the name pig iron. The following are the important raw materials used in its manufacture.

1. **Iron Ore.** Iron ores have been classified in several ways, e.g., according to colour (e.g., red ore), gangue (e.g., siliceous ore), element associated with it (e.g., phosphoric and non-phosphoric) or according

to principal iron oxide present. The last mode of classification is the most popular one. Iron ore reserves of India are estimated to be about 13,500 million tonnes, which is about one fifth of the total world reserves. Accordingly the principal iron ores are:-

(i) *Hematite* Fe_2O_3 : This is the principal iron ore of India. Theoretically hematite contains 70% Fe. Indian iron ores are some of the best iron ores of the world, containg 60–64% Fe.

(ii) *Magnetite* Fe_3O_4 : It is not as easily reducible as hematite. Good deposits of magnetite occur in Salem (Tamil Nadu) and Karnataka.

(iii) *Limonite* $2F_2O_3. 3H_2O$: This seems to be the ore from which iron was made in ancient India. In India, at present, its deposits are not of commercial significance. Important deposits occur on the German-French-Belgian border.

(iv) *Siderite or Spathic Iron ore* $FeCO_3$: This is not an important Indian ore.

So far Orissa has been the major iron ore producing State of India, followed by Bihar, Goa, M. P. and Karnataka. The ore at Kudremukha in Karnataka is comparatively of poor quality and for the first time in India iron ore is being concentrated.

The lumps of iron ore charged in a blast furnace vary from 1 cm to 10 cm size. Generally the ores are not concentrated but there is an increasing tendency to use "sinter" and pellet. *Sintering* means agglomerating smaller particles into bigger pieces. Fine ore mixed with fine coke and water is spread as a layer a few centimetres thick on an endless belt through which air is sucked (water gives a porous structure to the sinters). Sometimes limestone is also mixed with it. Combustion starts as the belt moves under a combustion hood. The heat is enough to remove much of the sulphur (its presence is undesirable) and to decompose carbonates (thus saving heat in the blast furnace). Sintering also makes the charge more uniform in size and texture and enables the use of powdered ore and coke which are otherwise wasted. Pelletization is also a process of agglomeration and utilizes finer ores/dust to agglomerate in the form of balls.

2. **Fuel.** Fuel needs to serve two purposes in a blast furnace (i) to supply heat for the reactions as well as for melting the metal and slag and (ii) to reduce the oxide to metallic iron. Charcoal had been used as fuel

by Bhadravati Iron and Steel Works Ltd., but coke is the most extensively used fuel. All integrated steel plants have their own coke oven plants. The fuel should have the following properties:

(i) *Strength.* It should be strong and have resistance to abrasion. The fuel is subject to rough handling and it should not disintegrate easily into fine particles. Fine particles are undesirable because they prevent the free movement of gases, which are responsible for the main reactions, and they may be easily carried away along with blast furnace gases.

(ii) *Porosity.* The more porous the fuel is the more readily it comes in contact with the oxygen of the blast and thus supplies more heat per unit volume of the total charge.

(iii) *Combustibility.* The fuel should be as combustible as possible, i.e., the rate of reaction with oxygen should be as high as possible. A more reactive fuel increases the intensity of combustion thereby increases the productivity of the furnace.

(iv) *Purity.* The impurities present in the fuel can be transferred to the pig iron. Sulphur is a common impurity and is highly undesirable as most of it is taken up by pig iron. The ash content should also be low as it means not only waste of heat but also that additional flux is needed for its removal in the form of slag.

Coke is stronger, more porous and more combustible than coal. Coke is therefore more suitable than coal in spite of the fact that coke contains a higher percentage of ash than the coal from which it is made. The most serious disadvantage of coal is its poor mechanical strength and sulphur content. Coal, if used, will be crushed to smaller pieces or powder under the weight of the charge in the blast furnace, thereby flow of gases will be obstructed. Coke is hard and strong and is not crushed in the blast furnace. Although India has vast deposits of coal, the metallurgical coal for producing high quality, coke is not present in sufficient quantity. Indian coke contains about 25% which is much higher than the ash content of the coke of other leading steel-producing countries.

3. **Flux.** Indian iron ores mostly contain silica as the gangue. Some alumina is also present in the ore and in the coke ash. If the amount of alumina present is small it may form aluminium silicate. Limestone is commonly used a flux to remove the gangue materials. Sometimes dolomite ($CaCO_3$ and $MgCO_3$) is used as flux. It is more expensive than limestone, but the slag produced has lower melting point.

23.2.2 The Principles of Extraction

The Principles of extraction of iron from ore are simple. The reactions taking place in different parts of the furnace are shown in Figure. 23.1 The furnace is known as blast furnace. The raw materials, which are solid, are introduced from the top. These slowly move downwards and react with the upward moving hot gases, which are produced in the lower part of the blast furnace. As the solid charge consisting of ore, coke and limestone descends two or three metres in the furnace, the moisture and combined water are removed. It meets upward moving gases consisting mainly of nitrogen, carbon monoxide and carbon dioxide. Step by step carbon monoxide reduces the heated Fe_2O_3 progressively to Fe_3O_4, FeO and Fe as indicated by equations 1 to 4 as given in Fig. 23.1. These reactions take place mostly at 200 to 800° and within less than 10 m blow the stock line. As CO_2 cannot exist in the presence of free carbon above 1000° it is not formed in the lower part of the furnace. Thus the reduction of iron ore takes place in the upper part of the furnace and CO is the main reducing agent. The spongy iron thus produced absorbs carbon and melts in the lower part of the furnace. The molten iron trickles down over the white hot coke and dissolves carbon until it is practically saturated. Limestone decomposes to give calcium oxide which reacts with silica present in the gangue, and forms calcium silicate. Calcium silicate is in molten state at the temperature prevailing in the furnace. A material like this is known as slag. It also helps in the elimination of sulphur. Manganese oxide, phosphorus oxide and a part of the silica are reduced in the lower part of the furnace (equations 7, 9 and 10) and the elements are absorbed by the molten metal. The general principle "all reduced elements join the metal and the that are oxidized go to or form a slag" may help to remember the distribution of the materials between the molten iron and the slag. Iron, which is already practically saturated with carbon, dissolves silicon, mangenese, sulphur and phosphorus. The slag contains CaO, SiO_2. Al_2O_3, MgO, CaS and some MnO. Composition of pig iron can be controlled to some extent by controlling the temperature. Pig iron produced in a furnace working at a high temperature contains less sulphur (which is a distinct advantage) but more silicon and manganese.

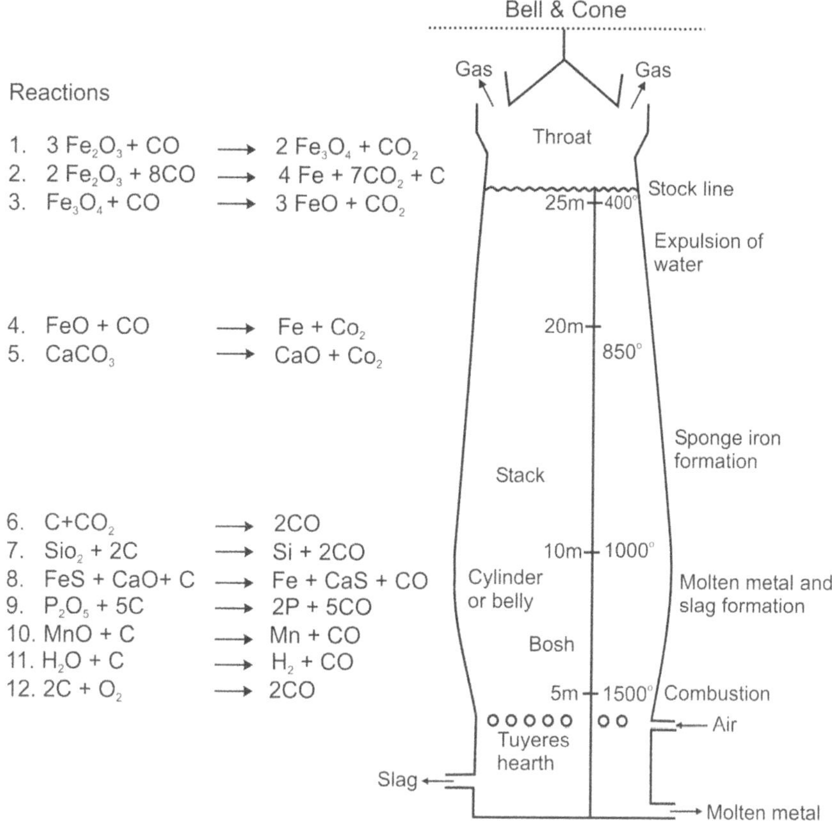

Fig. 23.1 Diagram of a blast furnace showing the chemical reactions.

Most of the coke burns in the bosh in the hot blast of air supplied through the tuyeres. As very hot coke burns in hot air the temperature attained is higher than calculated from the heat of reaction. Any moisture present in the air is reduced to H_2 and CO (reaction. 11). The reaction being endothermic helps to prevent excessively high temperature in the bosh. The nitrogen of the air moving upwards helps to transfer heat to the stack portion of the furnace.

23.2.3 Blast Furnace Plant

Blast furnace plant Fig. 23.1 gives a diagram of a modern blast furnace showing the furnace dimensions and the chemical reactions. Fig. 23.2 is a complete flow diagram showing the various components of a modern blast furnace plant. The blast furnace is a tall vertical furnace which is designed with great skill. Its dimensions and profile are designed so as to take care of the volumetric changes and to allow proper time for the chemical reactions to be completed and for the separation of the metal from the slag. The outer structure is made of steel plates lined with fireclay refractories. Refractories of different qualities are used in different parts. Carbon blocks are used for lining the hearth. High alumina fireclay bricks, carbon mass, diatomite bricks, etc. have been used in these furnaces for refractory linings.

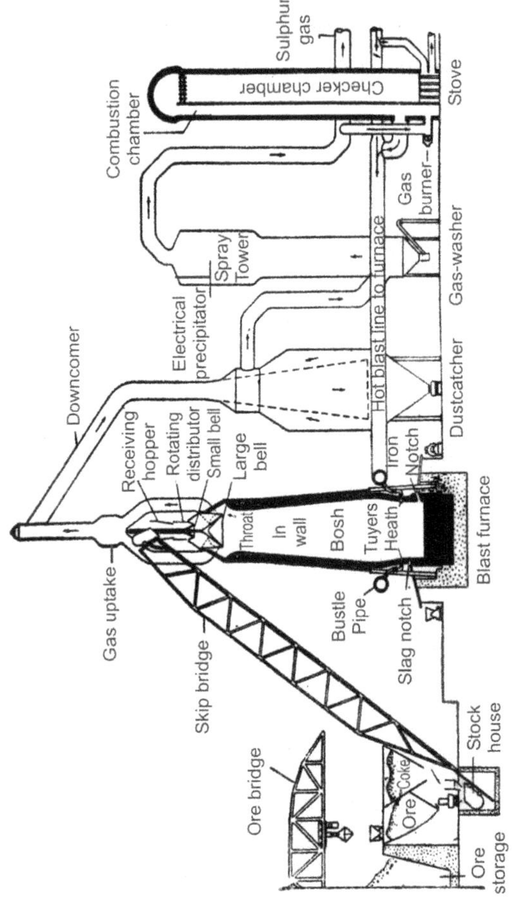

Fig. 23.2 Components of a blast furnace plant.

(*Courtesy, the American society of metals*).

Three blast furnaces of the Bhilai Steel Plant have a rated capacity of 1738 tonnes each per day. Each was constructed at a cost of about Rs. 10 crores. The blast furnaces of Bokaro Steel Ltd., have useful volume of 2000 m^3 with a daily production of 2650 tonnes of pig iron. These are the biggest blast furnaces in the country. In a few countries blast furnaces of more than 10,000 tonnes per day capacity are existing. The latest blast furnaces in Japan have a capacity of 13,000 tonnes a day. The Cost of production of pig iron in such large furnaces is much lower than in the smaller furnaces, but require very sophisticated control instruments.

The proportions of the raw materials charged (i.e., the charge) in the furnace depends upon the composition of the raw materials. For Indian conditions it may be taken that for 1,000 tonnes of pig iron about 1,700 tonnes of ore, 1,000 tonnes of coke and 400 tonnes of limestone are needed. The charge is dropped through a double cup and cone arrangement, known as double bell charging system which prevents the escape of the blast furnace gas while introduce the charge. During the first few metres of its descent, there is little the volume of the charge; consequently the throat walls of the furnace are almost parallel. On further descent the volume increases, the charge becomes sticky and is likely to cling to the walls and the carbon decomposition may cause 'hangings' on the walls. To overcome all these difficulties and to facilitate the smooth descent of the charge the furnace is made progressively broader. By the time the charge has completed two thirds of its total descent, most of the reduction is complete. As the volume of charge does not increase at this point the furnace is made cylindrical and this portion is known as the cylinder or belly. (A common mistake in sketching a blast furnace is not to show the bosh parallel shown in Fig. 23.1). On further descent the material enters the bosh. Here the reduction is complete and iron and slag are in a molten state. Also, due to the burning of coke there is a rapid decrease in volume. To compensate for decrease in volume and to facilitate the collection of molten metal and slag, the bosh sharply narrows down leading the material to the cylindrical well or hearth of the furnace. The molten metal and above it the slag collect in the well. Just below thc bosh are situated a number of watercooled tuyeres or *nozzles* made of bronze or copper through which air, preheated to 850° to 1200°, is blown at a pressure of about 1 kg/sq. cm. The volume of air needed is about 2,000 m^3 per minute for a 1000 tonne furnace. Just below the tuyeres is provided a cinder or slag notch through which the slag is removed periodically. An iron notch is made below the slag notch through which molten iron is tapped out every six hours. This is cast in to pigs or sent to a hot metal mixer for storing and is later converted into steel.

The main body of the furnace is made of thick steel plates lined with fireclay refractory bricks. The temperature in the hearth may be 1800°, hence carbon blocks are used for lining the hearth. The refractory lining of the hearth and bosh are very thoroughly cooled by circulating water through pipes built in the lining.

Near the top of the furnace is an outlet for the gases. This gas contains about 25% CO, 5% H_2, 15% CO_2, and 55% N_2. A blast furnace producing 1000 tonnes of pig iron per day gives out about 4 million cubic meters of the gas which carries in suspension from 50 to 250 tonnes of dust. Heating value of this gas is about 40% of the heating value of the coke charged in the furnace. Calorific value of the blast furnace gas is 800 Kcal/m^3. The gas contains a considerable amount of dust and is passed through a dust catcher. It is then washed with a spray of water, cooled and passed through an electrostatic precipitator (working at 45,000 volts D. C.) to remove fine dust particles (see Fig. 23.2). About 25% of the gas is needed for heating the blast of air for the furnace and the balance is available for other consumers in the steel plant. As most pig iron plants have steel plants associated with them, much of the gas is used for heating the coke ovens.

The blast of air sent through tuyers is heated in hot blast stoves which are tall cylindrical shells made of steel plates lined with fireclay bricks. The interior of the stove is divided into two parts. The first part has open space in which the blast furnace gas and air mixture burns and heats, the second part has a checker-work made of fireclay bricks. After heating a stove, the gas and air supply are turned on to the second stove. An air blast is forced through the heated stove which absorbs the heat of the brick-work. This hot blast of air is sent through the bustle pipe and enters the blast furnace through tuyeres. The present trend is to have four stoves for each blast furnace; when three are being heated, air is blown through the fourth. This gives air at a higher temperature than the older practice of having only three stoves. About 4 tonnes of air are needed per tonne of pig iron produced.

The gas inside the furnace contains about 55% nitrogen which does not take part in the chemical reactions. It acts as a carrier of heat. By using oxygen along with air, i.e., using *oxygen enriched blast*, the proportion of the nitrogen in the furnace is decreased. The gas thus becomes more reactive and takes less time to reduce iron ore. This reduces the height of the furnace which in turn enables ore and coke of smaller particle size to be used. Thus some smaller size ore and coke which are otherwise discarded can be used by employing oxygen enriched blast.

Another improvement in blast furnace operation is the use of *high top pressure*. In the common furnace the gas leaves the furnace at a pressure of about 0.1 to 0.2 kg/cm^2. The pressure of the outgoing gases is increased to 0.5 kg/cm^2 above the atmospheric pressure and hot air blast is also blown in the furnace at a higher pressure. This increases the density of gas in the furnace and reduces its velocity. Thus time available for the reaction becomes longer. The use of high top pressure is reported to increase the output of pig iron.

The moisture in the blast also contributes to the increase in productivity of the furnace. The moisture present in the blast, decomposes at the tuyeres ($H_2O + C = CO + H_2$) and absorbs heat. To avoid this loss of heat air blast is dried by cooling (to condense moisture) and then heating in the stoves. On the other hand, blast furnaces in Russia reported to heat the blast to a higher temperature (upto 1200°) and to add moisture to the blast of air. Under these conditions the moisture prevents undue temperature rise in the combustion zone but makes the extra heat available in the stack by the burning of carbon monoxide and hydrogen.

About 25 million litres of water per day are needed for a blast furnace producing 1,000 tonnes of pig iron for cooling the bosh, hearth and tuyeres, for cleaning the gas and for steam making.

A blast furnace has to be operated continuously till the refractory lining wears to such an extent that it fails to serve the purpose. Generally the life of a furnace lining is 3 to 4 years.

23.3 Sponge Iron-Direct Reduction Process

In the iron and steel industry iron ore is first reduced in a sophisticated blast furnace giving pig iron. In efforts to increase productivity the blast furnace technology has made good advances. Pig iron contains a large amount of carbon and other elements. These elements are then oxidized by the various steel making processes to obtain saleable steel. For quite some time the technologists have been paying serious attention to this fallacy in steel making and trying to obtain at the first stage itself a product more or less having the composition desired in steel. This is not a new concept, as it was used in the early days. Once again commercially viable units have come up to produce *sponge iron* which is a product obtained by reduction of iron ore in the solid state resulting in a porous mass that consists mostly of metallic iron along with the original unreduced minerals. The various elements present in sponge iron are in much smaller amounts than in pig iron and its composition is similar to that of wrought iron. It is the purest form of commercial iron.

All the methods by which iron of such low carbon content are produced directly from ore are called *direct reduction processes*. In these processes iron ore is brought in intimate contact with solid or gaseous reductant. The reactions take place at much lower temperature (below 1300°) and heat absorption is also much less than in the blast furnace practice. Methods differ from one another mainly in the manner of bringing the ore and the fuel in intimate contact with each other and are classified as under:

1. Furnaces using solid fuel, like coke or coal, are of two types: (i) rotary furnace and (ii) vertical retort or vertical shaft furnace.
2. Furnaces using gaseous fuels are of three types: (i) Shaft furnace, (ii) static bed furnace and (iii) fluidized bed furnace.

The direct reduction processes attempt to produce :

(i) Low carbon iron to be used as a melting stock (sometimes referred to as synthetic scrap) for making steel by the existing process,

(ii) Iron of a quality comparable to that produced by the blast furnace,

(iii) Steel directly from the ore–this is quite an ambitious objective.

Comparison with blast furnace technology. The modern blast furnace technology requires very high capital investment and uses highly sophisticated instruments and the working of the larger furnaces has to be computerized. The raw materials needed are very selective, e.g., good quality coking coal is needed to produce coke of the required quality. This means that the plants can be located only in selected areas.

The direct reduction processes can handle raw materials of a wide variety, which are distributed in smaller and larger quantities in the various parts of the country, for example, ores of small particle size (the so called "fines"), non-coking coal (which is abundant in India), charcoal, and a variety of gaseous reducing agents can be used. Many raw materials which are considered substandard for the blast furnace can be used for the manufacture of sponge iron. In the areas which do not have coking coals, the sponge iron manufacture is the only alternative route for steel making with certain technological improvements, which are being tackled, the sponge iron manufacture requiring much less capital investment may prove in the future to be a competitor of the blast furnace process.

23.4 Pig Iron or Cast Iron

The iron obtained from the blast furnace is known as pig iron. Its composition varies widely, but on an average it contains besides iron, 3 to 4.5% carbon; 1 to 4% silicon; 0.2 to 1.5% manganese; 0.04 to 0.1% sulphur; 0.1 to 2% phosphorus and small amounts of the other elements. Only a very small amount of pig iron is directly converted into useful shapes. Most of it is converted into steel. A quantity of it is remelted for making castings. It is then called *cast iron*.

The usefulness of cast iron lies in its cheapness and the ease with which it can be converted into useful shapes by casting. Its mechanical properties vary very greatly because of a wide range of compositions but in general it is low melting, wear and corrosion resistant, and has high compressive strength. By proper adjustments, varieties of cast iron ranging from the soft and machinable to the very hard and unmachinable can be obtained. Various types of cast iron are marketed. The more important ones are: 1. Grey, 2. White, 3. Malleable and 4. Spheroidal graphite or nodular cast iron. This classification is based on the form in which carbon is present and the structure of the matrix in which the carbon is embedded. The presence of carbon as graphite makes iron soft and machinable (giving grey cast iron), whereas cementite makes it hard and unmachinable.

Grey cast iron. Other things being equal, the presence of silicon and a slow rate of cooling promote decompositon of Fe_3C giving graphite resulting in grey-coloured metal; hence it is called grey cast iron. Due to the low density of graphite it occupies more volume and hence there is less shrinkage on cooling. Graphite is present as flake or thin plate-like crystals. These flakes are weak and act as gaps within the strong material. Thus the actual effective cross-section area is reduced and the whole metal becomes weak. Therefore, grey cast iron has low tensile strength, low ductility and low toughness.

A large amount of grey cast iron is used for making castings because (i) cost is low; (ii) melting point is low, hence it melts easily; (iii) it is soft and hence can be easily machined; (iv) compressive strength is high and (v) because of the presence of flakes of graphite the vibrations of machines are not easily transferred; hence it is useful for making bases for heavy machines.

Its tensile strength is 1,200 to 3,000 kg/sq. cm. In compression it can stand 6,000 to 7,500 kg/sq. cm and in shear 1,500 to 2,200 kg/sq. cm. Because of low tensile strength and high brittleness it is not suitable for making bolts which are subjected to tension. It is very useful for casting the bodies of electrical machines, the cylinders and beds of engines, bases for heavy machines, etc.

White cast iron. This is produced when silicon content is low and the metal is cooled rapidly. These conditions help to retain carbon as Fe_3C which gives the metal its white colour. White cast iron does not find much use. It is used where abrasion resistance is needed as in railway carriage brakes. The reason is that the presence of a large amount of Fe_3C makes it very hard, brittle and very difficult to machine.

Malleable cast iron The properties of grey and white cast irons are quite opposite to those of each other. Each has certain very useful properties and certain serious drawbacks. To combine the advantages of both these materials, the hard and brittle white cast iron, after being formed in the desired shape, is heated for sufficient time to dissociate Fe_3C giving graphite in the form of nodules (small, round particles). Because of the absence of Fe_3C the new material is not hard and brittle. Graphite is not present as large flakes, therefore its weakening effects are reduced. The resulting casting is therefore more ductile, tough and shock-resistant. Hence it is called malleable, though it cannot be made into sheets by hammering or drawn into wires.

23.4.1 Spheroidal Graphite or Nodular Cast Iron

This type of iron is also known as *ductile iron*. The starting material is the molten pig iron of almost the same composition as the grey cast iron. Small percentages of elements, like magnesium and cerium, which are introduced in the molten iron which cause most of the carbon present to form small spheres or spheroids instead of flakes.

The chemical composition of the S. G. iron or ductile cast iron as it is popularly called, and grey cast iron is the same but the shape of the graphite particles is different although the matrix in which graphite is embedded may be similar. Ductile cast irons form a family of alloys that combine the valuable properties of the grey cast iron (e.g., low m. p., good fluidity and castability, good machinability and wear resistance) with the engineering advantages of steel (e.g., high strength, toughness, ductility, hot workability and hardenability). It is possible to produce ductile irons of a wide range of properties. For example the tensile strength may be 60,000 p. s. i. with 25% elongation or may be as high as 150,000 p. s. i. with 1 to 4% elongation. Such control is achieved by adjusting the composition of the starting pig iron, rate of cooling and the subsequent heat treatment.

These irons find applications in heavy duty machinery, gears, dies, pressure castings. valves and pump bodies, rolls for wear resistance, parts for shock resistance, pinions, cams, crank shafts, guides and track rollers.

23.4.2 Constituents of Cast Iron and Pig Iron

The most important feature of cast iron (and pig iron) is the presence of a large amount (3 to 4.5%) of carbon and sometimes of silicon also. Carbon can remain along with iron either in a chemically combined form or as free carbon (graphite). The properties of cast iron depend to a great extent not only upon the relative amounts of these but also on their physical state.

(i) Carbon. An increase in total carbon content lowers the melting point but increases the fluidity of cast iron. Generally, it decreases the strength because along with the increase in carbon content the content of graphite usually also increases.

Free carbon or graphite reduces the hardness and strength. It breaks up the continuity of the metallic mass, especially when it is present as large flakes; hence the strength is reduced. Graphite is soft and acts as a lubricant for the working tool, so its presence reduces the hardness and improves the machinability.

Combined carbon is present as iron carbide, Fe_3C, called cementite. Cementite is very hard and brittle. The combined carbon may not be present wholly as cementite but in other forms also. These forms, their properties and the factors affecting their formations will be discussed in greater detail while discussing iron carbon alloys in a subsequent chapter. Briefly, increase in combined carbon up to about 1% increases the strength. The strength slightly decreases with further increase in combined carbon, but brittleness and hardness continue to increase and the metal becomes difficult to machine as the combined carbon exceeds 1%.

(ii) *Silicon.* It is generally present from 1 to 3.5% but may be as high as 17% when the metal becomes extremely brittle and hard, and breaks with a glass-like fracture. It is called glazy pig iron. Silicon dissolves in iron, forming iron silicide (FeSi). Its chief effect is the decomposition of combined carbon (Fe_3C) into free carbon or graphite. Thus it induces softness, machinability and fluidity. It reduces hardness, strength and thermal and electrical conductivities but increases the volume (i.e., there is less of contraction on cooling). Due to the affinity of silicon for oxygen the casting produced is freer from blow holes; therefore, white cast iron of low silicon content does not make good castings due to blow holes, and also being viscous, it dose not flow easily.

Thus an increasing proportion of silicon changes extremely hard white cast iron into soft grey cast iron. For getting very hard cast iron Si content should be less than 0.8% and for getting soft and machinable metal it should be at least 2.5%.

(iii) *Sulphur.* It is mostly present as MnS, which has very little effect on the properties of iron. In the absence of manganese it is present as FeS whose action is the opposite to that of silicon i.e., it acts as a carbide stabiliser. Therefore, it makes the iron more hard, brittle and difficult to machine and its content should be kept below 0.1 %.

(iv) *Manganese.* It may be present partly or wholly as MnS. Its chief function is to remove sulphur from iron by forming MnS, which has very little effect on the properties of iron. Thus the indirect effect (i.e., due to the removal of sulphur) is to promote graphitization (formation of graphite). MnS is much lighter than iron, therefore in the molten mass, it rises to the top. The direct effect of manganese (i.e., of manganese present as uncombined manganese) is to stabilise the carbide; thus it increases pearlite, which results in increased hardness and tensile strength. By stabilising the carbide it makes the metal difficult to machine. It improves the soundness of castings.

(v) *Phosphorus.* It may be present from traces to 2 to 3%. It has little effect on free or combined carbon. It increases the fluidity of the metal and reduces melting point, shrinkage and strength. Therefore, for making intricate castings, where sharp impression is the main consideration and not the strength, pig iron containing about 1% phosphorus is used. For iron having good strength, its proportion should not exceed 0.3%; above 0.5% it reduces tensile strength.

In the light of the preceding discussion it is clear that the properties of cast iron can be controlled by :-

1. *Control of chemical composition.* It can be adjusted (a) by the proper selection of the pig iron taken for melting, (b) by the selection and adjustment of the proportions of other additions made in the cupola, like cast iron scrap, steel scrap, etc., and (c) by the additions made to the molten metal in the ladle. Some elements, which are oxidized more easily, may be conveniently added in the ladle, to prevent their oxidation while in the cupola. Certain elements affect the graphite-cementite ratio, which affects the properties very greatly, and the control of their amounts is an important consideration.

2. *Control of rate of cooling.* It has been already stated that in the molten metal, carbon is mostly in the from of Fe_3C and on solidification it decomposes into graphite, the quantity of which depends to a great extent on the rate of cooling (the effects of the presence of other elements has already been discussed). A high rate of cooling prevents the decomposition of Fe_3C with the result that the sample has a small amount of graphite and white cast iron is obtained which is very hard and difficult to machine. Slow cooling results in decomposition of Fe_3C giving grey cast iron which is soft.

3. *Proper heat treatment.* This will be discussed in greater detail in a subsequent chapter. The sample or the casting, is heated within the critical range, held at the particular temperature for some time and its rate of cooling is controlled. By proper adjustment of this treatment, the micro constituents present can be controlled, which profoundly govern the properties of the product.

23.4.3 Engineering Uses of Pig Iron or Cast Iron

Most of the pig iron is converted into steel. The balance is remelted (then it is called cast iron) and used for making shapes by pouring the molten metal into moulds and allowing it to solidify. This is an inexpensive, easy and convenient way of manufacturing products of desired shapes but the product is weak and brittle. Ofcourse, its properties can be improved by heat treatment and alloying, but this is generally not done. The major drawback of cast iron is that it cannot be shaped by various mechanical processes like hammering, pressing or rolling.

23.5 Steel

As explained earlier, cast iron does not have the properties necessary for many engineering uses. Steel differs from it mainly in having less than 2% carbon. It has minor amounts of other elements like silicon, manganese, sulphur, phosphorus, etc. The older methods of steel-making, viz. cementation and crucible processes are almost obsolete. In ancient India, steel was made by *cementation* process. The process consisted of heating bars of wrought iron in iron boxes packed with charcoal. Carbon very slowly diffuses formg cementite. After about 10 days of heating the boxes were opened, the bars removed and hammered to give steel. Greater uniformity of composition is obtained in the *crucible* process in which wrought iron is melted with a definite mount of charcoal in crucibles made of fireclay. This produces a much superior steel which is used for making cutlery.

The production technology of steel for last twenty years in India has seen a major shift in favour of oxygen steel making processes. However, till today there are a number of countries where the proportion of steel produced by open hearth process continues to be substantial. In India also, even though the future integrated steel plants are planned based on Basic Oxygen Furnace (BOF or LD), a considerable share of steel from open hearth furnaces are coming from Bhilai Steel Plant, TISCO, ISCO, Rourkela Steel Plant (partly) and Durgapur Steel Plant. As such, it is imperative to make some references of open hearth steel making and its recent developments prior to projecting the Basic Oxygen Furnace method. Special steels are made in electric furnaces.

In these methods pig iron or a mixture of pig iron and steel scrap is the starting material from which carbon and impurities are removed by oxidation. The removal of the impurities depends upon the kind of slag produced which in its turn decides the furnace lining to be used.

Acid steel is the steel made in a furnace lined with acid refractory material and using acid (siliceous) slag, i.e., slag having an excess of silica.

Basic steel is the steel made in a furnace lined with basic refractory material and using basic (limey) slag, i.e., slag having an excess of lime.

This distinction results in the elimination of different elements. Hence the process that is to be adopted depends upon the impurities present in the starting material, viz. in the pig iron. Acid slag removes, by oxidation, silicon, manganese and carbon. Phosphorus and sulphur are not removed, so their amounts should not be more than what are permissible in the finished product. As the acid slag will react with basic lining, it is necessary to use acid lining when an acid slag is made.

The characteristic feature of the basic process is that phosphorus and sulphur are removed. It also removes silicon, manganese and carbon. It is desirable that the silicon content is not very high because it also reacts with the basic lining. The metal has to be oxidized to a greater degree (to ensure oxidation of phosphorus), hence after making steel, deoxidation becomes necessary. There is a difference in the method of slag preparation in acid and basic processes.

In the acid process, deoxidation can be done in the presence of slag in the furnace. In the basic process, if deoxidation is done in the presence of slag, phosphorus would return to the metal, therefore deoxidation is done in the ladle.

In both acid and basic processes the elimination of carbon is a carbon monoxide.

23.5.1 Bessemer Process

The process is named after Sir Henry Bessemer who obtained a patent for it in 1855. It is based on the fact that when air is blown through molten pig iron a large amount of heat evolves on account of the oxidation of silicon, manganese and carbon. The heat evolved actually raises the temperature of the metal. It is necessary that the metal should contain enough of these elements because, as the purification proceeds, the melting point of the metal rises and if sufficient heat is not evolved, the metal will tend to solidify in the converter itself.

The furnace, known as a converter, is a pear-shaped steel shell consisting of three detachable parts: bottom, body and nose. A cross section is shown in Fig. 23.3. The nose narrows toward the top so that there may not be excessive loss of heat, metal and slag by ejection during blowing. The main body is cylindrical. The bottom has a wind box with a number of tuyeres through which air is blown in the converter. The wind box is connected to a wind pipe which passes through one of the trunnions. The greatest wear is in the bottom portion which needs to be replaced more often than the nose and the main shell. The converter is supported on two trunnions and can be tilted.

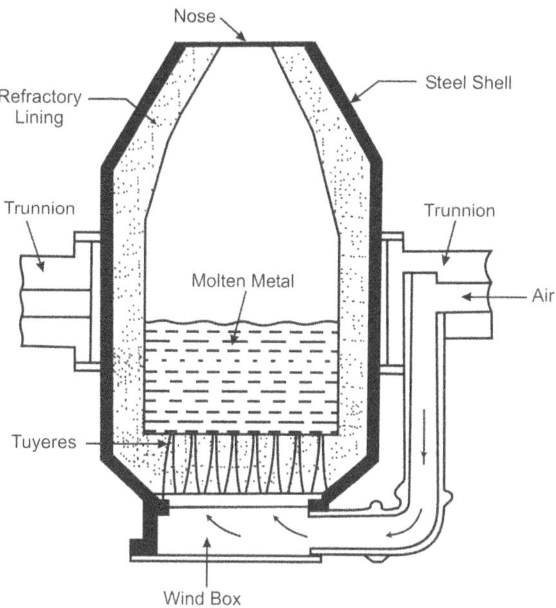

Fig. 23.3 Bessemer converter.

Acid Bessemer Process is adopted in India by TISCO at jamshedpur. The capacity of a common converter is 25 tonnes. It is lined with a highly refractory silica lining. The converter is tilted to the horizontal position and molten pig iron from a hot metal mixer (of capacity from 1,000 to 1,500 tonnes in which molten pig iron from the blast furnace is stored) is introduced by means of ladles. The temperature of the metal is about 1,200° to 1,300°. The pig iron must contain about 2% silicon whose oxidation supplies heat. Phosphorus and sulphur contents should be less than 0.04%. Air at a pressure of about 1 to 2 kg per cm^2 is blown. After starting blowing air, the converter is brought to the vertical position when a shower of sparks and a small flame appear. In about 3 minutes the oxidation of most of the silicon and manganese is complete and oxidation of carbon begins. This is marked by increase in the size (to about 10 m) and luminosity of the flame but decrease in the amount of sparks. In about 8 minutes most of the carbon gets oxidized and the flame drops and becomes less luminous. If air blowing is continued after this stage, iron would get oxidized. The presence of iron oxide has a highly detrimental effect on the properties of steel. At this stage the converter is rotated to the horizontal position and then the air blow is stopped. The metal is poured out in a ladle and then, if necessary, finishing and alloying additions are made.

The metal thus produced is low in carbon but very often contains iron oxide and dissolved gases. The remedy is deoxidation and recarburization. To accomplish this alloys like ferromanganese or spiegeleisen (containing manganese, iron and carbon) or ferro-silicon are added while transferring steel from the converter to the ladle. Silicon and manganese deoxidize iron oxide and form slag. Manganese helps in the removal of sulphur (which, if left as FeS causes brittleness at red heat). The carbon present in the alloy added helps to raise the carbon content of steel.

The reactions which take place may be summarized as follows :

Iron is the first to be oxidized.

$$2Fe+O_2 \rightarrow 2FeO$$

FeO oxidizes the impurities in the order of Si, Mn and C :

$$Si+2FeO \rightarrow 2Fe+SiO_2$$

$$Mn+FeO \rightarrow Fe+MnO$$

All these are exothermic reactions and provide heat. The SiO_2, MnO and some FeO combine and form the slag. This takes about 3 minutes. Then oxidation of carbon to carbon monoxide begins which burns with a long flame

$$C+ FeO \rightarrow Fe+CO$$

$$2CO+O_2 \rightarrow 2CO_2$$

This takes about 8 minutes. At the completion of oxidation of carbon, the flame drops and the converter is made horizontal and then air blow is stopped. If air is blown for a longer time excessive oxidation of iron will take place. Alloy containing iron, manganese and carbon is added which reduces iron oxide present and introduces the desired amount of carbon.

It is to be noted that neither flux is added nor is the air preheated, and no fuel is needed. In fact the heat of reaction is so great that temperature rises by about 300°.

About 10% of the metal is lost in the process due to the elimination of carbon, silicon and manganese. Some iron also is lost due to oxidation.

Basic Bessemer (or Thomas-Gilchrist) Process. This process is not adopted to such a large extent. The pig iron used should be low in silicon content and there should be at least 2% phosphorus (otherwise enough heat of reaction will not be liberated to keep the metal hot). The converter is lined with magnesite or dolomite lining. Lime is used as flux.

Open hearth or Siemens-Martin Process. The open-hearth process essentially differs from the Bessemer process in that the oxidation reactions take place so slowly that the heat of the reactions is not sufficient to keep the metal in a molten state, hence external heating is necessary. The time taken is 6 to 14 hours; this enables a better control over the purifying reactions. The basic process is more widely used.

An important step in the steel-making process is the use of the regenerative principle. This involves the extraction of heat from hot waste gases and the use of this heat in pre-beating the fuel gas and air before these enter the furnace. This increases the furnace temperature besides causing saving in heat.

23.5.2 Basic Open-hearth Process

Furnace of capacities from 10 to 900 tonnes are in operation. The 500 tonne open hearth furnaces at Bhilai are the largest furnaces in operation in India, while 900 tonne furnaces are working in the U S S R As the name suggests, the hearth or the furnace proper is open or exposed to the action of the flames. It has a large surface area and is about 60 cm deep. On the outside it is made of steel plates and is lined with magnesite bricks, above which there is a layer of magnesite or dolomite mixed with tar. At the level where slag collects, chromite refractory bricks are used and the side and roof are lined with chrome-magnesite refractory. Several charging doors are provided on one side of the furnace and on the opposite side, in the middle, is a tapping hole for the steel and also a slag hole, see Fig. 23.4 and 23.5. Both stationary and tiliting furnaces are in use.

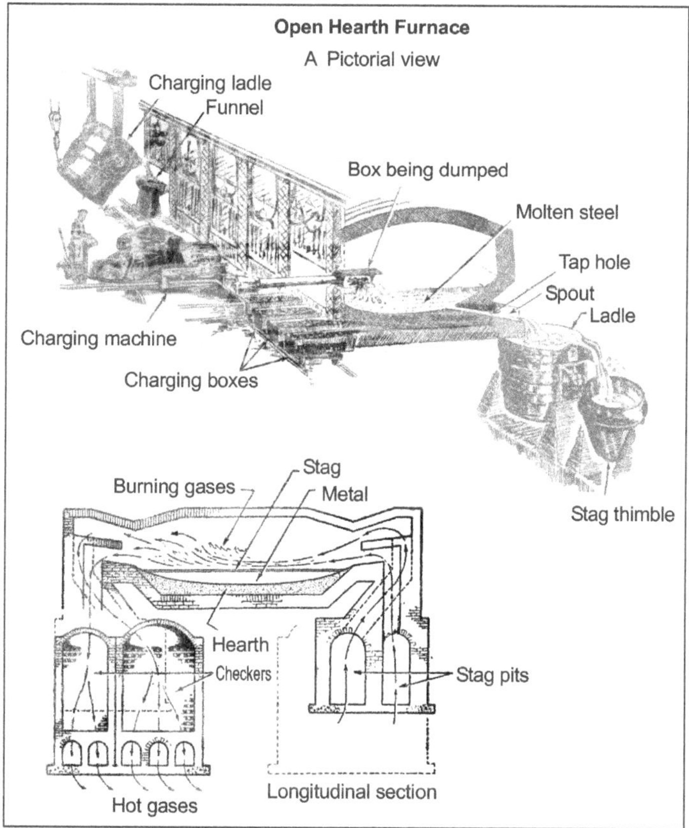

Fig. 23.4 A pictorial view of an open hearth furnace showing (a) charging of molten metal and solid material (top), (b) tapping of molten metal in ladles (center, right) and (c) direction of gas flow and a section through the entire plant (bottom).

(From J, Wulff, H. F. Taylor and A. J. Shaler, Metallurgy for Engineers courtesy, john Wiley & sons, New York).

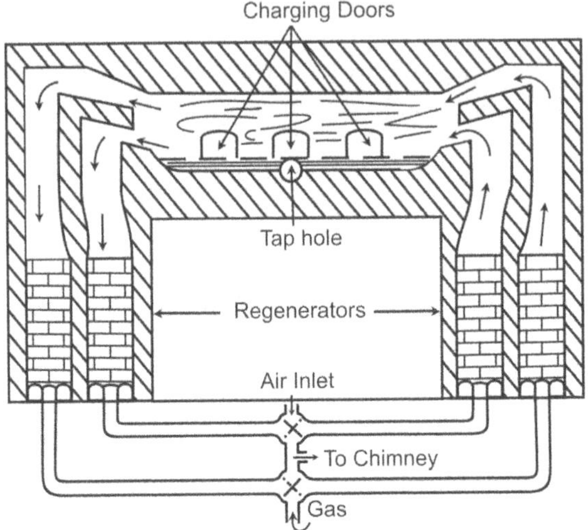

Fig. 23.5 Open-hearth furnace.

At each end of the furnace there are two openings or 'ports' for fuel gas and air. They are connected to regenerators or checker work chambers built underneath, two on each side. These are filled with refractory brickwork leaving enough space for the gases to pass through them. The hot gases leaving the furnace from one end pass through the pair of regenerators on their way to the chimney and heat them. After about 20 minutes the directions are reversed and the in-going fuel gas and air are heated as they pass up the heated regenerators. Thus there is not only a saving in heat but a much higher temperature is also produced.

In the past, producer gas was commonly, used for heating. This has now been largely replaced by a mixture of blast furnace gas (which is waste product in pig iron making) and coke oven gas obtained from coking plants (which are almost invariably associated with and steel works). Often the calorific value of the gas mixture is enriched by injecting tar or oil. In the U S A and Russia, natural gas, obtained from the petroleum industry, is also used.

After removing the metal from a previous charge, the furnace lining is repaired with dolomite. Limestone is placed on the hearth, then iron ore (which supplies oxygen for oxidation of impurities) on which is placed steel scrap and solid pig iron. The heat is turned on fully. After two hours, when the charge has been heated up, molten pig iron from hot metal mixers is poured in. As the temperature rises, the iron oxide first oxidizes silicon and manganese and then phosphorus and sulphur. When carbon begins to be oxidized to carbon monoxide, the bath gives the appearance of boiling and a part of the slag rises to the slag hole and is allowed to run off. A little later when the temperature becomes higher in the lower part also, the decomposition of limestone begins with the evolution of carbon dioxide. At this stage the furnace appears as if it is boiling vigorously. Lime comes to the top and forms slag. The evolving gases stir the mass and thereby facilitate the elimination of impurities. The reactions are similar to those described for the Acid Bessemer process, except that the phosphorus is also oxidized and is eliminated by the basic slag.

$$2P + 5FeO \rightleftharpoons 5Fe + P_2O_5$$

After the boiling is over, comes the working period during which the remaining phosphorus and carbon are oxidized and removed. Composition of bath is so adjusted that the steel is ready for tapping.

Molten steel is removed through the tap hole and deoxidation and recarburization are done in the ladle, as in the Bessemer process, by adding ferromanganese and ferrosilicon. Aluminium shots are generally thrown in the ingot moulds while pouring the metal. Aluminium reacts with oxygen and reduces traces of any oxides left.

Acid Open-hearth Process is not adopted on such a large scale, principally due to the fact that sulphur and phosphorus cannot be eliminated. In general, it is similar to the basic process, an important difference being that silica lining is used and limestone is not used as flux. The slag consists of SiO_2, MnO and FeO.

23.5.3 Duplex Process

The name 'duplex process' does not refer to a new process but to any combination of two steel-making processes. For making high grade steel or alloys, the 'triplex-process' is also in use. Ordinarily the duplex process, in Indian conditions, refers to combination of acid Bessemer and basic open-hearth processes; thus it combines the advantages of both, though the refining losses may be higher.

The molten pig iron is blown in an acid Bessemer converter eliminating silicon and manganese and part of the carbon. The metal is then taken to a basic open-hearth furnace and treated in the usual manner for the elimination of the remaining carbon and phosphorus. Thus metal from ores having both high silicon and high phosphorus can be converted into steel. If high silicon and phosphorus metal is treated by acid process alone, phosphorus will not be eliminated. If it is treated by basic process only, both will be eliminated but the basic lining will be damaged because of silicon. The duplex process markedly decreases the time taken in the open-hearth process.

23.5.4 Recent Developments in the Technology of Open Hearth Process

The recent developments in open hearth steel making technology is the use of twin hearth technology. It has been developed with the objective of substantially improving the productivity of the open hearth furnace shops with reduced consumption of fuel and refractories. It employs the same basic principles as in the case of tandem furnace practice.

In the *tandem* furnace two identical tilting open hearth furnace type hearths are interconnected through a throat, each of which has one lance for blowing oxygen during refining at their outer ends and one oxygen lance near the throat for burning CO evolved during refining in the other furnace. While refining is in progress in one furnace, the CO evolved in that chamber is burnt at the throat of the other chamber by supplying oxygen through the combustion lance to preheat the charge therein. It is obvious that this results in considerable heat economy.

Twin hearth furnaces are constructed on stationary open hearth furnace by building a separating wall, midway along the furnace in order to create two hearths within the original single hearth. Both the hearths are under a common roof and the regenerators are eliminated. The two hearths of the twin hearth furnaces are operated alternately as refining and pre-heat furnaces, which are also known as hot and cold chambers. The essential principle of twin hearth process is to use the potential heat value of carbon-monoxide evolving from the hot chamber to pre-heat the cold charge in the other chamber. Twin hearth furnaces have been able to attain an annual production of 0.9-1.65 million tonnes of steel and thus their productivity is 2 to 4 times higher than that of an arch modern open hearth furnace of similar size operating with both intensive oxygen lancing. One of the important characteristics of the twin hearth furnaces is their higher roof life (600-970 heats) which is 3 to 6 times more than that of open hearth furnaces. Twin hearth furnaces operate with 62-65% hot metal and no iron ore is used in the charge. Both lime and limestone are used as fluxing material.

Bhilai Steel Plant has undertaken a modernisation programme for increasing the productivity of open hearth shop up to 3.5 mt/yr. The open hearth shop at Bhilai Steel Plant was commissioned in the year 1959 and comprises ten stationary open hearth furnaces of which five are of 500 t capacity each and the remaining five are of 250 t capacity each. Detailed analysis of the measures to increase the productivity of these furnaces revealed that the productivity could be increased otherwise by only 10-15%. In view of this, it has been planned to adopt twin hearth process by converting one of the 500 t open hearth furnace to a 2 × 250 t twin hearth furnace.

23.5.5 L-D Process

This process is a modification of the Bessemer process and was first practised in 1953 in two towns named Linz and Donawintz, Austria, from which the process derives its name. It is being worked by the Steel Authority of India Ltd., at Rourkela and Bokaro. The converters have a capacity of 100-130 tonnes. Two converters of 250 tonnes have now been installed at Bokaro. In some countries converters of 300 tonnes capacity are in use.

This has been found particularly suitable for pig irons of low phosphorus content (less than 0.25%). The main advantages are:

(i) High rate of production as in the Bessemer process.

(ii) High quality steel as by basic open-hearth process.

(iii) Low nitrogen content of finished steel.

The converter, Fig. 23.6, is similar in shape to the Bessemer converter but is closed at the bottom. A hood is provided for the escape of the fumes. It is lined with magnesite and dolomite refractory lining. The converter is tilted and iron scrap followed by hot molten metal is introduced. It is then brought to the vertical position and oxygen at a pressure of about 10 kg/sq. cm is introduced through a water cooled tube or lance which is lowered in the converter, keeping it about a metre from the molten metal. Limestone, which

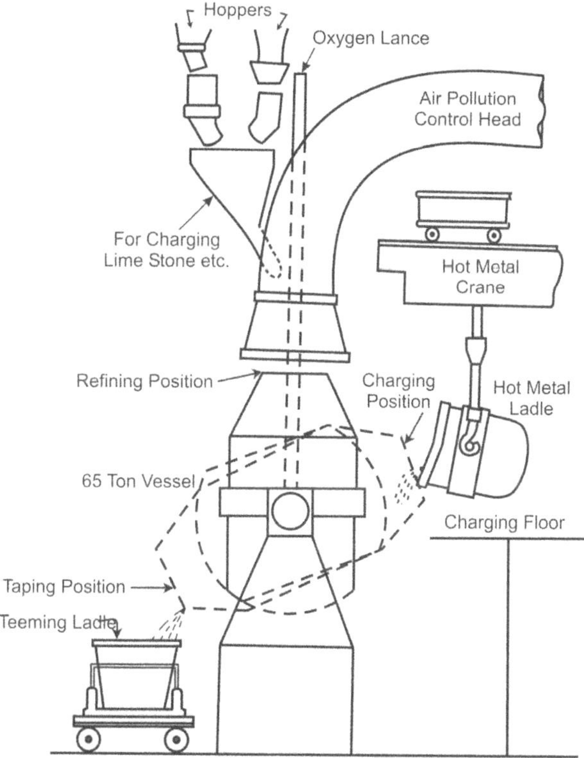

Fig. 23.6 L.D. Converter

(*Elements of metallurgy courtesy, D. Swarup*).

is used as a flux, is then added. Oxygen oxidizes Fe to FeO which oxidizes silicon, manganese, carbon and phosphorus. A high temperature and basic slag also help to remove sulphur. Carbon monoxide escapes and metallic oxides form slag with lime. The reactions being exothermic, the metal becomes very hot and some iron oxide volatalizes. As the specific gravity of the refined metal is higher, it travels downwards and unrefined metal takes its place. Thus the entire body of the metal gets purified. In the beginning a short flame appears which becomes longer due to the burning of carbon monoxide. The flame drops towards the completion of the refining. The time taken is about 20 to 60 minutes depending upon the capacity.

On the completion of the reactions, oxygen supply is stopped, the lance removed and the converter is tilted. The slag is poured out followed by the removal of the metal.

23.5.6 Basic Oxygen Furnace-Process

For the manufacture of steel by basic oxygen furnace (BOF) process, the inputs demands may be stringent quality control. The present world trend is to aim for hot metal with 0.5-0.6% Si, 0.8 to 1% Mn and minimum temperature of 1320° at the time of charging in converter. Similarly the lime used for oxygen steel making should be highly reactive and closely sized with CaO more than 86% and as low as SiO_2 content as possible. In fact with external desulphurisation and low phosphorus hot metals being supplied to BOF shops the tendency is to focus slag formation more towards: (i) Prolonged lining life, (ii) low rate of slopping at blowing speed and, (iii) minimising iron losses with slag.

There are three major inputs in converters: (i) Hot metal (ii) lime and other fluxes and (iii) oxygen. Out of these, first two have direct bearing on quality of raw materials. Lime constitutes the most important flux in basic oxygen converter. Bauxite is used as fluidiser and Mn ore primarily is used as coolant, but it also helps in slag formation and some increase in residual bath.

Bokaro Steel Plant has two steel melting shops. Shop-I which is operating has 5 LD converters of 100 tonnes capacity each and is rated to produce 2.5 mt to be subsequently increased to 3 mt by actually improving working parameters. The converters are lined with tar bonded dolomite bricks produced within the plant. The working volume is 91 m^3. Steel melting Shop-II has two converters each of 300 t capacity.

23.5.7 Developments in Basic Oxygen Steel Making

Top-blown and bottom-blown processes. The top-blown oxygen process, since its advent in the 1950s has registered a spectacular growth and today accounts for over 55% of the world steel production. Top blown oxygen process is the conventional LD process, where the oxygen is blown from the top, whereas the bottom blown process envisages the blowing of oxygen from bottom.

The investment cost for a bottom blown converter shop compared to a top blown LD shop was originally claimed to be lower by about 10% mainly because of the elimination of oxygen lances and the arched flux fins and the consequent reduction in building height. However, under Indian conditions, the recent estimates indicate that, the equipment cost for the bottom blown converters will be about 30% higher than that of top-blown converters on account of the additional requirements such as lime grinding, storage and injection facilities, media blowing system, dog-house and secondary fume emission control systems, etc.

Combined blowing and other development. To achieve the best of both top and bottom blowing processes, several middle-of-the-round techniques have been developed recently employing combined blowing.

KMS/OBM-S process. In the KMS (Klockner Maxhutte Scrap) or the OBM-S (Oxygen Bottom Maxhutte Scrap) processes, oxygen is blown into the converter simultaneously from both the top and the bottom. Oil and oxygen can be blown through the bottom tuyeres for scrap preheating. Some plants in Europe have already adopted this process

ID-OB process. In Japan, Nippon Steel has developed a similar process called the LD-OB process, for blowing oxygen into LD converters not only through the top lance, but also through bottom tuyeres. This is expected to increase the yield, reduce the amount of ferro alloys used and also reduce the oxygen and nitrogen levels in steel.

LD-KG process. Kawasaki Steel has reported the development of a new process called the LD-KG, which permits the LD converter to enjoy some of the advantages of the bottom-blown processes by injection of argon or nitrogen through the bottom of the converters.

STB Process. A similar process, called STB (Sumitomo Top and Bottom Blowing) process has been developed by Sumitomo Metal Industries. In this process, agitation gas is injected through the bottom tuyers in the conventional top blown LD-converters. It is also reported that this process is suitable for making stainless steel and for refining high manganese steel in conventional LD converters.

LBE process. Another combined blowing process has been developed by ARBED and IRSID, called the LBE (Lance Bubbling Equilibrium) process in which neutral gas is blown through permeable elements set in the bottom of the top blown converter. A special multi-hole, double-flow oxygen lance is used to improve the post combustion of CO into CO_2 in the converter.

LD-CL process. Another development for improving LD steel making is the use of a circling lance, an innovation by Nippon Kokal in Japan and named as the LD-CL process. It was developed to improve the efficiency of large size B. O. F.s. Circling is reported to have the advantage of shorter blowing time, improved product quality, better dephosphorisation and better desulphurisation.

23.5.8 Electric Furnace Processes

In electric furnace processes the source of heat is electric current. The most commonly used furnace is the Heroult furnace, Figure. 23.7, in which the arc is struck directly from the positive electrode to the metal bath and then the current is conducted away to the negative electrode. The capacity ranges from 1 to 100 tones. Indirect arc (i.e., arc between the electrodes) furnaces are also made. High frequency induction furnaces with a capacity of 1 to 5 tones are also in use.

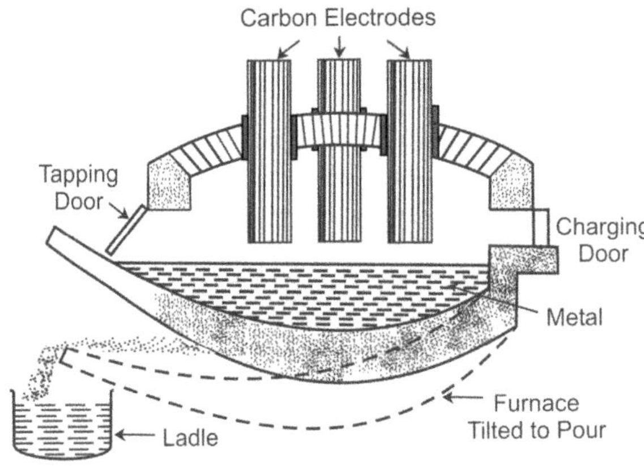

Fig. 23.7 Electric steel furnace.

In 1980, in India there were 157 electric arc furnace units with a total capacity 3.3 million tonnes per annum which produced 1.7 m. t. of steel. Such *mini steel* plants can be located in far-flung areas where scrap steel is readily available and also the market is nearby.

In the basic process, which is more widely used, the hearth is lined with magnesite bricks and the sides and roof are of silica or sillimanite bricks. Generally, pig iron is not the starting material. Usually selected steel scrap along with molten steel from an open hearth furnace is taken and refined under basic slag. The first slag made is oxidizing (by using iron oxide) as well as highly basic (by using a large amount of lime with less of silica). This helps to remove phosphorus. After removing the first slag a new slag which is less basic and is reducing (made by the use of coke) is produced. This helps to remove sulphur. Gaseous oxygen is also being used as a refining agent in the electric furnace processes.

The electric process is used for making high grade steel for special purposes and for making alloy steels. It is more expensive because of higher heating costs. The main advantages are: (i) the charge can be quickly heated to a high temperature and the temperature can be accurately maintained for the desired time; (ii) no impurities are introduced from the fuel, (iii) either hot metal or cold scrap can be the starting material, and (iv) it is possible to change the slags, produce oxidizing, netural or reducing slags as and when needed. Because of these advantages alloy steels are often made in the electric furnaces. Some alloy steels can be manufactured only in an electric furnace. All these give better control over the manufacturing process, thus enabling the production of high quality steels.

23.5.9 Future Trends

Bulk of the world steel production will continue to come by the Blast Furnace-Basic Oxygen Furnace (BF-BOF) route. However, with the commercialisation of the direct-reduction (DR) processes and the developments in Electric arc-furnace (EF) technology, electric steel making now strongly rivals the BF-BOF route. The DR-EF route is eminently suited for developing countries, as small size plants can be installed to serve the limited market and to keep down investment. It also offers scope for expansion in small investmental steps, in line with the limited financial resources of developing countries like ours. Compared to BF-BOF route, the investment cost of a plant based on the DR-EF route will be about 75 to 80% only up to the stage of continuously cast semis. Today even, large integrated steel plants, in some developed countries are favouring electric steel making because of the lower capital costs, environmental conditions and potential energy savings. But its success depends upon regular and cheap electric supply.

23.5.10 Carbon Steel

When one refers to 'steel' ordinarily what is technically called 'carbon steel' or 'ordinary steel' or 'straight carbon steel' or 'plain carbon steel' is referred to. Such a steel owes its properties chiefly to the presence of carbon without substantial amounts of other alloying elements. The carbon content of steel is less than 2%. One of the main reasons for the importance of steel as an engineering material is that its properties can be altered to cover a wide range. Its properties can be altered by the control of 1. chemical composition, 2. heat treatment, and 3. mechanical work (e.g., hot or cold rolling).

Classification and properties of carbon steels are discussed in the next chapter on Ferrous Alloys.

23.5.11 Effect of Elements on the Properties of Steel

In plain carbon steels the effect of carbon on the properties may be considered apart from the effect of other elements, as carbon is not to be regarded as an impurity. The mechanical properties of carbon steels are mainly controlled by varying the proportion of carbon. Other elements are regarded as impurities.

The effect of carbon content on properties and consequently on the use of steel is shown in Fig. 23.2, in the next chapter.

1. *Effect of carbon*. Carbon is the principal hardening and strengthening element in plain carbon steels. Up to 0.8% carbon the tensile strength goes on increasing with the increase in carbon content. Up to this limit with the addition of every 0.01 % carbon, the strength increases by almost 70 kg per cm^2. This shows the importance of carbon in steel up to 0.8%. Ductility and malleability decrease with increase in carbon content. Carbon makes iron less permeable to magnetic forces. Hardness increases with progressive increase in carbon content.

2. *Effect of impurities*

(i) *Silicon*. Most of the silicon present in pig iron gets eliminated while making steel, but it comes in during the deoxidation operation in which silicon is added for the removal of oxygen. Usually up to 0.2% of silicon is present in carbon steels. A small amount of silicon increases strength without decreasing ductility. In high carbon steels it promotes decomposition of cementite to graphite, thus reducing hardness and brittleness.

In general the presence of silicon has beneficial effects on the properties of steel.

(ii) *Sulphur*. Sulphur may be present as FeS or MnS. Presence of FeS is undesirable as it surrounds the metallic grains and melts during hot working of steel. This makes steel *red short* or brittle at red heat. MnS forms small globules and is less harmful. The presence of sulphides is desirable for machining operations as these break the continuity of the alloy. Up to 0.3% of sulphur is helpful in cutting thread on screws.

(iii) *Manganese*. The manganese present in steels comes mostly from ferromanganese added during the finishing operation of manufacture. It improves the properties by removing sulphur. In smaller amounts, in a free state it increases strength but when it is more than 1% it increases hardness and brittleness; therefore, it is undesirable in high carbon steels. In hardening operations it adds to the depth of hardening (i.e., steel becomes harder even in the interior portions).

(iv) *Phosphorus*. Its presence has been regarded as very undesirable because it makes steel *cold short*, i.e., brittle at room temperature and shock resistance becomes poor. One of the reasons for the bad effects of phosphorus is its tendency to segregate instead of being uniformly distributed throughout the mass. Up to 0.1% it adds to the strength and hardness of steel. It also increases corrosion resistance.

(v) *Other elements*. Oxygen is present as FeO which remains as an inclusion and may cause fracture under load. Only traces of hydrogen are present in steel. It causes brittleness. Nitrogen is present as nitride which is very hard and is undesirable for working operations.

Besides the chemical composition, the properties of steels may be greatly modified by heat treatment, the details of which are discussed separately. It is noteworthy that because carbon steels respond to heat treatment, if they are used at elevated temperatures (e.g., 350° to 550°) the constituents which are responsible for their mechanical properties at room temperature begin to change. The result is that the carbon steels loose most of their strength at elevated temperatures. For working at elevated temperatures special steels have to be prepared.

The properties may be further modified by mechanical working like rolling, forging, etc. Mechanical work breaks the grains. Naturally, the weaker bonds in the grains are broken first. The result is that cold work increases tensile strength and hardness, and reduces ductility. If the working is at an elevated temperature, there is greater mobility of atoms and new grains can form. In other words, at higher temperature steel anneals itself thus eliminating the hardness caused by working.

Chapter 24

Ferrous Alloys

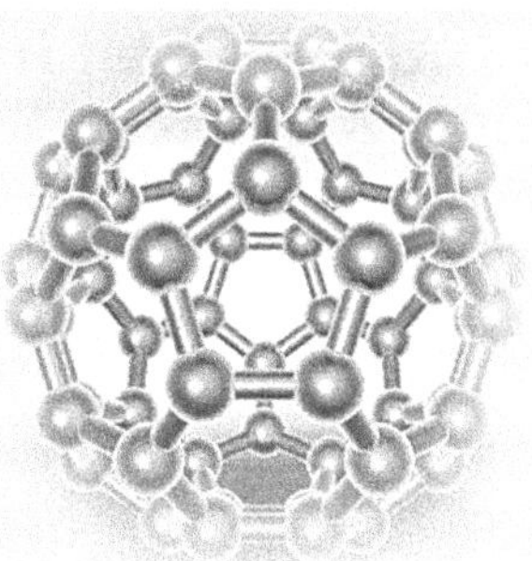

24.1 Allotropes of Iron

Pure iron is of academic interest only. The most widely used material is a combination of iron with carbon in one form or another. Like sulphur, carbon, oxygen etc., iron also exists in several allotropic modifications. Most of its industrial applications depend upon the properties produced as a result of the allotropic changes. These allotropes differ widely in their properties but of greatest practical significance is the difference in the solubility of carbon. The important properties of the allotropes are summarized in Table 24.1.

Table 24.1

Temperature range	Phase	Magnetic state	Solubility of carbon	Crystal lattice
Up to 768°	a	Magnetic	Almost nil (0.025%)	bcc
768° to 910°	b	Non-magnetic	Almost nil (0.025%)	bcc
910° to 1400°	g	-do-	Up to 2%	fcc
1400° to 1539°	d	-do-	0.1%	bcc

24.1.1 Micro Constituents

The engineer has not to deal with the allotropes in their pure forms but with their combinations–physical or chemical or both–of carbon (generally along with other elements also). These combinations, when seen under a microscope, show their characteristic shapes and reveal much of the history and properties of the material. The properties of the more common and important micro-constituents are indicated in the Table 24.2.

Table 24.2

Constituent	State	Occurrence	Remarks
Ferrite	α-iron	With cementite in low carbon steel and in wrought iron.	It is almost pure iron with a very small amount of dissolved carbon. Soft, ductile and weak.
Austentite	γ-iron with (up to 2%) carbon in solid solution	In carbon steels it is not present at room temperature.	Very soft and ductile at temperatures. at which it exists (i.e., 723° to 1400°). For heat treatments steels are brought in this state. Hot forging and rolling are done in this state.

Contd... Table 24.2

Constituent	State	Occurrence	Remarks
Cementite	Fe_3C	Along with ferrite in steels. Important constituent of white cast iron and tool steels.	Hard and brittle. Its presence in steels along with ferrite makes them tough, hard and strong.
Pearlite	Alternate layers of ferrite and cementite	Chief constituent of annealed, normalized and eutectoid steels	Properties are in between those of ferrite and cementite. Softer and more ductile than cementite, but harder and stronger than ferrite
Martensite	Forced solid solution of carbon in iron.	In hardened steels.	Hard and brittle. Has very little ductility.

24.1.2 Critical Points

The critical points of iron can be easily understood by comparing them with those of water. Let us consider the cooling of water at a uniform rate from above 100° to below 0°. The temperature of steam drops at a uniform rate till it comes to 100° when steam begins to condense as water. Because it gives up latent heat, the drop in temperature is arrested for a time. When steam has been completely converted to water the temperature again drops at a uniform rate till it comes to 0°. The drop in temperature is again arrested till the formation of ice is complete. On further cooling the temperature again drops uniformly. Thus on the cooling curve of water there are two distinct points, at 100° and 0°, when the drop in temperature is arrested and there are breaks in the cooling curve. These points are marked by certain internal changes– from gas (steam) to liquid (water) to solid (ice).

In a similar manner when iron is cooled, the drop in temperature is arrested at several points. These arrests take place in the solid state due to allotropic changes and the temperatures of arrests are known as *critical points* or 'A' *points*. In the solid state, during cooling, the first critical point comes at 1,400° when δ-iron changes to γ-form and this point is known as A_4 point. Similarly A_3 point is at 910° when γ-iron changes to β-iron and at 768° is A_2 point when β to α change takes place. The difference between α and β is only in magnetic properties. Therefore, in general usage the term α is used to include both α and β forms. A_1 point may also be mentioned which while cooling is at 695°, when austenite decomposes to pearlite.

During heating, similar changes take place in the reverse order but at slightly higher temperatures due to thermal lag. Heating and cooling temperatures are distinguished by adding the letters 'c' and 'r' respectively to the symbol A. For example, Ar_1, (695°) indicates the temperature at which austenite-pearlite change takes place during cooling and Ac_1 (723°) indicates the same change during heating. In the case of cooling of about 0.8% carbon steel, at Ar_1 so much heat is evolved that the temperature rises and the steel begins to glow and becomes visible in the dark. This phenomenon is known 'recalescence'.

24.1.3 Iron-Cementite Equilibrium Diagram

The addition of carbon to iron alters the critical points, just as the addition of salt alters boiling and melting points of water. Fig. 24.1 illustrates the important constituents present in the iron-cementite system at different temperatures and concentrations under equilibrium conditions i.e., when enough time has been allowed for the changes to be completed and equilibrium established. In Fig. 24.1 the delta phase has been omitted as the engineer has seldom to work with that phase. Solubility of carbon in ferrite (0.025%) has also not been shown.

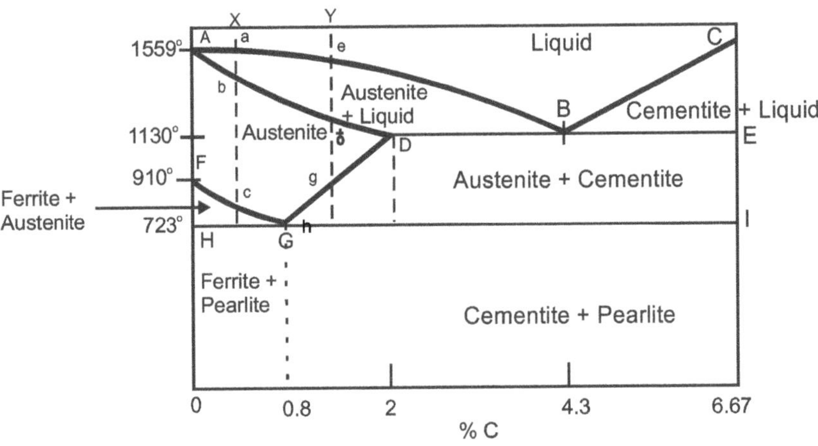

Fig. 24.1 A simplified, iron-cementitie equilibrium diagram.

Pure iron melts at 1539°. As carbon is dissolved in it the melting point drops, along the line AB in the Figure, till the lowest melting point (eutectic temperature) of 1130° is arrived at with 4.3% carbon. As the carbon content further increases up to 6.67% (when the system contains 100% Fe_3C) the melting point again rises. The melting point of an iron-carbon system of any carbon content can be found with the help of the line ABC. The region above ABC indicates the liquid state. This behaviour is similar to the cooling of sodium chloride solution in water. As the concentration is increased, the freezing point is depressed down to the eutectic temperature. At a temperature between the eutectic and 0°, ice and solution may exist together. A similar condition is indicated by the region ADBA in which crystals of austenite and the liquid exist. In the same manner crystals of cementite and the liquid exist in the region CBEC. However, solidification is complete along the line ADBE (which is known as *solidus*, as the line ABC is known as *liquidus*).

It should be noted that the maximum amount of carbon that can be held in solid solution is 2% at 1130° indicated by point D. That marks the limit (2%) of carbon content in steel.

The steel section of the diagram is of greater practical utility. The region ADGFA represents the condition in which a solid solution of carbon in γ-iron–known as austenite–exists, FG represents the beginning of the change of γ-iron to α-iron which completes along HG. The beginning of the separation of cementite from austenite is represented by DG. Its completion is represented by GI.

It will be noticed that the transformation temperature of 910° for γ to α change is lowered down to 723° as the carbon content in iron goes up to 0.8%. The point G is similar to B (eutectic point) but it represents changes taking place in the solid state, hence is known as *eutectoid* point. Thus FGD represents the beginning of certain changes and HGI represents their completion. The region between these is called the *critical range* (the critical range should be noted well as it will be referred to often while discussing heat treatment).

Changes taking place on cooling hypo-eutectoid steels (i.e., steel containing less than O.8% carbon) can be illustrated by discussing cooling of steel 'X' in the diagram. As it is cooled from the liquid state, the solid austenite begins to separate when it reaches the temparature '*a*' at the intersection with the liquidus AB. As the carbon content of the crystals is less than that or the original mixture, the melt becomes richer in carbon and follows down the liquidus to the right. As the temperature drops from '*a*' to '*b*' more austenite, richer in carbon, separates and the material consists of austenite crystrls plus molten iron containing dissolved carbon. The solidification is complete at '*b*'. On slow cooling, carbon diffuses uniformly throughout the solid austenite. When the temperature drops to '*c*' at the intersection with FG the precipitation of ferrite begins. Ferrite being almost pure iron, the carbon content of the remaining austenite increases along the line FG till it becomes 0.8% at point G. At 730° austenite decomposes into a eutectoid mixture of cementite and ferrite which is known as *pearlite*. No further phase changes take place on further cooling and the steel consists of a mixture of ferrite and pearlite.

In the case of eutectoid steel (i.e., steel containing 0.8% carbon) similar changes take place. There is no separation of primary ferrite. As the temperature drops to point G, the entire austenite decomposes to pearlite.

Cooling of steel 'Y' is illustrative of hyper-eutectoid steels (i.e., steel containing more than 0.8% carbon). As the temperature drops to '*e*' at the intersection with liquidus AB, crystals of austenite begin to separate from the molten alloy and at temperature '*f*' the entire alloy freezes.

As the temperature drops slowly from '*f*' to '*g*' through the austenite region, carbon diffuses within the solid resulting in a solid solution of uniform composition. At the temperature '*g*', which touches DG, precipitation of cementite begins. As cementite is richer in carbon content (6.67% carbon) the remaining austenite contains progressively lower amounts of carbon on further cooling till the point G is reached. As in hypo-eutectoid steel, austenite decomposes at this point giving pearlite. Thus at any lower temperature the alloy consists of cementite plus pearlite.

24.2 Classification of Carbon Steels

Carbon steels have been classified in several ways. The most important method is on the basis of their carbon content. This method will be discussed here. The other important method is according to the process used in its making e.g., acid Bessemer steel, electric steel, and so on. Yet another classification based on uses is also not uncommon, e.g., tool steel, spring steel, etc.

Steel with only a very small amount of elements other than iron has few engineering uses, as besides the cost factor, it is soft, ductile and weak. By increasing the carbon content its hardness increases and ductility decreases. Tensile strength increases to its maximum value of about 10,000 kg/sq. cm when the carbon content is about 1%, but this steel is quite brittle. Therefore, for structural purposes lower carbon steels are used, and if it is to be welded, the carbon content should be below 0.25%. The effect of carbon content on the properties of steel is more marked in hypo-eutectoid steels (i.e., steels containing less than 0.8% carbon) than in hypereutectoid steels (i.e. containing more than 0.8% carbon). The reason is that up to 0.8% carbon the proportion of pearlite increases rapidly to 100%. On further increase in carbon content the cementite proportion, which affects the subsequent properties, does not increase as rapidly.

A common system of classification of steel on the basis of carbon content is as follows, (also see Fig. 24.2) ;

1. *Very mild and mild steels* (up to 0.25% carbon). These are used for making sheets, boiler plates, various engineering structures, rivets, bolts, soft wires, nails, etc.

2. *Low and medium carbon steels* (0.25 to 0.6% carbon). These are used for making machine parts, railway axles, rails, springs, shaftings, etc.

3. *High carbon steels* (0.6 to 2.0% carbon). These are used for making tools, files, knives, saws, wires and cables, ball bearings, etc.

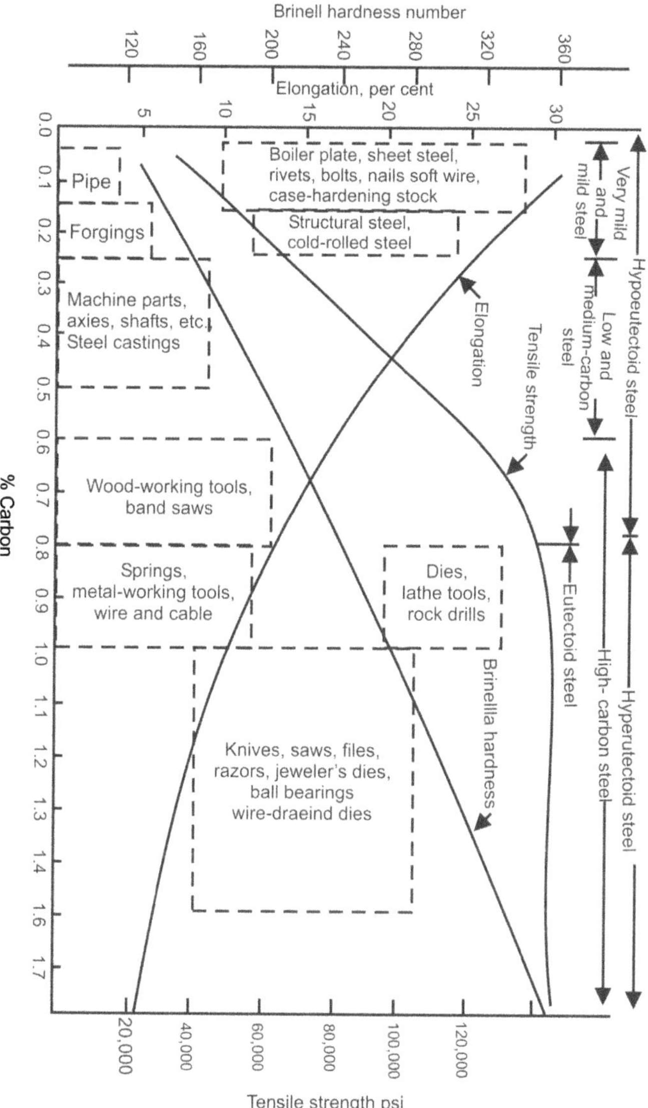

Fig. 24.2 Effect of carbon content on elongation (formability or ductility, hardness and tensile strength of steel).

(*J. Newton, an introduction to metallurgy, courtesy, john wiley & sons*).

24.3 Alloy Steels

The ordinary steel is an alloy of iron and carbon. It contains minor amounts of impurities which influence the properties. The effects of these have already been discussed. When some elements. e.g., Ni, Cr, Mn, Si, V, etc., are added to steel in such amounts that their distinguishing effects are produced, the product is known as an alloy steel. The American Iron and Steel Institute has laid down a certain minimum percentage for each or the alloying elements to be present in an alloy steel, e.g., some of the minimum percentages are: Cr, 0.25; Cu, 0.60; Mn, 1.65; Ni, 0.25; Si, 0.60; etc. The percentages of these actually present are, however, much higher.

Alloy steels are used after heat treatment. Without heat treatment there is little improvement in the properties as compared to carbon steels containing equivalent amounts of carbon.

The need for making alloy steels arises from several drawbacks present in carbon steels. Some of the important reasons for making alloy steels are :

1. Ferrite is weak. When it is made strong by the addition of carbon, its ductility falls rapidly; it becomes brittle.
2. If heat treatment given for hardening carbon steel, its surface gets hardened but the interior remains soft, i.e., carbon steels are not deep hardening.
3. Carbon steels are unsatisfactory for use at elevated temperature as the mechanical properties, in general, deteriorate. Their hardness is very badly affected at elevated temperatures.
4. Their corrosion resistance is poor.
5. Even by heat treatment it has not been possible to increase the strength of carbon steels above 10,000 kg/cm^2. There is a rapid fall in ductility, beyond this value. Heat-treated alloy steels not only have high strength but also have good ductility and high yield point even in large sections.

The Alloy Steel Project of Hindustan Steel Ltd., situated at Durgapur, has a capacity of 1,00,000 tonnes per year and the capacity is being increased to 3,00,000 tonnes per year. It produces high speed steels, carbon tool steels, alloy tool and die steels, constructional steel and stainless steels.

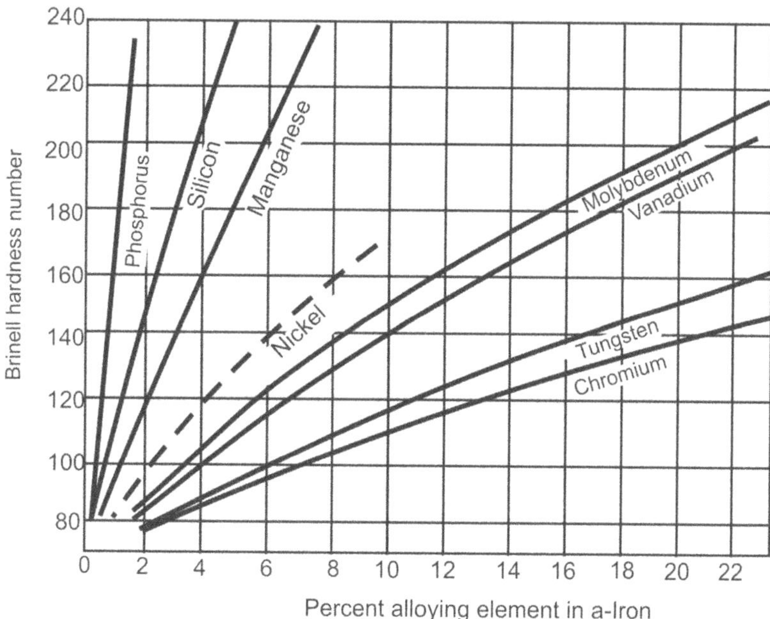

Fig. 24.3 The relative effectiveness of the alloying elements as ferrite strengtheners.

(*Courtesy, U.S. steel corporation*)

24.3.1 Principles of Alloying

Unless large amounts of the alloying elements are added, the micro-constituents of alloy steels remain the same as in carbon steels. In general, carbon remains the principal element affecting the properties of alloy steels. Although the alloying elements have specific effects on the properties, they mostly alter the properties by their influence on the amount and distribution of carbon, and on the critical points. These effects can best be studied by considering the various equilibrium diagrams. Such a study being beyond the scope of this book, the following elementary considerations are offered. The relative effectiveness of some of the common alloying elements on the properties of ferrite are shown in Fig. 24.3. γ-iron has face-centred cubic lattice. The elements (e.g., Cu. Ni and Co) that have this type of crystal lattice are more soluble in it and stabilize γ-iron or austenite. Manganese, which exists in several allotropic forms may also be included in this group. The result is that the transformation temperatures are depressed and for hardening, the alloy steels have not to be quenched from such high temperatures as the carbon steels. They retard the separation of carbides. Nickel and manganese depress the critical points of steel to such an extent that the austenitic condition may exist even at room temperature.

Body-centred cubic elements (e.g., Cr. Mo, W, Si, etc.) are more soluble in ferrite which itself has bcc crystal lattice. These elements, therefore, stabilize, strengthen and harden ferrite. They raise A_3 point (temperature at which ferrite changes to austenite) and lower A_4 point (δ-iron to γ-iron change temperature). This may even cause the disappearance of austenite, as happens in 18% chromium steel. Consequently, the steel will need to be heated to a higher temperature for bringing it to the austenitic condition. These elements form hard carbides with carbon.

The presence of alloying elements makes the alloy structures more complex than plain carbon-iron alloys. Therefore, the transformation rates are reduced. The practical result is that large crosssections made from alloy steels can be hardened even in the interior portions where the rate of cooling is lower, i.e., they are more deep hardening. Sometimes even air cooling may retain the martensitic structure. Mo, Cr, W, V and Mn have a marked effect of this kind.

24.3.2 Effects of Alloying Elements

In general, the effects of adding the alloying elements may be one or more of the following :-

1. ***Carbide formation.*** Just as carbon forms hard carbide with iron, Fe_3C, some elements, like chromium, tungsten, titanium, vanadium, molybdenum and manganese form hard carbides. These carbides are usually associated with Fe_3C and are often collectively referred to as cementite. Cobalt, nickel and silicon do not form carbides. They promote the formation of graphite and counteract the effect of cementite. The carbides contribute hardness, brittleness and resistance to wear and abrasion. In general, the carbides of the alloying elements retain their characteristic properties at higher temperatures than iron carbide.

2. ***Solid solution formation.*** The formation of solid solution, which depends primarily on the structures of the crystals, had been discussed under Principles of Alloying. By forming solid solution an element stabilizes ferrite or austenite, i.e., alters the critical temperatures. The formation of solid solution also influences the eutectoid point (point G in Fig. 24.1). The combined effect of these is that the critical range is altered

3. ***Transformation rate.*** As the alloying elements make the alloy structure more complex, the transformation rate of austenite becomes slower. This greatly affects heat treatment which in turn affects the properties and consequently the uses of the alloys.

4. ***Grain size and deoxidation***. Oxides of aluminium, silicon, titanium and vanadium have high melting points and slow down the rate of grain growth. In other words, the grain growth takes place at higher temperatures. These elements (because of their affinity for oxygen) act as deoxidizers, i.e., reduce iron oxide and remove dissolved oxygen.

When two or more alloying elements are added, usually the characteristic properties of each are imparted to the steel. In some cases an element may counteract the properties of another element. The combined effect of more than one element on hardenability of a steel is usually more than the algebrical sum of the effects of the same elements added separately. For example, chromium-nickel alloy steels have good hardening properties as well as excellent ductility. Chromium-molybdenum steels have good hardening properties, satisfactory ductility and heat resistance. It is for such reasons that in making alloy steels the alloying elements and their proportion are carefully selected.

The question of quantitatively comparing alloy steels has drawn attention of workers for some time. Before the Second World War the comparison was by trial and error and was based on physical properties such as tensile strength, ductility and toughness. M. A Grossman and his co-workers developed the concept of *hardenability* for comparing quanitatively the engineering properties of alloy steels. They maintained that the alloying elements improved the properties of steel basically by increasing hardenability of the alloy. In spite of the fact that carbon content governs the basic properties of steel, the alloying elements influence the hardenability or the depth to which the steel becomes hard, of the alloy steels. Grossman and his associates tackled the problem theoretically and developed a "multiplying factor" for indicating the increase in hardenability that could be expected from the various alloying elements. The practical aspects and a test for it have been developed by Jominy.

The alloy steels are named after the element (or elements) that have the greatest influence on the properties of the alloy.

24.3.3 Nickel Steels

0.5 to 78.5% of nickel may be present in nickel steels. In most cases its amount is less than 5%. These alloys contain lesser amounts of carbon (usually 0.2 to 0.5%) because the eutectoid point is shifted to the left in the equilibrium diagram. Nickel is soluble in γ-iron in all proportions and up to 30% in α-iron. Hence it may be present as solid solution in austenite as well as in ferrite.

In low alloy steels nickel strengthens ferrite and makes steel tough and deep-hardening.Presence of nickel decreases grain size, therefore it increases strength without proportionate decrease in ductility. By hardening, a low alloy nickel steel can be made as strong as, but less brittle than, the hardened carbon steel. Because of lowering of critical points the nickel steel can be quenched from lower temperatures, hence there is less danger of cracking and distortion. Nickel steels, when case-hardened, produce an outer case which is more tough and resistant to wear and fatigue. The core or the inner portion remains stronger than in carbon steels. In structural steels, nickel is very widely used as the alloying element because of these beneficial effects. Low-alloy nickel steels are also used for heavy duty machine parts, automobile parts, gears, armour plates, etc.

Because of the solubility of nickel in γ-iron, when its amount is more than 22%, the alloy remains in the austenitic condition even on air cooling. Such alloys are not used for structural purposes because of the high cost. These have some very useful properties. One of such widely used alloys is *invar*. It contains 36% nickel (with 0.2% carbon and 0.5% manganese) and has very low coefficient of thermal expansion at room temperature and is used where dimensional accuracy at room temperatures is needed. e.g., in surveyor's tapes. 40% nickel has almost the same coefficient of thermal expansion as glass and is known as *Platinite*. It is used as lead wire passing through glass in electric light bulbs. Both nickel and iron are ferromagnetic. An alloy with 78% nickel has very high magnetic permeability after it has been air quenched and is known as *permalloy*. It is used in making telephone circuits in which the electric impulse should be sharp and short, otherwise the impulses will be superimposed on each other and the sound waves will get mixed up.

Permalloy has high permeability at low field strength. Mention may be made of *hipernik* (50% Ni; 50% Fe) which has high permeability at higher field strengths and of *perminvar* (45% Ni; 25% Co; 30% Fe) which has a constant permeability over a range of flux densities.

24.3.4 Chromium Steels

Chromium has a body centred cubic lattice, hence is soluble in α-iron and raises the critical points. Chromium steels have finer grain structure, therefore, are much tougher, less brittle, and more resistant to shocks than carbon steels of similar hardness. This results in higher elastic limits and tensile strengths.

Chromium not only affects the critical points but retards the rate of transformation of austenite to ferrite and cementite. Therefore, chromium steels with less than 2% Cr–have been widely used because heat treatment becomes easier and better. As stated in 'Principles of Alloying', chromium forms hard carbide, therefore, it increases the strength, hardness and wear resistance of the steel. As it slows down the reactions, it makes steel deep-hardening, i.e., the interior section of steel, which cools slowly, also gets hardened. To take a concrete example, a low chromium steel needs to be heated to a higher temperature than a carbon steel to bring it into austenitic state (because the critical point has been raised). Then for hardening it may be cooled more slowly, say by oil quenching, than carbon steel which may need water quenching. The advantage is that less internal stress is produced and also that it becomes harder in the interior portion as well.

Chromium shifts the eutectoid point to the left in the equilibrium diagram, i.e., pearlite is formed with lesser carbon content. The obvious result of this is that in high carbon steel more of free cementite, which is the hardest constituent, can be produced by quenching. In addition to this, fine pearlite is produced which is in itself a strengthening constituent. Therefore, chromium steels are used where great hardness is needed, as in ball and roller bearings, certain parts of automobiles, dies, files and tools.

Besides low chromium steels, high chromium steel (i.e., containing over 11% Cr) is also used. A passive oxide film is formed on the surface which makes these steels highly resistant to corrosion and oxidation at higher temperatures. Some of these are stainless steels. Heat-resisting steels contain still higher amounts (18 to 28%) of chromium in addition to a small amount of carbon.

24.3.5 Nickel-Chromium Steels

Nickel-chromium steels are the best commercial steels as they combine the advantage of both nickel and chromium and have the disadvantages of neither. Nickel gives ductility, strength and toughness to ferrite and chromium gives to cementite hardness and resistance to wear. The combination thus give all-round steels which have strength and ductility as well as hardness and also resistance to wear and corrosion.

Because of the depression of the eutectoid points, the nickel chromium steels are austenitic even at room temperature and do not harden by heat treatment. For hardening, nickel-chromium steels are cold worked. On air cooling from tempering temperatures (about 600°) these steels become brittle. This is known as "temper brittleness" and is supposed to be due to the precipitation of nitrides, carbides and phosphides which make steel brittle. The brittleness can be removed by re-heating the sample above 600°, when the nitrides, etc., dissolve. The sample is then cooled rapidly.

As these steels are resistant to corrosion and surface remains bright, they are also called *stainless steels*. In the atmosphere or in contact with oxidizing solutions, a passive film of Cr_2O_3 is formed on the surface which protects the underlying metal. The most common steel of this class is known as 18-8 stainless steel and is widely used for making household utensils. It contains about 18% Cr and 8% Ni, the remaining being iron with less than 0.2% carbon and less than 0.1% of silicon and manganese. Stainless steels are non-magnetic.

The stainless steel remains austenitic at room temperature, therefore its strength does not improve by heat treatment. Its mechanical properties considerably improve by cold working. For example, the tensile strength of a sample of stainless steel increased from 9,000 kg/cm^2 to 11,000 kg/cm^2 on cold rolling.

Stainless steels are very widely used, not only for the household utensils, but also for handling corrosive materials in chemical plants, foodstuff manufacture, dairies, dyeing, photographic work, etc. Because of good mechanical properties nickel-chromium steels find use in the manufacture of plates for battleships and tanks, for making guns, automobile parts, etc.

A number of alloys of iron with considerable amounts of nickel and chromium have been developed for giving good mechanical properties at elevated temperatures, say 1000°, at the same time, resisting oxidation. Some of the alloys, which are quite expensive for ordinary usage, find valuable applications in the electrical industry. *Chromel* (62% Ni; 15% Cr; 23% Fe) and *nichrome* (80% Ni; 2% Cr) are widely used in the electrical industry for making heating coils for furnaces, stoves and other household appliances. These have high electrical and oxidation resistance and can be used up to 1000° without any noticeable oxidation or deterioration.

24.4 Special Steels

24.4.1 Tool Steels

The high-speed tools, dies, valves, ball and roller bearings, etc., have to be made of steel which retains strength and hardness at high temperatures and has toughness (resistance to shocks), and resistance to wear. The tool steel should not distort during hardening. At the same time it is desirable that it is machinable and grindable. It is not easy to have a steel which has all these properties to the maximum extent and often one property has to be sacrificed to obtain the other.

The cheapest tool steels are the high carbon steels containing 0.5 to 1.5% carbon. These have carbon in the form of Fe_3C which exists as tempered martensite. To harden the tool it is heated within the critical range and quenched. This has to be done carefully because the tool has to be hard, at the same time excessive brittleness is undesirable. Brittleness is removed by tempering after hardening. If a high carbon tool steel is used at high temperatures, as encountered in high speed work, its hardness is lost.

For retaining its properties at the high temperatures, certain alloying elements are added to steel. The most widely used *high speed steel* contains. 0.7% carbon, 18% tungsten, 4% chromium, and 1% vanadium. Heat treatment of high-speed steel has to be done very carefully. It is first heated carefully to about 1,300° and then quenched in oil or in a blast of air. Heating is generally done in two stages. It is slowly heated to 850° and then the temperature is quickly raised to the hardening temperature. At the high temperature the steel is exposed to the danger of grain growth and oxidation, hence it is not "soaked" at that temperature. To avoid the development of severe stresses and distortion due to rapid temperature drop, quenching is also usually done in stages. It is cooled to 600° in a salt bath and after the temperature has become uniform it is quenched in oil to 450° and then air cooled. Its properties are further improved by tempering. Super high speed steel contains cobalt and has to be hardened from still higher temperatures.

24.4.2 Boron Steels

These have been developed during recent years. They contain less than 0.005% boron in killed (fully deoxidized) steel. The addition of boron considerably increases the hardenability of steel with the result that much smaller amounts or other expensive alloying elements need to be added to the alloy. The steel is fully hardened and then tempered. Boron becomes less effective as carbon content increases. Therefore boron is added in low carbon steels (< 0.6% carbon). Advent of boron steels has saved a considerable amount of other valuable alloying elements.

24.4.3 Heat-Resisting Steels

Equipment like gas turbines, aeroengine valves, retorts and machinery used near furnaces, are exposed to high temperatures. Use of steel at such high temperatures causes (a) oxidation and scale formation of the metal, (b) creep (i.e., slow deformation at steady load) and loss of strength and (c) changes in the micro-constituents as, for instance, precipitation or carbides, etc., which make the steel brittle.

It has been explained in the chapter on Corrosion that if the oxide film is continuous, non-porous and adherent, the metal underneath is protected. Such films are formed by chromium, aluminium and silicon. Therefore, incorporation of these metals in steel increases corrosion resistance at elevated temperatures. On heating mild steel in contact with aluminium powder, an alloy of iron and aluminium is formed on the surface and mild steel becomes resistant to oxidation.

Molybdenum is an important alloying element which gives strength and creep resistance at high temperatures. Most steels for high temperature use are highly alloyed and contain more than, say, 12% chromium, and are quite expensive. Nichrome, for instance, consists of 10% nickel and 20% chromium and can be used at 1,000 to 1,100°.

It is interesting to note that those steel which are ferritic show sudden loss of ductility at temperatures around 0° with the result that their toughness falls rapidly. The austenitic steels remain tough even at low temperatures.

24.4.4 Wear-Resisting Steels

In many moving parts there is the problem of wear and abrasion of the surface. One method of dealing with the problem is surface-modification i.e., hardening of the surface while the core remains soft. This has been considered in the chapter on Heat Treatment. In certain cases like the ball mills and jaw crushers, the wear is due to the combined action of abrasion and impact. In such cases an alloy steel is required. *Hadfield's manganese steel*, which contains about 1% carbon and 12% manganese, is the most important steel of this class. Manganese reduces the transformation rate and this steel remains austenitic at room temperature upon water quenching after heating. Hadfield's steel becomes still harder by cold work, consequently its hardness increases during use. Because of this it is difficult to machine and is generally used in the form of a casting.

Chapter 25

Heat Treatment

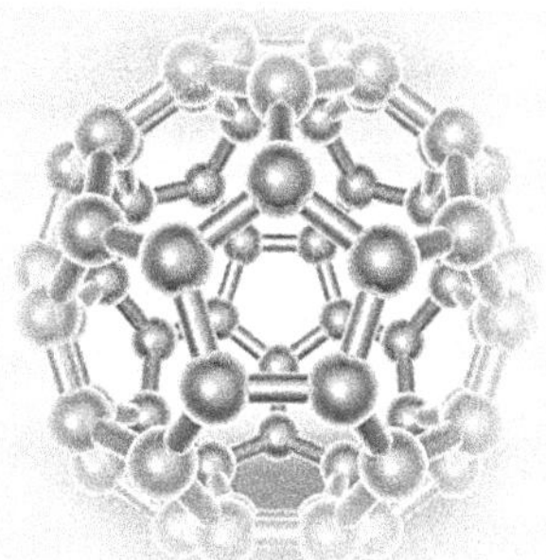

25.1 Heat Treatment

Heat treatment may be defined as 'an operation or combination of operations of heating and cooling of a metal or an alloy in the solid state for obtaining certain desirable conditions or properties'. Heating and cooling for the sole purpose–mechanical working (e.g., hot rolling) is excluded from the meaning of this definition.

The main objective of heat treatment processes for steel are: 1. to increase strength, toughness, hardness and ductility; 2. to soften steel; 3. to relieve internal stresses and strains; 4. to refine the grain; 5. to normalize steel which has already been su bjected to mechanical or heat treatment operation.

By suitable selection of a process or a combination of processes it is possible to achieve any of these objectives or a combination of them.

In general, the properties of high carbon steels are altered by heat treatment to a much greater exent than those of low carbon steels. The reason is that changes in properties are mostly brought about by changes in the form and distribution of carbon, which is obviously low in low carbon steels.

25.1.1 Introductory Principles

It was known even to primitive blacksmiths that if a red hot steel tool is plunged in water it becomes hard, but if it were left to cool slowly it would remain soft. To put it in other words, a sample of steel can be made soft, ductile and tough by one heat treatment and by another treatment it can be made hard and strong. Thus it is seen that the properties of a sample of an iron carbon alloy, e.g., steel, can be largely controlled by heating to a suitable temperature, keeping it for some time (i.e., 'soaking') at that temperature and then cooling at a certain rate. In brief these are the three steps in heat treatment by which the properties are controlled. An obvious use of this kind of treatment is to produce in an alloy the desired properties needed for different uses. Besides this, it is helpful in various operations of production engineering. For example, while making a gear the metal must be soft and machinable so that the proper shape may be given, and then it has to be made tough so that during use the shape does not change and it does not wear out. Or, when a plate is to be repeatedly pressed cold, for instance to make a shell or a tumbler, it becomes hard and brittle after a few pressings and it cannot be pressed further to give the desired shape. So it has to be softened by annealing and then further worked to get the final shape.

In Chapter XV iron-cementite equilibrium diagram has been discussed and it has been made clear that the diagram deals with *equilibrium conditions*. Heat treatment deals with non-equilibrium conditions. By heat treatment either a metastable (semistable) condition is produced in the alloy or a metastable condition which already exists (due to previous heat treatment or mechanical work) is removed. In many cases of heat treatment allotropic changes take place. There is also precipitation from the solid solution.

In the previous Chapter we have seen that at ordinary temperature carbon steel consists of Fe_3C (i.e., cementite) distributed within iron (i.e., ferrite) in different forms. As it is heated to a certain temperature, known as the *lower critical point* (or lower transformation point), α-iron *begins* to change to γ-iron. Carbon is practically insoluble in α-iron. With the appearance of γ-iron the carbon of Fe_3C begins to dissolve in it producing austenite. When heating is continued for some time at a certain temperature (which can be found from the equilibrium diagram) austenite formation is *complete*. This temperature is the upper critical point. The temperature range between the upper critical point and the lower critical point is known as the *critical range*. By heating steel past the critical range all the carbon present in different forms can be brought in a homogeneous solid solution, thus all the previous effects of the original microconstituents can be wiped off. By controlled cooling new constituents can be produced that will have the desired properties. Heating can be compared to rubbing off the old figures from the blackboard and controlled cooling is like making new figures on the blackboard.

If a sample of carbon steel, which is in the austenitic state, is *quenched* (i.e., suddenly cooled) by plunging in water or brine, the γ-iron which has face-centred cubic lattice and held carbon in solid solution, being unstable at the lower temperature, is changed to α-iron which is the stable form at that temperature. Carbon is insoluble in α-iron which has body-centred cubic lattice. Quenching lowers the temperature so rapidly that the atomic readjustments cannot fully take place and there is little time for carbon to diffuse. Thus a new constituent, known as *martensite*, results. It is very hard and brittle and is an essential constituent of the steel used for cutting tools. Obviously, if the sample is of low carbon steel the amount of martensite produced is small and the sample does not get hardened to any appreciable extent. In a sample of very high carbon steel, a very large amount of martensite is produced with the result that it becomes very brittle and may break easily. To remove brittleness the hardened sample is re-heated below the critical range. At this elevated temperature there is greater atomic mobility, consequently highly strained atoms readjust to new positions, thereby removing some hardness and brittleness. The sample is then cooled in air, water, oil, etc., to get the desired properties. This is the principle of *tempering*. It removes some of the internal stresses, yet the metal is quite hard and tough.

If steel is brought in the form of a solid solution, i.e., it is in the austenitic state and is cooled very slowly, there is enough time for equilibrium conditions to be established giving stress-free metal. γ-iron changes to α-iron and therefore the carbon which was dissolved in γ-iron diffuses and rearranges itself. This induces softness in steels and is the basis of *annealing and normalizing* operations.

The importance of proper heating cannot be over emphasised. First, the temperature to which steel is to be heated has to be decided. It depends upon the chemical composition of steel and on the structure and properties desired in the final product. If it is heated to a higher temperature or held there for a longer period larger grains will be produced. On cooling, the reverse reaction, i.e., reduction in grain size, does not take place. Other things being equal, the larger the grain size the weaker is the material. In general, this is undesirable except when greater machinability is required or for wire drawing. Hypo-eutectoid steels (i.e., $< 0.8\%$ C) or mild and medium carbon steels are generally heated to just above the *upper* critical range. Hyper-eutectoid steels (i.e., $> 0.8\%$ C) are heated just above the *lower* critical range. This enables the carbon present in the form of pearlite (but not cementite) to diffuse into the austenitic state. Higher heating temperatures cause large size grains to be produced. After heating to the required temperature the steel is held at that temperature for some time depending upon the cross section of the sample. In general, the time allowed is about half an hour per centimeter cross section. This is known as *soaking*. The aim of soaking is to allow sufficient time for the changes (or dissolution of carbon to be completed and equilibrium to be established). Heating steel to a high temperature exposes it to the danger of oxidation on the surface giving rise to oxide film and causing removal of carbon. The oxide film acts as an insulator and prevents rapid cooling. This is undesirable especially in hardening operations. Moreover, the oxide scale may not uniformly cover the entire surface, consequently soft spots may be produced at certain places. Therefore before cooling the scale needs to be removed. Generally this is done mechanically.

The mechanical properties of carbon steel depend upon its microstructure. The micro-structure depends upon *carbon content* and *allotropic changes* that take place in steel during heat treatment.

25.1.2 Constant Temperature Transformation

E. C. Bain and E. S. Devenport investigated constant temperature (isothermal) transformations of austenite at temperatures below the critical temperature. The investigations resulted in time-temperature-transformation (TTT) curves. Because of their shapes like the letter 'S' these curves are also known as 'S-curves' or as 'Bain curves' after the name of the principal investigator, E. C. Bain. It has already been explained that austenite consists of carbon atoms occupying interstices of face-centred cubic lattic of γ-iron. Gamma iron is stable at higher temperatures, and on cooling body centred cubic lattice of α-iron is formed in which carbon atoms do not dissolve because of the absence of suitable interstices. For this reason the carbon atoms segregate and form cementite. This involves movement or diffusion of carbon atoms which depends on both time and temperature. That is why the study of TTT curves is important in heat treatment considerations. At higher temperatures movement of atoms is rapid and at lower temperatures it is slower. Therefore when a carbon steel is heated to form austenite and quenched and kept at a *constant* temperature the austenite transforms at a rate which is characteristic of that particular temperature.

Bain and Devenport heated a number of small specimens of carbon steels until austenite was formed. The specimens were quenched in a bath maintained at a constant temperature for different times varying from 0.5 second to 1,00,000 seconds (over a day). During this period a portion of the austenite transformed to ferrite and cementite. The specimens were withdrawn and quenched in ice-cold brine. This rapid quenching transformed the unchanged austenite into martensite. The specimens were examined under a microscope and the extent to which decomposition had taken place and the products formed were determined. The experiment was repeated with quenching done at various temperatures.

In the TTT curves, temperature is plotted as ordinate and time in seconds, on a logarithmic scale as the bascissa. Logarithmic scale is necessary for time so that both the short periods in which the transformation reactions begin and the longer periods taken for their completion can be depicted. These curves are very helpful in determining the transformation of austenite under different conditions They show what micro-structures to expect in a particular heat treatment operation. As the micro-structures decide the mechanical properties the curves help to decide the treatment to be given for obtaining the desired properties. The curves show micro-structures under non-equilibrium conditions just as the equilibrium diagrams show under equilibrium conditions.

In Fig. 25.1 is shown 'S-curve' for a 0.6 per cent carbon steel. The lines Ae_1 and Ae_3 represent the transformation temperatures under equilibrium conditions. Austenite is stable above Ae_3. It is unstable below Ae_1 and transforms in time. It is partially stable between Ae_1 and Ae_3. Ms and Mf represent temperatures at which transformation of austenite to martensite starts and finishes respectively. It will be seen from the curve that in order to obtain a completely martensitic structure, the sample has to be cooled at such a rapid rate that the time-temperature curve passes to the left of the knee of the curve (at about 550°). The constituents formed at different rates of cooling can also be known by referring to the Figure.

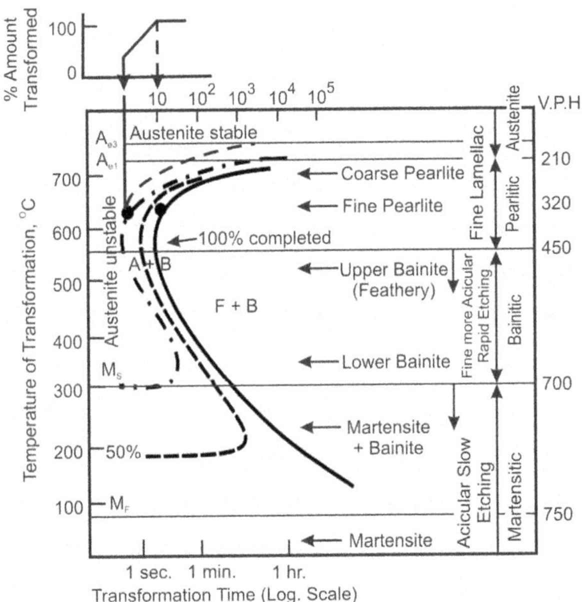

Fig.25.1 Idealised 'S-curve' for 0.6 percent carbon steel depecting time interval required for beginning (broken curve at the left), 50 per cent (dotted curve) and 100 per cent (continuous curve at the right) transformation of austenite at a constant temperature. In the right hand column the hardness (on Vickers Pyramid Number) is given.

A-Austenite F-Ferrite, p-Pearlite, B-Bainite. (E.C Rollason, *Metallurgy for Engineers*, Courtesy, Edward Arnold, London).

25.1.3 Annealing

The following description of full annealing has been given by the American Society for Testing Materials:

"A softening process in which an iron-base alloy is heated to a temperature above transformation range and after being held for a proper time as this temperature is cooled slowly to a temperature below the transformation range. The objects are ordinarily allowed to cool slowly in the furnace although they may be removed from the furnace and cooled in some medium which assures a slow rate of cooling." In simple words, annealing is a process in which steel is heated, held at that temperature for sometime to dissolve carbon in γ-iron, and then slowly cooled. The annealing process, therefore, consists of three stages, heating to the right temperature, keeping (soaking) at that temperature and slowly cooling.

The objective of annealing may be one or more of the following:

1. to soften the steel (to improve machinability);
2. to remove internal stresses (produced by mechanical working like rolling or forging or by uneven heating);
3. to change ductility, toughness, electrical and magnetic Properties ;
4. to refine the grain structure and
5. to remove gases entrapped within the mass.

A plain carbon steel with less than 0.8% carbon (hypo-eutectoid steel) is heated to just above the *upper* critical point. Steel with more than 0.8% carbon (hyper-eutectoid steel) is heated to just above the *lower* critical point–this lower heating temperature brings the pearlite in solid solution as austenite but cementite remains unchanged. Therefore, hyper-eutectoid steel is not softened as much as the hypo-eutectoid steel, but it becomes soft enough for machining. This is because heating hyper-eutectoid steel above the upper critical range will produce grains of larger size and moreover these steels are used for purposes where softness (as in mild steels) is undesirable. If the cross-section of the piece is large or the shape is complicated or carbon content is high, the rate of heating should be slow. It is then held at the annealing temperature (technically called *soaked*) for half an hour for each centimetre of cross-section.

After soaking, the steel is allowed to cool very slowly. This is generally done by leaving it in the furnace itself and shutting off the heat. Sometimes it may be buried in non-conducting material like sand, ash or lime.

The process just described is generally known as full annealing to distinguish it from *Process* or *sub-critical* annealing. When a piece is subjected to cold working (like rolling or forging steel at room temperature) some stress is produced in it and it becomes harder and more brittle. To remove this defect the piece is heated to a temperature below the lower critical point and cooled slowly. This operation is known as process annealing. This is commonly done in the sheet and wire industry and the object is to soften the metal for further cold working.

By full annealing, machinability and ductility are improved but hardness, strength and abrasion resistance become less.

25.1.4 Normalizing

Both normalizing and annealing treatments are given for making steel soft and homogeneous. Often these terms are confused with each other. There are two main points in which normalizing differs from annealing:

(i) hypo-euctectoid as well as hyper-eutectoid steels are heated above the upper critical range (in annealing hyper-eutectoid steels are heated just above the lower critical range), and

(ii) cooling is done in still air and not in a furnace. Therefore, the rate of cooling is faster. More rapid cooling becomes necessary in order to prevent the formation of coarse grains at the higher temperature.

As the name normalizing suggests, it is meant to bring back the structure and properties that are normal for the particular sample. It is in fact a corrective treatment for grain refinement in steel after it has been subjected to mechanical working like rolling, forging, etc. By normalizing finer and stronger pearlite is produced than is produced by full annealing. As compared to annealing, normalizing produces higher tensile and yield strengths and greater impact resistance though the ductility is slightly less. The hyper-eutectoid steels are rarely normalized because due to the much higher temperature employed there is a tendency to form coarser grains which reduce the strength.

25.1.5 Spheroidizing

Spheroidizing is also a softening treatment. It is applied to high carbon steels. Due to the presence of cementite high carbon steel is very hard and difficult to machine. It is heated to just below the lower critical point (e.g. 680° to 700°) when relatively larger particles of cementite are formed. These become like tiny balls or spheroids (hence the process is called spheroidizing). Thus the continuity of cementite plates is broken and the steel becomes machinable. After keeping it at the elevated temperature for some time steel is cooled slowly.

By this treatment the tool becomes less liable to crack or distort when it is subsequently hardened. On hardening, the tool will be more uniform in hardness. It makes steel more machinable, and thus saves cost of machining and increases the life of the tool.

25.1.6 Hardening

Hardening is just the opposite of annealing. The heating temperatures are similar in both cases, i.e., for hypo-eutectoid steels just above the upper critical temperatures and for hyper-eutectoid steels just above the lower critical temperatures. Hyper-eutectoid steels are not heated above the upper critical range because cementite is a very hard material and need not be completely dissolved to give austenite and then quenched to produce martensite again. The lower temperatures are sufficient to dissolve pearlite which is subsequently converted to martensite by quenching. As in annealing, the steel is soaked at the elevated temperature to allow enough time for equilibrium to be established. It is then quenched (cooled rapidly) by immersing in water, brine, oil or molten lead. Due to a quick drop in temperature there is not enough time for equilibrium to be established with the result that a forced solid solution of carbon, known as martensite, is produced. It is a very hard material. Unless it is a small and thin piece the interior does not become as hard as the outside because of slower cooling of the inner portion.

Cold water is used for rapid quenching. If the steel is quenched in water from high temperatures, there is a danger of steam bubbles forming around the metal which act as insulators and reduce the rate of cooling. Caustic soda and salt solutions have higher boiling points and better heat conductivities. Therefore they are better quenching media.

It may be noted that hardness is accompanied by brittleness. Therefore, a fully hardened piece is not so tough and may not be suitable for most commercial purposes. Naturally the hardness of low carbon steels cannot be increased greatly. Because of the lack of carbon, little cementite and martensite are produced in very low carbon steel.

When a piece of steel is heated and quenched, the temperature drops rapidly at the surface and slowly in the interior. This causes the surface to become hard but the interior is not so hard. Surface hardness obtained after quenching depends on the carbon content of steel. The depth to which the object becomes hard depends on the carbon content, the total content of the alloying elements, and the grain size. Therefore it is necessary to add alloying elements for hardness to penetrate to a certain depth. Of course, in many cases it is not desirable that hardness should penetrate deep into the interior.

25.1.7 Tempering

It has just been explained that on hardening, a forced solid solution of carbon in α-iron is produced, This is a metastable (semi-stable) state and the steel is in a stressed condition. It is brittle and likely to crack or break in use. Therefore, it is desirable that the brittleness and internal stress should be removed, though to a slight extent hardness has to be sacrificed in achieving this.

After hardening, steel is heated to a temperature *below* the lower critical range. This permits some mobility of atoms and carbide begins to be precipitated from the forced solid solution of carbon in α-iron. Steel is then slowly cooled. The temperature to which steel is to be heated and soaked and the rate of cooling depend upon the initial condition of the steel and the properties desired in the final product. Heating below 400° gives good strength and hardness. On heating from 400 to 600° good ductility and toughness are, produced but hardness is reduced. This process of removing some brittleness and relieving internal stresses, which is accompanied by a slight reduction in hardness, is known as *tempering*. If the temperature is high the carbon may diffuse, may even dissolve, and the hardness will be reduced sufficiently.

25.1.8 Surface Modification Methods

These are often called case hardening methods. In many uses the surface of the equipment subjected to intense wear, shock, corrosion and fatigue. These affect the surface and it is desirable that the interior should be tough and ductile. It has just been seen that ductility, toughness, and hardness are generally not attainable by the same heat treatment; one is attained at the loss of the other. Therefore, steel which has the properties desired in the interior is selected and the composition of the surface is altered after the final shape has been more or less obtained.

25.1.9 Case Hardening or Case Carburizing

Case hardening or case carburizing is the most widely used technique of surface modification. A widely accepted definition is that carburizing is 'a process that introduces carbon into a solid ferrous alloy by heating the metal in contact with a carbonaceous material solid, liquid or gas to a temperature above the transformation range and holding at that temperature. Carburizing is generally followed by quenching to produce a hardened surface'. The article is usually made of low carbon steel (say $< 0.35\%$ carbon) or wrought iron or malleable cast iron. It is packed in carburizing cases (made of steel castings) with a layer, about 1cm thick, of the packing material surrounding the article *allover* the surface. The packing material selected is such that it can give its carbon to the steel. Commonly charcoal, crushed bone or horn, charred leather, a mixture of barium carbonate and charcoal etc., are used. The box is sealed and heated to 900 to 950° when α-iron is converted to γ-iron which absorbs carbon at the surface. The time of heating depends upon the required depth of the case; usually the total time taken is about eight hours, but may vary from 3 to 70 hours. As the time of heating increases there is increased danger of grain growth. Depth of case increases as time or temperature increases, usually it is 0.05 to 0.5 cm. The article is then quenched to give a hard surface.

Sometimes the properties obtained by direct quenching are not satisfactory for the intended use. If it is desired to further modify the properties, heat treatments directed to affect the structure or the *core* (the interior, the unaffected portion) or of the *case* (the exterior, the modified portion) are given, based on the principles previously discussed. The process just described is sometimes called *pack carburizing* because the article is packed with carburizing materials.

In *liquid carburizing* the article is heated in a bath of molten cyanides, carbonates and chlorides at 900 to 950°. This gives a better control over the process than pack carburizing. Less time is also taken.

Gas *carburizing* is becoming more popular, especially for small articles, because of less time taken (1 to 8 hours), closer control over the process and because the process can be made semi-automatic. The equipment, however, is more expensive. The article is heated to 900 to 950° in an atmosphere of gases that can part with their carbon, and then quenched. The gas composition has to be carefully controlled. Usually a mixture of hydrocarbons with carbon monoxide, carbon dioxide and nitrogen (for dilution) is used.

25.1.10 Nitriding

Nitriding is defined as 'a process of case hardening in which a ferrous alloy, usually of special composition, is heated in an atmosphere of ammonia or in contact with nitrogenous material to produce surface hardening by absorption of nitrogen, without quenching'.

In nitriding, hardness is produced by the formation of nitrides. For nitriding, the steel must contain alloying elements like Al, Cr, Mo, Va, etc., with which nitrogen reacts to form hard and stable nitrides. The articles are packed loosely in a gas tight box and ammonia gas is circulated. Heating temperature is below the lower critical range (500° to 600°) and the time taken is two to sixty hours. The parts are slowly cooled and no further heat treatment is needed.

Nitriding makes the surface very hard and resistant to wear and to atmospheric corrosion. As the hardening does not depend on the allotropic changes of iron, the hardness in *not lost* on exposure to higher temperatures. In other methods of hardening the hardness is lost on again exposing the hardened sample to high temperatures. This is an important distinction. The process is more expensive and also more expensive steel has to be used as the starting material.

25.1.11 Cyaniding

Cyaniding is done in a manner similar to liquid carburizing. The article is packed in a bath containing sodium cyanide, sodium carbonate and sodium chloride. The bath contains a much higher percentage of cyanide than in liquid carburizing (30 to 100% cyanide as against 5 to 20% in liquid carburizing). The temperature empoyed for cyaniding is lower, viz. 700 to 900°. In cyaniding both carbon and nitrogen are available and both cause hardening:

$$4NaCN + 4O_2 = 2Na_2CO_3 + 2CO + 2N_2$$

After heating, the article is quenched. Because of the lower temperature employed and slower cooling there is less distortion. Cyaniding is a quick and cheap method for producing a high degree of surface hardness. It is specially suited for small machine parts in which distortion is a serious problem.

25.1.12 Flame and Induction Hardening

When only a small area on the surface of a large equipment is to be hardened it is necessary to heat it quickly so that other areas do not get heated. Gas flame is directed over the surface to be heated and the temperature is quickly raised above the upper critical temperature and then quickly quenched by spraying water or by a blast of air. In induction hardening the surface is heated by eddy currents induced on the surface.

These processes produce a very hard surface. Danger of distortion, decarburization, or scale formation are absent. Medium carbon steel is used in these processes.

Chapter 26

Non-Ferrous Metals and Alloys

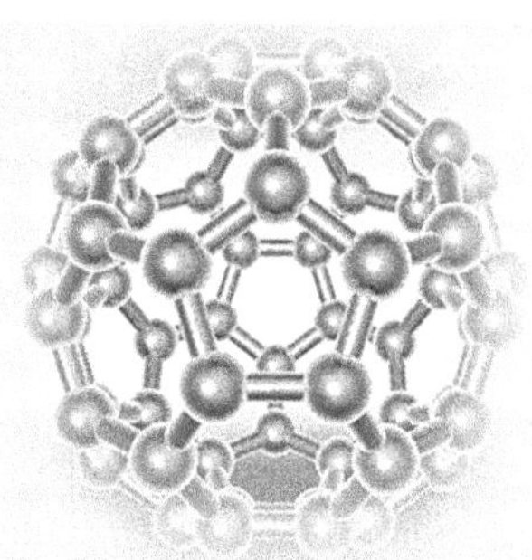

On the basis of weight the production of non-ferrous metals and alloys is very much less than that of the ferrous. Non-ferrous metals and alloys, however, are several times more expensive than the ferrous materials. They find applications because of certain very valuable properties like good corrosion-resistance, high electrical conductivity and in some cases, good strength combined with low specific gravity.

Some of the non-ferrous metals like aluminium, copper, zinc and lead are used in large quantities. Other metals like chromium, molybdenum, nickel, tungsten and vanadium are produced in much smaller quantities but are valuable alloying metals.

In India due to the abundance of aluminium ores and shortage of other non-ferrous metals attempts are being made to use aluminium as much as possible in place of other non-ferrous metals.

In some cases a pure metal is required, for example, copper used as an electric conductor must be very pure. In many cases few other materials, generally metals are intentionally added to the metal in order to improve its properties. Such a combination is called an *alloy*. An alloy is defined as any combination of elements that results in a substance possessing metallic properties. The structures of alloys are complex. Their properties are significantly different from those of the constituent elements. For example, lead is a soft metal. Addition of 1 to 2% antimony makes it quite hard. Wood's metal is an alloy of Pb, Sn, Bi and Cd. Its melting point, 70.5 °C, is much lower than that of any of its constituent metals.

26.1 Aluminium

26.1.1 Extraction

Aluminium is more abundant in the earth's crust than iron but its production is much less. This is because rich ores of aluminium are fewer and chemical activity of aluminium is high which makes its extraction difficult. The principal ore of aluminium is bauxite, $Al_2O_3.2H_2O$. The chief deposits of bauxite are in Orissa and Andhra Pradesh, although it is found scattered in various other parts of India. 574 million tonnes of bauxite deposits have been proved and about 1450 million tonnes are inferred deposits.

Production of each tonne of aluminium requires 5 tonnes of bauxite, 1.5 tonnes of coal, 0.7 tonne of carbon electrodes, 0.2 tonne of caustic soda, 0.2 tonne of pitch, 0.03 tonne of cryolite and 17,000 KWH of electric power. Cheap electric power and production of electrodes have been the chief hurdles in increasing the production of aluminium in India. The alluminium production has remained fairly constant at about 2 lakh tonnes per annum from the year 1976. The biggest plant is located at Renukoot (U. P.) with an installed capacity of 1,20,000 tonnes. Capacity of the plant at Belgaum (Maharashtra) is 80,000 tonnes. Other smaller plants are located at Hirakund (A. P.), Alwaye (Kerala) and Mettur (Tamil Nadu). Of the new plants the one located at Korba has a capacity of 1,00,000 tonnes. One of the largest plants is being constructed at Koraput (Orissa) with installed capacity of 2,18,000 tpa. It will have open cast mine 2 million tonnes per year. Capacities of some of the existing plants will be increased. Shortage of electricity is the main hurdle in the expansion work.

Most of the bauxite deposits contain 40-60% Al_2O_3, with about 25% water. The chief impurities are silica, iron oxide and titanium oxide. The ore is thoroughly purified by Bayer process before extraction, because removal of impurities from the ore is easier than from the metal. The crushed ore is relieved of water and organic matter so that it becomes easier to grind. The ore is then finely ground and digested for 1 to 2 hours in a 20% caustic soda solution at 200 to 240° under 30-40 kg/cm^2 pressure. Alumina forms sodium aluminate and dissolves.

$$Al_2O_3 + 2NaOH = 2NaAlO_2 + H_2O$$

The ferric oxide, titanium oxide and most of the silica do not dissolve and remain behind as mud. Silica forms insoluble sodium aluminium silicate. In this way each part of silica present in bauxite ties up an equivalent weight of each of alumina and soda. This remains in the red mud. Therefore, if the ore contains more than 5% silica the Bayer process becomes uneconomical.

The solution is diluted, filtered and agitated with freshly precipitated aluminium hydroxide from a previous operation and in about 40 hours, sodium aluminate slowly hydrolyses, precipitating aluminium hydroxide.

$$NaAlO_2 + 2H_2O = Al\,(OH)_3 + NaOH$$

The liquor is evaporated to give concentrated caustic soda solution which is again used, The precipitate is washed, filtered dried and calcined to give purified alumina.

$$2Al\,(OH)_3 = Al2O_3 + 3H_2O$$

Because of the high melting point of alumina it is dissolved in a fused mixture of cryolite ($AlF_3.3NaF$) and fluorspar (CaF_2) at 940-970 ^{o}C and electrolysed. The cell for electrolysis is made of steel plates lined with firebricks covered with carbon blocks which act as a cathode. There is an opening for the removal of molten aluminum from the bottom. Anode consists of blocks of carbon. Electricity heats the cell as well as electrolyses alumina. See Fig. 26.1.

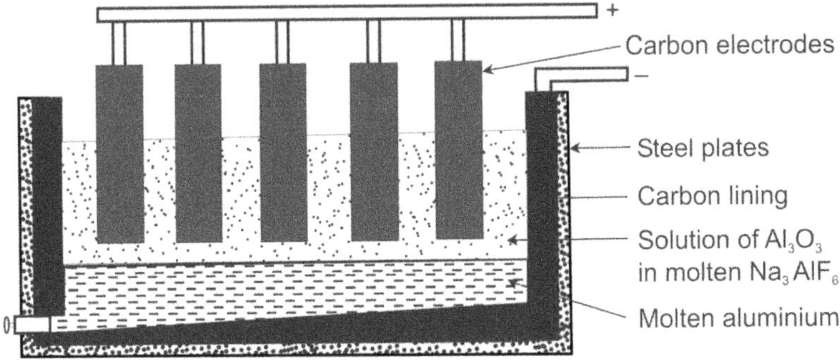

Fig. 26.1 Electrolysis cell for the extraction of aluminium.

Cryolite and fluorspar are melted by passing electricity and alumina is added at intervals. It dissolves in the molten bath. The aluminium fluoride, present in cryolite, ionizes thus:

$$AlF_3 \rightleftharpoons Al^{+++} + 3F^-$$

Aluminium is liberated at cathode and is tapped periodically. Fluorine reacts with alumina at the anode with the result that oxygen is liberated :

$$2Al_2O_3 + 12F = 4AlF_3 + 3O_2$$

As electrolysis proceeds the resistance of the cell increases and is indicated by a lamp placed in parallel which begins to glow, then more alumina is added. A crust of electrolyte and alumina is formed at the top and acts as insulator. Oxygen attacks the carbon anode with the result that the consumption of the anodes is high.

Purity of the metal so obtained is 99.6%. By using very pure alumina and electrodes, up to 99.85% pure aluminium is obtained. Generally, the metal is not purified. Only in some cases it is further purified by electrolysis in a cell having three layers of different materials. The bottom layer of molten impure aluminium acts as anode. Above it is the middle layer of fused electrolyte (mixture of cryolite, aluminium fluoride, barium fluoride and alumina) and its specific gravity is carefully adjusted. The pure aluminium acts as the cathode. On passing electricity aluminium dissolves in the electrolyte and the pure metal is liberated at the cathode.

26.1.2 Properties and uses of Aluminium

Of the common metals aluminium and magnesium have very low specific gravities (2.7 and 1.7 respectively) as compared to those of zinc (7.2), steel (7.8), brass (8.4), copper (8.8) and lead (11.3). Hence the reason for the engineer's interest in these metals and their alloys is obvious.

Aluminium is ductile and highly malleable; it can be made into 0.002 mm thick sheets. It is a good conductor of heat and electricity. Its electrical conductivity is about 61% that of copper but due to its low specific gravity an aluminium bar has twice the conductivity of a copper bar of the same weight and length (due to its increased cross-section). Its coefficient of expansion is rather high, twice that of steel. The great disadvantage of aluminium is that it is mechanically weak, having a tensile strength of about 1,000 kg/sq. cm which is about one fourth of the strength of mild steel and about one tenth that of tool steels. This disadvantage can be overcome by alloying small amounts of other elements like Cu, Mg, Mn, Fe, Si, Zn, etc., thus retaining the advantage of low specific gravity, but adding strength and thereby making these alloys very suitable for making pistons, cylinders, crank cases, body, etc., for automobiles and aeroplanes and for use in a number of structures as well.

Pure aluminium has poor machining properties, for example, it tears badly, hence it is difficult to produce threads on aluminium parts. Due to the formation of a thin film of oxide on the surface, it has high resistance to atmospheric corrosion and to concentrated nitric acid. It has poor resistance to alkalis and common salt solution. The tendency to form oxide film on the surface easily makes soldering, welding and electro-plating of aluminium difficult. The corrosion-resistance of aluminium and its alloys can be greatly improved by anodizing. Anodizing is done by making the article an anode with a carbon cathode in a dilute sulphuric or chromic acid bath. The oxygen liberated at the anode forms a thin and adherent oxide film on the surface which is resistant to corrosion.

Aluminium melts at 660°, therefore the use of aluminium and its alloys at elevated temperatures is restricted. Use of aluminium in India is increasing because of the availability of its ore and scarcity of other non-ferrous metals.

Besides being used for making alloys, aluminium metal is rolled, pressed or cast into useful shapes. It is also used as aluminium foil. Extruded aluminium powder is used in making paints and explosives, in Thermite process of welding and in printing.

26.1.3 Alloys

Alloys of Aluminium. Aluminium alloys are of two types: 1. casting alloys; and 2. wrought alloys.

1. ***Casting Aluminium Alloys*** In these alloys the chief alloying elements are copper and silicon. Zinc, magnesium, nickel and iron may also be present. These elements increase strength and hardness but reduce ductility. Alloys with copper have good machining properties and good mechanical strength after heat treatment, but have high contraction on solidification. When about 12% silicon is present in the aluminium alloy the contraction is small and corrosion-resistance is better. Aluminium-silicon alloys have high fluidity and are used for casting intricate designs. Ductility of aluminium-silicon alloys is increased by adding a trace (0.05%) of sodium or calcium.

   Aluminium-magnesium alloys have good properties but are difficult to cast. Of all the aluminium casting alloys an alloy with 10% magnesium, after heat treatment, has the best combination of mechanical properties. They have good corrosion-resistance.

2. ***Wrought Aluminium Alloys***. Aluminium alloys of this type are more widely used than the cast alloys. Their shapes are changed by mechanical work which further improves their properties. They are divided into (a) non heat-treatable alloys; and (b) heat-treatable wrought alloys.

   (a) *Non-heat-treatable wrought aluminium alloys* have manganese and magnesium as the chief alloying elements. These alloys are not so expensive and can be easily formed and welded. They have low strength but high ductility and corrosion resistance. Their strength can be increased by cold work. Addition of manganese (1.5%) increases the strength (from 1,000 kg/cm^2 to 1,700 kg/cm^2). Magnesium increases the strength as well as resistance to attack by sea water.

   (b) *Heat-treatable wrought aluminium alloys* can be strengthened by heat treatment as well as by cold work. They exhibit a property known as "age-hardening". The process consists of first heating the alloy to about 500° when a homogeneous solid solution is obtained. It is then rapidly cooled by quenching in water or oil. Thus the high temperature structure is retained, but this is not a stable state. The result is that in the course of several days intermetallic compounds are precipitated which give hardness and strength to the alloy. The latter process, known as *aging*, can be hastened by heating the alloy for a shorter period. This mechanism of hardening gives an engineering advantage. After quenching, the alloy is soft and can easily be given the desired shape which in course of time hardens. Sometimes, after quenching, the alloy is kept in a refrigerator so as to delay hardening and later made into the required shape which gets hardened in course of time.

Duralumin is one of the best known alloys of this group. It contains about 4% Cu, 0.5% Mn, 0.5% Mg and small amounts of Fe and Si.

After age hardening, its tensile strength (4,000 kg/sq. cm) approaches that of mild steel whereas its specific gravity is only one third. Duralumin has been superseded by 'Super duralumin', which has slightly higher percentages of the alloying elements. This has a still higher tensile strength (5.000 kg/sq. cm) which is equal to that of mild steel. Duralumin type alloys have very good mechanical properties, but their corrosion-resistance is lower than that of aluminium. This difficulty is overcome by *'cladding'*. This is done by rolling a thin sheet of aluminium on both the sides of the alloy sheet, thus forming a protective coating. Due to their strength and low specific gravity, they are widely used for making various parts of aircraft and automobiles.

γ-alloy is similar to duralumin but has also about 2% nickel. It is mostly used as a casting but can be hot worked. It is superior to duralumin in corrosion-resistance and in mechanical properties above 200°.

M/s. Rolls-Royce Ltd. have developed a series of alloys known as R. R. *alloys* which have better properties than duralumin. These have smaller percentages of copper but have some iron and titanium. A typical, average composition is 2% Cu, 0.1% Ti and about 1% each of Ni, Mg, Fe and Si, the rest being AI. Some R. R. alloys have higher strength to weight ratios than other alloys.

26.2 Copper

India's position in respect to copper production is not very satisfactory. Demand for copper in 1973-74 was about 83,000 tonnes and it rose to 1,05,000 tonnes by the end of the Fifth Plan. The Hindustan Copper Corporation is developing the main deposits of the country, which are at Rekha in Bihar, Khetri in Rajasthan and Malanjkhand in Madhya Pradesh, as well as minor deposits found at Dariba and Chandmari (Rajasthan) and Agnigundala (Andhra Pradesh). The copper smelter at Ghathila (Bihar) has a capacity of 25,000 tonnes per annum whereas, due to the shortage of the ore, the production in 1979-80 was 22,500 tonnes as compared to 8,400 tonnes in 1971-72. Khetri copper complex is being developed to produce 31,000 tonnes of copper using ores from Khetri and Kolihan. It will produce super phosphate fertiliser using sulphuric acid obtained as a by-product from the oxidation of the sulphide ore. Known deposits of copper are estimated at about 450 million tonnes with a total metal content of about 5.6 mt. It is planned to import concentrated ores and produce copper metal in the country in addition to importing the metal itself.

26.2.1 Extraction

Native copper is found in several parts of the world but about 90% of the metal is obtained from sulphide ores. The principal sulphide ores are chalcocite or copper glance, Cu_2S, and chalco pyrites, $CuFeS_2$. Of the oxidized copper ores malachite, $CuCO_3.Cu(OH)_2$, and cuprite, Cu_2O, are important. Most of the ores being worked at present contain less than 2% copper. Before smelting, the ores are concentrated by the flotation process and the copper content is raised to about 25%. Extraction of copper is complicated and consists of the following five steps, principles of which have been discussed in the chapter on Extraction of Metals:

1. Concentration of ore by froth flotation.
2. Roasting of ore (to eliminate part of the sulphur).
3. Smelting (to eliminate gangue and to produce matte).
4. Conversion of matte to blister copper.
5. Electrolytic refining of copper.

1. ***Concentration.*** The ore is crushed and then powdered in ball mills and mixed with creosote oil or pine oil and xanthates, which act as frother and collector, and is finely ground with water. It is then transferred to a flotation machine into which air is blown. The oil-coated sulphide particles collect at the top in the froth and are removed.

2. ***Roasting.*** The object of roasting is to oxidize a part of the sulphur so that the matte produced later contains about 40% copper. The ore is roasted in a verticle furnace in an oxidizing atmosphere. The ore is charged at the top and is gradually moved down by a rotating rake. Arsenic and antimony volatalize and part of the iron sulphide is oxidized to iron oxide. Sulphur dioxide is produced and used for making sulphuric acid.

3. ***Smelting.*** By smelting, slag and matte are produced. Slag contains the gangue. Matte is a molten mixture of sulphides of iron and copper. It is essentially a simple melting operation. Noble metals present in traces also go with the matte. Formerly a blast furnace was used for smelting but as the ores are now obtained in a powdered state, due to flotation process, they are heated on the sloping bed of a reverberatory furnace. Some silica is mixed with the ore, if not already present. Care is taken that copper sulphide is not oxidized at this stage. Iron oxide forms slag with silica and is removed. The molten, heavy, sulphide layer is called matte and contains about 40% copper. Oxidation to a small extent only takes place according to the following reaction:

$$2CuFeS_2 + 4O_2 = Cu_2S + 2FeO + 3SO_2$$

$$FeO + SiO_2 = FeSiO_3$$

4. ***Converting***, The molten matte, with calculated amount of silica, is transferred to a converter similar to the Bessemer converter for steel, the essential difference being that in the converter for copper air is blown from the side and not from the bottom. This is necessary because the copper produced is heavier than the matte and sinks below the tuyeres and escapes oxidation. The converter for copper is much larger in size and cold air is blown for several hours. This is because in the case of copper smelting a much larger amount of impurities are oxidized. Horizontal converters (shaped like the common barrels or drums) which are up to 10 m long and 4 m in diameter are being used to a greater extent than the Bessemer type converters.

Copper gets separated from iron because of the fact that FeS is oxidized more easily than Cu_2S is. Oxidation of iron sulphide is a highly exothermic reaction and supplies heat for keeping the charge in a molten state. On blowing air, iron sulphide is oxidized and forms slag with silica. Some copper sulphide is also oxidized which oxidizes the remaining copper sulphide.

$$2Cu_2S + 3O_2 = 2Cu_2O + 2SO_2$$

$$Cu_2S + 2Cu_2O\ 6Cu + SO_2$$

The molten copper sinks below the tuyere level and escapes oxidation. After the completion of the reactions the slag is poured out and molten copper is cast into blocks. The escaping sulphur dioxide from the molten copper gives a blister-like appearance to the solidified metal, hence it is called blister copper. It is about 99% pure.

5. ***Refining.*** There are two methods of refining copper, fire refining and electrolytic refining. In fire refining blister copper is melted on the bed of a reverberatory furnace in an oxidizing atmosphere when sulphur and arsenic oxidize and are removed. Baser metals, like iron oxidize and a layer of oxide forms on the molten metal and is removed and the surface of the metal is again exposed to oxygen. Some cuprous oxide is formed which, besides the loss of copper, makes copper weak and brittle. At this stage the molten metal is covered with charcoal and stirred with poles of green wood from which reducing gases, like the hydro-carbons, evolve and reduce cuprous oxide. The metal is then cast into blocks. This gives commercially pure copper, (99.5%) and is called 'tough pitch' copper.

Since even traces of impurities adversely affect the properties of copper the metal obtained by fire refining is subjected to further refining by electrolytic process. Anodes are made of fire refined copper and thin sheets of pure copper serves as cathodes. These are alternately suspended in a solution containing about 10% copper sulphate and 20% sulphuric acid. When electric current is passed, anodes dissolve and pure copper is deposited at the cathodes. The current is so adjusted that the metals baser than copper remain in solution and the nobler metals deposit under the anodes giving anode mud. This anode mud contains noble metals like gold and silver. The concentration of soluble metals like iron, zinc, etc., tend to increase in the electrolyte and reduce the efficiency of the process. Therefore, it is desirable to eliminate them as far as possible from blister copper by fire refining and then to electrolyse for final purification. The purity of electrolytic copper is 99.96 to 99.99%.

26.2.2 Properties and uses of Copper

Next to iron and aluminium, copper is the most widely used metal. Its wide applications are due to several important engineering properties. Its electrial and thermal conductivities are next to silver. The electrical conductivity falls even if traces of impurities are present. For example, 0.1% of phosphorus (or 0.5% of arsenic) reduce its electrical conductivity to about 50%. Copper has high thermal conductivity which also falls with alloying. Another reason for its wide use is that copper has resistance to atmospheric corrosion because of the formation of a uniform layer of oxide on the surface of the metal. It is resistant to most acids and alkalis except strong mineral acids and ammonia. Copper has good mechanical properties. The cast metal has a tensile strength of about 1,600 kg/cm^2 which increases to 2,400 kg/cm^2 on hot working and the strength becomes three times as great when the section is reduced to 90% by cold work. Copper also has good malleability and formability which makes fabrication easy. It can be easily welded and soldered. Melting point of copper is quite high, 1083 °C, as compared to those of other common non-ferrous, metals (Al, 660°; Zn, 419°; Pb, 328°; Sn, 232°). Copper and its alloys have pleasant colours. Several types of "coppers" are commercially produced. These are given names according to the refining procedure and the impurities present. Some of the important grades are the following.

Tough pitch copper, has oxygen as the chief impurity, its electrical conductivity is less but thermal conductivity is high. *High conductivity copper* is over 99.9% pure and is used in the electrical industry. *Deoxidized copper* is especially suited for welding and brazing because of its better response to hot reducing gases. *Oxygen-free-high-conductivity* copper is superior to deoxidized copper because deoxidized copper contains traces of phosphorus which very adversely affect electrical conductivity. It is the best grade of copper for welding, brazing and cold work. *Arsenical copper* has up to 0.5% arsenic which increases strength, toughness and melting point.

Alloys of copper are extensively used because of the wide range of engineering properties that can be produced in them by varying the composition and by subjecting them to various mechanical working and heat treating operations. The result is that the alloys are harder and have better mechanical and other special properties. Brasses (copper and zinc alloys) and bronzes (copper and tin alloys) are the most important alloys of copper. Zinc is cheaper than copper, therefore, brass is used as a substitute for copper wherever possible. Tin is much more expensive.

26.2.3 Alloys of Copper

Brasses

Brasses are alloys of copper and zinc. These have melting points in between those of copper and zinc. The ductility increases as the percentage of zinc goes up to 30% and strength increases up to 46% zinc. At room temperature with up to 35% zinc a solid solution of zinc in copper, called alpha brass, is produced. When the zinc content is 46.6 to 50.5% the solvent is CuZn and a different solid solution, called beta phase, is obtained. Between 30% and 46.6% zinc, a mixture of alpha and beta phases is obtained. Above 50.5% zinc, the gamma phase exists at room temperature. It is very brittle, hence is not useful for engineering purposes.

The brasses are divided into alpha and beta brasses. This division is based on the fact that as the alpha phase is highly ductile the alpha brasses can be cold worked (i.e., shapes changed in the cold condition) without any fracture. The beta phase is harder than the alpha phase, hence beta brasses cannot be cold worked to any appreciable extent. Its softening begins at 470° and at 800° it becomes softer than the alpha phase. Consequently, the beta brasses are more suitable for hot work.

Among the alpha brasses the most widely used is 70/80 brass (70% Cu, 30% Zn). This is also called *cartridge brass*. It has a very good combination of ductility and strength. It can be easily rolled, pressed and spun and is used for making cartridge cases, condensers and other shapes by drawing or stamping. In highly worked alpha brasses some internal stresses are produced due to which they suddenly crack in service. This is known as 'season cracking' and the defect can be removed by heating around 300° for about half an hour when the stresses are removed without loss of strength or hardness.

60/40 brass (60% Cu and 40% Zn) is known as *muntz metal* contains varying proportions of alpha and beta phases depending upon the rate of cooling from the liquid state. It is harder and stronger than cartridge brass and has to be hot worked.

The properties of the brasses, especially corrosion-resistance, can be improved by the addition of about 1% tin. Addition of 1% tin to 70/30 brass gives *admiralty brass*, which is resistant to sea water corrosion. Similarly 1% tin in 60/40 brass gives *naval brass* which is used for marine and engineering castings. Addition of up to 3% of lead to 60/40 brass improves its machining property and is known as *leaded brass*. It is used for hot stamping and for hot extrusion into rods. On these rods threads are cut on lathes to make screws, nuts, etc. When in a brass a part of the zinc is replaced by nickel the resulting alloy is known as '*nickel silver*' or '*German silver*'. It has very good mechanical properties and corrosion-resistance. Because of its white colour and moderate cost nickel silver has found wide use in making household utensils. This use is now being replaced by stainless steel.

Bronzes

Bronzes are essentially alloys of copper and tin. Tin being expensive, it is sometimes partly or wholly replaced by zinc or by other elements.

As the tin content increases up to 20% the tensile strength increases and then falls rapidly. Alloy of copper with 20% tin is known as *bell metal* and is used for casting bells and gongs. Due to the high cost of tin and also fall in ductility the common bronzes do not contain more than 12% of tin. When tin is less than 8% the alloys, called wrought bronzes, can be cold worked. Bronzes with higher percentages of tin must be hot worked. Bronzes are harder and stronger than the brasses.

Tin bronzes have very good casting properties, hence are used in cast forms just as brasses are mostly used in wrought forms. Due to a well-defined crystalline structure produced during solidification, the tin bronzes have anti-friction properties and are used for making bearings.

When some zinc is added to bronze, the resulting alloy is known as *gun metal*. A common composition is 88% Cu, 10% Sn and 2% Zn. Zinc being more electropositive than the other two metals it is preferentially oxidized, hence acts as a deoxidizer and gives better castings and also makes the alloy more corrosion-resistant. Gun metal is harder than bronze. Its ductility decreases but water resistance increases with increase in tin content. By annealing at about 700° the ductility can be increased and if it is quenched from about 700° its tensile strength is raised. Up to 2% of lead is sometimes incorporated so that the alloy becomes easier to machine.

The chief engineering use of gun metal is in making castings which are used for handling liquids, steam or air under pressure.

The origin of *phosphor bronze* came from the realization of the fact that oxides of copper and tin, which are formed in the preparation of a bronze are sources of weakness. The phosphor bronzes are of two types:

1. those containing less than 0.4% phosphorus together with up to 7% tin. In these phosphorus acts chiefly as a deoxidizing agent and thereby improves the tensile strength and resistance to corrosion by sea water. They are malleable and can be easily cold worked and are used for making spindles for valves and pumps, boiler fittings, sheets, etc. Low phosphorus bronzes are characterized by strength, ductility and resistance to corrosion;
2. those containing 0.4 to 1% phosphorus and 10 to 13% tin. In these bronzes the excess of phosphorus forms Cu_3P, which being hard and brittle, resists abrasion. It also has a low coefficient of friction. Because of abrasion-resistance and a low coefficient of friction these bronzes are used for making bearings and gears. It may be added that the ideal condition for a bearing metal is that there should be a hard constituent, which resists wear, uniformly embedded in a soft but tough matrix which provides strength and antifriction properties.

Aluminium bronzes contain aluminium and copper. The addition of aluminium rnakes these alloys more resistant to corrosion. Alloys with more than 10 per cent aluminium can be hardened by heat treatment. Aluminium bronzes have a good combination of strength, hardness, toughness and corrosion-resistance at moderately high temperatures and are used for vaious fittings for ships, valves, pumps, etc.

Copper mixes in all proportion with nickel, to give another series of useful alloys which, though expensive, have high electrical and corrosion resistance and maintain good mechanical properties at elevated temperatures. The most widely used of these alloys is *Monel metal* which has about 1 part of copper alloyed with 2 parts of nickel along with small quantities of iron, manganese. It is not prepared by the usual method of preparing alloys (i.e., by melting the pure metals) but is prepared by directly reducing ores of copper and nickel mixed together. In addition to high resistance to many corrosive solutions, acids, mine waters, etc., it has very good mechanical properties like malleability and toughness and maintains strength at higher temperatures than steel.

Monel metals find wide applications in chemical and food industries and in fittings for ships. As it maintains good mechanical properties at elevated temperatures, it is used for making fire box stays, turbine blades, gaskets and other parts of steam plants.

26.2.4 Lead

Extraction

Lead metal has been in use since the early days because it can be easily produced from the ore and it has a low melting point. It is almost exclusively extracted from galena, PbS. It also occurs as carbonate and sulphate.

Lead and zinc ores are mined from the Zawar Mines, Udaipur Dist., Rajasthan. The total reserves of lead-zinc ores are estimated at 350 mt. About 5,000 tonnes of lead ore concentrates and about 3,000 tonnes of lead are produced annually in India.

Galena is almost always associated with zinc sulphide, therefore, the powdered ore is concentrated by differential flotation, as explained in the chapter on Extraction of Metals. The concentrated ore is then roasted in air when the sulphide is oxidized to oxide and partly to sulphate. Arsenic and antimony oxide volatalize but iron and copper oxides remain behind. The oxides mixed with coke are smelted in a blast furnace when crude lead is obtained. The crude metal is heated in a reverberatory furnace in an oxidizing atmosphere when some of the impurities oxidize and the oxide layer floating above the molten metal is skimmed off. This process is known as softening of lead. Lead so obtained contains traces of precious metals like silver and gold which are removed by Parkes's process. It depends on the fact that the noble metals are more soluble in molten zinc than in molten lead. The metal is melted and zinc is added and mixed. Zinc carrying noble metals floats at the top and solidifies. The solid metal is removed, and zinc is distilled leaving behind the noble metals. If necessary, lead which forms the lower layer is further purified by electrolytic process

Properties and uses

Lead metal is used because of its corrosion resistance, malleability, plasticity and high density. It has very poor ductility due to which it cannot be drawn into thin wires. Due to high malleability, it can be rolled into sheets and foils. Lead has good casting properties. It has low tensile strength (100 kg/cm^2).

Due to its low melting-point (328°), it cannot be used at elevated temperatures. On exposure to air a thin layer of oxide is formed on the surface which protects the metal underneath.

Nearly one third of the lead produced is used in storage batteries. Other principal uses are cable-sheathing, alloys for bearings solder and type metals, chemicals and pigments, tetra ethyl lead and building materials.

26.2.5 Alleys of Lead

As pure lead does not have good mechanical properties, it is mostly used after adding some alloying elements. These add hardness and strength to the metal and also lower the melting point. Most of the lead alloys contain antimony. The alloys are of two types, the high-lead alloys and low-lead alloys.

26.2.6 Grid Metal

The largest single use of lead is in making grid metal which is used in storage batteries. It contains about 10% antimony and less than 1% of tin and traces of arsenic and copper. These elements improve hardness and casting properties. The alloy is cast into perforated plates or grids for storage batteries.

Quantitatively next in importance comes *cable-sheathing metal.* It has a smaller amount of antimony than the grid metal. It has 95 to 98% lead and 5 to 2% antimony. Much smaller amount of other metals are also incorporated in the alloy.

26.2.7 Solders

For joining two metals an alloy whose melting point is lower than the melting points of the metals to be joined is used. The process is known as *soldering or brazing*. The distinction between these two is not sharp. When the alloy used has lower m. p. the process is known as soldering and with a high m.p. alloy the same process is called brazing. In welding the metal being joined is melted.

The most common solders are tinman's solder (2 parts Sn and 1 part Pb) and *Plumber's solder* (1 part Sn and 2 parts Pb). Tinman's solder contains almost an eutectic mixture and melts at 180° and is very convenient for ordinary joining of metals. The plumber's solder begins to solidify at 240° and passing through a pasty stage solidifies at 180°. This range of solidification enables the plumber to make his 'wiped' joint.

In brazing alloys of copper and zinc are used. These alloys have higher melting points and the joint has better mechanical properties.

26.2.8 Type Metal

Type Metal should certainly be low melting but more important requirements are strength and hardness, ability to cast well and make good impressions. These are attained by adding antimony and tin to lead (12%, 3% and 85%). Antimony not only contributes to strength and hardness but, as it expands on cooling, it fills the mould and an accurate shape is obtained. Here, mention may be made of lead shots which contain 1% arsenic. Lead shots for long range use need to be hard, and therefore contain 2 to 6 per cent of antimony.

26.2.9 Bearing Metals

Bearing Metals have still different requirements. These alloys have two constituents one of which is hard and resists wear, but is not so hard as to cause wear of the journal. Its particles are embedded in a soft matrix. The hard particles support the load but the soft matrix provides plasticity and ductility and thus enables the harder particles to adjust themselves to the differing surface requirements. This latter property becomes necessary as it is very difficult in actual practice to make perfect alignment of a journal and a bearing. The most important group of alloys for the purpose are known as *Babbit metals*. These contain tin, antimony, copper and also lead. The compositions of these alloys differ widely. They include alloys with 90% or more tin in without lead, to 5% tin with 80% lead. A composition commonly used contains 91% Sn, 4.5% Sb and 4.5%Cu.

Tin being expensive, cheaper substitute of Babbit metals have been made, e.g., using about 65% Cu, 30% Pb and 5% Sn. These have a structure opposite to that of a rabbit metal, i.e., there is a soft constituent in hard matrix. These alloys, known as *leaded bronzes*, are inferior as they are not so plastic and do not adjust so well to the load.

26.3 Magnesium

26.3.1 Extraction

Magnesium is obtained either by electrolysis of fused magnesium chloride or by reduction of the oxide by carbon at high temperatures.

Magnesia, MgO, is mixed with carbon and heated in a current of chlorine when magnesium chloride is produced. Large quantities of it are being obtained from sea water. Sea water is treated with slaked lime when magnesium hydroxide is precipitated which is dissolved in hydrochloric acid and the solution is concentrated. Magnesium chloride, mixed with sodium chloride and potassium chloride, is electrolysed at 70° in a cell using a graphite anode and steel cathode. Chlorine is liberated at the anode and is used to produce hydrochloric acid. Magnesium collects at the cathode under the surface of a thin layer of electrolyte and is periodically removed.

Another method of production is by the reduction of magnesium oxide by carbon at $2{,}000^{\circ}$. Magnesium volatalizes and is immediately cooled so that it may not oxidize. The metal is purified by distillation in vacuum.

26.3.2 Alloys of Magnesium

Of all the metallic materials used by the engineers, magnesium (sp. gr. 1.74) is the lightest and its alloys, which have good engineering properties, also have low specific gravity (1.8). Although on a weight basis, it has good mechanical properties, it is not used in the pure condition for engineering purposes, because on alloying, its properties are greatly improved without losing thc advantage of lightness. Importance of low specific gravity in engineering uses is so obvious that it need not be emphasised.

After heat treatment and mechanical working the strength of magnesium alloys may be as high as 3000 kg/cm^2. Thus their strength-weight ratio is far superior to that of steel and is better than even duralumin. A great engineering drawback of magnesium alloys is low modulus of elasticity. They also do not have good high temperature properties. Their corrosion-resistance to acids and to salt water is poor, hence they do not give satisfactory service near a sea coast unless protective coatings are used. Just like duralumin they can be cast or wrought and also heat treated. Machining properties are also good.

The chief use of magnesium alloys lies where a saving in weight is required. Hence, they are widely used in aircraft construction for crank cases, undercarriage parts, wheels, housings, engine parts, interior fittings, etc.

Dow metal (61% Al, 3% Zn and 0.5% Mn with Mg) is a widely used alloy of magnesium. After heating to 350° it is quenched in air and then age-hardened.

26.4 Nickel

26.4.1 Extraction, Properties and Uses

Nickel is commonly extracted from sulphide ores which also contain sulphides of iron and copper. The roasted ore is smelted when matte containing sulphides of nickel, iron and copper is produced. On Bessemerization iron is removed. The molten matte is also used for giving a mixture of metallic copper and nickel which are used for making some alloys. The matte containing sulphides of copper and nickel is heated with sodium sulphate and coke when sodium sulphide is produced which dissolves copper sulphide and floats on the nickel sulphide layer. Nickel sulphide is roasted to give the oxide which is reduced by coke, and nickel is obtained.

Pure nickel is obtained by heating the oxide in producer gas at 50° to 80° when nickel carbonyl, $Ni(CO)_4$, volatalizes. It is then heated to 180° when it decomposes to give pure nickel and carbon monoxide.

Tensile strength of nickel is 3,000 km/cm^2 and on cold working is increased to 10,000 kg/cm^2. Thus it is stronger than other common metals. It is hard, ductile and malleable. It is ferro-magnetic. Many of the uses of nickel are due to its high corrosion-resistance, therefore, it is used for nickel plating.

26.4.2 Alloys of Nickel

Because of high resistance to corrosion and good mechanical properties, nickel finds important applications in making ferrous and non-ferrous alloys. These alloys have been described under 'Ferrous Alloys' and 'Alloys of Copper'.

26.5 Zinc

26.5.1 Extraction

India has to depend on the import of zinc to meet her requirements. A plant has been built at Alwaye, Kerala State, for smelting 20,000 tonnes of zinc from imported concentrated ore. The plant, at Vishakhapatanam produced 25,000 tonnes of zinc ingots in 1981 against an installed capacity of 30,000 tonnes. It uses concentrates from Peru, Mexico and Australia. Zinc ore is available in India only from the Zawar mines, Udaipur Dist. Zinc smelter at Debari, Rajasthan, has an annual capacity of 18,000 tonnes of zinc and 29,000 tonnes of sulphuric acid (which is a by-product).

Most of the zinc is obtained from the sulphide ore, zinc blende, ZnS. It also occurs as carbonate (calamine) and as silicate. The ore is first converted into oxide. This is done by roasting the sulphide ore or by calcining the carbonate. The oxide is mixed with an equal quantity of powdered coal and heated in fire clay retorts. Zinc oxide is reduced to zinc and the vapours are condensed above the melting point. The molten zinc obtained is cast into moulds and is known as 'spelter' and is sufficiently pure for commercial use.

26.5.2 Properties and uses

Zinc is a weak and brittle metal. Tensile strength in the cast condition is about 500 kg/cm^2 which becomes 3 to 4 times on rolling. Due to its brittleness, it is rolled at, 100 to 150°. Melting point of zinc is quite low, 419°.

Zinc is readily attacked by acids and alkalis. Being more electropositive than iron, zinc coatings protect iron and steel structures from corrosion, therefore almost half of the zinc produced is used for galvanizing. Its other important uses are in making die-casting alloys and brasses.

26.5.3 Alloys of Zinc

Large quantities of zinc are used in making die-casting alloys. In die-casting the molten alloy is taken in a closed cylinder and forced by a piston through a die. These alloys contain 3 to 40% aluminium and smaller amounts of copper. Less than 1% of other metals are also present. Zinc die-casting alloys find wide applications because of their low melting points, dimensional accuracy, (due to low shrinkage on cooling), light weight and better strength than metallic zinc.

Questions

1. Write the importance of
   (a) Underground mining (b) Hydrometallurgy
   (c) Electrometallurgy (d) Pyrometallurgy
2. Write about the driving force for locating steel industries in Jharkhand state instead of a port city like Mumbai.
3. Explain the term alloy. State purpose of alloying with suitable examples.
4. What is the effect of sulphur, silicon, carbon and manganese on properties of steel.

5. Explain the role of Brasses and Bronzes as engineering materials.
6. Differentiate the Pig iron and Cast iron. Discuss their properties.
7. Differentiate calcinations, Reacting and Smelting with suitable examples
8. Discuss the principles and extraction of iron from any ore. Give diagram of the blast furnace showing the chemical reactions at different stages.
9. Give an account of Bessemer process for the manufacture of steel.
10. Discuss the Basic and Acid open-hearth processes. Give their advantages and limitations.
11. Write an account of Basic-oxygen furnace Process.
12. Classify carbon steels. Discuss the effect of carbon content and applications of different steels.
13. Write principles of alloying and discuss effects of alloying elements.
14. Write a note on
    (a) nickel steels
    (b) chromium steels
    (c) nickel-chromium steels
15. What are special steels ? Write about Boron steels, Heat resisting steels and wear resisting steels.
16. Define Heat Treatment. Discuss with suitable examples – Annealing, Normalizing and Spheroidizing
17. Write a note on _________
    (a) hardening (b) case carburizing
    (c) nitriding (d) cyaniding

26. "Non-ferrous metals and alloys, being more expensive, find a place as engineering materials" – Discuss

19. Write about metallurgy and properties of aluminium Discuss about casting Aluminium alloys and wrought Aluminium alloys.
20. Write about metallurgy of copper and give an account of brasses.
21. "Lead is considered as environment pollutant but find extensive use in day to day life" – Discuss
22. Write a note on alloys of magnesium.
23. Write about extraction, properties and uses of Nickel and Zinc.

Fill in the Blanks

1. 75% of the elements in the periodic table are _________
2. 75% of the earth crust consists of _________
   (A) oxygen and nitrogen (B) oxygen and iron
   (C) oxygen and silicon (D) nitrogen and iron
3. Metals found in the free state _________
4. The iron obtained from blast furnace in known as _________
5. The white color of white cast iron is due to _________
   (A) decomposition of Fe_3 C, (B) retention of Fe_3 C
   (C) mix of aluminium with iron (D) all the three factors
6. Nodular cast iron is also known as _________ iron
7. Free carbon in iron reduces the _________
   (A) hardness and ductility
   (B) strength and ductility
   (C) hardncss and melality
   (D) hardness and strength
8. Iron and steel differ mainly in their _________ content
9. In ancient India steel was made by _________ process
10. Presence of phosphorus in steel _________
    (A) brittle at high temperatures and shock resistant
    (B) brittle at low temperatures and shock resistant
    (C) brittle at low temperatures and shock prone
    (D) brittle at high temperatures and shock prone
11. Presence of sulphur in steel makes it _________
12. Smaller concentration (< 1%) of manganese in free state _________
    (A) increases strength (B) decreases strength
    (C) increases ductility (D) decreases ductility
13. Concentrations of M_n > 1% decreases _________.
14. If red hot steel tool is plunged into water, it becomes _________.

15. Wood metal is an alloy of ________.
    (A) Pb, Fe, Sn, and Bi
    (B) Fe, Pb, Cn and Sn
    (C) Fe, Mn, Sn and Cn
    (D) Pb, Sn, Bi and Cd
16. In the electrolysis of alumina, Al is liberated at ________.
17. Brasses are alloys of ________.
    (A) Cu and Zn (B) Cu and Sn
    (C) Cu, Sn and Zn (D) Cu, Cd and Sn
26. Gun metal is the alloy of ________.
    (A) Cu and Zn (B) Cu and Sn
    (C) Cu, Sn and Zn (D) Cu, Cd and Sn
19. Gelena is the ore of ________.
20. ________ is the widely used alloy of magnesium.

Chapter 27

Advanced Materials

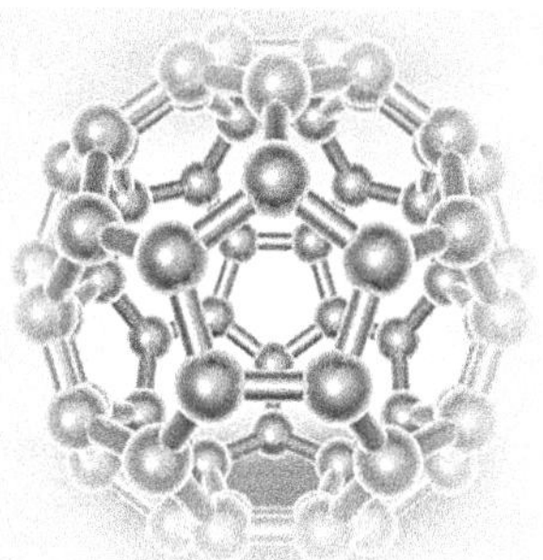

27.1 Advanced Materials

In chemical terms the world of materials is restricted to 100 odd elements shown in the periodic table. In terms of functionally there are numerous materials in this physical universe. An account of the chemistry of these materials with which we are well acquainted is already covered in the earlier chapters and in this chapter we are going to discuss some special materials that are carving their niche and are of interest to scientists and engineers. The unique functionalities are attributable to their chemical bonding, compositional changes and arrangements which were not recognized 20-30 years back and basically calcium phosphates but differ completely in their chemistry is still obscure to engineers. Thus calcium phosphate mineral tooth and bone are all their physical properties. Similarly polyethylene and candle wax differ only in the number of ($-CH_2-$) links in the molecular chains but are quite different mechanically. The same polymeric materials once considered to be perfect insulators and biologically immune are now transformed into good conductors and biologically degradable with some minor changes. Lack of understanding of materials often lead to drastic ends and there are historical accounts of loosing battles on account of wrong selection of materials. Soldiers of Nepolean's army marched into Russia wearing tin buttons and lost their war on account of this silly mistake. Metallic β-tin recrystallises to grey α-powdery tin and as a result the buttons disintegrated and the army froze to death in the Russian cold. So an understanding of materials and their chemistry is a must for the future engineers for whom this books is targeted. In this chapter we are discussing.

- Liquid crystals
- Conducting polymers
- Biodegradable polymers
- Current status of battery materials
- Hydrogen-oxygen fuel cells,
- Role of nano materials,
- A birds eye view of green chemicals.

27.2 Liquid Crystals

When the temperature of a gas is lowered, the kinetic energy of the gas molecules reduces considerably and the intermolecular attractive forces begin to play a vital role resulting in the liquefaction of gases. Any slight increase in temperature of the system pushes the molecules into gaseous state. On further cooling, the liquids acquire solid state, known for its orderliness. Therefore, a liquid can be thought to occupy an intermediate position between gas and solid states.

$$\text{Solid} \xrightarrow{+\text{ heat}} \text{Liquid} \xrightarrow{+\text{ heat}} \text{Gas}$$

The liquid and solid states, known as condensed state have comparable densities and are incompressible and differ only in the orderly arrangement, with liquids being more disorderly. The liquid and gaseous forms, in turn, differ largely in their densities and differ mainly in the magnitude of total intermolecular forces. All these characteristics of liquids are listed below.

(i) *Liquids have extremely small compressibilities and expandabilities.* This property is equally shared even by solids. The relative incompressibility of liquids is due to compactness of gas molecules.

(ii) *Liquids have definite volume but have no definite shape and always take the shape of its container. Solids are characterized by their definite volume and shape.*

This definite volume indicates that the intermolecular forces in liquids are strong enough to maintain volumes but not enough to maintain the crystalline shapes of solid.

(iii) *Liquids flow. Gases share this property with liquids, but the rate of flow of gases is higher than the liquid flow rates.*

(iv) *Liquids have much higher densities in relation to their gaseous states.* Generally liquids have slightly lower densities than solids but exception like water fulfilling reverse condition are also known.

(v) *Liquids show equal pressures in all directions. This property is common with gases.* Equal pressures indicate equal number of collisions of the molecules on the walls of the container.

(vi) *Liquids mix with other liquids forming both homogenous and heterogeneous systems. Gas mixtures are always homogeneous.*

Since school, we destinguish only three states of matter: solids, liquids, and gases. Solids, namely crystals, are highly ordered, the constituting atoms or molecules possessing positional as well as orientational order. On contrary, simple liquids like water or gases exhibit no order at all. The arising question was: How can biological systems, polymers, etc., be classified. Obviously they are highly ordered, but no crystalline solids. This was the beginning for O. Lehmann, F. Reinitzer, L. Gattermann, D. Vorländer, and others to search for new states of matter which are soft and ordered at the same time. The study of liquid crystals began in 1888 when an Austrian botanist named Friedrich Reinitzer observed that a material known as cholesteryl benzoate had two distinct melting points. In his experiments, Reinitzer increased the temperature of a solid sample and watched the crystal change into a hazy liquid. As he increased the temperature further, the material changed again into a clear, transparent liquid. Further Reinitzer discovered that cholesteryl acetate exhibits optical properties known only from crystals so far. Because of this early work, Reinitzer is often credited with discovering a new phase of matter – the liquid crystal phase.

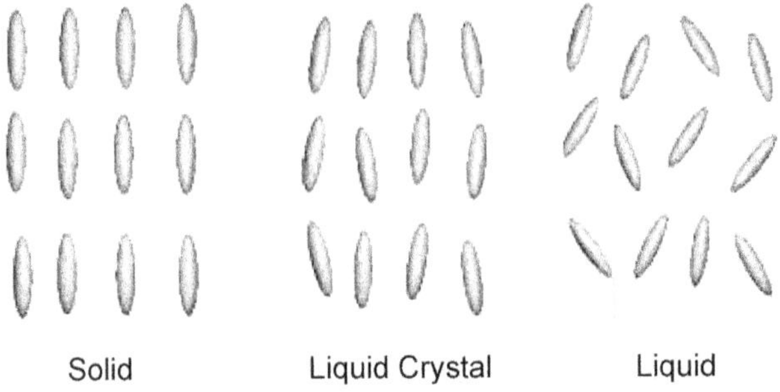

Fig. 27.1 Solid, Liquid and Liquid Crystal molecular arrangements.

Condensed matter which exhibit intermediate thermodynamic phases between the crystalline solid and simple liquid state are now called liquid crystals or mesophases ('middle' phases) and their molecular components are called mesogens (generators of mesophases) (Fig. 27.1).

27.2.1 Types of Mesophases

Liquid crystal materials are unique in their properties and uses. Liquid crystal materials generally have several common characteristics. Among these are a rod-like/disk like molecular structure, rigidness of the long axis, and strong dipoles and/or easily polarizable substituents (Fig. 27.2).

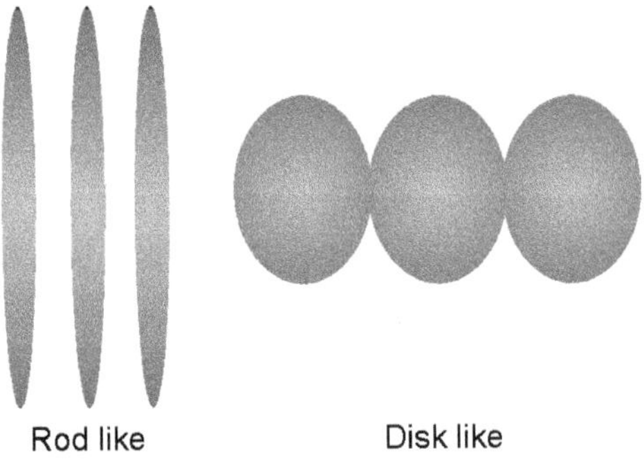

Fig. 27.2 Molecular shapes of Liquid Crystals.

They align themselves with a long range order due to anisotropic intermolecular forces. This fourth state of matter generally possess orientational or weak positional order and thus reveals several physical properties of crystals but flow like liquids and exhibit liquid properties like capillary rise, surface tension, etc.

If transitions between the phases are given ***by temperature***, they are called **thermotropic**. The temperature at which solid turns to mesostate is called **transition temperature** and the temperature at which mesomorphic state (mesophase) disappears to give liquid state is called the **melting point**. Thus for p-azoxy anisole the transition point is 118.3 ^{o}C and the melting point is 135.9 ^{0}C

In blends of different components phase transitions may also depend ***on concentration***, and these liquid crystals are called **lyotropic**. Lyotropic mesophases occur in concentrated solutions of rod-like molecules in an isotropic solvent, usually water, eg., sodium laurate in water, phosphotidyl choline in water. While thermotropics are presently mostly used for special technical applications, lyotropics are important for biological systems, e.g. membranes. In industry also we find their use as soaps, gels and colloids.

There are two types of thermotropic liquid crystals: discotics and rod-shaped molecules, Fig. 27.3.

Discotics are flat disc-like molecules consisting of a core of adjacent aromatic rings. Their molecules arrangement is akin to a pile of coins. Discotic liquid crystals can form columnar phases in which the coins are stacked into loosely defined columns. Disc-shaped molecules also form an orientationally ordered nematic phase which we are going to discuss shortly.

The best demonstration could be made through stacked and orientational arrangement of Rs. 5 coins.

Rod-shaped molecules have an elongated, anisotropic geometry which allows for preferential alignment along one spatial direction. They form both nematic and smectic phases. The best demonstration could be made through match sticks arranged randomly and in columns.

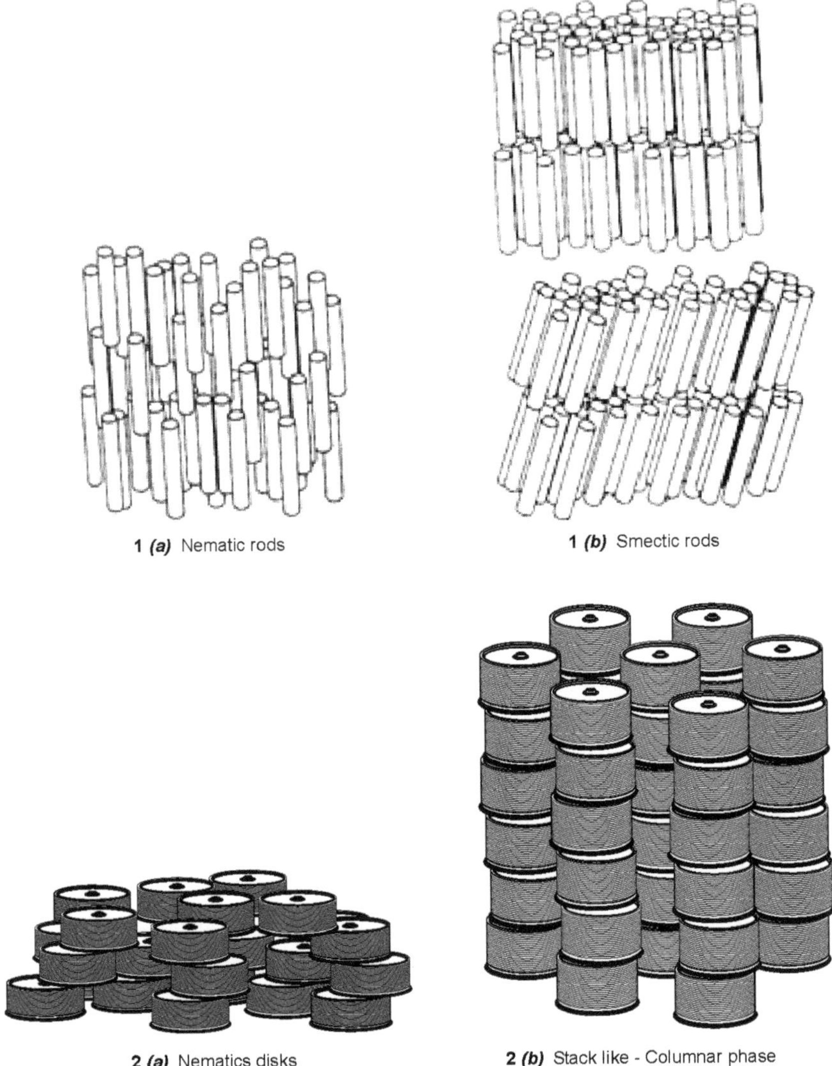

Fig. 27.3 Liquid-crystal mesophases.

1. ***Rod-like molecules :*** **1(a)** Form nematic phase – have orientational order but no postional order **1(b)** Form smectic phase – have orientational and positional order in one dimension (that is, ordery in lines). **2.** Disc-shaped molecules **2(a)** Nematicdisks **2(b)** Stack-like columnar phase.

The distinguishing characteristic of the liquid crystalline state is the tendency of the molecules (*mesogens*) to point along a common axis, called the director.

This is in contrast to molecules in the *liquid phase, which are isotropic and have no intrinsic order.*

In the solid state, molecules have high long range order and have little *translational freedom*.

The characteristic *orientational* order of the **liquid-crystal state** is between the traditional solid and liquid phases (*middle*) state and this is the origin of the term **mesogenic.**

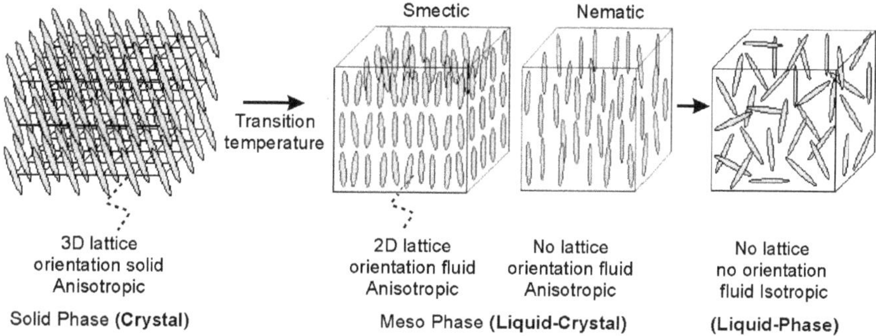

Fig. 27.4 Liquid crystalline mesophases between the solid and isotropic liquid phase.

To quantify just how much order is present in a material, an order parameter (S) is defined (Fig. 27.5).

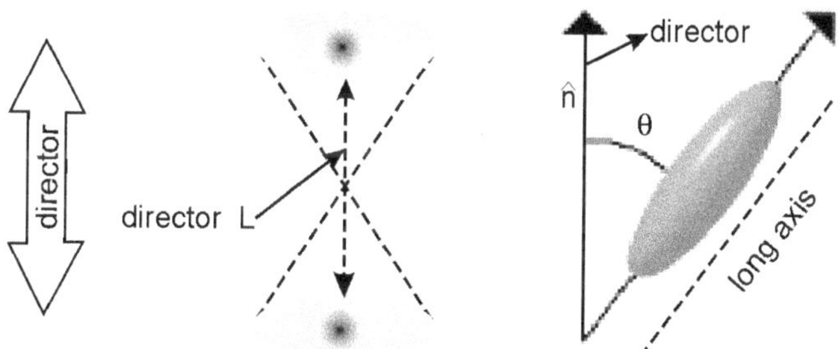

Fig. 27.5 Angle between director and long axis of a molecule.

Traditionally, the order parameter is given as follows :

Order parameter $S = \Sigma \cos \theta$

where theta (θ) is the angle between **the director** and the *long axis* of each molecule.

- In an isotropic liquid, the average of the cosine terms is zero, and therefore the order parameter is equal to zero.

- For a perfect crystal, the order parameter evaluates to one.
- Typical values for the order parameter of a liquid crystal range between 0.3 and 0.9, with the exact value depending on temperature as a result of kinetic molecular motion.

27.2.2 Characterization of Liquid Crystal Orders

The following order parameters characterize the liquid crystals :

1. Positional Order, 2. Orientational Order, 3. Bond Orientational Order

Each of these parameters describes the extent to which the liquid crystal sample is ordered.

1. **Positional order** refers to the extent to which an average molecule or group of molecules shows *translational symmetry* (as crystalline material show).
2. **Orientational order** represents a measure of the tendency of the *molecules to align along the director* on a long-range basis.
3. **Bond Orientational Order** describes *a line joining the centers of molecules* without requiring a regular spacing along that line.

27.2.3 Physical Properties of Liquid Crystals

1. **Melting Point :** Liquid crystal posess a definite melting point at which they become true liquid.
2. **Polymorphism :** Most liquid crystal compounds exhibit polymorphism, or a condition where *more than one phase is observed* in the liquid crystalline state.
3. **Anisotropy :** The tendency of the liquid crystal molecules to point along the director leads to a condition known as **anisotropy**. This term means that the properties of a material depend on the direction in which they are measured. For example, it is easier to cut a piece of wood along the grain or it is easy to tear mica sheet along a direction, than against it. The *anisotropic nature of liquid crystals is responsible for the unique optical properties* exploited by scientists and engineers in a variety of applications.
4. **The effect of magnetic field** on crystals is another interesting property. Substances in mesomorphic state are turbid, but when a magnetic field is applied they become clear, in the direction of lines

of force. This is once again attributable to orientation order affected by magnetic field (i.e., *all the molecules are bundled in the same direction along the lines of force*) and the liquid clears in this direction.

5. **The viscosity** of liquid crystals is *small.* Where as that of liquids with similar molecular size is large.

27.2.4 Classification of Liquid Crystals

Meso phases are mainly of two types; 1. Nematic Phase, 2. Smectic phase. Each class could be further split into sub-classes.

27.2.5 Nematic Phases

In the simplest case *the molecules possess only orientational but no positional long range order.* Liquid crystals of that type are called nematic (from ***greek*** *nemos = thread*), the name of which has been given with respect to *thread-like textures* observed under polarizing microscope. The nematic liquid crystal phase is characterized by molecules that have no positional order but tend to point in the same direction (along the director). In the following diagram, notice that the molecules point vertically but are arranged

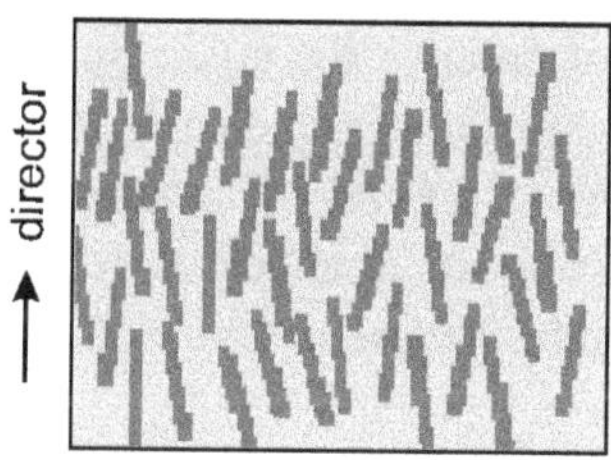

1. Orientational order
2. No positional order
3. No lattice
4. Uniaxial

with no particular order. Nematic phases are composed of rod-shaped molecular aggregates that are arranged with parallel but not lateral order. For example, imagine a large number of toothpicks put into a rectangular box and shaken. When you open the box, the toothpicks will be facing in about the same direction, but will have no definite spatial organization. They are free to move, but like to line up in about the same direction. This is a primitive model for nematic liquid crystals.

In nematics the distribution function is rotationally symmetric around the director, i.e. they are uniaxial.

27.2.6 The Smectic Phases

The word "smectic" is derived from the Greek word for soap. This seemingly ambiguous origin is explained by the fact that the thick, slippery substance often found at the bottom of a soap dish is actually a type of smectic liquid crystal. Molecules in this phase show a degree of translational order not present in the nematic.

In the smectic state, the molecules maintain the general orientational order of nematics, but also tend to align themselves in layers or planes. Motion is restricted to within these planes, and separate planes are observed to flow past each other. The increased order means that the smectic state is more "solid-like" than the nematic.

Many compounds are observed to form more than one type of smectic phase. As many as 12 of these variations have been identified, however only the most distinct phases are discussed here.

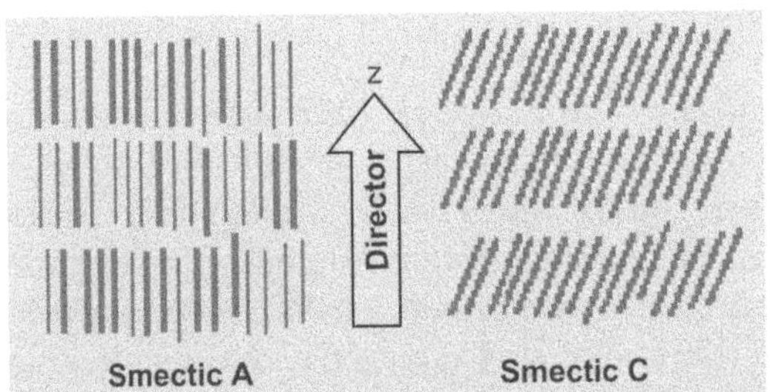

1. Orientational order
2. Positional (translational) order
3. 2D-symmetry
4. Uniaxial
5. More than one smectic phases possible

In the smectic-A mesophase the director is perpendicular to the smectic plane, and there is no particular positional order in the layer. Similarly, in the smectic-B, mesophase orients with the director perpendicular to the smectic plane, but the molecules are arranged into a network of hexagons within the layer. In the smectic-C mesophase, molecules are arranged as in the smectic-A mesophase, but the director is at a constant tilt angle measured normally to the smectic plane.

27.2.7 Chemical Nature of Liquid Crystals

The factors that control the physical structures and their aggregations are :

1. **Linear elongated molecules with molecular interactions lead to rod shaped structures :** Molecular interactions brought about by dipole-dipole and dipole-induced dipole interactions, hydrogen bonding and dispersion forces, are insufficient to bring about total 3D lattice symmetry of crystals but enough to bind them together leading to mesophases. In Fig. 27.6 the chemical structure for two of the most prominent representatives of rod-like molecules which exhibit liquid crystalline order are shown. 4-Methoxy bezyliden 4'-butylaniline (MBBA) was the first room temperature liquid crystal synthesized in 1969, whereas 4-pentyl 4'cyanobiphenyl (5CB) was the first member of the optically

and chemically stable cyanobiphenyls, one of the most applicable family of liquid crystals, discovered in 1973. This is due to the interconnection of two rigid cyclic units. *The interconnecting group should cause the resulting compound to have a linear planar conformation.* Linking units containing multiple bonds such as –(CH=N)–, –N=N–, –(CH=CH)$_n$–, –CH=N–N=CH–, etc., are used since they restrict the freedom of rotation.

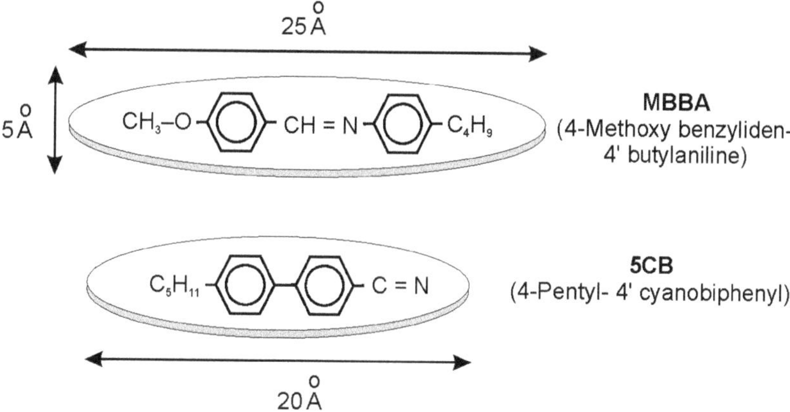

Fig. 27.6 Typical building blocks of calamitic (Rod like) liquid crystals.

2. **Presence of unsaturated bonds** : The unsaturated bonds enhance the polarizability, dipole character and overall linearity, simultaneously preventing the laminar bezene rings from twisting. For example, trans-p-n-alkoxy cinnamic acids are mesomorphic but its cis analogue is not.

trans-p-n-alkoxy cinnamic acid   cis-p-n-alkoxy cinnamic acid

Similarly if the unsaturation is removed, the mesocharacter is destroyed due to twisting of chains leading to isotropy.

3. **Absence of bulky groups :** Bulky atoms like Br and bulky groups present in the chain other than as end groups do not allow aggregation

$C_{10}H_{21}O$–(Br-substituted benzene ring)–COOH

of mesogens. For example, inclusion of bromine atom on the benzene ring of 4-decyclo-3-benzoic acid prevent mesomorphic structure.

4. **H-Bonding :** Hydrogen bonding often destroys the mesomorphic character (i) by lengthening of molecular chains (ii) very strong bonds bringing crystallinity and (iii) by allowing crosslinking of chains and preventing formation of linear chains.

In case of carboxylic acids the mesomorphic character is destroyed due to H-bonding and lengthening of chains.

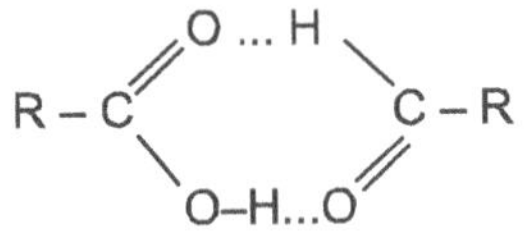

R – C(=O ... H – O)(–O–H ... O=)C – R

Fatty acids are not mesomorphic but their salts called soaps are strongly mesomorphic in nature.

Similarly 4-amino-4″-nitro-p-terphenyl (a), is not mesomorphic and crystalline with a melting point of 300 ^{0}C but becomes mesomorphic on substitution of a nitro group in ortho position to amino group (b).

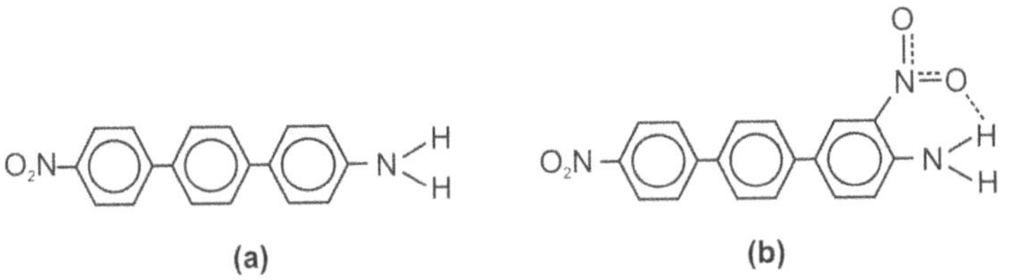

O_2N–(terphenyl)–NH_2

(a)

O_2N–(terphenyl, with NO_2 ortho to amino)–NH_2

(b)

In some cases mesomorphism is brought about due to H-bonding. Thus cholesterol is not mesomorphic but it's ester shows liquid crystal property. The columnar structure is facilitated by the ester group.

27.2.8 Applications of Liquid Crystals

Liquid crystal technology has had a major effect on many areas of science and engineering, as well as device technology. Applications for this special kind of material are still being discovered and continue to provide effective solutions to many different problems.

1. Liquid Crystal Display Devices : The most common application of liquid crystal technology is in liquid crystal displays (LCDs). From the ubiquitous wrist watch and pocket calculator to an advanced VGA computer screen, this type of display has evolved into an important and versatile interface.

A liquid crystal display consists of an array of tiny segments (called pixels) that can be manipulated to present information. This basic idea is common to all displays, ranging from simple calculators to a full color LCD television.

A quartz watch stops it's display by cooling it due to formation of crystallites and turns blank on slightly increasing the temperature due to turning into liquid. This provides an easy way to demonstrate existence of liquid crystals in a quartz watch.

A schematic diagram of functioning of liquid-crystal display is given in Fig. 27.7.

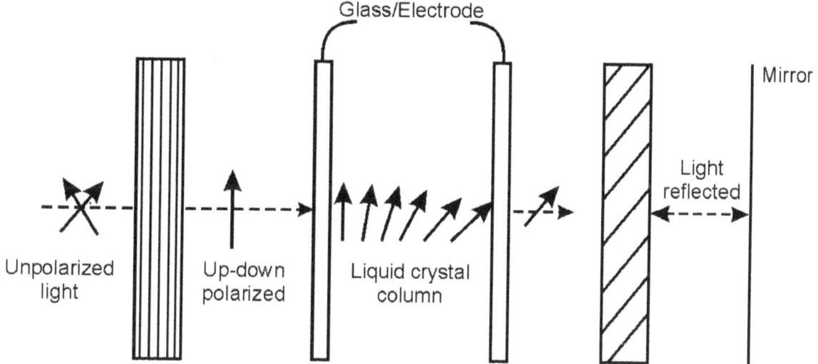

In the off voltage the display looks bright

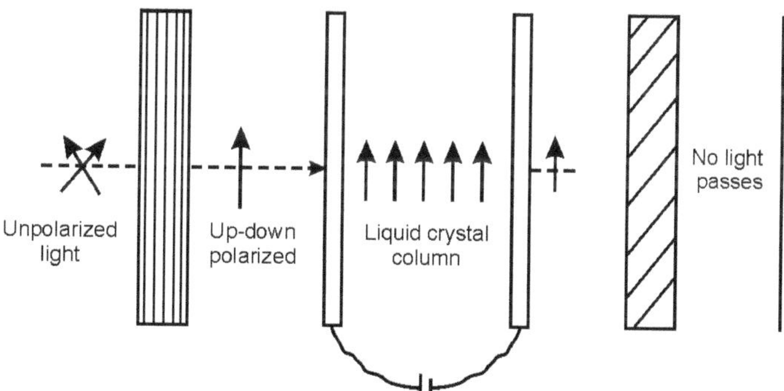

In the on voltage the display looks dark

Fig. 27.7 Schematic Functioning of LCD.

Light travels through two polarisers twisted 90^0 and separated by a column of liquid crystal encased between two conducting glass sheets and falls on a mirror. In the journey of light, it is polarized by first polariser, say updown, progressively changes its direction in the liquid crystal solution and will be able to enter the rear polarizer, twisted by 90^0 or at an angle, and fall on the mirror to retrace it's path. When a small voltage is applied across the conducting glass sheets, the liquid crystal molecules line up and donot rotate the light. As a result it does not make through, back polarizer and the mirror gives *a dark* display. Thus we can generate any shape or number by selectively switching power across the display pixels. Colours can be introduced by including dyes in the liquid crystal. Normal nematic liquid crystals align their fields in few milli seconds but goback to their twisted state rather slow. Ferro electric liquid crystals meet this criteria and are widely used.

Why are liquid crystal displays important? The first factor is size. An LCD consists primarily of two glass plates with some liquid crystal material between them. There is no bulky picture tube. This makes LCDs practical for applications where size (as well as weight) are important.

In general, LCDs use *much less power* than their cathode-ray tube (CRT) counterparts. Many LCDs are reflective, meaning that *they use only ambient light to illuminate the display*. Even displays that do require an external light source (i.e. computer displays) consume much less power than Cathode ray tube(CRT) devices.

Liquid crystal displays have currently some drawbacks and could be solved with the tremendous research in-puts it is receiving. They are

(i) problems with viewing angle, (ii) contrast ratio, and

(iii) response time

2. The Role of Liquid Crystals in the digestion and assimilation : The partially digested acidic slurry of food that leaves the stomach is called chyme. The chyme enters the first part of the small intestine which is called the duodenum. In the duodenum secretions from various secretion from digestive organs are added to the chyme. The pancreas produces various digestive or hydrolytic enzymes called *insulin*. Another secretion produced by the pancreas is sodium bicarbonate. This gas is the source of some of our burbs after a meal. Another secretion that enters the duodenum is Bile. Bile is classified as a amphiphilic molecule. This means that it has both a hydrophobic and a hydrophilic region. Amphiphilic molecules such as bile act as emulsifing agents. When these molecules are in solution they form liquid crystal aggregates known as micelles and vescicles. Although oil does not mix with either water or acid, its solution composed of water or acid and an amphiphile has no difficulty incorporating the oil. When the oil is added to the mixture it accumulates in the hydrophobic region of the micelles and forms an emulsion. When fats in the chyme enter the duodenum they are mixed with bile and form a special type of liquid crystal known as an emulsion. The function of the SDS is to simulate the addition of bile to the digestive tract. Thus this function of bile is facilitated by external addition of Sodium dodecyl (laurel) sulfate (SDS).

3. Liquid Crystals as Lubricants : Slippery Substances are called lubricants. They are very important in modern technology. Cars, trucks, airplanes, and machines, all have parts that rub against one another. These parts would heat up, wear down, and stop working if we did not have lubricants. Lubricants reduce the amount of friction between two surfaces that move against each other. The fact that liquid crystals have lower viscosity than corresponding liquids is used in designing many lubricants, currently used in automobile engines and heavy machinery. For this purpose both the mesophases are suitable but smectic phase is preffered. Even in the biological systems, several cholesteric liquid crystals act as natural lubricants in the bone joints.

4. Liquid Crystal Thermometers : Chiral nematic (cholesteric) liquid crystals reflect light with a wavelength corresponding to the order, which inturn is dependent upon temperature. Liquid crystals make it possible to accurately gauge temperature just by looking at the color of the thermometer. By mixing different compounds, a device for practically any temperature range can be built. Physical problems, such as tumors, have a different temperature than the surrounding tissue and therefore diagnosed by administering an LC. Liquid crystal temperature sensors can also be used to find bad connections on a circuit board by detecting the characteristic higher temperature.

5. Other Liquid Crystal Applications : Liquid crystals have a multitude of other uses. They are used for nondestructive mechanical testing of materials under stress. This technique is also used for the visualization of RF (radio frequency) waves in waveguides. They are used in medical applications where, for example, transient pressure transmitted by a walking foot on the ground is measured. Low molar mass (LMM) liquid crystals have applications including erasable optical disks, full color "electronic slides" for computer-aided drawing (CAD), and light modulators for color electronic imaging.

27.2.9 Future of Liquid Crystals

The future of liquid crystals has just begun. There are still unsolved physical problems in this area, posing interesting challenges for the pure scientist. The need for liquid crystal applications grew, requiring engineers specialized in this field. Statistics show that cathode ray tube displays in TVs and computer monitors will be overtaken by liquid crystal displays in the near future. More recent applications involve switchable windows that can be changed from clear and opaque with the flip of a switch, liquid crystal displays with wider viewing angles, and displays that can remain in use for years without the need of power.

Knowledge in the field of liquid crystals is crucial for an understanding of biological membranes, thus biologists, medical researchers and pharmacists are also interested in liquid crystal research.

27.3 Battery Chemistry

When two or more electrochemical cells are electrically interconnected, each of which containing two electrodes and an electrolyte is called a Battery. In everyday usage, 'battery' is also used to refer to a single cell. No one battery design is perfect for every application. There are many parameters like cost, voltage, duty cycle, dimension, stability with time and temperature, shelf life, etc., on which a battery is selected for a particular operation. It is important to prioritize our list of requirements before choosing one and requires compromise. Here *duty cycle* refers to conditions the battery experiences during use like, type of discharge and current drain, e.g., continuous, intermittent, continuous with pulses, etc.

27.3.1 Classification

Batteries are classified into two categories depending on their recharging capabilities.

(I) **Primary Batteries** - *These are non-rechargeable and are meant for a single use and meant to be discarded after use.*

(II) **Secondary Batteries** - *These are rechargeable and are meant for a multi cycle use.* After every use the electrochemical reaction could be reversed by external application of voltage. The cycle is reversed till the capacity fades or lost due to leakage or internal short circuit.

27.3.2 Primary Batteries (Non-Rechargeable)

These are non-rechargeable and are less expensive and are often used in ordinary gadgets like torch lights, watches and toys. Commercially many kinds of primary batteries are available but we are going to restrict our discussion to Leclanché cell, Alkaline cell and Lithium cell.

27.3.2.1 Dry Cell (Leclanché Cells or Zinc-Carbon Cell)

Anode : Zinc, **Cathode:** Manganese Dioxide (MnO_2),

Electrolyte : Ammonium chloride or zinc chloride dissolved in water

Applications : Flashlights, toys, moderate drain use equipment

The basic design of the Leclanché cell dates to 1860s, and until World War II, was the only one in wide use. It is still the most commonly used of all primary battery designs because of its low cost, availability, and applicability in various situations.

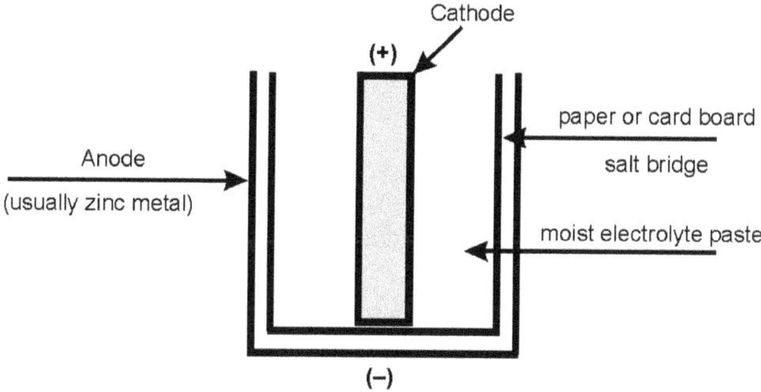

Fig. 27.8 A Cut view of Dry Cell.

In an ordinary Leclanché cell the electrolyte consists (in percent of atomic weight) of 26% NH_4Cl (ammonium chloride), 8.8% $ZnCl_2$ (zinc chloride), and 65.2% water.

At the cathode, the reduction reaction occurs within the moist paste comprised of ammomium chloride (NH_4Cl) and manganese dioxide (MnO_2) :

$$2\,MnO_2 + 2\,H_2O + 2e^- \longrightarrow 2MnOOH + 2\,OH^-$$

A thin zinc cylinder serves as the anode and it undergoes oxidation :

$$Zn\,(s) + 2NH_4Cl\,(aq) + 2OH^- \longrightarrow Zn(NH_3)_2\,Cl_2 + 2\,H_2O + 2e^-$$

The overall cell reaction can be expressed as

$$Zn + 2\,MnO_2 + 2\,NH_4Cl \longrightarrow 2\,MnOOH + Zn(NH_3)_2\,Cl_2$$

The anode Zinc is normally present as the cylindrical container (which is often surrounded by steel to prevent leakage of electrolyte). MnO_2, is not a good conductor of electricity and hence mixed with graphite to form a paste, with a central graphite rod acting as the current collector. The cell voltage increases by using synthetically produced manganese dioxide instead of that found naturally (called pyrolusite). This does drive the cost up a bit, but it is still inexpensive and environmentally friendly, making it a popular cathode. The aqueous ammonium chloride acting as electrolyte is usually thickened with polymeric material such as starch or methyl cellulose. The resultant cells are reasonably stable against discharging. The self discharging is via evolution hydrogen, a non-electrochemical reaction.

$$Zn + 2H_2O \longrightarrow Zn\,(OH)_2 + H_2$$

Zinc-chloride cell systems have been found to perform better than the Leelanche under heavier current drain. The electrolyte in a typical zinc chloride cell consists of 15-40% $ZnCl_2$ and 60-85% water, sometimes with a small amount of NH_4Cl for optimal performance. The overall cell reaction of the zinc chloride as the electrolyte can be expressed :

$$Zn + 2\,MnO_2 + 2\,H_2O + ZnCl_2 \longrightarrow 2\,MnOOH + 2\,Zn(OH)Cl$$

These cells are the cheapest ones in wide use, but they also have the lowest energy density and perform poorly under high-current applications.

27.3.2.2 Alkaline Cells

Anode : Zinc powder, **Cathode:** Manganese dioxide (MnO_2) powder

Electrolyte : Potassium hydroxide (KOH)

Applications : Radios, toys, photo-flash applications, watches, high-drain applications

This cell design gets its name from its use of alkaline aqueous solutions as electrolytes. Alkaline battery chemistry was first introduced in the early '60s. Alkaline cells have many advantages over zinc-carbon, including a higher energy density, longer shelf life, superior leakage resistance, better performance in both continuous and intermittent duty cycles, and lower internal resistance, which allows it to operate at high discharge rates over a wider temperature range.

Zinc in a powdered form increases the surface area of the anode, allowing more particle interaction. This lowers the internal resistance and increases the power density. Just as in the zinc carbon cell, graphite is added to the cathode to increase conductivity. The electrolyte, KOH, allows high ionic conductivity. Zinc oxide is often added to slow down corrosion of the zinc anode. Addition of some cellulose derivatives and gelling agents improve performance makes it more cost effective, especially in high drain situations where the alkaline cell's energy density is much higher.

The half-reactions are :

$$Zn + 2\,OH^- \longrightarrow ZnO + H_2O + 2\,e^-$$

$$2\,MnO_2 + H_2O + 2\,e^- \longrightarrow Mn_2O_3 + 2\,OH^-$$

The overall reaction is :

$$Zn + 2MnO_2 \longrightarrow ZnO + Mn_2O_3 \quad E = 1.5\text{ V}$$

There are other cell designs that fit into the alkaline cell category, including the mercury oxide, silver oxide, and zinc air cells. Mercury and silver give even higher energy densities, but cost a lot more and are being phased out through government regulations because of their high toxicity as heavy metals.

27.3.2.3 Lithium and Other Metal Cells (Non-aqueous cells)

Applications : Pacemakers, defibrillators, watches, meters, cameras, calculators, portable, low-power use

Lithium battery chemistry comprises a number of cell designs that use **lithium as the anode**. Lithium is gaining a lot of popularity as an anode over other metals for a number of reasons as given in Table 27.1.

Table 27.1

Anode	Standard potential (V)	Density g/cm^3	Electrochemical Equivalence (Ah/g)
Li	**3.05**	**0.54**	**3.86**
Na	2.7	0.97	1.16
Mg	2.4	1.74	2.20
Al	1.7	2.7	2.98
Ca	2.87	1.54	1.34
Fe	0.44	7.85	0.96
Zn	0.76	7.1	0.82
Cd	0.40	8.65	0.48
Pb	0.13	11.3	0.26

Note that lithium, the lightest of the metals, also has the highest standard potential (>3V) of all the metals and some of the lithium cell designs have a voltage of nearly 4 V. This means that *lithium has the highest energy density.* The dry cells and alkaline cells discussed earlier have aqueous electrolyte while the lithium cells have non-aqueous medium. Aqueous systems can go upto energy densities of 700 W kg^{-1} but **Lithium non-aqueous systems** with far higher potential power densities have been reported.

Lithium cells have only recently become commercially viable because *lithium reacts violently with water, as well as nitrogen in air.* This requires sealed cells. High-rate lithium cells can build up pressure if they are short circuited and cause the temperature and pressure to rise. Thus, the cell design needs to include weak points, or safety vents, which rupture at a certain pressure to prevent explosion.

Lithium cells can be grouped into three general categories: (a) liquid cathode, (b) solid cathode, and (c) solid electrolyte.

(a) Liquid cathode lithium Cells :

Anode : Lithium, ***Cathode :*** $SOCl_2$, ***Electrolyte :*** $LiAlCl_2$

These cells tend to offer *higher discharge rates* because the reactions occur at the cathode surface. In a solid cathode, the reactions take longer because the lithium ions must enter into the cathode for discharge to occur. The direct contact between the liquid cathode and the lithium forms a film over the lithium, called the *solid electrolyte interface (SEI)*. This prevents further chemical reaction when not in use, thus preserving the cell's shelf life. One drawback, though, is that if the film is too thick, it causes an initial voltage delay. Usually, water contamination is the reason for the thicker film.

This cell consists of a high-surface area carbon cathode, a non-woven glass separator. Thionyl chloride acts as the electrolyte solvent and the active cathode material.

The materials react as follows :

Location	**Reaction**
Anode	$Li \longrightarrow Li^+ + e^-$
Cathode	$4Li^+ + 4e^- + 2SOCl_2 \longrightarrow 4LiCl + SO_2 + S$
Overall	$4Li + 2SOCl_2 \longrightarrow 4LiCl + SO_2 + S$

During discharge the anode gives off lithium ions. On the carbon surface, the thionyl chloride reduces to chloride ions, sulfur dioxide, and sulfur. The lithium and chloride ions then form lithium chloride. Once the lithium chloride has deposited at a site on the carbon surface, that site is rendered inactive. The sulfur and sulfur dioxide dissolve in the electrolyte, but at higher-rate discharges SO_2 will increase the cell pressure leading to explosion.This system has a very high energy density (about 500 Wh/kg) and an operating voltage of 3.3–3.5 V.

In high-rate discharge, the voltage delay is more pronounced and the pressure increases as mentioned before. Low-rate cells are used commercially for small electronics and memory backup. High-rate cells are used mainly for military applications.

(b) Solid Cathode Lithium Cells ($LiMnO_2$ Cell)

These cells cannot be used in high-drain applications and don't perform well under low temperature conditions. However, they don't have the same voltage delay as liquid cells and don't require pressurization. They are used generally for memory backup, watches, portable electronic devices, etc.

These account for about 80% of all primary lithium cells, one reason being their low cost. The cathode used is a heat-treated MnO_2 and the electrolyte a mixture of propylene carbonate and 1, 2-dimethoxyethane.

The half reactions are :

Anode	$Li \longrightarrow Li^+ + e$
Cathode	$Mn^{IV}O_2 + Li^+ + e \longrightarrow Mn^{III}O_2(Li^+)$
Overall	$Li + Mn^{IV}O_2 \longrightarrow Mn^{III}O_2(Li^+)$

$Li(CF)_n$ Lithium polycarbon monofluoride also falls in this category.

The cathode in this cell is carbon monofluoride. A typical electrolyte is lithium tetrafluorobate ($LiBF_4$) salt in a solution of propylene carbonate (PC) and dimethoxyethane (DME).

These cells also have a high voltage (about 3.0 V) and a high energy density (around 250 Wh/kg). All this and a 7-year shelf life makes them very suitable for low- to moderate-drain use, e.g., watches, calculators, and memory applications.

(c) Solid Electrolyte Lithium Cells

They perform best in low-current applications and have a *very long* service life. For this reason, they are used in pacemakers

Lithium iodine cells use solid LiI as their electrolyte and also produce LiI as the cell discharges. The cathode is poly-2-vinylpyridine (P2VP) with the following reactions :

Anode	$2Li \longrightarrow 2Li^+ + 2e$
Cathode	$2Li^+ + 2e + P2VP\cdot nI_2 \longrightarrow P2VP\cdot (n-1)I_2 + 2LiI$
Overall	$2Li + P2VP\cdot nI_2 \longrightarrow P2VP\cdot (n-1)I_2 + 2LiI$

LiI is formed *in situ* by direct reaction of the electrodes.

27.3.3 Secondary Batteries

These cells are rechargeable and reusable. A combination of all reversible electrochemical cells give secondary batteries. Many kinds of secondary batteries are available in the market but we are restricting our discussion to Lead acid cell, Ni/Cd cells, Ni-metal hydride cells and Lithium ion cells

27.3.3.1 Lead–acid Cells

Anode : Sponge metallic lead,

Cathode: Lead dioxide (PbO_2),

Electrolyte : Dilute mixture of aqueous sulfuric acid

Applications : Automobliles and construction equipment, standby/backup systems

Used mainly for engine batteries, these cells represent over half of all battery sales. Some advantages are their low cost, long life cycle, and ability to withstand mistreatment. They also perform well in high and low temperatures and in high-drain applications.

The half-cell reactions are :

Half-Reaction	V vs SHE
$Pb + SO_4^{2-} \longrightarrow PbSO_4 + 2e^-$	0.356
$PbO_2 + SO_4^{2-} + 4H^+ + 2e^- \longrightarrow PbSO_4 + 2H_2O$	1.685

There are a few problems with this design. If the cell voltages exceed 2.39 V, the water breaks down into hydrogen and oxygen and may lead to explosion. Another problem arising from this system is that fumes from the acid solution may have a corrosive effect on the area surrounding the battery.

These problems are mostly solved by sealed cells, made commercially available in the 1970s. This sealing in reality is only valve regulated and allowing hydrogen and oxygen recombination using catalytic recombiners. They convert these gases back into water, achieving about 85% efficiency at best. In these batteries the water lost becomes so insignificant that no refill is needed for the life of the battery. For this reason , these cells are often referred to as maintenance-free batteries. Also, this cell design prevents corrosive fumes from escaping.

These cells have a low cycle life, a quick self discharge, and low energy densities (normally between 30 and 40 Wh/kg). However, with a nominal voltage of 2 V and power densities of up to 600 W/kg, the lead-acid cell is an adequate, if not perfect, design for car batteries.

27.3.3.2 Nickel/Cadmium Cells

Anode: Cadmium,

Cathode: Nickel oxyhydroxide NiOOH,

Electrolyte : Aqueous potassium hydroxide (KOH)

Applications : Calculators, digital cameras, pagers, lap tops, tape recorders, flashlights, medical devices (e.g., defibrillators), electric vehicles, space applications

The cathode is nickel-plated, woven mesh, and the anode is a cadmium-plated net. The electrolyte, KOH, acts only as an ion conductor and does not contribute significantly to the cell's reaction. That's why not much electrolyte is needed, so this keeps the weight down (NaOH is sometimes used as an electrolyte, which doesn't conduct as well, but also doesn't tend to leak out of the seal as much). The cell reactions are as follows :

Reaction	V vs SHE
$Cd + 2OH^- \longrightarrow Cd(OH)_2 + 2e^-$	0.81
$NiOOH + 2H_2O + 2e^- \longrightarrow Ni(OH)_2.H_2O + OH^-$	0.45
$Cd + NiOOH + 2H_2O + OH \longrightarrow Cd(OH)_2 + Ni(OH)_2.H_2O$	1.26

Advantages include good performance in high-discharge and low-temperature applications. They also have long shelf and use life. Disadvantages are that they cost more than the lead-acid battery and have lower power densities. Possibly its most well-known limitation is a *memory effect*, where the cell retains the characteristics of the previous cycle. This term refers to a temporary loss of cell capacity, which occurs when a cell is recharged without being fully discharged. This can cause cadmium hydroxide to passivate the electrode, or the battery to wear out. In the former case, a few cycles of discharging and charging the cell will help correct the problem, but may shorten the lifetime of the battery.

27.3.3.3 Nickel/Metal Hydride (NiMH) Cells

Anode: Rare-earth or nickel alloys with many metals,

Cathode: Nickel oxyhydroxide,

Electrolyte : Potassium hydroxide

Applications : Cellular phones, camcorders, emergency backup lighting, power tools, laptops, portable, electric vehicles

Except for the anode, the NiMH cell very closely resembles the NiCd cell in construction. Even the voltage is virtually identical, at 1.2 volts, making the cells interchangeable in many applications. Here are the cell reactions :

Location	Reactions	Voltage
Anode	$MH + OH^- \longrightarrow M + H_2O + e^-$	0.83
Cathode	$NiOOH + H_2O + e^- \longrightarrow Ni(OH)_2 + OH^-$	0.52
Overall	$NiOOH + MH \longrightarrow Ni(OH)_2 + M$	1.35

The anodes used in these cells are complex alloys containing many metals, such as an alloy of V, Ti, Zr, Ni, Cr, Co, and Fe. The underlying chemistry of these alloys and reasons for superior performance are not clearly understood, and the compositions are determined by empirical testing methods.

A very interesting fact about these alloys is that some metals absorb heat when absorbing hydrogen, and some give off heat when absorbing hydrogen. Both of these are bad for a battery, since we would like the hydrogen to move easily in and out without any energy transfer. The successful alloys are all mixtures of exothermic and endothermic metals to achieve this.

Hydrogen Storage Metals Comparison :

Material	**Density**	**H_2 Storage Capacity g/cc**
$LaNi_5$	8.3	0.11
FeTi	6.2	0.11
Mg_2Ni	4.1	0.15
Mg	1.74	0.13
MgNi Eutectic	2.54	0.16
liquid H_2	0.07	0.07

The density of hydrogen stored in a metal hydride is higher than that of pure liquid hydrogen. Commercial NiMH batteries are mostly of the rare earth-nickel type, of which $LaNi_5$ is a representative. These alloys can store six hydrogen atoms per unit cell such as $LaNi_5H_6$. Even mixed metal nickel alloys are used to save the cost of separation.

The electrolyte of commercial NiMH batteries is typically 6 M KOH

The NiMH cell does cost more and has half the service life of the NiCd cell, but it also has 30% more capacity, increased power density (theoretically 50% more, practically 25% more). The memory effect, which was at one time thought to be absent from NiMH cells, is present if the cells are treated just right. To avoid the memory effect fully discharge once every 30 or so cycles. There is no clear winner between the two. The better battery depends on what characteristics are more crucial for a specific application.

27.3.3.4 Lithium Ion Cells

Anode : Carbon compound, graphite,

Cathode : Lithium oxide.

Applications : Laptops, cellular phones, electric vehicles

As discussed earlier Lithium metal batteries have safety disadvantages. These are overcome in *Lithium ion Batteries.*

Cathodes consist of a layered crystal (graphite) into which the lithium is intercalated. Experimental cells have also used lithiated metal oxide such as $LiCoO_2$, $NiNi_{0.3}Co_{0.7}O_2$, $LiNiO_2$, LiV_2O_5, LiV_6O_{13}, $LiMn_4O_9$, $LiMn_2O_4$, $LiNiO_{0.2}CoO_2$.

Electrolytes are usually $LiPF_6$, although this has a problem with aluminum corrosion, and so alternatives are being sought. One such is $LiBF_4$. The electrolyte in current production batteries is a patented liquid, and uses an organic solvent.

Membranes are necessary to separate the electrons from the ions. Currently the batteries in wide use have microporous polyethylene membranes.

Intercalation keeps the small ions (such as lithium, sodium, and the other alkali metals) into the interstitial spaces in a graphite crystal. This makes the graphite is conductive, dilutes the lithium for safety, is reasonably cheap, and does not allow dendrites or other unwanted crystal structures of Li to form.

27.4 Fuel Cells

A cell in which one or both of the reactants are not permanently contained in the cell, but are continuously supplied from a source external to the cell and the reaction products continuously removed is called a fuel cell. Unlike the metal anodes typically used in batteries, the fuels in a fuel cell are usually gas or liquid, with oxygen as the oxidant.

Around 1900 Nernst and Haber attempted for a direct carbon oxidizing fuel cells but met with a limited success. Later in 1933 Baur, a German scientist, proposed the first fuel cell that can use hydrogen as a fuel and can work at room temperature. This idea was forwarded by Bacon, a British engineer and developed a porous nickel electrode that can withstand high pressures and can give current in kilowatt range at moderate temperatures. This technology found a ready customer in Apollo program of NASA (USA). This fuel cell harnesses the chemical energy of hydrogen and oxygen to generate electricity without combustion or pollution. Hydrogen/oxygen fuel cell systems work well in space travel applications because of their *high efficiency, high power-to-weight and volume ratios, and usable reaction product (water).* In the near future, our vehicles may also be powered by fuel cells.

There are many kinds of H_2/O_2 fuel cells and are categorized on the basis of electrolyte used.

1. Proton Exchange Membrane Fuel Cells (PEMFC)
2. Alkaline Fuel Cells (AFC)
3. Molten Carbonate Feuel Cells
4. Phosphoric acid fuel cells
5. Solid oxide fuels

We are going to have a discussion of the *first two* kinds of fuel cells.

27.4.1 Proton Exchange Membrane Fuel Cells (PEMFC)

The type of fuel cell typically used in automobiles is a *Proton Exchange Membrane* (PEM)Fuel Cell , also called a *Polymer Electrolyte Membrane fuel cell* (Fig. 27.9). A fuel cell harnesses the chemical energy of hydrogen and oxygen to generate electricity without combustion or pollution.

At the anode, a platinum catalyst causes the hydrogen to split into positive hydrogen ions (protons) and negatively charged electrons.

Half-Reactions		V vs SHE
At Anode	$2H_2 = 4H^+ + 4e^-$	0
At Cathode	$4H^+ + O_2 + 4e^- = 2H_2O$	1.2

The Proton Exchange Membrane (PEM) allows only the positively charged ions to pass through it to the cathode. The negatively charged electrons must travel along an external circuit to reach the cathode, creating an electrical current.

At the cathode, the negatively charged electrons and positively charged hydrogen ions combine with oxygen to form water, which flows out of the cell.

Thus we get electricity from environmentally friendly reactants and products.

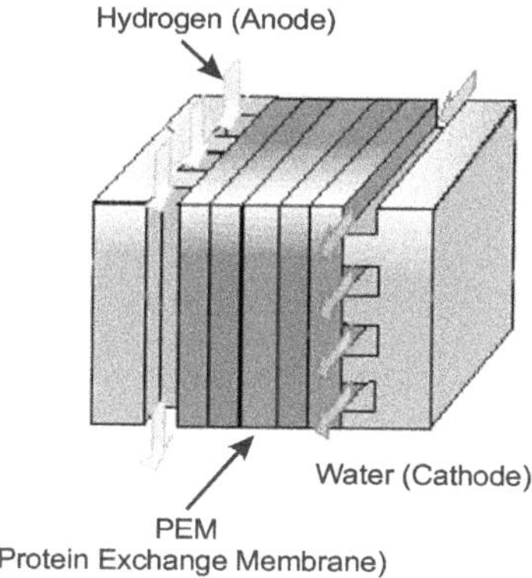

Fig. 27.9 PEM fuel cell.

27.4.2 Alkaline Fuel Cells (AFC)

In these cells, 30-45% aqueous solutions of potassium hydroxide is used as electrolyte (Fig. 27.10). So acidic impurities such as CO_2 should be avoided. Generally this is done by passing the compressed air into lime solution before allowing it to enter the fuel cell. These fuel cells yield highest voltages at comparable current densities in comparison to other classes of fuel cells.

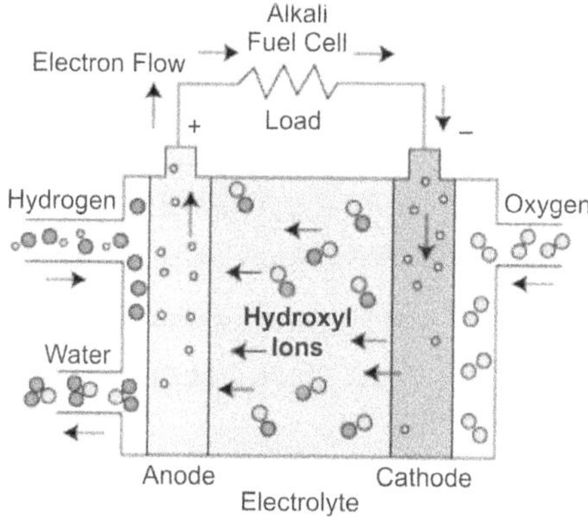

Fig. 27.10 Alkaline fuel cell.

The following half reactions take place in these cells

At Cathode : Oxygen passed is converted into hydroxide ions by consuming 4 electrons

$$O_2 + 2H_2O + 4\,e^- \longrightarrow 4\,OH^-$$

In fact, this is a two step reaction with intermediate peroxy ion formation.

$$O_2 + H_2O + 2\,e^- \longrightarrow HO_2^- + OH^-$$

$$HO_2^- \longrightarrow OH^- + \tfrac{1}{2}O_2$$

At Anode : Anode involves oxidation of hydrogen in to water by reacting with hydroxide anions from the electrolyte.

$$2\,H_2 + 4\,OH^- \longrightarrow 4H_2O + +4\,e^-$$

Over All Reaction :

$$2H_2 + O_2 \longrightarrow 2H_2O$$

Bacon fuel cell consists of an anode made of porous nickel and a cathode made of lithiated nickel oxide separated by 30% aqueous KOH serving as electrolyte. This cell operates at 200^0 C and 5 Mpa (about 50 atm) pressure. Bacon's double porocity fuel cell was used in 1960 NASA Apollo space program by lowering the operational pressure to 0.33 Mpa by activating the electrodes with a catalyst. The space shuttle system in 1974 replaced the heavy bipolar electrode plates with magnesium foils plated with silver. The anode now is a carbon bonded with polytetrafluoroethylene and consisting Pt-Pd catalyst and cathode consists of 90% gold and 10% platinum plated on to nickel screen. This cell operates at a pressure of 0.42 Mpa and 92^0 C and works for 2000 h.

27.4.3 Advantages of Fuel Cells

1. No emission of toxic gases and chemical wastes are in safe limits. The reactants and products are environment friendly and only we have to bother about disposal of cell material
2. High efficiency (75-85%) of energy conversion from chemical energy to electrical energy. So offer an excellent use of our renewable energy resources.
3. No noise pollution like in generators and low thermal pollution.
4. Low maintenance costs, fuel transportation costs, cell parts are modular and exchangeable.
5. The fuels and electrolyte materials are available in plenty and inexhaustible unlike fossil fuels.
6. Unlike solar cells, fuel cells are compact and transportable.
7. Unlike acid cells used in automotives the fuel cells are far less corrosive.
8. Unlike nuclear energy, fuel cell energy is economical and safe.
9. Fuel cells are operatable to 200^0C and so find application in high temperature systems.
10. Fuel cells hold promise as good storage devises for future nuclear and solar energy economies.

27.4.4 Limitations of Fuel Cells

1. The main limitation of fuel cells lie in high initial costs associated with electrode material and design costs. The later could be brought down with mass production and suitable R&D in this direction. The low maintenance and longetivity of cells could compensate some of these negative factors.
2. Large weight and volume of H_2 gas-fuel storage system.
3. Lack of infrastructure for distribution and marketing of hydrogen gas. The developing natural gas distribution net-work could pave way for this in the future.
4. High cost of H_2 gas. Technological advances should bring the costs down.
5. Most alkaline fuel cells suffer from carbon dioxide leakages and should be properly filtered and prevented from entering the cell and reacting with the electrolyte.

27.5 Conducting Polymers

Inherently, organic polymers with all carbon backbones are *insulators* and can be used as encapsulating materials for providing high resistivity to articles such as coatings for cables and electrical wirings and in electrical gadgets. It has been found that this resistivity decreases if composite of the polymer with carbon black or finely devided metal is fabricated, but the conduction in these cases takes place via the filler and not through the polymer which merely acts as a supporting matrix. Incorporation of a filler can also reduce the mechanical strength of the polymer. This scenario changed in 1970 due to pioneering work of Alan Heeger of the University of California, Santa Barbara, Alan MacDiarmid of the University of Pennsylvania, and Hideki Shirakawa of the University of Tsukuba in Japan – who shared the 2000 Nobel Prize in chemistry for this discovery. Polyacetylene was first synthesized, in 1958 by polymerization of acetylene, by Natta et al. as a *black powder*. This was found to be a semi-conductor with a conductivity between 7×10^{-11} to 7×10^{-3} Sm^{-1}, depending upon preparation process. This polymer was found to be not of much use due to its air instability. In 1967, when a postgraduate student of Hideki Shirakawa at the Tokyo Institute of Technology was attempting to synthesize polyacetylene, by mistake he took 1000 times more Natta catalyst and a *silvery shining thin film* was produced as a result of this mistake. Exposure of this form of polyacetylene to halogens increased its conductivity a billion fold. When it was weakly oxidized by compounds such as iodine it turned a golden colour and its conductivity increased to about 10^4 Sm^{-1}.

In the 1980's polyheterocycles (e.g., polythiophene) were first developed which were found to be much more air stable than polyacetylene, although their conductivities were not so high, typically about 10^3 Sm^{-1}. Since then it has been found that about a dozen different polymers and polymer derivatives (Fig. 27.11) exhibit conducting properties when doped with a weak oxidation agent or reducing agent.

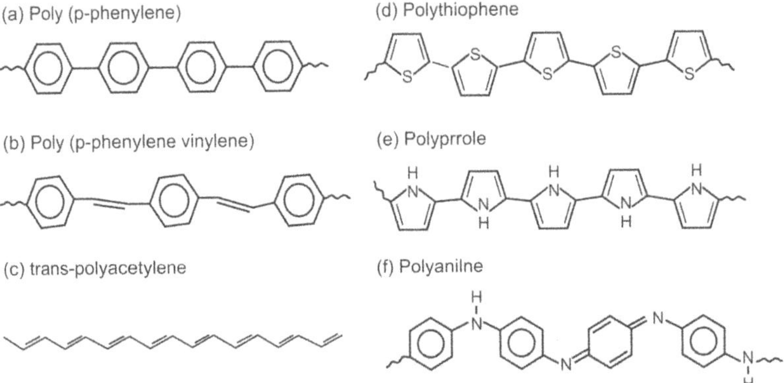

Fig. 27.11 Some Conducting Polymers-(*a-e*) are "stiff" linearly unsubstituted conducting polymers and (*f*) is a semi-flexible conducting polymer.

Conductive polymers include poly(acetylene)s, poly(pyrrole)s, poly(thiophene)s, poly(aniline)s, poly(fluorene)s, polynaphthalenes, poly(p-phenylene sulfide), and poly(para-phenylene vinylene)s. Classically, these compounds are also known as "blacks" or "melanins". The melanin pigment in animals is generally a mixed copolymer of polyacetylene, polypyrrole, and polyaniline.

27.5.1 Doping

Doping in silicon semiconductors is achieved by replacing some of the silicon atoms by electron rich (e.g., phosphorus) or electron-poor (e.g. boron) atoms to create *n*-type and *p*-type semiconductors, respectively. In contrast, in conducting polymers doping is achieved, through an oxidation or reduction process. The doping methods can be classified into (i) Chemical doping and electrochemical doping.

- ***Chemical doping**, involves exposing a polymer, to an oxidant (typically iodine or bromine) or reductant (far less common, but typically involves alkali metals).*
- **Electrochemical doping** *involves a polymer-coated working electrode suspended in an electrolyte solution in which the polymer is insoluble along with separate counter and reference electrodes. An electric potential difference created causes a charge from the electrolyte to enter the polymer in the form of electron addition (n-doping) or removal (p-doping).*

The different dopings of polymers is given in Table 27.1.

Table 27.1 Structures and Conductivity of doped conjugated polymers.
(Adapted from D.O. Cowan and F.M. Wiygul (1986))

Polymer	Structure	Typical methods of doping	Typical conductivity $(Scm)^{-1}$
Polyacetylene		Electrochemical, chemical (AsF_5, I_2, Li, K)	500–1.5×10^5
Polyphenylene		Chemical (AsF_5, Li, K)	500
Poly(phenylene Sulphide)	S	Chemical (AsF_5)	1
Polypyrrole	N H	Electrochemical	600
Polythiophene	S	Electrochemical	100
Poly(phenyl-quinoline)	C_6H_5 N	Electrochemical, chemical (sodium naphthalide)	50

27.5.2 Conjugation

Conduction in conducting polymers appears to be due to movement of electrons through the polymer and so referred to as *electroactive polymers.* Fully saturated chain such as polyetylene is 5 eV and in conjugated system like polyacetylene this gap decreases to 1.5 eV. Respective conductivities of these polymers are $\sim 10^{-17}$ S cm^{-1} and $\sim 10^{-8}$ S cm^{-1}. Ususally polyconjugated structures are either insulators or at the maximum semiconductors in their pure state but when treated with an oxidizing or a reducing agent can be converted into polymer salts with electrical conductivities comparable to metals. The generalized oxidation reaction that can be writen is

$$P_n \underset{\text{Red}}{\overset{\text{Ox}/A^-}{\rightleftharpoons}} [P_n^+ A^-] \underset{\text{Red}}{\overset{\text{Ox}/A^-}{\rightleftharpoons}} [P_n^{2+}\ 2A^-]$$

where P_n represent a section of polymer chain.

Alternatively, after the first redox reaction, charge transfer complexes may form between charged and neutral segments of polymer

$$P_n \rightleftharpoons [P_n^+ A^-] \overset{P_m}{\rightleftharpoons} [(P_n P_m)^+ A^-]$$

The extended conjugation of a conductive polymer tends to give rise to fluorescence which has lead to the rapid development of polymer-based light emitting devices (OLEDs) and organic photovoltaic devices.

27.5.3 Conductivity

Conductive polymers are present in most mammal tissues where electrical conduction or transduction from light or sound are necessary, including the skin, eye, inner ear, and brain. Its electronic conductivity seems to be the underlying mechanism for absorption of light, and electron-photon interactions are exploited in hearing. A Logarithmic conductivity ladder of some of these polymers in relation to conductivity of Cu and Hg metals are shown below Fig. 27.12.

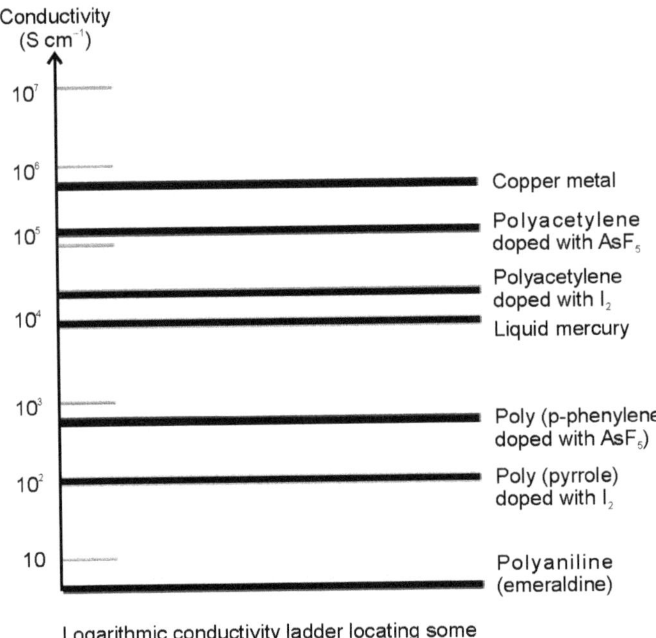

Logarithmic conductivity ladder locating some metals and conducting polymers

Fig. 27.12

The conductivity is attributable to an extended conjugated backbone of continuous overlapping orbitals, for example, alternating single and double carbon-carbon bonds, which leaves a continuous path of overlapping *p*-orbitals (Fig. 27.13).

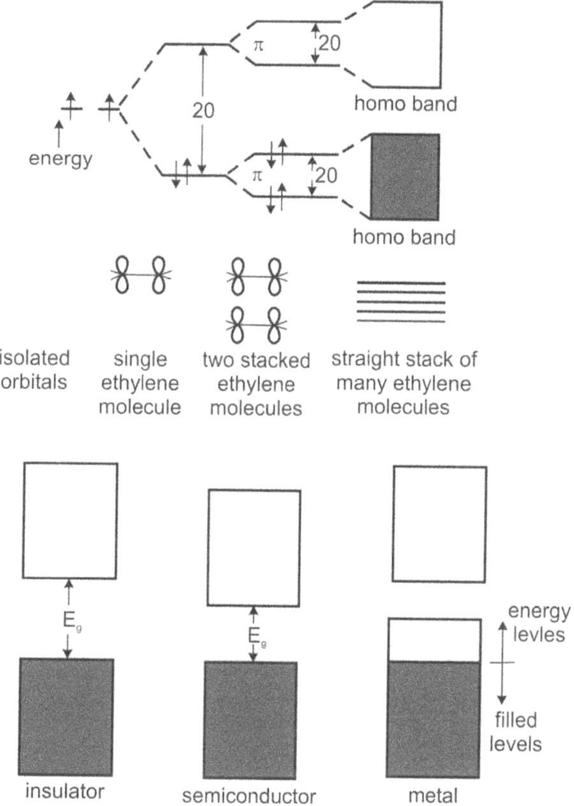

Fig. 27.13 Schematic diagram showing the principles of band theory as described in the text. The dark regions represent the bands filled with electrons, and the light regions the bands that are available for conduction. The energy gap between filled and empty states is E_g.

(Adapted from D.O. Cowan and F. M. Wiygul (1986))

This extended π-systems of conjugated polymer is highly susceptible to chemical or electrochemical oxidation or reduction. These alter the electrical and optical properties of the polymer, and by controlling this oxidation and reduction, it is possible to precisely control these properties. Since these reactions are often reversible, it is possible to systematically control the electrical and optical properties with a great deal of precision.

27.5.4 Preparation of Conductive Polymers

The normal methods of polymer synthesis like free radical initiated polymerisation, etc., do not give a polymer with suitable doping and required conjugation. So special methods are adopted to get required conductive polymer, some of which are presented here.

1. *Polyacetylene :*

Polymerisation of acetylene over Ziegler catalysts give polyacetylene which is predominately in *cis* form. On raising the temperature of this film gives more stable *trans* form. This polymer is infusible, insoluble and tends if become brittle on exposure to air. These problems are solved by Feast, adopting a two state process.

R R
$(CH_3)_4 S_n$
$W Cl_6$
n
Δ
320 - 390 K
+
n
n
tans-polyacetylene

Conductivity is magnified by doping. Exposure of the film to dry ammonia gives a polymer with conductivity of 10^3 S cm^{-1}. Controlled addition of p-doping agent like AgF_5, Br_2, I_2 or $HClO_4$ could boost to still higher conductivities.

2. *Poly (p-phynylene) :*

Polycondensation of p-dibromobenzene in the presence of Mg yields 'insoluble' oligomeric polymer material

n Mg + n Br—C₆H₄—Br → —(C₆H₄)— + n Mg Br_2

This problem is solved by preparing first a precursor polymer which could be processed to suitable shape prior to use

free radial initiation
Heat
RCOO OCOR
n
RCOO OCOR
n

5, 6 - dihydroxycyclohex - 1, 3 - diene Precursor polymer Poly (p-phenyla)

Doping is done as given for polyacetylene.

3. Poly pyrrole :

Electrolysis of a solution of pyrrole (0.06 M) and $(Et)_4 N^+ B F_4^-$ (0.1 M) in acetonitrile having 1% moisture leads to deposition of blue-black film of polymer.

The polymer has a conductivity of 10^2 S cm^{-1} and contains BF_4^- as do punt. The composition is given below

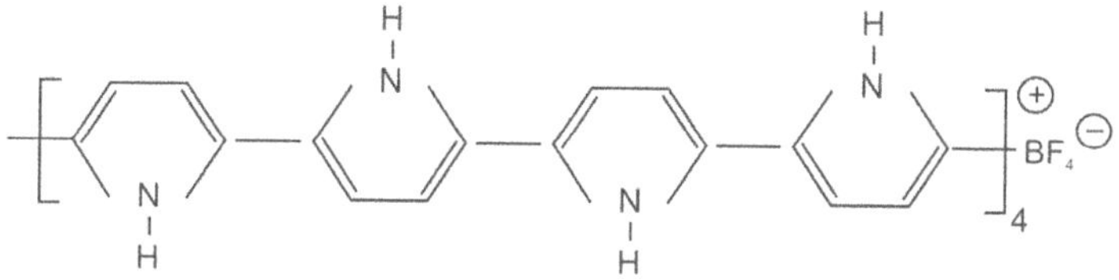

In the presence of Ag ClO_4 it gives a copper-bronze colored poly pynole per chlorate with following

Stoichiometry

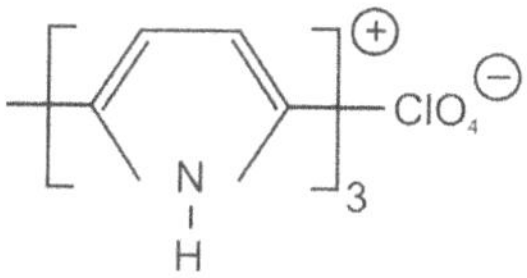

Polyamiline :

Reaction of aniline with ammonium persulphate in aqueous HCl produces polyaniline as a dark blue powder with a conductivity of 5 S cm^{-1}

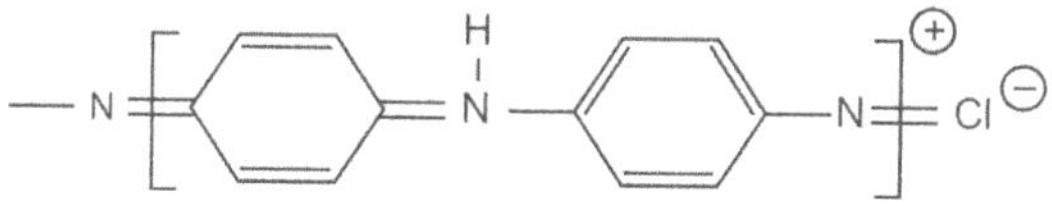

27.5.5 Classification of Conductive Polymers

The associated properties of conductivity and redox nature, group these polymers into two classes. The first group utilizes their *conductivity* as its main property. The second group utilizes their *electroactivity* as its main property.

The applications of these two groups are shown below :

Group - 1 (conductivity)	**Group - 2 (electroactivity)**
Electrostatic materials	Molecular electronics
Conducting adhesives	Electrical displays
Electromagnetic shielding	Chemical, biochemical and thermal sensors
Printed circuit boards	Rechargeable batteries and solid electrolytes
Artificial nerves	Drug release systems
Antistatic clothing	Optical computers
Piezoceramics	Ion exchange membranes

Active electronics (diodes, transistors)	Electromechanical actuators
	'Smart' structures
Aircraft structures	Switches

Most of these applications are in the development stage. However, conducting polymers have made it into a few small-scale commercial applications, such as antistatic coatings on photographic film and light emitting diodes in a display of maintenance information on an electric razor.

27.5.6 Application of Conductive Polymers

Group – I (*Conductivity*) Applications :

1. In the computer industry the sudden discharge of static electricity can damage microcircuits. Similarly in industries handling inflammable gases and explosives accumulation of static charge and a sudden discharge could lead to explosion. By coating an insulating surface with a very thin layer of conducting polymer it is possible to prevent the buildup of static electricity on its surface and therefore the damage.
2. By placing monomer between two conducting joints/surfaces and allowing it to polymerise it is possible to stick them together. This conducting adhesive eliminates the development of bad joints due to oxide films formed at the exposed surface.
3. Computer devices and cell phones often emit electromagnetic radiation in the microwave region. This leads to malfunctioning of neighbouring delicate electrical devises and LCD instruments. Normal plastic cases are transparent to these radiations while the conducting polymer coatings readily absorb these unwanted radiations.
4. Many electrical appliances use printed circuit boards. These are copper coated epoxy-resins. It is an expensive multistage process and the adhesion is generally poor. Conductive polymer coating on a polymer sheet has no complications and also has a very good adhesion.
5. It is possible to synthesize biocompatible conducting polymers. Due to the biocompatability these polymers they may be used to transport small electrical signals through the body, i.e. act as artificial nerves. Perhaps this could even pave way for brain implants.
6. Polymers have one tenth of density of metals and their maneuverability is also comparatively more. We are not far away from designing and using conducting polymers in our household wiring and electrical gadgets. This will be an added advantage in aircrafts where the load matters.

Group - 2 (*Electroactive*) Applications :

In the quest for improved performance, better processibility and novel applications a wealth of newer compounds with specific chemical architectures and electroactive properties have been synthesized. Better maneuverability is achieved through three parameters viz.,

- ✓ *greater main chain flexibility* (e.g., polyaniline),
- ✓ *side chain substitution* [e.g., poly(3-alkylthiophenes)],
- ✓ *fabrication from a soluble precursor polymer* [e.g., poly (*p*-xylene-dimethylsufonium chloride) to yield poly (*p*-phenylene vinylene)].

For example, side chain substitutions are now routinely used to enhance solvent solubility and fusibility so that various conventional polymer processing methods like cating and moulding may be done easily.

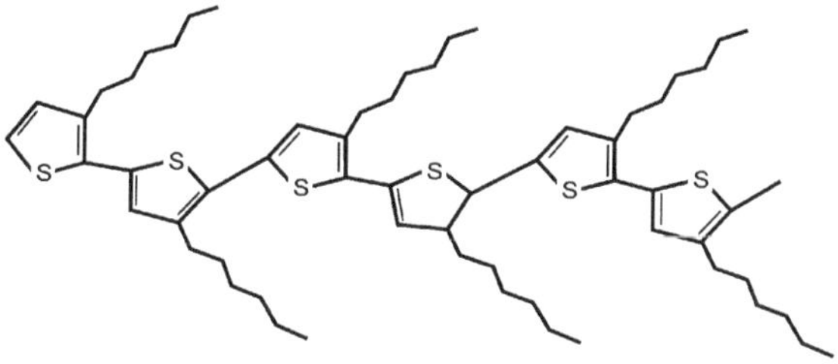

Further this may create new materials that exhibit enhanced electronic properties as compared to the linearly unsubstituted parent polymer. These side chain substituted conducting polymers also exhibit properties which do not exist in the unsubstituted hosts like pronounced *thermo- and solvato-chromism*, i.e. change of colour with temperature and solubility.

- **Smart/Intelligent Materials :** Smart materials are materials that 'remember' configurations and can conform to them when given a specific stimulus. These materials can respond to changes in electricity, heat, or magnetic waves. The electrical sensitivities make the conducting polymers used as smart or intelligent polymers. Changes in current values bring changes in the configuration of the conducting polymer. Taking polyacetylene as a typical example, normal preparation methods yield a random configuration while the one obtained by Shirakawa technique is a highly orderly transform exhibiting conductivity.

- **Gas Sensors :** The chemical properties of conducting polymers change their properties during reaction with redox reagents acting as dopants. Thus polypyrrole behaves as a quasi '*p*' type material. Its resistance increases in the presence of a reducing gas such as ammonia, and decreases in the presence of an oxidizing gas such as nitrogen dioxide. This is due to changes in electron holes caused by their reaction with adsorbed oxygen ions.
- **Biosensors :** The ability of triiodide (active ingredient is iodine) to oxidize polyacetylene is used as a means to measure glucose concentration. Glucose is oxidized with oxygen with the help of glucose oxidase in the solution phase. This produces hydrogen peroxide which oxidizes iodide ions present in the medium to form triiodide ions. These triiodide ions inturn bring changes in polyacetylene and its conductivity. Hence, conductivity exhibited by polymer is proportional to the peroxide concentration which in return is proportional to the glucose concentration of the solution phase.

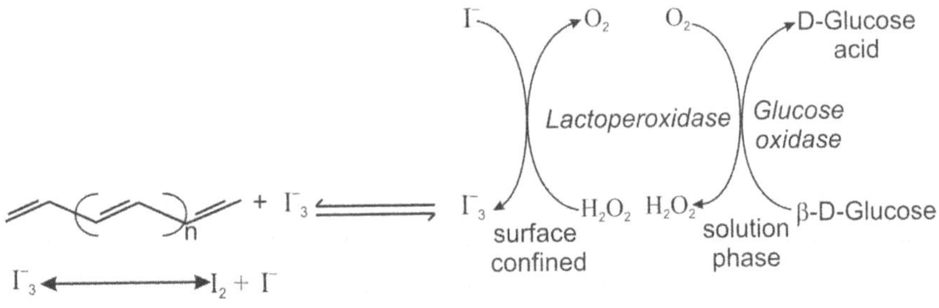

$[\text{polyacetylene}]_n + I_3^- \rightleftharpoons$

$I_3^- \longleftrightarrow I_2 + I^-$

I^-, O_2, O_2, D-Glucose acid

Lactoperoxidase, Glucose oxidase

I_3^-, H_2O_2, H_2O_2, β-D-Glucose

surface confined, solution phase

- **Polymer Rechargeable Batteries :** The polymer battery, such as a polypyrrole-lithium cell operates by the oxidation and reduction of the polymer backbone. During charging the polymer oxidizes anions in the electrolyte enter the porous polymer to balance the charge created. Simultaneously, lithium ions in electrolyte are electrodeposited at the lithium surface. During discharging electrons are removed from the lithium, causing lithium ions to reenter the electrolyte and to pass through the load and into the oxidized polymer. The positive sites on the polymer are reduced, releasing the charge-balancing anions back to the electrolyte. This process can be repeated about as often as a typical secondary battery cell. These prototype cells were found to be comparable to, or better than nickel-cadmium cells available now on the market.
- **Ion-exchage Membranes :** The charge residing on the conducting polymers make them useful as ion-exchange membranes.

27.6 Biodegradable Polymers

Introduction

It is just half a century that the polymers were commercially synthesized but have found to be ubiquitous and indispensable. Materials such as the polyolefines performed their function well and displaced most of the cellulosic fibers. Synthetic polymers were developed for durability and resistance to all forms of degradation. These characteristics and others such as rigidity, permeability, and transparency can be controlled by changing the molecular weight and functionality of the polymer.

Polymers are widely accepted for the properties mentioned above and they are relatively inexpensive to make, compared to the other products that polymers replace such as glass and paper cardboard. These conveniences enhance the quality and comfort of life in modern society.

One major drawback to polymers is the problem with disposal. Since they are somewhat resistive to degradation, polymers tend to accumulate in what is today's most popular and ultimate disposal system, *the landfill.* In these landfills, except plastics, other municipal and construction wastes decompose and disintegrate to materials acceptable to nature in due course of time. This brings about a series of questions, what effects polymers have on the environment ? whether they are biodegrade at all ? and if so what effect the products of biodegradation have on the environment ? why is mother nature that could accommodate so many constituents is showing step-motherly treatment to polymers?

There are several different types of degradation that can occur in the environment. These include biodegradation, photodegradation, oxidation, and hydrolysis. Often, people will put all these processes together and call them biodegradation. However, most chemists and biologists will agree that *biodegradation involves enzymatically promoted break down caused by living organisms, usually micro-organisms such as bacteria, fungi and algae.*

The challenge posed to chemists and biochemists is to determine to what extent polymers biodegrade. During millions of years of development, Mother Nature has synthesized different compounds and developed routes for their degradation. All animals and micro-organism feed on different kinds of the biopolymers produced by nature. The problem comes only with man made plastics which are not attacked by any organism.

27.6.1 Standardization and Definitions

Several chemists, biochemists, polymer manufacturers and independent testing groups have taken up this challenge. However, there is some disagreement as to what constitutes a biodegradable polymer. Polymer manufacturers are willing to accept break down of a polymer structure into smaller fragments is degradation, but that still contain the polymer fragments. Conversely, environmentalists will not stop short of complete degradation into the monomer units or mineralization into products *usable by organisms*. This conflict has created a need for a standard system for determining the degree to which a polymer can biodegrade. One proposed system includes (i) measuring physicomechanical changes (changes in morphology and physical properties), (ii) measuring chemical changes, (iii) finding products formed and weight loss due to volatile products. In July 2000 the US Composting Council and the Biodegradable Products Institute announced a labeling (certification) programme based on the ***ASTM Standard*** D 6400-99 for materials suitable for composting applications. Some of the definitions/concepts used in this document are

1. **Biodegradable Plastic :** a degradable plastic in which the degradation results from the action of naturally occurring microorganisms such as bacteria, fungi and algae.
2. **Compostable Plastic :** a plastic that undergoes degradation by biological processes during composting to yield CO_2, water, inorganic compounds and biomass at a rate consistent with other compostable materials and leaves no visible, distinguishable or toxic residue.
3. **Degradable :** A material is called degradable with respect to specific environmental conditions if it undergoes a degradation to a specific extent within a given time measured by specific standard test methods.
4. **Degradation :** An irreversible process leading to a significant change of the structure of a material, typically characterized by a loss of properties (e.g. integrity, molecular weight, structure or mechanical strength) and/or fragmentation. Degradation is affected by environmental conditions and proceeds over a period of time comprising one or more steps.
5. **Disintegration :** The falling apart into very small fragments of packaging or packaging material caused by degradation mechanisms.

When comparing the degree to which different polymers biodegrade, several factors must be taken into consideration. The first of these factors is the environment. Polymers may be tested in a natural or simulated environment. Further the simulation conditions could be natural or accelerated. These are utilized to determine whether the polymer begins to degrade following disposal or while it is still in its intended use. The next aspect of biodegradable polymers which is considered is polymer concentration. The polymer may have a high or low mass weight compared to its volume. In addition, the form of the polymer may range from a thin film to a thick building structure. Finally, the environmental effects of the polymer must be considered. A polymer that biodegrades is of little value if the products that form are found to contaminate water bodies or be toxic to living organisms in the environment.

27.6.2 Origin of Polymers and Biodegradability

Discussion of the chemistry and biodegradability starts with the classification of polymers and their origin. To understand the environmental impact of a polymer, the question of biodegradability should not restrict only to it's degradation from its polymeric state but also to the associated factors in deriving/obtaining the polymer.

Natural Polymers: The natural polymers are invariably biodegradable and fall into four broad groups:

- ***Polysaccharides :*** Starch, Cellulose
- ***Proteins :*** Gelatin, Casein, Silk, Wool
- ***Polyesters :*** Polyhydroxyalkanoates
- ***Others :*** Lignin, Shellac, Natural Rubber

These natural polymers are generally derived from renewable resources and are plant derivatives. These are easily degradable but many problems are associated in their harvesting and processing. For example; the growing of agricultural crops may involve the application of fertilizers, herbicides and pesticides which may leave a deep environmental footprint.

Synthetic Polymers : These could be derived from renewable resources (agro based products) or from non-renewable resources (petrochemicals). Some of the synthetic biodegradable polymers have very special market niches as for example "dissolving" suture material used in the medical field. Certain groups of synthetic polymers are biodegradable since they were found to be attacked by microorganism. They include,

- *Polyalkylene esters*
- *Polylactic acid and its copolymers*
- *Polyamide esters*
- *Polyvinyl esters*
- *Polyvinyl alcohol*
- *Polyanhydrides*

Each of these has its particular properties and special applications. Polyvinyl alcohol is widely used because of its solubility in water. Polyesters were found to compete with the standard resins in their performance but failed to make significant market advances due to their high prices. Polylactic acid (PLA) resin can be used for many of the applications currently utilizing polyethylene resins. PLA is based on lactic acid which uses dextrose from corn or sugar beets as a raw materials. The flexibility of PLA has attracted big companies such as Chronopol which currently operates a 2 million lb/year PLA facility in Colorado, U.S.A and has plans for 100 million lb/year in the future. Chronopol technology allowed to cut the cost from $6.00 to $1.50-$2.00/lb, but, even with this serious reduction, PLA can not compete with polyethylene which sells for $0.40-$0.50/lb ? In the future, with petroleum products becoming more expensive, PLA should have a bright future.

Semi-synthetic polymers : Many synthetic polymers are produced and utilized because they are resistant to chemical and physical degradation. These polymers resistant to degradation present disposal problems when their usefulness ceases. Research has shown that substitution of natural monomers into non-degradable synthetic polymers produce polymers that are more easily biodegraded; however such polymers lack properties, such as water resistance, that make current polymers so useful. Future biodegradable polymers must overcome such physical problems as well as lower their costs of production, which currently limit the economic feasibility of such biodegradable polymers

27.6.3 Additives

Plastics in their pure form are called virgin polymers or virgin resins. Majority of the polymers, except few like polyethylene are not suitable for direct usage. For example natural rubber is lacking hardness, too stretchy and sticky. It is brought to use by vulcanization and by addition of additives like carbon black. These additives not only enhance the required properties, processability but also help in cutting costs. These additives include Plasticizers, antioxidants, pigments and stabilizers. The requirement of the ASTM definition of Compostable Plastic is that it leaves "*no visible, distinguishable or toxic residue*" after degradation and is equally applicable to additives used in biodegradable polymers.

27.6.4 Tests of Biodegradability

The tests of biodegradable polymers are carried out in *two types of environments* :

- aerobic (in the presence of air)

- anaerobic (in the absence of air)

The biodegradation that occurs in each is characterized by a total carbon conservation.

For the aerobic Environment :

Following biodegradation, the carbon from the polymer will appear in *one of three end products* :

1. CO_2, which is a product of the respiration of the micro-organisms
2. C_r which is any residue of the polymer that is left or any by-product that is formed
3. C_b which is the biomass produced by the micro-organisms through reproduction and growth.

The equation can be represented as

$$C_t = CO_2 + H_2O + C_r + C_b$$

where C_t is the total carbon contained in the polymer.

For the Aanaerobic Environment :

Under anerobic conditions, air is not supplied. In addition to the three products referred for aerobic oxidation, often we have methane as one of the products in anaerobic respiration.

The total carbon equation is the following:

$$C_t = CO_2 + CH_4 + H_2O + C_r + C_b$$

where C_t, C_r, and C_b represent the same forms of carbon as above.

The Bioreactors : These are made of cylindrical polyethylene tanks with gas-tight covers and contain either aerobic microbials or anaerobic microbials depending on the aerobic or the anaerobic conditions needed. In each case, the microbials were fed a 1:1 mixture of α-cellulose and field dried, milled sorghum twice per week for 115 days to establish (to develop microorganism) the bioreactors.

The aerobic reactor was operated at 37°C with humid air supplied at a rate of 1 ft^3/hr and recycled at 1.6 ft^3/min for sufficient oxygen transfer. Biogas was collected in gas collection bags and analyzed twice per week for CO_2 content and dissolved oxygen.

The anaerobic reactor was operated at 58° C *with no intake of air*. It was also equipped with biogas collection bags and analyzed twice per week for CO_2 and CH_4.

Both reactors were analyzed twice per week for temperature, pH, ammonia, nitrogen, and alkalinity. The conditions after which the bioreactors were operated were intended to provide accelerated, simulated environments that parallel those found in such disposal possibilities as landfills (anaerobic) and litter (aerobic).

In the aerobic reactor, the amount of oxygen supplied to the reactor, used by the microbials, and dissolved in the air within the bioreactor are important factors for the determination of the amount of biodegradation. By knowing the amount of polymer that is being tested, one can determine the amount of oxygen needed by the microbials to break down the polymer according to the following equation:

$$C_n H_a O_b N_C + \left(n + \frac{a}{4} - \frac{b}{2} - \frac{3c}{4}\right) O_2 \rightarrow n\,CO_2 + \left(\frac{a}{2} - \frac{3c}{2}\right) H_2O + cNH_3$$

If this reaction proceeds to completion, then the amount of oxygen needed is known as the theoretical Biochemical Oxygen Demand (BOD). By analyzing the amount of carbon dioxide and oxygen, the percentage of theoretical BOD can be determined and thus the extent of biodegradation can be calculated. The films were placed in the aerobic and anaerobic reactors simultaneously for four weeks using a hot water bath as the control, and the films were measured for mass lost.

Results of Experiment :

- Polyethylene films that contained starch showed no evidence of anything other than biodegradation of the starch component. In addition, based upon mass loss, most of the materials tested were affected identically by the aerobic and anaerobic bacteria.
- The polyhydroxybutyrate and polyhydroxyvalerate (PHB/PHV) film readily biodegraded. After four weeks of exposure in the bioreactors, the film had a mass loss of 90-100%. At the same time, little mass was lost after being exposed to the control, the hot water bath.

Conclusion of Experiments :

Various studies point to some conclusions about the biodegradability of polymers.

(i) Naturally-occurring polymers biodegrade and chemically modified natural polymers may biodegrade depending on the extent of modification.

(ii) Synthetic addition polymers, with carbon as the only atom in the backbone *do not biodegrade* at molecular weights *above 500.*

(iii) If an addition polymer contains atoms other than carbon in the backbone, it may biodegrade depending on any attached functionality groups.

(iv) Synthetic condensation polymers are generally biodegradable to different extents depending on chain coupling (ester > ether> amide > urethane), morphology (amorphous > crystalline), molecular weight (lower > higher), and hydrophilic is faster than hydrophobic.

(v) If a polymer is water soluble, that does not necessarily mean that it is biodegradable, since it only gives a solution of polymer in the original state.

27.7 Nano Chemistry

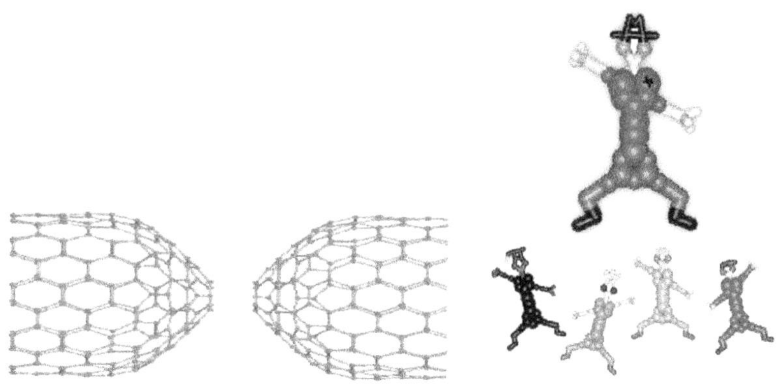

Nano tube welding

"NanoKids" consists of carbon, hydrogen and oxygen atoms

Introduction : Normal chemistry practices revolve around macroscopic particles and their mixing by mechanical means. So we could even call it as mechanochemistry. In mechanochemistry, individual chemical reactions at the molecular or atomic level are achieved by subjecting the reactants to high pressure conditions and pressing together to overcome the normal reaction barriers. This requires positioning of the reactants on a nanometer scale to precisely bring the reactants together. This possibility was first prophesied by Richard Feynman in 1959 in his talk entitled "*There Is Plenty Room At the Bottom*". He said: "The principles of physics, as far as I can see, do not speak against the possibility of maneuvering things atom by atom." His dream could not be realized for want of new class of miniaturized instruments that would allow manipulation and measurement of the properties of the small "nano" structures. In the 1980's with the invention of these instruments

Feynman's dream started realizing. The scanning tunneling microscope, atomic force microscope and the near-field microscopes provided the eyes and fingers required for nanostructure measurement and manipulation.

In nanochemistry, the ***rules of bonding*** *are extended from atoms and molecules (with dimensions < 1 nm) to larger objects having the dimension of from 1 to 100 nanometers*, where one nanometer is one-billionth of a meter or 10^{-9} meter. In this regime, new chemical and physical properties appear that depend *on the size and shape of the particles*. Nanochemistry undertakes the synthesis of precisely defined nanoparticles to achieve novel materials for specific applications in such fields as advanced catalysis, ultra-thin films and membranes (separation and adsorption), storage devices, and optical and electronic devices. The nanometer regime is of particular interest because it is in this region where phenomena associated with atomic and *molecular interactions begin to be strongly influenced by the macroscopic properties of materials*.

The Nanorevolution is new but nanochemistry is not. Biological processes are essentially nanochemical processes that occur with low energy consumption and high efficiency. For instance, photosynthesis is a molecular-scale mechanical operation and enzymes are nanosize factories. The challenge is to learn how to control such processes. Once we can do this, there will be advances in everything from microelectronics to chemical engineering.

27.7.1 Applications of Nano Materials

Nanochemistry has applications ranging from engineering to medicine. A summary of applications of the nano-materials produced is presented here.

1. **As Engineering Materias :** Applications of nanotechnology to materials and manufacturing include the following:
   - The fabrication of thin nanowires, is a critical technology for communications technology. Methods that produce wires 20 times thinner, longer, and better-defined than conventional ones have been based on nanotechnology.
   - In gas phase condensation, the vapor evolved from an appropriate precursor, say a metal, undergoes a reaction with metal oxide to produce nanometer-sized particles. *Indium tin oxide* particles made by this method are ***conductive***, and can be processed to form transparent thin films that are ***antistatic***. Similarly nano particles produced by combining yttria and some lanthanides phosphor (emit delayed emission) and therefore processed to form **light-emitting devices**.

- Surface scientists have been able to exploit atomic self-organization processes for the controlled fabrication of nanostructures at surfaces having well defined size and shape. This allows forming nanostructure metals and **ceramics** in exact shapes *without machining.*
- **Improved color printing** brought about by nanometer-scale particles that have the best properties of both dyes and pigments.
- Nanoscale cemented and plated carbide materials and **nanocoating for cutting tools**, and other electronic, chemical and structural application.
- **Nanofabrication on a chip** with high levels of complexity and functionality. Nanostructured chip devices prepared from nanoscale materials like complex oxides, carbides, borides and nitrides have been successfully *mass produced* in kilogram quantities.
- In **nanocomposites**, one of the constituent phases has dimensions–length, width, or thickness, in the nanometer range. Because the building blocks of the nanocomposite have the dimensions of nanometers, they have *enormous surface areas* and large interfaces with other phases. The properties of the nanocomposite result from interactions at these interfaces. In conventional composites where the building blocks have dimensions on the order of microns (one-millionth of a meter), there is less interfacial contact between phases, and consequently less effect by the interfaces on properties of the composite.

  One naturally-occurring nanocomposite is bone. Bone consists to a large extent of nanoscale, platelike crystals of hydroxyapatite, $Ca_{10}(PO_4)_6(OH)_2$ dispersed in a matrix of collagen fibers. Conventional attempts to synthesize hydroxyapatite for artificial bone or bone implant materials have many difficulties, but scientists have found that these problems can be largely overcome if hydroxapatite is synthesized as a nanostructured material.

  In organic and sol-inorganic nanocomposites, a nanoscale inorganic filler is typically dispersed in an organic polymer matrix. The filler carries the load, and its large surface area in contact with the matrix reduces the mobility of the polymer chains. Nylons are thermoplastic but *Nylon nanocomposites* containing small amounts of clay are capable of withstanding high temperature environments and have been designed into automobile air intake covers.

- Chemists have also been looking at *carbon* ***nanotubes*** (honeycomb **graphite** lattices rolled into cylinders having nanoscale thicknesses) as fillers in nanocomposites. The nanotubes, by virtue of their **electrical conductivity**, make the composite conductive as well. Nanotubes have been found by experiment to be stiffer than carbon fibers (which are thicker than 1 micron), but less brittle. Thus, nanocomposites containing carbon nanotubes are able to withstand much greater deformations before breaking than carbon fiber composites.

2. **In Chemical Industry :** In the field of **catalysis**, chemists have been searching for catalysts that are *stable at high temperatures and that can be used in smallest possible amounts*. The complex oxide barium hexaaluminate (BHA), or $BaO_6Al_2\ O_3$, has been of interest since it retains its catalytic activity at high temperatures. But conventional methods to synthesize BHA tend to reduce the material's surface area, and hence the activity. But when nanoparticles of BHA are prepared, a final grain size of 30 nanometers is achieved, and a *large surface area and catalytic activity* are achieved.

   The new nano-catalysts designed and new nanochemical routes could pave way for environment friendly and atom economical **Green chemistry routes.** The nano chemical routes catalyze chemical reactions, sometimes at rates exceeding a million reactions per second, without requiring elevated temperature, pressure, or dangerous solvents.

3. **In Biotechnology :** Enzymes are known to catalyze more than 5000 diverse chemical reactions. The possibilities for using enzymes in small scale reactors are limitless. Protein catalysts or enzymes have precise control over specific chemical reactions, reducing unwanted side-reactions. Most enzymes work by lowering the energy barrier for a specific reaction; thus the energy needed for conversions is reduced to ambient temperatures and pressures. Recombinant DNA technology enables the production of novel proteins with enhanced properties not found in nature. *The main drawback with enzymes is fragility.* Combining nano-scale materials with enzymes could dramatically improve the durability of enzymes, create localized high concentrations of protein and reduce costs by minimizing losses.

4. **In Medicine and Health :** Nanotechnology will provide new tools for medicine. It could radically change the way surgery is done. It

will make it possible to do molecular-scale surgery to replace defective cells, repair and rearrange cells. Since disease is the result of physical disorder, misarranged molecules and cells, medicine at this level should be able to cure most diseases. Mutations in DNA could be repaired and cancer cells, toxic chemicals, and viruses could be destroyed through use of medical nanodevices. There are numerous other potential applications of nanoscience to biology :

- Rapid, efficient genome sequencing, revolutionizing diagnostics, and therapeutics
- Effective and less expensive healthcare using remote and in-vivo devices
- New formulations and routes for drug delivery that enormously broaden their therapeutic potential by effecting delivery of new types of medicine to a previously inaccessible sites in the body
- More durable, rejection-resistant artificial tissues and organs
- Sensor systems that detect emerging disease in the body, which will shift the focus of patient care from disease treatment to early detection and prevention

5. **Nanotechnology for Environment and Energy :** Nanotechnology has the potential to significantly impact energy efficiency, storage, and production. Solar cells are prohibitively expensive and nanometer-sized solar cells could be developed to provide much of the energy needed around the world at a cheap price. This would stop the usage of fossil and nuclear fuels. Thus this would provide a cleaner and safer method than any of our current methods for making power.

   Other new nanoscale capabilities that illustrate this potential are :

   - The discovery of the ordered mesoporous material MCM-41 produced by Mobil Oil Co., with pore size in the range of 10 to 100 nm, is now widely applied in the removal of ultrafine contaminants.
   - Nanoparticle-reinforced polymeric material discussed above could replace structural metallic components in the auto industry. Widespread use of those nanocomposites could lead to a reduction of 1.5 billion liters of gasoline consumption over the life of one year's fleet of vehicles and reduce carbon dioxide emission by more than five billion kilograms annually,

- The replacement of carbon black in tires by a nanometer-scale particles of inorganic clays and polymers is a new technology that is leading to the production of environmentally friendly, wear-resistant tires.

6. **Nanoelectronics and Computer Technology :** Nanotechnology will enable the construction of nanoscale circuits and computers. Smaller circuits run faster and this enables far greater computing speeds. Nanomaterials can help by providing nanocrystalline materials of ultra-high purity and with better thermal conductivity and longer life.

   Other potential nanoelectronics and computer technology breakthroughs include :

   - Nanostructured microprocessor devices that continue the trend in declining energy use and cost per gate, thereby potentially improving the efficiency of computers by a factor of millions.
   - Higher transmission frequencies and more efficient utilization of the optical spectrum to provide at least ten times more bandwidth, with consequences in business education, entertainment, and defense.
   - Small mass storage devices with capacities at multi-terabit levels, a thousand times better than today.
   - Integrated nanosensor systems capable of collecting, processing and communicating massive amounts of data with minimal size, weight, and power consumption

6. **In Aeronautics and Space Exploration :** Birds have long inspired the development of aircraft, but our present man-made vehicles are primitive compared to Mother Nature's flying counterparts. To achieve the goal of flying like a bird, the challenges to be met are–material science advances in strength-to-weight ratios, reconfigurability, integration of sensing and actuation. Further in space crafts, the stringent fuel constraints of lifting payloads into earth orbit and beyond, and the desire to send spacecraft into other solar systems for extended missions, compel continued reduction in size, weight, and power consumption of payloads. Nano structured materials and devices promise solutions to these challenges. Further development of nanocomposites will be to make them tunable, adaptive, self-healing and stress-smart sensing systems. These materials will enable a revolution in air and space travel by increasing intelligence in aerospace systems and vehicles while reducing mass, size and power consumption.

27.8 Green Chemistry

Green Chemistry or environmentally benign chemistry is the design of chemical products and processes that reduce or eliminate the use and generation of hazardous substances. Thus instead of limiting Risk by controlling our Exposure to hazardous chemicals, green chemistry attempts to reduce and give alternate routes and preferentially eliminate the usage of Hazardous chemical thus negating the necessity to control Exposure. The bottom line is, if we don't use or produce hazardous substances then the *Risk is zero*, and we don't not have to worry about the treatment of hazardous substances or limiting our exposure to them. The stakes are high indeed. Cleaning up chemical messes is growing ever more costly. This year the DuPont company agreed to pay up to \$600 million in fines and settlement costs over environmental damage caused by production of Teflon and Gore-Tex. General Electric will spend time and tens of millions of dollars to clean up PCBs it discharged into the Hudson River in New York. Similar is the case in India where crores of rupees are spent from time to time in cleaning Ganga and other water bodies. Similarly many companies face costs in the hundreds of millions of dollars/rupees to clean up dioxins, perchlorate, mercury, chromium and asbestos. One-half of the paper and paperboard currently used in the USA is recycled, but adhesives, coatings, plastics and other materials on the old paper can produce spots and holes in the new paper, called "stickies," and they cost the industry \$500 million annually. New enzyme were put to use to turn stickies into a water-soluble, non-sticky material. The enzyme is produced by a bacteria and is completely bio degradable. Since 2002, many paper mills have changed their process to use this enzyme. Similar is the case with nitration reaction carried out in industries. The traditional route of nitration adopts the use of nitric acid. Nitration of cyclohexanol gives adipic acid, widely used in the synthesis of Nylons. This exposes the environment to toxic and corrosive acid fumes. This process is now gradually replaced by the Greener route using harmless chemicals.

and/or
HNO_3
Traditional route
Na_2WO_2, H_2O_2
$KHSO_4$, Aliquat 336
Greener route
Adipic acid

One ploy of green chemistry is to work more like Mother Nature. Plants, for example, have access only to air, a few trace minerals in the soil and energy from the sun, yet they carry out hugely complex chemical transformations to give us carbohydrates, proteins, vitamins and many complicate molecules. So green chemistry advocates procedures acceptable to mother nature like use of greener (environment friendly) oxidizing agents and chemicals, reactions run with minimum use of organic and other solvents, reuse of side-products obtained in other processes.

27.8.1 Principles of Green Chemistry

Anastas and Warner have developed the Twelve Principles of Green Chemistry to aid one in assessing how green a chemical, a reaction or a process is.

1. It is better to **prevent waste** than to treat or clean up waste after it is formed.
2. Synthetic methods should be designed to **maximize the incorporation of all materials** used in the process into the final product.
3. Wherever practicable, **synthetic methodologies** should be designed to use and generate substances that **possess little or no toxicity** to human health and the environment.
4. Chemical products should be designed to **preserve efficacy of function while reducing toxicity**.
5. The use of **auxiliary substances** (e.g. solvents, separation agents, etc.) **should be made unnecessary whenever possible** and, innocuous when used.
6. **Energy requirements** should be recognized for their environmental and economic impacts and **should be minimized**. Synthetic methods should be conducted at ambient temperature and pressure.
7. A raw material **feedstock should be renewable** rather than depleting whenever technically and economically practical.
8. **Unnecessary derivatization** (blocking group, protection/deprotection, temporary modification of physical/chemical processes) should be **avoided** whenever possible.
9. **Catalytic reagents** (as selective as possible) **are superior** to stoichiometric reagents.
10. Chemical **products should be designed** so that at the end of their function they **do not persist** in the environment and **break down into innocuous degradation products**.
11. Analytical methodologies need to be further developed to allow for **real-time in-process monitoring and contro**l prior to the formation of hazardous substances.
12. Substances and the **form of a substance** used in a chemical process **should be chosen so as to minimize the potential for chemical accidents, including releases, explosions, and fires**.

Answers to Objective Type Questions and Fill in the Blanks

Chapter - 1

Self Test Questions

1. (D) 2. (D) 3. (A) 4. (C)
5. (A) 6. (A) 7. (A) 8. (D)
9. (C) 10. (A)

Chapter - 2

Self Test Questions

1. (D) 2. (A) 3. (C) 4. (A)
5. (C) 6. (A)

Chapter - 3

Fill in the Blanks

1. permanent
2. sewage
3. bicarbonates
4. alkalinity
5. temporary
6. 100
7. ppm
8. 0.07
9. heating or bubbling N_2
10. lime
11. $Na_2 SO_4$
12. $Na_3 PO_4$
13. bacteria
14. sludge
15. scale
16. silicates
17. carbonate
18. acids
19. reverse osmosis
20. hydroxide
21. 100 ppm
22. silicate
23. sodium aluminate
24. disinfection
25. sterilization
26. Erio – T
27. ethylene diamine tetra
28. bacteria
29. chloromine
30. Cl_2
31. sterilization
32. reverse osmosis
33. disinfection

Chapter - 4

Numerical Problems

Thermodynamics

1. 2763.6 cal
2. w = –266.07 cal
3. w = 684.1 cal
4. w = 2897.22 cal
5. w = 1653 cal
6. $\Delta S = 20.18$ cal k^{-1} mol^{-1}
7. Pressure = 1.97 atm, work done = 4671 J
8. w = 25.96 kJ
9. $\Delta E = 99.92$
10. 4.82 K

Thermochemistry

1. $\Delta H = 3322$ kcal
2. $\Delta H = -116.4$ kcal
3. $\Delta H = 17.4$ k cal mol^{-1}
4. $\Delta H = 29.3$ k cal mol^{-1}
5. $\Delta H_2 = -1222.8$ kJ mol^{-1}
6. $\Delta H_f = -47.54$ k cal mol^{-1}
7. (i) 2773.5 cal (ii) 2833.11 cal
8. 525.72 kJ
9. Heat of reaction = 35.2 k cal mol^{-1}
10. O-H bond energy = 101 kcal
11. Heat of formation of CH_3Br = 183 kcal mol^{-1}
12. Heat of reaction = 21.1 k cal mol^{-1}
13. Heat of formation of C_2H_2 = 54.2 kcal
14. $\Delta E = -922.2$ k cals
15. $\Delta E = -664$ kcal
16. $\Delta E = 637.5$ kcal
17. $\Delta H = 14.14$ kcals

18. $\Delta S = 8.95$ cal mol^{-1} K^{-1}
19. $\Delta S = 88.3$ J mol^{-1} K^{-1}
20. $\Delta S = 6.4$ e.u.,
21. $\Delta S = 21.1$ cal mol^{-1} K^{-1}
22. $\Delta S = 13.4$ J mol^{-1} K^{-1}
23. 0.25
24. 0.20
25. $\phi = 0.37$, $Q_1 = 141$ kJ
26. w = 175.5 $\phi = 21\%$
27. $\phi = 33\%$ w = 100 kJ $Q_1 = 200$ kJ
28. $\phi = 0.3$, $Q_1 = 560$ J cycle^{-1}

Self Test Questions

1. (C)	2. (D)	3. (C)	4. (C)
5. (A)	6. (D)	7. (B)	8. (C)
9. (B)	10. (D)	11. (C)	12. (A)
13. (D)	14. (C)	15. (B)	16. (B)
17. (D)	18. (A)	19. (D)	20. (D)
21. (C)	22. (B)	23. (B)	24. (A)

Chapter - 5, 6, 7, 8

Self Test Questions

1. (A)	2. (C)	3. (D)	4. (C)
5. (B)	6. (C)	7. (C)	8. (A)
9. (B)	10. (A)	11. (C)	12. (C)
13. (B)	14. (D)		

Chapter - 9

Objective Questions

1. (C)	2. (C)	3. (A)	4. (C)
5. (B)	6. (C)	7. (D)	8. (A)
9. (C)	10. (D)	11. (A)	

Chapter - 10

Objective Questions

1. 107.9
2. 0.1737g
3. 0.029 L
4. 0.0176 L
5. 0.010682 mho cm^{-1}, 106.82 mho cm^2 eq^{-1}
6. 0.124
8. b = 307.5, Λ_4 = 164.45
9. 81g of Ag, 23.8g of Cu
10. $t_{Cu^{2+}} = 0.212$, $= 1 - 0.212 = 0.788$

    $t_{H+} = 0.742$ $= 1 - 0.742 = 0.258$
12. = 0.45 = 0.55
13. $t_+ = 0.271$, $t_- = 0.729$
14. $t_+ = 0.473$, $t_- = 0.527$
15. $t_+ = 0.473$, $t_- = 0.527$
16. $\alpha = 0.12$, $K_a = 1.44 \times 10^{-3}$
17. $S = 7.15 \times 10^{-5}$ mol L^{-1}, $K_{sp} = 5.1 \times 10^{-9}$ mol^2 L^{-2}
18. (a) Non-spontaneous (b) Non-spontananeous

    (c) Spontaneous (d) Non-spontaneous

    (e) Spontaneous
19. Yes
20. Ni, Ca, Zn, Mg
21. (c) greater
22. (d) H^+
23. $I^- + Ie^{3+} \rightarrow Fe^{2+} + Y_2 I_2$; $\in^o_{cell} = 0.2355$ V
24. (a) Spontaneous (b) Spontaneous

    (c) Spontaneous (d) Non-spontaneous

    (e) Non-spontaneous

25. (a) 1.5052 V (b) 0.09 V (c) 0.839 V
27. 1.06, 1.2871 V, –0.189 V, 0.46 V
28. pH = 2.4
29. (a) 6.36×10^9 (b) 49
30. (a) $\in^o = 0.1$ V, $\in = 0.17$ V, $\Delta G = -33$ kJ
    (b) $\in^o = -4.37$ V, $\in = 0.693$ V, $\Delta G = 4013$ kJ
    (c) $\in^o = 0.02$ V, $\in = 0.693$ V, $\Delta G = 40.13$ kJ
31. $\in^o = 0.02$ V, Not feasible
32. $\in^o = \in = 0.066$ V
33. 0.925 V
34. 0.209 V
35. –0.7008 V
36. emf = 0.74 V
37. 1.0408 V
38. 1.0031 V
39. 0.76 V

Self-Test Questions

1. (C)
2. (D)
3. (B)
4. (A)
5. (B)
6. (A)
7. (D)
8. (B)
9. (A)
10. (D)
11. (B)
12. (D)
13. (C)
14. (A)
15. (D)
16. Ionic
17. Current,
18. Λ_4
19. Observed conduction
20. Anodic
21. Anodic.
22. (D)
23. (D)
24. (C)
25. (A)
26. (C)
27. (A)

Chapter - 11

Objective Questions

1. (D)
2. (B)
3. (A)
4. (A)
5. (D)
6. (B)
7. (B)
8. (B)
9. (D)

Fill in the Blanks

1. acidic
2. higher
3. corrosion of structures
4. increases
5. less
6. metallurgical
7. large
8. Zn
9. corrosion
10. pigment
11. decorative
12. adherence
13. phosphoric
14. Al
15. to prevent settling of havier pigment
16. chromate
17. linoleate or thinner
18. is as drying agent
19. chromate
20. lead
21. chromic acid
22. tinning
23. galvanizing
24. Zn
25. Ni
26. lac
27. varnish
28. paint
29. Zn

Chapter - 12

1. $I = 9.89$ g cm^2, $V = 1.69 \times 10^{-29}$ cm^{-1}
2. $B = 1.669 \times 10^{-28}$, Rotational spacing $= 3.338 \times 10^{-28}$
3. 9.6×10^{25} Nm^{-1}
4. 16.75×10^{-8} cm^{-1}
5. $I = 4.35 \times 10^{-29}$ g.cm^2

Self Test Questions

1. ionic
2. Linear
3. (C)
4. (D)
5. (B)
6. (A)
7. (A)
8. (D)
9. (A)
10. (B)
11. Atoms
12. molecules
13. micro wave
14. zero
15. UV and visible

Chapter - 13

1. 693 s
2. $E = 26.46$ kcal mol^{-1} or 110 kJ mol^{-1}
4. 3×10^{-3} min^{-1}
5. 118 kJ mol^{-1},
6. 1
7. 1
8. 103.45 kJ mol^{-1} s
9. 3.37×10^{-2} L mol^{-1} s^{-1}, 50 kJ mol^{-1}
10. 1.0×10^{-1} min^{-1}
11. 71 kJ mol^{-1}
12. Order in H_2O_2 is 1
13. 6.67×10^{-9} s
14. $E_a = 38.6$ kcal mol^{-1},
    $A = 1.41 \times 10^{11}$, E (HI decomposition) = 33.8 kcal mol^{-1}
15. A = 4.56×10^{16}, $k = 6.16$ min^{-1}
16. 9730 years
17. 3.46×10^{-5}
18. (a) Order in A = 1, Order in B = 1,
    (b) Rate law = k [A] [B],
    (c) Rate of formation of D = 9×10^{-3} mol L^{-1} s^{-1}
19. 47.45 days
20. 3200 years

Self Test Questions

21. (B)
22. (A)
23. (D)
24. (C)
25. (C)
26. (B)
27. (B)
28. (D)

Chapter - 14

Objective Questions

1. (A)	2. (B)	3. (C)	4. (D)
5. (C)	6. (C)	7. (C)	8. (D)
9. (C)	10. (A)	11. (C)	12. (B)

Fill in the Blanks

1. bakelite
2. thermo
3. thermosets
4. addition
5. organic dichlorides, inorganic polysulphides
6. butadiene
7. buna-N
8. buna-S
9. hexamethylene diamine+ adipic acid
10. isoprene
11. sulphur
12. butadiene
13. silicone rubber
14. butyl
15. silicone, oxygen
16. hardening
17. polyester
18. silicones
19. fibres
20. 24, 75
21. bakelite
22. silicone
23. benzolyl peroxide

Chapter - 15, 16, 17

Objective Questions

1. (B)	2. (D)	3. (D)	4. (A)
5. (B)	6. (A)	7. (B)	

Fill in the Blanks

1. porasity
2. magnesite
3. more
4. all gases
5. temperature
6. insulator

Chapter - 19

Self Test Questions

1. (A)	2. (A)	3. (A)	4. (C)
5. (A)	6. (A)	7. (B)	8. (C)
10. (B)	11. (B)	12. 14	

13. Covalent
14. Law of constancy of interfacial angle
15. 9
16. 13
17. Crystal
18. Polymorphism
19. Polymorphism
20. Lattice

Chapter - 21

Self Test Questions

1. (A)
2. (D)
3. (D)
4. (A)
5. (C)
6. (B)
7. (C)
8. (B)
9. (C)
10. Eutetic
11. Phase
12. Component
13. 303 ^{0}C
14. Temperature

Chapter - 23-26

Fill in the Blanks

1. metals
2. (C)
3. gold & platinum
4. pig iron
5. (B)
6. ductile iron
7. (D)
8. carbon
9. cementation
10. (C)
11. brittle
12. (A)
13. hardness
14. hard
15. (D)
16. cathode
17. (A)
18. (C)
19. alluminium
20. Dow metal

Index

A

B

C

E

F

G

H

I

J

K

L

M

N

O

P

Q

R

S

www.ingramcontent.com/pod-product-compliance
Ingram Content Group UK Ltd.
Pitfield, Milton Keynes, MK11 3LW, UK
UKHW061535190726
13845UKWH00005B/113

9 789385 433863